pi to two million places

π

pi to two million places

Kick Books
Tucson | 2006

The number pi itself is not copyrighted, of course.

ISBN 978-1-4116-7237-6

Also available:
pi to five million places
[softcover, ISBN 978-1-4116-9547-4]
[hardcover, no ISBN, available only via www.kickbooks.com]

Russ Kick
PO Box 30453
Tucson AZ 85751

www.kickbooks.com

π

3.14159265358979323846264338327950288419716939937510582097
4944592307816406286208998628034825342117067982148086513282
3066470938446095505822317253594081284811174502841027019385
2110555964462294895493038196442881097566593344612847564823
3786783165271201909145648566923460348610454326648213393607
2602491412737245870066063155881748815209209628292540917153
6436789259036001133053054882046652138414695194151160943305
7270365759591953092186117381932611793105118548074462379962
7495673518857527248912279381830119491298336733624406566430
8602139494639522473719070217986094370277053921717629317675
2384674818467669405132000568127145263560827785771342757789
6091736371787214684409012249534301465495853710507922796892
5892354201995611212902196086403441815981362977477130996051
8707211349999998372978049951059731732816096318595024459455
3469083026425223082533446850352619311881710100031378387528
8658753320838142061717766914730359825349042875546873115956
2863882353787593751957781857780532171226806613001927876611
1959092164201989380952572010654858632788659361533818279682
3030195203530185296899577362259941389124972177528347913151
5574857242454150695950829533116861727855889075098381754637
4649393192550604009277016711390098488240128583616035637076
6010471018194295559619894676783744944825537977472684710404
7534646208046684259069491293313677028989152104752162056966
0240580381501935112533824300355876402474964732639141992726
0426992279678235478163600934172164121992458631503028618297
4555706749838505494588586926995690927210797509302955321165
3449872027559602364806654991198818347977535663698074265425
2786255181841757467289097777279380008164706001614524919217
3217214772350141441973568548161361157352552133475741849468
4385233239073941433345477624168625189835694855620992192221
8427255025425688767179049460165346680498862723279178608578
4383827967976681454100953883786360950680064225125205117392
9848960841284886269456042419652850222106611863067442786220
3919494504712371378696095636437191728746776465757396241389
0865832645995813390478027590099465764078951269468398352595
7098258226205224894077267194782684826014769909026401363944
3745530506820349625245174939965143142980919065925093722169

6461515709858387410597885959772975498930161753928468138268683868942774155991855925245953959431049972524680845987273644695848653836736222626099124608051243884390451244136549762780797715691435997700129616089441694868555848406353422072225828488648158456028506016842739452267467678895252138522549954666727823986456596116354886230577456498035593634568174324112515076069479451096596094025228879710893145669136867228748940560101503308617928680920874760917824938589009714909675985261365549781893129784821682998948722658804857564014270477555132379641451523746234364542858444795265867821051141354735739523113427166102135969536231442952484937187110145765403590279934403742007310578539062198387447808478489683321445713868751943506430218453191048481005370614680674919278191197939952061419663428754440643745123718192179998391015919561814675142691239748940907186494231961567945208095146550225231603881930142093762137855956638937787083039069792077346722182562599661501421503068038447734549202605414665925201497442850732518666002132434088190710486331734649651453905796268561005508106658796998163574736384052571459102897064140110971206280439039759515677157700420337869936007230558763176359421873125147120532928191826186125867321579198414848829164470609575270695722091756711672291098169091528017350671274858322287183520935396572512108357915136988209144421006751033467110314126711136990865851639831501970165151168517143765761835155650884909989859982387345528331635507647918535893226185489632132933089857064204675259070915481416549859461637180270981994309924488957571282890592323326097299712084433573265489382391193259746366730583604142813883032038249037589852437441702913276561809377344403070746921120191302033038019762110110044929321516084244485963766983895228684783123552658213144957685726243344189303968642624341077322697802807318915441101044682325271620105265227211166039666557309254711055785376346682065310989652691862056476931257058635662018558100729360659876486117910453348850346113657686753249441668039626579787718556084552965412665408530614344431858676975145661406800700237877659134401712749470420562230538994561314071127000407854733269939081454664645880797270826683063432858785698305235808933065757406795457163775254202114955761581400250126228594130216471550979259230990796547376125517656751357517829666454779174501129961489030463994713296210734043751895735961458901938971311179042978285647503203198691514028708085990480109412147221317947647772622414254854540332157185306142288137585043063321751829798662237172159160771669254748738986654949450114654062843366393790039769265672146385306736096571209180763832716641627488880078692560290228472104031721186082041900042296617119637792133757511495950156604963186294726547364252308177036751590673502350728354056704038674351362222477158915049530984448933309634087807693259939780541934144737744184263129860809988868741326047215695162396586457302163159819319516735381297416772947867242292465436680098067692823828068996400482435403701416314965897940924323789690706977942236250822168895738379862300159377647165122893578601588161755782973523344604281512627203734314653197777416031990665541876397929334419521541341899485444734567383162499341913181480927777103863877343177207545654532207770921201905166096280490926360197598828161332316663652861932668633606273567630354477628035045077723554710585954870279081435624014517180624643626794561275318134078330336254232783944975382437205835311477119926063813346776879695970309833913077109870408591337464144282277263465947047458784778720192771528073176790770715721344473060570073349243693113835049316312840425121925651798069411352801314701304781643788518529092854520116583934196562134914341595625865865570552690496520985803385072242648293972858478316305777756068887644624824685792603953527734803048029005876075825104747091643961362676044925627420420832085661190625454337213153595845068772460290161876679524061634252257719542916299193064553779914037340432875262888963995879475729174642635745525407909145135711136941091193932519107602082520261879853188770584297259167781314969900901921169717372784768472686084900337702424291651300500516832336435038951702989392233451722013812806965011784408745196012122859937162313017114448464090389064495444006198690754851602632750529834918740786680881833851022833450850486082503930213321971551843063545500766828294930413776552793975175461395398468339363830474611996653858153842056853386218672523340283087112328278921250771262946322956398989893582116745627010218356462201349671518819097303811980049734072396103685406643193950979019069963955245300545058068550195673022921913933918568034490398205955100226353536192041994745538593810234395544959778377902374216172711172364343543947822181852862408514006660443325888569867054315470696574745855033232334210730154594051655379068662733379958511562578432298827372319898757141595781119635833005940873068121602876496286744604774649159950549737425626901049037781986835938146574126804925648798556145372347867330390468838343634655379498641927056387293174872332083760112302991136793862708943879936201629515413371424892830722012690147546684765357616477379467520049075715552781965362132392640616013635815590742202020318727760527721900556148425551879253034351398442532234157623361064250639049750086562710953591946589751413103482276930624743536325691607815478181152843667957061108615331504452127473924544945423682886061340841486377670096120715124914043027253860764823634143346235189757664521641376796903149501910857598442391986291642193994907236234646844117394032659184044378051333894525742399508296591228508555821572503107125701266830240292952522011872676756220415420516184163484756516999811614101002996078386909291603028840026910414079288621507842451670908700069928212066041837180653556725253256753286129104248776182582976515795984703562226293486003415872298053498965022629174878820273420922224533985626476691490556284250391275771028402799806636582548892648802545661017296702664076559042909945681506526530537182941270336931378517860904070866711496558343434769338578171138645587367812301458768712660348913909562009939361031029161615288138437909904231747336394804575931493140529763475748119356709110137751721008031559024853090669203767192203322909433467685142214477379393751703443661991040337511173547191855046449026365512816228824462575916333039107225383742182140883508657391771509682887478265699599574490661758344137522397096834080053559849175417381883999446974867626551658276584835884531427756879002909517028352971634456212964043523117600665101241200659755851276178583829204197484423608007193045761893234922927965019875187212726750798125547095890455635792122103334669749923563025494780249011419521238281530911407907386025152274299581807247162591668545133312394804947079119153267343028244186041426363954800044800267049624820179289647669758318327131425170296461515709858387410597885959772975498930161753928468138268683868942774155991855925245953959431049972524680845987273644695848653836736222626099124608051243884390451244136549762780797715691435997700129616089441694868555848406353422072225828488648158456028506016842739452267467678895252138522549954666727823986456596116354886230577456498035593634568174324112515076069479451096596094025228879710893145669136867228748940560101503308617928680920874760917824938589009714909675985261365549781893129784821682998948722658804857564014270477555132379641451523746234364542858444795265867821051141354735739523113427166102135969536231442952484937187110145765403590279934403742007310578539062198387447808478489683321445713868751943506430218453191048481005370614680674919278191197939952061419663428754440643745123718192179998391015919561814675142691239748940907186494231961567945208095146550225231603881930142093762137855956638937787083039069792077346722182562599661501421503068038447734549202605414665925201497442850732518666002132434088190710486331734649651453905796268561005508106658796998163574736384052571459102897064140110971206280439039759515677157700420337869936007230558763176359421873125147120532928191826186125867321579198414848829164470609575270695722091756711672291098169091528017350671274858322287183520935396572512108357915136988209144421006751033467110314126711136990865851639831501970165151168517143765761835155650884909989859982387345528331635507647918535893226185489632132933089857064204675259070915481416549859461637180270981994309924488957571282890592323326097299712084433573265489382391193259746366730583604142813883032038249037589852437441702913276561809377344403070746921120191302033038019762110110044929321516084244485963766983895228684783123552658213144957685726243344189303968642624341077322697802807318915441101044682325271620105265227211166039666557309254711055785376346682065310989652691862056476931257058635662018558100729360659876486117910453348850346113657686753249441668039626579787718556084552965412665408530614344431858676975145661406800700237877659134401712749470420562230538994561314071127000407854733269939081454664645880797270826683063432858785698305235808933065757406795457163775254202114955761581400250126228594130216471550979259230990796547376125517656751357517829666454779174501129961489030463994713296210734043751895735961458901938971311179042978285647503203198691514028708085990480109412147221317947647772622414254854540332157185306142288137585043063321751829798662237172159160771669254748738986654949450114654062843366393790039769265672146385306736096571209180763832716641627488880078692560290228472104031721186082041900042296617119637792133757511495950156604963186294726547364252308177036751590673502350728354056704038674351362222477158915049530984448933309634087807693259939780541934144737744184263129860809988868741326047215695162396586457302163159819319516735381297416772947867242292465436680098067692823828068996400482435403701416314965897940924323789690706977942236250822168895738379862300159377647165122893578601588161755782973523344604281512627203734314653197777416031990665541876397929334419521541341899485444734567383162499341913181480927777103863877343177207545654532207770921201905166096280490926360197598828161332316663652861932668633606273567630354477628035045077723554710585954870279081435624014517180624643626794561275318134078330336254232783944975382437205835311477119926063813346776879695970309833913077109870408591337464144282277263465947047458784778720192771528073176790770715721344473060570073349243693113835049316312840425121925651798069411352801314701304781643788518529092854520116583934196562134914341595625865865570552690496520985803385072242648293972858478316305777756068887644624824685792603953527734803048029005876075825104747091643961362676044925627420420832085661190625454337213153595845068772460290161876679524061634252257719542916299193064553779914037340432875262888963995879475729174642635745525407909145135711136941091193932519107602082520261879853188770584297259167781314969900901921169717372784768472686084900337702424291651300500516832336435038951702989392233451722013812806965011784408745196012122859937162313017114448464090389064495444006198690754851602632750529834918740786680881833851022833450850486082503930213321971551843063545500766828294930413776552793975175461395398468339363830474611996653858153842056853386218672523340283087112328278921250771262946322956398989893582116745627010218356462201349671518819097303811980049734072396103685406643193950979019069963955245300545058068550195673022921913933918568034490398205955100226353536192041994745538593810234395544959778377902374216172711172364343543947822181852862408514006660443325888569867054315470696574745855033232334210730154594051655379068662733379958511562578432298827372319898757141595781119635833005940873068121602876496286744604774649159950549737425626901049037781986835938146574126804925648798556145372347867330390468838343634655379498641927056387293174872332083760112302991136793862708943879936201629515413371424892830722012690147546684765357616477379467520049075715552781965362132392640616013635815590742202020318727760527721900556148425551879253034351398442532234157623361064250639049750086562710953591946589751413103482276930624743536325691607815478181152843667957061108615331504452127473924544945423682886061340841486377670096120715124914043027253860764823634143346235189757664521641376796903149501910857598442391986291642193994907236234646844117394032659184044378051333894525742399508296591228508555821572503107125701266830240292952522011872676756220415420516184163484756516999811614101002996078386909291603028840026910414079288621507842451670908700069928212066041837180653556725253256753286129104248776182582976515795984703562226293486003415872298053498965022629174878820273420922224533985626476691490556284250391275771028402799806636582548892648802545661017296702664076559042909945681506526530537182941270336931378517860904070866711496558343434769338578171138645587367812301458768712660348913909562009939361031029161615288138437909904231747336394804575931493140529763475748119356709110137751721008031559024853090669203767192203322909433467685142214477379393751703443661991040337511173547191855046449026365512816228824462575916333039107225383742182140883508657391771509682887478265699599574490661758344137522397096834080053559849175417381883999446974867626551658276584835884531427756879002909517028352971634456212964043523117600665101241200659755851276178583829204197484423608007193045761893234922927965019875187212726750798125547095890455635792122103334669749923563025494780249011419521238281530911407907386025152274299581807247162591668545133312394804947079119153267343028244186041426363954800044800267049624820179289647669758318327131425170299

6923488962766844032326092752496035799646925650493681836090032380929345958897069536534 94
0603402166544375589004563288225054525564056448246515187547119621844396582533754388856909
4113031509526179378002974120766514793942590298969594699556576121865619673378623625612 52
1632086286922210327488921865436480229678070576561514463204692790682120738837781423356 28
2360896320806822246801224826117718589638140918390367367222088832151375560037279839400 41
5297002878307667094447456013455641725437090697939612257142989467154357846878861444581 23
1459357198492252847160504922124247014121478057345510500801908699603302763478708108175 45
0119307141223390866393833952942578690507643100638351983438934159613185434754649556978 10
3829309716465143840700707360411237359984345225161050702705623526601276484830840761183 01
3052793205427462865403603674532865105706587488225698157936789766974220575059683440869 73
5020141020672358502007245225632651341055924019027421624843914035998953539459094407046 91
2091409387001264560016237428802109276457931065792295524988727584610126483699989225695 96
8815920560010165525637567856672279661988578279484885583439751874454551296563443480396 64
2055798293680435220277098429423253302257634180703947699415979159453006975214829336655 56
6156787364005366656416547321704390352132954352916941459904160875320186837937023488868 94
7915107163785290234529244077365949563051007421087142613497459561513849871375704710178 79
5731042296906667021449863746459528082436944578977233004876476524133907592043401963403 91
1473202338071509522201068256342747164602433544005152126693249341967397704159568375355 51
6673027390074972973635496453328886984406119649616277344951827369558822075735517665158 98
5519098666539354948106887320685990754079234240230092590070173196036225475647894064754 83
4664776041146323390565134330684495397907090302346046147096169688688501408347040546074 29
5869913829668246818571031887906528703665083243197440477185567893482308943106828702722 80
9736248093996270607472645539925399442808113736943388729406307926159599546262462970706 25
9484556903471197299640908941805953439325123623550813494900436427852713831591256898929 51
9642728757394691427253436694153236100453730488198551706594121735246258954873016760029 88
6592578662856124966552353382942878542534048308330701653722856355915253478445981831341 12
9001999205981352205117336585640782648494276441137639386692480311836445369858917544264 73
9988228462184490087776977631279572267265556259628254276531830013407092233436577916012 80
9317940171859859993384923549564005709955856113498025249906698423301735035804408116855 26
5311709957089942732870925848789443646005041089226691783525870785951298344172953519537 88
5534573742608590290817651557803905946408735061232261120093731080485485263572282576820 34
1605048466277504500312620080079980492548534694146977516493270950493463938243222718851 59
7405470214828971117779237612257887347718819682546298126868581705074027255026332904497 62
7789442362167411918626943965067151577958675648239939176042601763387045499017614364120 46
9218237076488783419689686118155815873606293860381017121585527266830082383404656475880 40
5138080163363887421637140643549556186896411228214075330265510042410489678352858829024 36
7090488711819090949453314421828766181031007354770549815968077200947469613436092861484 94
1785017180779306810854690009445899527942439813921350558642219648349151263901280383200 10
9773868066287792397180146134324457264009737425700735921003154150893679300816998053652 02
7600727749674584002836240534603726341655425902760183484030681138185510597970566400750 94
2608788573579603732451414678670368809880609716425849759513806930944940151542222194329 13
0217391253835591503100333032511174915696917450271494331515588540392216409722910112903 55
2181576282328318234254832611191280092825256190205263016391147724733148573910777587442 53
8761174657867116941477642144111126358355387136101102326798775641024682403226483464176 63
6980663785768134920453022408197278564719839630878154322116691224641591177673225326433 56
8614618654522268126887268445968442416107854016768142080885028005414361314623082102594 17
3756238994207571362751674573189189456283525704413354375857534269869947254703165661399 19
9968262824727064133622217892390317608542894373393561889165125042440400895271983787386 48
0584726895462438823437517885201439560057104811949884239060613695734231559079670346149 14
3447886360410318235073650277859089757827273130504889398900992391350337325085598265586 70
8924261242947367019390772713070686917092646254842324074855036608013604668951184009366 86
0954632500214585293095000090715105823626729326453738210493872499669933942468551648326 11
3414611068026744663733437534076429402668297386522093570162638464852851490362932019919 96
8828517183953669134522244470804592396602817156551565666111359823112250628905854914509 71
5755390024393153519090210711945730024388017661503527086260253788179751947806101371500 44
8991721002220133501310601639154158957803711779277522597874289191791552241718958536168 05
9474123419339842021874564925644346239253195313510331147639491199507285843065836193536 93
2969928983791494193940608572486396883690326556436421664425760791471086998431573374964 88
3529276932822076294728238153740996154559879825989109371712621828302584811238901196822 14
2945766758071865380650648702613389282299497257453033283896381843944770779402284359883 41
0035838542389735424395647555684095224844554139239410001620769363684677641301781965937 99
7155746854194633489374843912974239143365936041003523437770658886778113949861647874714 07
9326385873862473288964564359877466763847946650407411182565837887845485814896296127399 84
1344272608606187245545236064315371011274680977870446409475828034876975894832824123929 29
6058294861919667091895808983320121031843034012849511620353428014412761728583024355983 00
3204202451207287253558119584014918096925339507577840006746552603144616705082768277222 35
3419110263416315714740612385042584598841990761128725805911393568960143166828317632356 73
2541707342081733223046298799280490851409479036887868789493054695570307261900950207643 34
9335910602454508645362893545686295853131533718386826561786227363716975774183023986006 59
1481616404944965011732131389574706208847480236537103115089842799275442685327797431139 51
4357417221975979935968525228574526379628961269157235798662057340837576687388426640599 09
9350500081337543245463596750484423528487470144354541957625847356421619813407346854111 76
6883118654489377697956651727966232671481033864391375186594673002443450054499539974237 23
2871249483470604406347160632583064982979551010954183623503030945309733583446283947630 47
7564501500850757894954893139394489921612552559770143685894358587752637962559708167764 38
0012543650237141278346792610199558522471722017772370041780841942394872540680155603599 83
9054898572354674564239058585021671903139526294455439131663134530893906204678438778505 42
3939052473136201294769187497519101147231528932677253391814660730008902776896311481090 22

0972452075916729700785058071718638105496797310016787085069420709223290807038326345345203
8027860990556900134137182368370991949516489600755049341267876436746384902063964019766685
5923356546391383631857456981471962108410809618846054560390384553437291414465134749407
8488442377217515433426030669883176833100113310869042193903108014378433415137092435301
36

6574869535878967003192586625941051053358438465602339179674926784476370847497833365557900
7384191473198862713525954625181604342253729962863267496824058060296421146386436864224 7
2488728343417044157348248183330164056695966886676956349141632842641497453334999948000 26
6998758881593507357815195889900539512085351035726137364034367534714104836017546488300 40
7846416745216737190483109676711344349481926268111073994825060739495073503169019731852 11
9552635632584339099822498624067031076831844660729124874754031617969941139738776589986 85
5417031884778867592902607004321266617919223520938227878880988633599116081923535557046 46
3491132085918979613279131975649097600013996234445535014346426860464495862476909434704 82
9329414041114654092398834443515913320107739441118407410768498106634724104823935827401 94
4935665161088463125678529776973468430306146241803585293315973458303845541033701091676 77
6374276210213701354854450926307190114731848574923318167207213727935567952844392548156 09
1372812840633303937356242001604566455741458816605216660873874804724339121295587776390 69
6903707882852775389405246075849623157436917113176134783882719416860662572103685132156 64
7800147675231039357860689611125996028183930954870905907386135191459181951029732787557 10
4972901148717189718004696169777001791391961379141716270701895846921434369676292745910 99
4006008498356842520191559370370101104974733949387788598941743303178534870760322198297 05
7975119144051099423588303454635349234982688362404332726741554030161950568065418093940 99
8202060999414021689090070821330723089662119775530665918814111915778362729274615618571 037
2172471009521423696483086410259288745799932237495519122195190342445230753513380685680 73
5446499512720317448719540397610730806026990625807602029273145525207807991418429063884 43
7349968145827337207266391767020118300464819000241308350884658415214899127610651374153 94
3565721139032857491876909441370209051703148777346165287984823533829726013611098451484 18
2380812054099612527458088109948697221612852489742555551607637167505489617301680961380 38
1191436114399210638005083214098760459930932485102516829446726066613815174571255975495 35
8023998314698220361338082849935670557552471290274539776214049318201465800802156653606 77
6550878380430413431059180460680083459113664083488740800574127258670479225831912741573 90
8091438313845642415094084913391809684025116399193685322555733896695374902662092326131 88
5589158083245557194845387562878612885900410600607374650140262782402734696252821717494 15
8233174923968353013617865367376064216677813773995100658952887742766263684183068019080 46
0984980946976366733566228291513235278880615776827815958866918023894033307644191240341 20
2231636857786035727694154177882643523813190502808701857504704631293335375728538660588 89
0458311145077394293520199432197117164223500564404297989208159430716701985746927384865 38
3343614579463417592257389858800169801475742054299580124295810545651083104629728293758 41
6116253256251657249807849209989799062003593650993472158296517413579849104711166079158 74
3698654122234834188772292944633517865385673196255985202607294767407261676714557364981 21
0567771689348491766077170527718760119990814411305864557791052568430481144026193840232 24
7093924980293355073184589035539713308844617410795916251171486487446861124760542867343 67
0904667846867027409188101424971114965781772427934707021668829561087779440504843752844 33
7510882826477197854000650970403302186255614733211777117441335028160884035178145254196 43
2030957601869464908868154528562134698835544456024955666843660292221951248309106053772 019
8021831010327041783866544718126039719068846237085751808003532704718565949947612424811 09
9928867915896904956394762460842406593094862150769031498702067353384834955083636601784 87
7106080980426924713241000946401437360326564518456679245666955100150229833079849607994 98
8249706172367449361226222961790814311414660941234159359309585407913908720832273354957 20
8075716517187659944985693795623875551617575438091780528029464200447215396280746360211 32
9425591600257073562812638733106005891065245708024474937543184149401482119996276453106 80
0663118382376163966318093144467129861552759820145141027560068929750246304017351489194 57
6360789352855505317331416457050499644389093630843874484783961684051845273288403234520 24
7056851646571647713932377551729479512613239822960239454857975458651745878771331813875 29
5980941217422730035229650808917770506825924882232215493804837145478164721397682096332 05
0830564792048208592047549985732038887639160199524091893894557676874973085695595801065 95
2650303626615975066222508406742889826590751063756356996821151094966974458054728869363 10
2036782325018232370845979011154847208761821247781326633041207621658731297081123075815 98
2124863980721240786887811450165582513617890307086087019897588980745664395515741536319 31
9198107057533663373803827215279884935039748001589051942087971130805123393322190346624 99
1716915094854140187106035460379464337900589095772118080446574396280618671786101715674 09
6766208029576657705129120990794430463289294730615951043090222143937184956063405618934 25
1305726829146578329334052463502892917547087256484260034962961165413823007731332729830 50
0160256724014185152041890701154288579920812198449315699905918201181973350012618772803 68
1248199587707020753240636125931343859554254778196114293516356122349666152261473539967 40
5158499860355295332924575238881013620234762466905581643896786309762736550472434864307 12
1849437348530060638764456627218666170123812771562137974614986132874411771455244470899 71
4452288566294244023018479120547849857452163469644897389206240194351831008828348024924 90
8540307786387516591130287395878709810077271827187452901397283661484214287170553179654 30
7650453432460053636147261818096997693348626407743519992868632383508875668359509726557 48
1543194019557685043724800102041374983187225967738715495839971844490727914196584593008 39
4263702087563539821696205532480321226749891140267852859967340524203109179789990571882 19
4939132075343170798002373659098537552023891164346718558290685371189795262623449248339 24
9634244971465684659124891855662958932990903523923333364743520370770101084388003290759 83
4217018554228386161721041760301164591878053936744747205998502358289183369292233732399 94
8043710841965947316265482574809948250999183300697656936715968936449334886474421350084 07
0066088359723503953234017958255703601693699098867113210979889707051728075585519126993 06
7309925070407024556850778679069476612629808225163313639952117098452809263037592242674 25
7559989289278370474445218936320348941552104459726188380030067761793138139916205806270 16
5102445886924764924689192461212531027573139084047000714356136231699237169484813255420 09
1453041037135453296620639210547982439212517254013231490274058589206321758949434548906 84
6399313757091034633271415316223280552297297953801880162859073572955416278867649827418 61
6421878988574107164906919185116281528548679417363890665388576422915834250067361245384 91

pi to two million places

6067413734017357277995634104332688356950781493137800736235418007061918026732855119194 26
7609122103598746924117283749312616339500123959924050845437569850795704622266461900010 35
0049018303415354584283376437811198855631877779253720116671853954183598443830520376281 94
4076159410682071697030228515225057312609304689842343315273213136121658280807521263154 77
3060442377475350595228717440266638914881717308643611138906942027908814311944879941715 40
4210341219084709408025402393294294549387864023051292711909751353600092197110541209668 31
1151632870542302847007312065803262641711616595761327235156666253667271899853419989523 68
8483099930275741991646384142707798870887422927705389122717248632202889842512528721782 60
3050099451082478357290569198855546788607946280537122704246654319214528176074148240382 78
3582971930101788834567416781139895475044833931468963076339665722672704339321674542182 45
5706252479721997866854279897799233957905758189062252547358220523642485078340711014498 04
7872669199018643882293230538231855973286978092225352959101734140733488476100556401824 23
9219269506208318381454698392366461363989101210217709597670490830508185470419466437131 22
9969235889538493013635657618610606222870559942337163102127845744646398973818856674626 08
7948201864748767272722206267646533809980196688368099415907577685263986514625333631245 05
3640261056960551318381317426118442018908885319635698696279503673842431301133175330532 98
0201668881748134298868158557781034323175306478498321062971842518438553442762012823457 07
1698853051832617964117857960888815032960229070561447622091509473903594664691623539680 92
0139457817589108893199211226007392814916948161527384273626429809823406320024402449589 44
5612916704950823581248739179964864113348032475777521970893277226234948601504665268143 98
7705161531702669692970492831628550421289814670619533197026950721437823047687528028735 41
2616639170824592517001071418085480063692325946201900227808740985977192180515853214739 26
5325155903541020928466592529991435379182531454529059841581763705892790690989691116438 11
8780943537152133226144362531449012745477269573939348154691631162492887357471882407150 39
9500944673195431619385548520766573882513963916357672315100555603726339486720820780865 37
3494244011579966750736071115935133195919712094896471755302453136477094209463569698222 66
7377520994516845064362382421185353488798939567318780660610788544000550827657030558744 85
4180577889171920788142335113866292966717964346876007704799953788338787034871802184243 73
4211227394025571769081960309201824018842705704609262256417837526526335832424066125331 15
2942345796556950250681001831090041124537901533296615697052237921032570693705109083078 94
7999900499939532215362274847660361367769797856738658467093667958858378879562594646489 13
7665219958828693380183601193236857855855819555604215625088365020332202451376215820461 81
0670519533065306060650105488716724537794283133887163139559690583208341689847606560711 83
4713621812324622725884199028614208728495687963932546428534307530110528571382964370999 03
5694888528519040295604734613113826387889755178856042499874831638280404684861893818959 05
4203988987265069762020199554841265000539442820393012748163815853039643992547020167275 93
2857436666164411096256633730540921951967514832873480895747777527834422109107311135182 80
4603634719818565557295714474768255285786334934285842311874944000322969069775831590385 80
3935352135886007960034209754739229673331064939560181223781285458431760556173386112673 47
8074585067606304822940965304111830667108189303110887172816751957967534718853722930961 61
4320400638132246584111115775835858113501856904781536893813771847281475199835050478129 77
1859908470762197460588742325699582889253504193795826061621184236876851141831606831586 79
9460165205774052942305360178031335726326705479033840125730591233960188013782542192709 47
6733719198728738524805742124892118347087662966720727232565056512933312605950577772754 24
7124164831283298207236175057467387012820957554430596839555568686118839713552208445285 26
4008125202766555767749596962661260456524568408613923826576858338469849977872670655519 18
5446869846947849573462260629421962455708537127277652309895545019303773216664918257815 46
7729200521266714346320963789185232321501897612603437368406719419303774688099929687758 24
4104787812326625318184596045385354383911449677531286426092521153767325886672260404252 34
9108702695809964759580579466397341906401003636190404203311357933654242630356145700901 12
4480089002080147805660371015412232889146572239314507607167064355682743774396578906797 26
8743847307634645167756210309860409271709095128086309029738504452718289274968921210667 00
8164858339553773591913695015316201890888748421079870689911480466927065094076204650277 25
2865072890532854856143316081269300569378541786109696920253886503457718317668688592368 14
8847527649846882194973972970773718718840041432312763650481453112285099002074240925585 92
5292610302106736815434701525234878635164397623586041919412969769040526483234700991115 42
4260127343802208933109668636789869497799400126016422760926082349304118064382913834735 46
7972539926233879158299848645927173405922562074910530853153718291168163721939518870095 77
8818158685046450769934394098743351443162633031724774748689791820923948083314397084067 30
8407958935810896656477585990556376952523265361442478023082681183103773588708924061303 13
3647737101162821461466167940409051861526036009252194721889091810733587196414214447865 48
9952858234394705007983038853886083103571930600277119455802191194289992272235345870756 62
4692617766317885514435021828702668561066500353105021631820601760921798468493686316129 37
2795187307897263735371715025637873357977180818487845886650433582437700414771041493492 74
3845758710715973155943942641257027096512510811554824793940359768118811728247215825010 94
9609662539339538092219559191818885526780621499231727631632183398969380756168559117529 984
5013206712939240414459386239880938124045219148483164621014738918251010909677386906640 41
5897361047643650006807710565671848628149637111883219244566394581449148616550049567698 26
9030891118568798692947051352481609174324301538368470729289898284602223730145265567989 86
2776796809146979837826876431159883210904371561129976652153963546442086919756737000573 87
6497843768628768179249746943842746525631632300555130417422734164645512781278457777245 75
2038654375428282567141288583454443513256205446424101103795546419058116862305964476958 70
5407214198521210673433241075676757581845699069304604752277016700568454396923404171108 98
8899341635058515788735343081552081177207188037910404698306957868547393765643363197978 68
0367187307969392423632144845035477631567025539006542311792015346497792906624150832885 83
9529054263768766896880503331722780018588506973623240389470047189761934734430843744375 99
2503417880797223585913424581314404984770173236169471976571535319775499716278566311904 69
1260918259124989036765417697990362375528652637573376352696934435440047306719886890196 81

4742876779086697968852250163694985673021752313252926537589641517147955953878427849986645630287883196209983049451987439636907068276265748581043911223261879405994155406327013198989570376110532360629867480377915376751158304320849872092028092975264981256916342500052290887264692528466610466539217148208013050229805263783642695973370705392278915351056888393811324975707133102950443034671598944878684711643832805069250776627450012200352620370946602341464899839025258883014867816219677519458316771876275720050543979441245990077115205154619930509838698254284640725554092740313257163264079293418334214709041254253352324802193227707535554679587163835875018159338717423606155117101312352563348582036514614187004920570437201826173319471570086757853933607862273955818579758725874410254207710547536129404746010009409544495966288148691590389907186598056361713769222729076419775517772010427649694961105622059250242021770426962215495872645398922769766031052498085575947163107587013320886146326641259114863388122028444069416948826152957762532501987035987067438046982194205638125583343642194923227593722128905642094308235254408411086454536940496927149400331978286131818618881111840825786592875742638445005994422956858646048103301538891149948693543603022181094346676400002236255057363129462629609619876056425996394613869233083719626595473923462413459779574852464783798079569319865081597767535055391899115133525229873611277918274854200868953965835942196333150286956119201229888988700607999279541118826902307891310760361763477948943203210277335941690865007193280401716384064498787175375678118532132840821657110754952829497493621460821558320568723218557406516109627487437509809223021160998263303391546949464449100451528092508974507489676032409076898365294065792019831526541065813682379198409064571246894847020935776119313998024681340520039478194986620262400890215016616381353838151503773502296607462795291038406868556907015751662419298724448271942933100485482445458071889763300323252582158128032746796200281476243182862217105435289834820827345168018613171959332471107466222850871066611770346535283957762599774467218571581612641114327179434788599089280848669491413909771673690027775850268664654056595039486784111079011610400857274456293842549416759460548711723594642910585090995021495879311219613590831588262068233215615308683373083817327932819698387508708348388046388478441884003184712697454370937329836240287519792080232187874488287284372737801782700805878241074935751488997891173974612932035108143270325140903048746226294234432757126008664250833318768865075642927160552528954492153765175149219636718104943531785838345386525565664065725136357506435323650893679043170259787817719031486796384082881020946149007971513771709906195496964007086766710233004867263147551053723175711432231741141168062286420638890621019235522354671166213749969326932173704310598722503945657492461697826097025335947502091383667377289443869640002811034402608471289900074680776484408871134135250336787731679770937277868216611786534423173226463784769787514433209534000165069213054647689098505020301504488083426184520873053097318949291642532293361243151430657826407028389840984160295030924189712097160164926561341343342229882790992178604267981245728534580133826099587717811310216734025656274400729683406619848067661580502169183372368039902793160642043681207990031626444914619021945822969099212278855394878353830564686488165556229431567312827439082645061162894280350166133669782405177015521962652272545585073864058529983037918035043287670380925216790757120406123759632768567484507915114731344000183257034492090971243580944790046249431345502890068064870429353403743603262582053579011839564908935434510134296961754524957396062149028872893279252069653538639644322538832752249960598697475988232991626354597332444516375533437749292899058117578635555626937426910947117002165411718219750519831787137106051063795558588905568852887989084750915764639074693619881507814685262133252473837651192990156109189777922008705793396463827490680698769168197492365624226087154176100430608904377976678519661891404144925270480881971498801542057787006521594009289777601330756847966992955433656139847738060394368895887646054983871478968482805384701730871117761159663505039979343869339119789887109156541709133082607647406305711411098839388095481437828474528838368079418884342666222070438722887413947801017721392281911992365405516395893474263953824829609036900288359327745855060801317988407162446563997948275783650195514221551339281978226984278638391679715091262410548725700924070045488485692950448110738087996547481568913935380943474556972128919827177020766613602489581468119133614121258783895577357194986317210844398901423948496659251731388171602663261931065366535041473070804414939169363262373767777095850313255990095762731957308648042467701212327020533742667053142448208168130306397378736642483672539837487690980602182785786216512738563513290148903509883270617258932575363993979055729175160097615459044771692265806315111028038436017374742152476085152099016158582312571590733421736576267142390478279587281505095633092802668458937649649770232973641319060982740633531089792464242134583740901169391964250459128813403498810635400887596820054408364386516617880557608956896727531538081942077332597917278437625661184319891025007491829086475149794003160703845549465385946027452447466812314687943441610993338908992638411847425257044572517459325738989565185716575961481266020310797628254165590506042479114016957900338356574869252800743025623419498286467914476322774005529460903940177536335655471931000175430047504719144899841040015867946179241610016454716551337074073950260442769538553834397550548871099785205401175169747581344926079433689543783221172450687344231989878844128542064742809735625807066983106979935260693392135685881391214807354728463227784908087002467776303605551232386656295178853719673034634701222939581606792509153217489030840886516061119011498443412350124646928028805996134283511884715449771278473361766285062169778717743824362565711779450064477718370221999106695021656757644044997940765037999954845002710665987813603802314126836905783190460792765297277694043613023051787080546511542469395265127101052927070306673024447125973939950514628404767431363739978259184541176413327906460636584152927019030276017339474866960348694976541752429306040727005059039503148522921392575594845078867977925253931765156416197168443524369794447355964260633391055126826061595726217036698506473281266724521989060549880280782881429796336696744124805982192146339565745722102298677599746738126069367069134081559412016115960190237753525556300606247983261249881288192937343476862689219239777833910733106588256813777172328315329082525092733047850724977139448333892552081175608452966590553940965568541706001179857293813998258319293679100391844099286575605993598910002969864460974714718

7010153128376263114677420914557404181590880006494323785583930853082830547607679952435573
9163122188605754967383224319565065546085288120190236364471270374863442172725787950342845
8631294491631847534753143504139209610879605773098720135248407505763719925365047090858255
1393686346386336804289176710760211115982887553994012007601394703366179371539630613986366
5549221374159790511908358829009765664730073387931467891318146510931676157582135142486044
4229244530411316065270097433008849903467540551864067734260358340960860553374736276093566
5885310976099423834738222208729246449768456057956251676557408841032173134562773585605233
5823638953203853402484227337163912397321599544082842166663602329654569470357718487344200
3422770665383738750616921276801576618109542009770836360436111059240911788954033802142655
2394892968643980892611463541457153519434285072135345301831587562827573389826889852355777
9929572764522939156747756667605108788764845349363606827805056462281359888587925994094644
4604170520447004631513797543173718775603981596264750141090665886616218003826698996196555
8058720863972117699521946678985701179833244060181157565807428418291061519391763005919433
1443460515404771057005433900018245311773371895585760360718286050635647997900413976180899
5536366960316219311325022385179167205518065926351803625121457592623836934822266589557699
9466049193811248660909979812857182349400661555219611220720309227764620099931524427358944
8871057662389469388944649509396033045434084210246240104872332875008174917987554387938733
8143989423801176270083719605309438394006375611645856094312951759771393539607432279248922
2126704580818331376416581826956210587289244774003594700926866265965142205063007859200244
8829186083974373235384908396432614700053242354064704208949921025040472678105908364400744
6638002087012666420945718170294675227854007450855237772089058168391844659282941701828822
3301497155423523591177481862859296760504820386434310877956289292540563894662194826871100
4282816389397571175778691543016505860296521745958198887868040811032843273986719862130622
0555985526603640504628215230615459447448990883908199973874745296981077620148713400012255
3552224669540931521311533791579802697955571050850747387475075806876537644578252443263800
4614304288923593485296105826938210349800040524840708440356116781717051281337880570564344
5061611933042444079826037795119854869455915205196009304127100727784930155503889536033822
6192934379708187432094991415959339636811062755729527800425486306005452383915106899891355
7882001941178653568214911852820785213012551851849371150342215954224451190020739353962744
0020811046553020793286725474054365271759589350071633607632161472581540764205302004534011
8357233829266191530835409512022632916505442612361919705161383935732669376015691442994499
4374485680977569630312958871916112929468188493633864739274760122696415884890096571708611
6059814720446742866420876533479985822209061980217321161423041947775499073873856794118988
2466091309169177227420723336763503267834058630193019324299639720444517928812285447821199
5353089891012534297552472763573022628138209180743974867145359077863353016082155991131411
4420509144729353502223081719366350934686585865631485557586244781862010871188976065296988
9926932817870557643514338206014107732926106343152533718224338526352021773544071528189811
3769875515757454693972715048846979361950047772097056179391382898984532742622728864710888
8327017372325881824465843624958059256033810521560620615571329915608489206434030339526222
6345145428367869828807425142256745180618414956468611163540497189768215422772247947403355
7152743681940989205011365340012384671429655186734415374161504256325671343024765512521922
1803578016924032669954174608759240920700466934039651017813485783569444076047023254075555
5776472845075182689041829396611331016013111907739863246277821902365066037404160672496244
9013743321724645409741299557052914243820807609836482346597388669134991978401310801558133
4397919485283043673901248208244481412809544377389832005986490915950532285791457688496255
7866588599917986752055455809900455646117875524937012455321717019428288461740273664997844
7550829422802023290122163010230977215156944642790980219082668986883426307160920791408511
9769523555348865774342527753119724743087304361951139611908003025587838764420608504473066
3129927788894272918972716989057592524467966018970748296094919064876469370275077386643233
9191904225429023531892337729316673608699622803255718530891928440380507103006477684786322
4319100022392978525537237556621364474009676053943983823576460699246526008909062410590422
1545392790441152958034533450025624410100635953003959886446616959562635187806068851372344
6270799732723313469397145628554261546765063246567662027924520858134771760852169134094655
2030767339184114750414016892412131982688156866456148538028753933116023229255561894104299
9533564009578649534093511526645402441877594931693056044868642086275720117231952640502300
9977456764783848897346431721598062678767183800524769688408498918508614900343240347674266
8624595239589035858213500645099817824463608731775437885967767291952611121385919472545144
0030118050343787527766440276261894101757687268042817662386068047788524288743025914524700
7395054652513533945959878961977891104189029294381856720507096460626354173294464957661266
5195349570186001541262396228641389779673332907056737696215649818450684226369036784955599
7002607986799626101903933126376855696876702929537116252800554310078640872893922571451244
8113577862766490242516199027747109033593330930494838059785662884478744146984149906712377
6478958226329490467981208998485716357108783119184863025450162092980582920833481363840544
2172005612198935366937133673339246441612522319694347120641737549121635700857369439730599
7970971972666664226743111776217640306868131035189911227133972403688700099686292254646500
0638528862039380050477827691283560337254825579391298525150682996910775425764748832534144
1213280062671709400909822352965795799780301828242849022147074811112401860761341515038755
6983091865278065889668236252393784527263453042041880250844236319038331838455052236799233
5775292910692504326144695010986108889991465855188187358252816430252093928525807796973766
2084563748211443398816271003170315133440230952635192958868069082135585368016100021374088
5115448491268584126869589917414913382057849280069825519574020181810564129725083607035688
5105533178784082900004155251186577945396331753853209214972052660783126028196116485809866
8458752512999740409279768317663991465538610893758795221497173172813151793290443112181588
7102351874075722210012376872194474720934931232410706508061856237252673254073332487575444
8296757345001932190219911996079798937338367324257610393898534927877747398050808001554477
6406105352220232540944356771879456543040673589649101761077594836454082348613025471847644
8518957583667439979150851285802060782055446299172320202822291488695939972997429747115533
7185892423849385585859540743810488262464878805330427146301194158989632879267832732245611

pi to two million places

0385219701113046658710050008328517731177648973523092666123458887310288351562644602367199664455472760831011878838915114934093934475007302585581475619088139875235781233134227986650352272536717123075686104500454897036007956982762639234410714658489578024140815840522953693749971066559489445924628661996355635065262340533943914211127181069105229002465742360413009369188925586578466846121567955425660541600507127664176605687427420032957716064344860620123982169827172319782681662824993871499544913730205184366907672357740005393266262276032365975171892590180110429038427418550789488743883270306328327996300720069801224436511639408692222074532024462412115580435454206421512158505689615735641431306888344318528085397592773443365538418834030351782294625370201578215737326552318576355409895403323638231921989217117744946940367829618592080340386757583411151882417743914507736638407188048935825686854201164503135763335550944031923672034865101056104987272647213198654343545040913185951314518127643731043897250700498198705217627249406521461995923214231443977654670835171474936798618655279171582408065106379950018429593879915835017158075988378496225739851212981032637937621832245659423668537679911314010804313973233544909082491049914332584329882103398469814171575601082970658306521134707680368069532297199059990445120908727577622535104090239288877942463048328031913271049547859918019696783532146444118926063152661816744319355081708187547705080265402529410921826485821385752668815558411319856002213515888721036569608751506318753300294211868222189377554602722729129050429225978771066787384000061677215463844129237119352182849982435092089180168557279815642185819119749098573057033266764646072875743056537260276898237325974508447964954564803077159815395582777913937360171742299602735310276871944944491793978514463159731443535185049141394155732938204854212350817391254974981930871439661513294204591938010623142177419918406018034794988769105155790555480695387854006645337598186284641990522045280330626369562649091082762711590385699505124652999606285544383833032763859980079292284665950355121124528408751622906026201185777531374794936205549640107300134885315073548735390560290893352640071327473262196031177343394367338575912450814933573691166454128178817145402305475066713651825828489809951213919399563324133655677709800308191027204099714868741813466700609405102146269028044915964654533010775469541308871416531254481306119240782118869005602778182423502269618934435254763357353648561936325441775661398170393063287216690572225974520919291726219984440964615826945638023950283712168644656178523556516412771282691868861557271620147493405227694659571219831494338162211400693630743044417328478610177774383797703723179525543410722344551255558999864618387676490397246116795901810003509892864120419516355110876320426761297982652942588295114127584126273279079880755975185157684126474220947972184330935297266521001566251455299474512763155091763673025946213293019040283795424632325855030109670692272022707486341900543830265068121414213505715417505750863990767394633514620908288893493837643939925690060406731142209331219593620298297235116325938677224147791162957278075239505625158160313335938231150051862689053065836812998810866326327198061127154885879809348791291370749823057592909186293919501472119758606727009254771802575033773079939713453953264619526999659638565491759045833358579910201271320458390320085387888163363768518208372788513117522776960978796214237216254521459128183179821604411131167140691482717098101545778193920231156387195080502467972579249760577262591332855972637121120190572077140914864507409492671803581515757151405039761096384675556929897038354731410022380258346876735012977541327953206097115450648421218593649099791776687477448188287063231551586503289816422828823274686610659273219790716238464215348985247621678905026099804526648392954235728734397768049577409144953839157556548545905897649519851380100795801078375994577529919670054760225255203445398871253878017196071816407812484784725791240782454436168234523957068951427226975043187363326301110305342333582160933319121880660826834142891041517324721605335584999322454873077882290525232423486153152097693846104258284971496347534183756200301491570327968530186863157248840152663983568956363465743532178349319982554211730846774529708583950761645822963032442432823773745051702856069806788952176819815671078163340526675953942492628075696832610749532339053622309080708145591983735537774874202903901814293731152933464446815121294509759653430628421531944572711861490001765055817709530246887526325011970520947615941676872778447200019278913725184162285778379228443908430118112149636642465903363419454065718354477191244662125939265662030688852005559912123536371822692253178145879259375044144893398160865790087616502463519704582889548179375668104647461410514249887025213993687050937230544773411264135489280684105910771667782123833281026218558775131272117934444820144042574508306394473836379390628300897330624138061458941422769474793166571762318247216835067807648757342049155762821758397297513447899069658953254894033561561316740327647246921250575911625152965456854463349811431767025729566184477548746937846423373723898192066204851189437886822480727935202250179654534375727416391079197295295081294292220534771730418447791567399173841831171036252439571615271466900581470000263301045264354786590329073320546833887207873544476264792529769017091200787418373673508771337697768349634425241994995138831507487753743384945825976556099655595431804092017849718468549737069621208852437701385375768141663272241263442398215294164537800049250726276515078908507126599703670872669276430837722968598516912230503746274431085293430527307886528397733524601746352770320593817912539691562106363762588293757137384075440646896478310070458061344673127159119460843593582598778283526653115106504162329532904777217408355934972375855213804830509000964667608830154061282430874064559443185341375522016630581211103345312074508682433943215904359443031243122747138584203039010607094031523555617276799416002039397509989762933532585557562480899669182986422267750236019325797472674257821111973470940235745722227121252685238429587427350156366009318804549333898974157149054418255973808087156528143010267046028431681923039253529779576586241439270154974087927313105163611913757700892956482332364829826302460797587576774537716010249080462430185652416175665560016085912153455626760219268998285537787258314514408265458348440947846317877737479465358016996077940556870119232860804113090462935087182712593466871276669487389982459852778649956916546402945893506496433580982476596516514209098675520380830920323048734270346828875160407154665383461961122301375945157925269674364253192739003603860823645076269882749761872357547676288995075211480485252795084503395857083813047693788132112367428131948795022806632017002

pi to two million places

603319896719706491637411758548518784840120548446725888514015627250198217190669608126277
854859648183696214107217142149863619187747545096503089570994709343378569816744658282679
119406119560378453978558392407612763441057667510243075598145527861678159496570625597550
743065210853015979080733437360794328667578905334836695554868039134337201564988342208933
999716414797469386969054800891930671380571715058573071488156499207140867582596028760564
597824237702424698053280566327870419267684671162668794634869504645074202193739452592626
686135529406247813612062026364981999994984051438682852589563422643287076632993048917234
007254717641886853513723326678779217383475414800228033929973579361524127558295692768372
312347989894462743304545667900620324205163962825884430854383072014956721064605332385372
031432421126074244858450945804940818209276391400085404220235562602185643489941454399504
109805918179488826280520664410863190016885681551692294862030107388971810077092905904807
490924271410189335428184299959881696609938369616443815288772140852680887574882932587358
099056707558170179491619061140019085537448827262009366856044755965574764856740081773817
033073803054769736097865438593821872205839023444435088674998665060406458743460053318274
362961778625180818931443632512051070946908135864405192295129324500788333987884293393424
351263433652043858129128343452973086529097833006712617981303167943855357262969987403595
704584522308563900989131794759487521263970783759448611394519602867512105616389760088800
927461158608002078033415914517970730368351969777660763737853330120241201120469886092093
390853657732223924124490515327809509558664594776344822699860748132973026309750288121035
177231244650953496536930900186377640940943498373132513218620802148099226855029484546618
147155574447096695301776904342720318927706047177845279391604722815343798035396798614243
709566832214914654380145938292773933960327540480095522318166673803571839327570771420467
238386246178039762923771312095807893638414479298025880655221292620936239306373134966401
866195108115834711733120258058667276399927635790780638188130691563662741254312595899361
196476261014055635033995231403231138196562363271989618372548453337020625634642239527669
435683767613687119629218187545760816170530315907288287007123136663087227549186613957737
305460659974378109876498024140112421427736680827513909593134041558262667895108467761186
659576601659981780894149857549762843878561002637965431783136340251358141611519020964991
335487331311150227006819301359295959716401971960536250335584799809634887180391116128135
959685654788683258564378961731597620024196215528962979048198221994622694871374624447290
934564700285376949588595916067892824910544125159963007813683674902093749157328962700286
568293444313423473512392982591667395034259958689706972673325827359031212887466604514614
878503461428277659916080903986525757172630818334944418201935333850712923457743755793440
621787113300631060033240539916936826037461766385657588775802012293663532702671006812618
251729146082025418928859352444910701382062115538277935652969145765020486432828655579347
072096348073726921411868954673227677513356901901537236690368653891612916888878764075254
934942497334271811788927599315967193547589880979245252623636590363200708544407845447973
482918020820449266706344204375553250505275228337788870408040335319234076856301093477721
256390886404131010738178533383160381352808281190408325644018420537467929926220376987180
180611226244909092426419858208617511771137890516091403815750033664241560952163281971223
350231674226005679412814062172196418427057843289598028823350598282081966662490358577899
403331522748177769528436816300885317696947836905806710648280835980466988410981351586549
069333195223943632879239905348109878302745001720654336990661177845543646877236318444647
680691428280045510746866453928053994091087549391660957316197150331669683099294663491427
987808422572206971488755806374803088629951184731871247772919100702275888934869394562895
158029653721504096031077612898312635899648934102470360366450586872875890514068412381242
473863854279082827338279733268855049358743031602747490631295723497426112215174171531336
186224109138695006888358989623492763173164783400774608866555987333821138299287769114954
921841920877716060684728746736818861675072210172611038306717878566948129487850489430630
861699487987031605158841082823512741535385133658953329486294944950618685147791058046960
390693726626703865129052011378108586161888869479576074135855345851517680519733344334952
301203957707396237713160302428872005373209982530089776189731298178819446717311606472314
762484575519287327828251271824468078242152164695678192940982389262849437602488522790036
202193866964822156280936053731780408637272684266964219299468192149087017075333610947913
818040632873875938482695355830773957614479972700034728801827852813895032179863452161110
666088393140532269449054555278678944175792024400214507801920998044613825478058580484424
164047750315360549065914300781583724301231375115622840158386442708907182848167575271238
467824595343344496220100960710513706084618011875431207254913349942476171156333214089346
091565615506003173842187015702261031019166038870646614388977363187809407115275281746895
764015810470169652475577408916445686777171585005832699434016772021567677240681283665652
641229824394651331973591997094032759385026695574702318132032437164205861410336065245369
391600506449530601612678226489424373971667176612310489750318857321655549883421218028469
125290861014855278152776256237504563757694977343368460156077270355096290493924870884062
810679436224187047470083688426710225583024035998416459511224852726336326451140173952480
861946358407837535568856223171155209472230654370926067973510005655549381224575483728457
117973936157561676416928958052572975223385586113883221711073622658162188424431788574887
981090266537934266642169909140565364322493013348679881548866286650523469972355747384248
305904236771432787923164224038777643301926001922847783138376325361210253369358126240868
666997382759773656822279072158324788886423693463961643633087301398142114303060087306661
648036789840913359262934023043249749268878316436026810113095707161419128306865773235326
396536773903176613613159655535849993986005651559219367599777179330197446881483711032065
036931928945214026509154651843099365534933371834252984336799159394174662239003895276738
133306177476295749438687169784537672194935065908757119177208754771071899379608947745126
547575018711948707387367858902006173733210756933022163206284320656711920969505857611739
616323262177089454262146098584102378132158177276022227381334954104810030732751077999489
919779638835307344434575329759142637684054422647842160631227696469671564739990437159033
239065607266441164386054048388471619121090087010191307260710441141432419767968285478855
247794764818029597360494397004795960402927462992035720997619501403483153809477146010563

3344699882082212058728151072918297121191787642488035467231691654185225672923442918712 81
6323259696541354858957713320833991128877591722611527337901034136208561457799239877 83250
8355073019981845902595835598926055329967377049172245493532968330000223018151722657 57875
2405883224908582128008974790932610076257877042865600699617621217684547899644070506 62417
1021332748679623743022915535820078014116534806564748823061500339206898379476625503 65498
2280532966286211793062843017049240230198571997894883689718304380518217441914766042 97524
3725168343541121703863137941142209529588579806015293875275379903093887168357209576 07152
2190027937929278630363726876582268124199338480816602160372215471014300737753779269 90695
8712128928801905203160128586182549441335382078488346531163265040764242839087012101 51942
3196165226842200371123046430067344206474771802135307012409886035339915266792387110 17062
2186588357378121093517977560442563469499978725112544085452227481091487430725986960 20402
7594117894258128188215995235965897918114407765335432175759525553615812800116384672 03193
4650729680799079396371496177431211940202129757312516525376801735910155733815377200 19524
4454362007184847566341540744232862106099761324348754884743453966598133871746609302 05350
7027195298394327142537115576660002578442303107342955153394506048622276496668762407 93243
5319299263925373107689213535257232108088981933916866827894828117047262450194840970 09757
6092098372409007471797334078814182519584259809624174761013825264395513525931188504 56362
6418830033853965243599741693132289471987830842760040136807470390409723847394583489 61865
3979059411859931035616843686921948538205578039577388136067954990008512325944252972 44866
6676683464140218991594456530942344065066785194841776677947047204195882204329538032 63105
3749488312218039127967844610013972675389219511911783658766252808369005324900459741 09470
6877291232821430463533728351995364827432583311914445901780960778288358373011185754 36599
5898272453192531058811502630754257149394302445393187017992360816661130542625399583 38979
4297160207033876781503301028012009599725222228080142357109476035192554443492998676 78178
9104555906301595380976187592035893734197896235893112598390259831026719330418921510 96891
5622506965911982832345550305908173073519550372166587028805399213857603703537710517 80212
8012956684198414036287272562321442875430221090947272107347413497551419073704331827 66261
7727599688882602722524713368335345281669277959132886138176634985772893690096574956 22871
0302436259077241221909430087175569262575806570991201665962243608024287002454736203 63948
4125595488172727247365346778364720191830399871762703751572464992228946793232269361 91776
4161461879561395669956778306829031658969943076733350823499079062410020250613405734 43006
9574547468217569044165154063658468046369262127421107539904218871612761778701425886 48257
7522388918459952337629237791558574454947736129552595222657863646211837759847370034 79714
0820699414558071908021359073226923310083175951065901912129479540860364075735875020 58902
0870457967000705526250581142066390745921527330940682364944159089100922029668052332 52661
9891131184201629163107689408472356436680818216865721968826835840278550078280404345 37101
8365109695178233574303050485265373807353107418591770561039739506264035544227515610 11072
6177937063472380499066692216197119425912044508464174638358993823994651739550900085 94799
9013602667426149429006646711506717542217703877450767356374215478290591101261915755 58702
3895700140511782264698994491790830179547587676016809410013583761357859135692445564 77644
6417866711539195135769610486492249008344671548638305447791433009768048687834818467 27337
5843689272431044740680768527862558516509208826381323362314873333671476452045087662 76149
5038994950480956046098960432912335834885999029452640028499428087862403981181488476 73012
1675416110662999555366819312328742570206373835202008686369131173346973174121915363 32467
4532563087134730279217495622701468732586789173455837996435135880095935087755635624 88104
9385299900767513551352779241242927748856588856651324730251471021057535251651181485 09027
5047684551825209633189906852761443513821366215236889057878669943228881602837748203 55060
1602989400911971385017987168363374413927597364401700701476370665570350433812111357 64150
1845182141361982349515960106475271257593518530433287553778305750956742544268471221 96187
0917856078393614451138333564910325640573389866717812397223751931643061701385953947 43678
4339267098671245221118969084023632741149660124348309892994173803058841716661307304 00675
8838043211155537944060549772170594282151488616567277124090338772774562909711013488 51843
7411869565544974573684521806698291104505800429988795389902780438359628240942186055 62877
8842880212755388480372864001944161425749990427200959520465417059810498996750451193 64711
7277222043610261407975080968697517660023718774834801612031023468056711264476612374 76278
5219024120256994353471622666089367521983311181351114650385489502512065577263614547 36044
2685949807439693233129712737715734709971395229118265348515558713733662912024271430 25037
6326950135091161295299378586468130722648600827088133353819370368259886789332123832 70532
9762585738279009782646054559855513183668884462826513379849166783940976135376625179 82582
4966345877195012438404035914084920973375464247448817618407002356958017741017769692 50778
1489338667255789856458985105689196092439884156928069698335224022563457049731224526 93541
9383700484318335719651662672157552419340193309901831930919658292096965624766768365 96470
1959575473934551433741370876151732367720422738567427917069820454995309591887243493 95240
9444167899884631984550485239366297207977745281439941825678945779571255242682608994 08633
1737153889626288962940211210888442737656862452761213037101730078513571540453304150 79594
4777614359743780374243664697324713841049212431413890357909241603640631403814983148 19052
5172093710396402680899483257229795456404270175772290417323479607361878788991331830 58430
6939482596131871381642346721873084513387721908697510494284376932502498165667381626 06159
4176825250999374167288395174406693254965340310145222531618900923537648637848288134 42098
7004809622717122640748957193900291857330746010436072919094576799461492929042798168 77294
2648772995285843464777538690695014898413392454039414468026362540211861431703125111 75776
4282991464453340892097696169909837265236176874560589470496817013697490952307208268 28878
9073019001825342580534342170592871393173799314241085264739094828459641809361413847 58311
3613057610846236683723769591349261582451622155213487924414504175684806412063652017 03863
3012953277769902311864802006755690568229501635493199230591424639621702532974757311 40942
2018019936803502649563695586642590676268568737211033915679383989576556519317788300 02416
1353956243777784080174881937309502069990089089932808839743036773659552489130015663 32940
7790713961546453408879151030065132193448667324827590794680787981942501958262232039 51312

5201410996053126069655540424867054998678692302174698900954785072567297879476988883109348
8746442640071818316033165551153427615562240547447337804924621495213325852769884733626918
8264917433898782478927846891882805466998230368993978341374758702580571634941356843392939
9606819206177333179173820856243643363535986349449689078106401967407443658366707158692452
2118299789380407713750129085864657890577142683358276897855471768718442772612050926648610
0205153564284063236848180728794071712796682006072755955590404023317874944734645476062818
8954151213916291844429765106694796935401686601005519607768733539651161493093757096855455
5938151378956903925101495326562814701199832699220006639287537471313523642158926512620407
7288771657835840521964605410543544364216656224456504299901025658692727914275293117208279
9393775132610605288123537345106837293989358087124386938593438917571337630072031976081660
0446468393772580690923729752348670291691042636926209019960520412102407764819031601408586
6355842760953708655816427399534934654631450404019952853725200495780525465625115410925243
7991326262713609099402902262062836752132305065183934057450112099341464918433323646569371
7259144893241590062420206128857329261335968087265000456282845575745965921205303413101
1182750130696150983551563200431078460190656549380654252522916199181995960275232770224985
5738824899882707465936355768582560518068964285376850772012220347920993936179268206590142
1656159253067379445689490708532635681968318617722682499114726157320358076462981162440
1331673789278868922903259334986179702199498192573961767307583441709855922217017182571277
7534491508205278430904619460835217402005838672849709411023266953921445461066215006410674
7402070091899119513764669044812672536915371622907913854039375600778351533741677479421
0038400230895185099454877903934612222086506016050035177626483161115332558770507354127924
9909859373473787081194253055121436979749914951860535920403830235716352727630874693219622
1900642608861836761033460022554774778136410126919065696864950126883762969072339612762
8722304114181361006026404403003599698891994582739762411461374480405969706257676472376606
5541618574690527229238228275186799156983390747671146103022776606020061246876477728819096
7916133540198814027579921741676787992316039635694928515136336472195406111717673873725
5572852294005436178517650230754469386930787349911035218253292972604455321079788771144989
8870911511237250604238753734841257086064069052058452122754533848008205302450456517669518
5769132000428167580549248117805198326460324457928297301291053183856368212062155312886
6856495651261389226136706409395333457052698695969235035309422454386527867767302754040270
2246384483553239914751363441044050092330361271496081355490531539021002299595756583705381
2619656831442860579566966221547216956208700137277685369608407048333251327931122325071
4863020695124539500373572334680709465648308920980153487870563349109236605755405086411152
1441481434630437273271045027768661953107858323334857840297160925215326092558932655600672
1243594642550659967717703884453961816328796144608177892721718369088801267782074301064
2252463480745430047649288555340906218515365435547412547615276977266776977277058315801412
1856880117050283652755432148034880044429799980621579045641619572127845089284898064264
9742709057912906921780729876947797511244730599140605062994689428093103421641662993561482
8130998870745292716048433630818404126469637925843094185442216359084576146078558562473814
9314270782662151855416038702068769804617474008083243436653823545551094494984310934947
5994467267366535251766270677219418319197719637801570216993367508376005716345464367177672
3387588643405644871566964321041282595645349841388412890420682047007615596916843038999348
3667935425492103281133631847225923055543830582069416756299920133731754891220372303490
7268106853445403599356182357631283776764063101312533521214199461186935083317658785204711
2364331226765129964171325217513553261867681942338790365468908001827135283584888444111761
2341011799187092365071848578562210211040097769944531217950224795780695065329659403839
8736990724079767904082679400761872954783596349279390457697366164340535979221928587057495
7481696694062334272619733518136626063735982575552496509807260123668283605928341855848026
9584137725589708837899429105498003311138846034019391661221866960584915714857335682861
4950001909759112521880039641976216355937574371801148055944229873041819680808564726571354
7612831629200449880315402105530597076666362749328308916880932359290081787411985738317192
6167288349184024297212904349655269427264025596414635259143484006758676903503823205729
3413298159353304444649682944136732344215838076169483121933311981906109614295220153617029
8575105594326461468505452684975764807808009221335811378197749271768545075538328768874474
5915937311624706010912446098294248412875202244625944776387494919978404468292573609685
3454984326653686284448936570411181779380644161653122360021491876876946739840751717630751
6849856359201486892943105940202457969622924566644881967576294349535326382171613395757790
7663707645695702597388004384158058943361371065518599876007549241872117148892952217377
2114608115434498266547987258005667472405112200738345927157572771521858994694811794064446
6399432370044291140747218180224825837736017346685300744985564715420036123593397312914458
5915228874087195087086322188372882628222884631843717261903305777147651564143822306791847
3860391476831081413582757558536435977216500282778037134228696887873497950960311088991
9614338666406845069742078770028050936720338723262963785603865321643234881555755701846908
9074647879122436375556668678067610544955017260791142930831285761254481944449473244819093
7953690082063846316782250648095318104065702543276043857035059228189198780658654121842
9921727372095510324225107971807783304260908679427342895573555925272380551144043800123904
1687716445180226491681641927401106451622431101700056691121733189423400547959684669804298
0173625704067332821299621536848814041021944634246462207455756439604529853130714090846
0849965376780379320189914086581466217531933766597011433060862500982956691763884605676297
2931464911493704624469351984039534449135141193667933301936617663652555149174982307987072
2808608596261126605042892969665356525166888855721122768027727437089173896397722575648
9053340103885593112567999151658902501648696142720700591605616615970245198905183296927893
5550303934681219761582183980483960562523091462638447386296039848924386187298507775928792
7220685548072104978176532862101874767668972488411395603494803767270363169210073508340
7386526168450748249644859742813493648037242611670426687083192504099761531907685577032742
1785010006441984124207396400139603601583810565928413684574119102736420274163723488214524
1013477165296031284086584197879511165115298278146203791398550063999603265912485253084
9369031313010079997719136223086601109992914287124938854161203802041134018888721969347 79

0449752745428807280350930582875442075513481666092787935356665212556201399882496284787262
144323628536765025914504683776352825876521391564809721419296755493843755826002531685363
567313792624758780494459441834291727569883762262618463654527434976624111384513054814498
363117897844897320767195087841586188796929558197332506999514026015116755297505754378102
422389579257865621284327312022007167305740692868693639301867659582513264991459502609170
693475194089753574640168308117988464524736189560564794263580705625632811892696630264795
359510971276591362331808669215357886078127599105371714022045061860753748663063505914839
164676567232057145168861707909846959322367249467375830996070425892204815507991327520885
837811176852142693347869218952406226579210436203488529262679840139532164587911515790504
605797108389833718640380244175113472264725470107947939969535546696197267632552299146549
334996632341859514503609803440922122067125676987234279407088570704742931733291885238967
219713539244924261786411886377909628144869178694681775917171506691114800207594320120619
696377951032270890295660855622254526026104607361313688690092817210681986185537809820184
711541636303262656992834241550236009780464171085255376127289053350455061356841437758544
296779770146602943876872251153638011917581540281208182556064854107879335989210644272448
986189616294134180012951306836386092941000831366733721530083526962357371753307386533382
048421903081864491840937239440334052449095545580164064607615810103017674884750176619086
929460987692016912021816882910408707095609514704169211470274133900522533408348128703530
310239196999785974139085936054335996970756044601342424536824960987725813110247327985620
721265724990034682938868723048955622532044636026398542252584164643242716114198178024825
955635449072192265838636626637508359443148776351561457107455280161596770484427141944351
832756984075526779264112617652506159652354571879566731709133193587616282559207830801852
068901515047133403861003100559148178521103847545429333891884441205179439699701941126951
195265649195941899754183932346474242907027188752235343936736336632003072327470374071239
825620246626519740901997624520561985576257600087081730832883443818310700545144935458854
226785785519153722923795554943334101744201696000906964156127322977702212179518683763590
822551288164700219923488640439591530184640047143211863606225270115411222838027785389110
984902013427410141215597699654388771974853764311582298385331230717511329619045590079380
642766958190148426279912217929479873489018684716765038273285520590829845298062592503521
284519259279865935061329619467962523739725655841578537445675589980324054921869628884903
325608514553443916602262577755129162007727968526293879375304541810807292858919897153817
973434961872329276147478501926114504132748732429705834084711123337462746172746265824153
242710593225062553023147387592517247873228814914559156050363345754242337791603749525024
930223514819613811625639114156103268449580725082734317659440540982697652693445798634797
097431244982719331138638731596363612186234972614095560799206283169994200720548115253533
939460768500199098865538614334957816500899616490796781429011483876456821749140756237676
184537751440314754112067601607264605568592577993220703373333989163695043466906948284366
299800374145276277165476238255461708831898108688068478537055364804693509588180253605297
407935386765111950793732820831462689600710751755206144337841145499501364324463281933463
890509365457145069008644834401804283633905135781572739733345372842633721740657757710798
305175557210367959769018899584941301959995730179012401939086813565855396619413717944876
320798688003716073032205474235722668968018821234243918859841689722776521940324932273147
936692340048489760590379580946960417542796137825537812239476461478329269765451622902817
011004378460387565441517394339600489153188175766505009516974024156447712936566142539493
688842305174001299205568542898538979426699567770270891465137368922061044154816621568042
198384767308717875902792091759006952734566820265133731115180001814341209626016586298210
766635233617740078377834237091526440630540718078433580610729611055500204151316963730468
492133568372654003075098290893646120478911147530370498939528334578240828173864413227100
029683119402033234564208264732762338302946393789983758365545599193408662350909679611340
048670271231765266637107787251118603540375544874186935197336566217723592293967764632515
620234875701137957120962377234313702120310049651521119760131764194082034373485128526029
133349151250831198028501778557107253731491392157091051309650598859999315608636554774035
518981667335358800482146650997414337611827777233519107412175728415925808725913150746060
256349037772633739144613770380213183474473011130326702969173350477016321066162278300272
692833655840117914194478087482533607144032962522857750098085996090409363126356213281620
714534061042241120830100085872642521122624801426475194261843258533867538740547434910727
100497542811594660171361225904401589916002298278017960351940800465135347526987776095278
399843680869089891978396935321799801391354425527179102253970108106321430485113782914985
113819691430434975001899806816444121232733283071928243624067331965546926778511931527751
134464689055042481133614349846048490512583456832664415284897139723760403282126602535166
939140820499473204860216277597917712347510975024030789357599377150950217516935558270725
339118923340702238320775858021371747783787783910152341320984894234596136923404979982793
0414444631627072147961174569757196812392919137409829258055619552074342432959828989805292
333664154192563673806894942014712413405250722040617943552525552250087487900865683145428
351677505422948032747830440564385815919526667582829297052261276287110401348017872248017
896840524079243605827424674430767216452703134513541676496689012747868010102951338626986
497482121186290403376915685762406992963724930972016287072001898354236903641492702369619
385473724803298550451120891928798298744678641291594175316756025334353106267452545071141
814832398806072971402347255207134907983989823552687239509093656678789923837125789762487
559904432288953883773173489411227570714109597900479193010467407504114353817824646307959
895556389918847737813413470702467473621120489862269918885174562517325193413520381158633
501239130544419100736284475675141610504109735058527620444891909789019843154852805339857
778443139338839943104444656692445508859463140817512203313906815965925105468580131338381
521764182104334297888261196304431113887962587460902261309008499754303957712432306169062
629194039214397402708947776637024881554993224588259790206312574369109463932528062416424
768684954553249380176393716156368478598237159023854212658406153672286071317026747401311
452610637653833903159219434698176053583803106128878520515469336392410884676320095670897
183674905781630851581381619668822220475704375906143380407258538620835651769984267745231

pi to two million places

9582418268369827016023741493836349662935157685406139734274647089968561817016055110488097
1554859118617189668025973541705423985135560018720335079060946421271143993196046527424050
88222535977348151913543857125325854049394601086579379805862014336607882521971780902581
7370870916460452727977153509910340736425020386386718220522879694458387652947951048660717
39022932745542678566977686593992341683412227466301506215532050265534146099524935605085
49217565491348309589065361756938176374736441833789742297007035452066631709296075919896
2773242309025239744386101426309868773391388251868431650102796491149773758288891345034114
8865948670215492101084328080783428089417298008983297536940644969903125399863919581601468
99522088066228540841486427478628197554662927881462160717138188018084057208471586890683
691939338186427845453795671927239797236465166759201105799566396259853551276355876814021
340982901629687342985079247184605687482833138125916196247615690287590107273310329914062
3864608333378638257926302391590003557609032477281338887339178096966601469615031754226751
125993315529674213336300222964906480934582008181061802100227664580400278213336758573019
0113717546727630590443531313190360924890972464279284555499134900051802957070829190525
5678188991389962513866231938005361134622429461024895407240485712325662888893172211643294
78161905548680549434410340906807160880282279596869501336438142682521704728708630101373
01155236861416908375675747637239763185757038109443390564564468524183028148107998376918512
12720193504404180460472162693944578837709010597469321972055811407877598977207200968938
22493032368305158626572811146379969831375179376232151112523497343052406221052442343537
32905655163406669506165892878218707756794176080712973781335187117931650033155523822487730
65344417945341539520242444970341012087407218810938826816751204229940494817944947273289
47701115741394412284555218284249222406587526891722727806071167540469730080370396187877
96694882555614674384392570115829546661358678671897661297311267200072971553613027503556167
8177654422874421147298816148027052438068176535732755786025058470840132088379328160087
69081300492491473682517035382219619039014999523495387105997351143478292339499187936608692
30137559636853237380670359114424326856151210940425958263930167801712866923928323105765
8851714020211196957064799814031505633045141564414623163763809904402816256917576489142569
7141635984393174332702378123369380430128926263753826677950341693343236075002481757418
0875038847509493945489620974048544263563716499594992098088429479036366629752600324385635
29458447289445471662092974954966168774141208821304770228161164560440072363515811497297
39218966737382647204722642221242016560150284971306332795814302516013694825567014780935790
88965713492615816134690180696508955631012121849180584792272069187169631633004485802010
286065785859126997463766174146393415956953955420331462802651895116793807457331575984608
6173702687867602943677780500244673391332431669880354073232388281847501051641331189537036
48842269027047805274249060349208295475505400345716018407257453693814553117535421072655
78356154998744474804273234578800618731493415660463529797794550753593047956872093167245
3654720838168585560604380197703076424608348987610134570939487700294617579206195254925575
7109038525171488525265671045349813419803390641529876343695420256080277614421914318921393
908834543131769685101840103844472348948869520981943531906506555354617335814045544837884
752526253949665869992058417652780125341033896469818642430034146791380619028059607854888
010789705516946215228773090104467462497979992627120951684779568482583341402266477210843
3624375937416105367340419547389641978954253350363018614009515347669614762556518738232
9246854735693580289601153679178730355315937836308224861517777054157757656175935851201669
29431111388635821596676188303261041646517148469793854226216871614001223782137797741312
6897726671299202592201740877007695628347393220108815935628628192856357189338495885060385
315817976067947984087836097596014973342057270460352179060564760328556927627349518220323
61441125841824262477120120357763888959743182328278713146080535335744942976217967890345
6816988955351850447832561638070947695169908624710001974880920500952194363237871976487033
922381154036347548862684595615975519376541011501406700122692747439388858994385973024541
48010612359080362745852884935632515853843832424932526660875889083187007091002373771065
7698505643392885433765834259675065371500533351448990829388773735205145933304962653141514
13861244379358850709446880454869753581702129084907873478068143663233228194158273456713
5644317153796781805819585246484008403290998194378171817730231700398973305049538735611626
102399943325978012689343260558471027876490107092344388463401173555686590358524491937018
10416262085042992586974358170981338940459344719374938776242324098528327622666049423851
2970945324558625210360082928664972417491914198896612955807677097959479530601311915901177
39431042090490794244488685130868444937059090260061206494257447103535476578592427081304
1061854621988183009063458818703875585627491158737542106466795134648758677154383801852134
828191581246259933516019893559516796893285220582479942103451271587716334522299541883968
04488355297533612868372259353900792016669413390911687588039888288692160023732573615882
0716351627133281051818760210485218067552664867390890090719513805862673512431221569163790
22773287054108420378415256832887180469879525130732663402785190594173389203585403956770
3561132935448258562828761061069822972142096199350933131217118789107876687204454887608941
01747986471378824621539559333332755620094395804345379197822805903959599274369137937786
6494096404877784174833643268402628293240626008190808180439091455635193685606304508914228
96452199877988493474777291327972660276584016678901364905087411421268619698620441269652
8298108704547986155954533802120115564697997678573892018624359932677768945406050821883822
790983362716712449002676117849826437703300208184459000971723520433199470824209877151444
97510170556430295428218196700092025156158441742059336581481349026931115170938722600264
5863056132560579256092733226557934628080568344392137368840565043430739657406101777937014
1424615493070741360805442100295600095663588977899267630517718781943706761498217564186590
11616086540863539151303920131680576903417259645369235080641744656235152392905040947995
31840748621512105618338545661766526063937136588025216662235761322019417013726649660732
52010771947931265282763302413805164907174565964853748354669194523580315301969160480994606
81490403781982973236093008713576079862142542209641900436790547904993007837242158195453
541837112936865843055384271762803527912882112930835157565659994474178843838156514843422
985870424559243469329523282180350833372628379183021659183618155421715744846577842013432
998259456688455826617197901218084948033244878725818377480552226815101137174536841787702

8027445244290547451823467491956418855124442133778352142386597992598820328708510933838682990657199461490629025742768603885051103263854454041918495886653854504057132362968106914681484786965916686184275679846004186876229805556296304595322792305161672159196867584952363529893578850774608153732145464298479231051167635774949462295256949766035947396243099534331040499420967788382700271447849406903707324910644415169605325656058677875741747211082743577431519406075798356362914332639781221894628744779811980722564671466405485013100965678631488009030374933887536418316513498254669467331611812336485439764932502617954935720430540218297487125110740401161140589991109306249231281311634054926257135672181862893278613883371802853505650359195274140086951092616754147679266803210923746708721360627833292238641361959412133927803611827632410600474097111104814000362334271451448333464167546635469973149475664342365949349684588455152415075637660508663282742479413606287604129064491382851945640264315322585862404314183866959063324506300039221319264762596269151090445769530144405461803785750303668621246227863975274666787012100339298487337501447560032210062235802934377495503203701273846816306102657030087227546296679688089058712767636106622572235222973920644309352432722810085997309513252863060110549791564479184500461804676240892892568091293059296064235702106152464620502324896659398732493396737695202399176089847457184353193664652912584806448019652016283879518949933675924148562613699594530728725453246329152911012876377060557060953137752775186792329213495524513308986796916512907384130216757323863757582008036357572800275449032795307990079944254110872569318801466793559583467643286887696661009739574996783659339784634695994895061049038364740950469522606385804675807306991229047408987916687211714752764471160440195271816950828973353714853092893704638442089329977112585684084660833993404568902678751600877546126798801546585652206121095349079670736553970257619943137663996060606110640695933082817187642604357342536175694378484849525010826648839515970049059838081210522111109194332395113605144645983421079905808209371646452312770402316007213854372346126726099787038565709199850759563461324846018840985019428768790226873455650051912154654406382925385127631766392205093834520430077301702994036261543400132276391091298832786392041230044555168405488980908077917463609243933491264116424009388074635660726233669584276458369826873481588196105857183576746200965052606592926354829149904576830721089324585707370166071739819448502884260396366074603118478622583105658087087030556759586134170074540296568763477417643105175103673286924555858208237203860178173940517513043799486882232004437804310317092103426167499800007301609481458637448877852227307633049538394434538277060876076354209844500830624763025357278103278346176697054428715531534001649707665719598504174819908720149087568603778359199471934335277294728553792578768483230110185936580071729118696761765505377503029303383070644891281141202550615089641100762382457448865518258105814034532012475472326908754750707857765973254284445935304499207001453874894822655644222369636554419422544133821222547749753549462482768053333698328415613869236344335855386847111143049824839899180316545863828935379913053522283343013795337295401625762322808113849949187614414132293376710656349252881452823950620902235787668465011666009738275366040544694165342223905210831458584703552935221992827276057482126606529138553034554974455147034493948686342945965843102419078592368022456076393678416627051855517870290407355730462063969245330779578224594971042018804300018388142900817303945050734278701312446686009277858181104091151172937487362788787490746528556543474888683106411005102302087510776891878152562273525155037953244485778727761700196485370355516765520911933934376286628461984402629525218367852236747510880978150709897841308624588152266096355140187449583692691779904712072649490573726428600521140358123107600669951853612486274675637589622529911649606687650826173417848478933729505673900787861792535144062104536625064046372881569823231750059626108092195521115085930295565496753886261297233991462835847604862762702730973920200143224870758233735491524608560821032888297418390647886992327369136004883743661522351705843770554521081551336126214291181561530175888257359489250710887926212864139244330938379733386780613179523731526677382085802470143352700924380326695174211950767088432634644274912755890774686358216216604274131517021245858605623363149316464691394656249747174195835421860774871105733845843368993964591374060338215935224359475162623918868530782282176398323730618020424656047752794310479618972429953302979249748168405289379104494700459086499187272734541350810198388186467360939257193051196864560185578245021823106588943798652243205067737996619695547244058592241795300682045179537004347245176289356677050849021310773662575169733552746230294303120359626095342357439724965921101065781782610874531887480318743082357369919515634095716270099244492974910548985151965866474014822510633536794973714251022934188258511737199449911509758374613010550506419772153192935487537119163026203032858865852848019350922587577559742527658401172134232364808402714335636754204637518255252494432965704386138786590196573880286840189408767281671413703366173265012057865391578070308871426151907500149257611292767519309672845397116021360630309054224396632067432358279788933232440577919927848463333977773765590187057480682867834796562414610289950848739969297075043275302997287229732793444298864641272534816060377970729829917302929630869580199631241330493935049332541235507105446118259114111645453471032988104784406778013807713146540009938630648126661433085820681139583831916954555825942689576984142889374346708410794631893253910696395578070602124597489829356461356078898347241997947856436204209461341238761319886535235831299686226894860840845665560687695450127448663140505473535174687300980632278046891224682146080672762770840240226615548502400895289165711761743902033758487784291128962324705919187469104200584832614067733375102719565399469716251724831223063391932870798380074848572651612343493327335666447335855643023528088392434827876088616494328939916639921048830784777704804572849145630335326507002958890626591549850940797276756712979501009822947622896189159144152003228387877348513097908101912926722710377889805396415636236416915498576840839846886168437540706512103906250612810766379904790887967477806973847317047525344215639038720123880632368803701794930895490077633152306354837425681665336160664198003018828712376748189833024683637148830925928337590227894258806000872860388591688497306939480205112217663591382515242786700944069423551202015683777788518246700256517085092496237477268136942843500629388144299879053010562173754591826799732177350293689280652100253962688074980926434580116557158867004435039765053

2347828732736884086354000274067678382196352222653929093980736739136408289872201777674716
8111819585613372158311905468293608323697611345028175783020293484598292500089568263027126
3295866292147653142233351793093387951357095346377183684092444422096319331295620305575517
3400679737406141621079236334238056468500920371671526425563718538895714164197723874226
105966673969971731681694154350952831935564177056686222152179911513556397071433128936575
5384464832620120642433801695586269856102246064606933079384785881436740700059976970364901
927332882613532936311240365069865216063898725026723808740339674439783025829689425689674
18643361349794752455262914265228424192430833881035800537870239995421721136865502753413
6221169314069466951318692810257479598560514500502171591331775160995786555198188619321128
2110709442287240442481153406055895958355815232012184605820563592699303478851132068626627
5887714460359966561084307256965005630644891875994665967728471715395736121081808415472
7314266174893313417463266235422207260014601270120693463952056444554329166298666078308906
8118790090815295063626782075614388815781351134695366303878412092346942868730839320432333
87277549680521030282154432472338884521534372725012858974769146080831440412586818154004
9187772287869801853454537006526655649170915429522756709222174741120627206566229898060
32891672068743654948246108697367225547404812889242471854323605753411672850757552057131156
69795458488739874222813588798584078313506054829055148278529489112190538319562422871948
4759407859398047901094194070671764439032730712135887385049993638838205501683402777496070
27684488028191222063688863681104356952930065219552826152699127163727738841899328713056
34646882273982887631986457098363089177864870866761854856800476725526754147428510281458
07403152992197814557756843681110185317498167016426647884090262682824448258027532094549915
104518517716546311804904567985713257528117913656278158111288816562285876030875974963849
4352756766121689592614850307853620452745077529506310124803418045840594329260798544356
20093708091821523920371790678121992280496069738238743312626730306795943960954957189577217
9155973005886936468455766760924509060882022122357192545367151918348725874239194108904
44115959932760044506556206461164655665487594247369252336955993030355095817626176231849561
9064948396730020377638743693439998294302091470736189479326927624451865602395590537051289
78163455423320114975994896278424327483788032701418676952621180975006405149755889650293
0048676052080104915378854139094245316917199876289412772211294645682948602814931815602
496778879498137772162293594378110044480607976724292762495107841534464291508427645200020
427694706980417758322090970202916573472515829046309103590378429775726517208772447409522
6716630600546971638794317119687348468873818665675127929857501636341131462753049901913564
6823804329970695770150789337728658035712790913767420805655493624646412600243796845437773
39026472512819416320076848736251764065967540693621758879307855916478777274739272002910
3429495624476613082007292507345291707642266210476730378631699542374551174565220227833240
96803524667663190861011206745856287317413511162292078865132941244815471628182079877168
34634132236223411778823102765982510935889235916205510876329808799316517252893800123781
743489683215159056249334737020683223210011863739577056747386710217321237522432524162635
80343762536068086691635715945515278178039217743228234366337728111863905118930759016666507
4295275838400854463541931719053136365972490515840910658220181473479902235906713814690
511605192230126948231611341743994471483304086248426913950233671341242512386402665725813
0943967621939655407386524229897879782198637918299709557924747320303239116410445906907977
8623155183495930353059237898175158914576504080251094791234217584828418819501385461656803
0175503558005494489488487135160537559340234574897951660244233832140603009593710558845705
25157042662846003544028236787685509826781617655203757956554816778960389274983556087915
4117774942357340076416109329400389998219926725708695732606877497422480202330752518765
02559684207606932299885875798988964607443817881700815488952265167228340452772191069914157
6463948523112679473086580319507645519767562895742888179681209002638714525785831527761
510908863174024369568056787301523542780479341426649522383370711751126537550394237209878
466804913947344653071407962259728713050307725871487557050258257346686661380235142605611
6197405543436548698005444879295970287590352258409782683598666446586045694241390729095266
249932902973440568160683805726626057277088407073471496060064561454070734432782514087474
27550672230484535700609221439000299298160821171704791761450519100813267037521493074056
785331110605835291278100739174994919784511291591368110739405517520801963053935074024850
955377250036705466516233043042508744232426240463211507899733692998540704165626104197670
020241509489241185609240963760442961200236459070644977062720791901923596480704892363697
986019828308728422856475235316288279132429552481444750552190967204608068954518171220493
032185374062724742151974030576904360268636078079200477623242955182947352202724437639027
7213920877670657162416397517858592544269234285352743288563368507896519620725194165560618
703705502184628454342578503830000953745182929584404649188386857934839611512971605816657
45096703677495836666693121881763679644943617130416037243050658485131749264055855194018
0051809084752118682246169761492432383194864344159085580110730703112015022434160731579295
287529368358203970033891121141706852193665897894595031543895890153038271430019295890741
49943592894083097077078362875914484037045038618966975811201852319231868659968038583812
37032915620757883594878094168820553160512819015264759280757495815456422134145937816705
69928682998956119823538371578804804787045841753946654976901732203108900703033629117673084
48450372145669644401469545173857434157810158618783839278552609399130570255575559060947
05149809348777332007279757303824598946680968082222134848587382299928179409082566520958
16554724752445667436975944746863763324289042697761067919339109833004223102937282987989032
093910926828363061736101738781236798986451493117024371282858826304862988844922074156406
0714705913740552466575697187021735528724543942771480917936443765063786186132434863579
741125852086345992780368879249835436329845768765016506511534500869572123950754478568317
3631557153527046524235259737513408825461609661440746675514226836031959801072152463551069
1718713357316854856312808578344356236709596509499469688206611851180860342028213318012494
1099150260143545001743273079362511307029825049941799428445114647932915459955590958780
762163666859179106543596606525352532027365072598912125568684280207724648772201099663182
955955290339331228436486447597356085984076094729838954243393262315323991898185226418083
129633354635687482886346561850481063228880559673784456200094146560349928087940511531005

7587129552571964111506850340773710604380371259575596985949362058477512026354947347534 74
8189262254190352671614429284899857536740692165271630086060654373736823556588626486343 68
9153218095572204456777137368310458075584529612832832606319629728527966674362974800821 31
8627921869044284342630735760703999669430789508147269730253817375694922751795354326156 91
2040594832860949992366412287881226419148504856328072066418557059520375030322916894489 42
7578306090910852410601400683274205583969773823150734996108758763704255564964086855071 94
2256344966732430656259250474581762733281816017019698166542426378763601453035946538450 32
5476674999737340835665138186025156520283637389171016545414882674448009105704186162626 83
7971120886141357279611099088292970229692128180978798951391504270936786444983196420134 56
6833908775943006442485623012124614511697921939634409508083229281294270436599146482749 98
4375942113020418297308417178813090379558545603247170819195302771465794555475544754284 43
4408139388908609776017857389307518661906505018077165001840744325854024184360501118242 99
0702323417243674525365349594799063334540754371812699399833719218485418735979845348934 59
2268515068182662490078029335012658824974226241885352526636702827662499349829488748331 06
1764208429016923052899608978604130065109028179805040587107671179041130217482796682353 00
1960220253185576789843317586806378359968791601538922220236575765581586611409199394861 59
9209159917553341783033347643131635012705390697079326567812415906434284721360235218236 74
1214733124499944334155915274315931687477882533155092770336202901222597794809855392200 06
4527162280855398278906584233447552821276517650572663267691141075034845871896996434875 77
5138479148183635100621466818585096348887081456976722020167991199462417776688907917136 86
5945960726468538810778783002161368276697026223459418737476733537998884403427046803042 55
1694127158739320398444374604547816113056625176412759821181939661101850562880555942566 06
0032312116180994622129301002470913347150682268430458680300904242861682025562140946087 90
0065191099495570815816505828983340739466084457565780636690272843462018587328252924796 50
5286681408503538519837523637451925622795490290557907030283950104854835929834542814487 30
4358047053315081510503001521428117175393649133166172621235405527863308002083177055630 29
4963594201654333094094177196326234119387105161570101798053551679370860291366756986097 12
4120368583812957695307798141365700174761356966986146068491439699573837631695824602513 34
2108072621713601943018087209888551415024163818325975259593165531865833117126857941527 20
6612218422661411825154657484878312610347834546749258308729985447421206445095233245050 87
7431496166555251797168020991720026409374921907569936896330281391647208963581771735555 84
8592706524504862516419540550801343510323389813378302497701822754906381499964723334079 61
3041469739476372650869273347108415685608430921316240434629863920841660055904598506491 24
3505264766067600344441618186403670083774114101094320588955598658670077863671896944089 62
2321374034113597199133135946553685446692367652589012108413777432482191812747847892287 26
4892970032371873456157981599834839100412601050746964599430331978810634913923812490503 06
1433407918328004063907098672596197098311265960147473725330526853717742146554005873924 62
3727617364905198713368067723952570781360686683261395014329509474851594724667527201684 31
6586608807512768584755541184381169011622005552113484448896066825922743131900796301158 708
4670117654935393046563356225311244727796669005831190616101972663073970542531439818457 37
9449486780134618217875939076999602029083965677287846905736401564015047696448993947541 47
4608339918696889271156942345492651246645507792554028105037622035967530558601856492056 06
2879090769453339208808849477828894851122154743230191383245562993881020614490266876010 20
7753210915684977830740859649857967152617010039475494539917698791323546550106407355816 99
9409756248149967443278429202762644189793918158394562708173301582160225519659898769376 16
4019861207466755048861110855726764507052622446130222335852072273620485057289238815884 93
8754535229186399714380884061757286220950122506515863104258884134355431973729856217753 07
2022629475552483044445340434888878581170341345342522354319407877972846760181583227097 74
5180929342193189815812482832658950040704855206099893783900341914163044639163880549658 78
6501375046341695655156618298878630705842306967660254053024811471007899784211830489010 46
4056896539702885595530925558636052158957375114089564905844156774937105859648014315874 61
4491250549253191164653821585197370093280194530320572628452658046046337816631429933076 64
6646530760590548962888724189716060225882617577539922055131509377200624863085562820493 57
5752724995567089221634233983602565328731029194007041176919220850015116735670101958971 00
1797019578120892910969417754369904368202563024054822625401905696507710581574240721496 33
9560365270283334407305750073674562260584649886115101689612181111905847171446106871976 101
7456587373796740697137423238753839030317200200207205928488785123911746471673743737923 28
3881966201687622191346233893762599527025672138622112458980212130501407288904300322535 50
4095866818724139369938193069148744717186646183111942603161664070377316487001864799600 24
3044003242224180940227853330901150988087067826883531720076752255313800881878043169019 007
2804831799287414125476123089606833095828377667688287578688683092976001011974533898331 95
2588619630132917094385816615374171794496319177154312506959853481285684619377669894277 45
9170918802520012749905559407289696594793331672243621567896776966708035229039018485730 80
6275670867658627104769409203565593025352743418965927002227049233186829991560936413757 00
4988537304596396152734629396974951748062696451793018719986788537581415975799314806608 55
7232568374305282764175670050288040489429899580948103534833934144927885925262192415547 23
1997143385086637320926632728243514933640704589683852345624744361175256766987767597223 43
9206357507471552918102762614012992480422883990297879925418517499129630283990729635588 57
9890593317795908769073905646025623533567221552259468838298452882922966275137162422172 95
4678670715840924184084147555758253938524096330205134970474069539956789798172786092046 228
6839735779815111868152659884606949758965481314651150392626377749513761557248195116119 87
7250344564710738513435927355538712462375598193813214238441581929070046389771683887207 91
6361741432497079109658162746429717072871725142745898356897095534626820169085356108944 89
8407100581920302176945120771774588795519510473384184739980796306767885845167575729904 30
6971542642383498009870869933670912108394453506245922432312348278549660374657188014892 93
7945147870540607924575900601219622123928720017215588666345734971409533721151655985757 94
1724419889026167016101611557834315025460328781198424027484608510722406676777876085524 761
7773833089502610064388350550205456324346167859451941795669874968515244883847513618180 66

pi to two million places

7108316165564209369270520611898517292617141714434655508706306063551012949400309759167799158426049197120954322702678432654296572403272088714321999645313202587109677165128549669962552698607311763718207498827399770601991362093083230736838206455732563765982912578131492224220427971241441629951265945639792759380383804782623160424325399132851123032247037561942321733047854078576244013291717992979240783390715757981426816864655382946847399205888631655934919867896962840447344968024077092831376408103352255242717404107673565424441004483347440101726441052954787296345898640501203608024451190350994974493973617181575277093780209236668135841636268319263406714182797421342546220705415600050959674045616840451771747952790353254932589120483385746590096781730416000521088934610768754004241977803082885181200173369559127137714195011361304409753279190504891583246399143483531648681548579178632935123925552510211182788573696060276931301469661433449642302114382483705633532793858895267672076688971274435815632088106650149568143558796576909857765902768707453659276364975553449617308078160987103248013795136170367763457594975686208013996374551762425147780628722265971455482906769295713643572152674468987889418820751292225756509143552828874614195097862427527881571566400763721037803194043095844272549269987169234331890022141503113998765260688761566740210197201719602390861082974927639569541153032275460173870795625993579785302443476716399591462317931239989986928437975702492369551587297683854005227651495614447105971962889888157109415171701518114743513643854005116246202131174800791983749700100471363432523281578911355450453371905275068229156185003328469567926226208190442473340362503892792071585960039363153368842724375366799698647934741133198328619441460653922784099903143840354565047056789552024827176011874335643690243503085631309559055250390492731613311734922584644609024535079190184411299321699770451832853586480428556822208737213616490586303256368913084103760215679927020005322355439804653119339775459044045078568021398465009693429547310269249947586466058091669984160684646087293943808274308285817479694172872990311013192675573897984091364253479694943480377703364634958476862982590103470727861218623001986607987782684245933835638919570206853521603211635230064988744600200170413056985365154668752023859375183280372851143274811699683692849220447380570633496618711240947835915869626858643589141359854253577688774932743634514754488640868818030369652431755688300205860773256959716086485415834468432489963077011371344675156930244885482077124133557732306949458067267845235943631507872728157901573070033178796854436279525719023623274614262868732738009497741122856237663214904653294072026197539071740422259539242888164559796570030957141389106936845036268231053986743753240052701534745893325679514941854537808827063457295962169085383535370381418115573816378209032561519869745357646412125498076005156141707298046994813593483150568116642793219335279822714715767340186088721518799669350252700757556099719882863064285448128275139280694702750148163289727314347348528529504604883271673978981563678804780443602109007320727369749344630499731442571560433133690387618100948873120713482710815889857483265854207510077953118326861708037070935927614936782530858340482351003632166378957426202550350116861543407379504516482896755698358935522020173679548075781909502697981271148703431190363112246128295303820512870430929471974594690821025634788995431771524379696211281224503426066399268852133079196370277780448857920573046990800923440186638113252097123096476059989947925759851008173039606822219975327301606582628525758257669507854726034938298133582528178670608512656002268871781125359782933734779141273628418865617592083287944741096970387985473698402545806329483502235939354358748022398976091629625011047393116944910066690723063469313016971182063253526924404384009372428442820970936485690946892008737175325255703054353982872781230113980809386701547488580344563187131960267854879389331620500767526411204439023758334272429869965478636853410284885737025472550236566341868091903838867078790720840361940216467012153483797815183282647257862881520710108149955898033811896156944175676134071704653851217090212377788433364965187211990540758187739439752836414395304424591390317881300418879188711455314826746998705558793104024038888408385068734162507165727418513495208496367095554245043948394804597915622828248378793415272036226336956180555637107681488889361927574265993582355943153088793305276755874751236506584396947560429719200231986802435171993786810036110231256836425607959741057415362829718004649774857371837863903703901539737491165468549971645394161121641761071714540176519056505252066227788312904571969320599024137539598386198260320549583950167555250964413711822256149601400302303540789920969867750786720003807426797053030716793229601564862280851840335235017060858951291222324611783025316362894394607365277133651163164644619909902122492241231516899276785586373631552600250348848781323300191018939961670273141699962651194574263676196500243473717272902846220979839487106598227000995491887769618850543265321180221944428222842515255614118743401804194614139451471287252759239125596443735683397289633126767823491035633296129471910151571431157954909339032614119186547523762472153110207936911584874220582274734320173558507712243796985796549158062795027409771688611480761631516185530685669245717176922044366843312739893379411162972245169998546856221570241759471176995291655021168550010898576193463945590882627077531146577522388463435193765397349848024549760760244030808448901068387869726123709783578245166801171485983679405529046198262165669172027426285482393396001825459940925430816969103297841123402288560019054934275022318529471282960969397681373419770427812130014732867760571940596997927551246171843495698564171287248118346542064231871455182415286763056751311626771773506175112454633879942652912701057899567180572143655791835069177793070407573290439749499582241062381051491765023850418273009662017175094059080540895728375540635515221996582075735131570759236153986394592111558640009880975526105383825689927215847850417460651615113378833609760121148487005560165812492470682568442720454728963094203066504452986462235942260085549915891499536064984280345794927570094979594506023787750194706246323949549578230822830668408188025210766390742309737209162853371768062164469354323179178553058331714208479886303408465726426939557002685760575393478885870946005827232305191081175142349126873365859607998917329289158960018150918163374008060354752000515117510290122992487096154592802620607616982721810291673155489294237408519674330791660784990557821019357136624359908836138598085161564174769460547855400819535306708030896976304529468682332105328782374389441156851762717116363094014799096494563545929501307390036268210073263700823561506912696431833517162

pi to two million places

5439030469898931426154426359511363466057378654951244574752621678954703628904830484996804
0377225134319373734412366185869445880640185840731476337929403863404359194198723552630
1565460805186867606804316084512845916042441326987912538560299159967278766195195053176488
3134693257366894644382558139108486209663742674579831301222343872583124422033094571457541
4704792938758582389977385152135237238955966431223564326262860114748908681715928106687270
8400820337718692153523526926347226809082598988984002620815217828261122931311820866007
0996860365409818326807558247767069504109975861436243552161945353029200254667367996485043
3731334952082107511992589266389956475698587079018561237915788643744690378715095001125502
1003884531192365296559946190047484662064234794232967006052900370917557818870819352214
687142723527763255989808694872111384598001412384216382782441273654244674883338167971620
1128861914154019367129094789902646666443156098372961501968624228250672306166720943546571
4251493086424887785986827595887490650772602509518295367651811823686169447243607837642947
6246922631949892196464406831692876616150605081384631941511620257790786307180123115945860
3896562526554223346234454507394788690268159497513116885143694521021688319044616862976
3325229863851818850049286935727647668238555646365544964006317648285575785866610228551564
8599088209586894443625469867952382268611596991005636608292679153375381606611224786953132
6158531871763885989377929188902998793879810003697307848959270625410484859315854323395
6831042390299070263443797875691855434089764407601308444819786265079476440830134942435834
2818859152592934714363175337495897010728735012707889804816350456766676932075530518404324
4610074032167647183608370847506512693070766084982529900031785030585368213951273503863
8246056425103377755809864643398017186208142663074172592226000511091342681074670129014301
6541010649332122837908275150010035300156545975083237729654396973820477416265710657408216
4996062622749618795334790706598897487177956433406484174564574790692517014949981009535
3413548908754836327579522407206986291024671703579251441766703886609906985726260581240825
3362252189920004189757457653151230000644457159317017716886354833330519215820559461173577
1632113223393196532038619900511617817133400107057665268991970816920221946470432379535
6411866063920558609034457064151797782145054722278852987210197858846070047420028468873795
8442289499743336562718779917211379161644925413297156528795295326397595385359209501386333
8050756136953089954758488302426196275898594151378051580502576754040178579585244883117
2105089277089227273431973823884687307168230248788688585510108073522781405371406520758107
2708481672639770987314551626469114232861030369329843303003236761627142640675878067318839
7151500279816337477907877503830798675940459107392103458740421961703492580818990720596
1291586420202885734009114955238865107911371495334639763988183948804530075074740372280936
8205354304949519483328334700751619790086872854399629815756058916376247230691628711111376
7608648032375245966493041175394613646433780467116505550467067183622128579504806716563
0427626711429999113487698447050370637900181096888629721757951732433802780617470496302042
4929166191718862433555992820932439194457118863215563201616542470553759386966246563341215
4101403228699093015913288580883124124288287637387274283803859071029274863335150309044
5328052597795658920554562434297982794134891756382400771612173324736428540160610044337641
4572207859217155914010378320201321338330963807789040957238105588293927963743816606868351
9505927701951536160172215890428785678482068291944169871819286273082704441630396254713
0532843883379133747687358261221162583602728961624559041896770247453827583966522993712351
6304898330124214174557885915942560597924277218199085562798486056174536844789237969079755
9455515464685316302446232567403489584546225674485820204245739199425309426422450420268
9038150152683602412559807597523648162809304891274615119623154611400822056396780658535407
6686882275426503812259991620760170895567474465242344520176616503259456659129667863246213
7991922296145867142248249288064768032108647799410041006003390679275237362546027742960
0734788038356687522003482457694908456862696057715701919174892260635208129738797443835483
2861369395624503929768057832234021716765559177668403757234844094617629312884926899368713
8983882227106027903799001904558336007973927741092665573923314702590923389065438842235
1324115388018559234956139930223919645050450369352927011566305153351918641864823442499919
2720272953459599063048723608041595760029668121116831723660381105428035914457202482564561
0571405546242082134352094810841715828957244507206354681600230512014084805435874252617
1017681853883557558717415424775449772221419261315552526910917556333193232224321852542218
2729149159810583689702503522813002141192486014248068079753699647771939490680468355280834
7327610306049409733091690316783097934636611832784531868716462680738833656704566010423768
5058013950744364796392228411269794513477300492498786496563679490992913271252897765191
8175427962806084932375520815361113240339713165504391887960198382138585000773242461778849
1875814596426423378897933308194881600401131265256356932446593984006368903152547229239914
1447437706963389357619260391892479363178008310261141954854360515778716004955788656579
7066588551042882466363057207778902266777042512681571979533225107638903681976284402861025
8805392339329474672024088541276492386447602161162620824212991660362299184923782236300983
4781195229138218473263422857591209798054782852505918379833680178741124264474600225624
1498069140074097972102327853957561512834580616541111792671042799057939449713494632895045
6512868847841871758020504583283874853137369113510255062010277534580943910500102183397324
5650472889476879298925945019875076712236379187586472012149660611512804870964886305622
8440839369443872169212084920085155838125107074195518720809374694245973117281172105192890
3896370394235776862127668210931827636649840421249381440979598631142254364839654999834790
8430702176438555435125743682822815303222238083476795111355701480631820045322072379489
1863572149106242526993994671015366846234105153338142684770627585203524099207972086991453
7301095516415033176282001969164115460268207236692552751418429969920539853433073068057372
3805041671972211273740507892726634063885068673445856077326664838457802771891147580132
3105519878413365218519071460681389868867103147598264611293795439526672867275994833590259
7445878687684964626834844344141359177145877660880778453571839329371937393236408356337576
6884682111179935055410208556188490102016005056395416874510822060355541081766646052412
4966224422804545243216032036019464135609792001959024049792923673298924553990101980112140
2908686999205758917771880741461222050247285857153675307478143897305717872683663601576136
1007722863196388526462351255380773194595635679653823624999265518043307963596211067455

285214290262949826567553352731004687886573104724664933265679273313451229550591862329373
933260860774513507753090157444382948733977960532284935830136183795862648032129736847481
751647691366211036036950910666650517171150827820093278835872259839404630683763181180890
442362621998812368268078579526219721668720174551747262781803268305854880397097704793483
103543985590784355277667603313988460527150313885633246768892710459585193289513916782385
773577265810047982563935519352005520408002870596782497393747886052835649359149783803779
649600052124458347790017560424658666519980770288394385163809550430492196032443609034008
517466042962743097683871519459826447359402342482110447572911177795877313415536095275957
089861258677145625239945007593802060935502489200847673322930857422225502064556902391265
436635785242724290560532057540308210145123820902174669757976534751725014658374788480805
377351504222240429576036137543248619965589193922050469998210629316096756517907513229607
778575533102658584257608668676453552092774827556754517716995087894118059363052499449670
123759800655349987396663953944170170596981015127193331184076792327185395398097640485278
467438723164329100290654953086128333026640075801296184992070220025559721569575883761687
843643467927558635739722535648841330601192895746428093578580811323314331152874821797660
397125795289003640719892332813161164041693773662801325973822223742681891764895964227033
803905929596496964821331144731667650419767811084909664694257170694570078712640144865224
284694889761725674653522050616210730010192624831468212035516995015220073163840041320303
332423121670826854689317584366304307843507859281044784926639526523987186441733800856816
923213474297545832694021612533328379009606486277854941266795136740458774169455961407626
566250299006922672678760365871379327960418488393933934692635434154809518362332331752293
703521029146413312752037117166754872063473892329378510729029514462927415467619479427471
669160304978292889614745870264997970792063872408250230064255449959040119741085351678444
090188064629374835443961440035352331030404117845722890295818058103212374382589870274737
040106837777159251264535706508300921479258349892475127453622006105854575997369313529707
814374284134055195444672148941505745283917160371545308252555834320251254241662445752456
296445791076971715214709518505500355054390631688258105785074635656204791466768055698438
455202770996971988980723371486956356703177687763789743273492829343905145567060744607970
476931646278121417138182743785614621970880870210642110573778514713588373773882407652804
519142713748811055974471831009393751976598021002410125112308136826033847449108771613228
576602639388492849598982365657272042635720263748256494949126291419171306462805956698254
936032613201925280434617043902892602799314043613702658201213128514881585731117821041310
335728887181729526271120008147506402683046418988769747879173173703813999188824241699421
215277604518595671190941807373479331099709283155468165639527101046113762540664495861838
546389822089967783295501114314995936803982223037136329574232173574464734210974149174364
199473195884005263872695923183642325491845595504534377846709470450959420120211422086419
127904935994521373924871107432314951138042937936554363721726348190757113531270930795272
952211247953149896990808946657476955651243605611420086639905609900038030250612423607750
329341347289050131677280971316268349596340929224303119508487886710353352002371273020291
659297525265703921042149634952385708560572343462157695698513406830454833154590753647114
699682420910232143117176922773853477041779407644100130104859609270721132052318538222744
487024332710398781147912754608083611568779215131131045008366363100751751102590028086427
715020962713662397401075288445468331618211502789264307297635576105511246203324800531059
951115054314848295534329598305742724517378865271930007323217362375873273148909109455374
027048118555719905168393874535206797085921189640785489504109405699659887159886336207795
504521932156336124685303174705443940294182926355240155452316098682553138970188015397045
962501691796648125015559323114826730056338357972603286017784741496004569725783495620587
328730124514555763452302986481495441009078835298012070126541095251846066620176742045257
367994690771908453787482060802904825167017661982073061833123921935356900407052154989390
344659388090475077241695436518580750664904594431888629787235716030224813522046010906352
145082806397492755128476943549962033991644887919743790209571888632002475020791023790730
729637463263366745942755637845356913673455240148971259094803685662823210050039400731066
320752572831471151926332892852069672393471750982952602125494764330195357438350925828311
133911539063376617373077236302798898699857994501659237690675488379889294006051628261400
481504694828140330839164342486509396354589091328059511163345503656348245191505831794980
831827281347950507727173359496633718821491928378711646390356692577994245739435547304493
555939684803279020861419681508260648109246885433833298663907454780526362916156279880318
782827074516303278639075665336219750632242486457694597535966732006038982629300007612514
947980089567124525695598275854857690124636865949422422772717715184964175107159841635720
724122437196806720392706478942789421712842641334271183184794413346064724314115015509855
117124146682433123520628406572269260690474791964472975283227495698196327787281625954012
020538073295825004974459308097824095299129654233184987988007716816319860865120883158672
565065944140618446837496318929137459934216034848228831582897309421614736892558516992715
531155888876007217034102445874402084434282730046730979555566681150130033888958380231464
313829002600763228503475830780878895180313981020762788985174353478225120846759497430024
437895842895680752663203627696299460180834941994912706559130840005862656399639110406851
041282007153246256426371456355757694528492711263557719632506589654553648212545926335525
729259528149934158787765156922311915102337344071699165647639820008969846298439977593853
981121332181032819896994579261764935829748373387752352859464035138238230626945363458100
319367250206982807384333411752831573143426398964163471270530347756991558003118159180911
378802688385475769729233988828603230299770430666288695530121027270576339598976894102499
684794981684201199256134807564404065594623837087236888125489491487948734808614168105521
140018455170084444842948475507327366428272220633658240174549880829130188391401568090500
008495465737300032747797209917507461785951579953202237285235920400742515225638616675620
318839811761861196022162847431907970250367459282804678178536647393560035403827828184576
694782337457113822121932616729501042706940952026502805228985909350023944908745626205345
221731194095778301953605185038549614062182530618203651827337062111989390244889753863581
809944918157848783365288654365422483020278924170496896511041727594750178122678581439174

8649424357300909171264877160595920974458114629554223100220085120522589764778114827O3942
6776664278274625939511743807198618722265586504030028469146927864680031836034638172640 57
0270742262034297187555809938687124046562233389146465830554301315509528510972630050 80518
8265272685335372937338569182693717167730316118647494810424215127915910146065697953 33133
7740959367493264414637024275245393350301309928336485407069840343991212452492755802 99798
8240920664464042585966200888741916498773027540372920421581093781471313622628866669 45474
2124495528490914921933719362340294337125575569988652966236450353519202677763794248 20828
6056893623152152317885014521313214914698685483594470686585010981314205892676416115 16210
9405356780736810089734245872932705210853572676380564228840929665884477795279546710 73519
3295474713015079220840328232204428944678218396547110902117340725139724757357008555 31274
3219996751259582568063235880883884366203262266191414934740436498000247398332092411 83866
7429609269460701418388178110714282439657796388439864782313715424989472583041145149 52687
2423618996763058816820846327437441210390552765218710735564525713360114558045585684 55865
0432859917676519619327114349866540777451450047307271171479571222757201812886446440 77751
7460328242317338533765298981044232224046772463204795179809715760258008857689751340 594805
4826877288477629384645496040270370508539419092769937066804551719416040376351180185 51365
7545109524703460226002074174282384948178225490636599208474903758320574467795910675 56606
4077500934712981700581876940802799269046059498721176341519148822518670439557310017 93710
0046657292180372848797971569227888839704198254565706428908985827958625659901375968 75007
8569853420944399597152366767355991155709006141301885395600693305082611578831597901 88291
2877765396964067539208084858229047556190518637549059417647208090848523929966365377 74687
0985680142361370763704674236180292186795924769777652926292904179839275053432943384 47653
3398501228283627985150263745427966717714841975733906572871543054321575235449320534 65375
4238204844850884634590853386677292538520444984413136863751894117684862613603681937 36351
3393254080685226921474307329134467625293226408453308449386471515618139413634350364 81779
4755097633925598827869036963238633034257944529229237752032874489020040532668139354 75285
5017464531717214599508145561364692526650227115337381817597855795041988075485811336 28915
4900903908060775415757361373755988018757307536248737001291223826113438103923437231 35368
9889153374949378632498494176428141704528408296939917243232867725641504837657731144 93352
1553852300178110827616363037090205259503779092534110470570046565251977925679331410 88663
2640592623178893126031528575871642421190333798725775874290129037593626972723431489 35725
7241883794186276864566775868692027601439805016387143520477673880900578928363381779 73884
5734410014996643323582225792535171105948560789182401521998285226946509587631492471 27952
0164467647402704689545435103069826179991402234072854891546806842095743207506621154 48762
6644675798636443880232586360886918759442271521429650664161384963815027972173071265 92057
8266002784718140034209265693070309044570245964675764901852781393148131509203641049 84596
9060225314474822945707025270436304061114455142227669366501254252372074394018277525 08941
4329152151705997454593125946821214351062276330331850433948895127672063729151249368 19357
0319104693572905276288768782500485054800597323075326522779255241991315961791152206 94196
8547918734156699781096702562993993208164507174173490564339865219986639055709352119 85243
9067986150214486239284387398201876022854712303949459661572587509650320071247665759 38137
2124801134153550616754720369579105597461067112541711745369543014719141993731972279 71690
2116135726252431164722893666441426212438549813623694963571282116036854416071082317 75107
8012983042538141908922492085953646108213956481132053160737077720760559934981503424 06407
7512331512158999246297497845474385785595227089267102479199199645043040166005621762 96234
0149282181611520504643814051201017632797902693271222701259270816304579408695938850 30885
8577776769880577120277461858372818585997017721116037109827393241471979376638648431 60008
4157927253061164085015150016520300200142743376390418788622635274702258984849469077 69474
7613276391052599405660382382371636943555470658174827307182474182726362724046239944 02844
4473642458644475104690299765267497344356985708539057819159958599609675061283091019 47488
6565075126139713632927641583491304208300950851100414074557443784927898576072610576 97418
1963369679075518838322017344376439805368296268732851893953081597213840998753657746 63549
3253113936255978954300091119142674075385925496901579734191837104016999179009456783 596285
7322447147907320456964719786315490862841233325174812784828809848761022100974278347 51646
2790553938519668895696510876062872957459088920170238672074010602453894151954739328 14246
6223126892362650272056402643021776903189555520611271146314671703891577339006545286 92327
2080811157875737499103532444669361653517522124688660805939738054689486755602588706 87103
0811898922024217495293458219535300991561355360731590956734699069924874268001953821 75246
2105349862701061321590757260240804300827868356293198384271052198354727511764233027 99589
2687273053118355805687527612409197424447633568095687484441045467028352365141527656 27008
0436309747774537678098208734980384982599248810670297754949535228299516546559850687 428317
6285208571961393797828505779014996232139220462341524168238038894466242673730018965 43376
4765036341251828509512088864856294714398779566559280749164896256218592671541469217 67683
9605450082164216260561064231444357982306919657804705747148460072968182372287977560 49608
9158178686729363237902415792047283646970210313975180097841598550007055364938753212 57496
1674875872583259925957615074339186228437988301346044540880817809685491194541193470 26896
5059919860410997653211196581062966550051161836517062029288087760914984616731644268 64197
0892306484630567545738872024760165257760852937721093358445387107402729259191524626 76235
3817978693064215340131633701135735635111098141821129662210736726269615672674830775 24887
4448416766573702400485083937025583859101226694835806839154547916601645691486305239 35977
9324467255886717416048550387114903176075537321944728305822191558078807524536969327 44601
7473605242058646968697577061218677619720587491045165142715495423853920232526975123 49546
5463090613294600566507283098728033873735155375223563183570253700649409263808031737 46348
5403611466600048468762423108947237916500745179705248628467276633755173036873683856 44037O
4980661790920083171078821049818331552614850537354075035108223939247445630109692042 27884
4737169688950911185736926890336659718522537770329622016708106551812675800940852515 06847
7579219138932138092869611953122090503801810765874883683178827814252786261879667606 82197
7039093260067296151275571252786437069898354444096139173790354548518040397333137480 52358

pi to two million places

7910955583040481534804539187854038243236907304310274062641777762657301034703384021129669084818046162496487394734584412155302581522214994582224994194195472564103175021144228086523028022134240931939327276781959906081125986239673394589896190716797777802595116314775762640285882625148158216439944135061960811758904619511585390826133549603880323713522245169681180597512189590028591797390866524495280407827130270045377437267855532504850397463757394646098408565893018482234161498658315034660821862236058019481145549035154742662660612950268784097547798140726823956931472487609828034508118938340409615343148630112486764653154787584549465222275318773560890835043837081120882441759938586466309397048117253004020305813409044745051156377054103501416686191248525269493348297851018111472329874045396127540222219095844050872306623268888497042234567000119497518597964940991489713853622794588740760990432854228127730581830402494510870633698694686740089481097539710090849476830410711529550638887652490545659994260773886347394552511448972036104793757254472396602354774812749416069835101314764023641949146105980556375704465155667123652568282701574452847602207817539723371640969862649205576687615644577446446649254773467297255570538828590789231759706768639824966294555601938731527103627201242931201764252246448031819544683337639946131383614457041608883422253715587835807016115602717754142472333152781356694009898004445823899842006407489589238923892752289147329455312404247755208380523795101239384358587754549990012720682866599985790984293038460073296238426290797218233372747669464015269204881430422739438838386988072365034008809524512726001361525704157749789546427459286696216415427519072078965765676204708762910259298887712834058061317182068879509627355230802280366588530930270461940061446449186278566424494208162102038327611169622442138639731157130118991853169915158165025834281284874149275360507355014927516496556894986881445782807241540090116176936589862811374592790322578489093397688160867085700299534572157942098099722053214575142715411220939886987456280116533207925455196985191038428157268351201092367995242906867995456830838859301366721852113536417244228370492060364815444971779988618739061970126506684370640425124459951909006226082179845415139874086156189246593084402747014710167254716016686017397691997662011119989301553540628177813282386798739883185480936514175269040502739923269532293931036045698425205947108776022321016774679279356253076833772206929809952133275493410764068293696256538097982992215020076190656713323330719175311095376967431445827047452191856565617305618532166042594645538561688375993453276738278878122231537281113417355451707355320827604407745254423078545374811259665463557459604327036854215738696224447960925936750083098914000685383635881778748642710688257878740799283418251977140842230489497915517987678274684754084928993864763498391753924459329312913808073876500505220066666272734384454049896801183432553499976250119217678755809806723324167826178257089116301798088195583791075401180509621601093080422570180549297646784115387691430708824753121723137940372365928771043455446962665999926233933298641137100126804081160276969402287136507298106445252016551733860468650406212924578927147227426763861426823676408516411947662651437101393855680642700778296596804860775179492212156291738671635464988985383575153249743158354139913221365051551384109030902755433236441202253007704282111471419181475709618331378229434207254341031555828186693283866836607269163836967793201021420290468133704915343805924654711497083540122724100650394974216418866922744736899506252894502777189894691329634675858792642352116335464746864260548561315778403611431490269544275056480384788879432956556048443391840602027045146827824231514065070221048519592072312004933717673835237093088565264344841946773453824132968854306302477825543502819595717543326873583172827933774101026347172525800055108998087920427447783853642749720654309224796057214003306615979398156970613660983964055202876699917225472402063960609642994542705915460007353673154988077390830015813351603573011111410928015412280666670587855509270333850098311567628516164924255092928303908770988934946072349028658560205422067037156804635003826052763710823986597931848309367641656360790706605233434111377931216120205880951461437739476835388395047212945283498654808648378850194676769456232670199871331845545348373608451276718005678754235887195105895652797804537834484650468146951677538136951845103083239037496571621433079638601544816144955239351112121894430238269540578601164673736647956520658725081592753057131343835699200489996180432549502052195550206179277993056424583665872167535192817503344992391833256236162650208149035578612440518344040381599135827173843373404529744999640599186566641535612424308001626179337509214296580882832219570578431716979462845513309683824600036989961805929879506603760712432725597536508820386360958809040038001760475078669744332587723215438325998399864395011449541507700972822653695839438085091284110416290966370127424988176163441016674234005068361676482327103889422394820253086967222925243407506026512988576358781375008510056886874328274718732324289847733542581504162589550238544890684967676489282970728115843511676077617260489135585109814789508429849836055936593710532020599790443697353401662876453206371886938218978015732190762998103612568387648387269853601294481607317618658066805968373389411982650087326242669600240908832076226117839991574402105842789845063036014199339283624554027683509989720421859620902016210156519223584211948820209123783927557185605541656205455347196978661235058348962821286082084034973119988107725904545863376610850509582385030751284259642859749471596754259240349558609796434019664667217572372370707851846466383706717029954169832988691247281876802738125496293898760722340846570950989432016548760479339467946851343732630392230933179068730316994180074048000687251365978579585994780199496523427286889887178135161715505778391587138640405789565918232137081400587138088365230471671271822006018608811257260339862403542067521276908921081552260329300444101890637236591957119530302882485868478256488300525181260810354213518122471584004627510592444870583709540835318975215236103420408450764137674234730058822034323160474633043506281423210829487240902594764411891032233740497947408578277622048261821951428217981124372676625846895195106998673740227323002602615059706421527460232699949700615823592828222978328684019972903653781681600288411730673324496628384032435365041397536205509105219749095799860595726941384024267555967486377429308583140664803184453153290815321549434582880442937355680052766701800094788733588609136494945838526892791365594342881741864555941029617929958126080970645474650902342618403450108124033539000610734694120978386716277216137083614515110500772011704214057510295511491370255453350206814116524477

6917845869435403411879135071947286833389662476101183017004972618956118398981605390920 08
9117277245282732995868083801073781314001876067250126926454645097673374700236767820135 23
5673242624788804823436290099963301097657305710745086213218779682807434398964835524271 44
8757305830321802494521092319912041786298321106456189823450495054397161803039568512653 80
1492251694878479554724186382786275823278212993978207428675547109249821824468614795808 14
0835500466875596261579061717590219271869723784547241129855757317937479535182955842991 33
6928140588480421571538074685311302335494627214184400563239744587537727518071466016570 65
0353750000780005476100367863699111323985862132218224624643435010363223985967017289928 42
5234113154343262930390735953429144139338742821872148418613127907162685826684720595466 40
3565113327927292836704215333378156489787872434723165771081189058811592205341344776752 12
9774635506551109801811454708921701244106349239492424226738349439407865465836386859700 26
0199154168385586155789670127220023220031686195419702892475742166676680152480824022111 15
6190982909528829342278406490395339672008649956965447075211846134340977857777364263165 86
9169876274954188683133247514531590023354409517149140813592731911461920067757921585633 10
7612547070933961164415088007272939456368492532718589155168814720960114154056640038921 02
8118648545950411900558007928394716199676003018770007299166134878103899189799277933082 60
3333833340579193386012599266354350647100912606346252385743463526847492979065780017287 66
5968256219468541077987421844550471048251138993654279944593202443898985134425672669327 86
1329504851702042670416810423988787766282835019312545495101087037669638120603127617996 21
8893187778305204501948120474270520457321254873390393028666808539289855145395183070167 737
2533915679276903907336248590343351476117870517797664710107502450768161655725395482009 48
0911058631732989175311841603640219503463573219594755860083208292675123788495516672506 49
2207206097412031293135743537452185545498302580415651798622780164689374817239713381123 69
5363735811057393910536917973929343197751880325243525860808275537409997210154008004697 99
2794342234544768970758031314906549976457271996996280332692090891558381760321398926448 80
2376910082742090668080043739925045412236849719409774670467316737887852049416564473707 13
2543728313954096231813376473848894121827756876058275472115348406411192866091980614228 22
9552490758852587114072134140163523811998912747789131397574682809342472823110218984300 70
2443999642906444508447880276686539463578359786330143574307385522480118057855163003059 48
0351702305291761937668044897455190062298141740225468793859809142285837449414294668405 67
8447862996873037366863397510139100798455883197189398404205851783126255609907516425666 60
9144857660683679374480652972403709933396292834348332661041368713447259629441715366168 32
5692987460751934900436754871245012517388228959426432206171837705951665664903889623415 90
3428365924676238921543162109473965009869257089507504114157819718945799485168292399767 68
5260590940847692555560320947301798892618229473834688688478774214747821124629005048761 62
4209757229517860733959886964186053995691274261105379964864827288214729865447937270511 43
1036415399504302492489038987190473804812173705725663713465147154131222056319569952971 07
4484542325785409319607037480624328873057403741431323821583556267142756875575513618201 91
7633010862837972585511567417230504719060873616277083262964429580482797563630823764361 61
5455540616980045819644670667810243347845988069248477274895298262045169437003711201912 95
3531129197138017595577974532179706899810786979967116140647258355731385280378144794618 64
5821634745203985589751231713640797468385145592041450052177212291446699278647652010036 53
9788997094195677954229000414384548714348852855651763080299251676444247682186490621512 19
1723425686851600605859780896623668832012839653122703074654818211999482253881430040168 11
4450362116720244462048282967776160165637897576349795548725510809105781339420347277448 47
4876989841921828085630416492602991762303626322504418296296521543856287607037421868140 04
7386309450159109132542103032561351107575582873478656260809325645074346337233422408558 58
1633853715306945878269202052395067272475369001398011496431659458297164868632204841795 21
9642449832794880631346462010891393287053134556150378876921145927268505146771355995890 63
2238650764778282690168036013061708569828863363533982166411661335548040370382100344583 80
8150558303401797120822493909503856609585571395374634762832404217519342656686392559177 43
3783255482070386105633012623762876981734728224250946153189070215082050421810397748940 76
5721499083247852854595100246795973930841106272522541569649389236827358143460772759803 34
6264312598278889441818491738026870449603886707186477083156478758911780354308201318656 58
2034354073422928347455769651498683915039761412613360789480997559164824906255168553679 48
2474050984649608568188917203699873757964398001165295270277237226019357555720232631014 76
8692847626362851893048492690926409854724936481814128316893832831257956621359883554452 06
6740895840923148625755911051962200050308020425737002899660124136355648802803399956946 56
0958857632199260300046853975598028765558317107063997506660476148677763563226116127152 24
2671096736184025291082552446153885776660277960808983028370687781398492381254517178987 57
7906769165132460310875518147960012167620168554361388753511114464644596594898628685003 84
2938167759796191272999045913439604283622782145743849108066267372039815968331145831327 75
5737193964762139470369487134483796533672088650760949443106748938628101668608093548762 04
0629531426836790162232434421625009619198865282501847807500930929896168789351440485278 44
8521019497293149122933664283836109583591179266973210503286586371961913064985733208661 52
4319891775175613307253369060628944014036246735791686124190767973072153896099260914778 00
3921829096605678051574245394812705158278656086176628088767548528264353457929751091037 43
2431480490509972013400938712099679922667327456972199757397498352955663444532434557032 62
6027829313689388962967691490051117916415739641516223459624143879984997239721062591045 24
2665562829601459679012861764153524786433047855814962571113956032515036318374506194258 79
0732974799065403378129323435496477095994159702169181036814733833330641513877132215173 39
8409381746568333237521245212042635149480179573706485748255881296241114146469266177478 17
3860156155696776808063542808133926222268057358604395739162738771435084847701866265316 97
4888647386824309419601892875891202138727709615384880950656532073442058984978568214481 09
9344327143794129234072975479326476182962040361443641127465240436917542835856614059594 33
2610091323144864164204976494795520171710865170698122416084821707217101649482479807749 18
0166663180760457163952518386095827183272086570529825589266492312740506731234877203497 79
9829560941063603051658168190384801114703042390182045758372731652085922539947510938900 12

pi to two million places

112219426665445908677926913711549507896665766765460962882777751995705545072979236662085
235078168943400320475437404007621799091881351094993966943134279859921580629270421382675
621435340592467202350206425854109685955128295988801679474853488276232260898821426027966
949488339973538091153102615727526061516646757472311267311304563021016442756282782191487
924669897532097832652921682584330479085478336542697584330779557195200010120787240198813
494984438436763827041174210036951169011180168326999466120100860532094157901928897613978
403516511159934642044414827682054550634184830616197994602704896489524389702584341717731
903153309321479805420208961951250759293649016278147407732247725732201913504568055999785
692775430546578798428594684085867841341145382412407206567559826482625761903033834174251
848538540384703710069087650808535086402176210101567282914356736771103511643978363440428
302347807354566914381770474508945872117878391541665309247269795195268639282330037168506
787620787754817839108197321829047879932913960788741768330818653181999406597926782213227
134596324714095294630761973967499846349363609758067253661551807859814534953582160148026
023317625201506366399391351428775115353212411225150570657231152085376502843221015840618
982570047043917186490724120891714561202491730043799349994206586637985787346060480619228
119464331562925686710879697123496236406193738811218020737915981801097590801132722578430
025011137880349579204391899288300516242921760033764107933719681331920675829918260784852
475711775242016834934819414005391646393521827371048915003658047925976158343651355349438
431915092146293081995018359167094253026540329803249676158439634711435324714370392214861
784382826113866885521598461344505803302636914394174355991753787166688140045296893435198
765272300845846550156565989521130110485288169394156867063517831922185595530500029864832
544477477719955016508265889671396408898805679580669160658060940485139280102227697615613
826083190760332454846528661464942948396677330080707320067510426251414296244714536875097
068785066005939402651877861032765470280632572990619689759188738667230511012379493292597
649574826255195927394471764009255618521185772443088894589313045709752725867071455651423
603418198903151954572188621149171034530596578450826186807436497735831757700864758799643
227445489500780966711961621513676950853089233612386662834811029398046074355342727244281
049032807567670033772711209491284344874508135688221560330504388351754108148303753443420
841220816836058132623457677542793161986045430504448510555800411679433767132055814705872
720882536047310649679318479637352788447885205873182866006563349325602359088898353777250
797020050541440210559461072076492440913633722789739946639751234117883663125090061416232
276570285410485067974498127181467643084141030023752565373049527672754845459997871633253
310506190240215181468100146512628510397598394128823698621131831524776496795777441913323
947985528716530231998698023983984731981788171333103443398908379580000513196534523383390
109097044471434794265026285740315181520354650728231183851986580293621352243797543193801
983432914312502757766754316869888602865677013500372589696445868683417647387839066544421
819235857731078700231917445428714160030268283724049464360347876903573326188114310108132
1885527985897303450534403303722769151404531823618783217199888989055082908966241976585598
057834142873730648098529078214594112649499219651136125677730769946058020640723918086690
020175695641759552721135933758979116047598231558725356445682571437465856688982037370549
704529071584697376355587060928012017697805329357967838079502279220010520167689887325410
838931925090717428881081070862320755101848004176969682629039239983938116236638478713081
932018555926786589807098502295373949421754246962535470439547324133924764852103761177731
1231385001618713047106477839324875850063619911967787775326807139246898440388265936051085
465236922261927240349912103831622629724114408355686804998074604837135925207539017014469
373916409286486391905375739329455653677543563294895308547919735618116894346944344364303
087144425491060982948288158115956356299337794739220978511040672166448032053106709130375
948843457873439847370765374740479308090343824433970583053269585629984793830480817797508
901932397881964474728134854864856399736790769039302521285919509594533031379751852981866
262011761260953213926339182718256327583059118937210691577643838872278422852900912261251
408052315081202726247737066716153729796236517171183091817152280526537593373755812823486
429693226678471338695988769158095081150499363373569059008428920070548252546176895416471
077801175860714328662404483055236425937757985524486960807267305907650248851408147618917
999896292907954060691650986275070330910088661199318365347811068950055323212323104099431
566975712843211058927290756266529830683461268817435027634457348731308127878539668259480
450244508994538506262228156572066562590807106009071947415806434289617313151570605581139
989607656842772395481206246549279224664410867393017052678406522475041053604323508688152
543821884057815229519878956064995606982745328922732703853758452092709242946673468959337
778965806769512859044905739913079487625397989989468534486708427632847644098046534885512
094360642889373837105351559587950751036819995860092479405220515488077774998306131379026
412827371575710612817362497836474502072277561952126743273581685496119698882583112616695
052224021881146693062574953847086995865745998878927868473871986438379048046374622281612
687127634511309478316617599707595085332574602849374001043645034556580449442950345318338
129078508883338583786977108498206651020627957076698334451779345271803769114102075574774
315429329032629532114978826203515987412546422884395277795499289564754347105898585159005
508490056969036939946380541274407827207958812061095018266675052829100428644011596909156
026024587211745604551094076846979736827481459790404552190484180115456634783353438088153
414037239817881907757630647233836848076617187887525444073186583050118647563203017139833
900789875424411026277749259455787263151608748702504806203816260628415675429971100845723
607943683883177569711607177476019773629986084709225612419033443038680607516077836502789
166628360931767596955301493681279793546665239389865492208212613277637898202946799581624
398705936239170511750705049392442937122875207210047900369520353054174702688100313142753
117445624640735452001303351544161161284530636382206203182714120347105733305706095610419
997744129437897233361952936807116294649741746746061619442841957715064212449115406707312
213842064141269671545664389159471777969493519583468433678322141303743107342917437343763
544159073773807768335545452204160755832450014127128997410149470548864724341589912960922
829862407455160589149631021000595871347192097972398368312801101752643186861118351701735
867540649265791513740582162972420188375102977202769228078017323536585248661037355246360

pi to two million places

519741758718234908738197745199604135160468808608272559061048282225757674635819166290343
907054759703480904400430433331741346142345412745677987258923240909151087305920242790014
967370151347721514257148023878189727890993192321188518040030497628938731198868763397705
690319074145176297505582950790551571289772603435467222518751947227775034780629888158027
640883058588732114089935625654452632562629304285439933282550329502893699077705490294707
962200839029322144411265738208956854344785225355843731269337547937659943069910056990821
560314508198864943894886795977365202377638052694955587145427065851747444596468235269410
568519337370049144862376065979574374294931376284954237476962984236204040699032232862548
282233542016522829128844342157512706020215383178452185648411506693943643644633903294628
692150012003317372231594599370244046654644010709546377862736676904564259977586034142337
627592585363126437089730757955269968503132069091830679132654203064003148245598623926575
975731775912862530894654125166228407163370149790267384325301619010137297886469540342569
455726305220387629423264806499623816308550031265168054478855681997310896795755442683922
048513091902688240337712017786398604639800256037206069295346015367351300935166490475996
904153484422840649464357839627395979697011999599689705500713980267143153912391461161358
183406808760534667255305042239792809656622109111184778965033519003128193081404706478740
367155552114034070303989072232339159423512652971711214491591287469696454455709228043473
384101385887428050725149320183676549865442619068767503039795693902421343747525920284449
370703219824095085287439294127815958647543036695336546465043812295538569601870814630360
006810223193535677588422170662717787522895393749739449846068819589926057904266324281818
832976825708783089016435405464175367797521401491698161349930449104204274172990731837969
851312455958606399199659668996108005049400729639709895951757463495011315239540543638424
771576730579689978093511231012700060683156013470561688420818621059065843854685352265309
940809550686451819643104550056985286403697227264496407220910728050656517590053633194257
188261901685209110944462304938727622601300966509801815021611618931499175544866484510193
964089242453518586296685358807237025208629039637513544240841676796106254077453543971872
022038982925881504881746263214401932459126384677645387821490032187360528840161581467693
409724342496695967974551295215247541300383824175967755422515486890349846758461066315941
988121179713345250927530701408561426350301527147370879790229663568300178799028808419389
392249188448911767008038038758887801697701113453349115348021065850875700255173632568820
097595527487122535718255169787531509556908689854648379484303518706149231357340296313652
791276152623061043140923956535229749326101802357414494002010757529248895892932458035188
934833623226621107047222795178964311353322215133111281302699657056542366660712427360675
833767835191012510994437030462907634661496449559967303212585228400681288632060138439153
523932091157906047341393629732349275918089423365652609394831336481029064358631183082596
587859784715023449078747678799566742485205104010399975710394022063069173474202089699175
600528889873675939662936740172098219541833712823393286247731719643862566531455126992223
677667770819864349679984152604519464045901639578960492791392910434902756838172684047052
298140890671314915262504417545271011235786801298936828493391396383366078142291794554344
916809706649131884537812025796215523212883988829483096025415451583015564562313284503174
185769799799107895565679960825291655375861223383807006921957963941983742611767676910050
735750147104127391778359634794411592416074096491892386416231443150984337999995741238608
455687906501796604659040119009310649145976455087441691709361597805467174658993017041375
390468254441984930697739630336143300403226370441388424564853196000910240359148604341956
718891985856156554657750890144317771286164568621900128459460742160742957104583140046201
246390110210193236887462318746390639051846090824746612222586831716989063606402540489350
875060019353832358477797777718415669271122400445516770541931073038836943879788904624217
590046666091040163622700645067167256329813561916957758561333390398277599652250990403870
932789862215954943799706306480709649417708005812270559330916211844004635893785632435864
191061540068204787901621404457877173981029526071730009912179711375424333488226661867180
593453500359794062610169455894879852873823946192592738300586578623690127192963826592639
378195968777634491927813839152734685103171283501167754128969634017633688033476132425006
547944835516002423125664608010786702586037609939080045175626009065555413098404273574300
500668774331352820617072990338937053222546700420588964046523936142830791854041696667832
407095595877094232009415610955534334434913854388408610824862428985961974125657170424006
787671236859353722710915670406062194347260204099413954720131755244915834594427491919291
350235583440418720697434588605383370185897657206225466863899147406171384409111405424448
924181252805873784443759903370271443232078520464193147559475831429194169719062976979044
988213080192587590485875701028049890092764667431817419312738798791908667056460174141045
183654736392112018312726421352990750753167418604113908507917404417265800928896640035085
618299372472136843413149569295704019813001606087541279574664190317973325939240210741676
702423535174221182857151618329768142226073090296369487133088077855566632397283342252706
565073072518903090395022975145545081413444428165414364410492175062270643628610175717112
048366581497058246357800755045626453744628052593284156788579850690105804527975626285722
083047835436681313303172332381352647075257795233015289166395286543189995731745780167826
728146022264038189956693799484242109824897420088233111400134104409516093083130905465503
159555154739774802214624067611052716137579986283543969665783552456700793604975187679585
004347859444483487473455259996323925882010445287895767233391108520813799484234715318952
612818751089051213546549692460665576734518571774051139809005074932280070940569206554428
799289768091385328823923127426496379071979978524909030460958502032813011881918979875386
127705098311267968672117810060942886033416074080204485324414144579454721054698929166499
819415997508117083997585525312534793070723771948237383367606554185021133373575357116049
840886306962649890151155629827792230433349844936783915198562685043252008447985546296912
629978831302936306463370450331552637520404739224157072857779988029635328900698400767818
969756354021766194294424753725649845722655067873590934057238979378191460803198271139248
049794110049229814317594991993103280897957472533768146061745433132644892480370134626426
692631734244357427051774756506755634133360059178313763737592038902042651716918654224484
136094659863626775433322172909728067721218122945017766423271673310919275233772424505908

0858927565564344115484438889532132702856054060064352403401177439426383149269420366767731249233446046152792228711747373206820737950407753807024875139736974872208079418362724579267158605875684373825696601528845015936360579976487955466689796317276414458267140983930260584443731183219527357824299238053046097925321759529437646696717859956554797526010481116090019255987703160369289053546421969361790098547207855125725975325768780341842823941930116764712780200132449054170687194060874864250992490041243777990256723623875213448754686801805700267716590457174167509358537466327846614717022291277538164893581403754204131631796620462601683005835840278084250847061028563214676349216441559656539951244111527225209885576318078808628374443953733638662891394399045918693562979931691320743108491238782696708173798380276007328352371305703835388120192178074557053921312504821976693406394280234533509698054521919641649696642051922322293325224980990680943829860823933868677395445267363219444015986590406652867020651004940978671302450896050636311474978975431863273763793202279530010877176844026182180038590906077029345978640932969511233853261494565585967711754424619462086741789881434778027023789183865475073672507012316459542104230368253015499272923049706500688744552908893664655148193848056374554470319717186422720965870630049834366571830309458525285629892546132607901723746233601382089476404721767102107836353063283918528942630970835241842688066639724543519498745949527236882351106475938370531361494523326299006061354420207900800784435918514238109540624635928800749174723260142829150664735646949149075313040411948746127584251725422993376561913228344136159688451230509792290809347780610001202430793367546069567188678475890291671628151047910984819968795053757476103438392892981834755913533728342888492285393965950164594229684902163576984603656667706884979061493789586623897851395030195525207115947916243038057133910441235127977174258949971813208993972409457630504381765420237749372929236666408582635630470188942847136621796280779475814106472039686900573358837832383938515643676929109532126309530237234188776377595132558571988684156351143465444921346183625820017739111963565973620917480289510711911931216161504935661400198915406771914740604502008489007852104489840715587249131814241237453147390958592854919261955127528154045555489486053043951830551638652963511435855426789578843322470303223984629694037003638667059755189622821668494721551679940102372605276192061750456049663717076263845953005334443878994432854366301464142075150267652998714841483859242046843515052858926483541295999641906383622255006162029851790807999551716142892743322162806963512162902965050345455980024292038066113122499875748777814543334957813655800830045879054556552375964308994728294156584689806294311272597554693021887912731035300216864227633661031890511086335963986070974737419552934178507801365337868776791151473833252513002371910235875886798039385529704998321830389985333735331510345804434025730422758682609723983422315017640803327317622631967565989772971839422971652277619673408573444137475914779317933398924359940581396032281346592478558756550575159428611613176739552834815080851852071547951439266728751057441386769718902087776119592459359290863839696200576865099629303818145491280734180972040320363366766499443919936264145227145073503705907609385752440000947482297136237721592630836022139158855909461407406976301297086569690667624421861836355204727903533175309360779778798470802390218595874487896074523749056283747418931026806280481743381300198221277706092184700815252327145986723437854310497839043649305860764535756025983828162540984963998318112971881438639542654408280086193057291799568887988182572440923086077708626351313609463097677409970284727668296668542908454052022919003034322471898204993824806075163872794566894084973666516228133694882858339531350502170536111817502101016960937372882174683892618068718872117122032067336498312812545726927631056482587901068575208080633928751364817583758109695596993384841991062692241136561171129547967778123862393493414008547053784583782851512327987308840937357211004586036545449453018681073293761108679782828034366133339780538614863594371635008771193119559848021828992677797694091393084926892397161087516259813646427951911630339179612919001760953221655734911037471209457900418602899221551175915683036249471608650518632279742956136519830351897142876022375071605999046852270284824202570096299382791478180154066416917925969650230616749677247841947414200924297729713888325116655222730338964447305270490241477275647154092376806641652822611221660551903510495321695382999570174114651029984898251643905699093945986820498552583128764456838984342109365894310511407555838492789369974095013891988212479916209888392435587365554523753623890641072663136573386887150143766765743928298320734613140394952617095308448965800565584630952329631871245183661663042774985273620234906785915576936220534724261112532638914302528937667373926286064609916642583999874647434141639526777322469333134089892282135236716095628251349234859268040735518153607196567395027069135357634343767443117249084553776703266698841449114808884801324930258438177011878566635729353987811406465883694172838736570843375751044799123597365972434455742718384733620516409860393102195921211225720343651001396389064945296714205608908617698288316388283825229707658189611895457298258107339454017277497834540687764110778004407429296639807958026689312946890891500618618418921853041643322169492782133921182771902167520208059672627494630028053887779459621856830743842989256463024089063366760647389704968736267734714319346437826952783760286146583892789336232361603686965888599440717090143857650085623703570747288123004277647474703779463200055437274736584724026183902508185020399413109503970812481221077612832459563995640730378422829494179041799135365337060929535804128449019567717436326558733430284401481499075465103281813878210907214339837454150957218087233216393141184885404882476133156499354030311313197143885666833802176668360829503236040595136775927155165679680295859733803613440693078137573011613002657970242655917863431943626466230186872587963057556366078289969536349814522388660073014771879198641606614390805777255192448707082910976735549911200612317536184781317654395572950385452923653669413348562178789261547404561504523088531184389783350748069944802081583047802929139502742186788691980176555468181445443074191102292721931864440749790317259939771362180997111761468900013772407009234849963308320304297321967582789400084665235071155111834810320144835294744885188632413303960676395857662392727435386476553325926113916010589720694912160419438436276922502040836018348118271585534392555374582362825523726253314359699646366678255933821009175987444402718522512906425157453594796130527189599488884782435317225627541310959998504427747535263888711892649770

pi to two million places

17055022010568232530711543895647583031225511628763968835143262728629815240755887995962 0
9804394659688932951904490105741914199814985879000053348961206916311175468253485829007 68
3953766264145205273936078651380572241506897185582776538952151358565897640730148811113 37
3869245888909822760229733951244135075100387258948206647854453929055116049254655683017 92
2363526377554626840904947800372684716102649508826206935748463164396897896270083376303 74
3017561945738907884810424309285523631029835517451745446606529767081994743205995165291 59
6008715652115461296735413957327751678443484864598339137584856250554601750792209883587 71
9338601394896751483376380213903415290428362645376374580878281537904708544575967614210 37
7236123296196402292022895146968811447058289309803570015041036948417054728692275197044 69
4697292034951969766217664143620321098749717299081440003715739281891528408319742294373 36
7747782587092187172252984069305556935225836786253877699037108329877979505169169629957 60
7026624877256111532102452287179703093738005633540592931089018740049575133056457554686 45
8819253443722717048411076045834505257242917684423561001394566824566042880004740729261 91
6575440953365000544583331333348665819727492760012423238225171846893069408572324615542 39
2813887422027661693797935663441045037115598236757714312491279623141150164528880440942 39
0051734645637803629395327787801636044104275918642025167711823591081249478448954880725 85
5125745496079956918012971317205403842490960873636213513109752862098622426241874243578 37
6772912641791880137675200320563326189241018616513034810244006856808606110481129427409 00
9375274857858586840922973157793668860861996442907361574905330345107467921392139357341 46
9378267840533131992354433034607195155205700661001169301061146455648916805009145005561 37
8494045436931348591061841933018954548638521986008082004028633222685779286594749900693 60
7510578023474201503531773730083649498916728935090935421209752074727250436661880061334 95
9093061140711046459924475971754240199654073065408416973523925041563550039026440029227 51
5519108629413672360002694120712781130876365316152244335162266919012310171362950133169 80
7700976703122592334099542352764898442089109243902756481617004072173927256602429583715 57
1067168541418713003571310230544725937706454206437302838147841910218688218466567138326 12
8263780697530988608290663303966984683962467476885114506491315186155246294792481115987 31
1090797711529805880920928581622770652756777195393112035731943345934347337295189941572 14
7227619036780043087959047999266424281962016300988837148845043980122446245560266040861 61
3319972328497876159317126814004045056189666869098470823941370855181326199637688970124 15
2313781873313001560112199565703541410653535638452439655642672721743450531708970862034 76
5475867412814640719792280574469540684927959945904582090187317658661625178733129693572 62
4876301827480530656002946241810151431318690240874938411242158215083730741283373100322 26
6839576907350698817682754848177304995391310318465327838386562671747600180627887580492 54
0088784039279856464964515527892734502001541003131905096271080962889523768982872651391 40
8194511671959166631818853373127982279478096384186330812324934368273270884716848408230 65
1068049840198996614184868219292371243226293284424830236101798391004269904077447991908 37
6102111123960672507192997931365177067316450479862322551979702992566531510159660459669 01
5088706888298252728640598951425047655646438613971390930202571945855825271392719811327 75
8886955446292060520268767522136796682746877587452876077863449138269956348008254414413 18
2534720494801421265432982967846686054377906133891020607653897467837990904198226428291 71
3569800434724666969930157511495371520437403191810795486894324062290945862326245220966 75
7449528566016465787368842465402656045769732900129583787420171151005742659749253328682 58
6625702458378115218220127757607872278753625441767851681849197994949764910003693490955 08
1945020556381122496479676624965078580232712368944622866979631971539024990109911767205 32
9659201024327415836646285103519400541457190714863882469446903822456883885007824410392 50
1635971537484905694524560531254037917603310165347750019986495805835536614207169974117 33
1055254042055211077318581045894610462735580709471466835287835222442439519510959648019 33
9972822544123729119753352333978820050032094830778066283364606324667100180087066628897 71
5761318039445308517785997967916175623642457991318747995295187367560206724336078627831 64
4655047133342557745622032970583706520846148146180327955657231128913791506107878236724 17
0631574279086027582680483282048253059594486535530533557360894366837877887790883577331 65
8156656404633363117896557755386745135965474379288244327761776652997753788443212262675 87
8961266383306843849005800577613730946043245733141597876165553722630161642335345100237 46
3536829894247824255806480766433618052377415631403789337126999008115460840814240586928 44
6408742389124577519366466994637359158441193177950085848065280520451386178972329910964 61
1770976297169880547414864040358883927950040568096688268252678332587535835160050579458 53
1484837770296761832636064913660564711850804916359111816805735686256767574836279625954 23
1444084268694441780846545900109830083247012732767325186296528101198756674251237185471 91
7419644610996381436922527648768652429643328488026710488044888015591064476982918336443 25
6383798347892249924247347347492558557293151861103453413733567227462578276718755128522 96
1571935018632517217599994227794412512769491665964117645331130767839435875570151126833 97
8807782308932767292196739065650167909884959899971836201837724669791646815888400401508 32
6413390170244028639070088310664906834976762880088097131577264334164705251536471773066 13
9272240563257100139729989909559374773055963634856006159849612535183107450428280599101 13
5615276461371873230740548644387095103762391293174413926799644747323618213633118585804 06
9936583777606558414953328326602877854696894300229268531019343019873705871735821809800 66
9389125076625708474659506289918468346949911962050562881006235243400502407512125659762 18
3568345522576684049165251570758414614413289520970093068729022716370563859061059216969 45
7351312296992925835675318834452109537570173563261816644245918307191732592805373518481 83
0987229456262172540444189864039750384513606111006210718088689290538855653803212319776 64
5007978808922913907197183215533766071468815888614665937080218111848640949124415780158 696
4737239095958580311735493963934232398121883858322269062273043691547964773290362031023 15
8462282118660828589608169094090000618964421346173446825214338630608641076491303096303 860
6156122694775672705661641983226612829559405418526700993894418145266998151219653967190 51
3843135365503213138068242451734889475925031241924841752557403818235113902616355370936 86
4688471015259866820062966604332671588470284672528273675136369158934985721514957696957 39
3793129333387868587015586438472121908813119471337087338232750005623992374477172103479 21

6899958700150469805895623651885426829398566671272305833174739467989387917984475726396699
7256515093350494496239329894118380951152202738593619916208931559373521319380127029848188
2968245692466401589102452240833407352947237676601871908356625734394683547048362244546
1993712921994552160770522653798347510666769463255511566494911680705230528173086908826823
80129412541814673058459343581273433407463471098101697337845113700036146661477797375667
6762211878253942360370654923722566475192700250824888860406223409811545113334223901773684
113599153372373184676634050715689668193810358548079907399613453888826857567243504591899
740069104470411162878652679201061613247199984482371523349978363752301431351328269553952
901086494205818604396159053058259754001573475299749827230953870577210005396418869704874
528973591568792707994416581049442687939022278200261738842463895922113926387495411419594
330027084671423706812813778229848743892258019606732295576646222560726500832043734636892
069742573101488777832814597005506211252970943955134820697067809204578900900556359931930
300746710425701791847467995201645098538151015396961735455277804306267757948771097991362
593662234937064837059816841914409008692838417581369607702626526373984217927518655855340
018024947388424795076359369625165815980054901107970726953244886137434993884408366146859
209020138746292972909384539568930915524703254564948483425584353927500268398089195124385
714727889228818004727979105941649371664175676509443374654097289014406328130189143863392
806334434942400260228810471699972553393957076410706789505905241632902212917615707202813
379630649859882322426710298264264548227932715480458764992124132126818276723090347559579
303115848248948301417317193431046635619982934226608452419772743000894757519064436425070
401157381317794809599533926915579884005478289536523956367416572964880463463057367371172
158099902098944537332554066492445556570479772079145812304506188806693477316115492135285
980811109640356420103206503138783298144430856387206579389407056232795868744608528406980
628390128319940403175369817291011930274216487446018619632159446845380755709872212964758
426105804371014414489107488133766721383545414247871366665387182071284847617070028023077
139862001523285284674980517160094177008483060781630740674129158570458579809143541609290
613494597096889257105679100556752900747504379946338211192119990091221539655631726332987
359358386665001897021037681056553912581127425650365892142910191935677400796661271382307
140818828418649325456700504789023579983462966520539034526722973679711222964757638427953
370703079415632893117466348996286910518604722726888778758797953654811330971852577488362
549950780896238311682394650511685470862613640217820445276222621850946877145846667658899
94793710284570278582886494557819210247088409805488404942892027586325135120327683691655093
3375756877423110361610668383215808025643334645427172202495621806059358605778368398254
618223644983354199190818175492396216871052804951422463758912011361597998438034553898743
686379416430030513037889583127924845498683990658600640789933528127851940984016719729727
069932213390718420955178247520680268463616539771651234574340304432466147817711996108553
728243091711263519501191538103322617009607819792294603552601878766923621248636248851290
354428397379232513895550640142391307665467538114524402470683765280641424872089134513796
385999449351608677107460143274772385102847494666363346194172301607736297628897725128300
258084687726530151682029250873001346219923156538719904106055074193036339018444239787442
338499826069676057020535368456464272727034894392366484590024597949473948604166711335717
028120922680527815688335313264331759029946538574852184710972047718248056721561923131996
627637828206706279778643822558087274035538875576372258299905067359154147149474372649839
787057663311505334211612174534089654152155497746247888629118303526040368732822025070893
530843523458081507195695889241260528757183964963055076628600911167261753007281738884588
123735985372692992626426660021729769040932291664578008028615731050138340599605215180202
337467493294109576913999967663852175374648850721464227683648609183197332363921592490390
006967888121011297463583734052586878544570222146208736858727966414530176263354155887940
590732122253946707378265467560810746496041804339579538721133064646799286122948571393385
632976161785089115582766119790233799986635770474963796822399350957954508205505111893034
779357024430352830442834702410590461224680811375399707428743435120724179827100082933191
371419288771409898637054627113614217060316038877158734107562660346260346932057574636326
530612059614741096786436632812848924621727699060440035648313727017184326110762869070629
628767824833725218167849520870187388883526681880688561553821029179384688125975922387175
756873776636521727918293598889124812904849996547644596555459515319230198673421439626905
245337463449860371792720542796816889295558794575534131465881283310245574792805020086669
571693957780153414390677074688444371997229473142096243098464505318539652190602671100605
662171450565239616762915821410039307333892918625670333714472417092407944822081957934969
811552492557325408808831648195199484925188597971817916507188649753536943195763660262426
172292425480056059572174815355934092538283243334477794234508946594682954801561640088402
355037323496549878662171076680106251027447234054777387228233706324422346571309983353563
617904512966453592077279387939270095466014610509180273269755513571365490940517098691433
383437353862239566253167050813221234673687814427618547883058500581078515556788076939732
421220873066182620090830504150607987867207800863874831471046796221804394757556309908624
424438280907075163603921360973967193494081982005189308463418418513775869421385970025192
357210352359781475656283700649893580619427747837673671656860440124253539425460837473462
224960829827247240218753734151044388427140893032903966317059852727435757224919804396406
893690833070460690336340376113566927200801720601652587016620924656531832178359034831846
684963362317735446303933793489237958382338014835246620707688841775646825727171361914835
528944036115796246825347099957785414816484667357356113380319206582213549678296294583794
899259090657150858589924036877724709560225206030410594547223573430761992020038703424402
223490949671809511947981181231766216132812657418887926717804023857800559852923256156882
467651635904834058800448384582302419984176242039750282144203323781364695612918160908807
052269274478502357943715614285496103309997013947721460617450078824754170067917818813373
073553878679601012421924341739873289763228098676229374534372899811725930082223246243759
854001837266087383264712072554491306433644995100194782544525542561198544468963386192334
108861190236636252006167177340726844487670870786339928851878574886890695595205756080655
359723625548665768065997300269614499791386394913764334395117818656169724575011955527139

8766633102419936496159367342333676593518995108210508545558659024045243950149586570975168
8017729800819922259725289161528326432871330191207202622505599302105520059364272067206 74
3658081959198389468624150753802751656622826042558448723696342315273704964736012472937 47
0582351894637772876085862713952359990692232587035991070927536077178731275481509403512 70
1381707948704002794636433688427716924012826404447538300216806055597399111532756743042 50
7916896649365346106649030339264547982624507527529703551170293895493926050261167328050 80
6361911350415038722553548052495030725922083212991676993938578960521919040226593296893 20
1528053855848832676736575685837994286855431488484598780431993710784840893374197790800 33
8636966596327000480075341073313028695828601359287661350885694130726895270622119444657 09
0135000285078170081732969360699447808011650899774698383275335446223117890041424456125 65
9236190671377822188309901262050387138637461107547138243333426066111911249960431197487 30
0355784675385580931940536564143872408715930702800223362024034209266924841036541392460 32
5281513910602580690086792469478464151377425304908113337485925659032521084378705836901 80
3059338532970010969600087425044814184589259856965345569808272371276255400483792707641 01
7020870006758524443257752645790361826803605262387899668756268684575871134948261702787 26
4207740532779178396605930246805376125287836224216318147642047643334565869242915619461 74
1479293032672745331987962759051058255643906412796059960516294105583577003536563242856 71
3972433093599861784854409718181725544777914093209591840501649984386128073788718816754 78
8756505663196319767304705864894624045949269766453285109198744337351215664488814513250 97
8299799856828301830292718126587579974915259421426066384493476182366943613010007783474 50
4544383940594638253164174696216796375403949152116008355340458730070341674476885386353 72
4175911919125730296287576998669830602845055125425577813049196573537081097538898051449 82
8195851720963288792497596687858557626872836385771428233523466567958926948548919544874 24
1952228540275810132725725884846046549518516222527214858969727263289515266100741919597 17
8328836594559768570577262847855954488372407579162883631484906477914553487265755850111 94
2264869962439109009594842195050118285457021898741034571838981790486364649082967773150 83
6776997335515074170081220258053885245195363983453187617781231829233845161492018946787 20
2174802452815919012422558651698747259021550762497491226737659456330761660211943484032 39
7991440702488173434330292725710928657389894240649561810909797654985118424771133900728 80
9299016301869411421261170343722496674766825988818337774891530013580023146024260472055 27
5799319899409643161144298528316114869897317486423082626493416316845278016222869452176 52
4878906995560991510960158794169103884595635966852936125991245729283769357494996000637 45
4051029331252353719421156501331547537362809071428157731851185927633100144847811577305 15
7277411636321762995559710643742787164040829830710463054191559937148153916125547811264 37
4389039745212073575767777507421150508298100857375238351838357539933202975989157824880 50
4120705904644073227688483087435345122645470626940975444513697572570891505730232425735 67
2721085168470673901113721018280580460322479166007383269145415931982924325433746486049 63
3965342248172938325475111403759384377805581002679023593384798958654860787419414168840 73
0434173496904240691742895829113388157394322771015619624776355190240217127468627824721 97
9967626629009191769556438105385692640185914761669543194077693489655590603130341579094 45
5197560296624875454879091117532699370937126380672256751463060740233445983148205780778 55
2538169343648053568079452045386888722145805202281371652698201162506165729737974807500 29
7233921909750122049474941700659392967296028738767195222550630864385036022864184376624 00
9174719032833908399536747468613101093275450853701032488164563575489558603679918936112 97
8761908356737312274823781827018531064263096134724871436054931890377026133291202074118 57
0203965492336859086327290034787222376598941863189523397575264482623228467814741678394 17
5878779284134112230198805548371919662085993112969783033886516758544622791340538447608 39
4235553449033163300012475799652761685263194532930959627313146726192661819832194566480 04
8912724042165730413638330422266489785155626456553219711447297326058213214861528010097 27
6801510402989652020638606860045981614285237499912085209347292307977350301533905977647 83
4289074748778151734815736268828728847309608641886645303294760075330935853802770492600 73
2884329411952086482971131759375253443889588142555483851732952836011377915329011335759 81
1747590808295590475065765845686049895198699305066250601709707832987606304681600952109 72
9778751360186320545895578180059587571719117232350915787137515399125515052606959476059 33
1576350909179773320833613680719455156407495330357188422563693171183439732516057365037 74
7321453500563595638262474936382247058368475214260727919955251074551304244338336405493 97
0033337134880029978594657544942765183034121119702102998733619911776479300476493264915 21
9099177723625580527127257799284186221272202594729578364241569518389042618629319498502 39
8088279018227708670870719353980183576382804721762702614902491846340252361295645126001 79
7544961312208728483738833193857402018908281767850298505262156335275083393941014555472 12
5637636854640790946476538165508050017967373743140990894748416914300650811821039900941 91
7142905544287434869177808284127163283349333738760189805192382363730219765007029919984 09
5395364081929393544844337867252570777295959613871007157921472183707580415005441349860 04
9290749969893790348810208279250693005742360174677126398250444798794775513836887538882 07
7572120635319586500300839106544714907549279711557218460901553945733251863898128582468 77
6959800827417655549925565263787184749887063291494390803074172603675896802548718373999 96
1968293266122412170677133971378809252017026230147178208006391621359820529738550555820 95
9403332647089156195552235663806264125747913463825374925991288014312614436201171810061 04
7225858418500286341562115688441856620282727660065536243416531861727054704601829523329 53
6489605733307453064730077394581740556221801029658695455429623213626808519358450025873 57
3058666519561744637181113447756361029316422929997712848473992479149975246975746167652 40
1333988711899350291995072594035417767887842758631733862021231433122324549995210226464 19
1705902063721563648704410426983363333138216958848319812093693683591904114931623478727 63
6627592154568410702417405297925694298191498174063952714478690511842343371955926123191 93
7579062117858090932058847948363057956121560105651820752164895293647504997836425968788 04
7609259997018653611113136048448104343137267327149325176406779591282704180928409930214 80
9574578663493779227121375537125494983646132191054790119508054816377823175531880540483 44
7456823482955282130638303595464797553133860371316576407833408859359457376719674086252 51

809781817880369860116613883471299915377112383415452874048995646902826030068854276345189
623573554618158222214071967866843102682655738115194937131618234925304365477918872773039
577291667603598906829849792753264457930205625004198215817833679758324582016703344016375
194361307936066877060596155074581873007405885541857077712937653954611123552017737452675
365027712360102626371140850249375451997238811849720048529540760537575504863349851760393
434036589529608607344805531229553356882145671180576047588419420583496338454216537702022
628873203281426271924191134698071535062364060488010610176139643065506664667145977479275
127501313346596764639606994405703118605608781228063289678165765372750629672835762639748
283846472950189179856048249198500799160923239976671964767833012638465080880428311109850
255466129686185565035001236106852974356644656198492092110126637583119546240112661948930
083828438659999992833337948765982135588393330975965394351687477025420380520337338231789
387828254304773685927377235747888656685873609256869105637744685113155947867365164849217
860389204692057392137396597629342661799387598861055713847390158695380014400337739425963
524869263689609087053952625109612729088737679822241074767848829902625921417206513544327
199164599833303338205097023670379189129777113900021796464556817013808941825846295987689
363592439379803701260437000195453753209475856566862618691377693236555385533736640181142
604011712634532053725124468808392504553806425476628093450603910861511948831424647739359
453611346263253979030531061551540475704318358069888911685088278357540626047008133489427
756419881104615033910819669743603856073267087156087766588589106089607208747158269701690
562668719926815848335174102410950604976133322103046832009316294819666417866410892596335
403862924130152476404151953276182470723527897727695774543149145720405417995318578837498
108505057157671110581585216705522011002403121471715798464585433248907341098761099295649
676156544718806244284933719422274740449837985964755841334894107426083336152120775019298
015129465672084211550763881645988966176464369760328924324510530298250512242680703731212
808393512202255408415132029479998057513768649336556761678499471334926956257739078483718
248283155679298197287786862903631056068580099072224021538766147136448019656148112453886
271654233428875619797920485573019299975004175881862203550843526937422418347754235756053
467255495615418988381785602926769085061313659528803917355560245687971772310603217457604
975950232256294196379063093795581044909587621359167786582953930306576530923070439867570
625760671427063852605547595952532130478006326107107680832162100145794640977692680069139
071937272531922852627428957389504137685477459296033592272625266668352170703189496282724
565284582414254606303728040777479798854129463539799924647469133552433723183045353848908
080893152518135768485272858917328591746450365612006882947050320471690415376867800192930
506366957785508855054236989012229808779126706105235629735806022201829431580735552190937
586577473626473699288881279782933393499869773523241375993155463631192982070653727478607
258997312069306272104015723943842608756039326387063929022190308589098777220198559385372
688147932288292236982590464309339788165229985971114388791916811255637498313161109319061
156325528926120586515985149397612705562408767671406059062759367897286320465589407531927
159129511701844375575853523697820603460308111408561622204290428905287093487193875366819
942119678716034475116563217044041605351341390173136689463887385553138636824336997598597
061645706267041304591264372849891483568904556090934810115809231807301845998408799090461
574931098614313315919784060635683188419505707596210326850840753951104607136774315063186
556811750456842910985936094863468695936722775807730607288379881424681003426858744195332
034222592225911315687185512988438399771818481775752765286872747867997560955981443326979
802322469251748008480437354026738684446482509456837198696619833088985878352579323281004
784980000165924072903146602815056472411034520315765276577171450510804603051297596390336
904878227083901331040053851493735374972951613489722639790211988963444866201881902957692
950434647230578452652005806799064539004955427487396033311151334342323939281539285755241
892542753368993670767360327076953407153977831769329985800290247380912222702470030149732
148309934933241880821118256958623294651857563689754163574689598660266517287106373178211
544073283084095822937176862803685645159152570329027569036857129883127811874734596074173
100978847315628386494861931043501661812266303769593726764588538380943049453023030268014
210975502503890721484246009339875439915383842137754597246409868737926602794166204708663
284387662736608782721500359892776517074454770653839619602834310285238409133872378563979
536825788370583048947266348134821317190888339633672412315363972952037995614054202652355
733182260536030151610767270161366775347202108995240601901907310716711572131531313991087
346049948558879305557329074866756924991779147777627525721533153059191543757640208556243
114944537254595680970256475764244423090474070144938720093148556612673864189942549493136
310475961893303490949930728432409009866042964776416063621289476951726567416922104126791
976202629175585305961605883598150943813988815546473953900221085978718592405964780276788
923924280477323241680115088099429075130067286149727378504160015538097278691011653816376
029956001998756771052874341796486349487590228434508102484523242850619456464928288338024
674531436007665393932531690693471534111025909155950980999607771081924043400817401909049
952241694593670841551263350446837423540829126465380354941695384687191594786448216907197
188279045374175897865653963543641749642113833239127266085382956774626422043748613750869
656038144115446781746318241578012548976258024056722181651902552564665510417840313993155
273497012827464078379677343103957500116764350123239218721736939561572561209629465861258
179225997122936015604832529324660590007467538289113588769660502304327546441577272041355
353431069230209904095882802842492545660922550473678663353597767011475477937895122163950
391748837006069208321431310565114032165914971605450331526087562443039751201627044475665
497445082910844914275328651257884320143371916195074243458542671276811026007996977327310
910874040713888398593020568547705681283700324106099488089120372337515691677129447677010
573628517526922673867332490411057618836343343739931740573619353690777058069918700110387
550682586512339634192984733096678757320329048370056903353621683728691586822484931645864
130995561280761354315839479796503645798442252939980325213460972862269536267247076289971
779632763346141120704154148305304401967545816235986063466572733057403342467568253998855
700384203956509771995410026837628297511970715692877805882319026171014758008973737834649
921004305707615859532250733610872957027150743122979203137211031512057869458182420174183

2056515117533812845799817329613000972285911308282090905331476011967818503836753034 70470
0057874836099759090912963034418276550511984294261174212501745310883761527721032091 62330
8833571020857721625950992529864364182068943965690856477512431829030183533951091146 75137
1853424630585177074434321613169130454456207295577914988904854785029494251869923015 64204
8236729967820854327708171399372971364728551623691028094394904980957111479873532633 61086
1544909213621071957862718265898464545958700906924924882052343511286873862691252933 56955
6562440153334475671624094781182657115595475669936842356249997922772333285678478624 52694
9813038295767158836825390348461671496801413859919405559791791785828197578481237247 80229
6273427132738070171213159345402254416861464162064185495562201758027171741932960403 07242
8557591403748752412558364868478265305790211293015046009300979113289391102092842221 26288
7439723987929998722171268024426957043640826917512394728858097663173521903477402078 30108
2500823068674816599291621420437855969070083963431749157040070491113309702304687661 58574
8313508014447599285202072786040624690986245818371056631825492066663392868941642231 68139
7853741745589835502398141347627568661622118636756113454018506123014505064146476620 02547
9372737016911509105700588058385528775155356834613555088814313744985636377736943347 30779
2236920232819512601988334853193084139129692103451156646155817184516091865304897119 53801
1024852574989315864723399926745372521914878779978880756267375063872378056469764352 68613
0677476116156403088981072299006136202913855386468368424583544342072490652694313192 63630
6455791910328174622465230508681145392237903469993576181922838411783111273426609317 17160
5472302748587000104786605983536876204234909356314679354437007086760444160809343038 89641
6912293846293502166110021076164054661453282613302509899295539192759629946278263263 21165
6587431955173359427872479954828722278107931497771103534255438166350502182004755984 571947
0764296782715877268483623611180659244515952829152301818089716722717634965228375068 07313
1741445335093301055862157197336759105167204885674541572816321725939792701826776592 78790
7269759586524444798627848766953949146101776057760360711075086603455755571296234540 66377
5844877314065805021814441457012161388944294254301272614399603975154880968417538877 87099
7710531568960577955363596700780699856501195536169958191091853337403661999066186774 58653
6593782895158619216835838537205517181966990029062252442971964776076579212083499798 14831
0842533800664605646546284410595975870105383783766951341441171157658015291972393283 18237
4190724182705562114292481259500862193482545185655397012584064777459094161077898448 66798
7879836035943067050826469850650965071424287984166501330336475959713294583569058759 69705
8365984023752645595142841527430934760028480597374451154823040085774538194414235491 87838
0929229783184414022384436112322168850562433541858843251154472064328496208456328119 41082
7058831893542884543650548453563300884266856935636428902027669230848663361182991429 87263
8798806829980861239497632951046359133826912525187946694508941539649332734549729944 89836
2994739917547441647197173177987268394360240105216610149815265541625403854517795215 84002
4958798797410495248004753558164544116079649674374767184221183581573767370489681657 61864
6684473995745738638952849566518957447866597778195075225888298702478900964065318520 47423
7695238933550121847859966007408965038385951470180407234577687838560758095616453392 16884
8975425983059917537610132320635432534424048860000309082261900373063418486886143876 37364
9417887401204826095051275986339050977024247252980175882639229387079367325221116705 79264
4140908543740148530459025037169637477458607191405425694381561170144378884418883091 59229
2719203584129871622866850532460389435650023073416708375186459536802527582405209237 44676
5733512706016011703490806822232723412140846959667332516156575806659024310130320641 15375
1168740775678740603592587886171973634936771114265430484708113330323186633985509494 31439
7480484078764776783277053488015967141016984435669780845487805182319957564073978831 77027
1135643924204452033300760976436796999004095854955620131358480587537494725693403309 09172
8323941836921932491518687235477393921275611794664018511800138075010277721713064204 25326
5553611432390788203509453770750843488923010206936485172849761293833257931632804024 02366
2247707358488505586196021481895075688961464986471085846445373294965523337264188383 26212
7117827240693226571570786417557289614533829164489186520495527295263300281049823109 85733
9430816022566981711150564218030749436110781361389682204877365185667020919787109427 22765
0347063385085500842117094040508256992457562828262781375133270805294552322160845405 76543
7854007179908127688366953749752286406714615345649011269387426711403621513820477587 54942
8565722785336658487290869174951010237587497660723016951857365090579491818691542049 51481
8950633136723233600179192443975940164167719835945106934272172934837133152708252285 87814
7644954066168266066328173859064681708480980195630954019100230303837721074832278139 01168
2082582389277936139561206216213391578640790409627777430623945887116813593241244337 10944
8308742299489657270496966891909767872956785683749182662280759470730876390942917918 46467
2898935038166571603238341300482214907355731011475604391076423070499714171792722498 89362
5118537718445653611243536680334158347109999781275045931072949201640004043873689108 489000
0220658968949509883554543303448063469068362642692622526048050382229656658564454638 17257
8720242239306031674501605397755165542460307432569145384140667700093348172625337857 83695
4968801819714207583047902504544932943440806547069667092081966871809574518223790333 11686
6601065885464616222513680755807281783990499382032540352222147912787357337924050581 70479
3436111604657520350964992030094306338515155701039654361560042502091754083680251075 69627
2405400706130739148399782154975269620067771746125375177474080770421469498072465669 21031
3803655901391446319337852495607651289588470395683600524056037732266484889767598647 22223
6870457260025131465330278949073668317542852793043641684491309014822977944414539776 70005
0476454539441997442534009022064970795065778667625625790416787951719322821604842790 42228
1457455555258501105051118532051282481704493408500651110585967966113480543157990100 27116
3704146255884514695315016137653098634679351398306442172125391421048484018069955555 89338
6469844709722072920441600174464574485789885219133254971330254820980219920946867055 13088
5041123215989403060607764070886215302252839630610614984492974704512812064392509526 83933
1630165354068929280565187157265787411940217478091727995418741181137373534823204924 02854
4437285424144786673531720397284099921075338521376852189920275476375155088032382034 51410
4490336878610551139745556445344133528058933149507241545365042536863587651146455776 38528
6184222500373544338608419457202578083624670516135441219360521249265478557979011265 81591

9933225542147336102522035640035827908575507305278835431594674179374264974074094794894477
7957316609623021732397288402601621550899074510246296718368591603789059816357439266727822
9502991817957028068636510124544515441318142965418452451978873052020028802043389552095212
2624250682073625164648296888315050959701000226437213534878582602533578984284992642598492
3826986555915745522772230447836700451292620325907284470070718264639429939710579650492402
2721513090902016322578929364662069079114189091709554858581709996939845824188862304346382
6468537094692019086644250014237049070605479440163636224484204946141454073340772056136752
3779947174346418696144163556429471591970959124572988939233815001041229439585288124290312
6381893911829364047567480132005483777642241308322733790168055134561187865263787390846022
9832484496777676526714460909842724092219442087290507772474227128491998627528840954536122
2442608122367302636241666463676956582340509347865011435452230172110431829674611812712472
7267475584183473918296468924243908358983041077861222164667413927458084410934467091407682
8908115480426990464476617903706913186431644872934811624753142709479512183711895430801602
6136867423308652068568392614804784456647494574832329837112783484945756818482357381296722
9860250944563100213870768049043011088410435606595632913551363659537905774508634658418372
9378550213855073066062032361892026534379655424091388667805176486602355686801024443819982
2174081868308063265793445013660695883116352765901963710912216830217994317817811597562562
9334811817590163704539548800254386919502939484296333878802324540268683115920771472660962
4081472974256413523770713265586567292609352131356326973863345139232379491272741604407162
5332837276663606992078289885158189007406817883560033839550249105442191369494384025928972
5768041647987388754419071010073882502600250529371571205988217997519052515481351289265072
0350312953887973951968071463129797393988552240677107478132966112514244409425462058656052
6386484117697376509322232005813738988859893022336308095219342652281506753067731168349922
0030749784495333173923562877249889011049829135380994323467387064792939183829847365091742
1599344224180136090702185376839482371972551488138816352825082378087561773037185933102372
6901551814895668026451066955667635627033163755042821846935526079312867717163008152297052
2501399440411109952375878216898707228324155404378594936488165971060194170111775308197792
6006102061075809541843822637717441589308934402454807763589859838646004481913063291821212
2522007280634089056273136156282514259729116909696211674082471631451891747360069596699142
2308087833837868659015986702232142869157014142480704589721910542004790420726183894565912
6757662433748165233431013197777875062648144789623796854491833393254452263282389839955212
4350864723998824618234678333412034969696346523102970980070312729811300298748758845155622
8443101315609908946158784058400383614543062750283843451683679399431155194067233688033262
1838130190651593168620191839636438811828697041164945876942211365769814951731860439447682
1922394006701455127928254056530324642352419083789115209165207534501147751337617613160302
3463500158304324119830345045973111548023529147267556528539615498251732218702811891475582
2192510975188147499627018320123866466554470962703221196735206682568834873759645072512072
9691451687396399872950892928615057450939183524898641711515633710772070437194298978525852
4106512202087219851152011968200668515495090775699216193168057612255084107995644735723622
1151384426059118785236111157667462461676058949088473218825118818916537294130184756365082
3622904096877270759063075951737344653812358167205699861544933744135511580828599979725072
0005425695844829042157032963296954183720611253277818507824353239187267379753901060421892
8213335680014917629276358973974915103361029448548755412659458830826273087297415813599872
8505897081564293241595652057224388601584207810475042628112904425526350548296613431983472
5578851932222671869303645667271026495994005116630866373172740445456949737487485211033172
7549364625380611334474310806832630846622039370773105244279995137450193526614235225514182
6805510400502143876778592990110859251867499131314500087258371166936982497699408416160622
4284063083328979971618705057651962404924316599951518966497547503900114739890318968783262
4557847453725180452235972687766876242850753816616792488000823409032034807146522890222302
8061496574270447722125026619237142356260929122601825058373181197103907517533857713780772
6213177245287947915831714843227314735068371778815798520230352800599998697766693700822672
0880420433042717610360443602119574053183239775082537624353359925874480669523131409508262
7297420082719591871616960153406545781475710124329470340498901172403145627070070858913552
5130659474830501092675331050476766851006872795324432368964938724349140188685802176697062
5515885025617415207031509272651458735885771669074118956676294168134057842406773388665292
8433582820992092796000256053731611957486517297171140435836830233310269244755634963018262
7857351110563974947335708175806329870766803421309668272612847950604361526544217036355402
6583290195474112632161794143686238782446810885100608798206571969473153168872765582925482
4100600262887084707264146369814546760230690648480001950891529208834752002948330118357072
1474860460032318036646630113783461481020801040824162464398628580275352540541481178772572
8449824401215358088326311157679388344399416742552671812706870485790500170018827661154022
5989664563822695284086125700000312015134146214627435881881137521596235509096186934825302
3819680850849675713080265221001754452150438824469635391354522294838227521939781610063082
1571394734757164331002885720115617471919226677195436928312826604396069925463721960291422
5377797398316744381208097218818831236226603387075326789425385591691829772833273126155082
4174849512359891579860193104630204088365812328283393282877527485978705364732951561411422
9853246103430255531301949643011670379286563766956985479637443740469514404752486274767382
0255896740849630272538858173832095777727044265967645023462419588725735933861552680812042
7751364027860596714899368123712011862123490548171292454815430238041036501487535674543112
1800604500426130787682215885144267302962084048226136949742620817609999350033446197688412
8790304159595153926411196546477482084960353618894576122048571862646143232749719188085842
1721650249255612284867044407945280918253914446987618136633194396064637822450816138177872
2928278397648591104634556227172221781769229741153867862146057242015889821754945547494862
3631767227436470898021546200732501302370572121626666252200530396135167883101300856801672
8771386008087441449608596103041041197485369831113671070824797474197170808243016916661772
0771312763331363815453158913375254168398408478643177506675039488466367772146792112185362
1223631672188803806610698593702379096318692240259119146345846149741712192550199254747962
0048460063345981864608011593744703731663195351890879205648107281187772402039744024602122

973911013499269664898978223364655365129497329341543406894694337381826637786050347493433
270290837561801105493469017933942873990566379697634781069552896198764618985072208634587
475775355868446872335724917904765480775103923736396185466753334959708917470501031396943
809023634045799030707248529632851430888786688074249816358563633931419476252306615252056
589630703714209157446786673768335155822444226371755529054939532882366689615332633149358
392812822458493254055594107195071379970356374234009731613098646213937953087094716536125
650803315785044573000094141394600147452544140381692099336041159658380050630368254566308
062825009488020034180021455841755463480187653567764411516477104384366900853706116905032
530314683543713358180929240076805095818888803131922996604986651192355333442715995130769
082085266296774031025947302259177682013259107773158578447731207588645093398775618726625
393836235757625158805620309231213866578072162611618127003756053446226349498386252566652
422923443651396972082378259957626108099849375422735675122410923244793072428280291762353
753386370876387351815527482111244800245912464051115111499664462619843390057925463539496
228889243623252186402524810490595955408365028689357489054200091253386743431340734226519
599814488762644831855273277494122878561306225821878120011628573521338086043652520123507
908301505963245468281892247598913287169435985142267573258150924982124899051846590727823
763964923211904205643849172556431873441622962006044719016116127860806915970507233831799
024001062116474775843902375746789131695701182264621770289457119136412685871868635824932
717465627067280751367431597507565774758376406338044944820668352178332133327896776383657
446746201728839572367211098154016213270068168740231366194833250104464856464603641253174
133332379607567293733052122974579333525661685589200437596251342030638342943060971584740
953801974115495300102821650559592594591948533482273271554448735213653447294239495596453
047880531794558629341890107779349027602218084991851412571653165137450875031401466774251
976476204616693113326045387896451657290843861519443114016151423070224716393990100437906
864103416236790741850646376825660389550334773489673113343136294285431488760312473133541
967098000845264274014209763136958762258591009311129973793600135533529207482985367204276
126984764006676698661053455207287218738180679105816290748701076736965216687344878743827
719973271864925542480668423833027410696091855007115354892417444079433704231825456068386
702420523393305803173064778859332292996554662168705712818066315810759698803795419028671
051589682183998617226456523727215921272699856166884308596839602871715385266941479317328
935458449531502185930086689117971366494924105395301740136078588915471340850039768036453
811115720861295639470964557427082387312687498873097059005337318346168969341709300000861
680278005895674152284436630022965265070138562656843588862975858927122897312250450193975
398801959992958594667444885279234641037247334135383902594807739551764067414764658014533
037551258783915206002730545980582800834158675087820218298029124179773152353857706406771
166845213368665010906443991846647291438415228435595778052417869221343902620970359030350
252703283979867654871112971641506576891539350909404216300292126234234712852108395421664
911751887684890160163507949908725145944284090769519699618037712827929233063139463215096
579366488528671853658985428232404638733828178481530209203088315697267343925583364321632
066089888458071136277639996649570648133324300804430706922817962968328613163949834158178
871426219665499051404499949051322758329020397338902854257513664074283771983895137584603
568593319676365422978795979675682839983101815254236665985727857888868064851894597071620 3
467370351680456789741083210206877691531050566876687732933492002389350574436954451602342
979457806030671893157679519089580811282704868678565179494942531798989854558463511016629
241506701611762219757292557732222995795702695142731341258703602132593747642947677233855
393949608034943296308145907993381594311461023743648260905274892609114997817599242523396
972869525241668731500923820412128542613616353249136625137866287441728736927773266853389
990509144288059316961768257728559277785548891224880886696290222200907105319867273320350
125608327618654686069004612176551141034532831271204435229510016794790313350534253556783
869192234312490521332794361256904680330454064259314334859893529878822549531857424881037
641375414844998295227489027969508981498646907616443895752343566506497982594152503242632
552944116596940559895866507612153399297486410528083098879197123728761697290730295301586
338095431940182026691046931393035266362835832196293419502205582156281151008278370219142
231861577528944307401251206982236257041351162127934474793737507085853449040251894677691
474206491390247315240473922375703568331255397444736369775913101672485564252270498558713
299184758438211851524915321086608709389477465558909768150090915524531843711016797043942
272006065934727864923765594695847171642902578632718343604387060615267993199251780719606
018199788961891441329681532735536565531782787898770454849256568315404843368663589348279
115378499601462943301785359189222687135602115638066888736024524286151770771110671285143
9717394625668407770725858919518657200283026878274880646248625804514333334454133086163786
823325729625795380067350910605339652325575968241504827951961974945905100821796236567014
770564590274789801810063095188896213790376936533729872681282088478870106308255415850421
334101495828542771806949463381388168245190344480504922435510003314142920894225768313480
195104195395648342838316899469970689361239529933647736059673795630161780318422618261992
081634867619660275866447118087603253007087453508535754908948331667080132534824971180676
522815802360708233390414281170229413525360033063302611245516864922753389765333275088373
087354659141118979834197708121109080471374423563241997436195814232767405600444674915694
945578714935547922254176429822307573665159603939567872952083076212995729056463332797905
608736019668380684152160053409822871768205430304948296407143779589677891785265134420901
479656996958603321761028398322325242090918749756952825023624449423568735010347018741990
530029380969860908761494567287112680687195992424006465327711570046123469550672596301566
722909054455688966949036381979374684658665340679559719446297756316458243438624037934898
047300575709839515821613921444041889422681665534895414328206155392681993338132341431398
790872065564411761005197910307921159446412482298695403958669789629636022480766326311185
609381709075532259658171492545809500486428193072375865331093474102684608835101765523297
927925886429690577225713908291190907196417085384594544335991896296182581379576619525337
770939593093755869597915058546959060081600343557079220572841848585599616477156190633768
504329365545474742979308228403401042147794004948180654572922448342610480152048933259789

3682357594775848939079653986132009777388783890023066496506731865265056828395821962580 33
8070209708988714146215856544262375254313938425321275734074533191162955171187913699270 35
3917235081499866237794428418843345714929271033322663099327159181177798427378975014789 43
3268497205154307237560639987729616687253234709907174640540240739876530764999282725555 73
3397102244685228197440635674154423398952240404254833976955371473159903911519958160949 59
8512103745365994424396455866218951207314020177355678185319574500159138619106408997869 32
8313648390096137571062723478005228242118426427552831612858697601566046431833533610397 23
3746019991538893157302858826916092049488454130092262588377714048796551601554359374511 07
8984718088470096060778907622069368407378496336096342509584708257256336812670064291029 82
2279991576193941230501066561932438529131227088307156747196820218627201948474469147750 99
5873774866029631262112393626268432315339171935691378989196606671277097343228082519847 50
6195406203449333070378426798379941771882384778573049239862558566116335286152795713435 31
4524810391638351705507787722297623979208407088711586623991923319336495574109949375410 06
6796880142650207310666332190372968824698040807054186317885193804782714122565417999942 52
0847288328203476858489725525747181941141110041741566799999641975328403240933119063192 10
4713467023378515181682298661343846179559222892272724792951269711902324963913804404399 57
4050092712081861325429437494680803495274028786638624393417108857657456509859476694892 18
4500640546563007857601863379039611427130965704638609176346038756811696167424770017570 12
0962241599529760603853488570014814031370011280296945431637235112508802119138585426221 05
6899489951830180914171906159263693473649530715417590666788072282014882919882051557077 63
5832956721911220357704249516850618829530889889133774280092605574823119088319103131939 29
9334559231342822908244952580052392312035468409591811803767004110412429520600416749760 55
5822753840278557228994429097079220373479880867350017022354028870748724156877915062146 52
4891733255247701844863336042379174274985534336281951376593862764032817426362481472009 65
7057617273393219713701624994376072232561327874249377778589269330335964016213344136498 40
2711391338427470757769543778601175664910861942707182917441242654445981363785943440204 32
2865897546386434827291483675790906124620843234390391923443343496772773556111421320014 39
4443227320381369085729795736326744778943865774890385918099259886296977925891374705285 77
9546130320543303677522033550855052641852468351949293468352432860294168994575328382103 07
0059714264453901409018029918233366474407788470720216230623856055975822134483772962995 98
8321194341336945834461478359693702832682714104848145288290526166403281494081840243768 27
9808314945204633401314793187522373778064144956575621060530337373631466749971428199074 23
9705585981535036662090465058448358290370627882179517010954976396032910465540606926458 63
0212687402703333762870900863607757172312759161950765391337763291958221560239574342934 46
8871298084612180268971042434170908330991098588888835254085942276917768288120756179439 69
0119075663452417006163200810141847533290811300309310975867707303631842545293345309766 61
5291752366323656474216904228061697515605333059925079176825022364645999570337747610841 47
5018859988302655204068322532391058724489413214920420150763661972890004060592720424962 76
0719992999765156898504788208519098035733115741544655500524131490124398995076737791147 97
1421276661555365700029980643522358559463340291519655744773725774525517368467724114822 87
6372680019635844862426037986498657582130805125486775367180449618700159104473879342430 41
8785461787045785436649442843850304116481926667184975252670736583993025400618865946300 44
2593498642188736746677914010289921935190341984732576022585319484839338206114648070364 89
9786708653140531734815143246518534005640853019289907636016009140767076874864987866144 72
4164384262549228598167910812920521882291519447434704103619261982249688650183287881228 65
5261494487243355986406705534886676421607699601535508232824182707156181963143431092962 80
4052569380172100643874560935856366533754096152099368441090064234555949678992586527173 74
9829803717638644155408339933247328130954900909116944267647099606051366703401744118303 66
2250489910202822410449800530639392246517643281963200444786431071064518182924901554707 46
6301366585027750507967666944709231116950742845792691986465479689769857442471202502619 93
6276904918689385379697748241302056076304338922473675747538314713417546782974962447706 65
4093819818294053395278667728983884828299114239277363245716014373375263048026324942165 45
5657671976751934720546499442516009891508526537508002510756054326553772723422307196967 94
5272246615973866021741689039122722547133825915532284525152266946972817303175525367108 51
9113588765425443579041298241035431744232764343271370654209963215706364060968713845246 24
5663352691301220789207803854120376020634119553945346946694930916207958199116593075741 98
2692987786665036590825853102107170150184413675291384847390819223564708665621950319865 19
8555690374767109471408761353154871815930278188382078139400086999967045174005890292947 20
4951246680739095172243055169301048038278147544641937702694249327243368125202460157153 48
6104406075905633203741788371475352143957277788274638618416087213432549823690048373821 82
6384009251021559976282492414832391100246927892536253848077699875241682775157981445345 59
2180912352016230923561872620356180637137437050124625681248863511622694756689681361908 73
9138611682781042246641848881377491637758235307175109336365159207832028507487817732945 79
5722880270292533039303562960965509081112345990450064098314626001133976600372981338813 16
1449862460738400410387389523346770156047656476774375309135303602773064948548181815798 55
5845871362783153768046482215248418050024360485920424819532883678403638789956319332163 18
3177839752991937552421965960896506553739404460898228965083088605090890249651264721191 09
6922940386605909137826663597944840783267636254438297382631612385127135883189511072580 99
4198572394263896590594982782417809235047599580728222879833836706610020415953764569075 93
6082090530464645605498351090377847790876519746000574937826856962268923656473689664002 37
6142139140408853022853014229240229342391847460728918244015840316196570370051165037483 28
3612130517927679202949584496107578731219312362792490708774704940277207668639512899595 81
0379182555275737019903563985512824029479351343047014985331634148827241470870511372210 73
2637816770795704244352542402658784910323099442185047657104762926221526379911773502945 54
0414719797361893916413646795825081052536221009567308707059953511023282255406880818424 24
6132905503411246368206259564429291757401920097057467517837870949738346200035152023509 65
8220513234951881288097417013828007277492706073842957867654565122328069601873592793842 25
0298239452654566153768909500376120416256518301073537003907029150204753714277894368805 91

7320302711857789917666634257164269537166959331831417686469920393292873147806545499610555
6358587858035598893825325627842779752774869590582901784353170386419677914076504812980943
3838768811633599534749783496325840425665564883523023097152638961085263284139935517370055
7015792433145571339263506491269103287457433680168470883210198318057258996356417499479991
4117646408783098587388760126224392915251351274316311424091659579854423119407426391419
9573700819368632439542888918921590733557117772516588695449446490515695732422360494291066
1139881878797814569230082568163508933768853608784915097614076727220176526327006304042988
2985323604100024040182990715058209534866786585491475231039176530439244445196651342514858
8659357253061878933173290841634035522164741041543526137582181818791279065282108056644
5817008188462120953276418823619373937158454165450046137634755726916725247610255780211198
2212191616769247994681485102211083546869776047065079702326979179446640582545878412351378
3915987868576580174717357584005545002169915662489343277570531623439857465121125566976
1595794165500430927839580643678562017610936953432274420323727782921264107279273315388054
2657187195231461476085124120711621453707052346098735285253526398555173759886215228325270
6233171771057646834420711218489697162632921249061416662418876041786839653352081340399
3199974584851636867649088685910452680787306216050214958591937822714926533322860968536505
0398614037995783932359268091077890548555861088592428224225927744773651178182700198138853
1607305680336579676271784577742916999791936962962907299726810304970969706175036178487
2804915714553234024897008651825057184139097089981443210863274307629534648301060291760317
3983162988558076971443395677290152947924948925730531036288092988571097742034339038942417
7496084967853115875752446072106263522179995794483282496498179688087770356049069740609
7558151120951620501327709107803913461147510049698677195780467282368221758850855512187378
8238435502397135356476753128488751114558439441307561669080219404705402509256163887305799
5935710070954215242402389738661449843026964361569759383503580008652520663448232509342
8912815946824688131107670648072715392133808549088932174463059788581127442534488131962175
5074539046922922607786828636587515668094475047862672273570769537148972648601362808015084
4226326597221147118721715445818774261586970793886955923103553477448442710277279181265
4193912554760484431809343679664633404282833273374185062986549946001209056686091094950352
0844183899163403069633435199713722340451018393656283949057157411991738814206864491885648
9681633355195066000928843332524806735584171337496171505509342637189402325303542599384
3941877187420881455435435616430348910314815205765886944478270644910995335212843251910491
2469054321738051067941859880544012894251232589909962312324053877398210144640584965597415
8659523205814498852510376930654974893135060329360744818149989820111827492778152011324
0464303834000930223108054725959755121674670665922944438571075829356865159801179019948045
3582471723450301763989149022144948902160198684151758737919168266109838573845376528041890
0933755032348767588757658350816808489804889946134638467583582758945004664802602247079
5960731123470870190122939638421992508876853711199854331293724294847578836115174083358437
5331090665942701325803295439815269206810548042155210247965114545433197115305740995493783
8369320017065641023993968520341513173309251386082983961034483756434854709456374110604
5616668328026369760559410786005301485403212528253223272517323249355788226593959508373340
0950598453008448615493760830772932369780539020694898436522867928580781581080858064953263
3173056468160917851471254000880722579371359859196020321176985166181382057266644879714
5605056476417427368418914506734245675641604829030981897917595674479970441848154395604702
3378435681267617715798737487316524458821001641061928767152951977309612579504013279951251
2304460717376533044348897583775022006741467780169732280054567344994253724138458236775
9639957228554593078385191403950474413617589100741462268192976969498861286529855178802499
3319663563824838294192474319235584267635073198580303015343074861824378325227933579938356
8537811327556538647300247674306723758445557066433223967058378975019401109845845303120
3974160814952863365122483951151426513952136199492804776145672284844312856596154497313827
8595330767369601494158637070362175658670104303586961145791714834458205482295971166547021
1362772824935407946290706014037201690357789239932630327260725450600403646050283809296
1007600067621093582161548809682798180459087699075582797111496748587103659798177900559920
4619921086218833393864367675453578236336989088161935642182109559511009398375374775546586
0778655943306224849127897875450813558000955361863224778945578216728582156558348574169
2055782234361503253551913069451960052894986940468655864528839323919612404399594779057554
3519058225812702468257321602699353123762173162563897324757116285960699970829394959814645
4681242911928944932167578936358775236587083126261297689522140371213333713736365700974
9611146795473894021625486684146352498148656843712993256610369032098432452443637457892832
7453254010137873546087085784915339133018487965021588810929903714350114962119197243727036
3331890117992931009198972066058919499183852698678005809392309173781954298508516846681299
2334259466707617776755886208012614126146408861560637038675646128881437886181688406921
0573731007147127556028255238461049428731994983801419274943751006947906095976275704074256
0527920403735132256437205320096902712617878841958243923433165252466820942546627297934824
2095027327770295359815649824733818061639387154774919753504932179174320668434092062017
5808477830518875496124423952011896490704766018606357333213987937346739149080881235134551
3774071558682223545588457544686343337754031387130262607146224011717060240106522549119864
6843096415721944924460282817325253667035372300424249866064805311275019543565232256873
8263515606179781774903631475049573203258272282087901580037039472207847114408535302162674
0506650512561669500573990832732506899521269752616060272524728366524467699546935694759472
5756685581189425853777257680983976858806496441857538717290871623665842954600064572836
0531387581386362944104313462995374182776271853061591934261217713201031002114525657669028
7097455533109073858111289551403927166875224799849874991785025892689021482459578259051864
4548253086709605415246391648674819969569196375971963039809610580335461327894359818435
0869745259200054859003037296168307195762268543536417311817450957993316477690774640274025
9052553880918629368604586739521133104685547484403817107250663691014557314738282505357057
5661393917552606951868964925034468664749146526156085505037913942029819942219947354382317
8322303684713733024748559429826380406512984891971277312697939942446836813979719308945153
1301022820717602411322963912280618157095376184520228786336178426103531073417592039

pi to two million places

7829165437023953430922591085010805655918772527775548002700192946141441537672275682553 32
1417372014074713478884343631675916143872255943304949771961233842226616048964396242053 26
7797041430802411640119680891010920634290380279255156967953244161928348664108382860544 15
3699643659319691377778700593603048202291302651459229613455802971827243841968767243708 44
9263675507056334050268371994493547326526563206627438083699582633516760708235495298561 58
3433119524392297003987910675268314944224875870597119750135716880807708801638578432778 18
1513027786831168589194611430108958921839289713359413928885648854509163725939835974207 67
1570746071497524605986398966957462315777686048797201480357679564845898219702887616123 19
4701320955924214488355505762723234434842642601175322306182230356585080470110118991932 52
5717205499629266412977350428504370262289723585281626756378963020389847435594806121738 73
9066853543845303929312988193883304118342377836147805975750584066225413362933578093194 78
1966392974235039084805932006978991767883396869131974825886474708627997131325613717273 08
1653340613946256855059072754586450686465652776825553429721408833837278820102890293240 31
3242102002610635664244369661208304176869322010489934515597321174663009086712008355724 20
5292251062850302940669270580504400681819227351425634658435481109593207340127496949000 25
4472079736037916466970319503383284835516767605831036545270857655498002823947822313718 87
0396521642078414038632005016875592892442489164321079620031371107462606935918955818239 98
8365915310970042358174294600735961247432905721092909762924104106566209235037924431392 68
9030306220340787058475213684434981400664399682817772883283068082967474851072684228563 95
0311923967939970227828083290403918794270125640317319867054809038172901093826770327618 18
7333823329928735425179121467416968444384160995792173492547541151695503632929460672187 98
3817798488683627829099798430217204175362522299672743257163080332626794270088346679931 23
7227789280490726906343593863344827373494687180880694508882406899726165871343751874071 24
4353589993574950576391055026023488483193010977628751845555614279728428487603938721304 90
9025418488426977514011626937613955045856899047300398762225695695285227027007070022363 12
7827564720918907236614533831506450866015716672503044253134573076142482529934735508200 94
8111074026427032879613545589972387692438810975970444572797225595582148318579221168381 9
2022376660147053550332990566389961139502003559003953143148531999733956110064596295558 21
4961621580455163249615249846254913386661556613057471073066064947612592513473986724042 94
7052713945870057114461774359248919999779853985891554580117570754584198570746444171573 52
8708831815566490671161372052484212406756883333463263093946744059153928124346865274150 76
3671083329467993079601213226236297192288906112943956865890674688582258888398916501883 55
3307523319815790355358685515578206546821833215907429103474695675663392485415223645371 50
0388621789026343137853026622744881799998738533234152500505075994452916010384924296473 79
2314485199676400312042619311018390010745597693245743996519682211157017225000078018520 07
6909279952748195722352249009245510210083294350604709038217623401235278483873772731431 98
1235331216735074162478419546325344615208289122378046922908509386280752677373364891675 27
5108867186907485731511798719112758973717212220069790268627015397703337623539168573023 53
2778080515008525981753295550807877886672815650966691615839112721698699388759112688648 48
5453452898384500172007531788096127347744030045241675032393038367061707101305504380587 17
3067566833533745378303685599937759086951306218465528579235933917419171205417969987256 13
2453266577397569709321705621938004614828574998937523164351347470736588209810605778865 41
6514732489817870094630138507925592226072971522620388194374843914310594095992584334465 65
7681739689326110459870100372754352511637744161227299994101861956605142159694120635513 14
4859719545286080974868254874524459036260473138064839379734468186624970072155471060193 50
0238648389343756227635012792584941732643662372023278553594194930450011152493701147634 63
4157542640955747394306944563542362081212241176373576970867776359301935638364440288936 30
5078333228036674743943248657079895085250872741832683527199515779265271987637499790762 08
4389463472126203607830817381428047878554978289786227472441770030163255013397053724176 82
8153235161769069219970255699962054642437265357754725102403129943553864594831470194940 15
6026684943031837836936554661866566254708258607489483972825155891603855349506451384744 22
1188275629862063313569213435053541753254622942738570185142216047979181239135818570233 63
8135445357112771171943216604661431015474198215549290475621090189572080606234908802904 06
7854566746372417724868119007420655784822192950106596683535208679087585534490092713251 07
3537813112328600410529188355048282568212439318097857966641441641974384650359754316704 18
3852145907794335773149648457421486085488674529131457458931518483420505854272116027520 70
1053028812182044257185040797177351938264441514303400003896508354760695211261435151449 40
9699151517833258517247989474052420610045984073638435113829829335370285516415328184689 87
8043592175819760111037188260115715212198992803575460838874094737522040639123362898280 66
1873195323552920401422000951548088070610074538656397258970803032798551240570967529948 77
5250348381191484476396069023998008588751011612900600807691194381030260949480659847619 69
0480593217852139982865901636139729473334245297578429975902328892122887617453643431583 75
3143784957460887473734258795875821990193538981422942391794151561315397930251464137986 08
9598876754136943230404870285554197809229580446989901929045589068465978383379944925127 16
0494133779070648657858949675759940506175576329347568082892202911154918648820159214617 76
5449921182725498867656896225170636148321951406030844868842947490817141227669895297665 28
4671870107291933779292824435324313828520635706158076925928260322211940276877904292408 36
5323232151023540753423210947605321017167804788960416851071973969399186187946346189679 71
3546786722440290516444396478293266946358491866150401165503213795823884640354533706750 01
4682450893963508407963388339316440021557629487655451496229849457357045563984582865390 10
3120311995586329789859964274241654564022155269311761821934040528049770013958185699504 50
6269083218442209858065603603966505204050926529449163112247412243985455233459397360215 84
8895956457603560112394722600290911023235818327077603181928957893191200422829722719276 80
1057856446673340203186065797599897673630045155344121222746492117841921042993023304754 59
3408148695733885585311878925724359962470195810494083427130065971636437516574637102498 70
5362916929006099797979820814714713112899508518492003986404614653025099491414340358369 55
6884216151820006672539858532035670787534474101821344997039591787397534962114723677715 10
7506443419320670978481013906119468142996565946949980301501505043949916581936434061754 71

pi to two million places

2006023253305100568566199539885210969917968103065156627611400123939441274050406560022170
0985477796442468587486319694618955103513339164119590397189387610544264230246441278596663
2014847955273227434092863462009840598245338576386151933644320983391819574962950525271700
4159411032941605520070797004452742655032910680168291821505488657297908306573320056671100
4039316642894607397427613272069913773588877646840767260164503796909067376249123181569466
1325842146511243018088628385018727320829304932053488349080225799962089315382043458746200
6252362968122407755716676147083325307435180280156466152412335772666545968950153512096400
7409879933525112423683932555080407795389065135984869314857268748994901408530010625403699
8440243398573821267762945491927728192707271007595054194545037909180516361580836288870011
5386724394500702749898432185567647403247143923664843411160620931019601825031780683539855
7258391335713344930361449170866597972333881453092174031811747752032581674338945826496755
2752520361126273672109764543134023806587201125134514611723801638359472687522817638356655
5896188613216729989394014941251035646583363688776090658837696741829191920311945646978099
4249838609041603309637645292794234193002301740054343225854625094743545517096835436975600
3565019923851473718492670597233277579791173815247435316334241172984589412910750455504288
8587776273734066330416039180826874172606596159893360778633070199222318466648889304527155
2405511746120223016036619219393661579378673696158162597300582128112825576467959494028144
6674576045577470673902200197698318259700293819541492759081133733236058858777871610067255
8359623260496001605899148893422047361613271007545230048439431098999163722188623262572244
7230711979182304944435140335744766397083610698607144570069276396639734920292183462976444
3818601893766880534512777038481569085406140328036150280386094903353489323035792511739355
3041584113326547129056739888443593082228303303222116592985419196559797184885423887158088
9140369301617172570015570614836906812742295502793463522645006869343077482074666368747611
4762002275018155179697826673741459504387058872387338963291213957303994663054340289132777
4681687546695021614124655037009126591798302903887348417613972343394556935660083801609433
5561378553746892071454423377646719631384646526315701017132358397487466544236302779285411
9045015666457881859979478971251481140502377690261728979301308065657163121212079142907055
4215088898379545365916435512334174598794809276941751149031174605522455785458135586702155
3090077031955655899599746805741613338361641691140099233415564386836225866442807940336266
7010522666936192467472371364090542898520518835100369268187997465647052545068268393626400
6994422311791299733364106678173591597162898327417728872302052609804248757771006988196244
0372912716284558358478403404924364878183372432037161878814931836632132424242420147187988
6601290829544902098739959542872139066776982756308916794217401688235876539750420302448988
6418896369096316270120557681969929154992775142543788129467665083250351267168466448444544
7240410124528064217832732227760436916102880783588718437100518084017958014108352816351633
6038053463076389194761501869867367060501475565451912556348547440616202739383503562785611
5295889468170169994014332311095287212448270472060546025850066704075791114136827906978688
6587117792043561148929968719888032590349546258685078645156073721715399533910705457420844
4700489981128289942160212220926244947274540561035820926425126780398190545265944373751944
2813217137033612591057551698992847294695342429807232562902588836268426784470298363132944
9660544162538614748828347981673228810978487694132343671883348297513277555209811183566122
9984856870217344971594558142051676013631681044749870916364943156667001634124731526314666
4694470222860280711813992815888751637214266832121415092317205731891117328825980525201566
0900415547775952404089135009403651970848700749678332743233588694631268790098502313172066
6141132108607604861957063566245230487204929701679457814582009036192806782139458937433777
7693126987686811712481640849105253884239333690894635409235802310817255763499799694364599
7544489485664773280449886762357891730215026987964549842771233602523960136788902639127633
1673348690099465888102863102237495535995017161877940595427203256807509917230406040592444
7593475587819231150708603864364001669769358844410773687028457703794092834941402822129588
6407527063935399304447238508439688527577798355208317581070948268654551492346771164511888
5672238076006299878184487827005272031293884799820971943202275713635203988800756097935499
6850722217381909642757568466440784384976235954164378986071667348604995364292157690926966
1517095282542108602668881287621322828870123941112135608499848560226167435034883052115199
9522213094722311882454739260808544121534421043454311042835336107232244610950475490308233
8497623378772397985764670714850727011550335079176889428553256555785689141110539376812300
0764172733233555555695819795161678765161127158802381737058125843376445496393209033630888
8423831346474132541575834085328701621478467527366035329814219899900103996516639857816277
0835896248137581128520502746831434621865421002873579845306419721733111903252007346192988
1287229517898245111770328323475986403956270619085455073580791658971007776402290351977055
5165146315695428841424375579757096894223288731250154465913235682234856323088186214869177
5254442042503115517112520932667209352445385328575930785720519631126771596563353595646066
6381215699176134271050357968934692560977592291135575500449546809395941988076919375288866
5024897112468591619511911805736623365074921836732839574906693966899486388125465885588833
8303308642792235459716140863913280169686706096747793497025136967094921185182680683710399
3297681827934909048809926852979497855733371545681229119082889999649617367275829677225422
7182642232866400132724327309242950923056622134697756027497131137749640216045186933589599
9433451701314743167166992535535262519182296068551102552106617693913058993047044013055533
9478586631684376918286472343532485938877973370002374344405223057833850423369748670050166
0028663716354807214257274236347165982592000599527350286342941390667926697237987304373533
9379577587467043870950735671245544966030978961181945541702455921930096405938055229427699
2173509881950338542439019622355656650959811895084955834758326794413719433477706441743066
8760728732386031909376467452918921839273405652449125058695651761156206981250039315388455
8184406490819305513822068081023933630856535953832850815185286024907380888193971974192666
4563416144842651154131695628352519591124428382628810103084755454897369025403588236483144
2440595060433363721723113697973766253778983291481467685475411897102364658779329452455366
6084629871709766714521539359362956508416867938874745177684647019705601206291119659392711
6942878200104738422691208420374736338838627479266343817074600861816517701247380026891022
8324861454672894643770339434204648424196702561879164897251838674622230416516000185643122

9965411759820850056332395241632046675535001335372968491746764631941349922374424722463300
2202185954740646378821188234593940899689958667766370114295285312707935566323782561966 7
8213665709220602831025653913540119066214292393816411208069617216043809938799300327913 51
9396160545906725965724244388667309883949480405001995869954087761066913890684279935646 95
0245990878656104815262619488029162203772854404310761915233096761345657898664927602310 34
6708078390992627547645000231115988151525016675637395740190577034126134204163590448083 90
7653748582777525966628543162988331420747782612095040776043338836635802430892448403488 35
4102870614733963283464657835796697459258740113463457623216081042397622251689597347681 74
2851273772134888431264298689167069631623873420014694898521423020835510710710502557188 46
2778564403760534154873713405703530471604771067752932007990830085796335058913648697747 15
0937612256284483514279338752636745707222002567691274834223794366061319862676094406210 51
5237198485974737929740617724333077735380254302218943957667695095667279812485008486264 25
8848457967719356146646462601496495146347149006188672601302167481074660541119266891840 63
7883531105630441708083578019992232956434743032959791358894379380097270442582150692797 99
8883144672532976896720924331096787787043725470404969378268535332779967817629151867127 75
4136572668369349108729256606562281591526335044669497567922949764583960403124782609680 80
7632457291796313570638055301817950615589346192005525020421276892047265235195908441637 05
9762275805273353990572737729245898431134662089469356846280770879593423614342618357397 28
4121665260195438481774502442968737870447818458084566985918167574593630071250992994559 02
1579797126797928681418361794529381147438345911304949490625457775739657448250418936610 50
1567223911406337904426932767178357282347840242922904037674700397134683438554640642707 10
2611753009130847612735756388934449578014367197801389826534243776067204873056592069332 81
6973770772050673214000573675534498089554053868887867159112407602402876493610914648563 24
3513922828961692053847422089604660805903823099659189334258907900622370408006699792029 79
8194409277173502701273368468208673831027079479355302208227752154460927356207151719553 87
4896681908468028606626805266261730739559289324327665608205589264922811457207893258778 23
6808279305050030741774353514258764320918185432669406906760079190821342039636895309452 56
3340221307302098645862976896554724865262428461104736657509041771732052323741407565848 99
3239270868216794264326875694735191217476911115775407999719992668288850793903934061031 04
2132964682504077064770521769095572432685966471769863829141153779769760002581927239446 92
0104966004285085470014809180810817266504567967186688064620584788093007116714190784971 33
9391499399525524545209494650784349719810361428778184033220570694639515476946972767747 70
6424864607939235195654366350830702520798246537427425699694577564626119873862943453280 54
1508276209906622774358444862703767092488431396731265635680597853428581984460900825022 81
5051063672691418876039788319773186265729314218073290550935385624444888087051585120556 19
4413737032854054757221463407137369326552108693222709423907542994089944254445906867574 11
4322524261672343521912782585438845595167978299328323642737457452544546052939896806263 51
3733587214850808820205518659958034081488329701253781235056793050818818568505731233257 55
5542419605427358319447976432499228822660435558523349606680905502905216337784746351934 74
9713022329493965510415987839740167516618593605179338950392466205245511268837311120785 25
7244245799623294450168341713595140252095179264681156829820313618827396426623321676441 52
4695487558164084358212585044247670699693803758573003905790110515414779557179169312729 09
5998221364115981595201458613678920666663532183944579112942949372746424648239215477975 61
5733670895761840575322098504748583570891766352727809495354274482525113739382912378335 18
4147182784881809377625946725543342069023837559765846744498857297100153365702593838609 83
7888370559661656612261881245463878074036437755829259340136451738584462455407649029616 22
0229244791778901424327249245624610572832994427679678314481934670551757083502942567326 33
5264906514141210237861093296718863103717170461762893116167259029067712239858836596414 92
4553081207285708410066076168543516663530341382801133819677912289974126655244951348338 93
4636181282256499053411503179141167093830767768774232569803429140799802919107611396530 77
6180407622194451519402604063470356799353883274378588152011080406490885175270082056238 02
0512864218424823002632432055997998346926232665644701956357300679539057244150398164239 08
2136235132717714586191210328112357269933087662553440894151205179902731473868182626644 47
5280406727464085723801550389418912595893739926501687752743769741533748172422037707128 66
4490771162603154417119414108348606899529507444772203362744266818471196563615713772424 61
5456070479650878312900133434911136292975583609060175949453796861506817908507607566212 73
8100117918293076118629911635574502602021275654360951138569094815424476722607340061037 33
4261273608044855312147578890237559057711317455009411859748652962705885639173897159515 98
8987014175869648654185324863779433780506989345553880505233124949841887573046444733144 45
9850552473986539970734623381939800857730435695476169828265893810030602411218665685980 20
7253371656135335099218859560107881521955992984830737114161764839950330037898824790345 41
0532502054955693588016154598918936886572124748963613671862818854644786179243581710112 55
1851317871774504307353645029761507292301108330802551534951869294849716900991730394769 73
3378956502295614877878048366582834827540230192303690385819788534303828558273006721561 30
4247679650997367389863963084595330994467366005278953510077510623540518095062072959121 47
7879266263385428792589775958630580646504484526239133538342627050430867009467036220406 33
9767252991365187842306583966702262580562122210733541161850293635641616655779237766395 86
0494693244550805903617986442755741294983021046969861644931370103702775084860153961665 86
4512853545304815598296382985981545562592486591863288176301101499737206920153869877418 62
1655782087885028970856782970192695827695239408257958934666666883918358815549069436830 70
3532763207934945109365399450972042836730670351441963155288753214822189325967173707812 71
4051334747386080963694563512019018439160557338408051663829148862479351379403713197966 87
5856259482942074632416148196268288849800968875641317790265769105550802543228031258589 98
4582872083257358894763134926062496271832200731813542439536437705648192953995700144554 38
3910878449144193680471065163474031170374482458505185788180686628844170793566042698003 16
3236349120302919753700996010666193896217318762267071826314852284417279433406818103101 83
8417534997349697901352604608389864938417085293469279158345594247787414758186260672246 62
4811772249856862298974404384392184024560360919123698959782488064463195555559308328167 34

6023120406670072487747599806326845273202557015621687662840583268894930505193900504950 49
5870154003485427760246248588466667342385974454567161141984303835706397426666703855560 96
4523903570201073652328352769206772213666358574608076159948257589026155644286649673725 69
2080468511746267024678766860322879651197857616442650025536622079972039998656146915511 99
6591892609987569195721982755095064759786156264742355786450113897041993509976406676557 12
0850295842115591494729075235534992741008512949193855962594032638202524988224921444475 58
8270029003679518705235762764423558418333071204601246299399154841958135512551467709344 71
4433092476373215011861279838185602557163141744264421039231841248615613047098148024733 88
1256960519677269438321490104652409981501183394145060084222913194160995009964496196330 76
6171680279966145964908485717408237805713129439661036877272697904349031896749323216657 23
3190372154146103647188424635680197125709771242045599277189401630807555791531803886385 22
6329349122868944587124407187398513109807299600005402969139086326671417923649756297192 50
2128839909708484680439071763198298386258976031273818102754934261012824458351039724617 26
0027124726441028393060367775439840384623746557117766042747940447110253227526070881915 25
9623881035944912100259215675509990359849028736639465333622278560198785244807812000092 26
7255630431187021878325473868804409188331048255150339506237035345911575694871584408122 25
3546614612133683291417713871207911325632996961058630638814550382930706507642500409597 83
7720091354284328731106694070419993253056831695331854406218096083461319779933817165917 06
5487955211443993463691039132585349777380538014249409345036276165813689500309512610570 84
1234456296013280703948714675890101664115170393932146989030266726605846735059647527480 56
1780786795393551032684912986766565426312732988529192700824708774002213743115658696907 60
6589908547798087756486559413089027045689772974195596550109221935693238497816225875176 46
5524209255740925717695468860519010003160801289728987052861085422973909396815077500965 97
1737146008611522092622608527082988364373624387798127745117082236808061077077413663347 95
5743533547250663440979289899184082181502006262900581367815452848577595273359535974840 87
2450053882741039998701952126233169862828034388497269141695862950362027229748868984900 39
7414716167457511413346027344974235505878072186655258735064125308324573880356085157662 65
9100847907204770453688975071997435665063066316758761134751644189050994953044117199851 49
9167397662294269445166214080877491355367345306518299977582014657530815794081675035725 63
1308268975276869491317516603141962741227162095782997451259507368949976478651309830445 53
9167618793163664040969778731171580041226555288637091406258178846923903643987679438994 41
9596332277331510624171111117589582042138226824715855862315936615312894321916548928211 95
9762276658143596743190469318970709546254984802349550186923112936640292909966700863878 40
0428904420862483661779064302063305933920322434365160794325702465868466897715343280772 17
0987980118148551579281644492135430015252996137723601077292108595131459952461659422716 41
5747632365702571880611706348762926273236008312525699654343218937450779674452915427894 71
2722894704464813147441242211665900810057217233044387008737360533164683029287005557200 19
0699431998706454465506242821727117124592068124294810555040470592410528835740065648454 7
2456074875624763472596201955416308086991308656786967875539700812791176866919496838135 15
0988085209582767929487854818158433903895764802898509257246086253006148886286503065719 86
5793656157955982572991894328947716189620569354672805441856350184626344267485715560888 44
3376776775181119587963168418536391233749766123771258705575367714255354528010236191288 24
6608468567360849341333119579933540423335773588963780531839093444280492270352162230871 49
4436067300423117979682863905171951575052097655902730996709989020051300226332647381845 20
2399769112952460615572933669965418267875614644743693887290887894259922714756326206666 73
2908094698629295343111076243281643273608630864133864864668368334034117417243361379086 04
7880568004597543289332721406080344475032843441146117190967017625398428226686468388170 61
0025364990074317384700086148176164319642146091993738188776548270699793984153938974909 46
1030806089521056237233733955299064854565477711132351150583518723974869707635229334354 97
2561003011215891267832849264645292657116115146530034496144130407078693714179233116662 47
6964087635487399017477537102018211428142144824621320489013665523144244134042877529811 83
5667348556593691796255853153675107980671452796637458994210311881547454807524651853170 21
8249967058200928170347143305649061103029660077886218643958620309126219537459319155011 16
9133155954733941172086135358840520458592736046321982702240715420614033109614862990759 08
1033133065914757495394365387018430653038342790401430598298810968662879396068421340105 86
6813687700862550410669655362243076097486920666744068427555947082594037595454393281264 61
5197860109409221003946623938100024885780821530539641236263035680445023302479433432134 41
8808431469281618392368201869189398393330782579391518765988615852658830313054820647419 92
3861166216919045975656253336318446768950752992586777308978113220554526893234119637741 58
0704297917296184933765166937562151464881384172662171532362271080278418137745960978655 72
4521653492678760880091880707514451795591893207464840761990517355858488913303280635078 79
7052313167676931577373187959490721237263799259715349422416504918609591639298051537541 53
0560108354141244266350884117088954264409770274228232128737818584813773935509749335551 14
0624466436894204535523793022955699025688892472476482856987927771770439584369247240062 22
0941325554943292326806265100656067112487799788039988221458634529591716662481653228741 15
5327526411248966562365362717905170820153100267353958824702235281639972401534641220320 25
7977808273135512050193684281552081855499751491511016991412711608445400907620830051416 46
1882552963462036087371679019058251894683895404682662971748666008393029512607766929924 06
9435225777438631812759679506943700156062505627855914341512413394030327712953253107118 61
7480257722349489292521980974308952122314619576662075923556359726690767986661233291059 52
7961018343106907702032228716252083611956494811752997132749730597835528521285785478442 86
1816852571073997916448507379463019479486010937938364054003035089249948913801089313227 03
0643660409213615225175033647591255299336234508746206252211615213453346405907315273240 79
5593956003274879097386942606636143145093479579364252820760576736682245561277978857985 09
0507465575999523325768019785164732223573444661249477999064293351032029241706181476957 10
5077280191727166542272028024548065568292656244457107484434380924735583240595727928137 00
9317949584280200678166703023483010740547421926860540197880276706177331169854901005325 26
5807003919332218325517622195004956023295431880724870989224933073759045534887851895773 42

8251250967651971856799652910171995101746478143027813335716956422319340757137678346086 96
7122438121730798969383121704204911241451586221205738198926028132533616506332709612681 12
7354457645034386271837391993894379695856116712668383393759855826461542797813317912057 82
9123789989227627725615951258427540001446320445791065468667324140533658619184280422625 82
1688273710321538212290016053895557458048149707951428828742756657075814826054824202210 61
2037688341073437044616953135658473158464995233288974086138926037436545571031357309787 03
0515797674186488333083334683061776199649533323434059168678388645247041275531439579402 78
8421613759684918282328600669289116050761181598098057229676116423560905478275530999022 83
6011825568757238788125858293421121206435351362342333545480003763735392284413374664475 46
4899727153248706234324739394940743678490417272542657426758951827960203343622606184340 65
4829291096947327758106315005802505689492133983705710619553810369925100616045006231958 95
6852776338414533870921568787580321274603112849248871469759238966121654100784516651875 99
9266072990204583456279634204397156524456500393326975841615274186890528103396227728680 14
5702700318962787707751372895137492853891601345118147909121245542835114507476620614502 07
8740552719831064913195084319393794051393560862448712063282330972563106568067159358712 03
9921409666332251191044508321653543621993777585843281227230971764972700282533520236033 46
9451608228728472752281846877737507229883391318768369026382449348885643460614702141015 93
3553708337592611935438143753258368050686926516021319638590042494502607779328982929749 31
0257474851915475821236842755637397781015150277718846739673439741825642715865300921367 33
8800791231126616041891722846890638267871722469773414700303377094862942467836229172179 12
5739785889572304938003585912363996896312161385831046483707963766267992976156682119846 59
3415593391674446886200355689651840618965020995787949475034213451064682941891357624099 49
5577188337647484493614890337338736408448766512857990600569180355802175743282237209829 64
0541398491768614254113578019232843223566220125337569971038210371450536113521580875443 25
8875177314981234159790077484154852468747186982843716427756796612188225898363586461233 72
7087316163958782993815527341580288062228960322744791973151341958948838419529290567529 13
5828470288997290468242178115881254450027577348975661069369938306002844248830408556897 56
4911615693828782862045901715920661835559705573502183092691196050687113637921989163882 64
7003032398559982585297372067596850212232594796092137013315436900473475800526697316366 28
0876754686843154412005445181096396331779963270733270078424261594328719836710018530522 11
0004993585898093472727826132452225547446633652346902607995201882984865792935643341058 61
9206357658021349497123815423332633081824963302038636180607430078936284804945727476555 96
8976904796307725843589609723556268852771769509575485467415631893654443452682522268733 16
5858336717410453518601689739003705114038721607492565728669414463434281934220107879944 47
9315289080704416783720859140380787192020468714895404296577827423277263006754826839257 20
4742895691916760052052321538211408873240679725588369972297703978174778655445133936952 80
4673097919875464405405015355984214901764939708993368236817978618263713774776189924213 96
4754681521802356570084650624621258009338239375893985352553247370307268761318693261257 73
3337290274919695015084048179987728373665255064027193614777325988089081494639422730754 62
1137974252544785730656206162327584566336871604105456555821963228444258001613092292561 16
9521705856174292971169937298798552686573679816223076859491733218637615077351715337805 33
6399472531737904670385755272237382781358856453237660838981202294975179584990141689663 45
2187860835838411893138472832576864873474621953538997800875424150586749780156015931136 54
0552070950803525500481212312377181521072980032310175918378625405659625399485447107620 23
8523408341501421890183896302766908646062889973158305000605416610521126183324563088749 42
3761321117383235991026715443333980903010767519215606860915099297579489847091340484776 03
7253316486633273997745741707870588584989036478250500607565276677666730181427983462997 86
3115472471904638130827026950271552434583777132888840113322856123276424758054914145334 00
4307351368200167103048967407913220417329365588638081990240250424758979906199739449424 06
1393859002043745081712616036278391241147268209085690526837422506891099193767722077768 73
7126770152907129682261584375714966534629615403528980698498199023815881324900728284203 16
6454586451867787181771772779283212526968322976641245496739715278796804347658957612653 38
5245739151343813784500518738591532963414053689484439722550801196079269028116229367043 43
7115837195386577860034194671309665344253552356135039263743335590248778009316758556650 20
2614245175520231051803797924160186816532721349074474187926304637935701957254656870769 64
9025628311394908306598139258771657534329051829883074422093153945626718913650932778527 58
5614188691505843112818062116453383461456499861027990887831599923320834970349900964482 89
7361997260841303050161383437573503350269679199100394576485031398899804053476207979951 03
5562800942711980771413862537468942006711292903794021105099312881767863557121288220584 52
5902329882784488972855767643376551320983720845365197273566294540752078683774825993769 50
8547458537785440151866870321270325108378857553525327422467456165530175294697049286034 93
5237663193775815312691121571250545649366284046131575493234361611438689415519179552116 04
0327941387040597365968287723555493695367249260335274498928882204488684436527515895468 95
5885890718317292891292345774428441927250527684755038702706328297997585388259938790678 99
6367663472636799709113700050045191515070502085744705320311342837530396450683734947465 15
2543161640695883965696047762481007698125762324027656324714558678116653563357384133203 75
6328577111457947736117758910978449597487134549945405008974943123702669160022779621516 01
6443144632155674657986969134302091737537932953736310293482594184851531345770064943739 09
7642085895731774231457672882967906750299223152573286983302634123352631634902064904297 08
2100632648815676763242544468703921333767894896001251362652354725651702225595569986284 25
1088668968471078726001673324221562512429272130805593262213072140936864354996898787430 35
2676884922123183449241907637471574744625215974576466357242752795222891504064277678656 61
1519193319181783056716465304813810106667342459156864174457688390624192018654102252669 70
6153890999072549984285484195668192454519747093061422753151298445309182757715136118167 30
3580932146032258472352811825504706062154262243245514468964572693823166555250959895041 09
3425374308599979713700425885834030449726710962996976323360777674373479878835673010286 47
1384545928791637490145406647519394899352212362474366131747830486884631516036592243576 76
2762344666539589796468790552923902702010757218919138214831626852490495848675432931241 82

pi to two million places

6413466272282093653277283976675572672897319381293419430572396207232920071863867466703063646013311111642546802512289430533112509853860120123607044969978521095859875329303277162267982305510767692680002207418849030165005038534475971018301673782681943612416569639252294741035743185176583656034123276433900956511863260791733899126277207213516175222255241829612433962825182328696862544411862381233064034533155601640695747232038365145663557498734411685994161655182496042597983926781613148318090253450716466664426702627611859764913247682952727805703223834351506367217706637637402490304659096285960271979725537800141820199810181398125950423486624834404392113648723666292020639396288453144837489010260840361484073120067415622915966963669408360326433414963712098545475250177366960197146178464515555994167263739708586495987795324215832821841093916405283567907068642107803466075719798914815540054200510730096279623472724997011221778165679844919433222633415033856753082446773410455032742856115745538742140007192843017744731423009836576075155127779628101472205306681742035059679410509804665631363778251724709140992555247103681267051382467521172005284942952197488628489852778783562106004878127114406349908816459244518980104429357083290472201607269660461984260772247831071714390934928973795075056471053802916187491886994635301357293501873206687311501731531091296794862549795815121682207571231891909138338353447102365979448080471233882744403505346799952913354609413927284465139119083607622657983981564246382915999044162845276818935327913567474032273515068909877547218155749984883466946222711943435139572756093318776721574285783033018304222517049632971612296836752748983294723150697497887414002192670067206567729213108493572940763592895618281290011097847243283038465197437585736859123098178360303140528230081302663133041399253399217941576479853470817863611377200140857083863943770352918349740374183511622370040173188269393263087505487566452909326566503024394453622727917080038157851325290236510568096258917942400163871496121216946992544239867472620600571311538783883388307801653788387752115931194493595694917939405788488606239594441849728928723084955792607211329771213723886969863602368291622254647108806210944791323990154066781602893469428215506272126054179829178173824891997338295301682667906177801353365047188633978427353358562327918535797779226627038024456968296862549118748685305497857965989184862186237485563935321563048992834865561541540649512210466103765481806025067654913403327386294169117762638138347851183641056999660949202044895026294461684668551060662420883145374012687794781398595777699039877073994170325653193556130005281435994560612616083223589903248956595675275769853245744035607612886079681857619771788765561985237527447223599272020602389167187908147088670683027939897683780233375796837484716792042056118846144350842383697378594825884978259521414316768984959213193894128750695961491932711470358745336608149437546971429195290310193894356371835937491483074230449340295962811166523899581106000938622216226555252976610650745271358949733447407281673926348622262971347155532936244457994650823197908769074458525370345709007744081534167863870148099674124003837808523427397788174691080535330509119443313873040838043067506030686269532124522901667503856318585859317437769494157410810572749444384001399152295292401680667458424696605569107697596987310950784518251857689800093942863712191016698078851710571144669507031273706962004730035675368235205815249186823908397408809264852704581680063915340134937352547150923527044161926572110042345848085322398093081970121586417329130532589871715885516842060650340556996859371591562193954595558557009347711681179835995842798195564356365309389050941964641889243417661217711754573714429402729377177659183107443058151531596094826350633655723861413920813075414610740512741348138890687520896517547286443489020150187201836613841728079882729582018977486126338360371109414086804414638189975514419051152014024187628978688233866528874956474011072459905537992175155647819809184955876752778280803822629818043941563979561725694090929518577447883651594478720682678596369454763706238206696202396620665921081278183219127468081453031421779867353368489380826681896912999835199422321272638771597576428521321515883717146485428812423122468402839056157796819989785562510271070628379399431907357979753622873719947845210383168668521420822019266723155810117372442375609151489386343666526579426037168289281580693159057152379480256619126870887647506950850111370257880233381801903021002975975592681821635953507064188571900594974446797417420252130947246191950277213237247025702961631681464762184643644651799535877759048091724695567396794553734971032219369455962778937791938340697253788415502062958387483096195420461546990222684347461767711319737486600087935443607302433663280865368473350668707408900184703067698214753133731542862215155131814095414979724670676343697696458309286795212019941406654043266683440819686918622917654410364920807857292423388755061809836591222653797288411120130691018576030498329532694214188425942866214695276880632082571964867134224698526419419022236241186339130284171844724822755723379969707482002437580371792180734202080536935740618765664169607739120909813494702120725197213699642344209305478465069237446490420888732630226156357919606309236991602782364930003449747123779455951240858239709946570275366759813304777505050536634574715516558372773100785781787153031613276848925357607846211478860351804029765696058486717567636659308748016099927950787178913104203849478943286084797051504283326524571886423198399932856342268607883443745309272893146092544299060787111736766959849633062177514884899337787867859785265280570548661217379213552124702395325608190678852803832422968075544717437748950143023150146961225494895338362756944869304674198022922555065087429772758076095106879827109193837142290968268728596321942836727242477443909060036804852784543854819955828743344189095523099265929588482897771967505439205771668938552397736092582090693430578986742357295312051485090384652493140068996173731735816222294455416149357871477506270376192498036384400160913611713729557661808926386467940279365670385305779912988573944783757639092679443336505496770742285963808721870399582714758000440222404214003303590360960548004718847304678286807740989832225262453168032034084435109374319499380299081241792110895423927096542582195848586679924115788447815219557498322258335674226798960098032009354865108549467676713405310343499864349758000215286835835721365978208435732604661260570464409200520643748808684041999585408697477316017505390253064903620449458476440882040053860571525182217793518019414711660086532948281060021915944692783460379829268818677784883782713148160668128480874790430234200377130896464781785599461837510620688441358628450630346441913942893762354742777586769014678228907006092

3252250324639953337566728997660254246597951963260902742615157481865278192977983681101 33
1339651625793318419407026964988951386923961261275369592060229690087420834720840833188 41
5826838019383358973224335136412244321174979404766824167809635203566415433254150964501 97
7910541460943749815990445792838028880133562481814072611423275972894824141887025957454 93
4254722746989976877162316099322885042028070838100814091887352633318358420740748446573 39
7838429805347106002374219987211762683334909209073865337959074928089283030107207550472 45
0851183334676304759820661789998004462744803370196555021320441396423674506953708781697 37
9969379061637848201169796270721270358480479488580658308963231288673402963848241128765 95
2185362411256969749199057478280326299861231724793050323637705845698785774531610386670 67
5558440682408910511818429025803298514096157331538756311438547792152983663838215871358 82
4082012778384097362326475844352630281664756079993221483927156321249990837098930946329 55
9859928728433521252427433494379023824949445785164936127032642339094544808620028353526 26
1752981835525297880465028135399112847161281153414460389703165467739525876538384445746 11
0351561641809273346254142217903310714720310599294953895958436885773489495225982103831 59
6420623273071483716691798967444541841890372511272835300592982739374737571099277652356 37
0360647348724784839684203742309758998874387876542841593565973588345060936129924492587 46
7691542804598132815825872999110300780631592481722052213206010771492336601003182710066 72
7266488949550942336897935481055796423771544954137177407995177501466695465574100801557 93
4179598301318715461713838220333287263136997808093756281698575352925390236568114358655 39
8284283241700051641990051764383512005756933430421802931523685411424059868057389737717 20
9032816486238395498050084360235358582554618855942442612928921434147898267094176760452 22
5134929872997435333820627622406331004883774527381188722581819982219428293676666000040 37
9914870018568674554412319573518712379305995148215954867701054047820258539083335640618 26
2225208028648666809763156107137641890902360603954124553570380667535676524726680367517 67
3845564696693596022634258001555720896238403647713214296692193472420798729629861675796 74
6082959712748570679034670391578658581112738257432519039782995445767430575299283028634 18
6242545022496241979163982734904359411395898404348957578332464822165249725331811930555 85
6140503105076548589915525542652886287528895545773678742029770378468475636642494767084 85
4307354132840361634913471074468329858980951110201242544884643071227174886586964367223 75
1257407663807577968593851823215803790138851324567042252785387661013519568286523394604 02
0035673386025205513475307900746894526143616381246602094339688182998572534654355285404 63
6104131214993769216302614831518234694209627911549417194660720665528440044356575326641 43
8934277220905575184236912080347379886707969228398693750888161460738382464200081539367 40
0188625730736953499730836725281014943043645634975213545319519500350764823703618453849 75
6361633974429430988638719898808188086747495831760222984672501959183717870015464719437 74
4024587964419343305273778617450245249707149907000518726929283458717863091748438785506 39
7547781397976147102974805258930692216662225235373449011353986266028141926476293709768 01
3187204140666876255459542229249384946271177550175862021378876760029805157411237809551 92
7818159082066363654035686833244556620095160463375225688255854582920019306381533873656 51
7945743702588756264732107732276466231522699379582538162507411935992575434703207518963 92
7929216230912990259044551217209318966179934694954150218683377015220759113008886890238 57
9915282639867824654608874627852622681424733188588572416651261959000329224404728408961 96
0264923773072793032869835071995091733622206904266211379357378789633982192711117924375 18
6838175762134729227304841109052893127397566546440159910892056359539550622684903481778 34
0163888074775859106047358664560729940094200063120435623081164981455149655513005854611 73
5524052167155666041333475875987920445592175677563283767227690115416492112246422360395 40
3368455011342654247448989545996792036442429665248273506879964950157404621482511116740 16
3812882370549267667972800574632906194561799730944487237467063062834619376926372843710 25
0629430239838747180411277944515182108640001558475791284640128739950977629770826226345 88
2505207818345760505308157127681646160126745615131039107176973845578732241330030005534 71
9511669012581135208015630373046908093097927353658564913574711350904412759076490299193 88
2008262173939592861233365729706646410270587838551318934657962685933047956026011545035 96
7710140057993336889004022075384825139930863716343366007923712406457617650036410612205 43
5688688177406253057006023018982911091534071177517124423703643637158902201162317102635 65
0130243991215404270127303916604348528921717678005443537960268144769874794055715993778 35
6399662100669274192714681089620407361116347202589862464744081961204033687520897010880 63
3542844369252180174251211967856991105833494449991683094494698470780636754666776782538 372
3040528489291173054802989310613282285243013974421278401082297992256374991861619095395 09
2292352403872656334962447446903480575135659465046250309625011185996363024036541878244 57
0740245894880605074168390715058032424183755862679604489403118420715618426638993005968 35
1960880991550054081911609426156177996494555738936233509560216938453029407415354220170 08
8505934108021537744168969765523900070011310946928000344435606360766131030272873892742 26
6524989909815901237651570432773192185028448811193320110357105719444387121835232255486 77
2644086673404544135367403990104641792881141327732957052332339987800916026700289290467 00
3455063211355182259645456365580270462153147060321476780387345442039887757315364197294 37
4656782763362311198646746083171624959380516317910160217431600363721351355065556811627 6
7164832287962390037143316348095868924384711690483078965100591104965015992831438312018 93
2525166768955897310518020709156128212794785768231503099654870137801420342350862188944 51
1309174155201212503779765726305117588445579181661243191479349987937188974667677782724 33
2922702482645480284999856755494526946870327503783940036651442685682081309020949057899 62
2100814077366965566279789587599381603739294081898326023119790605145978038449412185507 34
7234440464136333171482978197669866965514005181845419763310556350448849713422360339130 05
8979717346782373472329230517388505004636025681998062728258112455591586015018439090409 86
4180971710075461884773934911273571127107533095079036197946170873344664805241788806067 73
1106455884142874312055368645075413123789205016418245598529170285529823491756815198174 95
3565040453735880040973693100210161974099408857233681398906852305802152257830798584444 98
8490026722154928888612925028852813527173780318207628086658198702133918612113360246187 36
2649128598385704246054788599442082401809197362711751540474656341180486288643987511052 60

pi to two million places

1860076320866403208005880981246682872769158288851453555992972145134318817716645564502 66
6336275157142261212702829023587031467862427302335998951338331069080367912289759223209 00
5353398361052808487974347050510512429799469695877329008120707972879653583923242657673 39
2144380470361706529595672993234416869309201866257158203504592227460113349178476867831 06
3630236724355370932562694982307261863131091050164320612674246086791670377930940669607 13
5447772041240171387152541478713374566022914274536828100929205588900795084837232678718 65
9556212837654930431227464459773811156396674092749919903096783157044379273964166675109 78
9264093117468241878846539287943914280719137228194506211199604942014167567514155226569 32
8596939900541011164776752925649440428795835710036845090703458019087499993092734233237 90
6647410746289811710104027788338214509831606137185058427903895394961345986945534332173 38
8380442292218684824710117148515834710609975786976196816012437330230684469271055789326 16
6001295993498597491718450334461056240840010952490311291513102073536606699142509744167 10
8918044279263850255766220625664347056888812091343129654781619845396751548210810244160 62
4449318587351214286010858155871519419397655261062478092540814247596466270191943785507 18
6983496876926575171350176402003599383530178302781767102202449288655654620105595674157 71
1590472858301654225614200548268513719162768982527266000770336835926768927117466145886 44
3256295441705121686083735716597610278238848606701446329636821363730331746487176320142 78
8006742493485684457268867825525550925006154697582885492108122224766822902775116822369 50
2543987324561861209996738050145752145346770108025915298160421223116328760264578489208 81
4442541782351787729463684916863787103355988029352879751316600965034502135008786148165 27
5693425491575825447858789779004210159280113548097158154932538649021151389857756639270 58
2004783308103193586172095928503098371977956384664987334554901336566062958993312667035 42
5517958589534255685222167057206373166820932241554656528706208202685332600866580058396 60
9069504970302254534936941843479918148540317521615318893601698982971238272732961881513 54
0418704927348526265666408136486378871680299743419921840452670036155802038750040963721 88
6553766105646252585967623112091455580614923744622486559052594146783412301336488120864 51
3178145054641794164567238577509045217705499758332360916182468663731199597425637392431 93
6836066334687888366489399770870992397517694293270431571634050583519899477212598612465 95
6758031364020077933287978651130119476790122849334559372745446777306994245626020238875 49
3090223357398303966428565992346239434307543557661485851861284466173143979975977684470 92
9792773827647093562794945093757497580940229719554370143859221216058081004239743853304 54
3467119143871226627091401261538446277366108865182715566402048997387185384279740871780 39
8587857487216892636293407937055160183714050877149628160787383362335559788371360809666 31
5218932287510522740371018412548297128568954164194927943850639454838617154528632987007 43
4474646146503414460256193649389255719342320962385728409362207205517646982530400643228 75
6038069773146999660101861018409083474528089280983391290914925830365117302996765473925 15
1845027724484495376804763886401906348729677479902124856127316639984427361862308855173 18
2399678817158183206309699648514729573723694647944254825014483727864303542669964431539 81
5277168679844685777773176724214993063597651813595392768068710323045802519156036464184 55
2722886148251459740929971994529105998334724104185420272085136054307357487622738407920 01
6763466151090614719108133008769243989050542838285871745960020088457644825190313755480 86
0179403410944189883726523194071831370537998352344375954898132153424084287482442809898 88
0471971054529233998476551717751441096350331443841574283608079013413016396157944559087 36
6278909144275984522976305439340866678264314016375717056188134506536372888736845773001 89
7543538641536393817376290182296333049441891940659730575385121339862756462498470327918 41
5111491211352501046851190089611707902188891880624882538422836411906558748088381207312 32
3141344233353144433609656271921082476403927206088886262852588519928301333058905765272 82
9571426194979164995894363177324749580959841491639960872405594058974095185184537010842 39
1107823544795389772207975226175997379931801766025841678345852154531357858420969913069 95
2099187860988612444010607411986374471530993510334286163756809485035927570474426589679 56
6193382876884746673876270357798755596549401466289989209986971648540723033988839367611 01
3303784045113078379970433116053326219954425770307103968439752796919730812802511262236 00
7775400051308597498304645404951309704803426138354091344540564134101462193716056552804 44
8400880453039649492973826865022745282299484577467343378675502800997560510091528866646 58
7902625776895712418793158394872973887714835384248129311916831606013543029978483686352 77
3120290302971077830277473895813465194275616066742843607020400238768610459207769656762 67
8781970656061203397304722965481373446191321988589232186743912322415257741929078225709 14
1401815695728457383362291885079486832949330533593193572091676364595581367992386963556 74
9298651132482713946073162855012413231173726487739829651492342674132247288632846021041 36
6966644267728104149594302767238763428606644807904842677191598564512608618704025727442 77
4514307901736151561773151575005988399640141880497306997550669101292407530374958155784 62
7683114837351610082642105686878635684085892011926825243703903525176669009238408264675 26
1709260269710407047148153102057397997681579182981289235304146491987593615632212451682 74
6172277968157330253255735223022968339827799416034826498569382639736059056232139294855 07
4276485329426710589699458926421441196000845353331145040686537313195714843485415041517 23
4706871596658893468794776160506525205325518877942762000677929174286295148036393715562 49
2149219289945067840972054346001956298474409674862465363711302087381417548333816616561 51
8511191134684732365538248531987858181814501053869413158042894105310850262582815712311 11
4555123885490445347986700257077621741380291892762345238939140280529309686455602087074 75
0296305685666872397749985911356208348594264702238540331396655512294052067722982107716 98
8749068312321865667889253484373928939824303927063104601678559287530601787022213306811 29
9142564872664971680328549439089540115982149377017032767620929876361529476102296386400 39
0994665174286052716065117214013250959270559294839736129981798102567138533177106675033 13
1782787325121501327837748650207033135506227558130481080029460517179886418646938301427 22
2915794350397991778016490528277130295624570910278494445900502500126475623251401612039 82
0325502752696951967074235168421191098209014734553452473851605450203448865261194845794 67
3910324946017546065949133473564878481268188018707352591838083903679072721987136512691 12
8737953771599527413426674052986058267277760841997469706419659029959562956055600022176 37

8828180964658424294431160434101540324161837112411833413369086040738182186785929660060 15
2609792043026905143222568143657469655420071610492607105516213629879300555912132665254 33
4472375154831796125640786777424307070087662202918140655021360191663843859998861232751 52
9035298703495321075289690611404016598021882803768053487149020830871917804775313608584 14
1065967519604324017985891535324434233629910033903677261899140466810276148578721543032 75
7524359320503117165170343024273760823271009896214950638496910029025774167136585044898 20
7551353446944194185199121456681506843573075876541271316654235668122273722233387587767 39
3622874603410636815651866493228134230424204001730539139580450340596804482575134055549 04
6416335784381605886860227991791515677699184838574581978981362012970538786726606889518 85
0722007442610297073712835694266977933820878091270526740228190344887810682804959591079 43
0888100469561835872093432323304409697612377196896298211991687870983396113452620176959 40
0345867383397823414732191382499749914065896773447434028358031537479848996761924969985 18
2240176819305210022455108578603846905687663641289089751554366506561650619221818558606 39
5256352043478945916167981232360574968374823438905559643350429275477260719321983082538 07
1553885217730924299413144190263558021105357988665362641510146460711985598292895490755 14
8136944093607176494262910340381718216143960417698528173282088210369061259770468314059 87
6589814268531700310662742574082809102331168159759586548561781614606434476519302871734 30
0921800100587395483454403762764601382427634053294655808884773746683625616908343097270 06
9974817824245741942626077709798914229003500843321192397735909559457468156646947811010 10
3696356946867890309337571102762086607087821065555377926088164133752939155961539106238 15
3813148131766276032319888049792167796110491023883227506396471007696524744146098226259 44
7672975844810138808410145215932988735385180730698100946012616786893086024378498607208 28
0266924451098153916959736018213287944079223067584129848490365630343690870142583141989 91
0541398522693075466950199902772011993438099809657194828591987672414559171595955750006 02
4391473464999909496228073208901853177416621570733389283886391644137593974177979961906 45
2774096579769282536534878288646972253655754525228031688471039269429944017744150565410 83
9248185380972246265514689030002078212754945052791543698175496661875134783191865741255 83
5539740773733416015611445281501716175117999639940611911008630470479571340953158279114 96
9755061425259661879019404745287573343889299080029769877893086987339902732472293618765 19
3297280946392158801058120917330356069608825523217970057600041590448817939229844535803 79
7460712947076082006515168336564512412811294002079114608274243109520336002850578510317 81
2969011196732086099490036742606578833236759989031746784188273762112822615043735782612 82
3923583260235062150253881205038087610757792341102006638835964616931815750428606621212 40
2530812757970025787253844905787402407677517611828280220570076803317314373322249720895 32
2231769914830922918525246749051877165871928339145127222439107688038146764776825605199 16
2428994666415868570033589710058172746117001313172720152645395750670172388733144385271 94
9699753724585045185112124532376009243047239954398933276325847419966021261252980566184 97
6823050412057268302788959012984790370101236347773267203908107116393032892689698589802 76
0428530981257919573240805314535999506802816476376786162049908722050571792632644701380 21
0327447578509598153769279437353995599069201108684572761587374741497813219922100979463 61
6836883769800680193267246356331393619802284466029082574970887611612619391798898914147 20
3605599369088389305805359369339114503166658376790682533810154946336850527021605286589 89
6942257096353454924087953244983450152302310368334930834082351682915189641667157504762 90
1953467655050454331891572657051498776384149079126728380317905379403906551343242579313 30
4132494807608810469731249545345457856264329245753975443631106604365289403443842934131 02
9921856386196903953622936190101639935285350105729932771839446878649027719241196947766 79
6743216916617401837190656046390007652119611483507207555929101785378770569542074600725 34
7546329875918008302027150297749789152839894533255407195166657532309264951394211425540 45
1153778645696623468005001055766568622256559753200069485364386223037984856936822387490 31
9549004916657833974366986091833919987237194725884528872540128464505663054723627109926 42
7857024582922373042200103989251437607418119767998004961158488903136574404814727769349 79
3351969079124128680495050177445358305674042673285789757264025168112914401728938893906 00
7862203398066619650785808534824907943715105931869232064049673865635312813040791072222 13
5766548218780519858530019883207194602635121427993700694070856559587246813655434167121 60
0702677482923620401452985056021224418548337825955416419100110698441606111936134157284 38
5573768224370273680210549049859651658297294455519182415160406551183970720272020846402 04
3930729863001390554348608057272087118125877938449849043705292103749701001663998151949 47
6294999864284937367525363175218833130871088807978839241770462788936077376914701380205 78
8950494781158875639904502685755056174160558994625034600921021093521309476759343508224 22
8736527388837423211347106010920493956173174885378022731466288416038867881534237539156 00
3774078668328693984834808067071923600158571920292311134173510221745594119959835444561 37
5619179631107040180474480380943839754826744551977505936659329500786951398347929873388 78
1017794560817544738135591808299812498231500373506626543377645218316617295923566550503 62
9887119356012041679383725200771593141903519272458016944939389396886129000119117055885 15
1579807832197586364396223411559124784518708290402202070552688856767767572084330196215 79
0085294712798233970767046678343101904313793909567418493179487559919905514096196893922 55
7331938718224016540438942429761659128259606455657678962695006754576610574970349472098 54
9641722192264151810279891105903306539154665967022021495454292522568011997322331862993 01
2889772600548883051801907365617848724489615573216482574547538161434670840182571136375 33
9420168400511442960030308232427627244024934395610559393307378279093954401080585108538 11
4412665516154280952868117050960782891078997195299893421677946200201699898496514405533 69
4909314415663747898278927807417117097798317152252276910176290637536782986869272805898 81
5004823069790734921618995536707907033479375433604945307207964877016335038661247716798 89
4046172089012433709581716007094122493621549649575492391338905213792848132560065407087 21
9295204117451146578362108110624285418078166194941845801094848226063578640095318038055 31
7525908824674419440837169751263837722218999035166081850378401036964191489110716279202 49
7884078503577014056163478742264000063558174689957745981617176542473582133634390557463 36
7004182631314361141816033296676296760016799420553340364035181666054990897216378910193 31

4935297881620939541968206581906428364266241323705903925680464546366588202705763264918291
58713636046873638450544974888393625563446902584799733972437826866792004894254402223869
48917200466558257328802332499435398108946466293862106782615852789957371136482549194966
69990046514833047867362138961073979915734993372656791738205067948903357851703480156848
71190573315030732364815754371927707826788849883018486465398211832288477459906823397462156
12585382662372283196986025204337862773557515550720101605978712174650697369997748850582
3686182397405962823899061718458168239669939404234038027152149341645806650609425516633060
4931071671973308360331180912226728261653649177815454713336139513693578970790381291008172
08674970566963962750528372743979162876486455768107697904387778853689843730036014312948
6649262310772749037059562142587514293266878284788078762847818624595678168662583026820364
45978850982950892525944172113535585941142399515351241488610058114813371447620574893722
41692219063913137828116201787876086438886050658708324086986394651945116888379527746353597
78005996422511881275601601791590224752139769106798632092833840600861021846713981186612
05103771796785864715889119197808206511049872092732936744466455522783331556127984651988
34836197608015831317906745512410020867752382206554614559397489786921632321555311929060258
85276633670028108500403677602024625577852652669770340069592157761564764259634743356471
60218551276752648922167599801547259118353017790110214847022673250795852527548423162615
89296749128014980057541492894372407464438111610968912792553348665043947776470166897046624
97021733473368079477321086193443422642160858013035024637111089416681026503673752140328
24342933691937372637891689833875137555791026545295231372878571956727250352327250114908
85240111221231522395356681417563608279688942019932324527491150005680661906710073056211312
56401829504993342817836112004413870830454117779908282985239103165217553221739083863377
24070260851601865097547228451197539120397627596019905383829494982268416069423707468528
51185966687687972298646027571845029912469206489946948977483050501351974592227789428049458
88936266175575585016684911380345406553384045092547795648368531413220672805295322177576
79973297500007720903025851044324639934067451094331243587323598628587936132286240082781
50765608155394815807462635927191062286452912195489912738899347689840630693501530580839565
13944024323250666224298765923968269410311308341511967555361378398542211792194114495619
16538849185979570764326736974593593810668775110459390561587596349661795677581631290731
43952002117162416023638780963299013324703785938655291405189485402021747743494118074693780
36532613139946208945557490603976970949550080264968790833922207730630331527199447948997
81981828956639787262359965053845084022616071288706919179555348341272915563450783929095549
62176378094144760392676202991209797852610126161587127075355867886070133229374148009805
34901978765117235013926032028283193813753046174386185487073152287896043801626021519321
72781250101536609553959394826629785009647214769630414209285608367553697145767446582513779
26893003498187673095366561540940181812138145157189348521141376323974759124935970455118
11135126263852182268414222905761672336221741638596959428555841526603258173569687347871
52825385468661708317156789798692079668579385203867327936313110970175090915363971577478546
69238984188000859849849311591372836081341437393598408772681388157631645299040033170061
43555158713210484908604086309955458293249851269415089658896620196441030335698575787971913
71874747941989023985576445427531759127821820679194009597944889802165519947491076702194
96596100713430683605884398062502933339291805043787495845783955975219184993991566568420
7644401577026373497360798043894373233710357794629495956975286424169076671625974311702801
30433527213920989591016293150314406167603304487131088893032925209532460424387158071374
11333349675218946784204352012005101715588810140727903370901390608296232573654349022515857
15973438396432549682824253927771242235747636514746268630421600367373905728097415180263
32536477880629778365031696247887663049465890413981453683496483767637972403013154653988
0841739693614258671793819140481431262221184901047710700206513634019865582459549151893608
86020775433744258793923503783385025011677270527206099193802611795015938187100439454786426
11784251461725602723804763286997966113111614717459878855420378960304851193694719210877
49508553862899585037777990447987976819174494142900098035975024431445721638798732593334748
43304573940812551247937720838298860465524871445206812031443287202182491713332412389413
03220359557278051440495605813651409861600790510074644557432653939453916473024455409044
9359189950093007018061723337167982022317326195486552606537659624069393912796408926084161
14880330646499405407464597314824039656863432159312994393707782160027095587125926393806
26530906372271590302144540827890353670167098152389832704238400163481012749869965938331877
19507936973083569436728856506110250945352047041905954246820198982796106997322621366281
67363122989056247377762923755761011654006559385237273915066906457679463920589716522864
97743450228414035088808309344618168366673715725141638260562684009201088413783889207601230
00864027233105874037792368077723633760999633454914229514026340740695000357192016355410
39052403529072732698238914645854980607121971683395177468457273270578300165043438522673
28906604188841138250834136137996198101697052736772471979593261177622344539819532047244073
39455212937548499646212828779894759263556471124809844788635485768440400796454086534311
78446986669963155071535534675331827894217490529540847002365155937191300707653206101646
2428460575445913627408024944264302474203872311368140350116491673880339796281288237086263
14777371095092521166966772840005669652355331235727344712250584933421000654380467153351
52491823178461651025808818016446049994058848908523540861518389404360719340671292367260694
48064599978077724922090290386244074458697014205902219888068460965151709482306000964657
01436440766806637279687569407135156782799064907906327720904452450324857579280708225203
2623968335485158606931459783852836169456336186314956895751252054275834084337654387735755
81133233224458741691304462333288530403416818532765079629533256719680304662622269392924
2338176147406217694381301816446836250602887868826388486228440768649870505438229258065848
70404035562573256749520111431937547754077091962079537181488250616630692548318888716236
8525155484810807574535347566506910899890257900674711653610446354848625845329601116605524
81236658133484434023033838766947308531146822952900928520339707297247845950326397304709
6928772755667410787921270676969147191629646778484264375541857569851081658718271514749550
36156500371320738054521273860737134932832995067948381046672169613166745564983864106655
38519571147797989780405313153130469535595601250122307301351228235903089741315318724606

pi to two million places

7676506927312075405356228053969566709404554331001699200934016030151096700708703308962866
5869548111711644722242956459262470228438373631682761826381454527381901157150895220166955
5589525546220679207427677769230811522647511068244339134165002524040693302585989456339270
3641944075397812008218213585045547352157480438705096537744614781134687165655588835197279
2131850455068355394307605036900936296238783640866701294439893854444187850881588375087629
0001144469012888129558556775237521659868493432422064333126591574887295399533862251753806
8215345051124474409469110451163239958515057551231258294012367477151579026785466343783299
7684375181671324663954643020788768384514133131120043836272094728966779743948889034039333
9347714916556098784406261235812235129549256259585831916362448449112395365765871050788073
9670881097669121051336720210884909249834814108085147197931198325601491340711816926537717
6694886325677385081130992939242797311269677458349060895901467971121975453286879491003849
6940607005645672377105349189064450652802955744757018578593533302534373141442053035818947
2502565974199223900855033384792966237076723346362903759950087543075025796397415020398849
5903758576829100173441003016390174848563306422011752017788579842745795912502607278147035
7311880310825407223370561840398146430866701326413991073500244188772556351318020125185808
1489754097369730814873054088717373474428409192916434244021024496426392873573982403810584
2003734586953107032797985022930946626806901792724178709806252382975492674271740109313390
8440060317715039883971756517156642506614086351975457345974732854603525770816379080538731
5806922805330698661071761704723189417223854132675676864108506936197728508808999120593229
4787017236599579112547404902304217356116439548935384403866196678322738363099111005285837
0782496250614551882569385163576399303075590707409177917689600909421662686399459309871665
1875276280612167765599179102906097798876029111312938589553501801828282284251271767414232
7943772490834468421570946790104913429397381565935133360706191251839663489878904947087264
4134458081024139652538971443908917775225241178022016887498243016237329016542389588802987
5750062831044553948727701249308315249497974712676481104179236903256497908621475914072338
5699856848993207228126831503709899193130760222768091775960194196631360653375426514541707
8972656421694991277672019356518712974238974200774227060081833146868926602940980858395534
5298132643374294839713715782663489388817585285996432152468492022170504236438629715317037
8612025782854723968550109472648686527339361327053170918496084286797306300436165421346267
6610101700359875797906998622320548802641853248629251096168796598076953897654536145457445
5400165223914248148929729381427906255885970122387283489024057385524642344391199345027206
5771715210499127908992116992426409704094162072318039496941688985426561530328072246825542
4581111427009573232719015598853789575571161924596312339001389238727215278612420381681489
6467821416667587669182854585244394137306771464037343309404136447692935783257567547224604
9237725453066312261405501756381159994319702788365614699745356186625199217747587896680220
4666776259774383389956603904036282986148270213861905360663668457915145149129662414918969
0080815398786558385378115703426603443048225501319786604767627111519141329606396129679567
5148556053596642717648733877548421668073267934468273745356610801508605743399198621529578
7876111855924472352713169009007276022927785720407394928408102800388985665402155563337562
2291458982640581718488090352195923229845591916946392957967530091549871090141039887383479
2493628931057971150462061769010546893013669125649607645519105336273179156006459648274765
4805723188947139841098601302864866561626662957625009817839445743520393794916318616232450
8410436164555398170233396828075408160676789235051027647052040995697148193078321599322556
2257913369017793709375425041782576570705962239705424120671641874246415575661781751832110
0918462648717765091190334572307787317880484937765442539452471494240914793370735134878763
1457698510024967498296725718389578378464947863985440231214546407023160932103605559461954
7608318410781549758552449473221438932052337349758294772936978542447331921658533831755522
4946584875939746120313681769249128790055178403707516106150863284334456738495665891503494
2405200789741383221214924671798085284634286820447027578369882865730447370179875498833918
2164436320438360275261130900446100374779902794907612459938112405161919960596513900779096
3429358311903430562435671573409505616362874827820587615489988136228400631951119520178080
9674906704976589428203193245913242559617114164316694416180155240661886331173995068796437
7815538097222929745986867436434377244602250082170131493699181402454209157676839550131681
8107403428304112686525498680326457932318450295097745201389805535819414101913198483386798
5554819940171661584883611814850491864675639177286330583746651251909951762786218221778373
9216042843812365855035987768316798876917667864037696003396728064045734525975201932854039
0384327847518556147531022636373359384639799945119773456856654467428389581911035087502447
5420750117547265579343040641644816400018548961723573698765002091463124400680506656182113
2029336859567547226664660246868542008039260740566298299662282788667330645065503288416293
8829562588755409694680707065812198050857892405667820730191320067060364216836491652563135
3776254830895948420536098722955582752459315009433419009078154671122572710798521222737556
1583261030239205316792788614118329822645975576534045423461049029572239587533119579619502
2894605030835697403198482479375038973982795389920689718912962709702681601799948823685154
4102490634503793304638505980500068936885092159609249345461701869664702253261938819301682
1226484368777952396181368777350178398760142879720848366410773287642886925358714697839726
1888103384503337118151711484715753572829111377136313197782120245684609697774921963796847
3749691094564421462352746752726128530180078957354303275355058900276072417108242779772273
2735900626663866696013522302560850972996315375782433792507162172074406333137963875171443
9266238114553939004388678518424717587537903066366609326888319312103232072351409070336054
1657508220906033720166811388503146846445191695043655886615212595069382843445815287087122
8293140727555933699681210990351591016421256071107576563443063511705673167435289581947549
5216114251931100258904134528900750775831818122670748616937051137474051447945461979530174
7600610679345348377394055213592998818346545867982587586043173404055460118233649355330635
9083224391666806121029285929309962757245031274894390964296332087307746715007773300833934
3158859637012443376957769454826077160976708615481666794238910350609046146044130396871368
9488679983508780406806438162177240634791781191620062957777013993709343944321724972218231
9521253794132602753367456855860884410590851123027060655379689486119033343118293391087619
61856541457096

pi to two million places

8938743695706123428019773355738962407681631584433584877073360720706401263672416841255509
8300951381957688615124648660191044190004053873335671201528782626114531444001949050115564
1718014623553003346080217675891561479950371467332745815027281127211826466922555444131888
3985950933419623985945561184947674786532207149201414043587348389120710525816364490692033
9881387927289928539884606794699973386287843222510037432806665926419930608469361675677488
4717995553822497857465567250556748949309600038811165025990005993740173866604706262123888
5284817010946710138768352200253700490944667104755790008627486998607580100559897397752744
8532074183466193993789997610753993025114426156892048551972307840758227848383123586478166
8286347239705070337701551080372168639415071758912025235200309364453816100089088130502033
9169341159108232754929969978413544833229671875424182246552003796227904310697702416765488
2939497616404950028309838939426022430461690484355804747722403871786693491539288578630222
9892431436841730470301570109023060675037024472003326413487285601003219723656520159094922
9341482621229998231733207306487960120379727647315563630376092938373423468209183320342088
8037583199689240992749363529073564984723927517964835646038113180744527184622345859794499
7228431805274062500057844200483005238238751084855426648661840587880464120681035919898399
6098727131150641081845490455579927609435421840067176453548615105282475682962685981806022
9377282987924425294387085412073102529404983278917912774900315217552148825260347141601811
9538454176711806252183687581941540703676815615766181720477998692331464033613803346520400
1842615803902641825361857224684486061288736869992720274162680637666211206929034619695455
8113643447415987140189211660466226582661590542069763943593123662045528756034216500347366
0119434225614914032015794171185171542756396517256864538467095452483713059489858255974522
7756437837209390373760644875780538089666661399183963055434635153154818588677926291272533
6342628898525685446469814497461892414958636636719814006506858886086022426733798812768799
6940649702991545245272132575428195324917311506620858665207749095296510075340404922735655
4828295702569062935888169041465106971777242095544613025854381786304850806058990637380900
5430695026138424222705375354598590993266967332165151948172534532747333602744725852724533
9478578704905484758633115718366332359132347588259340641521039072871936326796375928473311
3161233978154985650774595742630192501361344218177865732684594980392574196969999876459822
4956709409559549064514319299753296990292901811334684918939731673374047376102153497902800
1317223379127998639147101057364580882496403779366914426022522432918220359694796522963244
1504625930376366432840865616023121610990271779794048244237437724217545327436903074926266
1725888065223326141060338165320932320266991087084758681985639904985750117619963690596999
2541043687532918190720041575982623454667270157369711333570414032093793451266060707990655
5868796161579984934109540903212436544310873161586375727374817450178665573793984866922911
1759920434224760064859760054978280629418739147496645660197689263616598289656557445804099
9142689094724970673522047011619153600945273632530666440201002320187322781976148686634899
8913273470144482032429311784100915283333037691119705125251189029708294297488398137149977
7805278164934370560432600535269810691898658868981610889926992043454781553574046529382255
5475792396512578169849867342182185312407343115296082141120919999406160101582191273730166
5017695811861903668977927569046785710181059393731438811929147494435652218962602863258666
3651901745365921218638768107774209158364690909165182739807531030664998062448492774756188
8452973294727139148997268407785897786865604872330575242285717247344366674181812327084155
9179147861681978003287524219464801193159379351523941740904190984125780990903889407794200
4672704915345000248604274553073068364722207589308219944523472145218428256209291437681433
8780213969363997860221263221098220773571441294126406537652642854348297069365511680672833
0614815535006779927342874671740836666020537292204848408702523012585717914569665795239633
5968627064590372072680587943981340066767014117658125222334810483868677174058797368961559
9621784654973073934095466043145860154049561577645616734472126427468746140830206387935988
0428496622472232550466095231768172458636261128483387407650758288956774589588736109519744
2072321262333556151933824717602181868389530067975021207690438838128356258145012500711900
2715651443462706481495059019699613904056079067372607241129244719947702838483531863329811
7619994731283314491596877750429032797703476193806295512384013842291003576769296985959055
4439828926666087801034405967055905207668370210159521951394544773118741072352794560844355
4946676079288268193567646665891613621404247856427926815625448506316685268327524656002744
7765241279427053419348280546267028159924539124873937558094012591977834653365573562593766
6808769975257460270216696492529983775381969384625470851886151047522647513648918833573811
8916971219583268100419576537702119211827256650889668246750486669965988504204119161533200
8988567238092601814281176234479558294288085813698860737275497491213068743742194301486677
6625969169519536885691174938231969755955794021793887372251515997140544728970839551036544
8666281965036848684661039274556841436359017774403875651208077145636487772844383315635722
4968367563148103943896809211928233741450761863056577773779414533302578207887933713552322
8193066778103474487881453302276711598246301467739131806469901714532318195729640171382999
3445866528104239029204404205030108503722889707226673204164075838352116255472935420977066
7186526207629467204363364327350566320411252128722589414997628046891485550759736146505111
7641257930283697453203604650615692381197213251056180613452134381768173276284141810216444
1341702491755843656811454797777519582806628442497503791747723620420424505736306091154844
0242699564336003530143666597511828032803953421054049191030567314210754130235793487723422
3907395938930043689132814854688219705348268064061334476035746590906564609800971717699577
5204375563545216850442243908278083480721568003121380513744755738663358066128982569525355
5723266307395195406369794691392739195092970992913488014867607271497851682702690507107677
8965765304404633626268887226274293120287517384975542143150499890745646121569554253570155
2014746265873588291544869367168210972638770366929876270333269267640511235659178773632477
7046116118752837864088038280136349041450851318295587380336571592375540437403660094312666
2493744735018369408835012318057025510894184693666883038062360436168998681815046281454399
9370843946723795281953032886060996315478054110539798317391048191798793376309918240071633
6953592567835866999085256828346179992044849215828682554266066694806590558837547674778900
0630307639773201191626441931231373328223641811934383050586585544982998689911466813117111
9421891601793736025759632185321048687499204734702994267871276133342376683428225657565011

574897202803431803206244849572309739005715093145389184344943828397373451579980515625916
432262701418620623694475930214105085028120360491099390536847805602166270463685525727413
290604228995635102522847519070308253273849955553304495780330902592753163522180988988262
911598033711257017217676690454568064922305157474689155717101756724035418935061128887302
404314431986958652186676073303854903602774609635450195252965340703015970324098511502529
305886567190112508328471496806489438130078371862392446817902162173569122887239480216484
647377517681894212220710356555965079794844990082719355281479143404037138717208176090319
885645874610810991059369177437947128793689503247741864858064819679956146436708248708993
683951315072030565303007886859820360720669916737616414756654287719353591044521169167692
816396364562978340737064788274061834054357077412161383202873787903785858932745536149564
4612550505547826687454455090886889469271859889494234449507483821850184341303236200466807
001917504592628408385053643126768698034026811580709810345898604130841735509956944151799
175432334806330732613391399797880003821091327660145496111570428580281606751623381353863
052429563833309503201928781641324922300476117953582495605914530004246447880621302469786
559692629257635740287994013569211675539040026996645560256829723695049590996027170973816
6616867800483272929905942810296816157469630606101062267170212539667479513893813748538
953517578329451364260069361377774400056699301740201136677317877044694506012922607496911
061576207637893345041373365119560032907835235964653265747497435286721116298075675851083
850160699896935867152259646305700913908762874164925417081690968473095488602332982888001
016529628089769877231969074210270093488809344555121881883365197843185355960674763507221
611787287356394656554342761406855941242591214170781163030801019258762199380989589430509
396825182771303303349266488532956187952664463190634938964927974779630581823776065803540
227966910818093226725142614287685085502340363959705621412515137620467224442897997855454
131901372056002945670634038242991780751257244844635771525893472233684526800050490579407
659261183404627646998353219893312711732710511293877462720081317263208671247247831036953
250571166846693391999383834165301330924772942935847074388263424007013217132097282749479
611635667826150715662825021252062763897515665851340604552901092611263816622423668992710
649041886270014421128735923939982500565800643250607509458935850072180729322308246108258
185874671334067334432680515475652761122900942154655618310341268701722865416226907574466
637402578505819839033726685391283425113889776103701554705969784284381211041667496423619
368984063561387622795954521514689288132823943963661294501387816765909292508975324716691
236834827515460782728044518243453759655049256806485323099928145147534550927555277962794
142455744052552188253239556250857211799635301956758117744367625509731975115565148451372
854240749048380819955880915161179392191042619092848578161390514823147445531015161591784
145999471223905165969441039727295741158339745939062050077580709596765898924961437343487
815637744208374999146183451341178572135657651002882165343788389069722998862946226868519
562883232057053789421789593678448999728380822512958573742956531487043040321954330164722
045813979057452880207472858810311408587987472118632864898447340531146044004800111896474
216571999311588903720590031466418126179209119408878500667780109991854619093489266918509
118998532817580156149634425272742021323083262625378297537478084964862057473772980595049
021455501089693480645109743330059979855532033131826917519159195006558488436551656335235
079497441486995545934682666761749841931923239198584931429070339724907410433531550716185
771284452704556151487208217879010758099459907819034076848455908034852561211244638832388
776007725640595058576145061729291017634210620163227631335708698416113338433519159552199
342391045655659731008231699479978456319598341143056370588841358572324394599937160845154
432247333257474326902932111340553605260907241688375335436141287023423782527089538235765
851659641732811004236702247942658948954200246277977930989345076021362417137031065132759
759613489924270047047172533326422774262347566320225179842495249821276559511394568141270
076623154851582573596335382671792266993633391806685509480289663970163358030635777920615
129299586818160233278268287561386953489833858078404867756502916232810223212628623703647
116725909111622074499444703371546717193557960340649572183991324315717586890563578201193
562277776342162367247566768761722061015213389428844275546380391429829340970637684665116
996845497749248421623454987901900590122307110248805880499208202673415213345261948227180
404524659228844295541908154492145808904570493927298321688341342995368258674641083672846
833132874321981230074513031444007683574024462780977013002142187123511884439078142864571
486713031627438140533885760388529469050776911243964028442753921160112468485588590871112
043313698335512983177044754395273511809456262736507213423867548781623509663987589890781
058433372203975442049861899214534394570010766993023334045070623558228321979391210716339
044784574064959683736835999610270559810934327545718226581916273764083264919446339254141
257047278930012181409950288177663218498545308566679007037626577058382731749561209175232
476012728488544222989867998826628395086789457714388158096755058011481632951013417845494
953248474350246166826049086054349862607076952206845769308881019936388602056908101664505
187218150057434727469240045662801352442429043237968476832649959785995418767960988824064
974266522958440697297761914626489783151028740066979236203320604044535479441981998947525
319571705208553161777916410772764613921571177402991355550151670979661965240622229141609
972270298654087146907919329111017460401206520811903347933507409339673354381066776442516
210910640978277403934724092269713998641932401833676257879516616692114857084403473110985
771860414380657030581140896522799033707067927777173240196063669059902396600617475966027
436477233417132112564069573043007387189697608262537784076168904565036964121017260551105
063180587830717710457167891720931555100513126248850749712570882776081846297351566606413
813185775692321942216169819818338615910911121296406834764541498748925967693915520889997
434834825109771717334774849042415704476573657457752887031370873919072182505772993177204
212596617789094627810737489396727533366949775977576140133909640059949599124774240582260
227674347914043659775007117490687276269345637528762791538610334809869295742898490087041
375521603759467876439843626959137289972377200731453366733526996836629260856570583903882
359383118961476361331704366529764436074940164969717395660023248475312782613511627451693
498591497284257801156635401125748838344957820547565134867525353393094929877785668247383
224713634127815663824590502307339832536083003873024839639954184028662980876689960054360

674637478175973859320010970384009432908482521486078558007203028392624814842107356768943
650847829172394313537730783382862045256079371434798902477544800015738911657789491143660
036793543639320677626302110521521013592154694544997058887628365793340606132391102338123
478911651339606482327343027611585343257078225967456692639944506549281999054027881507062
990362135050452317325013670624394106180766761378661414563785766044207492422926977034984
501265149216717696331051676726784883729549005678657239784427631134732497719890600761875
914089607306658215138449134561555571151084513215973829110011142873895994616239385058360
323234420828814504339150737808186319372078380311364175837348751976507933520535426106483
964680022831803234766267824389703438282856767409938801224558418282506165887184191773971
348424955753553615428351062448328211410756609679699510482522794215870693151868682289065
990544542897676834078428468663516950735929005944652517186518412045443271163745245912494
051031774974327164330478610425203579403271210384808446436333013715290149884275237794983
457479987528151196515943440298872149170655559184939637362023534636888218131437208134936
589804188526181461668564638974283915059558370394712383217345273482136156961283263085165
038215295650842082471083485563684152892779777775636659828621565022125074611675903211654
209654147012294283854575672141738992979980052464181684733481802173252198821195035184670
582808429271159259997015375097479007950930331880565580101398081229845654681618715385799
281286103766006844083084983972797637004166030652461482660428311779328935587420559345188
132647605029917983382716373959860235887851346092673231951588729729246677097635098946770
140282292245909364931925193174285969415236656465995111925468588648001037779316307387944
437871151634358086838550980361673411774679552505415696325551756593001109871034383335920
075203177187132116942973736665048161622813974369194360219793318911371477579284604851057
454283287481793287227871096158968260415191032160358867794747925929885099949904921648497
138544125533916737983269837424487412745470962063371390474048360794157529363363904481795
414111734936202604851201230536055436888793862914480787910052555989328252773354359247067
399972692273992775653397256696715240967730735176129922142025550653142954908742617053358
503331777460258915289378865430766613971869474350204696966875056766378250013366544341201
497803953340948430588040808713869531589891754432305576148445992971287256207647002380883
308255786375448067354538598763808584763650695267362809861199004026798113020461010194459
803592743530574492962422737733955201691580053320669755130197311337039125611913305432979
416991929482939055726795795281772696879329114257773205002147601996981522020615245179423
898458185268279789797440541656480562100833028605473010316050320145740518887619845350296
999926465795650019044827824173860954017910370254638143624452508870292266392336640435196
780035665454436425036250795296489213906564841401827369714502145264316466210037380995041
855887489630824657407334263002530994978155914212875197052710011571017163774988337879565
686539344266119544077414439924874232060422661597714780065489643349623583962211353125728
982013559848156570204574821091737378662802539307044655436889747490658477949409598122447
874322186601530097345480246961726712947787951944151344017301188323674487204478062366370
035242586176568380848573688569023709229088212227208341700897912925654354194078268993055
731519169901180705018958859647404813904626097073419544942270677540335371029648360620327
556173102159143468441553909565909746549927915376332947359960200898026407946829253612778
983205853605856186118945140611322242712861116686725025703363488631585717397116032882123
866374918745662604295318985208791769279853205968708273206077204908756249762398506391873
788063610737211590757336298970566574688377351598523062078348566726739945019724570616041
702865614312668507975571689508067386196133576130779338566529426117899557612822039627861
630366976572927463176005226248813002620159344695993323013044439085141440696129490951435
113240437281803780305270161474596669731823396165575509100944772011231301699270821104372
848353646580227984353176487604395208073622595864078734581915310594558252403107537721574
321457134477818430984474999189198857234881539219045201566171925554299875334540023811097
485202700537114712389797470101585475386268028132317028620129038658514423644864928655235
817137202938801752941010225202856911871860796450108765298425329716311744186507520187692
927464210613131762030850679066588256061616245123298333856021225199613207285816407020906
231271834484822300789404092439163274524087456678136735570039755142780739200343533132128
228732868954206188188329141890423929582934929305948139194192101543032435674476992430684
895495202395456455030650470971258953086410011569687327280575346288820928949980376942767
152372469074527687494117376112489070191987929642324749483718639132922040334047627285290
480570200588772263597678817471579821421764176643490054851532223351305020047426608426027
581114308115877357853681413656836671339174031607056503790852843226782164643593015192070
991234338444761974897013637518951684986306334712439357199345332511924340248672268509671
224442280954862096706420714603892359340106844004753696386497513597358331093783260019092
515744319203761229377089055745436284773845335137564981164123368922952288866851999159162
996178672081918371734730700682281270186065027530298339014669995747944610702671611160278
717065002344548526553180461527980301358894310966438222475439771623046764635318028199644
937566237115116051978758708342921467498000152971410916725705796936154875615717823583111
771013590125353955687127457997201759260654619005937960897846190272218145072358795484271
499133150396203051241109165044196697558251321818617835694275540614559727053523826735711
023180819720853948601822054826898663602869566818348668524544614408406952182633280487604
446919008266967629645490754572236923327441664913195564864399458899338775870098805433186
399855404932306776159182859287438960578046409842089740550683296113972392222269790364697
767875517303966644741574726584654780659639564895358195570035797166891226946992715128448
647722739811741814886633192899465947060089211318989429677196570486185276861342368815000
041723800282976700527792276554084855433348616889849738718678861898732323800424009638640
679843517162511269725924658678721107053801531949577164948506298157989469417142820421641
655866599072861984938491754802695846196422947793149812238364153855703808978900761390103
234971796963254719656491227455826354132341424364357459474929792785696077635914847280121
218205712372291254433245566053407484951814467690589598069520034923001249866193762108500
512364425478264357338213296609669731653535425624730809028817761133739720629836430505408
619406221838502449854756687212600676339743731532578383548748244097809739336148731020239

pi to two million places

0453380947415977664560313768110629892140901661232700390050502294761351885912410647065 60
3129801460888994927862354781233707563735243212718006105308551717034053603307376301871 13
6693532176984282601761121860063584896534143606709141997792464097211427449546989146355 54
8286434011040147223008474005897193942555677557844039936570126377009233040177201570971 40
2261897254902499639625668908485897750415713042927159289338014627628178042451243346115 67
2917087218116986695871312610665581097155155569633481984422493772789984900169133406501 39
2255837452536445871131537749642845154005364042185976909329790072008362962024673223965 88
3741317529058662626944673521044260379193152110356061561327177958332423894101137830862 45
4629514095817189416538188260985813625507028147207441010322838569770121266793214647280 11
5968243771101640588292038122298228255056488502000903159908409669801425015995974256102 20
2631836172557111491392144361101853384556828607031576644860576825091946215850695082849 40
8530180664499120714286759898668841279064653948357153197916296821916428762691256721694 72
2877436722690746310853419544061133608487163008322078153715481543546458370259485036167 47
0707302758499672313280960162936123357408505167586704703228598724739808647326794039491 39
3791308497363241413902943928457663835198421867634664301808689606921433831960422100947 20
9843976665228225443683042220547014325651194268703971992934248063911831147159679288276 59
0160658481161372294419943326376901682224343559260799088203400234350990585913929771576 04
9472707702830957584270709136977043713575902026722712135534497330430506741037054407734 59
5843092276937484039563820488592647043863621119935422550000256114497085052705009162892 96
0498473983059770893141204198377018700063858078442617736127875809955159503846662074848 15
7250182123540825433798202752568074574177955302589468194577646065374993269932888205395 15
1847408514744363568210924250171052586349534579159028715221299536595915807697373406864 93
8135614834686259397425293493723922991126095352789014741520946191693752339607918005888 18
3785066885788221740893022376707892649055901783573302904986104756624088404630174420916 46
0792765924033500521369750536665803340501312807124279897000627372915259070166674643724 56
6787432560019555424209445336334393919567680459087133098344796212966325811637654664808 90
0711247402815156434346310814453273397154323345449268619307448373532903424290227063234 45
0908155379512377571610492712179818535358509941553871821944942708611496926208222831105 45
0526017646944214984796916249383350864389979794372572585034947332112364201564544595832 32
1025867338212082720110236171813291628134769361336231630005551854169945123703087471769 34
7530638094914888282318114601484735917650196830571470471397168054171207608541527790614 840
8154352770547111386619355191825554967376875375655901891589207674352614885293763210787 31
2387572064840823737530328846402348829287586745751741196425955473254712998878441337697 17
4478289514806062975015981041096517924873725424158606070926343451122180515157680503952 32
0798390703844559080248976751242811186831122353594495362332805156428450911638253284682 76
9037798940560973049659821677479429060522906427115409075963049075004578669480644240791 41
6042498973639622383553426568824892490309401353227598292267618021399446801899603206758 03
4293979479500581123563398697279023670197762898407331986142970899455340903726057682234 47
4886920102976488719461875862934213170932727776916518710024989966658550838113356789617 48
1139240080692044146256653829452233915516045948248702775240265603080241608406335831049 91
5931308539474269073234720884119182024847075732911507212454468985268553115998511179393 81
0802097734738174933999888973856536994038759525336217482394715478348005894603936659188 92
8996217521047163046603844441237345091038293248389456001298364934173204224321656427582 86
2696462987054943708647463427711852938248043935821601986070062171119659183259180757449 365
5660319057421033069753707322301404429391360729974231482186205712797240843229742225306 47
8470222877289889172044541364168301862059266947810650015772730346839982955110599642751 98
3442090994298239413615583265388406852983750196100267432796083226766089824373992304583 81
9359944552036471095908251899166291593196398638609984425245591944423740980765055611750 57
8961464359199255642675963119627783751287465379652386652568169570032696428785072018471 66
0573792272520952321099076112702419491396280748169651494320431930648996967523523778013 36
0171535579415272267440438354774783461541713610807197104730886023745031213890081325624 31
7204976886802282925502433493959917827467695941155443098503616431316394425732596746141 75
8066342449250604023220280294698737518293127065541377829886195848109350266364520750931 96
1525082015023951281069618830208378779175314247378006661366107125561380956875490976302 24
8182547336957774425013633346505272962419515030700162982343399091000615550249467371288 43
2197235339945293806722341962170161634329247937574459146657715626682851107920980138236 02
7790852490861406132382047029603361421239678916949923423217152888329799391129466525384 72
6864053187319793226777027601787151571127131902216836416945329045195204615033553473446 97
8714152244708770708456141208314980110666716520746555367187872873746904987162762468053 08
5758352280419199560732766591756957730028792070628730635622829071093172254124410289965 62
1943930339359793127298249018850599820753028058182687345362620769884288389289635517726 99
7755270892807119838327126416498135850566092974668194143322037636016031702386454010380 41
9337568874559351238398276056529937969463111573623676580351575643680207979098069735892 89
5033363102750947847813570006470316765317984374893859319928446797050501465842227782696 06
7591977743142284898346229638095752245329234359223513044120345909610127448589140505827 47
6775911036197187481126259602724586676557147275493941205551336228916904263820835739952 06
1572394645174449899129702110657095949855647567105391013640465590616591177906243645957 3
9346071857771170611184517001545450809988500405931558750951206165545631462007387394433 27
4391565408222265516671149813613507373995489339174808637419664809327817100263955025914 04
7775193472052693611721552919289489112851231256310527709723493867708930988562479735893 27
5980846342321851624985305036273164555086002448011287948708902187528734539294131614662 08
8148268086141620159155491220419860259848860991001008935521986004274340573101214273402 94
7594356726977628542777276759795406783215499987080260583813286902818386210006100337642 37
9198001944237044206333199893214651697433457691282231826114090708986144154081991473747 4
3368164498243252660816669669733619695332127776912977263579843015097108158562779524101 40
3121972539950098548370069915726381749334231984170867848596330912936697738356288707840 80
0239223578231036122923133431387087137560726479553068785676787614086797853884183975085 04
6835092841447196834356692453914539697267038657031729624183897923545368707062910535840 95

pi to two million places

8625288172928169247104639591376543397751033038618690541627854067967188564523146253468346
6430020306636243007728041839150504839977463906523527004766822033701569158523770139905412
6383476484663841179107634163945096265761345248340913898753793488871084408225150247944719876888399200357379260736576854930155343026843848388931402721966820387276849040650786414983954838623991443271003548414285714663675814108685756849749642492058856984374594686574480184934228027958235637756438826823262418776221622607090451984593267734773501828543606939352416589601174507376114064068959882929944645381868606647472888991909822960178929780473579124796321869153687036559529444334995425160580908304927359408810125128045910650504766496267582224133933158027094209234354824137545730590871565675676701092095054671117837732104759766979436357024999172477640990996184223422593896684699154418877209465300703443718311572870573206739875957821407940736334236084963832333591790227712682630327694877532004680184757705394343007951196667752439615916630827808395905383229511272338207550744153078897776286771662518810911875351973300986377174786188154764116022239030319559678981537333332582983600451889734131385796009297774588061424010459150147820679739436362919935582276023675103478275564866271218825552853178253586010351082258145026112047470924017186025646900684617317679905734901100727287689261945275883358758221947384320834578633052807557454938289523900598456827291413464323488171485846067830594826014535968762159670812495532305763781564934456525782552293824962657511707494988668765441038270533740898921040368677365604542858451695031402232363375702641796577852081754489246557099240813236655786985264535388801118891932862519259222135566768155760927630615575930662647392608983278347680214605557131593915751341973623814379444978888581196334372829232196632033578012611307701085721598982028012452701419440551082110912628261670708062742710620807375623747391179018806303915191898251718666137575727597103898114628132094118243212011578288178755591908312841658981012959939724582828850347090530282922251789724313294878973743215073413895319923645094033008944417799497855061146956152948565531122294623526306051560149924641646941653581717906597534647747475189449033886376945491013384753795701243435328321449298275732064946005703587974137284730855685002414057806994944420965645454206704067126917703420558935459214751399465865637953244989394807296345359598977325035224253669755202625966190223911374464701515637749468173440379453288595394926777638755908647972478088680068172355853131435632975431766043925383854057568997429850951712778248854066092032655982812701957135642813925406613098387251913882874323038507006601992157068887156531369864586671792836573552565862718814401443171436793748105969137093212168062424716237296617153934399533680541456986793897178488910159860309541293529874616910919252380974294455148855831849949859479590242636642554865556331493468935615034148837488463129465300565980746769773601945598152863062761262676635577359767581138421612373314597087298604713841740124888891879713327363626251131133467653762929984089037789520000853939976358474428197985767807210489634599077801542669717456754261723840220327733647639755175491665338447273368535249669156297692483436265047461988233594559385423887390136417704694539579811875271215977684425117995807169454668617498200389141367574252951992353630286409958477380806675941697155808335426239966791371918119745650950142157414450246559428230866854834547550417532809049243969757733834069200365962698238063210584211683115360806396000302983486986250141895196159770985309893415906918264474621242983607543464462434641837589126993823538014384957364332358908803400365035629450989571731821193875360604033750257331182525704669344702757566572460202434358051761798043050015766122068591071191644783678775528755576514955386462900314778433524201822322818628898360499556710094713073625117504656828874933451108004370570085720732275083196736841092108726460308626987012176044090402806979491118396436605218431492307351631588904445480567978322555962857732503063907386237033313393794864907355954685179650429666182553105264598245173537563250283739435510180592720530901051429092433527312786900151513055874053955359526490302813127589266878502683802775398087841969542444707598265277147636801344512270464696586160140500059813554266004555304125645385648993160312805596469709727717647751362379318695356428514304456301276104282649403679748029330190134439986092840586881002688832877860690700575222916378845954295112712364162904172492592870335181217772457206347444140710457987011683486538940860879346428858140532372205220333882782491606550751428498895026730473386894146657715191706700715462849334130545953261782576247455778605289070556812093921363602079484001904534822717501991998503517218198291555373940524474881608401142868298542198154173994615194467566539911086257026617289157211616708612807864422596878065405524084076980926229798908974388648718812412152858621071441713146827394151245402315271846416021045875096948375943180736202192937400119027475027176567343594247367066180398677673056064018589905357512300230660837087116869147579133638408623253843492671860661390466843378796516790070950960770044545355631362405135816161738439970087207800572797374766973300019518157166514482156340621818656955569523059973209613527960436084916464544009853402960861674533438096899823943693824949555094716954167064128394245171412601916247382708669769311806960971015572581478531946257461453503976260554651254350214524143432982492413812177132034032371867152062735928163374410315471046396311177083453501544825945760866597757173914576609587753667249840517245700825958212454852196164378931310988248170038422332063288500432542084921594423734706930124693333918887639562414254068324429613965686240316412840305878256465234281833656651895855036990943259530942506080824684142553143703739066510989676539808735874993770334589530194951951702175226452073203602754639413113711611622289900457080284754036140638147708904136189639814679360905829482054185806765374568452152011414513584740042491714183497910852675578692364608405102964056854716291147960516605129860116421892358470677447836979163436922030219888963849015241726483510873673134583953442150224626155336633614834786432335613805300749667620842875753074484767612212952046402684256157402198984963409036109218476043128882969266380818247741624321491805174576051652767148595672360585448148544021329075323119337897054213766925989414624437580625161532179973469637179645547335354924900140560743663106474176679985725698630252839944434637993051824259841257983586495906278906953651854959181602825231129648862454662474149878023489619904934728196665830714757725320158683554707783672825756239924355463937745447797338296029223953910136392484245799089520465639804512178911186836846111736749565952373215883192224769664295

pi to two million places

9496940073685050284037768906265808500246717295897639915288552871279469207921220672013205
5280939645226322708682212314758006603178586184106934504552579097773956618012334187417944
1362215786050078339034591252524540485754363277089363734813762506583880204845701532917288
7753658510284304303738094645827944631703476961344868288054938747420783607209181966956077
7978078659557407096044302978609309079720730749607010815586850159480975343530527293411611
7148317473396610051752170023014799010869034522977048775984057286298941810215296900745555
6597243348361850377715055086033011150531761641696187723661381272278710986517345203857377
8730216612722258062394563423895118271063899938281939468089089171426867874887032369742366
9825435588883246084582028402348362623588364349326648017559046034282151721956396349530433
0084841662178271514494595409011794488525950494726556911945792036793654037611386749376199
1352883897654886121318115303047160619686704364428843498822552331296875029163545865426866
6088946046929370596495128448974048567800358527040399356260489828221525557199699352579455
4526174074327085989113001429067140022594327469021898299517955344274871811164211742934344
6618581259577501754135341180155914001252399483939017661752161192300632039269350307440800
0556485321736481168158330203170589764678232920238182476764490923997157484669006626848700
7926979745436105026659791718725654581872259956718471895396896356398273911694540082867722
2083973564852019605960672645551934292523068186375946771657471639851037580102664513149055
8946532011702593902980926721553261881121687059584164729322719691537323315514987881303477
3988948962542717046310852002030893497607474809695231438144859523362875963931570347642900
0052519090875253165740792449294631761411281605006043323678149217024471810409000252357299
0745094008754190944850132342773779028203768777598838891022428946586307188578363258401144
0040295820151572777505522049767174180652296812814453596307746839919743615077556084901488
3048815266226168875496803462833040684967248845831448961089841116418534724679049542984233
3687929502850535622738086930362906434961063890273423981644437129896752462213499807179099
8325385375182451431822981701498047147440520820677173482930757424460717784725245946185744
8904930509650979539054225306921236070170383791085714698302257724852517384591456079105588
8537047066293052686115149625701677756605098715221893119931908608040958932727146520031599
9118043637406495544985222116311079240253084120858743275083025736046735590502428762009600
5821782540707253588194242748229061261156506709906900009646228666919350261335698044849900
3060697708791796420344947066473435831304985932397059589076521205938976976179995460990255
5750129252950517564633281937784817982728921626883979150390284154892848405010183293430166
9393085976918820760983272888921135516982344564447333253072962398579235645767684446557400
7881847532800320620409124850379079033696967998575698548117548118386688492826248933731344
6365620962364360176047562884825574687983523166892032758120831192672738707762830879194411
6406020746280318221576402945658339747608798691752555031704962919619171215072124527733133
6375472863049900375024593485960032115144992840662158257436742274475501063912224218890399
1206885714990281250332229301019625987938312748207951457466369086901110213105305738750611
0287625824804729782975970378866527021744112460837370072764091503713333614971740090516022
1354287018659906055371259092989698875726878000679158690910845740780273990101872583402500
2706752349279084556458472338387936948393212193705663102735811096309442346293573358743955
4610171509748417603259483536217516712490048287878693443178634077789561343147653044727100
1587308359186544227533506094500454270943829595234500617950815499122260677053695403470877
2316703773580038588820185360607740785920309100130736686153320514309483297128610836602522
4555926973266001032976141119174374276782789747510302954650308104060484212662927492258711
3195804378325583814279728206104671644545396678275066337611956154718081141066372890445088
6071121650660339892385555376753205387993468505034921858653621561162560773785078336813944
8450925056970346054311689014345656230772431780451284414990211879930904828898966619644777
4295261486897574573204681701313930905117056129681336246565767032752997884256367261046844
5139557876174554261407949927885159419323455730645855363767666277904565194687523590507077
0290264235937689217411258335714398554727179693347166990452457385765734636323402095801122
2354476444172330199688759484111588591938802652082412625415775923953557139009940619257888
5762438343967082535985086771745203064771259716871629271981108722640716731620311995057499
5353350785579058055228056768709400358862145084193945110212966418030102507190041435180266
2583918416963342871083924470112172842730327747013437984117330124469137759748817280837800
8632835848060410924220865767728752209963240080429944929304986884989845824998371385891666
9131411594805379770420015970689347111831573389010474647987808156521926441124175366266822
1681770769432381466336419486790863825847134143907867852662542025507987500598344208643355
3203403385407169700485895423819416463202364499218696935197625148758953644751634449406411
6198941671134104435014824843798746391600097858007148865413513572346046623479297272831422
4155920800251034678954542752194241325704026306976946540161354854687985714429486803039100
1844108638904144811237544371285330823993732836681962313029569185695856627411377033885855
3664622747193167250336110407331565700765207124240797569995015171682190064511788702874633
5229298088187710072903397299225664211305601375775977190139941236326728084538940031909611
5421499319261336411225553601183662732783852674019875478187635393573394928471029582528711
0380997565439732567129487558224783626807452739034903745390658115194195726455858788269611
8859947491839526549635447571365041228605931178327747043171702175554273381131644612205777
7914607365779146307623015698777942799470800066693339080866312852037258042871394552756933
4418643828321630754249357674340668984292481754076245634843585997894799507358408972112722
6010980185913187269858204360215449353734228209983215127996754771510867255688821989769066
7943231991850034564659754689420908586518688541565005301770434777447943867270393095250800
7174811438806676944040880337002762289229403949546456869467365627651215744427275561854722
7297092316077100833032761204644001950108825543666118384017506433079878960184957256409222
7026453638384878282644377846764678894521861373535543656377606476678170898450543551146911
2314274141648367976459749610075175159580739164799319511126936601648584829390733187973911
6908788195618674358831373515539061314098627551521295444877104580897721910582763398947488
4910278339169952254377681439407418668237567124423323514823465867649964519452476330870533
6440687141456832606639769454801930943710086795751239891190608598079561897702861704671400
2026390040755211696790396797200717133559771467791845713611149407966712462922999331477633

pi to two million places

421654122778357482586275349900679211197806030788574954697328419646448724815492388450416
874884403265627747095296067748119527859251481070284907091501865228753429418363140611237
085873262940243390981983868089791860127462081893950209887488319592020920204199143110243
288618404386721469844721805882747761188553143345477594991570811215472430488110982675308
501879271223606726542472255495116778349951376070493031367598922164576741761560889488299
509311427656818795044573907260386601558121381695557713584204354934789536420233897464954
927668535362013175286570550588094409771668261877850356569324883700619168688132576989889
215377016429415477035280056119484224729851874877749535465998647312837668102316847838642
080357150661043760802759009924176268410207319104116864752349250603645638677729618049536
610561814538747453647356295576007682838500265390338220423592553982932841939449420500089
988724891806211287049039130029485565147495434457524482287158340654542307467849427493096
347001514326312416129821109765779780864620896364347208805915536423226483126451502160519
650265847206670613012049338696992206072120550748469831325044567903619793751145105610994
059722706102455316434184231593135390530727736431563672645801322676377266861863347929609
941243270017744953982304320400254498544641258218149205601492188788848500428184182953837
747170367919128937630870104272072105279340761590955605690287884154359370412944870677376
427126382152837911463086145996881385154958969349477753009109089564506288798749499879189
773300553955499696722311303299622373574385677800288472896421335832669747258346061528037
126274322172432529339257592492447411545058597603142539540190273271795348245344711818326
753377256883131935700208316178318579346955506250987414808507383726201448358464003533691
270556211958906298935556277917839908857649562403797143088390971110264225897462931766896
722340401278924999440030146467922020383042161988077173466473516468109820565844567718498
749974291629569527774417095631284596102690145422333953644324790898827529451163209927530
7499379381898314756794441245959637262578848779824592170128005557580271614757921687730287
677241427839045907391273929426788438577192396805229940340533470735733345183672515542658
263961999930983673079503724486864611493730495761294157070706620328918110723891552753833
666381708043033055670727526167741946306060870155565744523087465583961240503014805791061
458163090131489886188210793827475130476412486827801601908496784421890111018392559678015
888440508539397683873601419123096060086888484095875909397988754571250989027721154014401
926221727965465649895992143756911429020204111157924873126508075595972847278699682789162
682786949115247674585192281108652677009819279435337913553500514689857938237138827353572
717178938301422164857171410290069972823532328848462192811289408170797402442190544369303
801749299703208434011087332111453684235999932090895156690859649152277667229639694224242
341883180352109911640280643484735443798320000362808190976855849256117523929774165820418
448109089512683647084624741173961271488101932945586738194325086126535883736855992025907
815396082128790730606533354490598834747010168086963731773732582913940201499232920969270
678316759681476696675208077482680711116784929358461984186261721109253221606852538815918
559880627768721643818351604955385279364210903131036078162438914712543316537004929543573
131191804284196503216157809042520133124856053203608591448176717054976690739276489514055
215206092541329165702491817166443720366613685787673325110828590381921544766528622524635
921917501303428439331954912297279269763953174862232078789202684450835204124961260947859
764764050513446164577995797419209394138731127672405794054104620755970157418271819156561
183209665877740814676916977483766418386081752547859685152469480775281790629399270652523
129308948664501479045471076151553997996389545735499561642096460474745985377240519452042
446951551105760149422144598507856289800520015014421723603039442834244487088873811175695
484586881015884730508202936051430137010450699026815819894595150453748572348798071613236
899889286204993413477844596482623886893449564130688693968016240225696097259058369035914
478304640969184582654758153494500776516075819566412400743361172866660578064097906850684
383908655013660341715661217741365352466629240385273163158429247510718207376199742570052
663628734248527697091241463260439116468257193844749765274127912883327647534004587978756
721972850802515877654905655892204714962052521040120166987644538174938761539621762099818
669564349377520407867045502757495420828343826527279906640404630855560157935415801838870
887714052300577774791013360758346370816037414033581662171052377954778031082878931138989
447958611693922813772482097662231537416555187072430037844683873444610185876496624830235
355290685620889367050129145033747129770829859205524051878310562165213224612288003475473
255932571175091617659416648239497291335237054388627240848370552817002962980905865080479
495462458224013571415849006733084457388181196744514291152143427939269965562257198643919
606775303837468667222170355689084250348329476678415036783681989747562636076056173959495
904537872887088011965418789247653078202554123421527555465375278485116421930021040069601
544428550071743703540070335935650539865178570700658209980419574495123587644781248610414
755659045005537787454257327663137388250201726489696161291010481279326374567313473871166
387709984542067027002960111722637627272674521018118655910525861453388197012899119638473
502901740007068063146230130805397545776028724740990975069178826192831971647212849587379
180354955908545006179881321198211429825837530563509789912359405448601790248626076719483
518442349058719075337294433959240918535280295720103839420962334277562087477231701175275
687241012245302815387958450963485798391045192131863495690303912564113905964125401943629
294949192135307171534484096842071064228228981186402676350716079693504702100231878109856
135679106593066007897762553197898250066315156298335443916444925805700780106162803210220
195704757986565811680933242300056066489673694876242617241011990759620694346859404061641
090571155345453768092327128723539990744359195398397372110133494916569271429930280203147
745605054466184575323059017078615644303819701791495676539404594356060660507017393893132
134625850381073195038674483556019491822805167730576850268843254958908956081419950752565
189355402263661008481614620386544647710712187951630698336202140746124327385607330074172
908534137146764662082332653007577845780127550787262783016182508700842818774533207879778
942964858597581928420499648296204273540733853100546993954125461947347217039952702270357
79385851296836644067889837465656361354447806668384669765176614996185439896235023717671
102740890919399583145146872361287138497242069080950442236785510287628680425188163584026
162985180721152833223758097090574535691784019381600243965432578250454451969403488409243

4085789998277191512303094738055403082315560818607161583433955617281063733110606015264964
148156840431946235604363017431750920771304908868560472738551730953808475014486517604975
677360783315447229253433596384563021521710498738592531020397895814333663841559299255089
446027807017962274521055506711913163262652799366963598923830060969816001670881344300203
911771916307016377800383003711264182414601498704171480566260338497751756038495491915791
417929536849597062863297174522149452643640004011685127587387943486668837572288996134299
386640236405894482452912428201658004781669418478063660478483817618565155840174603728978
215856590831489306511779198572317164764724189304315319990881549971377420721011833196861
968494405184713480510376244887581817277233442721570874000852493919493398103083199952288
542626308518145514104967489642576817203145774197605540116651437193376372188186506524854
450399937660922677770747939980142258086621498719124701387469895676580981634240795135703
733486839946074001483818809109122785049874225639471028589036178924869455199904917116231
709297408195172163625062371420825261190713179187638003738209989415516736290310549402905
372537989577370008817865887049043920910261127811112935519422778192147007436373735190083
294733687322396549476329928275318351812624007118920345658855468095030263042519219879777
374432637260928911109005219868755372281053055764147614824639858637189772797761937308767
042649704411511412890062126113885306037367229595816117021395030074141766112848332886016
736716732035880470158024784864639807999576764707923310644566260337307361589922617875267
426013536007295278511447312992791452450623340290963971759232197958011191466992839606609
054620079823745212245036016910411563221953297269544555151828409453955078674040937185323
660387796280514312676627403643982398318817605978528134663946956055581184589398070501135
751682068069489438552913508828544734828151760971542385952213273086132204129233077656558
692790957004784303955681994015964223359594550228105654377995470862448559080037069501560
801930721464897267316393892553815995861702205913973027090063858884149534616276426044287
389125191847819850025498263035851006341034437633407334699031270355830801124354815866116
467990470409954792381287897186780109815204978806050476668203662967250936939607313669337
472036724300318966620499750652520175471357865734643836467637926850195846151069676536390
244748995432199723190611693296228778946122656683064351895616389957450772210945127991885
324444937648778904054422423347097958726521769377763707960407327897024558295386961641262
465754921044812238089336380764949967196348615540609091415949767808774619122770828684460
532572718019232463199946676789260303597957811827900047139409217193998740760009997069581
634479233605106166460077875593954578581356191179734847242756439544469001853703038899472
199830805883405836720473010593371645853777333738261881997860740674509062922554889908344
358447071868346283907929470801169686394851185081164381346028123477126650093708643498081
061192511699839069112411278849225010740467001002498755498035240563673715644048348352246
102196750403342268090196091918367697539184488898103076931360368464907518519208998079674
849520642528150398821992945016312244419290790582175500212464156438780114736353486032318
176976992568862342243651161413783584860345729184564597310145801442339103436619091211751
414322400433814714649570280368174781573399255278767581363359753605595393840176867674304
746201122791642138790162717829363320257354064982943159500554714114406637280225906499077
297215318483539621059207509481870223099482546729701848378246112123226391943500731777620
066954059960374037654410628719241786662420882146482782794381136197446758843595640054338
403532921278595748814000737497083250483278575562787738665283977888843093724219594555535
436816532937198863634368179075180196127350934580410944655172588496802612101534669727381
161498560713593318421251782086976441366280313195926629980080049993801799153449183914186
001491765200955505742943903063235405129132722852895331838612800900242018592068301245576
885159986036670882144036179949757693010158168404896369505707194161819229869985461811066
431624215434388574483433888071994766742309255414252204646132633302192541233265542251790
039623700118772248801218997398674540232783101115581388876253275234665649792845609117583
939724769569229581647917057753460630170075274314017131409347082620758055636513836577977
785668667231433899715854051283817539517288672679243351932427944640427484557220427137038
338428243358563142551037569101814745835641467895345086856290184867609167706199880065923
053443639571728250888425996639289712477574013093554392473324079182045553817663723533222
093512540132481590298902642974259278789454750065499469212124666206626082143493200208055
224057223984383044936328281922845488932895874022579320820777499212636085911890296466098
389588810029238820492462297133229297037751993136100980352670524486853226374478046384325
646301000814486291219945715644790099446084293682847965274461276533324488110332420251747
386222344906972367553788033995966640307214375360233103696573323152377229874426905668961
779628232189871890962510009667381607994856169911256164664511017559716379949948745173586
586770840471684474770849909769979447071072216462801394627545923362533618584457602386869
037340402342997958662188971415880457537565078504261021206974369709692144094113468994263
087437856931812699404387145948959983500248233904352960214104347987103068353429770829833
207751807152848088283370145995157366923408952207467531109020004087766773452083041072554
159390052576034609120814028417742037713070084243715867293605934806649256089060062140073
524622034669434043776297371417295657738443594005475357783364787173858831995725380639809
960264540532720562873151122978157160862957374684566542360082900430576533154127168989746
228513385107670664275111061263511316160924243466569700709329952178094757112910514811469
846816362099491211915737590934654666176085925249964296490499530084958760781963600471026
739400740732084362442003461449470324239517678532424534809937752802805259240486576284671
412186729974077030839307426174014500322104972620715170167184912813438981955969766521920
190813700036727820825480844407705488949773529007409396411521592813098524327606824759522
533080282483140911455034964805588336314363788086926850984022754356109530387770701212000
289803251790104923250778850259083263414354851687722009982947719960230524388558044873833
160358617226920300671018787426069417860407069990250750049323362692993214094658099646531
428589402950507599383742467138926990433088969689317122152250932961528322314040152247200
696225193121876948328674200291736644356806635821053328041767261515623021242888554325019
347747162804383419392847933486122205735240128775008941153319230976508281811189635502495
846710684529264484555742771213474285886128655316929049767845662596418247266927783440025

6607279474144408370664575851237670920004831219403483729208560022902779773086485414746 18
5881115702402873149042337471022051225421059966070353839506482334592609712442501847780 80
9771726739247319401655039607867012304739576533280658508348412526008894658467311915729 17
4574224177949335497989190078206213086108038655162237581246483554065072432929291154509 26
6713185929676930903243805005047057980295747163465461868596663348164737563758253029528 62
9237343678949466010883529136131466638328071198371749145500064298208172745061751153316 43
1785068605599580414470505145134434661354349132377733690004775758504046993549528650821 23
1068855209618997124393434111680987953962481708529338557609002702279497412417974047571 78
3589151301019483114208868324943314817802337650953324299552859443683435519727318517804 94
6558350991943093702256819318024684488318677087309347278038059155250231204064223609470 98
8530160758370543677837414644116053421760045932648901251010943688839493750108078235517 84
9509082358995407264746204374288810508396455884266548692765048403283121820919225254911 47
2664063622030793455066039883857248600510852807898750922954152053685929261987791689221 59
2205220551985320147926147108591884318686574115963789931121609989761109796335654449027 34
3590983289478678839749661090310579656789888481463377937809720633905840102515148001880 401
2308861312975542072153696492705801455275021614678086913660749812251621634354851006334 43
4910353420533934694524974769550862656936276212610915726891851277303763839795715936693 06
7949298648386633844544863127608551005181894581104146470845401536291458227457290153410 57
9881693540552831867360382508835230473920214607034933191067272651277179914138372267050 30
0428895154913480292363239288242207957730865773544300787149940796063032576958992623920 59
6601385554180341966989324367028510494283621790242444777381925654599726442141645880183 32
4265738561878678959852363478423536809059927119995597854014483596606215632951638907300 42
7568254001923588832030019113497523286130065109787692945531139668955583476437080703315 69
2464770566831708735818394804538875307517147833711950768797640772884648664979182318450 23
1538105517587340718962538798537621437172436820710468814205249239365508704489135541655 62
6666713154192176366666445461341970658844048039586284011613603368548134550509359031416 57
2525760552586987870305119319075536363454402515452339582336587160381608528218500714103 54
9936084662784075061236838764462145891921779142593006497312418546365599567512958502030 82
2784103261371511351983906124159646333982390087897387993371672887203567894869612784691 80
2988400077024046141684357096461162379484580024369153446231929702763704581054668618648 92
0064818578844704613762942820614210703480363084411167800614815023913690139665758030939 65
9044159125120969529735812730979032583292747939969282438314920625902933092860103323917 40
3838342745711351398457747832024488386021968076716331653498673034745401399921598221116 71
1151254425853760916734327699904005844974347590556420630769469347283565064991077767958 92
6507983234810840182244891170749665104061875918474857697210367432681514842019569722074 73
4482208792869960836293587631653890478390048742747935151528419727078614321965828417069 39
0496815772568280501220207510923634655176931515723238105018022585517518247822341364317 88
1652068506613942262653446972893597805755792607832166738898066125192388000169534325226 20
0528899561086132879950759573554749552474243083287019180305647708273670143445393759376 31
5501750935185158798505069463905231958292523097514074052545956314007436631365318598857 57
7378878419026141186895445650421603370647862526272470854341759779500692649026951120275 02
6309543674058016472131592679453794369497526101844660858000783127191316021235091382970 42
5870233641604644846128470918634315198653654660902676182247471801225355996813235223866 79
5597369258068959753037126480244820458137018203432869086371659833775710858330103687434 65
7741106218143176556339621006385831674674809188717077376535589866220294998738274442554 79
2411634654679406036502474602456189525909349966735951227127320524011111391852302706761 42
7723092229472881013319296333997818550318293443263220646980101295170455704300632501562 50
4315083657628326741230020495063971736784188910051425253052496886256738738477796628662 16
4973624028836930961735437337313667758358357017944714574708124906980229776815688257225 70
8949890899120660554025206080132245048251208137567437661986796426412986059431128300054 08
8818812331334729747312126744443118675943209106681841781026557335326213877374737553457 82
7904151159313218518285887774417722966204691386602093496942560536450662133317325961987 68
2588594298794267093801141054153993918960836361924874187916930646357844807215658399312 03
7940118324480201556714142859863727859846970740475543618234815879601141366235245480672 03
4561475693978012289565852162256169671298369008539004383451036299591188765554858501038 24
2000728257383191539907527071272412720139581995535663735512890906976251070304892027945 25
9155650048553046678249437412341710845111272716022109419314554015799975257597062357119 65
0920779653514558284969464124712726275202638568898807683516345882147443822121862964661 26
2708267422635057195047392386350754121166483026200971276710048982671613245191035278362 07
0128544446701113052325226930150870306338451974189270634069245458809137680372957533468 06
7793504686478745877073762192530528748136198511385757845429291076654292834071343502250 88
2818892915244527783987472190081314380720430207245467931758778466626656846528861507688 40
3122321218733296072442684943391394823440048794731810015672345261770257670886790779488 43
5759761351817722330324699911665759663561548880404973201297560095265988234960223329496 04
2355117872785529240802858776678063370228800022181033547073381937382622675719292593356 37
0954061572778057244148311164042802001173781045377356971226702822063493126905454981671 84
9950281156890107968802900112565891790555923454722921378528723282181212503451820595480 96
7722776607131017786034388295733182064235872369934100981061723496382468683009133256534 49
2346840680693901747353201504440611383475519042887754060775819161781154130039925740378 64
4359130190784611315598122874333833098537839728020727286892320298943973394819817757139 24
7491694646666789612342910937033879325123377397138802549835070664655016435659685314582 65
0756107057247029983740904860172437641981422704314803738452936874360053639126726507615 68
0781314200412241195949403929770163428124078720808521666306459230567032070981617225290 26
9602306308129260227978177435716138102919298062250972902342140272119169380327132983200 83
7285066796162815923582866865760999044159915872039418368224730903669496907177457012177 71
4467268351119457992357643789469356535420860629730104673071982276607743930230946886154 82
8109515202159705255023615647835579419688175556091385778521922259647769941023057003836 67
4762355069788231896556982414650286786318921131224060618093860488326451730840169501420 82

1573260642922288691115250254993602180051041932609690738474883239140241558533260115285070430660932244424825241441746074448844534185528241403762246430116084929486645829985520542671517405454420206280703433294106933727264297066891081535848401490945693824621684799297070687378774062892832545134226040883718721990383555267472209412312676596338155725024484611269592746975238144932164044468989516224006928370958627380688833629163724004187595864552046374627569583050798453815347152306641100725692362934927639367109131351270403054576969966845535847145591819283771246258352141244581202749660784771810569945704336050816850958945374630795183950034717251948443599494854468526812973590190690653598903356956031006447968263120457319101154449655511226841732515227738630308135975821722905976613762764176616347690912507016199955375964321155753439662168827108403675119299960008862463219998754180943600530874133962692998207629668015488090523558718680761698556060434175392524049679996465176000269269165516987526525067739508513040805388445060926010964368810894202685615907389985099082501549463918220970064845536636689968589228638215971107671797678975011542808723039128878886996186789307198286754991974326270934100026851032592963268818524148957912443276487218014529821857497714292943666893783643822296585911141794051849420735706452832598845912259394927444557146677651511456929031395253758046337585796415811636218722766151898241220535132242248825616774245967654125403317844988384111376073500772008569727762648736527833891636142496421063088349308903320848779428644049722268286185994668934379485133485590582820696461239464975741223628313977338190874558033167517272238664736033632294221653127763584773063153429737277492314973942733916813987517165017810326131745506377657709788695976757422149371635288456874197560695360039333733202065164936970814922958343311631037274090986625747050514702272309629622876768936937738712390204651672955925396375704229682701372159784961803623480433790639703585582422108823529872394968853577050451828950991937634747320351938281114420925467393367822925849653220080015263180291408646888248451312773062679808155954764828504172711392640315538504058354882187206324982256830837811447496727888333200287176700343696901126333771406814924856099222370955185473844571790541768739279171785165931986828875641818677687684101914918079399611289070751588545524699476508217675677211646636427681998522410096521305046508340727574484966197616146084059060741022144497622366799448483777754612007723490483468047995536374920022287112718513560661682291232392180055827814185815990440624052625463677109244383173398476328514706535000797831849158240117595659005350270640384087071233412549043005892552683369109471018856806309748638940766073009576591462156505019760734488992170674595990100871084361834433277876536825877230822334434543292752645035082751036885278769677924752905435163484717783060043423068025104169270984042942326154928313761218792561598055496179934882234043222837656531886125750264078045152954197194195477977406155462426767143364705108681554032166445251623107122601230893710475506825086032404639086714436907732441343984928414678828471479332996215272656768340104003538027202527541843522537228344897065811560857105993366677451985660472200841601908600007880595268182669131384173418199055901058235929020135405146103729259840892444057996292612098298196317512145353321039055651854357550155063302303939832174241638876581418283754957382481357953537641564860019910209763533920754849348388692816112409042860516889724156259375795831898950072924596046798855441909311233298527184462335677735993334804074703720532803470697637002715728202307305769614148719664204255727504919503663230046513954181407251089307094439783121107332766875928037765071725136087557193764856367448558888257887661339184449302119188229270901950014684234363699275319954611567138685704507441432163390407802999234905814465652044962335154357201055418206044026298913181811835652148013865548465039391744227689983209599473453261485153380908818440673726935784293194085081800541112501203457445303331397380446948852770989980569600051914114142279477629690392241524461598136812557513984940301589089128390396906823715967622826543755905616030020457274561527945965191825007133487156545101752772344850870016671622088134714555957733125187680564852910540739280222112003286278105640243980038070190569773965297877909880031620066898403721551962433429799928100956086160201491088191150979379290036944991206217653774823719553245806361501300980865044242529347835698514942884633840196508628261926818181331453522710221617868856397961804255727094019690419527822189723711576046001639909088872571825104688428346189131812029696413348931760054804610495092798975176311147402264219675764519848872453240032533025182483077491232525668847561695423934889778461915757944227708558205785410061634879826139855509192493376344196938647024154343282328276845684142699438372354025409393730738046343438282071965531403267158470169414833139149967410908125309984316312073108505016906329319101821105395558556703982419629169520765719992723071763730813642389401482326265470550903841961663957458368747626735252558119907191836237358436871834305909221799544996922233003428708226933056544676836776755877960837571832810682555956854316804574768968447920124443974874700573757245740874921782756424733258593382718360118550637271168238142462458904590760296921428081805778560188655269099927922154712708959240179476507844514414645517172724548476941607662478326065735413894461988583756749847678053698392957632660226472399204136521623736361466640323155185415154811185810844339855150436734807098016306252475101971504669545974914141011308661681636860421033880745651994932458611604871116486279983802481253941899013637727311133834266778532592324533344853759664716208338537412263555308937443901951541575679942453238033408307216841994789968124268884616597060833397337177144364986798767187972512615063197288554063191512610386496203137400515441345721044904467051945156922739366732497685739160310543114816892225159257668780953462048818191351201284162581472102780965979991944160344384182326231448081673218912379946597469447371994734475461447290698588542010162523064196805972372284233732451140654028375526303970784269804597321019493457002525055959469814542210702836906765111338008271970430451924780925117856272910352627916974258068402627307584190214628440917898442561629867390910053700297885506985823190928855440993457717507874925479171378543226314655666153587021169160431720984265232313966065488930853830196834019241368671420697276336719758147787236143301726125555205583002475666557171155769559720731113661647649212410007432536721117181902698647349658130101367113732229307682146982330958626017552721672584247759443215834483251667717953813744964440656738138326645751744905748004650568402111858982706460025498422293207274982206974769800

6226082660110961848556935211806622929940380157261010384282960889874606563046798502292990
0209291932461771606160384801422508869123734248586441731267184746634512149080855321275810
1894748251785940175844722914259381996647903390875103666661541646865470072022022545975340
8098423501364848574947885547459213375096376782165427914135474670062812767442222991188270
9981357269073785469091101556810429717305778842476385374026797415237094624639434605121630
9245148210507976032685986609400873234156423535667666375312276348010107704548905103240570
9565966741308311579279748096695434726072474106920153392090933458277744733961650093575240
7112417075063503208867717412193495146667638712637372846116040877063015837600111533649210
2190203318068500324666379221737979268046636376195583620471957455881725511200080419911460
1363395552011300394239159753566741136702232551901934176455731469482202225229548333297740
5560513730774251770977444659807607594546606593103127348715553264389083383702773822074310
4568944586437175412106604933774542490470014903959465745943771237897280058429396210552670
5039566321492376729814066350061202360793595072500261188229881702440932306066540097229820
4335376524377994159185149339041722501469974253353780504362141093577235490088186907390260
9514016697214836261672332385111763897273755722811689919980399572937460322172819284534090
3325844116963040623830035426887429021182429856451245795207347984627346195104552425613730
6299433014983937291110870529969519183533650534844242848046310476522083420859365173096580
4496130285833654819543598186756220294161384321163257685982562967201828906781124186868780
4597313387140418729700711919867701293106293096949899354813921875928804839845138740758640
3027656157147335183544073506255312343664960053109148813032384462652043513629026265933300
6812051158885104358569945649938869570706111923572757110294196928994187655436884256928060
3861435130310638444334313396196979733561523242666417066390240362365593574944251038356820
4938543066369565628381349757831909024277049909445141111241230206043422328892597491438230
1575907984423517784541128724259493533330985710978755386839817548252121751810308218176500
5155563452429948453904577562674465785517591489162251178070532881170459740597598645830080
0056103223325384300750981134310120450833190422865468587799440122965616563890815959223940
0343922260010197221765736217116359902575752900033866022749259421935596129475518513699390
3249692786635065238826768941167638908982314619095968338612311858671495757270575686399740
5427113036591818384717808180841752010065566213047632675249466820599746393688064999076410
3425808930820409023632383517830632217617206433632011040344099409158405146878326260477560
8040712615484260346833880268094044793089193734239036386440254924481157390763168026646790
9679017556718706413633240288705087457165871395916429253614402597840290871343774441758950
6655811307376296889347527173711313779000310082729387122487986914124280084502725122546350
2721992018762423508027844796784337023680736143992859048111112541931101350959312727661120
4888954689194535563621879329974884586783481448626980470399009162895288368911374616731560
1702410045157757001603778370339572539261354053250114657419224801218243169853035363945980
8575041132622240054109705194916292574379281916876204926756847774860345138396946799094350
3055861500427944483112063090593443247000937536710056044926556034727477807907836395045180
7624483926979873135352843484638881332304397927748022015296192355486000473427309507867800
6253076736433322824196132427056557794522660656950161892625626948238005695033649150471160
2428579689682908969058363758278283892895520363223930960420462287810637658111782742762530
2340505564585343287393682881055582302015544340256660561199160833124527637674938321952330
4956317862584221341665748262887744717933870329083625039959451591113255050506004055193310
0678540221401392893077471031783341055948291837130769245697952206414022795846705868857750
7846872528947007508518500453068757078397466638450736019535737073785356700362585582066730
7243338623835066357323527255029874737157104957827619393563501675928367423085383857351270
6128826173200948780873129781099401372087532187962176750723431042040983005430754543089340
7560999967053006260089395875398030047816817127195093934191068331792429451595438511210900
9172267321832118162279570595447822536126829509548622172312363166507257218634531741072390
7650382647261206586272339002819509088574096005768787459135373146151589768102894467346080
6010997367803156129155718923796941378351231627407516945694908680544178759792003655469260
3444618177373227167003444266776046094924812058797064658497352880792415315639324344125780
9177935725901786375825693806342504430545882795813741075602287584447810965427651767062220
3706972419791802152291454833956256228460783866292668106052637667735843824873378258642880
5012123329247207096075960682756058028935698068009042477941448052246140801929827445359140
2613006731539974220398080445375011953972619448484495199114028190660032628026970954487160
5779231879915526503112323553639686684530243015961973492262352274323053604223375606417770
7316238560718050405591823508941421612604389373200349753263396931768396397408340876148960
6053467650020180180950179015247573815905144668404233204625195859647960289531559077547200
4187516804221048827397952482737496742588221229084184267327354050629747395625186097845800
7013227661151733251592801250066103078456552279339066119554676984670693114454341653858720
9999117725534075836267307205981923186415819683344077561735876011585410628859025148597060
6302566272781881934320443544059426417472874446384970887538924365482695425677051955005030
0485739671925936918313224995238290395190787500747741928834128339315143605456554992740030
4000514752860117649136881975405835181301064281918542772397988891155654420613295214834030
9746559537469317679712605932622735788936943952814037155498125622918360289709884483519990
5909272476142927384761134273942707646596586099675123050324376092528374535447598558719270
9745135456040928446312383889192976387224509486946643316458586117087884072595663446488770
2838981448069759334634892092084747933664114769169483043759599830298448510466971167611800
3157567043074868319135015108662681481108066267644488187637341191046301486433951333169430
8648671912530511977122226051292720662888283651460221944708196954631978248180623064295910
9343803191070756728911303416669390683766514373340098291769177833502351137723780700801340
4937512480505007321973812385162506320810496669536517392868858139774907719078245297941950
6168869722316752104506671379192086538936749986314866410170044457629958945321955986639580
0917941851056008686137283520554464203327542845960428433290425439861323590988961610491100
4037053812955077468163875576813162991902508157027190764907757843602169772279041728321520
4831115467377986753486330097098407188529177293638378968670454493802201715742430599092270
7663302429744170450074589944751500765742782902593896377634935315947832002254243267251840

2300506882020825497212921405633725581581127104741570185389317829398787712798274886567150463224664385555348488786876641355612761048582420589186519723435336395723609948081681739740311903093045100043961806912344770855514924729667850895219861090165524898357700496192037386165583736313265234598245319951059160583790003979969517770697952469522983676074314990085485641433272200349890338404285312421123402100498162429185892422243495753608082603067624015469578806472961926684527511160095905378548316367124548486778820172520546398558650035823500202150851642366350078776071519856145256264951591055507477278537981540632716819511175410228892983782225453367566519025121682301691819839425989997594117695875215092794637661963360871925094446859022136218571040722049966386358712990530555270284104677262021274466745452154577237082364445335706452251497567870517194800500802567824607818185487583892258881641762049036112904312816748320766934852285776067934846458657669095206610087879064301681675339162115205681084467894159123374146368211388337577326140274624666451509892104406381808305608040137010089074171937662958444216912279089328319826772333212792007680916610268387238432454420686797560394826736582067345155042676308818158558039319161732754429557643651620151664466855759372594097115187836921732999384157882766022670925291474618234186636721931122698423608685452041883128019978033487875658827105870237096130274799132348225813769612170462274224961016733754788010634085514866228699569152716263350178980339827364487742823153814829001343256788832287607165110115958718714501057834762975288743392745818276860004527757177247880209896564385294238817387999187471297587411654133239155651089087752576862866168071921090343265030146354274920447493124652014732319757703612045011747939862821525354972026717908950519508590979562153891218056487899678573068849963903277813229401520730505202096076546883252517526410135506514502980213435336587555916168367494289441214804865998225611198797208669302128499501664795238170649564273515525846846289200141881385743510670806803934006493980278204084501559743111297779680235822043997189182740819520245230161759947970828058600328892886741062471328690946946684390420120577410669946215723851109156923759278558684493977360688001922934914717580163165254747272854041117825003154049416495321147450754966548023101855184474204933682149867867387892495751470690246623904739663020066157810935792046362771350394697843435917414264377803119021954752269208954796887825485645707135566698502327546123988230278689013117573219171678493016365013895529131590117422489607374946020294851279855924507737935690813338697470374157396558258857374134903582781434074625542355164396890460779583866499664945074756579699351493513597320013303372081166289167650986948072944032991761290723011140465891138242727263296851760942841879398026052395654066459330078965741576697086719209294525393155274373528280048234575091448533542404918205830211736693760497384227215913483552040425761699280642854714019368556141102765828085704034232587586569465009202308498308267843272774869277955040661110043597647686786538495502722558854224986136573631739614809989360847409717649547154334062370732658697527645921337429714124207051122564494990791441924448328393791380420636238002970028206239318075619226485397394933083251970873138458119857017100249136525227198684083184047846936429025111292751766993988620343878917517267572217442145385761100283487432190524333293999280948839061529406411018793753941134303171916852635531673529853639867761635066119682247234864034096810385850460672960147613107402400742653402436803792106844615341970808088122076807110643890588634215785282582287743675470140831003113842985349352824935850403623459239266883114480775467105910694065472373465877894192487323643702402024017517597046829540255079677309844431798635413464595390227657432035188431115597546352330452355055292850636311524081176784645471413279813406592746732083545282276634305953054734309381750377872317646489096491044255879694041170527766297023256952463671775845878822028460530530546618172209655098183967544401193567027827210833556064466575228098058628703330757873821082112921021038961652312284879203280376653872019582051856182501442409628366098463322528411015741736148823414602873774984914984590462988903284925608395739914059102412068455762480345904723595368839274619006353006025286844345466005785745687250028879740781269950143596964947641493185049625552169444528557452948072324330277365557456104296608835697425307361113031082380003158538736286884921091749180339054828505462786410001316947199894889420953086446502679610071504843926206331171186086120818715732734551370014186768692196704885901036242331516776017271742233050495306158524791962889675944421346235651931221942844071430978955055985886244201736282607350511638463571118784683471588349062334258834361427479382954834126341741688260288873554781665323832154076835909681205488454817189269206203670965360564089875928450815742747862194136716994322945869330495353722967997800487191020347824739491797171976726253884952817305136008922705243085564874504121603355151493103964186380892659519034350197950366229937107061098032834728693454151359671815062976637443099493236957049534959141935272715141873765869451223636710793801152421662449411980044655721259753580463999102393279770838718475347998543603066177468548167430253509114142253635633838835435446860011086384802147927631584793510710622722123289000304008555103460768066519552168656718918133485068069377003906870137625405524668497671014459934961842246898041701860235101964990133778650832343686944378216392021885085996581311126588948291773489736110174739245996397882377084643593256464055446635456450627816507891000060357895847141178811400147045565792468479338985904435609519395598304841190708596839082696964630103312919540548356633470754825597162240957245607899522751374317292043294421457175702111966492732950458924351154224989284182930968016279099901439100065160664654765723303846462439511383067770124161991030939928940398116061674207603382917593065660440087162242294864120927680054840263546046021286431296861760486768545873324590490445815446897600129176759371782877815746344749744317472245714178249593326589599012326316543180978557825775355049571779819902461257772849446353525217033433931343645138304258981514773623679199418168832859918715721722876519947031424824878759641163100819127666028856647310961164666767118936763472129096300869239623420708862933863125561287340653467909741993828816638506027873280048505449182019933787113082949339025444084544528311079067175578158009386753600386035576806398106423575109973318402806850770996662013993830215705749043949115438309056158300113014109913958329032586958376761970744891911326312164098184347256008766211159695133389046258905040210397168875297634115653016376140172417412754156775733851563340968924430140617974975371744

pi to two million places

2566473493454579246967993011026194337636448667537561087691613043613748330479441225952612415148981699807681018481895780038325763387804353118569719951740840404292157320717893877089593155702274272826581614974080133206343384008689023889391833314200451596140089861121562924364169472543135617018117920055959285439183334369451919451431715413500742006390506071676581905824216202897696844090552454470012818868581435495454205566898387862448320752121741094873647410915786040991025855586394013091551756002797658653840756823639735834723364825533520524772296088193054082105328131130488092625504406239557590391944317141526527540808952986572630716340732243950379987148815126244606312813852410741435979408576657852516943899274655621805886920662325044937029212885127459270213767483429277774127189221554062683300828600901792187538294862915016245407247778866998890070980505692506985818373901350763943654187232930311836581686333644779329007078574510611399148091109971279294384939136701584241896910938623802725764521667814022361322904680977248778268641881228779934329767071638705695119901896323092458999902179757131345174259360325136902088395860854675047773229290628464817203195072814911800959505125813447571570554678484436237769148191799840529066771547929117426693348858567088139473448486214381111595475449852804057422506951464720084246244722993215792264672178814520648208302769361192173852367856740962791085234229560906326667104802881922637139090737684448093184449896992931060368700953087272410262136788706972698569734002280363504755236880049558697359039020691931283070710617913556767421587725759822191294318645745477205132894587827375775210307119425545175288583633445935800552400893120991882570655486639231448086237441459530559003065809529244042617675254206771033553684955246546435770101789796416313038674952396108289032577244899186229965198423278274995958934083103129326766102724108073795462818807634269861761703310093468008306518347713287054254996232603250116192827302129934934139799634261512648506347348327591363178682472894449321466061847965057304067187482841519147416841690897229561606700728197766108311410618692743573355356462725414379409813463832286243364133478978890912641531873040605317247738103537082669410654440622661293766233632590613838197631940130398985582094820953918602481990989664862713566498757014979534006764478586820491076270969775705491508796197691944139015941256832679426395790182782545568969968032231781589778225657613871677013214802115527276731117341653118922038451473699023164776475938806016075537483382328545825950629102393600165447353942982876682073712192752999073697440811342372020131423404505259316972775590585657003132022465479086657583451266545682630554619314754718739127923972118469217783762006326246681456322496877656028010455006968281528657542598234834009264710128328430643025697652644694135027062145831911049179319414349301770348702633346016871029183608374125327587763154022303956288776844723280302142987294946474939503146491804133514202393881524875937371592838326600354064289535416985061992235303829218610486918255269025574988931977847206199198050179722622476371814459796413713846207982090839588211870480155631852081343051685237415842174781458011563058311767877089770942569153607878091136284354824022828394565804195220130311977227965959838840936358355901185742704482665056343842162536049834085438904028543049268993066135302956244002028267343672871926207940297350617969254101291753762660825396822180621641812462173118967334743094220887670606305467169963141821559029243513784364350632262093120680143169163849781012271071502167924720562634687067390887567539442448270838250878826535655819744166357784924176318814836216441222232363549938942990784092165099611235325120110314233835506549388909275961105369398083023861037130475667626055558309278494357851978563905347432674505604909407708591886510655453008198264452528174087746547899161511634875393067419302334275836504964156863538834492970269883021283850750117307225319474121886030606608665654771424490261810915095583074276082948581103520110703303058709257951235338922157247424785671676788358973767617721175130586934335536764903674373881704440543698738593926649447153503815259632505530020490676462445852260521393121962997057825503270457798044858956070209902153446695370684284521782144523866710469408107506167317478569718979427878871605283450429022122439834339069476764641314581807048184216581860366937640989433649381426799919717981523351084157064761650276045386329995224430338938448698694879649254150996947664654689769216486142392356827531850965408813353323603150184543091811589795300960882380182295483646650730834615875373909467531125542610804096592752714562722552735012176578243220128221736845401883409112563696058674173544188303029104229540931213916325166527162812140200393485246292677025335567801321211000746875104927292150677332824634054330514641101694580152392810648716511937484962847145205922860221643092270732911314850551462608765491036968970857811251399477489383288426598056121218316115303041915518697722069640705170655830745429556802055860647919856113720832021397337158971801009341952992408631804186229969441590014844534898620679645053077361183991361144007305930420574967863953731729127783584821562914800308429891763640342582197758590998861542291506488185489391946492442442707346065366282411043806874634642445248921727008002698896840977049906879061899434905998791232993931686542977873405363741086041168212232803214430184507966819142023711524639482201770931427298845543362729864739836067619735520646544148606206582731163150571772420887655624872226829266602037148511214624441496499951520297356963468957934911848613873235673727236286722956930702227282618936242091446834342646800266077199502216509365191242991298909496734886916099755708415385632669799928723106696812003873504357001390292512455612525900332219730742628557709051926829142078160159206846851894960268032416450232179784549350036868531120074399878675353188044734785139543177396060055361114406353642161812602126386573094690593255906397219492038056287678828743091909615406869152123189362789249870124538829074308381607486273896787745378236370946788911603417540455089169754059227394049266623912040646855819111208987613022144456972568625452950130411217199809106691154157468748584959986619908398378171263310681533641332040836168589505925150168276847524016347344155357829189499213859109112248318187172194868838657130404829477742444060109969015638352378276129095967481510722582918132998287427054987359677484220856206752067159942180911885102146438815360796234541509213403430061299786868257960293849765918712706805152707141861960573903188851282433872683766410207671757126610927458223352290512994852828161359254699066916018502759401681624498230386088013824249883069963917623093961348545994517800671078225519324205087390126795497844015698988750791231652774721289479153173128

pi to two million places

4579690612883700197518951091669085067985676051658730747998414456849249212797902881242410
088408847783731591936132045482630356747492775654138559987303215905407629514363788522298
669818514967083486052744652644981763753211029230625903686358567994891492488089604126511
073389096634435933844124958326982682427601020989659454604773652398711801751865136733933
944944267677542321191082309423728968680220347439593248623769437410866602820174765563194
951087225889202215249803245839271724279586677378380658016613729297771184466502501152581
240730709630180406940749655684993087263042183001570412013792245678731940258213109454610
936997492226185837465197323220368127821679785482540385357995868520967119563235803556744
705872326862792820603648717936660559441175390180898377902808434422479687088565952716127
883634043276080005804509481613762417573420339477986303223673828425719406058354743783890
446415406579099993185608124308463533967991722881263788799160033833788588455471982316793
889336283777320641146617025954208508851805129008431630715043310753954554199079646509023
164428834474068719718935466735649456812354304961298884537609297708931994214630482207660
235398243215761625487612842541210616113313364882245413244897545356626534914240822491340
206617500721126943061249663413321878554680294591249172174641038186713621514571649507321
984896957276872688364311053604427157154289136681571274428332133397633430005081273956718
748263504621039314324319975491958090961365625700243201665807095188730589919787067936829
524404853420238652758845077629278714634221930150056380457251317594012285611355436854201
081823715869013394106131783292979204109913274541054713071745330047233839565461871634457
034018556047181381578089815947036849425786821926363634477428221860596424745794721701500
258173333022732047386573215346948938772327005556946006475062041487331611568622852690741
688093499017469127606927198938505564840024337032655975293686023874269601591645095650888
549020898484988762500950696676359381231697790345117805540667349228798064769733991389207
356808640524705632218673824785071644611430980954948809247373180071966058926964767438019
702682241032886561113658492748669631268073768413367434448435491569475817285335944010063
322752305783993875303255072167800795954106798161051287012353251890481593561721703239862
529923914117857155601316674495414313623225134309093811713897244019512308210327892658362
850240295904940492941532776864235979783872458059395139056795560489022429600734317792826
184751933955564593715046875619975797431690153572144066875808686554524850048273571308531
731244135792320993581940084780355302666178999034001645004709491013833401772294629926564
233454788104605647714030170786181447811161624796255323924406800969011790922851352731479
550364501613013003828067884533563391135173141024267843977071244855993926430774456327190
903208839351852366033052369976131317334834526825772849605743916715539515852956333008194
184226563499761732066839099307221655802854054586971189012219660240669460829422781432154
324365761017502059129601843671261833291453458847496239270976581693940402863874677684859
859221382070348649830522504843916857753454819551137494718952124618141523990215336585113
352535414111656440333991014817160305596064373428680323910031407094111326232632399839699
537619227136973500148398583884971448168151714974590795901774492774451113062728420131635
843764179209332924316744005546123114291616263976070663560386006560837140438303004134347
463989548459451126970333775832905533641222535927524973455348333723258625709633843342035
966408972856443178958761351810094275452037400888010727884818895951672312621189125001920
873733861336501373434886404252252001770462072668820070446561847389682355647147107826826
300452149043241055363554525591842135419686511373978866630975008142564319847403775914587
895906098457426078857863202314402576460524637248392024315470427111900318904204033381700
009864228643418074777777983442555989308922906974570187202046818294167524913485599606198
009894844748916628760419800602597001273656939362975409320859454667562340804615013545821
550863207226603893401376730576253406555169815277785599299882419464266516768776119173622
702092278336052507704807075907180343633570756382836596813995390760727068181365657591986
683751054611521808378119196475540967095824956017828245672736856312185020980470362464176
198682717748478222463490327810885463141517371814329792883256249937115629715737390115836
310870448602510300496946914258386937065120377046630824216489443358000596868730214852492
879538242286100073642036496791486942425477306447281042550872919341960667052564506409608
790024404064247311413566099006514678880932791384938464806546101789056276456355644526787
973176600856459859045759450452936327322914034062409343851631402526002102085325002803141
809837523389639583076237367334254811893427718926930339828412036495177176010034675192081
583382936321282066313108914560201482252304552882944291740051438913118279809819848432290
298386962825148739445820391094065328018875407720949074786117915770017190387912806376236
617440144045207022924523204540576280696579308502039812183784020672025012026675295531308
349435347193634177273406360262579603136511978554856693728464042046848927715778043458677
610085289607369314413346487377352501592452119765975459087695020605617578193591077403625
835765360080893765328137084369439022722986532221828843740013882581116297155345756740321
498609755428688657987436900949705097986093770278357223388331453980493989210171433582618
967400312252799730336457106160728496826402668234770455830154585574827171372435847099486
137265871302549402449573855889966053537090338925114540555812456929413788827165199000437
610796725728059987482047989567855938858499483469651949308978149972776347330585707179027
093568227576306393049702296633955287633799130785859314207811335111432012102601987304216
706260143575841179770790458083808849808816662618535883559242006305302464346289923082030
708064941073041567597710077523985586867594573174476709455684268903853112849498801814477
456650509614898991517629924164287800047413850804520329530539184097689946319969559127867
694931959273366205430918120556692462152740786651432352659207070867879558641686045277535
750207487671433377060119129403158574310767777795213590261308082898324883948320949988456
830767241759299430340209439932270827548357388507419917136940049879858619423446279608414
447356652037928295317016335118153029312723025435629105545863957777802211658866611269335
740729443614557490563720071282544811355783402901604851760524329698135502747147052635429
352648136623886958489819516790476124747446800847725887139455273671088784750842568825983
963683066764766451330823429953840637149396551260259641269166395532942221627797607874955
291748568842182486374632474778324492983235440257156760792867425952849433898967643436575
4823075754784033503696537687365498022398780119203544049128826835941953971843647255409

5314210556663207320463884838276837926105500380573953794021513641366249674935373241044043
3486238233624920495354428579053065452772650722034659290443202201716324235831378351252 10
9576415274124465776261675436094709743356400769041436221806829935515109138556573734119 48
9032184562204438771527004821101276120814078245264988636103832650848085252951495226355 42
6460671844543042653382666861006655771695171442956555905423681933938717532038641155224 28
8474087963872655996503545316017872842995906248975694314657253297995656442753810259566 67
2558761130308635459508684842081702309037760107313710623429337807454750823785605494798 76
9021390566558589286009199045602603206378272907615539703831101800844901121481192777967 48
3910272882057559782053508834615002190348376576463110568401425042106378331650979093472 59
4994266170452072326910171868068931598950080623997586948389705241612230171728940390466 99
8494272133929568126161004650902845621267573941439279503195865023504811047168563578354 04
2648572127540263881287194620920381325464811617031358676710643658766055165513311331702 27
1823215687736219584821685646528460697066190543954014065106309733365138119633316594903 03
9216427085354228049798026714911895636425174891344121426361554780892145283670822169402 59
8711263211438852993916963048048178929629882011238074901305294249294801611435330239008 06
7065721378167971985686130290301299399445124984690100198919360598279169730514759434649 60
2883328969660815056345056609378129236133490585780550945642103530907360195844637121650 73
1982015642422013268456687741832331024731921868515643412032717030573066078517538509706 91
7170791725285511743627871301600952208920242405030575640215372736959266799747810707279 37
2391235577709346828475601076301279131199539176281861594303820778398243261731966313336 20
6379349676875089524023642469231904541673862358360482837439278866547759485902892040201 93
9593770656732119490991043352855179871403502030760557820191483882880946496482084241766 99
2456758312262478070390557653141263260242922436203719532918554718091596443185685205788 23
5010309107612806044570442514799758960888028125997862387743549659904929673220844972443 45
8243503689780365184909951214229401566917453416838309035284779643067608611599763678720 49
5505795636516693834521021205712467189023635837908339119080206899596896990188122321855 25
2869348573651888630160452941028179736080689549524036066488944683485357371170607994305 47
1921648759431314126979595251661025229095753755095093371854490007290767612634676529166 464
5580371533060205534741620555668380872331011456706082197136019911669601177265351241440 51
0936203601001758405334468987565349002447580184990285112905603628154372796762883123816 57
7437517662456404578370496485690904281846741434107660754984114657421533437962825237739 35
1775877039942552131816901739901861642141354392779733470876597369481710103318186376892 72
8376366023019205919792959179148224416394031804147790028285712517764484105931564467536 33
0924157970212626481304280838933770672398228654341731736481424562966180793136953250911 28
7546949801550317994516691228413844646308741027987820955877346176667793320063616141299 83
6112387852698449676224949460162224198481882844175972508965043238838826776211538694490 72
2314080038640966747955659603365865500834501574668100371549812154559177082855269058782 74
6268018954840985480647767322593083364643266678951981323034384780554257118933244880337 10
2766080664261976800040145768192614123421421090837882603488039871589674691868127595035 41
9040689672781395132198842118325610948747352764866436713359368373719071671361534428920 72
5273057077805616065916154423589107846465547369563439707372217818591230109443692313952 20
3010113674073457059526133029367437932120406159970890681203507862354127805416826582353 74
2593856966435762710973540865230333395749249771995346662569428121211926674888665256315 16
9706607240021939626848282515447561496357933365845237724099687357953227591900979741551 72
1334845333578681422873993851902093678274021559991420456446438381600099906505371881484 93
8160865503572270641774386629751678966655499987889572179026230908454480646518569309255 69
6453172241089451645426796761819728832958413935133844596041672854573991415080495944661 35
3439845014276180542209659848671099440825081513239252136069510626733736792233221425995 23
0222936409047664596154505594842048813114413172046469267049759749059935116920439027605 15
7446677396870803247804063437778416725021988849435409828211600072772915050759869365684 72
2016941046189444582618551160041549451062815887248514034519005556346661524473749607661 13
5778748374003886293884886101950281280781792745034958405752928452983890915764913247310 10
5633314781346402650462629156753779092137247828970031963259689125133021524656120543583 76
2268609282030777416870045904352635817494636724551789784931750675390464041603363847240 54
6498075003930024576610714660605719495109140248232735266912214960160708972207220546288 10
0387307622968906215262971114289273463392143785758381679957096512975121288247076229375 65
7213489062361860141899595000293934330117463300332972907834026382527837960530000473559 27
5468487189299720656136533751537477921962495517969220085573147944574288225924228767773 21
2885980653704654024619938729649935943563230213110848242495018006757189398611897262182 43
0778317833445857036118160941397634465162725658288616878213013425589073818405734222752 79
0944015079633506963068315858425959758344133931666799730480514710420516213562175409048 77
7330227396980656495900945695698536584320835620615934529254241892916173052220979352465 71
2270664005413539212620953741607025988131267956667461709323717405236296319608936529844 42
5074302280497664164038282925713716360306176259672499571761536958524866449317201096085 34
5723423625450385444144127163847672628333308189585593647600616352498590632887445032551 13
7768181305334664669950154774932420985686593504901062114129914177309980459978865399855 59
9720886527297388216508774800198668603163056123011444933193578407633418331385977273234 52
7021265265772962648846204405032377509270264409159921265248626771659965913245715413925 40
0153811699661401449792205985286546311988145874191873375518550958118710196924176642924 23
8937549451631594772453110198414508008761556264407882172093511259342618446830352107379 40
0041838289360585440706517264491688578728545265072810491172241294152234684844898973496 53
3155693932685540211665594490751531039708324623445957019685643267568038544519358687335 14
9681959769600820125379900840010546335233641891279605446876357037106514135683715512448 36
1849192509499414144624632178459676671911648776744489599464431583958487181884662742027 84
4189992880327512449666964867934589413298602330348292887626063713644580737134010172699 24
0031409996289875932823997324878713822652547419034882217749819545570796378004278014587 91
9441189077071435801103026624542936251505434616515198607934238562390664551545908689970 09
8727578338564769103346863889942896361916953313831063514443194692997895215042734302745 05

489128224046567516837384091737414843731819711882264119670295140010484497368688360489262
885407453712460157846887947781317083920277018500839599401350787510645356146154845035346
787490153402751409018346456754197604548330869216939024898067509229922940715506923777878
266699123015899093808133728505552990599347167842350786739058036553895201811147715527516
138372665668705503251456831582959065357006080657269902272143379149237524221958255515527
390476641515242308413093279355619405005324441453950610949163270387153037015281008875408
093329479098659178396540897411919871437341136512716438240524415842887697571497711414714
279508295887029927924683321337051526756439423113502628776890344646636321844459217157587
924113199632987541312018325222678696789964132934113176366538896832051191636223996373640
065062421869198223064419813515321973198591015636256986218174854708888378220216171014912
432492165323865576908527472547859682981249480686066444493519183037483665508175542257335
268514038987865030070402899334381972301961473408648283476126073019822268614411798984367
558389159008469991314054138319391816564308843497882991517174297486490963838967343065171
217360275453757834311352172150182695929149322787473742572457213602566263841389526262793
021300099661963003232522013138218844822215385253127676763048551870068314684039926818548
765384056384831921002472231916610091343950767855513837048214284915101698975339078975632
339921778903880476336813748465168922635716230718406415663249241086692396760121601081445
609232133742914578448806124786377388264102086180249513057338836941585087823197098151586
711709517388028679580151067880449339024806890990529195328446696828825455292078708090501
661485367533081336907004801338828585461654064133202506938355963174243658840647261575760
099347841140840629982366482357485543533590505361262742820018784805295304476986322636627
829632741637011531118234081786739876610728127325778513921138076815418944404176329463049
006186478075989126428325729987352871612774183368051756379419524402321288854911774150653
111681836226989531900495922925083762608050033174333856378486749582231058631889407398076
144969201791751393353298858853433644997913001657128680999951557636883579690344998472342
604194318599122046582749564413763677702161143127001434771612016464832132927118257132879
105841357861931189374595323631023912708890139129091665271923774586864170364801203295548
751612012917060959270907773561674019391174412447124601417849679728249366145899072550082
4349970890968096364168915925626719343047171456304432399815568869354337262302
614980035283716651359121693178382309796485222062854188473486939359438432529987537651192
492335099196668931068393430992917742911260879728304331663875840237022011217239456114473
365412763340270584541778577485248631649999170485476948432053120929273998661075263131976
433765380295221416374236023722185067791138872580576777554374253574423899796335819714032
277935613974470711946114165176151512388236279056488635894726860573344797283092570943913
779516563058538904168168987692580836506882500936119261078911242709881222693468531985170
663717420468096376655728936417132493864433405288735279025508689950937615120464984509308
482094646064177940775927273518750614934528181767517108450236520442367768151326743193251
095192005876749184930277695596540981225396357711694671126060236069439457213648076464990
166378437484109735730098749733872155726959760331137128831583803062490323833048619521149
826235866733363594360815330962043523180699058672531667967198977573967198505633203916276
929612784504325093027849365575704663665005042353807000210433790543654526769156321162307
081538667932875280399181022287966754927414138146006565485087779489944785505088949148052
882658768844456627293908196144006839830805240372569506411438993318116637701630751930445
002156616091239778765007387433986121377676316379949916035800294253941509361182877925648
901997063611511833437732287805137817006854618939770780027545047057465744096115186501688
721678158080161854864108089863222334099124749225808111853269987957362030363601202486339
705294671240192398688862699835431092004562279169984416988321201809559455053488532617542
549536385150631189625309217665291658243159004583496939706876586542432819456476478537355
251530689891097966688781002569839068711319214254198866419286668754537724431761446354156
657628714055353642878518017662196471266899624319482731009397417104259055243749687333036
596872146018899966928025823050002504947431957887361774398118194129480046029505427368669
231007938335527797500383591562081647286299516181415118243048649558704764203871577064527
587236770808180590408423523775775754008846877561116658139251193567319094020961428901109
648995715039720715905042478578296419139818646456980690038836467983679810123769463228883
609158544301049426148760350379903441776859995675993305022532347652546889955258062893178
117218231637950877845934372878641514463873034437300982480492495540033223434378058884644
265651671117254089024251656807434576029571481282240947846055854321075345638218480483756
258915751706013714681964216897177605495301992469530239019961482626017062848187963579912
396970015944414686775318564583127254717439445008216298283049375695975213397439120310652
126169622928117872149914975372547212930687038508756506150275226420337304121616234963978
809943527080416693322721835932242791116573630925466496712499296143990750000976357102050
913521721026748783818004596833106999955907782554648912836743394515824781528057461034151
134356381056637703548798094032652665484582486845829067556435863245918075346077580169583
997580648813770434341301059688439299179317417474286712514331325517759566739120611354673
648311519793542086154722283275856040177338917321114186236520276449176308259943093916713
016035555066396640063769917744823839840452727764172011229122906118559512750665064983546
029610296574754375906955550751018593507588378946923408088442421040443517845101844946977
660224325727576337213826673283184851041913722572799003023019519881457021217165722765189
190273755803239808560285417910869633038050283058254155207922154675509988166071263796669
062669622288041052194375553599347863223933308807429440316363392974318457429479645304848
126672429605547789372416592547542572729481830358524079897060138339019218816473115578505
264328106710783042538278625507357514417809437945152087696441803929450533710695800288060
095926155424042953849392236692825186545787154155054376817216194941624398023669701764585
168224184823494985612261220525940694686843355828800042360426716492051983003040063908216
049484182231793780018789714732654139165969414662854460720116633852755083190003281934916
500791575742541572642210731921314240334532607660005408832422430395364285170101907195968
996222168572420582700804488458661600852468547117664403374541716972573166435712993493887
181992917594663181328648661847971944939975673768896889421874125051812865226543689028478

9523945318907597818378429373869271123324522402704855011368071301499127539063765677761950
408247294577951614725178334058293631438937472774496789651364811407002313274619898292467
466885589843355545576042775737627609120580484802719998495539526723426269717512224621696
901278008109844442553591517508360950391542809560322940245012534448001359071329437426440
765715671941413957919227136541009016170299103199865705258409257998434141590790940476127
073830478178955109473090662575049936351422786265074676659604091872737225477947297521215
673568572609788566464575410138193018478212336515699772263615169997116230814735386874455
953454013866559999562536246834629631683409797550456405010277261083787851820349370533279
213894340837041728605351627431809719641851581334638617864058085289932616267479202822900
779806148737877411733583027613120796025607848818904686228962784390204998370368536881027
711277649164807893994443904092507590483328908728324142449335236463081679818407101984769
366305538954028736744903373984749075859503560607200358784504301688110214426498847297177
449594105845200382301253166078708859906630371280412152890785515342213121547113947843863
137802569372757622158576792891521263000669716993847263443082846306827831953212479757976
403014368143734615264691989502103421817625936597845326676025066744884671246096686617300
970662524501769227885205690673590084316041245928474805668946978182767794755884701037881
345716318817494428998929871672786857254665536773728311244461767304087523506505439172831
961983050563809001407118907712213292090837662204930496794578667884181003586373834270913
535280939425518198079349785464480249642736866843355216708089658368204954333674108689934
436229704144434396109886127019621236168294238935804471970209738919920823023231464235056
710649911360357731956388471918197698807100580670387057250153019061535855590146412669669
192362905950385486873376121913721744271446155552674522825740005803470748350308148551071
539938916838373110927502058170195123131177677851844877768726804213939286034213238995818
513277666936559818064557155854038012159297126776864384285371888091790995730806801022541
165164298547985978012005303253225752499741035431088380198432200235756740827330608396642
555936595175830516153497134438263679902877179428908266042597494911477071422970252511258
606039005296024025458262483257557575612161312795812812156853840855916278057092937246612
436720588868157679076930350517895756393400311125929797253577754420056996640220213834735
660376916954413189061916061446825181316017660986167723206634974560465097288265951275397
273336869417550848721788938864061839846003716868968509185229834465775516415629814905434
676268421144524839941312303852580518486801957088922618832522355364368736458347914690356
787225982557597559602168145222994919737926897880657277499602061831236191259842299974459
179319190700446995540584360693267316708926126170133984267188893421249875569902430595059
536766540275330755030949068933593376555732175383356648904107287803731742673096181407451
562072125983147245748301211066094797879629490438133242706116552697331798122204673427121
226641381929147327894366091827878882764146146976422205029114448418441381849237635277149
146906974336080814504292765761587542170524939409283863732947357784234240779548820953126
235340275503505710289039431336814819951535661092447742704699911672378989551639046374983
966032427419931139442903905760590741555533250655482151792922547642545087189622131135166
993304543120000747182980065063873898926256489054397396866794321270549392327444279957173
363591063334844185146953429812808968687408323754080262605943298620562918175441222290018
402100592584355705001162633413891116472241032935430679924686315539002795139232997222766
212995130994097950530207390559581191512433304040788524971092537241747430138830317970184
410857045135768151291536244294925037526161101183732100465189614678269724442617804346498
440708181946488570155664729124940018323157474892122721505485676173310551732867555555137
275257228070158444430690911684207944852719275167523884694520140584365412441900688299574
590054357430805615594652422881931272032923409249033976547146518111314592519057258049351
511243668916540226006275545801761174352814314894999161467189352381443368464042767295387
167526081339509875962077857277892498559828348232891772081177773383466334241884922791980
529683092637567318470970487223370762475836981471774211480432136309414872547814926308746
678478952455561005349890788889845921184102235978273731652128019748541341986977054395388
749739014574122119048053900855254751776918270607097967712659724884419682338105825994352
982382672363173424267155782964601050683100461379274890656030763632598102793661123570622
546093038459223095699447464899594352803595728122073590021484674876096284970187989807161
708718601131703969843543710968435176649247955442027422470637716000357522942761374328810
277374243763384654818232425458586522993701908884747767400268001009673197266849558645446
767079878771751353980883983207327701780462499327861888076713309254338928428954733998046
782679145981967469019830683989226343929035718573309596628538845031122658632570149517844
368139185358392042964337583894923848122175655021035540671058277268875751378435979790444
914526917905927035088146718778168140149009915546216901425697803503595872473914976161903
348045649916980438948284871605733097080720504665480348755712333122224862473301639986713
795127886798643810255425604253579275162413162454955297310236459199301134961429522185316
982957104068514803822139888376963907581425519571199359171118755087759625477377513592338
703229940139176365803706784400859562468763995140147145722468543401280785856430439394706
997121975940645921442901071293191405742653347134164128664510758564588123951140117795508
073216367810160373433760157315563492556593937367165619355004588107322335593024482569699
655583883053413186676168998085666828277132356870681226254846298210313176077180123905872
553472420474152001617666021805882461949664874606456387899631507124429153884042324507560
321404776242580365260920519148291010271577459241426287104455972926999565011286066885462
748757187665066969779602823041381108469308787168483579092546278034944264254586120459971
996080031663034795899289456325531254253443173998536945980583602867451850810533130476475
285376287457097777675413077142430023253474040930306942182911681643903916557470032165881
100062420718524757974698052765170972745153025094618659937285040116814957781425963612401
478096837868885112514712762231791533148704487120579377655030401664297015076738855047738
328881787312124600745276123541766657688170101149428992573491012356466763962580651127133
964984228245027305669370892373581646095356116434375056502996315167974534093823532899970
256271802165624362551179869721624983230956587196829602546680650000467162224023966538241
855057306581446063590559554964198201139696528443930157393105206283089149218063428715535

pi to two million places

3700590035047086460963541097884106096566034365354449061700707899583180561453390650477052743156641760972151791948928336481286604608353071881948053442177304236040841191576164461309136759315286399233963870540720547988479579886169962180644201535353179004476525582272767064593635057287804275734855898768296689233572428808680624632489791895841694457900295928632288829382800916032460635320238223124737736787406179409010413815330576102830088488649415922557543809460630258626769891731846156393679957705382514934153048349312243480633331688402699767024473274906189736334543327828082077440266709778178207831208572634456094708554852142658489101849573503186642174822985673406328525642088746642534053395045023102754493429019160008441503984952021532560016392766936647780958412656042906452568061709558520418712814813474344515393746475496344220538926109954944289146367538957876055842484190258311817288502158858378806898930594521539281374654088777536100423510149310846076543290946086796910018840384162250090888837411244830646123775233176545420148684333531525807526673468582177646255479877158003780737152412484086925690704928900666781140279791636537433395145161411107226456176282259912478849099284872988870529808306524599355173740824113367757972723356826803378116357549983476669908238378145543916215512856385576818804480343213210293942162408135480262308822241912311925661205255016134989977537211303331839809636216624718633004541360033833053057938607902112932032887174995145272045062395800542689317348183540808487347093887388619010213621756502595911351442127021011648093093037494167291593624568786003466022400339535225142449151606042997647942423498377961134986659102717829299312465284256303982440607897986973420603951700753187578630616126471450010032081159416960435788247184765644413513385091862089785746218053947270574914758885846342665182414683879858080482488208055020192887763490205355158562400189346433242686366341174030312234237172519433745140146909287147290589015061523895622846152031461702617566758586209515477682835625450643162643745807607141785780027587608048332596758878853180959596581481600806521298179195843985925295184458469272466477076091516851746385647187622149193215623010127131052844507481070026502446417858724356779922130399652183827914184826528162510461472116304790782242023484217623506439141529823005543625701476369958878450144051305748180118673087237596524652046435093431303966655856293925156810381694666620313639439917344896713982981562741198826272351095251221821704750686253399267537797846680999542042149911343540701022056926819994042516658933928498565667347332579493437311852048961480398907491103503139468334077786693687354754911077611269470298782112564765681491566691909552936732055957727929512982090715157730091788477963374300214580360314477890620077155490476480542427051316776689353266943735909722400831594023758388314763542140440860421299577016536729659902048363057779854577067028190587186483061382527527860075879070243587438121184097439290836223989125389316731842659559659548495881455735273119107469179828345743767858841443980936994532326065725996975882421929358791454369349104390422637817675573224987314756957013416041498872979606838574148390973894482340813010956583016594628071917844798263328455536359745121872603325357817924068700106861650620278236810634292874145081373060954890892474450662277330838526810902090813243131511849533596968989039103170864371432842682490648126662406926704213367160470574265531382424342208560979350065014126838867933302418654622307472025320717893805031719917873491122677621899708478236081116007980195606821356266409245951405941919888659605278045961898798897812460445075204391195108742576253503374285344359966802548772112938561910122074296511208884295468554439251909040934596997623225136684493959391431058220054977616511932363355784477387007275167088770183360897310533306167400940748760848572870502540396522062498438780791421232920380618831048172321093001720191485125537211230915157190161271374065368354640345909017516919077376312212625722658664549644959123749618371939551519665045731490348698140453817457369758982275113774563078672356216976682523924529815904675621852858455891453014822265840535952639098539371766197101587838732782383755847244288253637262108186657407440201307721398294034324598455034889846916089350460604602972075243052225050884840981344559999075868131547522204656145766233140452632696535279292379792937392919037486967196463071806072478474739655051709747509660626023429037646844450582247222021323863885571451323474982034996604066211777772978606443414673662111064805331614307054651648294660337643562092912703240902104351027595403970654729979272108961839673150847030362204984020805668699225246595358373749601331417900353700524645605869477674688973094413691010744202027223857514205225470819089634991421944317404112374851728895430801845740910319994611280556433442262289579340517365029531336028314329702493656220118807383796596866938563040365542449375719742102493740570193694513848256986001953215162568971401835116525496006360110312028199534967046160974620829177365729978564571306965165001622777885273834072598355967397082404631525917673804229743178528315649444954503695640110936985581798511502721191591763800482965813078985947039731206537802032276034416297329698012059066824690143029493995250158989047710508025383515715842805178152642640065745840279170365562834115519991778849951688578980881405096867648256081575871689213785261434824019867008595949183437996872749380392439900124089292676454523422192207578413785334248788335354749324685978019743073891678142961050444496628579719868493170449749155067552127911161583805706968196572594815298667213335258086767899959649515960610988893062288696613263121755757586483262796857114153350264721407950235985958527648203136398655628137447612938148477402022504450859121825095624779073889654945265020370785100531156965622029849937475086136190439762995990313119008043197524580940000914558217454526625805274039981316932570018276842172231805140682005023092966177270185446003704330143234525164686645610521720692906036720273335396157708284888237288135889404534490226994538768784657248578192875582017026334879541782535545557549068250069979739505950461342053430926981905725531230338713302912850319228135696396012838534552115943569140977460158198777412381988958667271114272958310827089863920462180345379455624339718562497127140957925961923078188404535552979777422106889367338855485881072236768784859382251540354409948225290592834847801360135009151581145329214423539629875548306801558853165966563394478186222393658069731261285120024220474114337596177699106333914637580520358154703062295277217035148212385304932657058446113196562254523852860644493309181674712723697272029500262669493867606739072357441326860714781306327962522521358345146175319707352809011354314677739453471024006089517815558307572118215079950

pi to two million places

5883774454731926129253830499817430402307022389098160761531109928886528725602179498866 33
9642061820921316043176801177815429663673507872419183405805978019413641380162857929818 32
0868424446974760959467546593513927192027086066670096446912642578969760050375281909661 53
7566185545806264352294921657908349843792574319715877064686820018182320486899825645633 40
5013849665576402970834480618786696893121635133966878313623511774979419930548228986619 04
0064715435959579227544277779439667263372974662779775357319608434724918119021011929439 25
9038026026484174844478200516856843034661441250061225441185536036696829948065721395351 33
4078869245327059129149828017411210718841342687878882980021071193184154769063232133035 66
4704280199834162572610516704131168493867700277509498844108513693169564448607593170835 46
7673690177738942973154551145922770111036084305577182412122340329282298744398644640191 95
6092300014394993453060442579969384917723978161494511312042048686379167525306349006652 39
5804402898435392555784845807220033202925034659744813261401733733484152208726498583672 36
4880564331283046930530487353905968489776941066248996816465510182556276908923306543747 47
7325157482346420761826937202001112884908374084156663787904917715791626174472533569211 02
7963136363961933383031690960585634786515836410409521854218925393845365190009456821882 35
1219678534912907472733457619087952770071453429642885777891979700517737331894256474677 87
0595141670950151254363254585850590927777223574413690610705925417965794073644894013368 46
2125974037769436292671078648069165694144947649627554797526997506112392906590555602998 06
1827757923211986904515905942490767601449443302144753811078861683941736268247379536204 85
7866736619434018375399507887357076956973633489060966234152033032736644168409155972675 06
0681869195428972955496780074208880873199984229331801642263918301140795970491267195672 66
1938762353423067783745037399215560497316196545379184136237601366609873437405615646163 45
9852384782852331973079137019825090585326929428640128896615562366533668086796762690219 33
8587009470620408502701789450516817868277031934278430701645193131391148579096169684416 06
6209283732083338786764148839135298925848184530866997588412889658670242875568773123590 03
4961649957608292377522689365557076354134082655772488902435754853975257909113420179830 26
1153474517489394228238827710449742344359228203662147297399136740367101215970943082487 53
4476980106697699031419407850208010006384516220354274895328569552580166987140127909455 46
5844685317297663885922327228023922957255162170439537798680918870851195550148345006535 42
0589588172819071594632777061363476090473165184177320017762749668619298300484784222251 66
2526812410603171436519456728348892810958904469510765410361898853483266943402184793134 76
3806133555152023602176365618271131545325315248318501600255035300235099811874568401397 84
1324504129248995106356188398860593998518606626698374306821560893536408037221056922170 62
1065402903346895715239006679969843981971994494884736379926562713791440855451262773768 03
3692487909647451106309430481047440825975290276493019099618286720668008381247708280425 34
8545154944826733517709915865139720744453559629062029789651482279964382284624100494925 38
0966317158494746496877324294171486011775792546480922293925634847344849734476876789725 51
8676844578041930104358838478744984719157546612527742106519834036887682177098564798974 96
6417963758532760889483399378980386935905003885915004182247692621391632221151170732974 07
5729995059216141953417954539564825806957558191410105474085836697638897485443567038088 77
7622534237252366858625286860701112073766444177034759239022054029211833635920768287468 19
1635734436212258468551849117378281498933173294328687866734127709419506140678430955963 46
6118300937723559315500840818820429901112536254951586587987779332016060230253996395820 88
8578524640683893060314881551188185106339280138068829477553386857687006287381871755096 20
2301678908295772993703948123553225117730651413774797053438937964519477623368044456610 37
7282423774640653174719122855087525701248555303495842547755111923104174126089604176453 04
7384486184959876882445549497422370796274252261378595615081527071735522514060924714662 87
7076575923280000636236617818518201466129968973204506701938791223390280196792848576421 51
7536050324280495374126017027574030305342271641863949578600006459952392766917288951034 78
3250731781367442373727643857425217753618604996157784516405621251200757112686925439127 05
4804174130629085265480196487981111437341575547549991708223661965071527972205089698520 51
3649053554727844829720710781457451456846086111572956596317579388647484246331637656494 48
1161619216046287370290404097536728861306625138090935098491430915201368594034990362979 31
3644033125901186964880120878666141208928978105070175592631593289475933353555853961535 73
7486473277884692900514837562696456927138301087192984228256114413262879693085543514821 95
9294806708059222682459066959870498056520226321886000458133848213383801071401193705031 99
9865589124946066131408979946504502916697723288306019518497855653243552234794761767925 76
5445820526605167994476072270550260402538069305498861065092174655658888730178883841289 59
9959659159322793045738084143789823577978122166391540010741213360957168963667005748458 95
2054792616196173666179668924730031816330776842912398177400293806904648200505930567210 97
0470723585762765597152866857405809911535605869057244722420834985942707651457804339879 34
1679576813796362083390395083293449558059536376048546223113625167923643813542477484194 80
4358933214598819421360579294167412261599983610855016724402649352902692947624134258256 38
7230319743507861606461769510121854106322030817166112414886764403388729757658443952202 21
9610073743454148508916871337442678359527056131521252620738693321830593565893062103049 29
2095355381454951160121464198439793903718964398693487841589092209093907042757821905935 70
9430767237024389005345312096700350896192224298816087686432982939714882496041244651328 08
2112489224188331444749268803634239182966671662822415678183967524379665967458116991492 81
3690145022797801359769313865394554720684577777174591385361784792669537103695089637722 790
6128123654791570888690766768908193749340610368410673860054110026278700874705947106586 45
1439196980559706950556505015123393666589057173313364476421307065675731991903938811893 82
5307521088285951614750680946883924574001113547027478698216862127432149766518830031960 33
3456223553442141736811700595766349367622329069184880323734519243491246566532971823841 73
9691986587133313412071051707368341724123447237186724941510842704961554950379550152873 82
2486053846067672039287586862626169843822805601587578668309251223040159989753848922836 00
9598075215949025807471715773537480669005001534985359036976722971043715921098469391204 90
5426246339608550824918232865843934026293826856371525962389474471783369996180201154788 2
4991332365221595665340485701123258277081886215013071577934600274395168927524355182396 24

0839150123473924668651002227673515415339817443806293681884318702195394674578806838733304
5026699348204740930850952940088706951818632548354824966260706650249956468194106470407 12
2731100421544185549123163407340196180907498723670338997499343924399165880651283667905 64
1695736521605028224488521757361133174294735635774830847833098492957430573060450412840 29
7114897365523337025293332859341544883137400581072624246477515535611897734274294259684 01
9406813333814161074909164044307805277729770093827687366826928083628346772591003284128 12
8935787611886533037994391571099173130876841482981473167438924150708123330466386665268 85
1715228773495808650709021102713080114880705251086164181615255674817104786219400105612 80
0134647100048960081618136339861314375953783573446347009738022397066733317884712174216 62
3797833950508494584056480430059112018417300502241962981137643255508625993667297921212 39
6833626254330792019745755537985778603339936113973147973585880467482663190555031714751 07
4082657545538452600524391171639479854543854640477445274442320820115805894011041094538 33
0920475598313012780340587673381134517847042396208849358293399476294892749263076151959 47
5808635230500390635269755089737196321432027975808869585297731621983594425668946663498 65
5664182852784053071039460383705531953070578143320616548372087637305421510498196398231 62
2394615554016244819227488988991548181616101412911142233274248071051202784972384904402 64
2968076980344031601013659880856422960754069569557133508005356595188841456265319836545 99
8149354703533396578804947612732090048553113590869747145353689015718969841577548117779 41
0760419019145652728884040518827950849008264536013242915228888937975199955998596828893 60
8911271091030997306757062186597296734639843061928429550429210546691609154182803517832 36
8415521818655679634071112430643855154156248975176397718812620984087360982992383637303 02
7505941877062626357378481368868368293339698892774446869696629499689660276916003698605 96
8904718417160001971841236265199793012787419139272279869802272459522940152432208404817 53
9812400825763713606703033447765411404274369962054125145048270021051517411890882549260 97
7472146205536677900755200879989232346535925261273772769545671215444999501484526854325 97
4087574445023559503175888094678670624279504968122317381953193469955615100420921536083 26
0321530691857272259929860039539254734318064647706444154409630859833209894281173897950 68
2749552487066282732805364792474107235461194594426537978749869795964322144950143520616 63
2736083387789574626900889609416213734426382002676575339103418924761056359888170083539 64
4955852192240769813052521731950132374196453428976165813540733130263030285478022485192 55
2078323237454832679632890712562747524612292929641409424850290594741557599275124250850 69
9663473537642940738391820741909162541609728445916256144727217059082938576767130454384 33
5990826160862846719632875160286080403534720853621937494941366551991508536892373512748 50
7429481445196541693822917931530918037820472872238945587726485845658358368883541761056 47
2125564386401518568769577988275174222434711115526234369721170763345046653272476633423 78
2119587911761578339722874708451279886599245183279164144598280214437270189462668035792 33
3375174806410499318521884550682118360856897563251181387504609424195574432639719843107 89
2775247235667228839552735523606622598788598539632877284454683694923676064844122735368 29
2191324554704074421666820674612656707964301200553515990474170854616373362027038659522 73
8064231262187406268909039981671086150219582650064497781735731850521577151608476313142 97
1006824824913916249289255612638476738621823334522830285701381554374765057846569058839 48
2531341998139629573625019198571058084667923649110592908805506833821771837094406269993 82
6919706824470532005794540835186100366116156094508242398650419987316238383574666454521 88
7359026121934531604858339212666377250620851905462362373373281142581664384249889798596 67
9541759200369130045370371962753606871830137648121274105614824487959863510850054873490 29
8447752975753493966553071505515441039093865374166652159183354997382566893264526862082 35
9621402396340971224268268700253395582693224380599832426908453047814199149893110710157 16
9474955576352194587457686296301838990582420127878675579693472318503074461060348391459 87
9497794871139135462902548628222497438944743868266164360683181248859872326879107661623 05
3800860292158338144328453224074623554187998881747238128535968382895006202252464635731 61
0826636036431485635531605049915414413088721541443317452537321065139611973306948781391 30
9687331861366696193940257200499308099995348219427655878113235218778124571834239764680 69
7306979340608138301890778537922762154846640882312642981642123941745697135666153982555 54
3529498172239014522235919257419431464218266477688866772502171232941522897844616291056 62
8542007145388341678148913456674450692912906319135618469153315855813507648754132185945 57
4577015648665261674662085801070023556816875810595967699890198547270641753128848749413 90
7086642080806276944501319670197033551810041719242513644274485219781537035777096179780 36
5547150100641818050275407358363121007584869042036045306374303438876152382898355737285 26
0279010965811358399008202510641500684886037785145170399362693832261888230189959127100 08
7907541662874832239445228502391848679138156920568635432170416335863703532435303860313 82
4517889442520080485301480423264006520996296009664177693761308208068702013088347209999 16
6455805747029726506424859031007184698953106901743228378475715367509849704348348205993 12
2322475198548535455451980842281450746416932517417116603293266776227819083489722751003 08
0897525205030246649351264612784890874118303858779985696639063450520022123934526686579 92
0442386148475724289010511432888381458370014867822833301640727611403694136211573716998 56
0584173544560803368889069252435338105397159315045259209401288682676195785113132383976 36
1566212857648264097235668922065045948831798586410105643818698894289927631481613114119 78
3948514389640201616144702822217564827342006461530161701567304518618437705212706257342 25
8715062895985488617620522961688656521583778424714947986648770706732481799084249744141 30
3077278122172757039389853107662656948276197633287446596045593501802123231361682946910 51
6536799613149656188142307979002819702075307204137744966745306251107845657924638003827 38
5686022458954061439361499404848428696980648326210846438074333655839868808998847151429 57
0101451205496684289667486610186687651297313962832144102683416586035991143893180762564 43
4460927475028253732472243963582314418661898353373236916808695029220481436237204812347 23
9217482268429106661178494353167259238504105377931104908157295800821890410996537405229 40
4620198482059471956546843007682203732839906146792078803130621080742887826745652227951 03
9753018549245010806671435652203409852097171275158839048613112946667339640992434926471 62
3122346056816004405457214628656625616892869974683821633904331762863708095828508190969 74

176212074055085110018153302081718136717685434872720891726067431810386875499096428742804
106528574447831399487497892443435373201883750977953460440052811978526927544248963257416
298794288255060763951651083871176646737663875269750883793989032883574021122956354320874
374141624949105157682271174177119322329453919740921365908600004762024328188811611636449
287155599082998145435840371709565305275679006103858220050257742242925697737322966093854
673369294496815432605685823408507390915045923150813226767810636325059586274551489228285
654069452221056355802862241676405576952603221833562396373988080142461187550546095550941
227200200166732907897800070963848283124462955965450221207433264365718697134514468968889
229510802001625680799512184131609833097021782768602448714433613144592320601473497481486
913207742459736433949200730885748206722411847926824053304598692754589707213154584154047
723222208392080363241272176311628918816433940368693171355826800063517320974505243783052
6334298205158572929372906514408225209314003833442860094371777696508292941118971208737078
409355275475946676077007395829963258388610673745079261804330672514053521161337678960365
385798503221845083723457258939744824492032873386027992042012799506284586176929349347486
450464919686353383914495693293534991399224536641881006310307109505687734454230964589543
614186308762414944474198666983477134053009577978720792914623896263908846392996906393101
638336383213195393793042804984875962794959779336483584800378416101863317414981336434273
879803025779702183088650364181062215710292820215871814456266435519128923907464494910427
359692766523114695665844581811447637737523501285931265925760750556501984335677543209175
069011358786716195493655202913909437717332015580807227333191001779316519681542280883705
765075988375970217541177380871379853389628968302629816280553011075577589785348238548129
828105014962885887182322180488826018274616448790291728418400029367469706652600862828438
830881335633580243510577635285933515444380101024499421747818965531884708737815522364407
145428295483213118953754081821607486597977762279102308233648436172336602917193399261678
828689800578911911569612861137304042229962444456452288117974715366357914615049438121796
999117195533929154679858885690828191103147183811299678656514530596798131908004221827090
056930448074830083456425150236093637525563831935174270343629268402347764138990390433623
518513502639300959664448987760417437860381723420607907143711944475395228435516778885709
093405909842628007555002487258306505575112593538963151718260658460012893268297421617913
081168782127210392175796983370070556004792659974323649525447559862348447404096171475662
882753250917375915460325080960114432704119873159696800796893604075977501803740941714699
818850466289826305034062817302823197991255309183628077992511330655577725895805497219375
476854503466072184115570530967474247232869920729495417686489051896687507736218429791511
575851590669815433469973569518765532300615728199574087984332687865483877037048388679696
344710448810366625529586283434384804626781221476425166019764227020362945087987909230689
977626220233634117711820153990501730714939961155484037152282685779983851093840045282636
757612605639843913529408739455742776484992414146858242913857917756232954020168247866760
021021513826664110130042178006234432179195236861776961388050065925090702529001557499487
526013723194269628376917390123280026029927806831063027349010110031406707790045792769827
931368494516308950918290483029650019089283364988372143961807932087423287680686490992044
495073598605289572116995190404083389840835633066438541288900714754386656707085018469537
900325716436271571357347225105929586511098406845689715658254557256333555945765777476180
201568378523640314978538098315662531858094257851058139046154652480727198329095772958055
300883576439354657145791359276746643254483845455827309139861671777995943499455031747714
365979668345645423457823495302203379392308962273442866717466798655682537325145377430092
237212027531693856659776107823598018863075075604083677127418714589669834570275384870360
304258428021277883107135113277277749920464830373404864279022583676007739008365615912219
662842690363666029854323536317994514703396394961386871671776305654609188565518443445168
903346245846829124143175358657498912475955608960402921553939774464288772795162230612464
779308826542357480107178091527740165487104682655357030989813108310430948667539415116316
766582814037008668655476438339770427571250447492214229690806227876085444048588139957127
475613019026375150738783118851441005371031418316347433675117492320531768258391503423545
058602263058687027792432296916175459534404574473997117121192077899158121806247094415506
947273514012345502046290677833781762141918901469088070798181006734859396797266934876965
238322019108208731369657998137760621433456083913079919949161884261517662011516330133652
402686506039217254034345528675180825801046290582083646811073518337297519614561225253713
567711887842519940235439622424775173733453533891507262882321321961008440631515479985995
215888847115997837510010625972439544429056478372428271127688613087297176674503877844706
050498179263263011217932857796302134172905845069384231278574470982819903689333323225245
658784864990971750320221672852149988277546583065590159584351618555205489782673613579627
957493305220111916108420802369961389004399328350663451816407683899879224132306415886725
004798621914872760913836555345671757845474556760412913322446095514831253536112871452107
430436355478982136977200927930137261704890828039390999835574082898005613032632804878407
025880190853877303188187127307461366420509108749612419017691344997077457528363397832315
611035648212470903595368161039384137568833771578279382941354623916270576056022717798353
211081825134419902677286104968703506073372089355308640081426599165087223597269462486048
366658911303958507510474494726993926452646055226651892497645308072131093520962315760848
831138260364791668246800841422358877741046115569322707887932687436128379575768538184074
677820984677627367244691118916546554304847971042152397792892067638966358963835664180589
442147453962262317532655797252681466311769780129937110982453405229270933003996295137144
383195135057317880566639610421225778725280132528384833082997288182507498851831471768420
564258524847512955453433867631293546440777340500072199683801405226959470919911895518840
537007723165819521053047311491449137585879146562388719464798452652848026896471317271515
010401260230712122287922468881136328743346672378205581117942027216479225653704224487763
296970129276913150042520225981080099979885590235689043290231478538738724020947448385394
177836794323325613093817720501734309796245324951033905554193731155678109600573604690408
793506212987488240528497439396469465846777436237028014143003044016617901165797602697905
113693090919413004043992505664956242468380203405089480290002637622005786405521562274015

pi to two million places

5388028003468955917543127621861314760525568998950949893238347287908253998423463975791200
0470117076564797800856626474611921493201815268567633485843782147331176775382064248945800
9658200413472695716237038372077935659762005227802251822529632003253692765319844596583810
3755079283425278582591745643500626084344163112970195537547485185666661090821301768897900
1760811867452945298815710766128602140319019653862395629151313915405878496630720194218580
0207054805464477547695567872089601597179073617241406114197727650252901385992813531379700
9256036120684983888215192251813217998093335978213175107048380602194223577108994591747050
1824708603819575788002816155616116658533162847263728589024423372208724629332702835317490
2841004434726304179900869416928737222783808028826637801392053292867729280423656591256450
1489454928932008400609684183178390593490237620500022433912426191828329180108477069574150
5867364119280128349645820683311575462623391664181390619885224573436830677398361010464230
1646413313553235227375266445338203000409551032192889761040287882571654340181783887410155
4129489492109750777460955977152478691288112448292842571974688204885654909561576851286100
5935267802656143938335859217886735660059653009853654620642543463791062988857389831835430
6967921926477578037976099151997763156973019113881696777038313617431644537854570073193600
5219700151735410076686960130543727588160704998778936501742224661743819326919286242930100
6198425416346048533061312244441005381190421705295021020394928499328022918752088000318240
0833795363864223783182669354546763785703606403992213306997783815919525955558887819155270
3311550013170879490166772728426303449497471356587672814394265492052630061908000591094490
5621854521757908484182984669384194955881207470010603768356344103320545001615165724072010
2529860495889470797218342597016438475984865828098086905490367726146202153986362941637710
7930476272183136596809685002918031141079704311381186418491415869758498660224546492765130
1028727586293867040021300059517816358644077874811780652256850653071350734245894460268340
6208345803133921453542233875444863038607087278502777777508677629920222402710743245913400
0402892872090265865447356210842527192228663131418011220286876581195927271283513242150880
1645019120899669021973736772018155467098939927354219911627316526245069499200284852366340
7161921013531794460181932739607212488739815750393461817020822302795639335431676555278850
0293516327345947265795104011499734482377420658757303463602505161561392726898206604832100
8554232208407202400309137515825417460421057455961533919845890027397157437538022426415560
7778759307934804245360144550080463441240544089726593996633007997972592129223109419470780
3842041830759662553943244560963589541422566136737969285423713764749526337556977162602190
9226734913942723609178467619658720575970336763996591575636399342505878894376683014289160
4175738503388810078600218625038800881754282047678427259567196048406057121007965616492690
1699693920855774499424977562114141333243237951509364096641340808447528614157971242508500
0592580508106462051245722168848201979344115322991552482048598975130100755988479695867640
9524880279423231241980738873590666649649151113962924488710570456434199781743710931692166
4762069094056656347912220876673531620143957944457621663868466016550053394793158087747100
7647644547839610999272348900285145797974434466044051117604558617700843387605314800885360
7357021156503610454879928403753217591482018829939970865269064553815917395418408308262960
9421422960361926568803854599914531553195305193918345501858845493495578823746509856082120
9842400795180417637267312953711599532987137948096818664554688514436258350366624408525580
5930624414700201882122046421925917234593398961925790163177529238655519765624923716284650
6853893447270690882441102813258419107757550548159667543119569943978415163324068894070430
9266081121561861902530322071390646769773268368150661123769280024896835575713723112884250
8276892878231095678118996696977409489348724146614973972794941266107636940295823628303480
2422371763319971843239990398128493939858682724464605407169925940845181835718891094351260
4176846914779092252837837532395601947453490905510164233807811599663448262071415872381350
4203240493175757533707935096946134828527465137714683420562623632197172961999170866638300
4381504998872606717267657248389966739493478186405997311990052966003532994139326485263280
0930082210836770501971287076699002478130085130527296684406061016437823071913630704942200
4938341960095350999398713515879252197394280615135656431340816156888490174782485783660680
0040392675996290863093790311150672767194187688246283075188795068973905489798893999883620
6631762238054216550097343843607589214242046806810224873073877809478772352739016455743060
7898417558609780598011596558666081808783531930027125973791335895789733400876344311886140
8697683788112151087712667757231018332511139198516665079680964485325566384175831166949380
8592161416674567555538237860442445571263339667721462244674158690129562054565276810473630
6826097898649630056827907376919863156012916142569306478100698919702022653867416260304960
3399712736676795742895556631401488118461544582007593167883566826708991945569858024090670
7654274445079856845354905475878108274277202197966003820209978659519699449293354316949160
4917055441294482092967609251479903225889004953954995722993018062179262112407110162209570
8552704888605381928423608529570140657674221348313326026342177054373095357010890780730220
1354982636285288782655869156856346625947313258519565913846892446244583440227704967066530
4993872354752090977064881709000110639125571874068247047136722570921712228672111917253220
0469145969221561229752437455506882056173695494066627332526041370446290865446151044256000
7632917948614510692258879443748157789125088591113417223303156749738192086397220651352010
5828008698246876537375323344666978377328101114526195995886665010821688159759495666029030
5275833214937286435874599676476525820860555244733912117129797672981403798909728650697640
9203220992792404020662929795608333348370973437110383136407411648476005069852015674472270
7835899960842557884272473530611705160202186456818014842377128587661659119901186881409510
6094024580682619905112961121275457411927534638229399475349910748595614107959401047161290
6588915916567992554444225077969298705769705847914304011825454535611172427337139999943260
7212771932319339130006890262678523004204477598136728474379012867170965176478255946009070
6012677038666595545097404684858912713323990972933798358653508097267094853161645453079080
7157352995507516924250275201254820633988339034478284894862532638437039404450543436312020
4049568194850386290285693371915547792962898845325057659768207643446294689814524433420550
8044994658582375053659570536224403818429935935441506813507092057744497999051667690538020
0962176636560683578158732037923102676052862595077654529057952721042041385246156860576560
2821874755389025205785080074069337294298773746305792569937750402685661586616804076154220

pi to two million places

1619388979463853633831125958244187970971954952240747959539519422976858251390076954836547899338635688061829056793407651975802467421791025033221696437760044128207799049032330812317611237997257214692833121847175212958011916642273935144376698375199845306457873664077738069965291870312319459500416451412022582697246981461488184628904987272836192381272041977916155713134479968757570339354019325461289360256544031228921782203409543443883693546040895445802479018084762970341979623021602945161124769165018600598116417274382667307289529784092440165695776572555572957613024253354790950676198191970142415166593574460395967361853255106100792443466532944181807012147646030124862963997533347247459834339008685160467240225216136263343752004066323425499308421418588260982094361574324084564710825443101883090772625857025112104655164462007203915374647393215292063379797098517074773038060536283898151196954606442017113929566885311883280626433431701717020517026485281318412516343924730717133554950605158671361101058050469783588061150722926127576939521408030298283970431417265709132950829112534317694308075468655132922242764990417404748134107636819495739774643796865183044465668398318484194824417605986382138378235313730452285949831010176224943940539018649427132324272489432022720850526670401611065492980272688629950320140783550068178326612442947337934708704036933947444883640146301430766548886919995190654063116647678194049730100375198570541672527647023085919922739984234149334998390978640343907692242729337668926604934944169471569715547059904217421925882575342592999565986426768451410612384687880654457917717504027762823606330079437584074366948131901830157091267666207989331700172020539549068640945149774097741094373452109428596847172702340645067100714287458635586867823699633907994488074376088353368203121137324451571040418642925879449637775145772351175866088174938782675387764130377951906050406476302680647426906879414474482693519539380970178496731946452891438922238632801012183431411809807504797024389919357272570527060547047281474647446208222454074714169406520696951676001055917701364897132278420170033566432055511462217933132322217119327373477715778865837696990791171131728353740193571986318244470476154481845770252344124845509681281166649709060432274718336680981167836515700468018768170956889324184407362310602566274757479697637472090635484029491734727487308120195052709637583607462545479352954234171272681981339902641902576337361165697385508744130241104699836432221001267728682180949890762368689263384417518856262832129659941213751796989052092475199878946684512057837358450431489084456881613840158803969872920098540862969471322598632313171273219428914135494323359122937242309660154616499698557962763275554577984454403223040904934659346663308857369596023285750180268502120007127142076555657726407983020729194806512936021776113609391135939353081217736284318944173962135545360500246138787960676040805316930025093441825713421613614440119152289530653007779053972003396628941794659641777983559253422287774656400034393526477351783515933055719947159812125038017559547577745951520713908433610011537904901507890705670881844574114508830512299700209969611309712189372853308359428809904927007467960128096971829283941281989605213233610705214431760149950199589535897183369078283138634236857956796563419293049769815209974528878312145973684607567476636060039687917435421385034895326295820544748769241330366118328689112778317546029127912955676007418733225542908244759673043008045896634533781987536994863350537722445159010541656658870698989013753888024327585151411858454353652987491572718865356408268327225110147009098986271677224152877689862267703698396995978281655794088825371837645374022181884118744813382879471964454937570020723469625336015922182028081797260491282924207267129800242376002246485008798898857231588422146946147491027891554652123059424861027260471269813241493814990176735748960941182680504771730131645894682345409075051861661015410701795541759759828042293862715551077779967459122466163847477146011178964858677219186104417530731906039973098127182191067244244194814711527834303920653681221424655022693020656758812764754343855734742632815492770562660066546711848765207267335654731669662177069833645835826830207766337778647489587312552194785990526293246984363824685738723284277789839925737597904038285297787397684663809624175477591483096429915528145165803728244231957520927726582693230597524612450564355755914053401429767508244334746624398583437614879389964144402514113002468869371952157830448693443840734559086690946448138235353204219253479168899801439178623237601755214594654274459179747606361665240186206188847558648738058308937960029471475549016794959042815618428725158529398298363206691979956738947880019958798034828312599253297599105020735073626602542431051328629403495541763991094763294485429044481026891525958952145732012266399103321680006863504862007357034858961061790277660718167706817936631804538237919125956198438857609518028943215796521205657611093989915496884031377549393540856953449908009012703840172626180182046673795994682954369719423257922064747097589525107002037741719426389346496757154272504046938718835727756344487694459508498638877616835505884970028685726928054212917958581319142283803871372827552868306454333148688092151697237876013755279811045966988291777962410970709137037883930450645012459886785895788637091048216529811520583322780817799156082638433465541976031504888419198608484062649003995878169329427061962297453127136532694441514636256973427524160873440269797879002146362696262012037753388367754969157208241546392424135057574943231795379554632385241993089406051134109851878055884092735591167161670187118155350582366098569193922700064682288358994486575226714800463165675219420196618307034521639282318251127783428328954142472172600810147875803021229475153440371272332167086572434181080477957841442651899778860114020271232948237623382689209641111630110322175485080943353504340776283415930030005339134180446423526240038073951095137011158509534732482237438035089352893521658778097600325712128354195312255082041269876035418900771365110860471819410808843111074253639171389881359753193233465157986477504684575869893919631188241144154303570400082640126777953841106665096407904987522171722641564904349596267065632899045783760113685132624683460213484557564626077735218836021662216832732675246600907786809348433879198156466120824848678050644296958460210293045372463348760259962242700993707358710745996846332376105029378740807906736268831432224002634254183494649749433787642510786294049435597088950968013376240248190955811349132979136137997670206146980435014907067331506378762402066239070271832790480285148295606115474069599725619492966830210211930502506226826388089654062984556197933891284197588113582007257435616857677236659365775185363740768701008248672732582705594735875949915271835

pi to two million places

6599156303913785716729854404027517173882658738499177463098812776933411133364070227666786
1050716885045092417768179812507691095090914411583370303122408506723054581245437271259392
0137937129895049768896410465834850748194540118577023591724690129665114821500743422992828
1222799787375669920071296284554728348592496576820307884679506470219473098014949774430994
9779104146628500906670090421960298663311030110011132782301805889845558966393460875372851
9074360022842303896215171619564180656270009999560134896928557393262657251636400204079984
7141233603888674683408729576146976265971507573970514521674038808385420153198494093562083
4941808448216518439071864548579354119886526224032719177493852757822503822693597054005494
5456324866881669370213670549848865275526796375925429531152739433899621680155290688351370
3290248458258150712763600548356859654294106517041963038966990987130227585345904165117509
7486133027113604353170371718416809873141819585031564870772149478854628003926277518456650
4309126034142218423042101331402856915028903541775090965422814083329824732595632824701886
8131890697734122400982457204778102124425786794196217970824549471516573880791213055942557
9829365910185929521458891989736410574890894887746130638684259775642325268634634396779838
3912119284815590463257268442980690380184893738786626776014667556930214317529099268057699
9578973176289166009235193383288594182243444483733745857062267504976827361392921256590388
5590939178962936286792195698567220771145167687115170363840099018030189287795668529739177
0541609614530699375915962278339171127981012984300277489093587489453981173533815768461395
0841856380458329867284936286928780127547239866389241352820450578035388536484709732754492
5680094275596516029103457751245135941932152689509014544595241696113698064708110231533338
8276662298702841007248866052876037540430218578429642231635313350340680402784441475200826
8900378635478415653636895673116883950433545631780101661551036438448861498271395607957291
3905790086466129824415228134778644226453349015863717448862610877742837337586277208196830
6963831805881107893895728376778301186999700894336560624571898381501767465512480893494477
0008373453061948200224387252360495475711739466291361825244605762600948866902565813293113
5067443779323202503802054350185299498077810525998530448875507249312550651003885719082485
4783899326602860337393568736445574675696601191916014388077529876643945135794760714470988
8932776605307411631153013911049232821931105880973670474323440589088285640026797594353464
2742107841490615409296240833879108156578990949881571269023662004328021528958961577522206
0621457988284049182338365735125431269368037987268684755236920077772138117964924003615902
3457700349411534068335573824873913135391914758158527625413375899842036188795513807317346
2244771235853672790530083551436917470851610091475448224060922455757610398609663567882894
5500333604333720846556725164894623272786930989509455863098301143787825883212299067132918
1939765220257245584008140241309321334209506896941907787020261535758021821051232908135283
1950915725992555006719879687514630234571315295363312886696129887510461508593563383915070
2622404505679511688694605110946627672376072375052884555332350474774899801582880457530050
4609169021596452383038262921063032258517321567528089135848835485487124126532742471657522
0157935604338198464883776605527026552678470835601024356860883266117880977031360613779093
6091130872340772096181023901578532920354712719691056747947044146433121314395609675166261
1289444050366388193402864204850380728609879198298048731334990048455219406573465415553780
1466230578392118511432796651242618604478370665694246510969746211106625572671764171963200
0060679461544447383009587849363123252352851899857198091600068141919186168389015181248046
4347190128400070704117380054024841599367102849240755930396348724030137535199115751180444
1626152220088089789683333245220004404297312612251617655694246030207535958932402085289249
2435270739656513885729183164947634488104460321439087648326551672987621999736837094585493
9343383325884620594015729844642978627786282332906904492327659329244991240461338339690152
4758560101968276123720343055108979231752867773878201326684509880718026820392202804849215
7434242568046730697177121873456085715203567062170083893698275361917197371079407399906036
8395206127922854971386002256544673985627696756447769925630094957092701311919298249557757
6375085942369144906834763454604394437131867956302588382681792897579073867577438197352081
3126806852841944365315510971335681257482834697901789014318277526180883313639973602829599
0100005197243770152558648191061236176764114583236934795506473752503406099278972307195082
7784970011960313162475551008728852173812918560876774785619606389351240925078580628668215
5371236246489693259203489219130671347171099770337362895959783509532523598271289649198711
0448156130108291858127392922110179269486428196832862469292108464389599692881655076958228
9873372386157832942848807562193230957406308380891462836413811692922908323367569040678765
3576163961754266645938024360197471995403483650856887128855292429351867994655684451320866
3020748995316987888798390898954565467592320605819226177883766122253354733863709677295930
2399118055662280926495004498775277334625217332638938949491465403821243218073615845361140
4602155525975366720223919881050284681073770437723293121597536227505157238764529436348387
9135289887879058215501111042335845183102276113343358047897141497715251577898482989361644
9182897931779032646722874941867545326918364087982826661910595314252866309267070173240595
0706227386670579287338827184951879760685322806148088186465932879942779746124529540952671
2496904410461310739091921123969406835184469699304716866174240505192229767529852966848673
4836811625766352554575224403184392246233255616522514104235923273201888336360890136050472
6337496205823706184846895299865267466363021958716136936786724835356109940373813996989004
4051538895903910328279457281511397587762747445318429305366136379516981727460520885464692
3204175911435001350395075348489615692495090020755605575438417029014607818186992492393911
9438241100257051821388230098035458587811640045530311277179266614743774837366237460202073
5384034823096876778936230561678938067317856631371635387047158740172890527249125208271689
3043739822965883776322658705827681347353299478638582774381654779360985613156779341800020
1475804245593773783036039563134913576340135030348407399275786595343912755160207530865236
5385274503788859712486671650039877419265374145011160495071001864930153582757306955302684
2287584834032104779305348522095490792573563977027573732045507991015818436798097731303516
4498978068477009241246589017179499863901523452423144640733146239704547452605043018742129
9438212570644915073546929709320638029037759118818739015791038279468492628606493557705792
7535699984474878454484558005063460310194327718272203125080977502995191976880246589639811
4153371350

pi to two million places

6843610011310996298251092340076886268173814285820414110607895701149506894308627965260288795353976116845807304352996562712209219685525228849169734490752782340903793436643175688240831929497835431473144939914844944463428965717965533285040094675939963594990733864266562187481072632478724304062904946792025384859153980266086302820683710681925863675616167488300314934301281052995998880618862997236896416520456806077951010091774657308145469192581327312967302559205871656588042033913153974415917287194875701426222147192789110660275076128944273224291366994665927065725104193631023859817549079869943892438890089393463412138281362194718079811145030002043390157920439312553995192226090889997125630923302714291250144039818705004202596087808708135868664901772444736952194467034906502238409693051825939968039421038583216016400769477891924212148954359374400999379643728561303628943116743708946744737971676182086415931683561842400484542707621856066439597880214216431665190096072817460641376846555288918618196034882585207484555661908969318905511599162690872569882176388463745955064737773914158066740766500977834149348421618440383883109376015393889363106315487134597252904837036883415016860751585799025411531953560707703814971227446960968940953052600980817492700921933267421388744897485958260981627811239347933827705619740666378971366211011150620641783273090943864210443410015684879462446967599935247049308499270573111248239759233598986452827704154827762908516503796081505783509823027587421577959779004592060888004354874347352982814703676657845392604783416704594176917489468082537728362160668568691135194523833890891891110912798514587414076502852590672091111527992591421284355439757589985220717590994624144692846326862634021582660249829211586948010765766605491619087786452758667194236834434314341938123350021289777131425000529224971673107771001716851188822998892084729467521062242455347160799120832204726268215968866450816735096164393399244775114623696703082302094264625367491347034074245876652408826191033625198046411371161227393497964475670145458128842701067719626329369178482286279120561849828090900732388615464457885782858409709604774411265916188951776629166252716872171876251651363170897961764648430969865502038327499036901395661677245334777359428410507910768933523879355498187846781230812595819765132635425639402346170289036070816642417750144015178987133940719540680368401437775382302327416865127331446717927706754152593737093421319759704093148149199488987228678226746965193156630529145957136183974955247480194339279493656351149799084354265713102019714434595201177875945803750787462518049950522139898147845950331786257848999775100918367587992192282605503802850781001785441580731247007138124481319901891828791517297480631535418402724973198667605030469892140012207784918654167062889461437398736216971466574033954035465704207132586866333627928321215614881054507326446439898380024170684921991297480094285081669435649295783434412365261698482303194094863341021660698828462310983387014189280782073748320543074954164289632481395094558993370449182618664265415081900582707235884189828099801915103517706388164396642793470437736480269578536810367986653193903768776582660850368718103572836664336505771722786667287930908099722192218881367130934805010525726415738134064117816210480678714302389166811566310728760536945753778478065058502995269117032333037167196011235234440747526812892835586828684026325716056974500194444951701806618108190644105455246621997460551083766582885812981445152674903802269567811500703052376160976160751647435839334104077980557722934775991343313672159786610550734227837116776183199485930275857261397058658470986866795339609462918867805571711368786100785500743881857106412692050966991492894227497425451085029268848939544748228630674993942734032486837449145564735830378788605984756918370045423301286627585792709510678969053335973429361420032882467148804583531441386307976265974048930871107688512323078607934245122017154558018416921640760895771328756451994313802327207500512875563762033975330060539152081158801054294431785558933982463564583104692415828661178849010046341407609499821446541460392837510247350589440734919649540370272245345722012446368700313848339842309183240249015795186593378599282947015985795919685983868598194490544129809748419955368963699635471720618181358552592623299539916937591709242861366421855846766772062305444148858262537053208245437448229030278027065751038001710881783281865145986558325871955464585253419757769782177292361207133364124028163497001437315008855271291758919608289738898020217608210487754500133361319171319762085641491481379640663040654035429084875144609304542879788815724796980056566079959438836482028316033894667904451136553073142332760578710820792676043953535466422536781850268292288474386393039583740097385870142651086344620470517584760549904572725750315085058585681974684292904170671661413318850438469807973560177710481176836641077939679436202684819117731482408997692555128826314527801264381391375696394891108464799849976779062995454495157317870080407124802833371117026427230154423913602066185340630711308281520749753719314684009710622503719592163865824282289408364927262824559605155252350591207841467076056758736733618941117427616791555301719943675684792853267322657804559329786442566694443389254568991712253772984042896325472825553367989035077223103164222255936866224029328970790077496213095713496292896492938787597564476663310010012085380491365778048694771004653842197087555332495654302279840491862319679062531769931473584528027778205886769223247796532393559859917992633825686079312971530659553948783277738880713698149746405066251751431410646697548078882506210803693030149523602476687170177830459449398213546668120189155589510932377910663974321717152861012900733351801133111413566846800791039962245315059620716207040285515702614335810300780277816502377517200967856240169078815127492674101606302319578673330966741953326317915486900877212517307235789809252253022563226541902339914103803099343538458027068033444271651927725823425369059249275644499599318933730202415613392172868821116886256585270998729060165657108807389487583616952170620956326824609039961837889787201822687023785356834418100121493461723067675286359901128088910089647377924597956882500739850238077509591298155530669248953573727641323856684678177814057638772348760403251294676471359436659413455310631406958562634633689731065163807435534312610069096745376501440519811834923531887194079939990955389289577987504772778032708466093615488559657996970259562328028461747441648263204872412829894947810298822837746185618582323346671858688349581842191943883222041292566218213616277944900410223801402421658236604363913231229544346569498272220679082328807024151373524162312274775730325612470426854433853224701279777859978351819954302148759477816076188527083820208483499747127552820257969469966553810

939593798289613778443007953018540036966166115762868120795797161535606620273576267247120
993482629114587258566100302026165460068481438714608372797925961459932392110170397337698
734954390474516654195473774411618809167007285249734723534800924594736914410232283436343
841037207710013462057946640670753098240855503576886997007848143675510401545336522149188
838320213279173749522508591040941264626158726646161917696317331416633744493848851485951
358679018700795817364758504070965634445300631086513401545686454609857793768228464230331
017327992712144695553139665054817276401972657906219538813253509632509474459898910878590
169692258198604873251616314872253320868115250387783903092109639731408701463262774873619
992516036903516401813228638410157789138345482086953949114165419212244103725882353735283
296622540405395995512541880469553470169279658180842216911467794957709031814719952242004
890588559374644161534909346821095273819069249340125854016212983888236067112990472785382
510932212676008635678872991110606474438540631307026915793291147168357489308607341762034
842422557974325100100386436881605524668277328014816697873496332099634172377923788303516
605487167179287400111914472624567124700249762245822402739770702702350323771241691314913
084480272640099449725957209235930230699273000522490736514197778118272030525906050536647
931018487082839762435976541024547190212964624407218786854291307199115602213435981092612
780992449298835321421151688044305242231335172036687591092061218171572150132304915038524
312760160267807102779059878363028490995133233325655424283233126970826048284109116462935
530797013471599928564268898897007674423346489650045114824894488438119052031620239561251
550801142934559384022589044523549160677057526417751973609044020448336788189156897027789
366044998316755033346099743453906679681237983313639844586220491826985980185062808365058
175856328286723970216470379261157543687834566404659060788858593615726073376479473628394
189492835760504833644940113498654912082100481325506837562819702568696995491876219763972
045027900446675639511937613156006454486485525074979942085002895444499533574504683662276
687208248316413599480307060161182230915617525952849002899529342873761735102674241881593
715994890969792225772144013909127224617888387436080519675153031479114143357320736585904
930427733798447129195444645430440095975183097418233761863378115291280151786636009016476
974544958957323146795549938933751394956436260145495926467373472160218853132654468600885
372077221345275103105952625307111023553788569164995911692208388877078051735438398567867
015096378088686475757669825323540442454284008826874777703285382619976254258199293420591
217977808277870511858452308972985638768651127507276341003599414660122279489574955920360
994603780484838552595991081712362827442040178498021103176278778875030363022636160976667
580106039555479978745699815797334233997424447588453139334536645917552581347550463442671
610948908179968959226704644021691768051005907184473526312354164424864777438787765173853
478975014025204069329911353255614813604353296831292899129535615290402759131767734127770
463653085221325754863079335478829963834069993715154224943808824064733263123350461388215
947951691759254509848097911089233113778953966407460834573009765117060752414362883466091
800384906356926853296400551653599785791306066404745571908486625041462763357204425087660
332076412002768371472025839577572548308176352281706577594153270832662553910968973058504
562259368984989756227021582652652806200251844164898191969095582120789897267176461338008
395664877220193204636718817239547049302092798661041184695704868470049638641259530673766
660389421761893875423752240187458159722844252290797737229255101808867399083988549214491
386356253886378916159188842405129981936517136925916957791998849494149771151943157558263
070599485758635347496397568559703862678054007200897450252705193979698125296883119164590
452097563052833730948602383296272132256900737675339111682947198127705742624375221375825
032008736375320464500057389179346592355770836220414528501390794640672466736018275379854
478140769212560856944810541661956752647507452390254517053109406626366824745757346070406
527575357774320102391334113813577503322561143900976099469521398177028414610884136085659
183932993931358081952707069270927607721717779098763854634460240516904994977475774828730
093397894064147369457198504829905134842867070991509330445796639195355891478010944345098
270173677240979049480592883846575269082140684093675224888424661256052695097801012605772
822873599379849977045731761175869419846747429048640123199674462226863633260076411070293
488972360126214984596841603187424531584352054891859045356941964460633379884931531169547
583611157497667740703154480578978170504573192258815494311439025793450499895500372704236
182645680415899700136977093646184318296660690731354763645120346808804485446394798810546
809849670131795494286681388276584446505851797712368142674585475571382290726634603164381
375014654291945359934360820628279072531886575179734245746612250274430190922429627769366
531587168094442427862498407494665563526804502768435782113627340694356164251537923224666
458237107932145904446111092210703976045651010128697605935565797997230393938683961799189
869179915938694708632424160101610309887378543956773148297234896594767142721341284004376
210660200550566202333951957558645128302417714282371577693287679784522757104237281724684
454616224472703666186405467249250144671204565478357269214470538798054274436506654919003
697852300703720061418372571130911168102172286721258959485746004353295033114695796564900
624462269539125112528680378263752083462010919392052994067353165017583782632764100489944
648907195082147841020836669964415554899731185444595312327881988435193646690756191744363
874898626362765030272068609251880972360884793801625295039225521028318591195209521603087
970922306318243049051361251785266783098964840762618711211938564523325880156358456632568
153659741400635166538527833027993327618187316429284343663516156632552808774054766967000
188148312992649497560617449945560484056516920660629441907471164740575561955183453874064
046466344652333203510377844767588566666090172875209822447115645045567243110919573466626
779195035111191248491406455543525772549656639267852219176238547538197108319832284839469
222949537374931725775542580964794985891026863594535761355146603751391496845122967455478
424717793060749999248715039173711331059841824004577425759178667121950517428961196738988
890457931260778363161297825694113684507888434449786640449589578177977537633175038686569
214328456590763877035085454922881774592134644790364096719378848001565990852627617509773
940164374064232153788341300502620171319759124864587923006850025338135489151319984710933
979884365446048272891549710760373174451260971717062154915230935580784562839217995093664
990440960915699421493871124291466312054691322386063352687404641869764975134600318428982

5747512025844598310398201406094846870769161838308623878910614682419728320052615038511O6
8199058351769563236569694142362610152155381725036939198820292713685444010629436726451 20
3334424480829528507786502235886684798729233633452675826546136476448809927763316550213OO
7449452989543396175266117490169121300581095542449122673852851223438369598397804839752 46
4510120588267428306399960890765573443634480149048829835718981640964510119289853987608 82
2969226435420824456841236875233258977838524384542275920567609260795846727938663976401 33
3752981741038008397136698750179251888900509616272530290638512244815629473486671807921 67
5743950233541879714637034135950706887960310150164644742239232867292371725653842901470 65
9068867294069484254271590520534093390143210813138545825970434858093148550633941870543 75
4820191702268823175109013294136171877005809113345796164220301259802204310738296686490 70
3750445386844284408790945807138389620954492075996155366557130684404695271299487830971 34
5078827572484489989734970396224651104987770041793712631986655042482645042713229161230 93
2461641353303916768945148358569270216603190656899930352729669425409329858504714285408 38
6710889273443108736345700269111924324963772982294009440207520246620664473839172044257 54
8340145394080354627340999096406907207562290737468413119858657879885591425214708035178 49
6735893693353500754467232461312522582687046375634396463905947907039289612597648667056 907
1815846029936042856641652378271137607745621090317980865717319993431281796912942962045 56
9520124331544608359576565078324467211258500776099689071429906214637225018370319985169 22
0050813910205378983915622992500646873567979540662782950224745926656176868862932116256 56
0500845429455903291737200982085881735374687831820644702268612764976090634339415664902 64
2621944959997262787988742068653774840012402712025274733443616681302272047545870270464 09
8643467573641494455604018046902564908532516727167095127900523934217027430328861413233 09
6169647556020183862159978723609621227570421099687370007122579942951869630041295418877 96
2872409327846179804864790769989057773388605583824911422831637486928785381481431462242 83
9849623378557735086051101050926129323099492474189267513161881862075325740386848069649 04
6982799912216113866566382032601913596840251564924790644398333003180789523696165184056 14
1033843479937617818210047435190907065480640320569117166432209921547124046169424710431 52
2220510488536638270472083522207228179230772437862146298606683003944318403189711959387 58
8071981150483977508624928452053766099357035113846959784295954406485751505062084971228 12
3360639541480452318303759039230419312935804702532239639115127937593601514087053514606 29
8995194074575535488686198226932469553821021947202124884904426365539530539435330040506 13
3599683201666987947720795258669143318990921186522605849608814962683654672349840481438 10
9431985740368377466370936580637339294587946644516826114833345980894813230142344973630 00
8177003029015416740179574436162018422076053577654231628456427644093755689715537872869 59
8722457407255862889162754227400139392489093720555456160784241539561883999055824797402 97
0741478814357270791492210435577454609349005737681596828196801977159064579660575494254 18
1445329924798199665711964551885655656812151334909804658037684761826601373717568372741 30
6170464438082924391653713296382616530723106068233388999951025530732744120942699413792 01
1010466320874971541058744582611599590687724094286978759462694298805477860776458005229 40
7570308406385160436059312555658553312311655463865541030072699344773118169078662801195 53
2248099561712788155639351901256077112884694732263635778302596564232697384744696747938 17
1918349844120102842351921959479411888645207796023134306321717435406924886237588135079 50
1301954212568538960669312542793294445046041439975110593068307589866631208970917867381 44
3106267328572913524733573733413996498162256056419741787984415602133012302960042148911 47
8485752643862518142961913689838222749415586395787408237809038341408792976445118888023 37
2009886401223617417643050637038107692783411890172940132946238563930853153224422276881 85
7172889388516865007590484415709736636539394113814143300708417983117234397432128982693 08
8281790845460835227616883923670670181937790638751868898267898592612248770725537077468 09
0917381219556815424678175824986359084577528221087337094710053447043162645326509875418 70
4826040850494328944967115111556404503759922183153150621894176807974004026780591637359 492
3058021891056162459013019130734230922011877489054176347523478720509575083012458009760 89
4490201422832151708178638297151787541023277937785407268363951803722727399072461328116 67
2620238536295804476874489455893556129416678771151444595215072801124528978389889243980 57
9261133951171633727247632203561341376655493609107607754288639415375091128355095389101 75
6992734810580205244749886065694140688059815714553496102116404129920711915738223964968 79
0061577179595751167559084819296904355934449028936964861900866118483640844547092298808 20
1869643501024165892806728934618921288399805886754723382266287467489827600721396917908 19
9418072364514382989430173553032251147891184679935694376925654535114088246264417878175 40
5852046275209119455171507153223229003178378639299726457950413622430738874742958842038 65
3796300887828497491015367140448727060522398327465443452935652807696047722150302406032 99
6082014428740664630902436955822017514819021276959194854806500251596627667211562658270 38
7072145811474469794369850676576882350350632790855260275245784521393118971219595602227 78
0105217056312396344461127002125867235737090090246740694925766696530479737142762657053 59
5744879087619929979754915470956745686626669331781602694992456378613345040128648268256 45
4678144257343138486006682679767270827808689640327625941678840571459411480462916488128 52
0010261056198452785431676734301941800287392809461519481854686631099079515044881336501 25
4612214135427912283913036947262589780027331680030312219807922272339302333969902493238 13
6301781213601131368310388677339001225252410376288392001146874739783019681516234484584 41
7084711540563067122150252400600956684502099654423577855478800658445306335458651974851 72
7110468692370821037423834451621446359164268326469767485119161178393419514216263434572 80
6939369671644034523111798133136747255810302330033564709193131184154932715418721453955 00
8962151995577648095322472817056846283525646275430576200001721499028964513843833117116 58
5841776451061858250821047245469890855260571543347364077660598021844622302303576492869 83
5354772286691167255092926423495359112977808670676400914176047251101842295428946972427 19
3139578251117713736466128861099385961544636685556640632424096551152712448740813863395 84
0329068301663750515937889447003244679444205006232380567807586140516179491567587848547 53
7980361332045886720131374684222777393383010159278738266403520793857372733187872211600 69
7886849105688767373546820717934420516918691207339060230403595314402484827552498849668 63

pi to two million places

9056151812913549133306891980509583195474524347688558773831041434453993397826437905600079
3257790326877238933597040892906379862359132688276912048235322685331113094927661454578891
1147071816782987954873977966493933404999580445111442637878055006181201942856580115116604
9659751470936084824478407151016568819151547630853708991777495003154677919940554302384555
5684329587122807033930365508052035902690222574326941773378330207785088164599774998435551
4703041899310927886362855595498666452151460832243270293986535471171461120978644646443325
6180023520004864169280901251955587967777071201415854032429272796513998214262542785238836
1824897440083764657199183395660336509497078523288974588149334773099486397383182019324446
8791919227598192870591900590754067357185234811868346434778505415782822449079858408700058
6120528447163358777233043808270913739688950568455535232618221110237312713688142058398838
5474704900053609386576177773048783078559721753881861727200525683687300867217691544126628
7754950323922211648722178615226209112425923774670550161255823546154420874458421496477760
2559572740291847164531588852588278842868558894965084270394298993544274755784884092534424
3675964612654381092025562998205154148019374700480357966774587410088413181187258238578854
4848881668277133032050914804990270633866946909935124879227100509277461414952914626105517
6485990692992381759902441618794970235657180951442585499576517929621056744600221413367737
5480798603596959468569833797726728528759250604441846435778240362377667029872832370604552
7442555068151809555888683589147533045701169217308190413682196197924446042649934384845527
3237461911371108244643350169480155025328418462873349226613456894413900247066335547078881
4160548567966865643266981303178362164646588208300503644548129228849644616410125023757768
2821894551184580595732090939462064867750938024379128960180082794047481742206332791471128
4209513701056194327176292868579090949321805556897906628628915473754867319869373846641195
8561889977082799001692464942519034737770398863014920111835283723279199650777155816340061
7619698407222318678270123540349943842916855650515174383773539203225611599297568796557748
5637139984984188981872386344774773555150135350791910818082698953601252478254704321409778
4693232943801472551258437608609981460219969863154847357629208099670223123077999976689926
1493709481313117612672502902025112501769538583175793324823747587160195989309689954294457
1523802778922356850487641824139102326914916557444484512460830514578741750738717674417735
5110130764553789788752222185205861330107591272012841581609901291829118566671573929980092
9179914916100046603332922577608675656663596534181854225605988431844272181094231810906631
0605967733278403950596066977210277736141182720643429853844246587728471851788363375281193
2542583005499614614837350414191861619862910670638791195176205055641054814853078236437720
1653999652095042389041167497643602902434195676224468514208945656803232777781828040235531
7172716161384745264342561703004038807168888842147579572813241630693977190486932101774484
9343952208897797746064132160659123542873066349856495138429915111937541568268804640344074
0031916487524228128990849604910887524818976629544394637536219830608530266297677622972282
3708841064030679839570455079864255624132956970690312606937272270497385849063548119466653
5336602643153554561276200224543650140928707010384502494229968246989481931106380584896638
3496684231546895857394178100478027431024343617063937322667141404764505532068525452777727
1337995989719129335109533521837623781448282123898331839226954036294419444779393386294478
2085653916027686547422160219745416250347948657330874869636214605079665721155856826477125
6401983236872218016273702740854612331531487465319393626134387278498409826789986196912220
2669754590120334748717152720342378485320199773941089399183239935343608383194483504077170
1605001971940795569291694412482684602905546024805566374925676902358565496305630549909947
3833868200225327275101476538201571896891764386176038466914712705020901016679314353889981
7992613894958205684315254271733398458496778421955422849693775382141987539836280415138853
2709515070677831761849268674924518788782565238807875593398741899239410646608428415951776
8002918996744604562347684174628436149052409961460913299078250776717299487379930049706609
5219029159187881565502948692499482638863968238495608685005799920812591698095762500632552
2742977572235324464631299191602952782838780863743794848781600798669184494495203389091118
1996840014360193709330547221904717861619952110929112791700197667202021639669184222413349
5352640541503434784937300948107500099230370915388201550820033012007604036740047502823381
2172353310546983710100965755488116614281741971422411114653964940881068713531679967489974
2793722635135810389988254516055367251021364169290032107312234300723995569142252464301126
2735666817822859016903399525558864708517542843979342327492533998925840392320798359932521
5743584252303743666138993863200080064574750880709379984627470966570809436929936107002273
4538156848014398180022649796165498392424721495385516649061338869947944876407062566016605
5171789113105157898124840674415404386343218080496035776369336965075024967546596535171150
0859975076400045595426370119626833504239694093247325407321746536577121897863354556824417
0391037818242656724415781843849453825620349781174947104658950823214082047820539992217708
3096379247191435705268927378829630172045984163967659793992468451202167315575940610855011
0840150149395848132431432648317063835229338983573286296250064539653232340901663455349976
1453977754354551018002272987816661057242312430623503991266927255939838704468224405690021
7527208905973140317194993937576065170443081784358468902322640906702558256315652710399919
8787449960056696531169420178903331930791287640450024529260777573554483085149912160462260
4079663570042929414152107851793951248929311310872340368754933321199716941558224225323345
2699165148427080749649824320910870913027192207360528239889033377648244024821643674489928
3893271787246301295213777584065676650342254844795273438929626352170692482957223372372260
5212148675590124375106886361686206848107532525519080870082393756679930005256400410568868
7321345774201100430212747964046267720796028868075453328446116396367029616763610612095564
0915903922675977256127708233691017979324027600947790504939059499035509762328552456920014
9233803895551145369453798964243907753154386610796172549357971644803446126662353804145555
7367642625144590571925802222930640330494317739911077459948051848434169030124710528400011
4530117015926417603100466879843400676366135754159381073949023384595997856649006331001192
5807617965927489021730881865451249156100845649219173398418493640078924253400528851274409
7826072818449936233443967778341643032861707465557447095887106612285979843832898882778860
8994982593444570625520846669336207364561329951753464990966070993431256359749029656718846
8035188787644371927343228496757534874380605868393873208710712341196033089306023521935502

3796475301514159372862118229525906701857585590486981033613106195370441077208602330006669
4355982298997200389420507124130963301247398988986501613446041636976412991855139856413334
8024401090382042098050981881637076560253542288520642504748958680899179466861171932503332
4823024105980558476638045521378932305723502009715574760259377207677608746814821345225331
6302087888239853556684084620188776333889382394005969382347555811966044053660170851554311
4092863357092159448116017532965834133347177302711059709051781158901708660299016051247944
5070243012323067026217970114151068200226813999750725832130356166794912610054201286453222
9800672689000948209707585410219884885295459601974730636132842969855385226523816130889666
5914509124886813125953536296057660319750429504118843939724705360578947986283171400396844
8076421190941427568120273245423319593111952395056290622611009964398948381646448745866833
0754857785328740819937572748521974137180942967777411722239364135603321190933440755678788
3811304519984514862898000608483869420621852719280187780424866808029951289703473294463177
0946003859512545338683557968905846517230067044888968406108630406121351552038742139284499
6202225754658582086698640604986554258859081455309948434938427338421786450513985427397422
9095857008561462561834952700228141732536765397946912752974701317006383541596544634244966
8352635059485344744721078056107810829649426478810025979318775639239043291785327634203755
2297565752743408295084547947015245260899313885783123911751269225566757288513340439769622
5403931174933713994495293568010603796944595685975249877267348079073267618245233552121622
1496802344929254288655145733756557659455570923953342814246290317278154039983415564198377
7180189821124760855955518999506207300714034520815503329814975070244267726436033873753977
3148431374070926654492295204231990073459463931199653505680733298148658411091994439462722
3284536771128484736224606331360285910596352371938716345986963644390685405322319315241355
4693248757673046338170302944798352260205181494445850496120326909233752716235513352623433
2072194300935881503393598974493352695787457278314039670396910017073414932531022063263011
6925237018012024422688492909819555117195612083815501448583657166510269086648717323819011
4860992469913154608200199270504730568876141893298108310235264828108102485450220875722122
8344134379484999727920258354341720442598469327409141782143949280179974565987369828742822
6826748442121371546823275112853416523703165307043258283372112371376096959399375495362233
2222197465961933252907404248760251381952426973910175637197534300447961782504311533150677
5825627353434762525391425152757047878437678852418719663462419927008025761083927497622633
6549001865320645499515580290839851327262732196728478302538522190479278689389538780368699
9318846603104363352437327154699898811186443670114028426202615047388235899747281493343255
7065474637022451887289061255030273791902640396577417606898976983456646470520471635921444
8307099584306471727507633718020714597445652514188505037163773818902969685284409258193177
4410055576089850292327101600898257223817394543698529654769490187384046465643730713195900
1140761317428220388331360297085261451234907307147624534050245423763666685575740120604055
9886955630114415443141696986074032207886003132790553917869674163555240250767653085632244
2737847149746037784064609346812991871902892759763070159840887781745192690147691030030233
4979458511000180886606216286801110915116104098327308808995433755118718317378655877648733
5885449036687907416918253831330636233820582015488544982788382174243758054923815972964066
1963105821516709190326318730933813850113091332770930342511221097215505610491387050426211
9805260247611607505971037943664771529353491718612506611636738330499564878779365697911299
3195878277774060107533372432790009732407256070388800871137309659634457096041175089444799
1191621745516970964844762046174128154536128422601551301589525862806957063924443547802399
0516861338549861926656762276035823498556919224890690116459866096156795541549215758128355
4092025393170807378146507883203526651345707065536856992811242656708448017786461497404544
9211843122762184522441088471507570190883495022437569885049454241057266246094583209596255
7287203099477654010739855806996619801953450050651450105492401886108913417306075923843944
6124656986160560909454177290736400939132195676836433329961999796542434802656940683698677
0617587413167650506027135882574433707216458819522613397054520523441280923173065051989955
4502858538352287378728698291727580782420982610606090150925211090898996438929786294301411
1140067628774992078172379474620916899838961894863077306027491026703883943352474243421233
6712177024610116024061118571687050824404916238943435047025081364915515510432725074794100
2473747819226520593355136255301812196424992488212992601774080103719884212547447216029699
5002774268507752175866917993801362584224173985607669916543275706123045286728070763189477
0589544061884213157138003398487408680941446184585423034495144852105399124893245548665933
0533558457827714423774338533920824127763216596753832665064306388069556915620262225946944
1429007998769834420914699989795683266709041413980686333251565256968787899225743279613966
4370265431446393337990008198575307978817613374360948392830223380327973203865364624249800
5511402246922647893159294491804038298364903488863869756441485560856053717722873714822822
3686427811365720394749690693723991561003682807514117847378256221277599591677581403853299
8688885136552323139384317422585356280701915440077630163476477812613243802484218656698552
7400764105119010954528983825634840704558080905221476972204287382202064451116558020215811
3722046634766333517561799050089778088560093208145417644390323524193973154721892092020244
7427040585791354379260097680763629875661478264433892612121345659445680104222761652418377
6401867345339148807900136360352843217478040168314672618806793649304686587364631148328266
9804920227876255454017850917749772829467569293545166609864194814583644478909603830458288
5369895608066566619036031004530472002422639388157206867469006192987308224849001681634899
2125543375444758038873615865885924859755874278385999295417099172251153402542222969223433
6617778192965331349674775722097843019381844505985075080564948318967922058343414789051022
5761749099758200123472441028507941231843760147142137829511021873589939170816868055988133
3546991379453783137711413549197124346581093112783326621759222758936993477049843525192100
7351757370200545734513058731322143846208760277518405348937132376629071120552694905300388
8822804204808210851928761076772814548927944032272362841247943192644192430796813703829488
2005869742015013235319872051992035387731421588221148597431823879098919932539341503787688
9959695804327251777689866125889395442013917130641886295106652689606241452084480340347411
1331488004313864731715844681575483653167051260174664076072809596721327294224536104266799
2809469773583638349822811476691695969557327702827912099792608638170176542101511158110922

6832874859506462623629925924969437818222916923432548536160415251420636925214774598094l9
8892681477644390537611191746302607417041449224194980017062386681694660798924751697185OO
0942874444662410058638035586306506360957270979734930709648890760630190792334617019745666
5624871288498673826785462689451209462290548275423302517321328278531651758693405411184557
1510483463283200810115252541311936795456126121610319742783210387954825391743140987706652
5700603767196838304786929100326748344206684066735150885860916629765722779969260038682733
3496418758332118080916861851713459115685089314940448199610725023378967798991887258268667
0537757435124650813798628134835066431324662709245236546983517407042651369882414330883555
2631945475425324845884093993784631867338133661031005811464551518773050141308156660180661
1322683529639711462401049831484646043952006163735425846982605212545385915053646201416659
3061524137743303724644103975985001568921027909378678603185176724046959915723405290038166
7607242477016978152148615194746270546142211256912722538159050203958305146858500360242000
5490945502682618509587542621097399132209502954895828917423298134315406925757058801573665
4535178927415271289413143182694769025888547865991968988359990438854227484694731187519774
8877121501915919088858132637699812159080896918233505907971487461333680397063500369555539
4225907345369332638969616238425410479462948379374638132461240825306902691465021517196555
2425678847127229752667920280384129661575021846811088693939925268794395032672870087581888
6104930085975019692809520482375713894543884424162175662637351803364327740173412594274555
8202675440278812140928688041203000227063180825931867664814904472579944670459423828111448
7861177124712994023435319347638977894844625604174547802052955308150254298015709039238221
4721457480550269684443809603801644372619530044039908184265874879532636017477343963474113
3302927558182000846488609818329170786212437034594042815016041477058185644132223188779339
2163988961273699240022516053512829372365573731931938983417538439708309035853330857183669
3113650370081905654504329842209938149004544540044861822140550605287168313414262789644166
9533398029687951753666784755096725673907734718169339975900118989111396246520616071885664
9119266426940160695906161044378014986669982843324652576088257101089919591111808350303665
0797428123070939519840254801694462620592363635107199014815677440001231307254102560560015
9316840532816997339070715372088090132654154084084869464851337622748284991612747470532222
5505791833719782998021737591424344714866343808459910139255494465966742373011912156910448
4733069805081712972731380662922967849316570700310830556788979123298130375103174678340111
2081353091466073367511876218722324789683703827950123954449539675885373844664134720247000
1588520176131215481437213347926567224469179562586330030699458362891038126489523388807663
8438188092118073979276831412308688094691717655267197134112902715814675294427625213295011
5430115336903892213841013378832753935920209051942093938979412893807659530870273550773004
2219114300432566846240722890340217115127316890533946785187382278485158224704841277392668
2805651726037644584400682846673735448259249691621423903388931811701138913150192228541007
3818145229259218514857525247618384807758382986666755676288831008601526358935863571274221
7772895154458312934709800847178289672734129037076071597089783899342792625573818895631994
0507275208732599945654560858685535826313370849340641393491374917907721822853116009498227
0366792789235878843160849183895336236157657683342228959270239391538681195815299606645220
8188618438969720018935678548949422165123610171847376046060382961644273311730402586998552
1537216477536320660720612329293239624078004742806780812454299279509350514397602119291113
0378021703950829509906994439391195500340721574148017759229971932827163039713393800349226
7913341650891139464247966132094725089819572375049113312916698427371148958662864933613339
7030362471630128321007084152026276899630681896952666694405167511329378690500341200428441
7394295414458444754124677185087103019497637306464566186349406566159555902220388135713661
7682347784631098854748031416075183130544034281127035071959901420305881492321488412935669
4452964317031319035779377391901656912577178069851218011520363784718232894953444655146226
7846820263039666301990692261062324086257493667440265813971924801331606041280430940301117
8183953384569734777875549650522671989203980646508809876595990402683545829084048874359001
2102485707190569558518229964816324570157037483350716256177878402602480795436431427740977
5204629600770595232159378392656058476318316453946086912913902970752932457528311220631665
3077945599709358801910179547846822202981379496733900870993483523659252153174780914929663
5957680364164531574198408294970424039023534026592213525985308033243178585774060569003995
1749662478311461547074710152410927099019069460555923968305614807726537399849961888020445
2613695722308907368225836972534673293043490796231583862612634442408631410349974036550886
9216050046280979394259573707752242127755962792858417731369309036459436384493982339291885
4895536764735608789899719778070386620927463185198508519268526245382587024957380501099388
1632387082022856447402189035037020515868245711255202315308780901619437053871768164482993
9449118658976812285822828630588986102960528325443860452415988790437047737737136850213553
9382265394902342171088978371005174981759702068682570272597123999047773133208906742112112
0082183986777870565629424693820042378134913066630287587520925647732731674636966474625770
6120034671564090347896314818682134598061272968256545576749859197286737975381067421738558
0194869393292329785455610079000547253305167081854955082957331645303896672805738782805882
5883386743956721320161223642719023999031850697809484729803770517200091305626967205284331
6649649046311031340090378152174924340671446624526247884447970326708092047966021525362229
7779841303529182491667754733386109757717520892570062709539841927672468102127146446516336
1829038678557363550700118367241542895404774834328434837475643045274450680844294328800332
6353481326604760172142269400349628563970257256283718828008954760238666306549747045349884
3766403859894645814480531950919925945149823361365390441231674071296632618704241988407886
5603468894409807356511171908841621313370383284752715801402333740874605509019733708360447
2009020165095766376037863721883200386390086318700252115943278393284792647056206906139440
3648830945520239437276001155644876784475408356160399848851377292303432353009793967833669
8300912777997949717046285314100435349333822674849658177523531271960159061728284621371440
1030533439202703451301949107034461745657717921913241437364969396904896573282575669132776
4531083445687159449900509202240475799421485141392454572532607827164713056995813723489662
2724101513581463393598573917624057344627157841431895806876744880080349010473159581907224
1464172911598289697025937777675006732038336759992089814111989857577054569008806673510553

4703721329348905545549720535301055732469661053806609950070275475222626763831652725279154
8331453619391671955103025151941717812391244827611145322188667771767341740645237899301 50
3710421102322388477693731425534798388888374132460169484945229905107108955879321620563 22
085156530074079505592494459743510349190441133791566366617724617723425057713516044266303
4312685087965410397347392745290514055124103029836258933623583426786553204778075390909 09
4601404380676438291147830296465182153861892014007114945518626929913247375729922061152 52
478291665457569566383251147867252560975782740644380460707154246177387337680719306797191
3031072597471709514887374746950349152024257187438661942079828728831057102815793640271 46
345327984734941390242082297386601437354856090802957134692265331182640679652544349899969
780862934703855361570010379699864618068028932857297252901428290197040501117093960401799
400772790730059574064535923916663881144597641874269200937829705225640556982994851451162
539665867422363083935081487782114697145156264218397232451972563939261864816182977438532
410643492525416862643522704313723804662598556860369392975077940069921092726870253264588
061666922557296352244204858671812045093427166631890519262450149084600688052235094443465
827402258290512950304979961401660772556448746146270978688973256721019292300982454765438
008643740010975723666092413723186800774476268629452639172489702676745679273486954263296
293307799383060695174258956537533033198251374669746258716171967671157859591293894641900
519594231162554265589036040426117249989093064050953199964610719970516938281588424916149
236396001113845325917165402764954766172956590312791649517360294213628187264592677678276
896029664972247633742345885325543235731917632165397243501468576239165650433876800210156
944358236086348457350102531746380709120581568187968385839451042367958767056686080265948
295232559420292102688752911543692228947178983633062840897730393136992191147148162894383
234242910144325546512377642992120827639297377159406413974052092305309640784163165543554
240237060821781822992158282156540176766976612089914480600595299065437623636120263059889
510077235755651659219714516783674355129084511136295559530875867083778001919961816556831
004557022827891916130240361835164122316270904542939796741282350843309452202160626684268
919718081496548307692214680927243772183273970929550675247929813115669882315933109698993
912254344793577456066112690464296564074054192882264785479797974455922998213530029315531
271761722186319708982235311610694068488157552894816208209181667248131289081046237699070
132293532844500814089415189311098796540721461827574805585802435381615139187720044458506
579847920254569441147796639922979053202771302349978644034421824961632018918118043264783
111515946811148164726454776176349787130866887569529762603031666841214634302624407946869
782859874183950317139429001355908772255770537385820548895316960180686323258791510705889
327168594220715607638907859760776550843037319077176693145523913518622363114271160854458
040847500024799875823005870882654914622078726726806363745375980674164794484814973892387
547284593438752792867664432725647955615698313724625104542888500733489513025043675009241
929034354360108566920829426185038883972921380359209785263348133849959272530132582108608 7
156279941191224265330135185468301475883617271648821814983502877885575396597889061068322
286186195298276402577225660572474465593403252865816014386518537361611416307557297037999
446651141140540794271475377811114255133972834961152875330388644328160811651900971960558
504031993456527982222342807299181713054322774875923028252948026550571740861992523706296
814981242042637768598819539599684750680120317039135076007001949208483688796385142405835
386068968037754420706427165194190884481774909538052581044824739134996228296437835613093
745711476150935895945792653518444582440734844891009395575602291056062448456125600438212
279003427907039318710752380856389211364039358354010582073711565980725630307728235851731
361937077949085709984740133668500208412981911223595969682225964545883214339357119483477
892683458897400760306939454021357274766342848666575966790937797646768163015848816770646
038970658327592363081409299550394583781385143772011353236289264203778348412135950821407
271208953373316878810000342084057618038903775566026792357942645825046342858264058246647
041364199474705466118335438791076331452420005378089995503474173982479708963230657788066
150878820932715957565448426168832058940287967211719805372836240656118907999138028832094
343311246912681414321820670235390665254965636176151324015679568910354880795582582532800
003740724316505915305906941655244026442226553465708270971438428418562650322627493196295
890247541653457612824093535406270144007091148209548333649386576628566250273021544259798
453829481913014999381683903264080298234887741471682495878481818005552066910156137025327
368239846402589188012033765009206477291158684977059128142539364633256149381380988193472
958921912237302958434157505102177387600274257006713072132715585027183263165673229741655
699387896932382888666604753463987363328505616254877385435688300572624974075468515505447
792066495159632292580229326617229622139077254749522646217975460868209939045420657122232
019376562938297808861803061952849313208496738723326039474869730793856785007594093947867
425498820242041547066719824068073770803660751254641737104593569506195610338555273220981
093912275466544307152856539660824821860992013074825707698853489469154319716569722476616
161766706519887344489662458628625152225384290156087408995434217695175410335336340373475
038196456961620860504864141458492224453185569392104609170949935123097068124336744992981
962743873931320970401669537573173363924066206733660664354609510478033905319817058757712
382737802524734990313350046117788777274760720342573148159707116345409091842317519827442
776377940258294921372461614244122992493589461434008809389144670032034652189837272897423
974379715319830729612428502696561588466394484055677766860581950531483370976868517186306
676495978906788870683150525108802156270017004232925652577553571474880717033038406465942
131322256028818375188180027781993016740998644106271998252745569596806164230149058371294
203283526086813465820686611208819994256087192395986575599068847944798957284718971157601
126635233211871997466584003580218134043653557895102240963230644670362396496910274014153
840261860400252104070107826599150449718939463765389742339594447809412596446793881059540
177061710349317906010285848179428692776802516063469818648761586233701525160236371817879
990343795955115086514540258797264232005725044441943317120137468254190708531072152085126
939846628017711102642445638079158824956674349287553422153367398007018414596890640693638
164153469373358041348621682362812485519628207372517116447100484339267712904709544982177
124599735379498959250181926670099192319815139687203924078131733288227406183489936891596

2588277943418125978223746823356556966161707037739424555386021304334936400753195339849108699633356133084982004622052379421261127386469773435298406141202933574769854283798507314230296086006458390203266832971066474859280280706222894119846852371184628817825233780181757362738656338576525114589769142517386579768133455331259610892167967599216200477777366017396981570051893055632093476805752082392470930750564865354014709692586332447535240023579514208088609935236904979271649529733073121762702277582805843309927781993804392172681117659365167746348586811523913603817239938043619137087041783245836870732687927591512421327930692493680672567164093795811543741008447431807984798184562295337831592094370585987930674314958421280917714693715998839338365967633328550845214487173498729828022872245972111003370678851957086364693267471590612320181128619220703781668982540615318307653843983956690719804851395299403331186528088123217077735132700970934377286450690552483201886405445912131117904412799490178046065119346213835182007919417118694779020864787750881187224477634397287600533491197235765406868201936808677188641543680800990685123604632696099408509903969483621715707186981568085994319275507836176310635228624435829757772397161712443900078527257874425455388265166580154504944151640169244989042683457345626173402057140117938253155396691786508532861602277977618284482128672946293714911091163510449320107800723707449634925691218904931357361767273707579489881987081480329456080270142457402799289156950749771877632590121855832116833245638324635426413663047442069533906874618048742695730825289285497755704465530725372653544909165505854490500908073608133007772880591762113407812702207157721799184691999647647655001009776601727964696655224452669933769931290031683521179052832728749544751298134025062913810954966857283877952905044288966151554228237601744913735138627397538983742878364570857778791777228294183786250926281394514140283818107124993708570349600679768563369686476720810794013433268992292032107971528543223093687838982497846997755897775841621288966611830971893956180558988873478822385862983606148281970638853766387023699719056887914200150058932781576679554126561532407775118370166626802333634507878413669795089586402858492504048493379617118032804557079693217083088711762591995943564413971495282622098711793930564122549381874381430808472755308389172693462948996786375493611546243336309926769946870192230881483903481460666095445247038837612083165142010076318337589349399962008248009986046283062432490492757335719180582605324935512741107358603022530576982049946002145986397578505832139101615820893810253664043283856221560504818745659654507487713704550258145886878709415841423636958253905004410260672794345106150247722806223509663315736125142667943414074665206353454658392226114632355783939937097026201916221347388562453983020165547146208476297265257464268723386362315313555355681663605190357189524466423758661980411448025219901302595893499907018346443127068059900432561387112104506890557506790767625630503842626225137290652256549426548273213499848416053950047182498565601847314490742691964574526581135702651071623877334688337258997786436725917411557016552924842063152187379159524644831952515980633946037421867350241980390342917288205856935267866581582011958760698252396492923489146794430847865749745083669623296079018930796244587484372517492298595716781255748423316730358061373238720790093363369257424578919538675866761134125824716030708943168749649125716727634443030554358794396727983916055021351483332405448295912401218783134030459294416212927504998450167094648495344091524718377825576002939794250148934214633867588345896387026112729779481845303444231471312440784982148467358627326536286255010397360301794632125535434232983465042878299269795250762117660416274969836359499603243495791210300046219201462117222885844133017429091519974561741221425481510127488263009469748351230072170223802516362654245731668291547444218100277761276072978923682586116206667022106584256981663821798412038784102644877355661996291918779924348976235918131207934212628580852454714318097767637213645840433599276725551587668729268189864664577371815671372712732706917446076837023027879242438214038304398781254700621379655600744129064710835313042028929121521463452256753566839681362781899954496762854591729513324623126862712073070316867408660159570991663158646522388745623449645159907497920994590343139214940679848302722545724305143530454907855046857565598503141863084834865010575124492875671828991276457246072592552964338464547294520764118443817902834258464092984568121756169389359659757404292955715291507682970780202397891696925219956916743579510728110135844878034761204586517986663278274801506364902275231920574521535572880248195865656936663054699706295114422061255504568691480327448602817268281027424886402616018934181365408817027519036327982203827038208983512455779035361653446984428513345483864352548108312849983289418172916120839071390684868789213810103679650443525120422214656143647220149894200214383442012409701466236343858272692961086252552511892129071483604566795595168653305310545101978321657417626828559346918714924724254791232715290930876743532838750563162595363919968592580225709413045416467857239737626687111622820934741373839530974538562283954012484780817487662218271515701356954660145145276861488219451341105980195908637689785102904745476657438215181134510250246512821307882573115182513573208273422069550416205631424449297757854181047996194442936394669739813593669111845157777992643506435444778907866885667636555675101330203920641099446748697238001890185771237393338518911517615291790335703578359795955444897859498954246647154327706708272651894378379480640334221433772404786786940630662072056502773700258523743934557949437175759482223948482130909053923116986640964837350683077249445046879290633705387963710117044090375203852640296257030044175089847800778547400690983159410928873587189533947358288473512024777514415286414333015237602025716546639313515169580288664731538340423443705796376486698042917498972943309837151215686192965419767908754359006588519119491014260755969465368844561880990145071492835593653328658780734465634881032988679671646781357384388603935675481267316383463260815222677433610342562799044580710243729086055639419844097513553481936898944385856295365499862588604857134920887479532609109770303465012940402286028361334223505478683117531048043788706828828150004082704666900724460766081328348616471638169678446051551761653711887635152359278589755531581273539168302801968872202551866090543478414559163466458713173467524042320513784665145484852108274337280013486724506825679412532619761659887652396688873925240799849352814127096654205069769479109091969184310175731322507064339087907860867329003063983535165159800074053082689602417037023267069764470297602604698157613952922096431184121990812443297453496

pi to two million places

9714035525042861389842059873557626554353472123109964180755739430498475835897786753408277
7752559453469072029492901386136911719381317360164647662549743551192372319033756640399955
4244778394429835182877424641232791662248726116615311955651531837930374483071056491908630
7692464303925333281823210265162984382479889408161531526098178358205420475757423919079060
6108906172633363527535883672852532356699956862701830812167405280180002553689929336938670
8811674407772991654278804467841356200936297554569151807671331296375531481657990907293100
4137962847599417790724890998839946581298870509883672688417262456006546811448063717429500
8458798293125338228957064263555895195747921004748780099067471349081771165970278740048480
5584618443621880455975536139781877110016001209738065852060227467398432198019506909533160
2304589829176162586011360218289353059467362871285557042048740358073801752241601053644920
0727003135873627446547077793526648446408067183202372794201434472347416804982143532190660
1854299254690278390239469367565855047152011417500088637443752809863355356663053144742780
0108559946161239902956286531638308517479794311770261662110750667259113679565731261079060
8742527132102089342043068644265621908898010268788167758633391987936068177968001520738640
5811186433222300276066097923728491881652219262768209018854780389443560320424330725687340
6061737023245268043661758961997444611691103048605905531997056360086339357467682549327300
2610292972652219997013784745668336927819603268412053872503967733874100943023310659498590
7575628944088749450232447104514115759283854319043656099794417786945650590867372794050180
1035891330078022518367047129478583932589452033737115409652517261432458213651094928649630
7279067566775702907881082452198737279403819995024928527363407436172234914601313374188260
1577753944998175824449373817062049505167229424285374716677617880644347567422409659208030
8034162784725694268290229252125238485853373479136715924694997360835010090841599916137800
4841580776617930919159471885690996102325600636796509775882954357381862985180328592847700
4137243664895146145050192069446910055451753162806533052264007677708933108986747592363110
4249211964737983859542557785447923057506806386126902090132506307939519222133860918595060
5125949662111567524770317261323632703634237172046692361752729643717179484510462385604250
8222738205746694163997921781159413559646469298483067092014257797423800935295307642131180
7963032500338498464286036340724983128555939809627486824431957818590388875500902591367550
0437589147202605836213776664200909010705930533871909593483383302999128166144933023488320
4286278094503744594199622771925912129739618715920208473155346980822917945574110692956220
7707464682922506407698844047590191734774146649892635236134717021650665605904732803649680
2439836851481673729696986157231072880503086554205465345998321799681376938521879386471370
4152993484829918880895775760669754960742573489972049994459247477655065588130988577915320
1253482542661842900833581533004255332405321421248360436128192217295783844480015460029890
5050118238964630978560709107880102649379376565094718793225833884460889554615294626640010
9608389113091285689074952441099295591850870862987749246293509843054163189206518901022100
8368738507686044955836700884497199414711807644132023195029043987298197405881795755592490
4324694154799405009578634649107893595827727866007415601127758927156974669185672088046150
8595689739292346460865182597348098762780327357831228242397149180793499521896489149874890
1195446601846485979393343279816952173481047300746483125306807551970638079106679558987660
4530004506852433458445014194380760573215972428933294095357769028928132730950373344137440
4842404965267264912307457602800339029746726007190696517893056846085704862419219218955290
5012064993829790452218907115449969937914434089777511702614082584014441535739125752687820
1120477956764332887539697262447318790329525014752686425937456143548799640235550255624120
8656418887114956390547794276858725041127991197852527863555532605490775416372818787942160
6756748578561416230433638240765037880156019880957723804879880877686818853155071552135670
5458248621074536528904229917221564124329654859604047603401089607930001970188160340187060
6876734930116806762537583410194493959959751799294529013819672487642943530349403840454890
0206908114517035801938191590053992538549904281164432095981042972560098939829781428018150
8679033616012883408918242656076965727547631996722453307756635651498892403328017852909670
1591215592207966059200516647151130840157471985733155950814931327443860967772896845628660
9479813650313566094452938685876216472384194726675264827978036461664616546009384320096380
6651058793835840874963614197063437324407440617596395040543023084204531168926186905477450
5123081429286713145478724936722627140194805807208956072954026603573000165918462286944290
7073535927977913514154572363668056135103373594160579869350944593076459253643501294918340
0746568221068024329544721198860420889704487725978522293551441435958666171084057619129640
8024968957296336190810647342929158012492334159507279086087473472977792831022117036007850
5457695093136478690591301928975081368206937476624409797390608581957036947956217339092240
2356878719544888707655404175386489494700125053265954906591158579312704816593456876477100
6138345165728635656730814481072241362520297715123011521636643617555415377731353126073670
3845116106084151358374964999144671831238478172661278322910119023056934267889747688453050
9278879053490310507761425542996293875484175196399791206714877105669967322253891393223400
1564458501503835897166317796965033760872633416256651521685707737530402247944039875456930
0997611847595036302112888839234495926700379042225710280651298054546955321149537582766490
7186706631970791692260507908921887407617766347269293430000285444517296016471890156412060
9943099861693668518983340105486629594899465690353605541570293708695036832258132391740010
1335789122838643818662142763911201676285091701580132394079716586787519933592260677409730
1015117422689023963220013016501117008964085660915964199602020603998595167599336164608720
5161053738306083013854701146917863353802341074177249771296949823893347460978146663308890
0623941931970797684886220394489665218553085537196766750963598805177974726179303269128310
8267486205780926525652903428019088270590140546637470013833025889361601715015488868935910
0569440449721720211392456484749888583660011930309334050424819094808467898760038276692180
1719086769660182978203559920718977804292361836630667945726051889605382347287593282301190
5158655518017828313356583098613731652593503080582396255908286572921386547219577931224330
5914732375240378690825360216462948132048845739870513672463824650599873114603762253497780
2210301014542987511382244821367352673721070466677797424225083858464887294909600521861470
7477151996602465522606962127062085416878723951599782686120792232886495599471294392000990
6760846261445296688181911180944889919558814713799314815293295825160760107116624951556590

pi to two million places

3720289303451396577612021582500179271658211474939898252164131398331629215911678786311149
0350370804672911346756645736776031670447118722869777749372049841155253422836893646994 86
5992816389029333329596705023331063186454715946198921174696044074793011211416649749227 69
6638793413605543628013565218859169191601265017034734921985779121559965545078845902800 24
1979283614688616696872940673265794485514121029331647678050432030238929162094969940946 26
8112688477619419298887283020453254835559564561049750752610524425580938627342063589277 68
2844228129581464734736707362255972829437263774877683992863763234539366344505846403222 02
7497341502487576599102151954483914148945282238326814839805199707655926889831544715769 51
1831555106187300676856804057123078324335686278130316986843408552081541485255677208828 95
1940358785837202454478168634608411970360599469204205022389505099905474622524732812866 75
5075605926657023819679854436973843036347551353010142231318596659835501729091226635606 83
9627244308404652362865284232381423499086766091580811529985673188270001310938215761913 29
6593145799781199823426305914119833894125626021521474556185978394081691455551300017513 10
2083825563173616215459035129106180015379921299481140928264215227003187999595833637241 43
2647124378051975244525862986609139765363120503485519687014261642737394425538704104093 40
1178690284831732444669643839274132175880980004949886150245568393213699722603951315999 09
5278370145480127595279256570668760339430121291197850593657439413065417846820219003381 05
0360714952969682719340824995408242161639655390093100714463887620313491340775627556205 87
7748079843945811949518820071320525940334421340012436887461101510266689010672516105842 32
4526485513982392400435721956683073657863463628471394771876051423987077753537119885446 76
5729346358469884781033914667847854708731515042908742459239461326108929195037101182549 23
1842072104379420062672244936264350959550337583817195105443116873279605675329829161312 57
5435682524564650081228052263681461159753211991706603643597448082885621932267393686560 62
5429382249006199560775333065246484430727348720887976729269681479748532637460821151712 27
2174344612137262433632401184329188986371123475830738974105916818188166895595371823995 83
1091589558651551079391105153715928239197272385189345950241777905515151572649757087242 87
6794409731453020409590769173875009613264837074559534153513313449000387528031013378564 41
9114742350224523620088655342053548595620977763485917378780621225544022592527384830776 51
7999637382300909619759380174752578307961563233626646373083895738467111670927506441574 76
3282421096818667021420776326837552607761110898894496467327753439014910896163618414435 14
0276458122289605060381554653157323103591573592275891134925690093560714797768870073195 41
0228342717487574907091871630476238872340969630253407824746509725002722414526033508279 17
0509524408937573333123198203021541635078677826550932171378171231371611421231812440806 33
0981536074376395026542557472387777479516037076025486348948281530335202194134646696375 14
3271528080291002816293441826410827591952495181736937113651514537697574630355039688155 77
0393898348715498840132387691533308860839615203879265983426272431774927686269635413106 84
6562448434931165089668748446934710340534822954844042494454802901500871209116791765860 65
2787248125797753474788031668900351087458568174549707049735995711339834554368894821610 05
3715261453005639912424453996258031328022857815556753370517961433593548312607199002585 11
1086675429079991703537006064368388657603432842134674936328479813459950952445941366688 58
8636535801656439672455889752138057636159021484215833508687176912138457600443756081874 54
8473053787916068543093840810194972061059938135377308743093608025437461836043691487411 27
2220989011476728779478953951733778941126562046748007712938053458407655616163671403281 36
4506738962178520907892224643916798604380401984876059535757297848080536465417964743416 32
0801881522629972791753618419010572539102352610066612857938612848287211511443312174816 44
0400561836018672863279325396928232426014000725989950970512646811985138070288116929747 03
6513882781618018114314687279332331454315102504802737697359069296971888893454531535369 51
5189875968892467578877943475069709594838918709199985678213907843263703165798784080761 34
7005954231533063135629197234763111237012637437169883645699391802040236017065484741044 82
7168499151119737619157080006875431889672294915351919561931247877010488364900343071220 2169
3271644873789074869716889055669789349557482685994538815934237990106692455380866035693 09
3474851627199700083726662594609163691191866834087603981605345718730448837878445994564 10
6619464928279029871448513829662277069771018679726274836234505524986843346217582535199 48
9358389264000821604257858150101486886919441391356686968831445986211992334972258493374 10
5532462372231784115404225791978377626057249768611808885686579798004500743655996168491 05
0849723073534559981156167551316684195733447513292296442913651579966325538347945076610 32
8968486987628172465341810383001694345217588085496074784900756730397047497994124351261 84
2137150277561668273690261688150889213493507321826932280660518231576305416072303671389 87
4837793340501584198071058315310913451714395671880125030598869506311654406196180647080 73
6123767536839502282439875286051051135181419197379569469363868127453165281063503834682 83
5738440551422934800799366821080449360686216460067663448303192249589315595404581694972 41
3680574963656793307919636715256980718618747311974596214344478745387556767923021361485 97
2863033894182374504989176974007142884239766806283172901190857403250338545222354033759 89
4400897932960374120546035438418200763258809362869789359558085094975650679953503345837 06
8005723967268132946214075401312828636882307453773754966448767066237058674528560526287 68
4371031208597830707742988942333196659933706724709689525249854458209814394921207939343 65
5680256945592260893704379680883059217354990280818723942350083166725862989891886521707 04
8580918439451741645600902259738411457762210305768869092600196199756364983572315537116 48
0078041903407349958308441455122942817760215210493581318239662767871573714998284140803 68
8197142273883020874501573985852397251066895699623496283549272734508268912027147525704 27
0506111845946471828532054542723345159408698624119906327074618266795601170127525835086 47
3884920460278571285023478467533903131878629664955207064550768958438391867679066785694 77
5621300377608024518082734525214320551606331504036994154860613067595336597168002359216 08
9478140468185331471060954683369162184806558707134105519361745103710159921613619950970 86
5013778053389759398836693787395482278968333613816287002510939802307692164016669599177 39
3122483908264479128099857806405324340913718621885323049102890113716108308680874867238 11
5859742181565370929674465518492928750040651580719202828052992244416273543223825649525 17
4246715774826896121618525639989365694450558351032818566413199762669743911116778094871 79

pi to two million places

3696573690376143192238347665546373088851777088449683238058186610490890214787876907311577
1261412345364963900842699447080255146246808750232039897736546071330432841705373441390388
9323484250318168830269138738845884794203557989165117728792068933863168270043214700842466
7578529416604696403753841191237643274386841834063376908656334246919941284466004865121777
8043276898551840227751809879011069676458190953231321128789090149768694698126335988541955
4556717571267667883715576312670276152597776992561643721030108499313035437189899247009399
5752824298729692241071111800967264157307261862392713715782806114143002017111426482584888
9272072257212203277130099829499920651648922455188366142159595891282375248360621720444955
9024857568099691558623443816516714631633100487392293226221943195365465767781209844862544
2039608301127767452831340676229436884135422093949607182598326001355753099690969163734966
6559745037380554178792938203007569843076695439031187270674422084002598679309241436629055
2273054873320374993181703899256416167449649356926665215348150771057177730318828452217536
3855090902876005207644455852172346692659704934706045532756296067906066323102672442059555
9496173835522058814667667226152539099092237347001406561397500633569448750968771520848322
3518979421232259292010069507200894275409808572680945059064154821843092352059880849315611
9983713486308631608177530993431356826119391131319755311787891831559922775024846159670666
4664052848809211928532569954957327013366289102018529597151020987318184639442800526951900
9043300223556760197787053090664264844723116541768542221191674862350821405552416336285611
9372915602656303748519633798312493712512584668544944179398939503613056358538146534974299
4520242828047303931963919665686795097812119450371518494839553461857076075329697620604211
2644505057272373623975657440000518850643825547281839696475595398026300097329759096766811
1250192035086291396430856380268780542422691206708466695667793808428879759066333385191833
1658073331288641077969980277388121632164618197229799382327922503439232071557065563693155
2921882935546190236507169076916658534114162027868814520633343518244458692779598046627977
0127434881999983671416159300403854909347712321623085392496688136772265723665048007428666
4506672319728135654420431444220544279556928924816510679063605431422762298205106039468399
0735987427441902527429805800573784246772140507586893275782389649330295393167543655826366
8293342278667996908620188089559616889819483366568308504888361764603121668763086594919833
6752609718937985086429615427801315738842262690972075493012383476933259946850644885035844
4843865983493006017943549795468471188058153183483885467009000362390165675090985197438266
0891762970517516264463210468224004535486063995436349987617932954392450932753221167185122
5536171072025658333543526968008511489632677184139489710157968223712323777998170453134977
7953440975537822404146878310311873204302877139841266742843083009963756515199425640823533
2499590528093660801770815480297746804698730184428172248913336690764257349831495301549177
9718603868488864164136682031527841507301603828760904785507046797051029723762400497326796
5404437574346149263092329939531021385733403946546431380393284754601668289457039852028988
9705619905005856633168102542479612249799416192380808171869561768335559308203432917111866
1194066142459302341685892832540228112237514487838380437533386005245983771521980857484744
0911596560510126102135739087758744852230707278675102693405952584004438769166088899837699
2971157675464446866868225821631593841533402184543020044556787861653538779006847964816111
5461477425676643466678616571432139560580510685249293020348631118769790675737306130865800
4980539962498308851629744832342091870932486556631685998882180763426336180380330937080688
5655211630835704265987849396277425742722126865260875537703432427972131624718656218150211
2721001533814352764044674384081792139852471765606453271666305344360658788502002162008322
2243397547438846569080848229761095989093615129181432469419017754297201682602926069362866
7759394413039980700485624722707461638170802726093244910407161694311272169324596134669922
5072649358332736922363720775270567953185986989393297451923406580320289778734716023174722
2746080557162944886163769004988028768178344620062203271019135123357703649008084555442555
7682671091481559673655385825943822195135801563771197167936702062224568634854690598778033
4658974565844785240881926883092277705460494635561584140974535542394776734275305008393600
5426533858342924088679140936814423285879009223052915279991590500906519247441417091004777
0849988903033856339419441048643921902022566287739846266971296286873757221618817127821233
5633490471549947075886804059463249806132450533895452386272552456440081303686346872684133
7921937957850774989027824315838273235033109670899330157066419524016404802598619582946666
1146570878969931266619486819119516889978571382608897201650221151500004591675918314137955
2842924016263075906928102591504769068722721155458369984724422211012783008421494898542755
9892875796268140828424620273448115637761588585666606287075979845389858885397359150678466
6677440735588962117649026388773046312288150082414864035251260700750590426855154587308199
9448361425175009691988121471776942905584636475578159388806623954201893749084032275102066
3489232830934102823789408898953172570727814669839254676150509011338562734748228265353599
0100895281786892385130548337080121637570618687637859548021488100672071140270517244681800
7607601299422257578378515262372980443754897446037591002939914877675347724241123099838144
6911879572544474496042009487586321049675622617109806570375457977845287555067068133301199
4037028368463183005066807530624706475224361857438269778398335421299357934225135203492355
5031032263242224496188486719775355803165214710619958410406221190835966907897963380812136
3989284373107431582309352969442335189333013437082432643853278990826149765002466223360133
4797174646420807983695465133294537755766227252446205078516996720198999934010133350720755
8077104038265583648170001882520487386269472797768681984124497130143674446177383603784500
5886773104155213802955114829413819681116530162355989101477594807594117715701165086921855
4114710049853076615445032636243434205052785612868371178869862422845371122780475373192144
0489569910861835758095498384445660130847465276512015225164046008638825408921010246158355
3618988556156795455594341539377450291006612155870640166529816814549150487091045363602333
6962679007420452451078747671110858160388335049073019845459657543115162722501328826413277
3272004586475040035975388411508587123667673305367165176672359234741082205566207389714588
9673612442860131844809900996619418551265914031244826821505050968211722386212311526523000
7131658654273609247852137350152630873645042097086869752775065195387346837523167170317933
8647078333812547030567421606759571812448172415490067795539503394974344065110140266883233
3981384072995257794268605098038571003884722484823787587024923392172716067232378375966822

pi to two million places

806558779053366211090973343419979784433485265324350722533603987194609870643016788954090
843383183273800955490568085092791321896161996636262009622696371104592305859857933213945
718121849704692397468471194090316286548267278160233466412458675531537220698657075888456
159200392763767615955391438114106410712803664182738485702131664766755240750094911376831
891968759459457677975505054359139588476862792044160738994386605970487557360320761893992
907847325701218938635124345040256111531106159816041062934721259033810339622107691013592
064184427275725636244992188419296928450488655716808147667896609363464271937555630095802
657166740606967045070055976452397530935669267213475656322310521709797267882011756733844
202585858563099582772237670124261950140214124151701706257482391993757315398056518447478
149788507156871414235362443327302122681085470827797568225700932570140588369636640436028
018657355839449184790336263501554324008794445528858414945115640942709224065884058157027
469906282629123540380245799757493270182391447626286719728006741510646157842608708921008
843375777396856008810813692857565080184359309963392248748227346979480784064776775774350
810648131138541202668180146216960634863869351822833302134661336932791685950307992469114
783325130820766910151785797169586601857927299671940055236690498805765228834054038295723
294956666106181631097892263994073011570024576283747735763081837981616381190797360315872
121983138196203449769816206423985075228327733732583324372882160597886109873551913778558
814037298650838175116326674547594083452969670926281699808448890104436399294171335504917
218530444136127553727404380196616706233382973802404906298258609513935027495898750528072
067318378647302489130579054815627366934203153365934803545221237593231989288843748037324
107862086056224621755669140664965603256270152469394700985147116009940891283519475682748
329672272176682993494445340679073655254171748865839010393192874133368705592913800277273
147754869775541198401388960452885541646155295967306253288304118555011888298415869115770
818537363775060713350543268713538259132850979585614230274360395764960678598010893795565
430875945538732132350076030077301079178964083799188708619632086790511588913741262048385
411578488220395923594332704524591411473507805455040335819033847064810142486891972563811
686319332164976298148495396460838748495439694585152579518091162665452367335174501636083
335297396558922109211569119783167291844257925757550040307490466642270906643224285132454
408076050309972583021790257921821690823779438834808656545165803224926641396247339605115
247595839356366074438299366773118971243814345071442634906683367462679651448757604216490
046578045668108976237352466304486513946482319660407267126392740537148060020288141194791
417421096450631305594241005639877803076831545495507157300582014792515959046024661151783
936785601612532406275242043711925061379648508919819095877705809924908420856001038860239
431557064092622965854619405990986964765074089090371020410737732283300019434413131898982
911460768793479475637926076608714832586479309319426065037650312207896059737211942645807
433393229464562171529928698775723744217452435105937015224486976130482977756125599017514
547595430957297867274007217741044287624195270575130818896669903902513418046187913963917
519996106880395794578140869943557929398861400711417249894605120231528627326923240980270
344039357213519034584028877988233814226609893318496674788926698874274514772695849546580
773589496632043196842272939054023420699365934655175946440193569680887573651356552737827
558456003178125785472579739016890446296702126248612178766264775483706533502859187059393
901778046471759424323624082047979583793498520455967990254051594049454774417608395779736
867376905855647918279488713678562976470361971927727368758174220619342121583922147169397
052507247831658368690921206606029778144229145154061195058652651412881871106491879697943
160479965128847403654083999670832363146792721324493626985707407047130584348263293188454
405105866194037558732502573156652753929021523622074153014465759307227508883267063705821
488431961565310418901791255484578746340538187505614044231461197845130970565453691268255
794448780425882950076968232189862623811110723687449921424691504040400313321866823059154
359549671890135590314696913736224077614431031226486037647919392677174935407534162571025
588711228300980455394033370487328235883976509095216726564747258020548090639892790959497
565422906184606794762581291478097037557429248089816541744855096727088920335142693606922
157665634684417142514188413007346342117709133884972815682697556483084091332633671724772
500819038332573981873743660364293416815416085287464587255653507621150730885068803456968
232915354060692724241361263834962430252227348220766743282446251953438168663298333351231
892641537539026898020923069024691880866165342597137031996465636551784185225207740037755
013360935861416370449092199891298081723596202953847731659451593254575127621826588269650
230832748779521264161760154839038458698458319371172093681990832889325916016831523789841
750955421456573542447703079868907388542365511940355299976861512185036911858118061965525
762964582474395494638849523052293904319208347004085959062477586926739003838567519596357
188381477340819136711013050785355589760197241318688735176938889609209708119960863932376
114367610703567567519545098566772609283078303333849478963026102904982778516773496566301
792404581418889016861103714921870755368544212842300294361846299784470990238920908702825
867759979329002406874041289587818001551262385066983520761015258102922206918505898797807
610317527440586023149271114853812525022120659086590178202587801042247426440305081493400
889355369942155412427940242883301635560084548061142519736000794683676743928913680375452
258525035643374923040611881347898649501791203952139272589351885032274797283045553778306
485639221525481817233208066725276659643162054180594480709373376273851917793186994730245
908116531708464778489793524433572038061469870092327006538735123781189355588073827614313
871047732133065852429217695200986522155832270762404786747201853835783437702235042508142
457128437565714796893103634180893732997220191856775283873360133944177263815622028490159
604474228410691468157779660193578485360756676916433461402954821501762780794026532754016
988343543419531698259725727032376723271043885021457551160781078490367273413756850547512
833318281681806879234839423579028304653257090996989339293195685942215104275316137785891
330262383995724924625652092380982163100220099385178387862912847546474997380846098849157
178387404461981374483906873405912842383139922683550617309537760077334415519711344778011
363671293049539990019837057605614721077458334412357522587981973201543708493212917688204
045074494910930517869705671177011419656169548746956995545794664902459880581820522678280
554400630551716201091706454966577280912312827303711793448888192757525099425680572045715

6003088351254835453182167003120209888175665602481564333985612103352180312220387245461764927719760756163989992555884482471253975444199575287553381607523828907495794340847840981905049954722349176780899897355542816911986602910505481900466341371555736993081687879278573666964628266282678449837678112704991616800588086998350764297274029251889063949221831873692325603991587304511774483342679125168538532935334055740695466163939922312566237332891308244660515960756818641377290495515788532889906154123437813616994804639083110947354161918944252795610239155036306202464858209675195205668630960889827424160266486843101853579201093414899591329401297008097920933872852450683677063522719375612107055299577005685216547395630934014599507090684943999526109950370114911485835658403569443924401813758729506682050712955420991850876507111812723830525509427097018083673324604796439163711962481671287816880866722863275435721997602795855021249960429342072378556610335213428835403247208840753622747707672265703221614598530848070143271133232795558962703353311330191389309
8742633572074810144260604122194998739520982043140940011739601500116374649373933374250093596798354959808168174942641161758900297445379024645100115732568115366055049218229071820849066730683734741354087686468803628105346366063639637050840060322594442368068407010904147914080743724325375420452945492320548854094472777330867728957284483519930624062333601489465701029537276931923980079137424612898806209404251071821687003541827178246935885289693683871233950818071126520323160694213943504416422118045865972847019874705877049174523614062166471826842276504309872213025329208062441013565190192429768319418422479532304887222761999981305934343540963408734149400983611379929010404902165198001696534510509320118652889500134985548157976486919975355400199623921830371557049566111407490698906146880316274565043092929685649204939142663256004909546724624506038670277906597782558864298724579515518278895492937913618643354949560747960629939433289895607101850074129740279027598329853445366547373825443050649218755849054725366313745370776589299887545318170758033901446976126120801219236484960131914537587384208870327736310454441031752042168170202492424685433938481832397201173367271437591403573249742190766283951872620948773642289035555070995680644813853912139494407699817657122176758776977413747693008038530924147025926730972434251473897943330822070919844429634936827345559992756543084921450058959498583889721949080480021103101077469457503027986720456029668290243300490195865656152338506601086085247051259251072368903914426044804191068856917115605546765577541313378467343572819135579215212524416342672879351457987262360684768794244924543295937593225680382412430804044151703233035468059302554598084018141938839130991313780395165866885640534425045976863068464703852930422812537128881240638371919682513155064045865821220270334452515002455569718520449427126617557097790765220163140992434562496582346327419996695919096303150772162295973794133044906912354662899976787063964427543970530720101567866765146256209664985206977071028331992535522210179071542554910668909890157143522523208824935483264058254433228993388181433246077620211927635355605540181246640118064490748656792493045753030540635899858103643050669178836044633760072521059307210192292241895322382915890934128228750510980146316239573671434576541180542227401565044111311050586337680869378330384195969453861337328924716025322293697313224937997292800599267556724797579095349634141700203866341961453441828590549528053409997763087934339975978910568604133901260650786125608232043804545448631332510663723910133243023185466526509293911805432574169027595407190841762886773529998642546058514480703245585088002374921985388998257635690469402036938827446681840139984177397803802254291492591957492773178379574501072916189699932883137541980767264918001306438990549637448406914582846426124204961678749190028707996979188534126248108795055420743526441794719404034479651091027244817919032055597758773842727381310581436032811962348249687706680871902988723417289527936418706406948463980108956431535233422900314798101668431369479196147206097308580136150705516651333914433244858694690120492477325571823058188574716991463603187793301384759759861120961707162675773157297613346857048140979691548600125124206029878855353376478465112312928998520040610540065834250345163468563030940509789836252773934123544691012936261701989971395671039397745310096305490105448365702167199187220346357563638244111617093788893854939355612500904793637171542240806103438118860330575561248673326845606941752793440949702599774350146701995027107914478321097845715562188832741071097653063416812979334593467427567072441346943145162167047337673582685317196054281712858870830159316048884149219032436305805033072196601828090094043271791799057699323544388105032406749186069157840628987344737670923422578224494543482808126567717321258040238692893611255653082045065306634056149010369686585620638108841621125072437468721892420926302653347648510064882401875311077523879905191475130170115551259365095027766386560475993267772695534780632704930394893489759809117789256537443265439374227850627165326171004913012764765858878130048084071184414269029062542006789764961699662037240749990183839249623080167971334236069975708749062665937334634933117472389156734441286767762850073286742893474298435416914069489814464185413445247285102226600079613809607527010407727596892661516744436606657917119518927091206931157006078461310486009590279729514654677233831975300392998202306775086147937831034092325166793058858094449717802012406075320584269045332099732793580656207789144551244404825526930353303513359014814445164701717409678054134220952991809060291260715683339276616892674456115532092800093355234193476248116875078333750521448641230046935912724551684094934356435920208400866397288745264476416812224319790057403767520411314593569082948636502886651386671870984019126520871938214610496291771605611241326229847291819735019232646934760673591792047346019502146850204222725499060500390527173983088239346961329546058235596168859614438551773255682572004086406671572614287421586562936534666765030539943726433777117552486333465468661747102947425707471145240840693574593305810664791058725770870394121595834972080143473202166732002959217783114654794156763235809015934498393436031139470196023680643647101552654833033229032494848874017487162555417843234598358313173685674119482164882832198303929378208900666864165635632999002589124253674659874275784244531350568116333665037981361610334287691399779093956537587469917820529512660654358748024953105462079082869238253173487093488509220298997492167707593046651158113338304783246584537799959642245110552052822598551357995433140780468738288309181716886504746473420060161594458175927648783175100615405715334080277700173393486915972544835849957031169089050234791

0041826961127743891011136824983651124352217329412770603813026341816523575148350077986682
6106893578108357868111581663802542863945488244743152171446531882122656033716864388552779
4908372296727150585998390200737043520451962130066826186389711245753979831674835802866609
0243753365970379553017428601709822852543466282260502978228192296797495070684694701411114
7182712377179454395424754558231763707200209395248030550248615291942583807464456124756663
0329362111940643853512617336383757851990930337789563507099848726475632881577185877829773
5370548456218406166516380521163355359561571365548830402308390484990534640227536350532221
3126285798474887120879742391971298065157156225145357376229649699578516189472586019304480
1887884140770642778982111550389340417299078428877913960319090947564277462823464504961885
5895718672708930505099175082590601623035608541002615065995829409418822317117668561034431
0940901519556035941976295271519144654657011462735647603416640733553010784066121706880048
7767658296034459262386475855747734122855909945512619726150335698046742946544652410747999
8946799784648927454814462927046102427724945172854272025170751372587973508902883565123773
0751402162131170264048251331975042822098951384552813626884177326700882525431724357988898
9275269871603988463330764027410207254286075391463205633419005178016954816417949317348887
8122444725186119410088351313755549049418167286424417218802638816562753985733360041105995
9943360044510938252578802766481442579095548425763256667686186591270514838980441597550220
2022308442316481714578201023087613689400621715918636966475668011505893950691794861793888
1106913573911192911947645716384243985067676092701560138762547387755513082161491786331110
7567699693263983636019984305639886793035036311014621259261823243292023050487397355510338
8061839630338392022445021877806341801390292508001654765599060390880691771852440750963551
5195819308548534943637526931428347260128632155695589347590375217333956986053558181334404
0468345871203940744923635469708013539670296892056415327057617850743694104216200281385997
4099445803948437171223780859161062547291289418501433733201139419237769927284987921679557
0485720848262117895374509959016057319145103301592195047799861639973513634301812707663661
9625642182961355714741419682519778451292399518094802761577905150529962456467685941080003
1965524777784431018487536837769304812085949347514957536351908366451038348117838304020008
7249543294359801839909611614425208081024615483437721590074746546970789568255917810215535
6444060139689315984451593321201982737877807460527984806318853393592553264056804758713992
7361712743904944420640017604659600972674199501118019442199470307638680820189152106960880
3373114792754504691807086633068304717712769938884349752036150838640309236770796659624007
4665303205885579543535898594777542084658544662131174173213638193761117452671984537710773
6595659498165968875953527256553052258677787436196995545270088885043127590941433023642995
1983092817046363671057700408235562634578787478523448239984694378073980038821035519714667
7936743948439795042261555530639372957759761213908282947761609069866159952443474072082554
8727474163790541203204376326497675788394491159485261755081482364352014490068449037552554
9504528711277590244026828966023991381226668728716943914242335690716721974852985490650028
6693853139002780062228105465949194965767734087173852622258531958476574101365001688354883
5623699235440122359693600605122970484808670655982833625461624988840580805683874768059224
1672148992546676970704279759470744399240192171358768929455771482444048370028276093844443
6672655795920533328636382118343620277464608717645601823692049975214261411194950914136559
3993959888496873253905456168986309629627745569315962711129138906875337142858168326558822
9153167341288902773433554934478868355341061282300218466236526025203082990557359962941221
2840361584876982844767216650605084309332357791634125986725241074116285556088741764834998
2071420906963904058285391826216228998268695975949380590488575368152351745149646614269665
8795620199766438100506150418006870765847045347714700596330723357790794376706421196119220
5824254444186413088962966896033391500132432796099227783533958918466257599319452669024221
4636598684615865059340714840086040303385526382246381589158118363359664373818562104058220
1328165698540316735563816301968056458734803967516057164490401683827820160310032606803226
6839604685589812913403117536801291255768900097036049925914526513977577298346853058553669
3635182475723337804400750475514350907561272195228462960672210621607461237715153711868885
0400371478628178842646139058053647502894690723928909472263625662125720569197736932903113
9341358756978228791242833507250272859563234780250407896120197892164132387436929916913997
7434727149780099649672978953914872704895812275014589904462389058696429492723035412933553
2387618921156458876442971363897816413221384394558034626557913144029141250116885199892228
7079988203332745885087873962019584284916999880962566639784614021609505972997287096124433
0457625312926815643291803738394819151464952919885361976689649877753470040989333379727715
9490519391803031244093812163606427205974993743009579616220470674611740857341097442874990
2407222407192008491185818151812427633852311408809193386990524737551796979153348369860777
8847341792375900020696454778980465442096165582454565757260109829279462120160358645909880
0121461108129748652676649377548555016380093639144038747044068074173071114912039559556447
6378636872521258664199651815527268261024910471618972792199637288140577295437189483001229
2061255825008809586482343503115842725044714417992408858316044363542631311998838150344774
7327397732657258291837424868253221336201914847369762675550760047847475071302633152791444
2464845831054261792732559597899502163649805680167217023986364221513849136789469665189559
9636981895289292091091581455804158302963877917869354121830040998688888707650560675784552
3488371448929958031397226925002634423933729377836121998946004608051929181573650714060552
1324366571174865186510958665531766993318173830344832523723928096067690523685146455827223
8435892090666957383546278011242910414205647458071394447904816658809815878347299839103310
2752287469474046967738211615109724712756091818216032132711544828799022091580995446717991
0239857757760075937066236993152851061780016222800130689503482824380598897428078097863337
3237536738751563996250020268891715608720568198038132159271334649860797832469882632505221
7246773232158505276772769080739518020633239202222893513074342659786059370251069263878950
4893955632192116661135515559813269057575409440166368942600926755204065333655395145959944
3033647298697252246130287398349730483019618694555657529791067787277547211347230810666551
2202661837023659008353118127529782410474176812005473285408824483885466837414233650591225
9942286879229483507726271457547046200616509400348912926039955431957832683200403542687118
2806825496525383158353257730798874142984638739305884324111675854532875489997195502300333

8352132642356527110701750793748806830785603325414601943320967706374935741539533003747883
3990990070253146296598041526455897799394876475410724850931927603294897917174136213784197
8103506849616403938713561098187853350649482250675345626451525297740329892753756169181748
8537555073371637048051131082092768493599453069558121008228531454181705533962376276768523
3646265893677373342803558785781280821115743061979155371243564354768881163180868339377582
2789315224641995493001697844790900079766476198783361464566192197575452830238998411280198
8621038498830157743708738410828080144737287666819032370967428941970934024336445816131807
7477228213377537599246894948856887259048714181460237646959950801386604347059435174986009
9052318312201394591848890753040173686996125439466721399672314030349362286270110183021106
6675111156974413093694485088430863920946963800556700634047876561037082409804867884265850
0559964776275293345172179481954550738493811330423859464446390168372344019907188086077474
4584650233245520572489711651503735461248395335503707166335469558335922089003314811093105
5035625241575155460739324444620243895162945071839767616987097469732773185008363285906286
6338132577347176797086008286365784771014243655708737137294057536068519961990142326153519
9121878183240382660104099322768038702518282689905013928749433754762826805592644380644635
5852915698379751024085994057155596201690611806063853047946278101163688371115018556420832
2409881625698054524196110805010759134257423116274388612649920868926439355212150847906167
7359649534179203357299319229870094573119991169784226885366510539372307341483362776594610
0820275072013548479905377197752110208021488139107284434838958337452396079131264461657388
8531821170465993665343126495903472419700891057207310514031003142001607836834277549263847
7812555726811479079790178690706587063474951441625253213465913541611593773542711274878442
2640103209138695354514175104568359401016226775468370908677917638329951341468046889569352
2868045362009755798588010754417592852429641027544394174983197584543691671545375831879858
8306467153427646260166170736520150241250941328917174724357727936423052842049153843136718
8688623786700688669902695498242234826535568866776437975758217353681724178526139621292352
2814651019033040202959808631994332812220298985891741331294125482553096868723311629218467
7821310026202656856968633389860311490682515184065358262028492036911080130045106582899768
8893986223020029873020266682395983433721483435941141868009441024239480597129516215285958
8031825836245884073891924717130756271362697428833359520054337402297168977565143850023979
9631220832296886854415180768757504850991986416003851929064901878184328260738036579415375
5088922473330912890232978391570165470989902590963377562583277115219769901272065427673634
4314435963386698378990691427314298771210280981354038990518196590257528717110177255359810
0918897180596906653462252559961087106029038506826103736595190365980945903875680234895812
2098378184566328475101226255811761539113972787869656636476038783309584586952129741360212
2303926230727583162017153270980917602947021388975447440476453541813844023239519271050083
3654112614498747762957664613152927304082624646701708792176731621559023521033971585954705
5802422838270279714940186022288724774495150192048406390897784706393683763842470276918437
7140113263995349055391609284364993786270814923084851585691045365720342141118382724192599
9609844030715132883908461395367071412105272205061025340510194029407497595745271749295390
0793858606386322716975883091315775480834273084500345820943756785117623829181332285007239
9565267328818090238219283414941449565542842602213790588610200418833919731786325472260696
6786349814689795481129245649195627574858991085116766023520108670357206241041911139896508
8056310177625446789940282116489206299309939504162691936328525050565907122368264291345975
0011438126624463961940292261249313966460082178386024222634029098826070714131013402251822
2925181145074532496117982780980909040598668887394654345337415292835273206845203742286706
6180187577441930845756845900830486689521818505462058364007276520648231602447922945765035
5027161024023604827609189292591418654431079730615857216897581301459977941667168583567014
4562797481377628779120199707733760091548850548543734919107244488782685079767274247498877
7503716950996456850662105235981331559735770965590640499957013762197929214384231902193401
1513373371463885699756025752609691992041679698230878351338934097212741361796713331802161
1065533514784012271805000560589962544108742917710596386148887121653420274202194001089823
3491632143341096645523645641574425476162806149948622628197947120995332656928835757076874
4231482565476213966576158701886088308735206342138180550809538710626433109792183401239101
1558732344978992864043400856643324403552063429457083508674597822201907204349182098165274
4154755619205328716377066988391265389325883009078593309732527980300713903254611166790612
2622091484958642463137460474292851212258409058847153194384311331074768044632952910144117
7885336084147241830788228795538892654286664484346740126017527830053237795047173946198949
9841265861788389973276677309259772363725112409369357153099344533436315957211000477806131
1956256649419026610029205275667024981564837479664097209386142874282180671772944466864229
9698980601045005527182047419353303659476484286197418817359912109181105178317173557233620
0487679773497979516425829722861089343501579983963113356714420777512245221594458881235393
3183178984277679077619574751252027257634592410599926915418595094605377094715366442336816
6034537749447820380314799452485419024158225473078010510922138304388873009741595897624392
2851682724173540249533525649788361744765198146214873797373350201389963174984048031417473
3311253576810877282054402753015794992122482281883158599032176421808576117958983050763104
4579394151675401359916459608896611203563607240992607138768703535308360231371618275879494
4370788026235451349994710057516165840831401814160964148485556955730484032393220524854208
8409177214991579667050539409497091309426035844241073566596751505941297650572681495317756
6547067231503130463608454835845721446246720883776265194604923072910857551718087040119262
2985996743739966703984299785629244915783679456050193823228919978420229143846192877103398
8117953279196400870648499992736416101929828283644198702283182353696013372952696400314320
0550427157165630034780171924642065185460756811038794804264588691923654859303362606440276
6948220974068354234243978019485317192026063360302189849987739570514319242794157428371466
6917177565362215383803955612588336253255619898881383941315190594078361441569787973390220
0266643667605661260341772385273381717007465432876226735779917344206401459575985605811985
5204360990748786201063309505039894971353174758183494361183335852575639212464655851461773
3314300998747082934936630501465316745749214912742258220888494609209423211433462825171607
7831824274822368063119758762681072277963874119144812076079613539844998783245877808558477

pi to two million places

0791403580403227933215701389593658177353967847577538591986059077025714985199792918862O7
175540665044143674061959756902461075245136349660724935824938152862368659264139236327584
459542351653026603370230664555840862306562445697110879197830061029764884611057424265295
474176486625207870400490901790467103598496470060348647617110294936726514970098727032847
990599934789281851306023690074930957379371813869516821395468129591464986234149183262075
502638768248950956748676320264693455175510292818249839119646790918239352418715552522863
268318942087699775967873611749834858899300898246311854478422410113101911458213306528058
112412300535896490363692652436919364069404865160756328368948571924613377198958925336526
525704820267206476980220983714151087480827271214552656540049463226137117556522557855785
438620484397274512811246989303953851327557208738586136332845154980999121622176081942298
329537528843084974815265989509596031707675498664537413763046783260728838516515898281905
983662442409841239767543381995641388773390255619104043407092540587331227195150043907332
570074022910892710639857026423394507230166256217803265052508088792039039830239056304093
083018130172614570730839500184286195290125738124421806436611596997022276933679377048967
651600229489255184171690301299072120129650133350627007142276635497411119992198196646987
095666400665324210039471451781291000017803245406453689450147394974900566906224257146068
056925494622647946704886636289350462532097847012868109030596278379131960109090781603725
759888909156680494319319589059697362378318104294372533961007287257463297767480226244825
115785530275005860141541908753722113152887672443495488939371268118235765079757375591862
260954758793900685505379226352071301751998848581413739120823909552910494808863207734526
534495606973773156538854783575430682309858090330634518463435242119359009917725193273291
229892982399848033143071342088986768649183176648276455164850978318312757196668594096546
739916866673803114287725605476721566676445897568217849958036979388003509182753585483735
102380350966032255255659914155544417369194496215692433112650812479498775233971600989640
432045163241566124325014550343166056753606443540198147107297747801155023230507765864292
355729797955055139760232195070145877926441473921211871557593117881085673494674367757908
697004868600761048553967400939668266925299485376913467099834065831062322136420749971036
766488090636658182890886783654476560523996116874664350388545496579393366782999422123905
754675789611320021463888751427704284851614103790853626728543299282619009124004269300184
230897419472337188277076536459963443767507305972448909468437335025368601750831720395152
360017879073227288852436370333004444092781290593453686631414701046593418834746892826299
882363013060137669269882177988517212454145733784882303824671916659511051746324312790315
608741488607081554831102132540133568685405588343101887088938761393732502340880796593820
148048303164481123176201540243450258972177670052598768575291107994887617033468123201993
231132192874341846612599870181746561179146118689268370252016529911989888749488292420616
964965430894423463417530646262066320412705247904652222594748526298821801665103773915209
569257176760513915129079083306308913138467076780713608298991899449053998432749402438897
106017627516486543243504174682174047720535790729788190300647621795656051593785317469975
436785042996228068593836058350652163718143758120359463898013538578900875386377999442527
513971642857645585388150959986542599611112126352521835373754089383829940714767194795565
653338103356092091651358796043175645214900210873745219407016607907421171463892092871847
601609024923191104226715102906017895674642383409519835911424086426457110707485300762498
022067363837798445988414775150716229321920310260500551509076978943194378348211223131719
769687308328746838329398680193191653702663820034824649888828009953080219176380419759462
730434237050498168626631463313819924499513504093368521326486221662614304563801554167029
975670181079914598371430134003203497652952164385778342024804974604813565562787670014116
764532765709159469878574710951707756175897195470146914052898762386344666075216918405152
920370643416714344581014881245904108836676936963016122140430307962334187927807074145543
096121950988033073232712251430746743792949084700011181578721760472562843687444029999034
907235233647795614826072754304750733835794169520854118581414211633663318843613930460864
044381203050087374740743035198125879556512101543796185401817683516395531429788921097933
506442189220638279260170808596615134092310144550959805004970933341826034628222661365245
786243689338287481808083116632140886018962793337969179670238926003951088492322226248791
469952469448221322207162281876337541174407176440825635977749100498441131586645655216934
794699385345895276480298615840226409999421000433420644939416446515860822749727905680465
910580231998140418166646897107038158991782599052443794164767665313637038164955688078417
197066690888187111892963554097089449350838086720874087385891678280578464638730133563290
056081755657051868983518288538558189418761846431885541883532205586551491960840135051091
304296438673726701769209462568404821695594224381628363176054907299398382901877071378648
219596279582737284384930210765170111412097127189513677811336345225119432564060929092039
892030311428693110299616289715741651353122650976566387254150218818945769606338265402520
174627484331378659366835358927288944137222712232373031899762712187563590305524059334406
804067165885499108922339510318522804003163077793139388788124263739945766173505780454864
709713365612269105452680323351709465782235132634711977541566480012164761891538394454382
707413571109880273825243581929452706387243028983887862373972699481019099564763398726779
743818824078646963720613457502040438651404814540848637229480891874953336845383329185692
611600136090526980748507788080971992207905493846449129981160445105124820157681342303697
584597931352507649917267189783592046244135583539680204390112988082084879270599398452081
514555277160455545091663910861465981010943648552995834158940501132217591891822740788585
450573707541719807935765713476422566400785520271235964984181478091852475405178298598358
719450900926456203221456793600320098036589140365924802970597042389340141784940898340588
942082813754108453271947659401578491808798841278671288346973044453463300113407842446974
676100521632523146960746179723522751888911366108472825044733387698808997882496174571432
653895931989193809453735620069779566007329207375987877395334012242628246381176046655295
493277601516544433987977965759642130364849538029733629734054095365660271566620956242040
189725410026903088730688596758323634848608003136749337880469881081792434870555858612604
433511134155068347210280388630798842486479599344269107098070530828950651392898724560947
408991150499159326607612639813504186421268398792438281063901902442716735076462402457681

7524129773437704721153408616804178294996765068580625127475299506559532498818661118722l6
9921472095560654755219761455049109990697568423675215339297355971225275715087665659664 50
2719171782052938851093694472103279129972997894959537217965414822046848471079713315292 42
2565810659649076885751212831515757008915688390775159233949705557155439612034287580617 51
8867039830867833408101348168394370339219341974243163346877167554010287905951855469702 44
1074836909988531592235768360501858756785573637458577148410634013348975908777490583345 53
9770579535213590168266477382725085565578135488763598832002785770631642240468395161657 16
9656331177116454122497180862165255308450802635618913260435962920096403238435463721295 15
9475370293413557820560910346503148279361410656603454420828537160023113668131909196410 28
7304920850041743703833744628104646209427769637855809342757875987841833340399660193542 02
6714882612819488625539504381515336088819835281749473542529613052058898947452978981927 65
3623021464927164086320292356592994191752454776144084306022379318567606483039434162187 53
6270414749761396384298639152870831145938517668485369224524791339701856796618981007020 47
2212331804541923079943922150899183397221294664854266918247805798787826538813387791747 99
9298627164543339304246091128474141007611054200897125853667283631486089863434645693410 24
1748675676488649993201676076913951174561630327374498044609078090640304676349444315588 69
8973721502306022408768962808996777200829540097286219693697999085628637818920687343124 34
2519125716658608488533132234261843260558367351735755946247449424091891352020742489171 84
4022318266702073646768610186247364849275801473588812961571164530777730991879061029888 62
8714930204679227525167103708071639439123743167928668219344462247657260407054599859682 87
8959481812296099664498418954355051269746222228405582160178156384893241562942941023547 24
4744065298275956508523080398810417675310953945082956686670098059680397238783071088873 09
9167083990986667030216146571722478408522623333842572081681007339653460321498432069726 63
9309186514925480137010383870547849580569239080907147014680319441188291677410010867607 14
6367034609701658774793861986557251491603212619971997380349016484226754491259673931239 79
9007483105538506866183048290644335568139253044901755675497722458655370131148854521455 75
2765003400128947427422375583403216774265860294150285405959573417873490709801590858265 30
2204657806921368634418238335855058044069078904876946952301682422689530301950384904574 09
4772378584130809424481263867625452617907185667849491594475752589043298597155625391687 06
6405003386911470252752877463230763947736622050212431711197669755407073331126759558114 30
7664350837766138393741882119872814024301959257723399249774565359917373704823455256901 74
6838618160590685025236871722925582045471781431991858074949168211910106141017546675307 62
0289154632134291872260156914532339244678353609292392595631799247736426558854142993028 94
5714297643673232226292360240155503056432028370518644027032070094133089307407897145934 11
3546630626365872857188977005569179639209408954049496757766916683128261519805386857951 63
8874569339612697366987222044985742652078573393450055218249597364838727810394612054451 56
3797961203029165947657469934154327101407474577289265442299660080219143075163201211471 22
3362886891100314198269762081161023720046209913211643260706919886802864097226678090238 07
4035935421449915746197968355714813677142010284368270041034431879942143613811977053870 57
0251577675008745353928774720196545049062159447237705651061967599990856948777593914911 59
4201505099136774196405319122353927497551027522621259329031592920206322743156316398835 59
8947694912780282598450835836799862035335202068546055921678655283576498156695323158588 57
2387298888221915594480378709089164856729907213738605360437121439616910385695176160284 75
7070741220885574454803861554929996011109008952930561509283466502880398315529188908659 02
8176649338550360211301004261404612185620272908635851705705207750060330829518090619335 03
3657336926887231145986400466223734847362980287798810214710192458549374877745311596289 79
2540550178074749196477840674655279039319556581389669254392861168127028607801649247175 79
4769004071383841871022921733518989407640808971431883089221639365968753798701420400378 49
1301275010036189355286464804238014072668778949947024252513956832936672012672774688760 32
2848694287301349973554634498410829039902461431124852884825552468148762739942714989089 89
6406588465382777488201549894005594865085108465819786193302486083380072550353705752672 61
6262712089574838570781071679039632140611479857589273165206651387414183990141524080694 27
1641531248414657507367162101437285661506728048482094590121411539705704846221539045505 32
0545140864908348169336750662852070850447616870476424706292519842182340567119317597738 50
7121384356616120054129148709109996813318550345675525027394805609455333242616500497427 3
6992368959557120323458164445061839809446368120108418926213314665672159947081981768665 91
4883268231854601655417288345341670449309166374846568976763423120189832643839103418758 41
3624196745799464920222197983459305656369275684935977767109310304141130731253956424863 85
0145550075794360426654494747022596898510266337438301815326070463610412035069829100774 02
4752336575842434925980678196106766125498936694745793203834801189180462399344020486054 74
0053972919887064890835327384625425978152377016539340906639616141813699362622724220637 33
8198430677526480387417719061345607086951288294213418894326141155983741984309650618079 92
4824859955747397586597917835001625124791176820566112456878979546722894411612072462218 21
5036111871960386759404634081534052093195489945280136392392045582070502328159177110790 86
3859943266252683370835162218627906963513461000189272878972239673342112248855253794962 33
4805017456457141696886360100538717492882149746928962534740324906591107947746995501662 90
2714298465088391795743901191544231663338727905054893157337140084303338771179398455028 81
0515225387855858852767867246546822526013941421263800251511052536202028508833681167117 91
3145351827458269079362143382873636714785502540618315074263817135131076739357650065187 22
5796621355848452519981400465049664429369446264325353422704810873584386515316574783693 49
4381756184393891019209933920793591730235133613433361740937889433243636766210205752064 04
9860033947626117730659790071733843508611904667283091919140548761824903540960361117175 87
3842829531071297887413006781572900718720285253473736830526838820885190065288899206711 41
4175614821804859030161269936302200424573036545063083444521271814048110646265502183349 18
0872813431700059389454647778071780075541159447956636875231302809685638497664674164239 79
4038097802400682239304397514877618551014680749244431304936842402797966380697010721859 44
4694667569526315883828526261340027805651395416472679784720187392873431743195634271468 61
2868703186802680513077833113336497051424345861943399376038313489195361652219857173400 60

2626816423331526275325615269986604467428210001630787133567564176057061036539724403434999
6407552391445970004248827807009018247852047697306068182728689501112304020259654646391688
8265344062451389438008685826309926370738304783630389808601099489941257512561401534463844
4237087490956244130195998756389104652096675458776600865903952152693072494759346376552499
9957398136870468238357822213502275156277174392239955413454901430780658888714513281337077
6148502576852323638293314742805966880964620998422476207439426900279429172375897478932799
8562424729659085321594720533236949043402796626630740273131643223047124289657816081090466
0225680448819724706799349489374391507550517355788273674663011336512806280676387389443511
0734047785428449458103240215302688926709289273432162228866530807917255253664825319222488
6046719040118814979669189723839048992144990637834224725829744875713871639376603835319588
2212583899500531756700955293648507888404290003623246079851080944704118776696569852700222
4236542148408230742496591289909650888536308725432732151415989181628756781130705162568511
0558151267135934483217802678350896047258005426171033289518836389103244737167483205917877
3365096282974559694346240925565281665664281336902593075874044002346731373767779248672611
0262584036880816938609418304354216051232899431137753391065117317425791903877442755577466
6603040662009904063042605149202987043184601327389509099815270306433694469041004457120222
3545117101132875640395937024233171029839349008207273903649597967324607011744165743432544
9961178069176467596474687979151557278151624730605833452636485128981677846980881899113211
0039395551118696836023267657819460839277758877356094075598291775428086114543301395004555
2465512429100491137288596606867189535571189037333006490897568335165004948243750201336855
1572849963696746425914953603739411549609823443143510932022180970935978032954975959889500
8110435013606216420030405425352518200915587623321754421758808594192994016616000363439100
1534009403986138161418529659189582746862217600400754022405234914487411541445060350425633
6232969603659720823649255942147652077137457479512200232533075772735440666725460638556600
0200246857044600372754039232960874325328139244892759626369997460819803076121586944368122
5434647600582345170986588687578964346022705480070837900413305141721926594157615687911500
1913402974858505171486081731560973989818711788963997543859385148127122856592027869352866
0760961001450046862821433081002880034237990803160388504060829762941823082783808603522722
4981023677059060464634773095240249025118717986424339190253045895732039085850787195225500
1777037652162664218528198174050734002666372515280934052081167101126969867793722598569333
4951943269320125902423076518277713527188447253277802055114483586444782301154711844183522
2932511493257269886174912260328402072777884330020182435128895262643485040180117669218944
0030138462303925595731289815372438169530773158947855646025489012335984452603058421107833
6641770498438042272775618146361497082205297894046841964210519595297634427944938008762377
5274587365404368603239256812039681539780620318441175173406354964644946886431290056599233
9710398026055271913444121764931576701250232821586829133991709434721860199061499472704199
3722324481136577736478433420225999696279855298823483581345198218142567592434988631331577
6435549852201618747009448486245729014154559189488870773043749586720792483838574340109822
5006289616607997109441836998747844395676792923888624160244369027154652760022493934903699
0547167448296577083073924210015283272337960935692399033882446560129800791917643031420211
9423739939643744425088139872031104733044683994406298819697937197577353254193649997033299
8030950573019449051768134116524453593299051529119861470957035374526557874245185688896011
3513044654670270758809946090330183569536601327917187944495410156034369228648022224704477
6758696090322096842256361340563483682971743439491345035015456271211307069128196826386733
3221318404444149770373845094446175483054536899360682058038898772474119523892924216374677
8456249854279850314493299533158554300276671540262962651669580914607881017471430699174411
9986584732904016553566585762630805024149558847753348985236467223893416365653247943645100
0590252258632136464125849984679616184355234035232472111052122663609157360271302132944822
0897661410378070919365580262218178495712207585119042287800087459286773627633230096904377
8031370895252076667175727182998614393655511837166922372541946679808216666811103956604399
3375037280755451484806816604367467894326404537115665863750531512081271327549205306822200
0052569298501430885879183833858882722616677568345546004203873216650375630854083599997388
3442031879253515109883833853900329096587405487398852972968379972293660129231230716020555
0973393093605034590395514435053077998616792471614432707476245085130197897386992709933255
7895246455475067636682646452715255225433388053548362739162623925296676645875489467344755
7727335601383827372900539389665659223059857104848277439804972058382111553820098920966133
6946893177199111474717037337482698105962706129131399606088218772148525578898249605715111
9740995507139928669201545658383431014260308085868849327192298415895092643571831409247100
4705184512875869988410928735902874312039343762798516411032441226292631100110969149554455
0309453357692140980331567654806421257727767562525366210180850636818295792871608398234022
1472035362598206364552008523128058003267168668344815110463737048499734839907210272119033
5800884324222116433444508002259779528179717226997323743864517946984457648063948949183344
3852518042878693263275290244789047593794042859845274992227797210002389112154893838239133
8287298993173119476173906115044782792876911023764755025225717321948181473706301308841788
8981959816299954108339024441069270673759595699711953593093849611028657407650636769449088
9301855864987037289727234334572249278915326092232477022877262964249176980803027823621722
3937988540050362571554887536100890114568649828243767815051248282055049206761472527146522
1896630049688579599767752259397406030511028985803962621881971282170519263223089517468155
8647724940066347625239985417319602616103692419571597760197169490239932872743974658804366
5659364968801685286397751552247599976494185950268040500640969843511307379711044119791800
0574646549307802152125298100873140604694735659064689241814839126360000736247105564819822
5893808874576453627742993768135876541917973572296127000892968471369649368367896352518233
0389131039926337585965257961649644990890955243550865890255302785990775532590127306002355
5311241372288339546404865777833161576829861517865092413747423720887013088054395225927888
5302394309216595649098407706095942612962824796778811133533262952874797540987883556678799
0042919545157674414867840448236392233509566007275479391401697107231858244127989233882000
2377940639757536572516250133516367264435915977475061192571301623009093734510047452761800
1638070967737009437680596671422941358960082475538324597480393206079604490501769207058511

2367261984589568309379680625434025095746216595188797550577965491955049492867123325133755
5673871605735638002894299024885121880124056867923618924755604824874955328263873146464164
42059885385147743343317259129731711974000426498722243810614211032749924136371337547432424
0629667251815657913864370256202430396479048900504429852624465756623621882085409494236855
0573272737762283655293864213194617852606260499906254796884745853044130593739472779307755
3507819355762734410692155894072757362859694496638890921585132706101710614979762053857088
52812095752763294985766777194759352152421676877868173437055674237402436509635179971533050
2057143111464013586402829024515173261076716920225250063376243107416178747624311018029011
3318097223112382400446652702557913433386482338478240836415091426303214665547366175962566
1696659433120665985127670461450435565056763231927238034514025354212809618536406586065955
6865009005429840500460935485306206267704765845643230355579621397040128414507155132958911
5455166928658278389403391529922388233290253888572605849243307420504807749658189661060911
81058545419793248020379655682803999261459692054638058771364903148774404809112742816548282
4191745721110244974231615616924754790847307516632660981952379567646387678825343150882202
81795667477147068016351059647568318898324971204609208556997133757414465046934783022432727
10038421476879322142485713565646283401032424128263276542081608944807016915495419078899090
8583899738707067015416665338419583507169714519341920374457438205104077772973273608393244
1637456285892241337653863674955049543056637708434508365177004646664638153286744482262904
9601846860503688344077608448239700256776213235721377269139239309592523794220566769837044
39260789034826737347545283328576599177610125695553501926205517993980215710312414311453030
2306985898701030358942128852315150644414206528495349362202442156032850944544546287414077
4018450857333734350776305942612250192525532512991863421476582140383079795273873761052733
0263924182242641542150906460098831844152564307260014686146011619491302403669382475017144
1894225920208067077454915759538454237813886087021786642478602868245538257060700785282733
3222651056334456649087436158229522645069096083169561726052652534915020704138021903400577
0178831183123741998178687238825101059747512023490654168401573350143178373352481938619822
8717997108611704819560792586428195619770249670042110009538004738803920047245467873090622
9279686005426820228388866840290831335207688650527791865629012892131240315114784046500077
5712617797115876960036259178899584553203528776418478397863166507073750966908836131673911
4766831068048300176113605941255839026184975476669621728534035859219034523767151164313377
2600671055941435933213580593431965154632317833809081857823319571680232256364543546573966
5389158512696172683566529545299336653616507398029873401838864612441636517466666989389244
8273782645426314272038650117553097076155873345431026760891681516242126487058077506359277
8820073571778056908881607984334565970610092424036098417826254172021527883071915797667422
8851450587738133761448400083912643956891713569322761335281604797325611648002436478133944
1949391998144463450338977304830790172218978761141526758491378276713640481452224170097633
8024592754166726985901420341115880451518379364707694489921651958233263828168333632551300
23426351694440084457734264891932074127715509564322610386891038570095852192162848418489898
8273268655470423667527507529843122908730541983950440894202141667808219680982797670774922
8984971242388093354144951082942562973278266923004101161806478685421633930127455892232422
4786749160761576954116883021345421715965846090848019719649487228542292491332269577189911
0652192682352097342852936279886092116791707629485847496149978359834308720070047710218566
8974412617231091035586226249946839024978024823106107738908043031729059847704522430321000
3304957569659557590980897187735513274829633988645718778469106403556448961252735144868233
1053002778188431067681436348836868815197935919480586451837858659731027120780587817682833
47642204580417485465272557925932127542209355067091521746074186345010479544484728043228787
5904278532798925864532242985233863325720785494344100713049181600750957198178380956000288
7475827557145959121423798241034420119904298000834846679847791736676339167559812330736044
4998178330002714620794715396260742401905177826968287930733427372635545596820513214757799
6885165521578563821506100375744210687869817590879723105471878597945093416353173097134277
5573684804654936846085893279519387805483535183845795512788897107538526481259181979522711
4467314889783066814412948090438764754172032883679315394873192784282061408378211112385511
8592573720264234464661698520633845340600855266871699988254685318368450116433542242467666
3197456134600849630856057453735903033205858460474211719831580072893001356115756207174233
1489330479644746806496381293164292352358113902966994946800145068838572950498800317429477
5562367674376499424361295901887816363422319493407258497317389718473874274935509850647266
9696844126520678050219420428761073628889385885038732456855643881657884628098866182032033
5782303338009930591300723334132345096025973746052004357099860029814550957846283200151355
7359254602735159675644163653011226471278640324482400773799691764665060238733966563567933
4039835656807221965404885119322488205427980912971100770045002277546120661716691559139800
9765659822716963173713238023308189464381281348664524959954457359960273403749531981034133
7354585996149549836091761262853953078738457075946329303714882251930381717511543835002677
0899582654526381103725254887743926013540602522145491981699579873716453513255099052087966
7799440782253080775816995602711127758544868440276052939451428880029095380284854110122611
5778414915814077499984149629224019889130831785966691538822900994694745024784490257136733
5697263979283040328606345468198590148086774140892108904010576575031104192216149418743144
5878476136714773918305353143832266954538329922394045613360601782141188650929220792949666
4091216003590511538805649216270544641912365189082065327758919973892229390120026823222233
6977367233003938217236746530526507439140683094747321260320884009899014802678019948268588
5535148065705391400576934547136732038757724513069759606056795390037265846113845113230644
5833725058053167934472599430552175008531778636339819472177438498394166462144855051887700
6616890278874741977750727859461678481964887923839242970123021952643848769171169294191366
7645398975302213189442746898644511952336113580869952565738499513227234485893231138679788
3119517843877135064823078704829980344715507014188205310414266822948160081609502468235977
8893323946769501559475750223592602042472263849410031136704409745365861030801205930892755
27610728526394257529284362186377642535427818993064800665696367275161697181990722601937575
7116892594797447612487628882179865013674750750063832347988396497740048841235756668657166
1421583110847360913934500027320051307981281570222561690655268330308366563814134700708199

4221664848221041593434919082040564085952240388003780734926165030023171799931482592911 80
0377474465950156599813862386928690626523820612336236745964072098350770110829907902806
9034109175096357314561823190444770495486618716069228030350137359522412331696418348799 08
0748080408689982217275513161958780967752165398983096203489409368385653942119612308102 11
0347105174241643465517192077927713852950602675186424396926553672334478410068145595114 90
3678283881757053538003894600276907056312702323014141306631680174679733509725414626095 99
5788159410727806596654228530160830980948279808777995415133063418519778723030126639225 39
9955941394962110041954607825206744250803288180503393892187565244451699554137647784167 16
3730755847972338659392635224018226080316927670846826907128840619197491176562869996908 49
7073082337564779768748466753052691929850792803668182143767960730508738080830144642975 98
2541700786439730496108341861969661959632201840359163563411843581859820514136319153091 25
1744066240493909245135885190762706889366270990559464689376680069204682836304625016402 10
2743791785448024851286182161251211457000357346740692536790368909502592398915481716225 41
8252452080600390600406155405892907732068730958006174997192036471209788419244666709204 44
9749874824088659052669358894877525751640135436742379201453072202353576834544681206865 95
1393272635592769965995737774411037909107156835865846562208586210713954935453912256732 80
6295275190075494090489639438880642545570726221159363124394916459725649101842575408220 04
7228888466345128030448319017840074011676477395615436471395235581999769359010841775219 73
3620308326257616596841193661458533114207532119529276697170420670518598424997628346041 23
1639081227908900560239147276254723044656137389193482911984549306094462945096159115367 55
2528626105912712122041446841774978186305011129740039411193508189083573332905511440743 04
4446758533039081986774585870646675310587332044486643819547370848098401901457110801511 11
4446629507460652330517345945257725758930786370071957679284954220239137265682599318384 96
3573717455405038735780540832235428668250983407424619172124106592840528111662009232829 60
3017213638492851047735852983920870989263169843588574220637445799561054144370524882233 58
0267567460992544022776840093593181777507857673345320731185308379736957382024604745009 60
4524055600641568355404686418106415591598692574489030347146086368684207141529519539968 88
6399441629850126219826547899506312921479605647184999313392444952972883378335522530665 60
8113911155759997907138289241837357409051932411808327532105758344340786287664294881133 59
5300781151425957827964092837812763167468852532329802856792473204532093854210158071474 01
8094794611604862776786734377575114375923330492549945720627684233643946932701733610844 40
1875653569316078803127015677432921109546037466986463058964329961957990839163885107358 36
5539735868580394756294040228635209634721703945047038525710853133624475454200105259671 21
7835787463335941659232356257039331280188197993487698085088538737901567888592495993380 41
0450709566819780689097913047531270144691199081713805793823536727157978743995647891549 06
4076938192367836672321819058213639903497314398119674217404866069196506586883151348340 18
7681346790426439557385900654837580715281812895107414409604501704396548535905382780434 81
3583077244516003786097373743147217940264953077294295524732164285858641931339046225557 31
4287679022533447878688563797709322070475438244713707210817280726216192203451676638569 85
4214600293710661317844674349494603427459097077940257119887375313991238132600095632063 68
2362858307898741532274462127591793546312149215856310068890958077806059372828374066045 17
3381475694066896879074464372890324045717468931622799152607670087495794636552981080060 56
3205359323461491322115081869171155006556665547745497875592906074227619249013128677758 01
3421084016288308729212261577650952210150804466374403297822650479584839492908809137834 91
1052701891586665978153952243363020381943077972210074929529190141757752516992977147935 00
1371899644889115206473629667612182183930848992602460041899154669973852196756729309643 42
1698898063419295331116601520268906755263925108101725929474115970572467020836237914457 65
7307631055046947966134150629498747616641845707823855574370474740572018709533927231232 50
0033657654418215016266023517184726721533121075096401585101898137749142654529986669208 70
8890369491023049304630341750898351467990256972876511504451026758356083496277333454143 95
3879619686112286271837770262649999543789756618638452242447394924921505485101221670755 52
4051021038733002845936131845844278673382314261769735636427084212028318843673819283471 31
9508717312219101203167211411093958999228846748017416567609937878196877076344759701878 70
1153635070426806203432221962481896791090562799268720631573443595078978529230696719511 04
3330556678384953840961252779583891054879684848620867971749300845214435942534620011241 08
4266566758689780827762768401346982941929580203305740047491397897105912264221040732557 89
1314047746709521633731095467107147882434746973225362089718434140168951520932937289579 61
7990097453132280763181828994331896956895304723703995389058396574835501508194701003364 94
6075415680939094827544998118100311514311243716206028508211677160529015030383998177874 98
6196350048908052208969068279491550381572239746651144204071213280056065362446019685738 57
3258138094507943474066036054359116810385474554139010052108568269641743645926975730111 23
1427615691640643829304144191258097001501476260450843029947397770443406025584831551837 09
8621043718244490932449990941239696807273557499097475439290255798479709348219032808505 91
0233185056595885693410369752178796616771042304942352351086300728128713214793278040206 64
6142623007856140840325983489255712085118985382238513620972879195187746506418610105011 00
0152392140198811550103331906715391496612736381353490620189880118606264888141694352927 51
3020120744485069394971565696370052810443645796540085580441624842571854483720866433386 65
7525228581094828921725783915819147691364603268447620225583378843070662682013656256706 04
2916609673993739633372559817540236901883535300799015939672492877457231001781338885062 94
2677684523610642620854720708060536737626847668768462104365662552545771558209684895512 56
0427094838699004537060236388671367910424911491996301475646726002794069393629208526804 15
9391655694283101370172150024613412555388032120174802466199405716025981142053849733099 09
5858647713112190057785168213546567692543695868395539592269791198151056786242787386355 96
9635159652578010088777516139485947653028933659176240229706578369853607110049534756757 22
8407933874696396982052754885413863809128046556795786738024779624558074935723887491817 20
1030089198899323795339275624929514306391754175652362056553753374784035475143499180169 62
4212773057517531727140899284179971054379976646930483998576569703889161802688948286649 83
9647382240305236838578791765498736162847160152275110555356422709303412906341214050374 70

pi to two million places

6538761104405763127767768795582839693606879749299247305575701450712864877603721671366639964795168121815089563593221450808534862645244138042319376365335527483533321558341288886647780139622494602435843022305917574155275447784716651515806015968314346993860224116703396103314344414215237812127052900415296832835814274572054807634173997685403211427870270994658214566961420493586005178320307495998499945367759639015443329837295987702158798404530424172368853956543113249128001668861432133590181459881534511564969308722687998154401637903625847449402767622314058383024632327835558970491228763755160993522863875948264709234548966040439552829693496327329619453926341254044358306491272796994144257715378660212159628384800807764860068442119512842811118606563381627587966850467909393030243819414713450444610996238141708045889385979634382447612009431475013914511029035345846423398665337750340328875117844562170190700826875371234894254845267952905967289911416216871720725278954130366253131216168718400290849140108824741929033100395853328090305689816195958414640350088183835447766161764083433565762829160365278550533429201734442399991298215606563923309683123260611349847459047534817572479352289989350094349507539637348289115471101729844079071163848822988417921854283174985756016443562226461225946402830864776638735945988424504709908678771675009139300382117519811184256499449961925019393804725337399459333773125252463064043429925100636277264404252212933536398388871255865028214839351953782912192325132955047942707747981757306909813981758364267491567563803402416350300899745884664455951052637730388753348734402177258548165703260033562049041773557909734759843947599584542976546367412107535150701385121261710170943881638681800325344560780113893154572325876316875914141839336568222962466009146205514597833791156464792666354362782330254858219782070997473109160351106470097487400073152228766473962912778621844683555002020430719142007284627901831863978702570277226878239103697244548641105888916691110592202944493294362703541330988052688008793461709563048460288276601190708940730028200664359866943099128838669523792986662817889842726970488860447376760942026153771779179096775127871974710440809155990679190772341720808085990428600454545675142277138473782341100531182443063238871528440862687566050697234784773621962023765844110337215904381189469829313069286115985645313989313994899983340440092422837912517555217462189128760513946898847267718746650478527066436257483191690849153712588054145403632674786953964910372400574613022029319950310198775060288023797500255215749964464245334988591590936954395808452804500493663983056378225410562624168321730113237466365081832155139049801939199625148203485233603520297989224377311115009165857010326003536444475177424698915893573470565875149762563268039695816969490397599461063976343230542272130876246685734670460622349378419919838013099392802365227419198605426424971179228205037053758742713666727148553094608077962908093583854654683498403635552168457034303500634102350285348776635304712506884408723266759056557933478459113321260789019286980993596367757831289572697042883799355130392695124058919984490604631927762990564603947687565277618898780750820211548536425379197075472907112634428136059928119173570992152555519802760560371809051890207185773505552327139139625015942725393023718644501766178359500536742452835334629660040084680727285331808352724863433160206496873873921616095592777074720918638361912857557193948445722793390984130659405999651263847999733328982724471352363001173194587972985469557496641486067831936412121572645340765807066896025318540147824874797280631201916673807223776392087232475421013217402193170524688311956613036567070305219123786177939256907627223847705052392706222283749423814306103440377982382108877414396313901510708203127545660795464337135345992806287196946972555924876234056085997602542338053560291986990956076137368277070442866704641224740569967492098598383612829365067744980245221678095700993792811010739323086789354647755651487750747946650087569269504913255612642806005988394995155162576408277981605727557443961201814749780045132178212978637437510997476733763131344406693221698979064814152245960579603637492938539058045809826035568195289395221669574156422043036643722991469676043864419421301367590016932422693491304924670270778248184552311346110345348927331506000123028538334230363824715502555136874563216693665604441464245556981823192741194508279688704694176702966019507450249851652980618680546381347525143843327899199609271085812200893576622256979935986992214992546471768210127651995959782470050142217475458619429603923945909288284181468774841913614189873828126483553432410169649346552629534556346170833510950168069402286750567763445714741321777516730620778218770649224444752082800098342557045778491919168841767717868863033212689549197645739407537009883714004870254292603296387792875437706956043733999010029485263815003262899728551130103698581932674488505222842191805455408227747527607465389890506437479816983471777490761037600628289061945763967812992880200275967294498615477065204732195418902967085837806056950268859915202286168317793181381113370084664828231642940194035016674892993924087057564794251319995861278330973526538433593841675247530968424697781918624343771115599252828273132951369787821407425451631268645232757808217439154214688943590977318156580220579640441942122537014799189278853790337774328734095511741378352019197915279650013938688485569374748216129271672957278563847386324693858405292467490402241343895188323801079314601834291621633196573700159794273900450006384165131451765590859700264700322130221985224974139157795298792909634972898511760181137446922094253103131383449613559931817883544164714503878554716659769824672479744031166060619891225041569090447646624571283638206166742756470277596897462784181051476701358042590385753062036572937840164916694827135928569273543067691788670004922027323162640407025502795620934962162273386194868110608449358956017870858831338441728876389093153740724000728025325627642840264865650196869797443042592258495804741792279253400552524744950234083926561723909309423006093663032348020210867886808965918168479273683301432714695684457049365421273852364197627489417603760297520161535938944876220235739135468342725946282950905765143194209595516072612741353598331918412357841964213428872566873897084383114104658560037688220324630865651541079929646904657770652379505345960246149402061560544843064378729945258226263609197006342345695812108101880429485136828673985213253451985198680652792016175389656184118252242529689346349832387862657383224882146718221239216145216332527567001704289399052465482587785124125185612578868945553166549754643047535031903559023214381285817927533984012460823890717054605835960586771990218346528305718682775107625066537094529888302119627302931858892708475701485689985296650573380

471038620599638943209413359577964476992214153786551124648537943925407362192752468482382
849973125718645765551508695824151349798157017437827336637993430650906064980929838630333
539425021822566381273209743546622445887643499407355388635877067206336811113229422983654
052688215612702459628857235482642183145461433191254583311812559791473648401291468622198
673758189771951882327852080933278280528503438813801952845546505139324690269115602676843
585443935076285667261266508394535898309320803700107893243658291550801322381298871464809
135644029247125244100245252544508023246157822063586871671105569328543801624683461967492
375522775131051001267760576438799157199485970651760213891464063173502233846434583948354
350279819027287973032028508468840198759498703781461796686462875466703998963042483342254
904924467013293924725832362315311997123989446217658842719338254666216103821400699023027
742644385714175745879439789795948158049729059777726218748279191542139015671040424987960
383839873080715504253039320113817262336691434188476626755032586344926729416355440616416
058126006897850489024654095367384485080441994812315226437778358928010770528723579813191
764225440790262977522329943162440568228240489795858620420959030167530770098412550414395
173705772057555508175512601790181200735133417723724762208120008604407951239514265989643
403764245060829599666156088903857106846402914127173765715134887944642689107694108953101
190999299956309309050352227772332629147014017864451463531187383784955438825008569273087
839474528749201768864473117831041101991600631498818929990610152778168708421621381839557
079184051198067597769599875313775772688789108864591654468983133474235479298051910921514
683071632385535103872718754467670829529749053459537652593192516594514793368506381679734
786636883270789543959667729832709668006279053959998294577731682383260738801806541025146
172162886788358706619093677297966422255933690824586710321214530157614065638488320462046
551157310033106271776366327253551051140113729479742423417996595373489421421400236584408
113388319761752550589009245453137756058842247628652387606272469903021267047078094512414
716294955702704018998666632017984230055070084407533279625699917718765426525703312549513
970864944719145272944883050946018415295562514740409525798009901463383797769021293940853
102488561567350606338634923684489507528233401007520258283062071137190594267815521410921
860570542096103071329372555368225794735587462567776516453310929822876028379225930251318
516581337706052092108657561743012334289084699223497351511631421745254267139789248050025
172232090821245741107761163535916860465237641185208310455600513909589498730970708723111
425470231216733203810854809201787391487934488837268545689214877830390016547741762281260
580728355415331369007931396300063769702007625350507261233441510111428070936819402236989
991308247424654012701940113222999932048332874671355383494579635836899288623290439722584
493817107725905803949716259506636916042428812825483869715966530554742543545597343320165
017471694261408641380380466595322388060995968930493981398914417781080440177680412631187
307038032840781365152378659505510087403583849737817232100166230527219947879907436057423
140992833458661530302659108802848943882627192860592688546252611811506554314391860473863
832014952014199240165101739767409226043254842945659258581776899771652026749864198907493
364258824303008229914088423037033492000321094764235749370825153883596128554028571511999
684121309513297601060622384467853304303605283324594771517521109132184692968901359920399
067517466637717540893162635269159223166758528381513309573351829442340194857599928875715
896113735250073352994468645177277810729355506620011166278640684583474212201535461842745
627781395631003503800901852220399726275905468272699143753600658655126345316534223994033
256987619903270018293229045380216469805315530988295337618967309534457130377128599254581
802272613746556905822595786920989804611674009391732335754451424181559427904164840501217
527511162224841376487939528948768911062083467875763236881995065081723493681850049201395
396931150450840631833169795650011516330083782711074977286046415193311497771862005817211
835717658891646355701844887330656741216711045991852850612219680110732254829518774076669
979602303847200725332760059467869526790514319525735477141111573062837948717238799010110
737197033795111387902442285766119513470938240551686729869870945885528098965550905005839
479776816362135995896454669367741167952365593301962543171459828163763773483041585352887
106282009286734513178670579055862422877697703803358671896644007604521050778010902637401
436327800462862893243121698489569692681269965570096116297810488083332264011584449865788
691989155116498775950082011654710794954761627253597443140698950143479155214870180524406
888053182445054861510557508245833483060153051527141034013461587176204932737682281179363
822377263676950899606005764576074349083808672495330340119364736422164031877350174262838
309181603371305308194700548145666334229294394379129613611797429979597898222018382043393
751513900818795675780849881967116995779814800468611110202998559769628419388687612327451
524627733080446573369546365493840400819777609706639132376542539186868203566854276619326
843902885919967881472483502319505887747564159106418991240691253094163125619541095435308
814642343408331609704950449309811673539831293735539341187320088670867106762928026623131
366609838364307561568243371003247612866087421391893567521305950626336204982646550082066
501877463331840481096537269399354992508460932223638918187900587249238610783215779790260
035562226643917254444606289329459454295831001567300550754372474262118465163712077024599
682774758902127077460823281087774656437622050892211762862594923337322306799176150246435
991356381620607408584397425133159389863383102724114385075320805389733801159125087956234
072913945303862707068178014681947724028939617221644175848630204516488795837610929850676
053716776401041227878179550018233197260574617611883779468454732039891838117019778662208
080181016483471431403292545031424952200821114330744664013624225319259875091575121739132
432965349401209539286534708463158821504955168014428706494848315638437272630481694795792
035566844577863829722889535344118520610069545041770445474492597086698863609934470061993
886472734499279127222316585283623292536482593421073552499528548442312732204674710780642
436699584238528637432273244201828343973400032241859019238030590058722292896105514993883
061413500649369104739021291543977494536051080648720801311190490223110707230770624283919
528937220911487783908790449596315222989682708220530489656016395945586075535222221595738
349596092864920413661120498768165209416326912589404845282229036070277275509104234760715
102608470372049953307356616520160803158835638796224312089007094192173450477878774094071
468706792259425905227518180949282295331821489040420843933377285890253665084263277258143

948601959376487549244711520859661665883085955336160717058520424797750579059521204948991346273733393517973537490955401805020862425294715561008799154147069653545729992240709325803842553897467763514808951876988364635894254942842207312036451005027160780398336131700022763357322058050472099901287776893533759857416645852007639216878048573675392949503384098229397970658314255553928295919229698068777227966397293907779082178517324761087355641896708494182323029269132494491340375769778800008521206899488519501187042430819704776776564770516660073820649884857171048322723457119259180652711670489692290985807515362751709550528429039224365004824880744813185743646566984452180536664675483798735679164220132961903517086414797337157754106151617424044495799580315561115791030873134720990353018949994658219299219647710568822829861410142019463905423285844381712408363322656243251238385947763567301206764410850147539814463428593104949686936382640046416259646951490319611104854477591917065843927067602400311452127176470833320094175688738759477706324099809206846305347743324194522002176300046622820238082778041977949338939188985224408550686669098726150999343275594219513614894603275485400282474638185574303867224848145712041289402200141526884770962461222114999228876439191910899400076450042673636033595644644270818978527417707745133958409576044621143276559892571212464070497606068894752177888846753657731308884831704130847083028117792594667012087718412865941990187787509632002811023755143635612304865546153298828299904617451774858147760123133434173138771090557709367065736550203175790043067229303144502419919774280967622124251992863274025837040075297281743548041063934506373267506843468818388748332354112166341880424123303403490967577941653779084125868798288610323527885733215533815198830588045853153469043130898096369406641813704154859314966671159813089944082545715355230065250822849617287239674650825190045324558235748687720066747949712163602820852354302783865361171124532148648794241321331700852315433727746076806637669961889512288049108911176595515736498473886069524716684752375144641521336548924672761225853936148416514385818691738416754348278131766314291169378556461817166096634029127205365302544476383065335504511464152247086512121312900999001968151695921524303910229496964390635521990651394321630365345397471515735014459156097003147953738250722286432679118022285454450510066868382649729074813258480871020887495051426964293739258136771841690654521561087615737802053527958004468491361369174682537172803536784350361890124577775833864677004871875515418115037141294549114272696877208861952903110006521480604793894304261211250474636222567539347681922221920063516876682582150682798801607357061108055578616867049474864042027000614400977944187614785497645823956249805444955125709106402708323908144600925117778765206380393713571176447632922162141265648373947107451322905073720554233262086352301211192300993282164753643692379063250335682531355433034789629153304492311538091995755532944987052801903351167407527636655981472206121804385730030729787921735685000125623318067425988724010996896981385239730619195592833694861190323949253594415836595816138391218541195151992655070437222451106336712668962567275866773882387907913364509386511722013962859447865442962183266784517002027818841924009364903662723574437287448563310872878958458482352508201562742207923922033920450828194622615291844607061975822852133877969632367864013313041099556453747740645777576849035137916732532182650044015006462416364046311782779660535756676037161374420266916172189142991639230484973825428942219985454786894825670545771208306069664015175470113984382899319336352288857298115482869135960324285116362192220766798190063884048427152608659071553704971859335226057118104150457934735396327639988723256063689608410315441364214878261192854183849957430442768683851459149189187424006619028314479859226445963347995310286302780183115008930007657628088707768885566711361061861689962499638799370291136195006215509591002663942980584231778964965066275652978344151497338825523633644652036220132162803213500493619597507026917278323701383715764323028881013296332873938245738746245096895082238330844176192408476051027246860191447430391510803777481923871052911279590551749882293907551274408036416932829212553780088491928702854675425466697357397053653624540072223989562013067604811339156349727760567144964090640451148094824868511796216404428068971957629753562236181688850027285694336528800131844121214112389838519527851194814679016652840688382186958683066129590397745990561487036122898098411382006158591424712862298600417189064530100820327940885803857608905122698760086424606482694850486186296517221848751835565288146631275237687067466752691724416729735456956733166771849284394313859957738504850610973138057829209449944463239430606875958119003860261904921098398719699336474633114066294511147152055694804027987358243091859738263977134041141601662377526935772236147756347905552752166482414609981486281328766311875241074174329874362753850457777425420575766273931569840438567729143837835901823517368770880480343742863236659074522885395228357786813472195376605008468619896263300503309360409978222914815147525771837952813848915680491921812383604122713582961164972471080261254592059524865114227833911564975466627448671852187561816294711964713806687771536085417869418386146607485365395525809016966237780000655836818847197698244458732298445308869902993378877952657197280899159793941934367522718663437829079368244403206246396018669904972319031502436250408130553533830653161092378952733391436979259369072869740426234042475870337917002214925834352410157186453983478454517589224123613673529136260171215544108493033216442300759697110588546958695711726320379851371929401448711950153715879163321253830793896944127468922739861011837208514286931971502864690987328481720738738152015911637945123010101966662036445412956291903554810519125343871316001524124785504522454804170858009744164360840375963801883860748953526626950353281648096816794488017615939929356306431457111483516674756546277594216725378229613379520048290422881659995670507607348704290859084996849095294910463268517636522463170134879989376887798420929485129627852301598830153361268342991776614639254947705193020500310556054936776339163283953895578869977697143135446101324961890191705701201820667021165776654146051368534345117330284374109752675183557592471851518890916898948657604164533212417028081148467590773013272547946094155097286789671879618012204433502968796219644048327886379960409838825362938230395839698373949610171235821842177439742703646911258107515945266355464675143678868797822902295599477157643067126525971855615110357476610420964178412476830158403397936011211878112008231745037140757004092710837435401089199345949837567061274097176992139541109521012508398136549562045150243668431133989738587361130651247452

4232154257150917098311414008602648905393707127744124406690768316708542405730036117869052432320544268235686500327030306501507748047387002402672924144800205153067732701911145489873963492924892062897129047830449268538002831487535810599705614838069273730096866109888637890295573247377321839296823726596409999947667955760530718216186939459927781644525369696581224500935645899443219172791686763985578925399696609882487227058690249200178490271214863539553280594682884699335785668968529914380341052833693838980810654163125074949460944740888646908365216571068529029379011400101710418758204392619824337261561113568584173076286520631003127497144699781910388199260901664617975406910297256584746904507192524494658335276417463995178861690567326593163341245854517825808244900718850178514288717667208311599555929267262117825611904459506934896175722832507114752044127677657505086893670980336668678679866809585451635645045670009828213046712442375802553358494967834550276310756161856761102420062290862217868011243459156477566136273107961732523468140907001105009794586345724419026620057736717904612327442085379760972626868770094642272586850071636459572360638163384749434997520654319048826875825350515691074271345868417945695582717709802752816862020313195443597491646865492287630804666214313185341739865035226451902580545172423793197133547989443130184302401189808568284287635611511359254505175900660902931753419723703731662676310455267705728410826619539539676802350046763923223819889539040992691678591566892197646205371386957697999908884131768915113743462702375576135602359532129295133933068804106622580955975301522759071143072612099804061120495954470252902245829810326046036661079078461456247718072122094537822437960892084783698815339985784383584762331111455294499316358956545170743494100864563756823652452255289027971719998672996381621378347132538829807661287146255347835297146301393797884322985295845439519676803864770811999621581708094439895803825070711358270798334780985661030080924836334010664378517170500867285605725665824900630381665925027603594014207243253350329071534064372099104955441721929617285189318787667540913587710997530683942283296170848065834349711181400922782530261593481394755173603558408942664474933095846199862068129248429990069049530956019916735927003422770580777725794298919248350750002625357538268748323634227672480871144139303257644516263630141577372991358552647618310615475055054350039788791534532770215960445663570306770661001920179321401714796971673846973333970705605859892283092531295264942795361836760792879940177086017608475303934791124788612396945329823362750327417646243217820505863121003280810253530905228112133576906734827893771929083668640350282799470624862476886704402085953853472413704692825937227528959641559749916775787268700961437933909121938699113630197318971094560373016111097666244244001817806505557246233985925686553861168261270433407009518008868971398949213194807654561609544651226431496693496974361698396168741124092692508791649510122518675248363561605712348638468927964566676498484647671650465612669908654854037052810502823254158231964824582861497900418034575969586571657893599120269240475446862562656707144162771143207057262320457058642548648538642871832592358827210050178192591032186210252542906106419649321973848229246721450880276773631002510614658987528184567259205007900609926331793502930263391497547899805591598387407228201211160318474460731309264223605720140683187407414368473566930281385968449636781635546690457525318548265991161791886475069922132683880788802178150798752627095916528280767673143687407626055402771583328466679056225244150316045686489418112599997950353070728399541880040621837690405286046378220683553744365548569477836150626359899363478702790903099746277218424110017648215901270567118208766878227576467699428541130542428469796771293655637190811243475249918894104423899887655814909816313382834439867883056142206994865643705634568169510209714342381265370529023114891741626975984689067549381513688231255317855349374611505045803566781944318476851348291726795304651549598075641168998237936862654522544768231938216555988168975654498984722336080402362152126378985700273204797098350733847558808852656004601113636641069535797344908686214328577769297713883866875491736835535914853057824571910998302983139513757052525620958095401789769541394615172026176496607952106330548645811890332327753556080420928807954813104430831425411756469379644937008805178843906465059869952993456228849781367900568424690668982348037672283914141463938344197050525552745661524307031168939958640952184680068901136191300908934268278288375705639519533012518018235004929310697272580570319664319775643414186491957095194411520225796015794217433299871249539848164321588401682318015676822058890334434057690623837262060541947083026988680883194001517792506775717517459372238471772205082093070415931173622020003881306078884009111773966418883673332044652964644593441976859642812624451256257788632315383190956542867923083451240276168883596525102882925470174588508778546743233541314358139890052342270388006077143178342526680299665251595805267396825662975785411273245999963482719371057021772767908830058490736336401316479936837878094275476160827790635398026354770894899217773445189729100846164905691264458400492070830852606455660554188794101768916027728337257152926339124809056000282337920177526406893511804497791997323780203805484515346421441121574092611717531775253425231265652789565479949524999661341866856113717265753576161246756393636346585290219883593583139219249139341864245413593442816603840579430340585830595161258412086641797040450055709015103142797901457995856719745613645353724475732597176241622166560981547765107924332845973035022141801910437784872406174681283719961462839166425348030966724024117848379051186988338391790267930764956491327966578197716956457475933313313426260774897136719700587905164110575608680390392686582634870634540551576313986167638107741445122594128550754494215952948573989863056847155351487711933227943103860066287606970726922388392110104220541823141878387002847488883890566750633122209205148078706136108428374406008904461466797371582627202911168422932474824789178968587778059776094181864431634002885026453644551355067121340118869078555749941020501202058436935943833843142118798496695796671231829694191171815804943525795240601837585099793437113088026402154288164434430672028630306124498537156718090967836741275202011134541349983917111725353851702142430673210003144137288710554407895824702376490475232030970596062076120274233173017656903200367749269422733032275776275170079415069130235233822952293804237429919553110013757008735740489604930149100131051482856386998429294173647557855294153337949392024402317194271602902312715943693646130477801570469751026061543356023532272713255237816494055253651889464983903451788574439635435801343497602714738438551l9

pi to two million places

8478108928668229457725753597842954543499095269077618698050126097324257566739651664 18609
4335038414961838737035093807038010169536630461609240729436221133735537225631799245 20868
1827167064196100450690001783517268153921786584748124069882994394469284753921204769 67040
0891751698004473501340113780055210563049882543493206799641734183811320822606619087 19833
6021714825656239371072770800442603025786437546914140646737998813330627498043404884 44339
3728585142092907140693132785150534696812734523204636356663009170263359763238861424 43801
8824049100851015825229325655672793009966171516757110370227909005772432264519348539 58153
3676201804307906634693859522827634852807373926695415340681286594346995691180472437 66089
3831562192438656410313134058911508072193286739168832381494477607199208103547538842 34538
9673265484968440806351067157523712030750328876853691661238860177353540400910880395 89275
1210025497166937079787186642922014014558824562342414467031323035012810324420163216 30576
5377138709905272596849408782981186152588849252721860328952218240262829832327008176 34556
1299714577465837854724296746182466438490297865300776315937919644254783948828268065 50176
3316234010146327094725972018823335388133530254565390463483105173008497042615373676 40620
3301379064787387371214645255533658355828253129869085396002636689072550568271410004 17352
2898482171759994026807464914188870306381471153296467895593180865122302663699720471 13174
7863490527766941427342272327958087000259828052580137853878320031180872150984623227 07431
6277615294128040427317387669976955481915380842773570932813737605617669370728806121 19558
0689008159939826487645120033217856486984320906376025629999288897307612477412283832 69616
1107516489152282506445462683064172271803383431737658197124639514478783200093513333 18655
2223389556602516470810190022446747793987501087746162698894095028131074856975128637 09006
5771914143770296754856231438325159503852517313664426550285981057641837070482406073 20787
1177055295453096981835197292941154182519583530833634749559978987631994251041743808 77385
6423077173325405197636112390635219894917439052550024775592393944610777614131867655 09240
6892223028644317156237562055325359475888341884110318326056616085707801241213297274 91660
0004899274741402158343701248157419129887709471169331110411304646457319878710695701 13548
7660168472395555808872908971746912203692511897246805917112574573943911456518061060 80563
7865308957847735391188780952243969780018458236443954482446565562789229023722984608 32554
7054940682218788034731789918341605402685996748721800813207788553382430527889525289 70977
0802608507881752684419747475030089442427002773032472938196957991276667626902535976 22952
4623326878831389484003936817044687218510139571324767540822323043782281750498121221 84923
8110607200441017372402920022571947628031499451547833447030334645316373131527491626 92772
8719658075797656249029624121342739474949499605804582887192221824373601628086164468 29421
7844866433081941654909805035061939353734418444988481855950496502632226450862252087 09839
4179516137292615631666411637908625996619584832695382064061026825170404948988385791 92421
6865098291704758395293260119443105729997009488200064628250641429808857846119843850 93174
3499331587540568461844308724816903828496944549149121128389917442698035544256672059 23759
4015085287584120562381867054311890166818097178919022122935188742790921955155180888 68903
1344770845777714549283578452961724874657333204316660641302224078094099240861486196 61646
1791573178124113520858151699182455410544125676216433441070173512032368369224702394 75223
1986468094665767008441477627112781820370673947730272527129920311438451352019946347 08981
0581466817328707877561024420480747530842041044301634872633267883450445024484231091 77551
6736707605284105175214522934932480284264838884846900209944528626464184203472216570 956864
3477981110366220426052840320341215341862403961367926539763057239176780823941973879 37396
9931185790454468787315002799164768258851740784400056127034280313124159400625600806 03168
9679607752678055851584447737573204098686615089568326179598719347191082425622188268 90023
3410539718917603630564713421885935991661004043119568998668547095296307607868634751 80887
6682139900467692737094847486632625785263229965264902904755726556426281710673612277 93268
1885116213824241463851419800618482560202162079647746012249996696722868524506028513 80061
7648263418596708376361601771787875847872113524257127483452764923176264625743670922 66211
3077335552056833860543070541587402449290035756111655557169985793131000681933285380 01828
7774723250914115819850576954509512870720057652266342717714995753519942375852010832 75572
5591013419830661178092084032701596302419735914966068109019295387886496291209154791 31840
9276234623131144410252780158536443513026311959243846145507833436871321090511487271 23095
8757797122207183607136023614162793363027620066151301431755842436447252712844828608 33476
7494112067999001847319319046961301417860432255267100830950296615161233914007232481 27406
9433748481319458570184419485195460907139625406959265556536231923829498572212861264 509463
9194954110726922190617568177212932823950981632942369731247240843462067641516583724 29522
3693017432684138741020941322315904311230900855917808980986398114724234312597727307 25874
9654507988460850364940355636064213602463672502975825882142397090696389475158521960 10056
7087576174342220068818401867828964079713421198798942420054262322639109160808328221 72062
3832181566009563766131150703525394313704384764067125707365986370470574729955770563 32924
9287066767157842639748416481898746442062732629180918634569514082111262111830778423 18805
5048390230182385596198689725866375383685485188089002756723914879675747587144770449 63905
9666476388405551398320851105180869467334421469643889365198742929450079635793367763 06583
5479134409437498487491781105259529349460886960396179237563635270568286632335693875 47828
4960491495594355811233762962794911089627284563066959031292373894987390646454823452 65301
1245970936916636819842940197570396110501809396370767769574613416736594918684071597 99409
7749212951447536435570510404907182268047753468921921922396655988926401383385425649 42877
5082404565581803397987528075009321325165955626489684326472095084572694267619620324 65242
5361138081285541080986138899323175276100059368274818919305726879270527066816509473 84411
2413522522462164058966977377662669472230604792593765890405451089087271966929602615 64605
6922347083077659724942225173449004958810325871173498512933407482889628228684514065 87059
5822088546356622237969268457727080006242862470839100012713227466932910775957841160 55523 2
5394275509600607608370533448080696135747986610034569429050487428865415852057182474 93430
2926645020129015228508576137455210972739110882540490195467902253732559437010933022 23343
5336792479155486508905610315092010532977003311490993319144158254039376710385615125 89708
4131515152832179811625094407838446359926982485147799826238367154228185069666716266 20176

1609705670948611250935942092576725027370408358338931609750567830354428400003970020286 42
3333532380030836727576947167063207271563288145135402665452805370561633756575655615604 56
8397358211272334930266571373761578088814841695460695189245089776145827730564471156036 72
3401373751239113342221350995201035617764307709080447304826899917797646876003448036444 14
8641346346899950784555102030298886333848328181072699200088982857193684138915398111676 35
2370135996047767943273521946249418339830341772752871973216535239747836661598830001870 13
5480072549967794815641222507319820777374949369340515926151214725240913312432853522660 95
0991788624206214707618144436516820590669370269728484430350602521914049775512614504465 72
5697123159579764295821796831320720497667628657047131171465971689941414194155855279213 29
1655341080358645394236043976346169533528984633901443709771031718852633529786009866930 86
6984352639341843697031885743906286547106851908002382479226597066957969129228277977600 81
1198961917051876577047154804076234201469014747012300723672006374794146524884880021869 72
5445463705686004723267422019698082214227784721393055099639306665883512057342095362327 06
8335190537432350964241692802434924637529131968240070383892099468207970820508553026506 08
4172906467894329248904226662371493914871852045333605092819272200266783545090067219293 44
8357187400486452584361950094552025533853015019360827545508148367762943173466687018398 96
6327368706671738897838170585143556166265753058934928379956883715661905117469340198535 87
5250672746157717542792976654112430661688579214663048265376236291786363447873206048116 85
6445133196337759745210891420642621077337168716290579109047378376399934031761013295865 66
0178686150841320259943918468802660091941207015673670527521044602542464766222553796856 22
1129982222136619496927805123461358938800937870064458889553009309678080220567122291173 24
4962194666336085015095949168091194431883188755727288072604920484225726950234777273625 66
8174264307924072732430554923883140548639459295216734612059347331081226707443585443063 90
5135486124584692352772955595950816339124034488084615574831651102805659691260473822009 62
4298786189595228730193832928596279934984472850958341618215438661086913654406425908911 58
9698129897442714708606373440889508112079256324343912160486709803726929822324855824995 70
8943110863765484037375213836977540372535093086840240831865921894754342546568199331920 28
6973641687638078055942726490940543186086883587062399303809324893776063958808816927882 17
2328401008007787446855284930095356168133698782188131987960915707800606587512041368105 51
5013899140721605454073209842445711240786902752955637176499692273209734533903274938422 46
9717756042912196379400439291393993696383431238105727123160165462381860803416937792323 60
8064841668516700884580070799053990191389869288678868501254679168252957297950907781400 77
5837291959525923077898529009537496931446488560420183072483668164534888900616418487213 14
0176197920673842538852829602398428554013089339238141175701208083015541888719101171220 35
4396041258813688804892939409662947667940562265648193922536371686301060079822869890537 18
4707195524863614424529839860549722667381399271232326979135267819475081542475158259570 71
8215174779333083805385422525935868790011201769715069484687232397756960219053027729134 41
7989533284571722569595139398450080818555050284617312093463323897674396686784116298253 664
4122223505420632363899786130116046064276425247211995387977652547098991387573337095875 44
4606881810333079159233516940026805099690092016813695028758939377149493311215724159021 23
6225154972698785804236244127487899865593145938269756931430554981795065314418668327328 81
9513027519956721753823705715225229178899738983906301739917299478596473288245148057315 76
6239727468127206928354685997204915820065493214263737853653777657763105425649156377280 01
8981099444126794727692115095607280236282848806019820301770557360355034313476907412760 28
4240702785624501867604936809217966663050628579192569032160921550382981573843650495683 33
8819914203756320762894376846024766103943520229448101014041168409635222224465266984042 96
4025300170064070372331719352207617831350035742568452310201544861847411294624049632889 83
6405515453802300065762431476684153339805267798198723759552846345594350875975308122540 78
3115285645913826698540619890759859202685372792300841475303461621845748488155548751802 80
7500870114860205663005110795262621816509997046246093820030562461035531104034228743366 87
8696989658957146052636013394740295502774288600482358997616032607498570471221845586671 32
2714100697884695817126271499385898285859214625368691960786535613356845007576646743331 08
6324711580071025086357042200427551027969222298288679505160390327842273887381118303297 73
2968241604268323692533434394805615130277475655642243538631283413939265597296620263849 69
4533758729394690259628738874014880709463380659799698316511192908802511925602858173049 91
0841216417996843252364020412026660339061356441401313893221202873062944532613198313335 65
4125205821195292932149405364882300337137813069933752626752573427205471825985193439712 28
4754974232282540407662617308559179774871262988700226674104747061146869802870273514819 88
7933069078040518521698287223191317558377555383293064193433276705871185727668714564964 75
0046792370667719907069138700087116303944297482298951815094107419158382849830008905156 33
7499709234458126841189055720390138093744926726325991473345522166651405838474159317937 74
7013883109292072972574807268927486362545010226180365457994969418363051852033485830613 88
6089153747576814506890483050474323104175672544456786775405653532462442638450112540158 81
1337628792279144576999324549020710057193890243323158180677444472667231766456076823483 86
2792139092387243041511088833740482602075644767577685231345785784346498458293362407464 80
1415979464521435092855444146566236726007057107442947148089930546252461505395466729457 45
4775223098859383492273211814012181993728972747836391302422295422832265812729939788697 58
1742933644005462333979847923661918522402262545620101262962274256238175300517763286375 53
9542768604195767584878682935658016484751646810850672530738693501865443186067821174807 06
7102386061328973257124478239794432392685146855870717593267869335876854716034489616776 11
7163412996673364505897921107546018301236333606911449641883879341218421949353787499665 23
7975153917630591596461034715212146564274723918539650213163259112308165283502948681956 57
1358874767723962901916931991677686458730587552887475970901591888584134023149445351637 15
8204611939295386500630901968781148725200246325005859550973265669759408809248852766267 44
4246672639849454940963335885449443855070827786604326376695588990942339213345324374586 92
3695535840844157854104867882308876591499258587761349378939338172395661267326458605886 09
9833130238817706692933483404804894481441182843465637528081665602947298876984895191128 97
2799505976303277305177693767829975992595734475281486283944418936299209900241358060024 23

pi to two million places

3268552032534364649358977527069034359880903887764835281141728985220615766577189745996931
126904994274595857748900715017710280050510218508746333748028182672386637611059367212151
178910066446038280924572444314173858083144173658768637249757501724180505564815645888269
442413625697379292255945902050613210392317227946862868956199674192340073901316879381804
182128317285099359198047059989085315255060611129029607455276541605554323462610167088128
515203185163523305468450163097876868862516095539218201938430432208578808474078482365115
168524579205376362858300472488949906232202771501038935424792067272180390323180854549352
300215703224048303575168346141950451837704613129450220640492325433720823309566895789820
800186133500506875378551738982296086457926134601229336427894129347237377241183471056719
947498813255663024174855112544613530731382508017759590020575525688293535410332940582476
334595489119148002630594112285629020991982970892267520245118222001125571579240733650574
612758838238123948024110404490718897939017921349177985750287517112124497071036628890526
745050625093926019965399546707668561456593004672173104967569904556248663389377247350889
687245530867838193979741498390808700712484440614848824896131026973307385912498790374187
062600105142158390408876123703653713520292948787650548885182853428248410038398553662313
376457670923704477054434480282490886232209658664498591943363119205831620545310144578482
233739950955681727341450850356502805930648292718529747212853238817532400077625654889901
114846834623218046070717582378163621157379393336563514361265578824300183870597868274534
862241033127098248694442254632051824297733063193782880524736277279010236575158387770445
736380232431010591330252239746032544228425304763924993687412259412525344274571914357904
261985598032355410755517179602225731007940379296322417785536177805088049496315873832128
675554297068865951555521682850849181846715119760879072356978603646204109197206493004123
150188837770458311397629187009238052681820978751665089521199378270494551399204447989433
788484231222487627259892997895482574664348575579264603758622423085401606220231239583793
967906186082020738486233385006386037167953946412827981719113119614782167000830808950785
445813266954818166386256950925231648219178221346241417591335437979028341423778353888920
595668461979810523192825766529383288091196687610481858896454424402205427489691120693335
345523517307201395279545216123690563455727025608582660648538391808817916070760661386419
533907534631042316314346145514954683921525457176523017627907134670298964119348104686248
731091290588286930561548550109865716994317948320400204615028922428325398720403509193836
454205680657981934923777973992361643392328441222912416131023681238412309400859555943921
238902610148002411843673257218566928685211711743895773556713694403655341882663032196737
122047784613967242423934382065575710146521083904110580254990574309270936597181213253729
453276129255716033811599327295120079560078112053314636112722208931150147192517953524282
097046383536281323733050395799214257143099602033906880057835187398188931411813030607371
891553698731169760498551140763775109513049911398393241826775067861293093343419446004710
929787141898483801103110274188437832920482839062201145924541976522126891453178721214513
312677987845560445615959524975452985753522215468171489841451113842508141055013085489623
424362849791041944000953541984782940988043650418337571950530816933625465486685067659486
021084403208883697917857456235888145994833927128325817566478148930763018318845031751877
025433815121721747380864741836917805118542713928305925515203696154426557927433409517408
790292846894847675128182899369369638005209662580963095224478411416233470638867515420150
413175353850008132877162808104700188346896555554432576568037656735020267265299682668844
678106657495635099985961788943244307515831054977176345323806390582047731811620890740012
401492008019238695449202136793024258024646517698310567148501978569587399755567875029661
732867079490827566132163830257936527271172357649371977298818387650160046960198076260717
297338719586498709909968724717494761114101478259018801182453610215220422049819761816353
003398859997872605603757975986803073314752247564007571518792201644858905540902200692157
100677635924985723995527632014441216849154357780055265954849998512479780901999090881425
545314201117411693915483314820573828041454362718046468662629137358234966482875220896995
836290036596212940685485088790079706097153615049303070971447935664014903112055308082409
502443841987582416470860586489907394068754413110174450351946487233176746260249006611578
886109716268850323904256421665793019714233077714453553878628432514336021250189909452614
454005268521377177876674943223919259355906514401227995323657387859633667188679821005112
248481754029247501366950050475967084418012637345880106622530400185066981810971904908813
0229187287636601125981634073025813256662620787942939377308834058169822940480343932468067
552000853043214053837036811967449736426643299033781535255022524682542776448272604906490
82697721388698375592442072377285780670194840754825024590313667177786794582809712007470
914103345397923040702124704789597241630631679390426246365331538246802784629998492955829
706777036736178091274486510961319478375542147347260977852214220697744173339302375889154
517446218414658881740633219251066399862636936880015612905397748917869227522718660000064
868860694384334169568976191047931550407379069948054142393140342772120564176101918465198
316227552484709378875008888303178635730728291876730356327135387492718627847836880270457
848570870734728184365507652351170625675594045614751081746288865138894890196169187358664
401789113965033124828628212773234149566257038143667039496496422017426965254125031386671
549257792425824718393040622119315529395247295659610883491077311846506610853782237227650
723300729864336816764066348271265966113355194054621502427959882603393437778513689665892
077747703809123061987976754485388849348861517902298951335515131781694479389844805510160
4475387293460208105642399995653132943818111817949168206643422986141748888624688910910401
578383578904647033850624225450915261785817209601549590419019808172498633323653239962035
299833581288184440031231668441282092157103650031779327671655485926788764291194438852182
338965094495108299292607756543518139988333500316594418749015738251515349117302529210911
90759492242574830715970710339463947729017711111795483875324543082191957109711163136558511
133616779926745002254567461727026192658228096930167652635367978562181504406855242502063
277431207892459280730831609544932281990578760422714125469899220372255809419313290739795
298469074966884973028286127683424450940254197334278386371332162982718868028770488590398
165426651312217466506883903438805406628761324151473649801132256835245396445917788087746
545702095388905755986962043372578858288766430401486028813478566777522316770085281567031

3517086993687228395979776142704304433875367404610340962659814007959537338037660820668O7
1019298809834646112535217262647461480906245084356091832662332861476571788736707043264449
6423755831118108613532063045273135501338932584698039650758945995278709868677366166565 27
2940966270953618992380834352325064535406536366700812380843142129160848939260647575042 43
6077337007697801048622686697327206131308771242883807951368232596049354626987547850697 60
8709062139444670987050435399717068285577732455618338596725841529584388095410036379769 07
4832639223226967631086303998106796892341608259466054119657825083379100517643071960066 91
4226778080310863589495491266601968216898487279809973651882678777310532850591912303608 39
0383406391113657529384738732739647586951190390961292144657800965627823085485257649831 92
3721384198581880491491408122029836664762563895359153095424421023811400942201054358733 60
2176560751537841989836183058408095095792328872313144825688115900293087041948841952225 15
5771425360196960007116347447439350056845634437690033574176028255491088852153239250608 78
4485978816727896690263172001067206472074045942489005080762823397203513834604101168558 95
1501894545922832586999099358318483888333495905712543219713999285566983682627054269275 9
8788149893889555327752195716198846758409266923712272429452524398252383641763866889502 39
3648767029318151507672585976784849233745844943089231935310651250470844594598229323390 44
9455206919519475305035974627883356984613202221299389749069342840233552962635602769933 22
9272508943020206397278510558905844300789720504880441292348947467288678389150262288852 91
2874295275043556502096294224736793464035255126954479331493143201237484386524665786633 81
0460290770251674067522547035447178168614852922458798618961723244658952819715140614101 21
9273275694847104783728576946092451816916879953558040498412411049197574392925110095931 42
8097659219158870146386551402977596012953584700768611682922644374888309035896186112660 68
4982818072956009457321865073950634395701253548259150353689577146691650957644504336265 12
5119916820408853488218399214903746883206389219583967718072126073278856195131355765752 43
4134536657084865934159844354645376945359090314079655081066049468224721451838346650131 78
9067411039280361890014419260045177269902946519122625302754503806032488518836047936463 22
4990548258049577601974085448238309028870415973120470316393483654892654723025152908040 77
2707730361851825196120166242493551339087586691042329502521962227426242566715929828930 40
4311900679570412010084094782659788365032760310302848431325642110534772075835031460386 17
5329382737617103295213812756699397374441853371632506355839050047644050431846555050730 40
8990389269936288620469402695844315063108897864338252987917006595528198783375541777917 62
8876197775025249067296380621847945014014371102715759436540408833825817964419430833308 59
9847759889176852252902745784216896900368437138924036903256321745803965288188275818677 75
0903504077856772821528037441782855392463930301793557599469619528141809981603585622724 55
4127597453696375361955709805352345132665746446826235548397593427568458345430580915983 82
5089473892876921029768874299871189541517320435806519158561327173516811149098354192211 82
8347036572886744881559980262163347287343430001461760387865900136880255812776602634600 14
4920817095907337266610360736242308598843852524693473596609937969756917834925603693389 61
5646024653120908008670272784774156103600087897533676639449245622950923406638335836132 51
4639323912202114104758373638153882739120159065947488481804948983636563622348542146894 40
0775250800230777863764248086281910281820347118317437303493344261779754369537719481282 15
6429854650561096435593803114776643035513388882948516981378638751307499107404631767287 59
5323655332912659476404873562844117507971390474005513323938961955512856978598020477697 60
0525206068054835205471581242947362801219088311512081693681709227961780504089455783599 13
0934763228080151499698891356968813611105140234198784170595077656404824631188859014808 33
7761115523768193830635714319409642574226714354981847395407794813508778895488544130013 31
7461251000430733254010753142425362432677798010605783400730462256735948868856191266923 56
0123298435577265000064743197514702226705282803889500570021520539080570268516881790966 53
5977581263688859147694197974829700113258576587857266967996800320896771899522050292034 90
7180214016910200562865477296008692638027291114694014711955067728439687724873271581274 80
4285978103024505132899744107575415357075206103691029094313706923000275848953212511978 46
8846423010680449037328919226058377388613091210711770587752870318746530455255581540388 78
5652417324957369774887607670119504928941832459182180707700622495028789984059966097284 35
3140332007498193707480320834245563618631659974150015818925437235841475658435711934943 34
8597563546491917525117456083081918043863357646830719241410750040948868501060599618501 42
0458616536896955114758739606606076195171626252502367814350592999274323654217871595338 03
1759236278898782949132690189601796158873699583536315303050304832586192093522214594208 20
7413039779873837909064062124779703439511416304880082178681941363928392496378920442456 71
0399995289756704861790288062734238454739784472819355128116073381263279174219928320972 18
9396961112814976972284264370275407503476407901345333133132456496302881157135538112815 49
9523882630906800982570403252224821418954573119471050493392745559351320420656215368929 49
3664686409056310958536598612765502965285049085025477459821566929512737113321745590753 15
7180290994325883405386994816409967232531224608321089931752489923443811948206643167694 32
0572801425335205262187828721489083506561087419272164437457408832701675994309456501459 95
3883255648240406611283333229346944738229814758202131797535505554577624296325512225636 13
9644685499789403443974936816101659194123565017327701212257496652664631173071157348366 69
9751314182591158461432810857875813421737613298382076751955597554195717239664222676454 74
6520021198627822878202403645259902635461770596464704118323000965795490248713711791563 18
9956210816754283526888091179891832700151008943093350226550988041726814890881229128708 29
5210976641544476183309811369127897747712298020213321274658985263467192662455538831927 71
4499914083410105952487256903065476732185557814100283844205889041468210966905433755520 65
3291342320411149444100772459855641528901216225842635255114707859555321360988093289506 96
4483761877632525942710870131467907675027632598732576679119043428262705147524536494231 16
9042157035922697585321023117194458974786817232156114440928449874481225597231616794441 50
3445868448206178121426891723482568747787662783616836774324940877324494898869106311260 03
8047886581813424621072084554473642151847287687459371382984280350920827521176899384594 91
0962462437545367749182746541846734745927049744842547339625754198994080811398880961788 70
1145201432449909586952926214021933973003160651735918939332358754137954654082138809127 47

3065435451735664490932758490061726388976458421845913459045197700840345225999096188879193
3984028895432356279060824936894154632361167752940382683462355519428633099565126369680011
5648880489294626619364652318434098834586027305854133387446270545353835020397571542522 1
2852521282183013029933266179041222692387061468284264574806844585557582486687213502541 7
3093712604179691242764813199746114026534138005020247181395569397702665136484009479262 01
4589759138181669048177231659380550527069076839774618455309912602312300839997984107604 78
2938514159887872479842239091437645430731876545189095326502310947742636863639679565102 79
9543304550196299510289841429099115580018885576177703373467445467848761475525096699445 06
5903373930150031316083832059972984457035801649539935756873406560221992121815670455920 86
5803054460810365368880519721305776033697791282130976053685806300795151760564762996410 62
4739834969589994916851870974989444093459835935232634681288025988542531965131635958944 31
4542006815724356053980185968094371494286556694414900349111556979734615304236866068945 04
7000453891921643594562912220420454155483397818361155128983312621025076967436929855168 32
0378541867742237633716951455530847260914293919256400680948773987875363904313284521095 77
8702457216803676542536975566933769168937215968264275753744949741800541699442708567746 86
2794334574127819345975089292370144992534455143533966206168461628694631883411638168732 63
5660966844266181685087835830220172682727951594919629889465471415030066994178790856944 50
2732408037778427150341385373556050507742961538376582005594222095940727751019318804644 24
5589129485936453837844050159592091691809920932208512267292474920514401517221610979212 21
0269157230236569349855124309230660305936577985029687173529133449972416296473482567376 40
7878330294277507365219596517539511732842538480999346666970118541311640518949526554950 08
1439314826648814446464020940783235393423612349538986248150164626287251406844154032324 10
3038129212028434275834077158714583810420168862545619529592250289375037943829260402097 40
0297659130668606595275532751069879445414339965551619498692094065915238201172767866616 28
3443831248166625203883657625412669984546456630562695630052414535081644907959147990964 77
8200244813028767072494468710070397159911375934701497108624407665152147198933063018602 13
0504808607034518380062795641871228536242328066541232425929709109770029297212199474442 62
2854330868412733791162698475208548230041569057210122026554698035671548157735251904691 40
4309378893340797835766389240794583105380933986171461332131500267957972581896229057661 51
1991624099919326321505998246815961203508619616218514095538350773709738165512965135202 63
1895293924356239514539031754422363313065739019184123145894299056068429253759897791824 57
3698694733042393692337352111541335847620381814269245862206816051705773722333299611929 88
8870521300396966800044633671809692370885305036087584320402755756772967005007093445629 21
9875590098517416117101049895879708446952195784811694719989150710844962635254670316057 91
7742116938220772209323526005718234236239343941116792458409152606168641300255914305681 11
1177970827773615183230788938566750909280970836933132742700410401902946144372591081348 35
4917074104732302975622169328016927704919328931914111490886972438904204922187582597997 06
7564392655969402138114267128981672957428040764685833698703542339906412061908925873490 11
6640570308240076491226128501018729351002460461822256831532382619806998211838401071948 22
8998960737201090757034014732667850030562583899804980719849773423190191747246598372272 07
4210855769543250590340703340504822422614308762808883420167299418449561628869589213472 34
1857679721081980546181583973331783147978139288506482582508818965761697179187141501706 09
1545003996639897044841309212546912320047584424537510041844647710493477384411257984961 96
6549943416890264371295126805623028208527087804399988117787752643841823140576165246619 31
1886069040513242323953173313951449370956869460930746931416427616166093185970588956635 25
9602377530748304313165254572812065791236728498218575347455538754768260270022662474815 46
7105317825391142737172074582716509627640099158475043363431826546354402640405973284371 16
5072254032651366553949818905099347937981985806978218655021813270393153600498153697632 98
6505536587743634152179587756124518580348309948337088602548909152582823779836683487395 57
0987066276329805725192187914536028557177376713782996157796051868521774624727977445894 56
1254974374283190481535080953603345176201862224601260462488022681448903020869954053406 81
4154385184939659903845911174586904095735814668336030323456446280634305244412220994971 11
3470241456023969311815207755289287487893639618928372948700793311171332579697602283983 17
7455775458612993110518107239694876579428203755884462077707173234133890415195555404647 557
6327027653380107926590450199139991114863195571322212459932729859118497350132241737752 76
7249833216157371180056909299700801812447186896657877268005455511536819797198826848078 94
5227542433487036723188421035203504853087241154459384754965862009222207569799320338036 93
7197011862068678541949525556332466091158655042170617676506664877202306733521287122957 28
6462859329268677061307692555141372493400192765224853943270059561964109070391620844032 83
2798986619398759246467689391149503182808347939986046658372506950213256921756490238876 56
1524018521701221162127468589481789850286474477872880522352356822852035937224412799366 37
9674282448396437806808494923017209468877774251866079593245371729080261399410969767654 66
3409431271434368806589206992388966339923778147478935028094373660275900364624161300918 73
4985959634280478473695673504965124278435885422229450338001132620039715846114551960415 42
0912023544705338201478038741000722524862318388819378019216582103744167097585735374113 40
8393955228578747068218429437450448918247514388781233455586345077626270217716576308624 20
9174705004017086640127559230703925990257096154954742574332757651801896883382854955713 22
8848591640290722819747439084329162549923360313093772591162447164874142268157619944324 61
6057295606007775572712147755502821957088598761030861652699743456773684934533075310785 50
9520368342153820268676367990480875512997682554833269000521950313002848256936888101150 09
0070833401009054160543970501044880124123142517279266978094246624616089531136154168053 09
7688007906225531274958649555698799918888536255300611324583354828575063694882255737304 0371
9252797800932401965754539426019497499777946963916736741873698798782505943711750613684 51
2558358007146559799183227867154283537193419549062224893595622487235001596551595820360 27
8287417452051357454840974327538257555219568073477789127242952528547537741631054371227 39
2203054066316539460792954280119497228598688626209597057713657674576946422985856740408 54
9931614689233548567863314418192212213688629970654031711152797921389343628299637884132 77
8293950989205495568478674730118373845505613147815705416301181460518125831815266099326 68

pi to two million places

565644749486427236111006399020153193412882598321073694949529160862702977939042236362515
914082470343247036819246398275189526910227950314933730899837799279245439411604715911973
315412320997178133722888601536303280053212834939228219427985955455416679258510853206403
209848850627815661621488128466767270416201689843688069485991007452575301982376453842056
047226797984563260150554934161892553326356677175294303411190187870568223554712015982950
289360956336836118560837692702315069334026594230419562045967244077011364731694969106130
497283217640846953353920648150058350182558510308549034880383374818319442188130958477422
037695226471444172459939593411907054416363079721417685593828641120947165620194101307656
939546975599640082976005354188661788231110201727155730225505952903350137402586928542771
974669844636030298141183451832193293466211910497206119736836295670334194277916762383415
463921665589720671002110390859686683905281737196527813921345015475687862913183328433454
758072201247546748768289818923468803249712157862218584652381817916033876220544502610535
277301691773664780882417390884452589138524041891705393013605620502532289090913146599752
230465464962648727050504279856355249995689435484656290533528389052407476828234694425124
220524321460496911515995027451192115682913619877757442997508199457149787740475436646671
218367186399859386477584804973795743103135561069226488933237941886210343030099671295756
250603303590322176674941553563372391168890323669273923796055594782223183502575769569048
499578355341992708004537102909673080487193192187349050957832790929334097901123050953551
778787974257302591266151473309719502128366443945796395942721434847918265638596525320195
043897241592694683925242421860380728712661664237383521768393446568942250005575813151840
308505972561946962356965072584850934813782928372217068228979426416542578445420092847486
454285138283727783813351358194082795357922839889739911377359314873156434126855773893828
279744037394038580890547585376469202656818166100037622459230782536902831976059742661345
582628952000747175198661393484522858606709892291398600737672164274502184189906875128210
516407142066063895878028324319775810884437261383554629612460612810338131101281474830649
250015655123906492637605631557241505944464475789719349309910266809570814842381048841857
925500513676842913511411251757939022262587400884536416338712775753025819623722715907425
554757340119363083529254662769457148113386654846390212303691690580495406784173105594659
185983035030658313990683169764838761011995193662536138889762266165084667337594784663042
039384465890384648834728228452379531706994116136410819446969967244407469283923641305679
339230152919108360753578089544072265784231508405883419547823568799736621842186804968764
3075313951636703666263754439135185429397386393035047322416695286543843954922710470012092
173801038357609531156944974648759568577239395946125495083017942186451249760690958553284
814130398283321957441569941230839326637951758977091254917797311845939926153452213996141
098349106184057736741482444413357246529150310050804462575025374527545985515392948604233
021458280713117375319139408935171215518043017046220554479737669667478931730962714817804
320524153739850169540905116883068125314868090439523232437602316225250438836689868570661
653066181904568527474665587161772658516507341550659451756136678425907159518526492326611
989419927697833028837779411117501447410422908089023660833919406521113932372676135978853
892342828890089773054733631268033251710912983926148497000542686160299429609638488644406
004910294550396652917023423634660650422229602109878588006944215698577053669844841086867
310506963029609061934231606638748990018828949512561559110240446282053159377096842656480
346033781731481405384504499075920355090644590990909144525972567149530282726452152739661
221826041346147501028649224457265765754529032405471437060123443121775206577764552719001
153953342505418075391219150598379852615733705162295441197030788874317399891457948253499
871589694378773437376751615663954647624352081865315975103666897077758328386505752358050
171402323620418208538777467983029508442338331103745806536419079447497062847706547679482
962188669259068217600673139160665816078583284251386246833062603366795467912492230323889
295687728947612151536396030940627653119766901091138048740549377875158608275732363545668
580674906271659721599029921867086138968951373706831731568009195548277920465055777750668
885211866929765211039077863184433695906723520283999964323963876867320361379164668679503
112179126791204942268631969522649415260318384601653704509259257880042440594434946762285
550854525566020694226877321894633390276715358202225470331223479856682702825245988012346
839537763551746891225367881238189199546378441270332418737376332561304223327118444507497
424223782661970461040111226532758895572384985057636480494092480401111627708715552798840
790361257278835912414460810333685676978349582569733383995616923989000850293535342999657
406963779502140851903006449547774735484116840551028776110864198567266226787228785939335
839702878835954512635880762654146340514808297800269911581141860547629512881994308386653
204842674055555102533227461342219931036194777204952433882117850322504622922142456955550
033137758428571026505201688448712793185637412607277411417869893971542267276322865848616
967758187395294670929434525373350650056960134607318025776790991551345034841583984675265
057374239747147481254009357819535459734584707650074470973254939310359524408780242766698
463084264446386736380162865119392598513341956303926559571167112304497992321173606848975
917558263162239022622489846228657935727962491037640382310408048197494285927024632127390
069050749887731732188546893642438942096492856644246175264232707272793308155715588664841
631953161411736909081320190273895284251549967224303489023763280543729161398227252990836
908892497388522995923775450126780887326119546163620900490104565727238121860734029185447
799535289959211945518805213829562617740804127330379420903678075311256760080426049507406
190564878080502343735589161553899274656330762008009969129076100632053732312693727962007
390543343746221025899572139788919097031755019735593786924046361116027206338338409032239
894189167224940658460386883599127252142008927927430986616350795728379051438551387295271
129836678961619181394864073685794226166711785990514038234729140090097922491302499787040
038054145640642906637903922523044921716095890520281781615990117043582891340913588943930
495096599605113324294482509836510749851489015128230584721525802518501199967290923019534
159104663497514809765758541954441938660301023932894081909479278480783045240452965721750
860347225859417577247107124427077986907209955806822251978998647932984732830779343237408
852481802412599360762807499552231046321991068299808118204157702024037867235236304158690
964281330364662533949494024268686257659183352224135410200488785056997471685240672865751

943425706257347366754736503125572083747217077731732426112803077590654692395022906758282
061276734984710635197644664427167209384615607255682037913752094903953881686272030842192
749330119689065021855424299174257922262045088519366428928772287755130094773646300064241
099299000459324667735313744405486684875877715788647542882513257102499583552305099040681
646341314573144849393869833134777427352150116669767950323492087026670633342677121526488
087953659299373518992420672251194807313265923343577207964529777754867438424768926160681
733916928124635913549460964501115658857292538244691171975572851078655298454278716583023
412111155300074663525550482452345410485731877910034762330588407257833498442549880584587
627134845326920402256838316423222370976454714050912364134406361959220500003871490019578
335980781008598886055972848479116532070874593435630802187162111489589626800897330821420
606706376919309905812237161050163381359939183593187130304492618018971089282042610031614
996598585966795900943374721233344389517882139213168961373456008847211834055370417390886
550221553897742480597895417234353658614901584283694749006043088952560611138755438584962
726915091086239755480590136238040358262600530735745124871077414263477512509773977309390
725282336321002648028822972706410952010528838078676946203715639576047661145005808047717
642728780013131277413244302723408851298968560383152465747436798237046012790757806089242
271283222644030799006153231410207971238005338205305121911738709301983853679081734700155
973475132833216025461225027248671258349531573900562069353295402359717153946924644274388
027596148649412924752773374134950714878033537786636311856230615348675776706325490186829
580038258788485176455630870877721920831838898622536779155140403649128193327234646410410
420881109312609839618401855846370235435946724857491530908048594183126096729940667484419
440084259832175480334598617400200431378508286183299756817765710311963158186599504906723
797917873672190751564065343776447630621705766985670874922707502808767246204225353894474
142413436311016643663969988087857335045346724540669218104268811566954384513391960569769
130745213394121929508769111217525555025022707891471738006854236503032373748677870255838
905989363050210087166469785711805166173156702947958588442627198410355094621510005467265
326668617442935116210475673902419773096614724220870452411170743338851786662622838975950
520312887201895235448217394782194244399758999835006078041238792193065679325724872870197
685892465232506696114739730088004771044338654372268843956825008857164813094684461270956
543139557547580507935142675631381936217024229731878481227768318957407048462335383621659
586843308579628537616031367647478472565866092667662925760874527312462223550441586597063
699892768170434994116812797928212969226927665070866724480088023309789282035593082888597
853534119785021182140046842505467402549771417333366895230491683144437629773482720505996
184279601382438264090054050050543899562203257694031141072966938554123861845341651357163
376862041548822631196037539535157965115615791687520081674536224127084386082271549504444
160687673507152786736164990684079102050438446012308262194961869244003083142930447840925
631186320777512643059419959491771706413089068677746385439177938197576164468111066389323
583618409159532116359788090633295173261028171014866694864396113101717553654033227890444
701001976312123410764636362458309773268932593277419195714992449855964697617304672915269
924740555769815934966307737547040710403479749755177108493121353924293299736757220293459
757588025764039177569051215528023145210293130499653339373023200920505409685661504058680
341637934252197386129365897185693312537414578482782700860598921654039560817318742975871
919106932757121799520873208269818635984600672764338764820681561136791228409737944035561
387140597390368679930474236923503106534214666419163614019702393423257526942990274693396
440815930157918795154429053346261410491553156884501255855500091332206111270101844350069
521746173065959864168810938163042497608770615457839544937918719179778214050225790496722
122077816023801492381294008546323714040529727769335214212268461859782117443585242059109
702894198462468999523242239816555904667713228763619601838487431946537298122963352740628
643961041759162471820635419743568571300716754368106285383585873611414883279233104600613
445141308647000247195647216798552823676900196362140435649853331976888064888620020461779
025580227094968342346108349257247535445213392688736939312898504859588223281475103811888
094604514397803692352923365109003413934584678504122917413504943706196058666194826656431
821665662964560637334750932895534325174972737913622427082393564558244327306600773155471
301780840169319607522344183992270011897688984844635055266072691642378379008047941617684
334383228458537684138118251167756383889258394020120613567446919670757527804706934458182
769157795686023249588411602895983784051152559797838818410290147847623450896702896672888
896476579778230699798819270883513493304458981708646817971112838182002044197499117721541
034205165427027031999846634914695207615986506848852203163287322590381506483550100084517
784375450743614505922194240241872628249035634126526431192655030406320257536605315091862
694230400933310248298609192470044115776935057387004285904866973642307055861273564020540
482677643880903258027885087690994809257329017331171653054678392387839859920844949442823
545511110755671417941567061971198641995640898197469650383598560001537763197865942684706
532072913018346587824492507819878301123504261299130804129377867117144174364941704130959
984277496152385231195216354157162041855426172577328975427489642524558369008052069719897
834053279524962832641569341561399857286730987422621402885181286001908460323175013330780
748312361641782613805117793371447041481211899591909802237467877728128904530357995934005
672043056440449354780269634646397191566000926368642173474117619390641454023179390631738
285189036523986290070998309862522489234371226060985810304095763885109243895097633649477
485841674766441412104426996776851984981650675379539545328708054990529826809162734102724
026108419569462737591630570306936049304955176029167714525673235033671309570781769610890
255690131338374511581734260872080919768962376158247232799696762659491769689035000181087
114738574349836409909279336945471815629034528700893317786462600184845835027509394656574
351979662469396278967133275300103362167058028060203271745531773549327571281289722301694
317565952958213137736407650325824120958394227661748987572818484763028751860746198159530
914517767537614228781538324691582039685564713187068248641469435543418642408986534022652
979350425310530745564744585987765188270858920942199651867033066055402432358530574124988
971883964658269498141403272968040098567409447220294441441026819291341109421874151328287
331079823292514506610782792101218010311197042776907694303544975283057095089336353333318

510969023890698062935716343038815074851498982790171065936441270747611298518681777313704
723445403745600286219167913221365782707024079509967914180334821962558626695197436112488
465946582728765644989771594444211823492755174989682744838464215544522543357771015711153
126469088386801039213792718885796109842454401760358927360092579861974341024609362898969
559726019190217380700975580237538083550621119923314740908846814351990335032977622554102
622476650244698571713535726528821856861123930221223530219051627735697412863498354083903
277354810508071163896557294476684579217411209448401996556143553068983086120849486986451
710667370769181572544604668202820526485233586992499908059551927828814586941368229897692
946966657523038260343104885807605511708300919710119849253648072836156976781877422150711
037399536927661524524496535091003452889597770297427788541533978588106607156894926213082
779492654943621073456967878550687385561133352065110009516722384444353315866717983535953
576451196680957452740004243401544004906266613796295747065274075783714609400233265567720
783707945010252698047934975995363121472488812065999504473245405295694916113543765127592
357126014280388914399713206284195468196885884175292922668864426214340633843235449035871
836990505347094758449581539405452988391465186318150939736233214054509690894535311623502
316264542428788803218861917253625798092179692510697663155348667212890255353427663211799
541889440802311171241078910639117913801116908415614710432641203243520194668781977732716
307048851095190043634508471386726771537892816556212509319933084617355227009185624196261
048152846404552681817698323603329448717990508744856244052584667057078202967384386993693
546543645644472300445362273752665314529921962230988117645680676713978543791658473209504
694430438488370771865018432569270115171349496173698078242694080406450092250682393379568
371761215390570823858548291004571562854052309447709233679921158488718861524422945904780
396105235477529451843291821776470401267419025562578477103968755033652996221448817935264
058678133326215920913490441894123125521539959779526570682574624076495486611392481467006
572277196024957345080584087022130787994543507255603602188951099248370331300766185308525
391274076920389698099676917560701484757577914252713802209415938774884940496468312025721
533555806488881999490244912487443870269648190456178567151821532142145868896572634255779
575626527006292978459645061748964253000210922479818088894625167573273500296771470503279
987579899629127928746607578196479484752392035222645594063126283248417539754817457010657
092602259317232087409327491400979781869702601437421478631793853130815571167180161841988
500770608288114464967535146217195521385239251104198410816414048220442494178279538017357
430963694436495063518748203960509107089209844314167840644619109367713615057721430047516
429278462760612587932649555864471865902984818220160210049977198361803778211895956516892
253934182635010371363211577812146023709366910632191188391695270536319994253584490860239
808314298724363262584106915366179150117100345827398845023001900255774774312143342156091
021425630047292952168051282215647987913148361230334611027752033829796984764875107530862
096284647678635312023068605098718087512822655052819891699944505458252874274597063402296
033568538869497993828280147109986008705956848412151964733434766337627435135600524136353
719011479120299222531533188660251537706833101548890999536990171736942984470666770482793
224482341820656742587914678111518974774907255848915546647004332922203989506494527707553
293229942626743329882814282556896573326678657561933719593908066429191475031113284467514
873963577287934298975641336042658143050921348136798528882925725159524636866705264718983
961963598492113415695319863804558148479071780907869529272791840452294301029428367110839
506348425063820855586539232765452862272664883418507487413938549307417630279555031029081
367706943028970120253177018494182091497923829631656022298361980348204057952319381326654
609721358839304485216628721435257511372368450592929824956114742268292183057942653346922
593252263720317019199984476757154104896916082005560929388043856153939388610800038893141
551425198807281346757093933154300525690378371445159973023442914505699864146759503031812
715272789425023834667263317333745588099435443950036775670310919094071611411540557340884
834283353599319214437106006859366026469934657200936031623444263821514244283917159614658
131742473170515632534823793147494492006979548016468218988433275915280709539882126839893
642318106548904480489276848921854710676978861350333524498113861637395731026802516423133
941154981062670497502013612853271679450641310008468510052112136399583903140226191080728
990981630702063541311154226660786954871773820214105474014024032631964563930096218551712
914326524427459441441827560214455153440454002874324957770584494080984033027258180963179
652491735070149484668415725465928994947623637005980489039657631382402769423666262269015
147894711690498025494902525878222068537260457330359730907758353931703088870854234953122
303180229287959906405017737334421306274314312861522090518089923909090287912051223118684
602413327860975002467507917126057740457529256436482325915999316644509890191461480286002
548234387196903373987053600376311002385030885515627394262315739071459324738698502972134
778226071168099600604407499093418534005721719934091308208823412305076123526962124545508
422240149861662157316029979040149669949172746237826730178940942495960688437738947431540
589473271470169619858250912821571405308578466191861322806103650342003901308140379286128
430225378811284489468830537103303123570043462316590768763089808395415916852054588094762
371192788990746785753669928142352470441171665252891003605377559735588435425423746517694
185144707314525476876928997005954613496803973823293623899801986205650562460668131022520
368054832484334969568965310038523754667587856799400946425372666150546757091298762361926
700835987626785550302863066879939226509937739884608390668208206623012461299730378011574
829644982605247059002013885624414633410581453363132825605320520059292648083031321203507
87766295846454418916953734242813903516292914225048390362644425573581074868348925185396
565314981963055712419995035132382195118696723815145833934908285975599196697217045343043
273161935718967626863978978868443886119041809441858258695796026345630375524775831115431
147620359386985073776707427651082481928976810133859699786143424430006768439420759554957
407609325016782399296004574035775833310658604115555081469085084839959586167662512905315
303224394011724448874034522407180535221879714738531321688962602853125348758585567251370
781868635211696331105720317223706046781195086114266665602658062908484881723155659726462
424913746676244614371952845693522156184400208981301595486002990251885850735730715293231
231813184991789363734145399134110220355654105101077243598605563174827784724013798347463

pi to two million places

0972453826070078510252638166027436710934976867796097426453923241963379897682484422209801
9133054859645286814246290873537053411643476969647296675258300786930930798498427116879O4
4083381863267759733803252682422433296807395064180671565602714268211230312342948551653 65
3215269457610646824700707679329111011960165468779920874150621869243525586056608143500 70
5463657282512899089865298827626568230074377036105806222664547996570037627144257157725 81
2686825466294687239956406509728324077614389187704249767623422145130917936877460629168 51
8390246803136439897416961358829185973655389064577513081193455198291814912845765152299 64
0726298338964187393802463224707363927685951291328962522675690312898982088924713495701 42
9446868226555957527745822828429972809777383212732530514169924655225066614655913575986 55
4748942601571238087794284727911437223671115303827477533414242407179488602028340577593 85
8604743995094332651136487754324221472663136485905777194353727858104371833790095179817 96
3113508959960557056842175321440798955923184808524878083675500683459483934247832856563 52
4764189493364922492977088581624231502049139804359044981004314932330105944902474718476 39
8931631684443474047505681172374643921613431838126854709402109279571879936931386000038 46
8514368458211668002182820129991748092639890989566539405829588860028274480269991698010 64
2404924265810759212724690630992456074216611694183897204028455317073315231511142056785 09
4965756120390076831567476272107045735661718784029626248721159176926327063591387678465 53
6199985983182304514554009697426850377501200832133206958860197305121870854184388061088 38
1433471343256668979262770244341992180768631001648573384040824075853991365579081178390 85
4206293306812294729335597163936794620335391527090637206946734233815138342694714102439 82
2438830866768989624415715064474790566932331808524642956292310160051118640956880068052 25
9396549137978135636662786584825404606873151675168565871305305211587055927383753478689 08
0814844722468403352770000119125467123303888615629669601587218133024728247879162197849 11
6433286067222446352883105320858360458051478498520944223549847131411318817352440769169 83
8170078031706833636151053133551533095846623655392600802125833620966795554446296223486 92
2443190847463919351749601595928620445379912792443954761912799230415844328912730521691 24
0813192340886616809185433824240294544610170882751608810427472192492094539469702686028 52
6307940310706909913092204311596450429819108579944260900152451424283341581548822767582 00
3641525105027903580048053428092302369982089414594633169359389907001555983506446908224 44
9861971704430762869320144595156484877701259485011514053979796339301004384053190679074 36
4712995847183321388711090998235667221263805486838741894010107540013788101839315914880 87
4786587879500151553964407616257750818635821793101172614404666888481821945604405137924 88
3197154571243841934389672475514442366589364718518226071883638750704592945198112364117 80
9723074340955219414052446422009022216668654835608101851443171555087452030706681358203 88
4352154987944960394947296065470806398245627394706397440640513789295335728586689990085 99
0869067039434348708008176071405043256903200881536947167076099967808378752608107329924 79
4981733321395318387783282359167370125831242890647354742150887825514984141423501301672 67
5082179111002557972517427600739373192142103926032970810769819058988038916181382579914 44
3432210529156905787884745821110382660549894669376430572820094574075292533611636575093 60
0636734985028438872658171072599410784930308130117586486115832906605732742706030699969 48
2400586086403325051074570771362035206886494623441190563168069892406954125655639819424 71
8889220352243226097231754888531728266773910375723710106739337663808981515842676046852 03
6203672407891490187927355618076617558781815383071457220388177297187593895662961497505 67
8534508134597659282837182367192431400974105437297844510162575439587624579038208902934 93
1501481048364513538889280304386036901642658923162207218966651935591627497955048009255 51
5951904618606791880818722697029796220208880024877871779125810211789425072761319795431 09
6246640197720972322610026863745530018619088598256565626527936608540782281152515511826 17
0482175695615416278513963778247995239435931246975243690592186779426383330733739209231 89
0347396557573966490982052618777270846169558736145592952198126893316443382397314238472 61
5945308852540887188657764022385105108797310172252284529976947617375824995263247141181 73
0160201833111918112228135088569821004818161146167604851873699956114717104896951452849 69
6758125834536129780347732313296535965223906335742142020499047977949714036872153280706 43
5798771212837235867186123284810142893866885703762072348447967592914421953864220121082 07
7988765511025031152937169329199651538878042516256025746324085042288449977623894692696 31
5286572937413430169360076035329369706417827787737793065010569739456116219240336676403 09
0783541455138316434889449979237557815875014455206486934713086498330073898088056894346 06
5207348759524887381452856331188696637177260653929012996245121362161821249758551092460 86
0670329278198602652523263228264835722732391838349258202127593894057153041127362818836 10
2715025364815499370099624982612448826291798672824976494874375237721882327023286078786 74
1446467607560018209489067346574075767645525563642243021209167963726132117855243254451 67
2661784549610873790374606632941769264634585023927554463003063866779511564310976617009 50
0805367466292149031922893154613447952205175673050782240665412164309774725130449522884 99
2542381654379640699109303726768644369877159674558387881782369993907291362951135658592 12
6247060512379920709844530680675102192359762580382091795994036229322416000214494830694 98
8847984025711575886642601488261724776092275527824931800612454680443270369494280519969 32
4740567656213986896794541045778067938189867874389603154930808754485087084139693776973 87
7043177428658169640187113187555513983743414948043673850691099689283408455486588315078 29
8044442921781351968165206545962885924899472173332920607370072548385216598658988907709 88
3546595044585146811251957872729831356143487873935778901650631075864406944483973846508 88
5441240799357171892723876043392675414211824046241390905429387061282390467449182904617 63
8581776736597386933106850735700688392261440049675224410943938679474310473017573522143 06
8707911389927519833743170803283017996889374427312521582135631706351219458024320861852 11
2521563609484108627492827688256250168628994059815893041995881642867887107620110956075 97
4128723052213171260507160053682802813777234299991773005009432736417487408931931722894 04
8033459985522641089298221544911814713634885180832952437115609605107576510978703876312 81
2439595878309461734903983475833694660579154522058959144056918656096427650852458957058 77
4828812041326123928589487352356033810431400035728912378562743250250463650074389009557 27
4063694944618836528666054372615245417220582983085096593648284195869616911572593241982 73

5832458538314401670681614873475086126487519897807761879174968270011712481891846613898879
7807611495726137823984153462245734739133766906177880605463474924802150100016241121217131
9454356212000027292737880906863891198335160811628709104432373070159783768052842819095211
5390089954681456048093477272389992107712708831199940831181902204147888292396454286509791
8811164285982169556628812903107065000093202373656256754844374179931102181762310080201401
4616668489913449894241224497019175557649285423992397248206504152537039228810588634119191
7203443941487016575998455338478687656777347709176671468038236731340163982840788020534951
4683295900144178046155394012922359304469985445894149744390975240122334235496542515475891
7214184893791435027789414016885281649814395529829554011283072905955477612250008820025121
2725613693737437692659490087374604434475687479171702479007356271678183376663649010405891
0049089190374645620178677383153699102840008228759819748535247250201437242104792848796751
9970557786331111404474698078827521309539908080311709187392213784736322633200766495163501
5108258978872205344514715051636392345586235874774466224273621281109257958287611916625411
5368465731854067726266281807464561236527057026139992916165305519804652650305545484991791
4037554110839891390539496691499562413086576574402038396516635578730617890614207972461751
6714788535806936170019844579843509116551020022736519845348479824964461874462583047674641
8261736123338632176788831250491277920092668848423818185480897273944618069227465490467321
1094749474591173096314265599310511814464984916465734889299095623605691807453782343016621
7431623595003450763718679986930718629506983311087556768767520957403143485900228811237821
3624226101946741693084249295818822909711597530124904150583934405280223506510308621844591
7274035293927698964695385469623728428517188700622734337169764549988286823553023276435031
7635204423303121918594735894560426683544961268990924013089983378701351245239487953845961
9039342470609246933983359445142833816687683906014541970792247310862821685927654723364211
2180787384299729325758725227419692110088899483495642276964023114275391244744324195895101
4686825138415711722663081780340834694446916054767932938894204339662086472722713405850341
6947698378481116592561577153522840427648249729162557314786408820040147197345096547770301
1284755681410163286181373244098590685626637160588907071996765696936955777186970008347821
6798465723983128062485664119583559466864633682410147066349690545555892651629622100768411
6618973454423248558753157654115752350824419774405089409900945155868296661067096709024211
8856726351302207694616474196101248350048832439909968585541868766773096136362855749905311
6728342790555208749417006678209583055699624947773343363507138316650091609944586030701991
1206362094726162816281543050557973001978755133788352283359301917015024377492382939563791
3589629549538010617350700506211537941161369245401779331106731368079062915857711534352901
6341740273374652512188369391593552057152548039141014116076073493255128653224200668287081
3493294702711238230448191080636701705555168166051617085257753897094270729668693022600641
8965741466207261504708613737237246453254378255759009256753842261930842019301417055445791
7375197530047041686118589733242245538937939958889725739655857866212042113770585010041531
6324733806348493087817055660615334112886073767434320714040687517684530102491894478822891
2467092476454535781147272683411981251579374793859013653631922285377259074594948916500891
0635310373293460482081275476655448033468975099417337722730080088351386612163960876950801
3327744632752021123833663890757017365868834071432495737637613189023179572011933183468401
7571300355600358910191442671794848640975070348915715578801216721948131742438746984840191
3683550642575388859536354064578459525802634894053684852456612071352028218360912107282441
8603570984859862149911072056038245919485982982354366474772443187443625938524431934988591
7695084769983030114427791664023536244069488936433666519120633002270696431428914178156511
5762413660679408644277393150876453278200005143114416938450213602453859527358241280822841
0514800724995059504215819802203211911703806022726228531527782250798281601848480446388451
4238187782171045347616593762607297088842176998044060036095071630593792353372760325679081
4296637714903184469807592422930615928875788093754556644160988810322733243261160955215201
5586557288084122820000875492308382839802493037018529206223877098770427810125622682864581
2582074893745065734680589573571270244699330407523545448638482611794977490652298806639691
1549161945675734171741466566677615589130228517249374989142315440420360083760231932689101
9975546265851442441306916999881628089074859690483068772415009109248226072211953662783561
2688880666917145514720406546986724461341839275466663949242227578726053861942662037939091
2649269513080814316765928250487647210486358393805609309137306920848388137756273064149151
5109110844125268842822447799765828045167088789610015596232545784169460313922987805452871
2155596749804085046227082625986099215606656776059196478661965350118610338730120681862691
8898317555385634933349620362545889058353970499641814143440022080602998861413468584239131
9340153564232472142530142859120565763856935627762470877172637820785953181433146684288611
4405778794843941854414553901189633756637537059024278787829869814914979048998218102052901
6879742490994284322732392927379229358949375634604471174134618118763775567553171353328691
4467886633685252997791338884062520535576727452268759094250269880262302902959198007187271
1543905075709204894742788081121617537057081777860657232739443689675971961324093497733801
7177485067680661100024970948317758504896131681710471576013281159979796697117964893749361
5801828293146885326670686356473845919381630061412473853884559004582927311830453194923751
0739495359713935327341573558516210856392947218980959393414820835849569334691831189479801
1443321814633528391077662888449608686963782651073755036361644921669180879049103339948481
2442207474508737919354745696156441020723233200904265039496704594794672172322314104619771
8179232279689115160333870859699947512842805010007507936666864488336267250377590252110261
2588133484892873136664049593470280206296700557077340622029755773349891842778084850329491
2415000656210226872992169408430237869071121487459408905047676091555973770140654311513641
1993910153221693197750605362646324562937860104085102597225325173092650965923479694992241
8331176811659480608800082618923704059498077040017396745821892429548950716877133100103971
8515335345314710685957094450605373359195365861719631497006018216015116781644587779937231
2745484370216719698324035215895168535513924782055912827961534916630151458051788841148091
7186706984272264732539627856885621860135755148025841191257861698154027609127050741951141
0396563792898288285227408601450395352112653895388386779399517295862927094784127391152431
3112925946400072613464924997268181094515144033606300528712881176345574744004968524233431

3988590299130642366501730923117254932235693646270826473653961470193614656488082366385 68
7127431181462802429670841610218758998609649692350312982446820228624101581183380869587 32
1577601822378961801576779989263352748408957310408461068537718696398441318584611504148 28
7305023118066959863417933895317469817706686635251177878445758531114952252892091504281 57
4922455591384711330427158913553411142349566732404385307357246228684692228408379373448 47
0602916936013993040950654596190030818380252728256308220937269752752989520766338895811 00
3934670501955027215254070784222031586900159951308263848419748777684991692392219653723 16
6544757440764101009965061712627959017809881925888477927081016545937332535284128020075 75
1048259116525137578340100895184768524910186221173555348636969841574019955602865124223 61
8740583409084476878971542125953035591327072076061508973819799481185638356745062572946 95
1343691678918263066507952431155826364371997603774446201879101613542447700627364765654 91
9626761644468077219092292908086622811295988492478114867153425715536760311952454756156 52
3696770385370476886604698021385248030561176382068757506758136412862878013776514061907 26
7002330978458776691822833391651290787076086693751711689174162769080955175727969805201 68
2102716262136069706041609156178870530540720979768493026637638968662736050113988988367 33
5913513960315474288974713854169693929318308680021232896329599381271653722118529514501 67
1084320746942049020976670074228274667222156833863413023286091104602401561230498455914 09
5239210460975654364834446150855037499705955336660221346260483996012273273641910694943 99
2246524570005239136227333832849727557393356495943155230901233230563434152351056962822 8
6778821093084612156866114443870760958284067151875349808971433102743967023207286830105 30
5585474061120710903593522245318973777489287212239388744316548020903900810810885266927 72
4619314382529097258631124615779550275982207061668708048769500542419875562046699274663 98
8639348486840142233528728214440395478422891745476582719454397372171674600060853686206 65
9734305152428332728939125107799228171714059172236020738527892642080373792927864355780 21
1583575735835906020737813547108144950490459183211755301545019836835859869923267993632 24
3604356246020284705006583453866926334314789829914678538149164332937337706472856119777 54
7178021057206562416780620492496050681014505544400083691585386399196405583027286332121 02
7159254780491989067089066871512194349441699015818121629691645296566728882261734646998 74
3806066636501254919352995156342670614836862978791948285828518127841377026940221675580 68
2835676215271251002722532175501175391676185708244100056978635520154571775096098390350 01
2825128730291388517597460971546374785158162912190028421493728409410165155706168620646 22
2073668955493931454351822662501773714797996892061206532527541841391447611392215213461 40
6928641935582165447967025369448854441712394333499860298027098502714181595416780254324 50
5451516922268832227680705841685710882489400810805982436388069304940140036054058825147 27
4558825823285297655543660060976141881745439506835189943064407677097341582749754334607 41
1032688819736325102959868912713918327323940654677652057012685602158373854658756251286 73
7197305200835452057115115637967152422854375612319367176856700900637354250309455101918 51
1962299514032597437920935437420911843155116834985132549158654381928399557383125945367 24
8250671147389629672990878885848882599701074227252504035481804962852257052082363485122 69
3831562806728367147435055983927973260668429507263513827510419746889767672591369017306 51
2365731719532431012869561072580528336315890095826698417353963848720165200490649064035 49
8544923722942741837841103308077738694039345046633322001742224053633940752750598293574 91
0869881854105227312991022233397661253324690730080270049025891197866095461412213341213 89
1962638826661786556146344210813263307717607866380636342431909183254194322747324115237 70
4877561380927960483576499695368748511308807062476127325750376659107831658785768550661 98
8397945830269086027019641122022573660586407392914387722598907407359876447331552706846 61
9167081990942108608514033280601787851430149967994736342366222291820690324453015250155 78
5760152288473671518234944926548288571797699263163527538455765576440677544890668248996 61
3455952180097814220867268863056429154013829653468573489967197242017176041456299913025 14
1444752488487885952192226843723645329643388078851106993397696127782691625463097323535 45
1462264137652139297506809924776597347261269093978722139280745585647099981589429668515 56
2704033660587623053403109834593229001269832269205022015814343028080002236580154632882 55
2904706825073630063078118956871980940767158512610334401938296271880860086116970723967 03
9431812613097424711171790198574301239907540354716506288051264920095605513453336202926 30
6649809154087581763402129960016194911103493751576558875248272142376797264256426977368 19
8890882172269926555304987247904798100065046843426551930595486463515965834193256019953 94
2725424637317151517635243951403807424423857724765596152456746699316286028264855040685 9
1224634076966397694270136693882918560767582008669438337563145932941714551996669772822 18
1470620906203497631712086466030605675614933109167359686926299803365159852130088802404 61
8168331860877929665002779221347601161748135200060259895150218963264153851311775210714 64
1070725779349718225272679005993370186596157932780066176911183167614715796464874106952 24
5376114568879231150886303623840717691959989003565807547270107199166447759227730658896 11
1939427586024665729164647782959517792521728407860744329511315658116636186290447363155 51
5655823373410205410970553981055118587774847255317701694561977327400076274662916015578 14
9893849915480067001588028727139647788555334050152578884671964802052467242449395518643 72
5584409996006306300289818578112735319400344482753400429606559508922535186015542228917 64
2946353097283905715525192200671822520448345695728302485815877783543719512685965426708 16
9607982346616792592607356189398955902895422982006901135011999278337057617540712583480 58
2780681238942856193111114132504933164260528023303981416900679889803737603209951643848 90
5869936960683071012317961526692453046492374632284627990824233479833037403445017864480 39
8812755922590204653660846910213794842368881452629696866213566476702014267253205644350 31
3117144413771351467367169726545226546510614403820274351610225460111162195775537690901 29
8148136883286248160366899277390833940514714338487476920116426194819239644946320044638 47
4688479111517044843919386765303110082423870498152452040557396205818316248694503295350 91
9858056458989087442568312292543450747373415195274236476418028199548731737721290125242 14
4675404294585540242390647170225700446424875936387663670677479620244276809437785482576 65
1284377690093142098666071018702933859331043773285303437168895184802547799127333139633 66
7250697624004443999842871548273540021362280036387835780861930328130999490596589188937 50

7538344269255703320890375832462846794994772016161849244806707260664239432321923207176007
3725231360026007938117888524025045773149253502497399425440513990997242778918511892389555
5672490883322532107988627081564000322527803151097668662283346777383494174612258945420988
0000290932969074377260138690910279140625985152505977668184401078167066934000751493482855
4055561430475539110533143757464229486209744679084475876468363178927730859885111355095799
4744954218629866774966961594744440927311320066978611380859054531762649459016781969714988
6139949226972233845270514380938951350464437551225486111708913966808936677797309943848800
8619331909104357860720849673073825937923963599328475100407272793862627167044754075719200
2503934191972545189483117279024614049668955179079986910253764890964845981670901713935122
0678283957996122631197331491337781834338419318523675386629019004633433285328343418249211
0719160927673080132353694726484892764429798706811256546280611726046073321917630474121500
6403020178932057956876051027752505304361222709604675845312738661652142419408683408375899
1400951141339295457700947317481168853640941835286109971034167235581786028234982020314999
8679401740417191402839362105194806048190182451800817013710421974592127984040242689630055
3355368549416400863921774567695344586508832862982165846016450780000359898521290123959000
7042026607449887850627176077622255060639074531947718892509905810093672989195409957663355
3865837809804479353779918771214297746197640947272121402353265517772361634196607437639155
3866019450177691400624068412731629586080635067629377512529343675960734271996751352015800
0873739548389734399012382565682907859114425882852136616920140299004131462136806323526500
8654112180847499875844111836290979089198341187537664960480936264893291043905977256329555
8138877674469546181579787041915975583350003772867246073518535088549546420293097158563699
4978576404837415055580991884020934333351281955145551857333488816491676944277824055543366
9773119201437225415419274505976013798414437359942028760525557189378041148926125798303611
8380107711005004239239264415692277900570279474013649422918479336925354324840487802317777
4451841779583558257547618425495753958785494583430848044173708798492674195289323031489588
9916006854228795789294815900068735373333333863813908420994872551172617108729440888684477
8508116345585269214154029967209886937158862322203314855817426826359506893623448134224733
7869394609232566007547212567414103421849013118963967654732916742471899342433028029092699
8507522949079709443009238672877439362551113136287615919738324134916722682612282631277933
8117508074089589439719002542641266494798017644201645415881317601897265962461441391319200
6785035246383065776401652297320470892011469171337228786303703845313419426441424966498444
7440486177385135293731080998594728151117365071913988126930748269061403928642276627849433
2955882859383721484750707783551198962249119558823704506458200561017020532484449021502299
4561765561859214574399009554822941375154413613291504304691417994122460933816513626279022
8278842138122077589424038840622943317279892594183668293679259604224184594177065301954866
4394817033552410287470447311715070347750834323733311632587672763683468584448656253918722
3946082732471350044895708680315032503867640173531507940034557285503700184560276996150611
5682914161170610461617408248462670335189152551882482127260072595765678899125678702014977
8667908471066310748764673098909147197987925985906257336497349823503126098309873468616277
3935805790735490824683049728409773238116708249151637346808705105219191762054169882625477
6054458177111799377677869654216992578557720426344244304207449548970339043450720611900766
9736351740195265632213928258312052824003746695459092804525968798460814087071953425476133
8836335351191121441431485055201235813806264923135383387675809188923797585515732036588311
7624241169167523814588592075164035236684372679175906370539219798112659770811394735167199
6299997052090170907589856916389066427042357007445027751201049039481483294574438097462600
3510578981954076398706078758177407249118450697884299413834206081242839014881487259985411
8148029294922787832435610554915641091748877067067820119859109589098386883951171838013499
1482549927491426985525951776536126242157266244889609614829797030842021051602796714785611
5940646136382477589020110251992342153210060175230257421223754210749591872867518955215533
2994532268942518840942282675774422227552820761560727710394718256680247106606773831206300
3146628474433862047432595685056892871626532908327878439965071672422061382945329166004633
8087256305952415328812093700992298960069813962796862795676351987418240185629319849226233
3364318993090170488199525882733880795326588514493938124115432732058916457864229452684111
7518808184050956904691381344307784890211487971211628397734430548461498079356243549671411
2947333266642248305004364454547027491723207839956508680187617327033190656579547539520699
2925201530706435047130688128903205385455639879921010956672982873047946610054316358462344
4874415554071338143234413117884241960190370286869624227624650240042081371350464015993477
2053275567950353390671272161779060302162399780418576334810054483887031730716255256479955
9999869353196813010156297097871167306964940274916657267494343208132314613886925533494922
9411831638778899032594010934111572742980133181457828168791351424395578734205915614587733
6898808079170711455115476266168208177757472484278797129815482385203689591652405437305244
6672873130301925829524987639320985731209161056135722017639271699598686108867606133843644
9616951865484860461684848240743838118074173422244839478939982780147342223057865410217722
9264820914094850350132241515599990163071478148527051614432258254780143440109533660239366
2642493138522940547536416836467041574300558372984704018145571002969623469850599778855733
6998021822303549238318797629894156977212363844088917892752775284714477621892057425453155
1037479977706862362421899063030962822117732197082303477454550602385859114038989691666222
0778768326012396174199976236550636326601699066116890886877413337809195161497705492141711
1819110149543478693820620980827878957313278990600208633911278471536843043814981509703044
7708868816359741925341341221913962874578109934248327141091782763176542096312507136692633
6588213575119987540115790280285078066983338418338266739455368319656571703194629336410922
7266727424499274732291380098357445091340799308568360484677866535421058668005211542648677
4972395379364215665358720818554125981945474275161118416936625563241935915856950889053531
4121833662399878022115042456600150236469108159974192025443787361022671294122829888805333
6794395032940480496167711072878810316146773037150218352890241464817543095490594488754100
4250168040699998196719379820827733464855815859107761788284975054685679139904011384562433
6040357440246191237694402200760421433573387260392537246640896649147146547300721952927377
4176796135652330678042682000242307412186674866946280587887873829952327805010706610138544

0288011913855730911380572755893681940309380718863793758215444351626559912030878482052 39
3749351895597158155182804814807831310535096823567161235509631615286658578590195287177 54
6425064129849451057303533243348909799720945729170095878408322140440501474114381709565 77
1252671577981709767638478642064876136793945784423858603564966446314013104866705613462 76
9017492832290043271120819229564158948152826558422192618902530513815911803423108851491 22
6653917944817433904322700506874288819967061960688500673328353614680498596081402879119 63
1127482546640437038478288400798157913445232101668673741121294496813451012237984294021 94
7946916648150312431873969119602156712114227943068201040876756809373898205468421478560 14
2776882018807171741412749367467817513227710904673001530946777306922329095349569449160 11
6904681818272347781619751257278515355698616292202918847453026983015663874917702194760 50
9417486684119727660444085483297504989078258061539020030103521802717920995081783912123 39
7007002783290619371336821833696154082652039271615180227915286407683150329974060928048 31
0506255157226287131327681824780116915493786248322401723837072716088641929421181718564 67
4811856940087052996113141059458293870330528629791882649324742017955124423294762903488 34
9527229283080401222637227611027520380012093752088562406453112503726401539964371233707 90
3743951203202853502547482779809303020219663250932909448819057451029852172830699622421 95
4710165193932976937065457633343243325643313639122108352913555770081265053979316468149 45
1341207512188350547396859555387809194737465662790318681617136416152155407745354380414 79
8454584063447474508568028086232412696939931122964056032731092793849169187933359614853 51
7001814969826164800344027822813985879014862852128674617538485034806838652016807476744 02
1147665569643873967579664429588640957755915419261507196653734108170649748222041913503 52
2393203692779239073695588057799755260190308143550849247598185779798029812641251926980 83
1075657465946935112856279759057800341932347600136961144729013111372932651887731467214 107
4127522101051515586571349391276885573006456556663593525354509457378968835800277072108 07
5468519790215677663559608589952449272204977529981545862535929556885865521908620348857 42
9443548834017661683055326602458359944543309529736205642829474539664101759387807875790 40
1431956726445025653896405200714870685370656271174667615719181064228046438268678064178 17
0171014557998785949298102867487747167901743550399316968352859211648148786128628911637 59
9276847108874366161776239309829498234064137828071121741237834222414406998486328823924 06
5637743898084631704339814190170417704585646785993494113767101851048288345847591129859 91
1326180631166520977109326025457776640478113979812027396279052178079656593348303196386 37
2636054817261754402626843572645147411143619747488532813525432313210406553747535609140 51
3357644845126503159949939847591233867625899886282674242141065631702826842027427753547 85
2784980625547796730956213501075135701081437671209736106569338981128770064395086822635 24
1957989987524453151326843577125269551165684260624979530056413423679403268694320221277 15
5284522945590601316630023329786184272936970542564089572247314796084227996064312115572 28
9888413406901570607916985539706270694986922613155059545816240665427626536098988246922 84
2402217051092088452264193800405064723514106544629397362950062687069185743503468907870 25
2668899031059030733202099770701955626708007842141750040248776436938270496675406914489 56
3400092985235499129699604654028321299512881718293173380258934336088863956188145305493 63
3210224670237123490797406812568724883352801820879686837674893286595156350015843734542 71
4861534249712522615178086471835019904993217819915535387378562345508714313905038432869 65
1387690329053401523842824396391325686729272248708477969680927365604215169040230075017 36
6194993585447144041522141064749724216152303724661255301796533454354088980086901630718 91
5880845552219843115742231294519670375862605877626333668485075275284790344250165376491 63
3179283207357204363149629332621719841204149693900215789875840550688363131897828032706 81
8206201147097078246477602762963513245135225192242786044766017845546358689564183440825 95
5325433408871016151070367468245292204678084853185295603026039336998791252960230732785 84
6937162703749144838891633750052621697332332152400708758098485646289779180984825342721 20
2481272565551505030570041811989518609216787809270937241480471120320927521638219693005 36
6775084683589386544528152235347557559076380237852151946018272456823639549056812593274 72
8954569893447114180981315364959748515410685460978633032486429877701863553136745081107 67
6386047334640295768821152889813566449309493104358903645141317552414780992535059148130 95
3147122623413540646090598039300729540969213694564498578781314849834710618588580258335 38
0086634822074587547577692067410067571778702148024906809935403404978086974752318213687 88
6719206942034418621529900468668535749778710147528458215215350982709863527703575629384 93
1393160423887168990739823956189887221884206119027709429903227573668718607977070772155 92
3264181360169717277740702884134571773206253029881617264064980178019269641550387293286 24
8745115669440090146910371466589528076907506309916580190402824912396345679839269565571 59
4761366943873933300683738308181230942044159295951718719249267979297129502949158118526 94
4654488816064489091085747384972252662561359833739178319495991506543998379398350900517 43
3878817182955161844732570694229375365255153471178576184977477973316771607149942145263 81
7047522727295570500173374401356029781216871886130729572008682704728150994493655761353 00
6838203007248723891860455582340567631876262362589043407960239552914408186887350391637 29
9347162815765862853465444767189779664009434202490164341084863053640849409128877119965 06
9804578807137977951591204253802570919860455967388601901634414829009972971392282778638 71
2630801766921370493497690438081956561880129790914650754406845318051694049624845495251 86
3363242222261458699500929790456268865970393098396956023564388968632992332376590811424 86
0358035451697820676948892486638330687917659426382009263778903165827050291137857297509 74
0049386753580516323216984736338584197352048763394553946427392885249745167489904651459 47
3884977874365758947077177943562707432572285536101492929675650265851006003218146153612 92
8591069343214878686374342977020598254510958054174455625081164881956344323355550915407 02
2873450617663489800492811148486033667358805224298027624506580641661646093309658322774 12
0168700461864708735460759460023553734363668584630008298911593217465432173511755315755 51
0186309658856027447328817436176453655926246176438777403799768860801841457644764303571 29
8399555691565594231174530659483625020603643822173544096553539509045743693865104312934 27
1665810521464773628560410427152372939781248710605985932017807577384367065631135068593 05
8886083620234778064987975853178401672905978765702813875802603233337988320303896853999 14

5202912912522642348966197633971826810602370959759662980918534819012504031589962520471 70
7676639986835315319660868752912542311537503475500515367406868469956767733057751079235 04
9034577877826339647387642905640164433316822456509039570066856800982518273305819042432 26
1128948518491882214253968613100337137224301868395596441734670841783204007571512416808 37
5541314824792400452430812536772968970444067964317974268159359203135821514639678923853 31
4691684801923785613665706360942213296256718222140853765807068346168583496375655648152 13
2788002082516961807984840474867542864812471892324572784880995178492002381321495055976 86
0978043922721365667914325583872939628855575318985133904712378084559407184483283611533 61
5780257058364385933712923478051058091607157923717358832712616456065679450348099079972 05
0842273054741907781106494021601666662950945932469680827885873826896084877445611837984 48
6573045432053689268403491423450881535101678575706668749563760666307293636497984091052 84
0084546233261781814979761412181872861268653460711665095513547783230627411811482071716 46
6526542709914847880777689282345795486110671504947364739973648066529953736779993402353 27
4964801034460008563976973633732229471469730277640940775396288390125824662939693898770 15
8620893799185516933084638159965395913392629245005644067609742316903078027155970074441 83
2783863920059150419266647287527897954927703415082639154789692688330816604795642091123 79
3125124523826289132391142592592488723531353414037167337993061711996480374076644403610 82
7781509371989266789958837505255476092942001843074443896493496814253542442287548263467 25
9939978795979725472244117287600771260884093950481192159674887395558626710654903021324 02
2477079265249776944955205108056410482206802172927635615223773851063602097203137894838 90
8760877304444339107037138713316374909282154921296609435599576542505065970239239800133 97
1925439883533581932604235533464070907285804865214153390352229706047122013049412345535 77
8954712486300047055628894550446086222664852384731073317223038564186531180217935120788 36
7893163243192580644850051145282276974702881297556723041938969669115168403425219126668 60
5613003845205389336911210081205025241087952203462366013906294095868063810486156703491 89
5288995944342180753633129570224126076566283785776310584133249700802109639513491608642 05
1940551379803738265147259899902124751359728276487048030240957692462436925511139223531 78
6948292842185557243747361826577956003518901587687598144266576265825837773537988011685 81
1161188545871990328237725765974792555501517035225204329594566895045977985931335460305 80
4628712105099200265391999022032802985513499055895296203508118519116239331768415451061 06
5516247882198069414695314021638542948465099305759018425477258768576414740917080050765 86
4163376935814660781872244505837942437576287085454033501105370926710216016889742519017 26
6477109118773401965876131923824401677866944731603643323591053046392692548647340576221 16
6484269938330305048174397311935005367616450508447630473256376421666139818172167153111 28
9102455851620731970070031125405052082382908931287575642187541678615691797009700938050 14
2984666852120015915113272059232133736245100927185727349413528948123231159923461589690 97
8967445413062762687256330468233531371596513000406153320056264213843223148270354735863 50
9198844653383568400484168465383514752987266852104759009105167315924626485357070819430 35
2359875504929773330753172203557347023651707931593278744454981644593247393837794365278 00
3109277535437349297011203088850811385110362985620594883469615636232082831084250287376 13
6412846726450367957498216344694599244023203866360957965273125048912012359996284808751 70
0271355908821282445921236597263623626193171749930602780067270385204438694420239626594 30
7173387544843964935739401654469425438918315939596750099329847553183201970958681114740 34
1146430774907873234732383184316977580069510475962534780392905312278972028317103077044 15
2040962027536066168775095387491356486548993528510813754662302301044164406078743379938 86
5943431973972762809887255359642183688625252868429209407669353101746760614361504022410 61
0065751838578184594309738166652742282235952326563156004363460606617992954342681278246 98
0231107831922175488282484434894215590509624968095516816141784081058068173009002631527 21
7960747234579744753439008266238408070488767662012566975422632476841443602996127063203 64
2476725566239584136334669634201833963782542254474306328053514644920802840116935071719 25
1343449175916344648154738160058964567608405798016590255748976711877225395270001783834 61
8076833574250619637793972907186641657756554922398951160504436181020594168505418765847 15
9029625940307677186938605462437938952917738297414846392692221118535479059909509729976 074
7387296264727605210459511282743563127096059502375573180619545259876531988819158370780 60
9460730104615314079672252575778609273091902469192988451739871631507685766351864937976 29
7380660316503198618172065118786733591976523956744318887953365229088652008014049981149 56
5766450135714869973922194603166589011920226137063954921500008096808049365156179044413 99
3654732119422707627168066232679931722181257457609152455167050976549279875792849661673 76
1936853435468604260079278210505029351060269911077292583322331389423980871654050463146 58
6178672199499749697290597915791846480370351873886438202706098049340045567860930765382 61
9265263181120218638592516569258659483274483164346145940369617235805971795888290583815 23
6978853316823013314522929991540592302274058550374062853590459765785964573806521600263 15
9177428881248018307666463770771998546947143246041543087106778821876907079694131367846 05
7701838283928309479160915776228351293308159945269760684026594233928858711054408683063 66
3858102047363209449653498877627578626990535391553279118811926335106919312733976667409 14
8613974561452545275688380518228526608713496328388418368484801362645828954007306507915 60
7410288908345465962538086962552626371159235948263845456087258368603761981706726337448 12
8763726829988777229445404077067899401643812500275542016983023069545116299831343107188 99
0986254101600293552441548625913761597478709185153402568888130198921884456611164910632 68
8832423448309628880496579742710340822373667382825577250767715305186061648306335598019 34
0710989716309884087467259104459888759596053032924889936385211939490968283710864571352 34
5601580438642971035384668305968987968083587349785762678556588838364951627235452070877 95
3346765867204916172087981314914169891375664645456974216663388028870931159655290890968 58
3246620942368725955209587658676396452581062760953955291491765181737672918987667798425 11
9216807552581600469267289283769650286492087987822244210714046467307485197437139760530 74
3223451552532121235870851616285214538008368612462951548039453329683773642486296457024 35
8189231686226499499218634983246812149467375119948195045074496597043271760401338355222 36
0815340004795981936852230925881457545979566969475623894957247325248486607065428825628 76

735928839761419619059132908399469163841005321889186943256572066776523238247316438079769
604445729694465028343061665221698708026175757051241641110920735772620397041168885940376
105843313029314448290659556842964872963771613219507321997164144170256546894610127030824
597238418283604031166034891537160762328639666165618560009467155491519746730842325586767
410973340298461696756421076322315367897998858068319892283085976337509294649038791074370
774008463493623624008300849500735581649669167597778658081118230477074246964360473279351
820384719889611703590083004556851252805095510667699602373878235376637278226774052051061
532018998380611491985952875002662945542273526058148948809979407952381788622443301332364
554325741038837042323980921416149632755395663617686752438176556625101337430464808207552
851564911671828345413047238795984888999156275919082152195945358943987391987885447878558
839301953170347124025007207286819666318891540923756298247737363589629303730274864925136
919788906763582461536907572381188900038683408930397937519930653817228736677538511417161
821464063005399340793672109509412332835057142652831949674694850212250586274045481109395
874053728880966522799491313504114857375818994915227341352040220177174269756103240539002
593855895784915494068759108510900445669877902963589995785804399594722284335544039138455
157545905101223677094900429072516327250643891141494031439293381604921411482986961512607
568119300488516042892565334373068623996218120083759114317308944198998029337723050652250
270984972059596353606069301554054235809356222199901761551336947568289739010093518988689
102305620334560674781595563737372431505131046388436846166022210580766055016338443951339
275390504719811126778937842527429307171428737605543790415357814619485598470604040101633
653118930683696593975950084585133272847308800484877403931718396492321212575915598639683
100503344056052789887661147243089207390677171334489995904995565286687043138974556386419
506525626338200725605325025449791258937286606936652747569545855493793654946690595626482
216601109072227166433627147859946459999267490495496151264230852976404043695030783887567
665223227853593755677217005441761522757890068849222464725810068548316761687053704714029
329379429575971294561345245527545656824943971433138580495097372060753941257415629101102
326051852161087629611346623796743611234251779934005960149443294796416460031088377668870
592004886201801963376976145058292176891376685475541938116074706436461835509504278367638
447274620716138197119177044448797513778892289958847382008296507046223127280621051269910
394458936078904429066128912164886732932771245059557649995316456888885393740496576571688
091303924234856937644501999120287840027354636310802804883903986446286631594078401029177
968777418982862222702853291914550355495243566744611953366897039280307633613422784813008
059024523229864536534759379247009771237251485597896223350550371408237388559899963572576
815282498573230291020966336552980975186916429280761927498322965219448169604371548847590
855272358644048020175814257540052934387946196431967349029347026961869732830560212662858
359414291417643270438839971403798486296507440722652641634634289829191540694582238400532
287197813012034284651150002141596938756059895846743378832151044581733491029314084926196
254430106947558632673613396013154936002702882265375501691319481721677272774887979071308
645912517689622702343482671737444762540360443612383426636852909793473026221774344093610
047343832594705561796242470147957759938396128282186704046594736713467400380288689067343
505606541962915439823233191663415772757772625566712728756776474035878251863240142935123
099265241861053652623906898057679590093782672125122055111315834782392514171893266684869
674104933862880589331412583648730855968523648709573522735515641731600057043568337512858
648679877762687891197017419892629750373901696681145531056396389133491947849112600288727
202243629554113302103288831252179038609153411678577623388013856364347123025313312758349
214962681684346180565506437686484432169160099329793588697132634594804765801673028762362
637540464177371171233375552907616845760984120314814906712654478813087496692652495072837
633958248312795542416998444914156090821234201446635611514398786928367644038199996073613
035658764068334110290878236852304517716218148049432624678420340375691218100204857133468
386031640918904933187028256511334225965139518362851792326534003162530311776858304390585
530314347000949954042899310620069093842985984946257642436427475502009295982199705371385
675402423995822493614688183578390529256276625781476282549052153118845167267925851062996
4112189047429032689829030019539195556490734818124668433899387652291244231198145746620093
780807965110820809202500372217656466226546285167806975616512298469040287638196536023135
636136497663890221979236172338854918198095735229701423204549139476002030983118265134708
319579082742177973229110014981104209291997270793913105416884305647668068288143263931229
248475280351556328279527326560834108816169796102396306652310043702682309153399901101851
784085347966645769639956438623259072566307560394705513589758512702593763532523454546071
157876567775171323140719849308707274172599038347990837752226623850161890001369665331022
529573865193690993625896333791041177586051878192070877765609994109805155176052433838149
857884415802148300377380729422205761142218734191201738097419160390896945708128691961868
771823441333977805975939917040852574024595279535895516836710461224141484882350704209894
946405334610298148181498384962874854695004074324843010042370237702693594977809093900955
516428125416837951222634081034024006017318562283671178791286350576512210523464875470892
286547579799198663491343367457317808000101592280324580415606204058814973369054688383056
831118856426927446668256282605693182711497110790802221674273725205608523980970519287617
281829491707968108495634209221568005222657659050270810105964689600419159413971344320797
487520061970362143653107681949231466574683330194256177258029656287105746356679361964293
350904175915476056437228482315205448832642676308711147460636034732834323009419042086748
123219633559808929813901374321723190294403380213835037846658247849282371602845709018509
399863087689746361213922287446437282223068981442710808767398474519868597356568324758502
379022904387433865650163543092049497513946517120423996380485152404382714004998736609221
595167943665520971624671902878497234306879414622093306540351514653335343817621214097499
529081334366887421960854630056294187185719897965490389226099400645813457698989029375252
603159479352051728738371667007215479619563757407012855926075633358043611785695703271510
860973326600211008827123199163759340997123032026138109889964222097317552741425536341306
159729484372048567305363345356618321960256972367743766404198553908458665477313344243061
721260086297605296720785932256229235584626284633318929283198361000259218711370399715055

pi to two million places

774932048759783197671362863201844282336539347163637945712771588409841768777705511446946
524042208853820282498929639508467046570193475974601082109002237981389393889173661995870
235322219629414991396048429927934116586704877857111337265349285665368928875089626084058
604973081180779650060100681097962304559548011897685661967624238931275652416559831945661
471675822057205150733187438649959207729217154115270702515737429327406030388997018176392
492844496144989211996008741455920119711061917813560083117939437002185678808012611811579
946414717303547906767039162644803691523001638097509549864785771530193320758707672807382
416272997985657219816684445193115121472153946260624154747884492531474522234289814443935
668965208172835201849104468501236353466594439101771286059078923203698177814414065765795
314169342756627442756479249572256177112244150820590572608481471404592395307959706042717
245927521655060051371539677724263182593431649599940848813489629443980192805044699030743
329586964202347799440608555298026016876274594816524700132047509569931582510565956554511
246812167410441892014971291705911320464869299521893607872571989543326870276421120797111
258063437179070285365686818813064931529035864912313662571902811608044992533173093318861
973891372118396348879203809214817739751512155506071421237847824951464949328585000899517
323133417944919571954937381022516204335585096404429995065494821817491829820980285016135
764697993067913222478498512958741989762340204826428706202051130484543207456436305468298
589979035884055646100674589972300286489589524145330379221860767572196030001998405346104
236967394846598198903322644713641778288583186156410899810633325669588014908754891185529
444754877770703325428165404870114529025916253519801238741401043603630920528139960818578
217505919920444071679901533844586401542054227660394950985253171464765665886681458842246
276815471774897706020161939694773855297795179206393425938196355238038842483958103927567
056400517745841545364779201707342682078953784241931871811251340019135490752230188536642
456151886403026504152062258752697464483059606000866480846739758746940310044305071100803
285756185260033706833217834100697307370364032129825039545915664516251801635854889250188
649041767130862347416451200097805267165914094586669529719432662613627554666276397479883
132420766002470825655516863465415732273037984633497435417605339118055495848517100306504
714657091740008220220333420842716210255069982951669623227920520679012499910597901723864
217585369064213618772371091338814995600185227362965360920637316839784737149841162314047
954571631028508309649294656142890777360309568914137573724220084319536767372075167234021
174327264670777570720566446538170861043049130199330474798596351648879646194492952890582
573097904095062772066683569442798805496198375127327465076018212152053349220808519176266
170836135723334251006968310417820797186065027973257761157165072402736511990861376703898
866503275577438942275815661730769721836436801942195847681141757576578022594122936937628
548376150113712094447772678839072637650618932744947410926405869319895905855996960097204
635671095586425942225071342281083161078785456208386552686224968775589567428400965170789
743998364521940287896992967688552250992454811316798119609584499945128301743956586455425
349246733192769929221149864428842742821896234377707149143857760805808485642730371713537
642453067937869457905752335076438815218106566079270751572528839918515710526005618833910
853622127568842615368679981807360170875673642170133324263530619346354543601726038855255
677458062131463820548899779437994865925280732477127701533566724305128745511130522707692
262151065261479813961301205482595539849829038221658540002787182817945275934575981606268
256693535910919702531758789585078642512381690228684408855504623386176809374387125500306
276779114460427175023690482605349682200346806279793975282372351973107085666232473255481
566912018466156539344860475403574057353264973565944918990736160811763133524918574724029
491422191055346694230286637390630837957952652291658230260997650740918733400133052343101
030377785078752063516583462778448353683665590270558349479612983505612394339024757396930
505229513081104670913249882048315378942261322355663430981141830986684768102825715502868
274838339757192587347401153664671682629370155268224421277110523909179693192391277379771
768801542246883496201400997322757593293917724086463489283646444353876181138341439888235
343530623717836037727092220679014913157494612352486498036806094484885719893841191968471
524836899608147222418754317335333742369490081762643689316367885345454285713963603275099
228957125803241463056342872931869971175945658363103616835157351924115291745856865483974
709206998158899997133603670512517832216587046941565206271294188303672861697327759547489
832700933761709103805515938607797518316479113151091832801275251593651504886079398848586
099308119349380127789670910847840847094315489455770550455036182181154602650233359866262
119075457284333340654655923652677967900972402255903766627701571787453010678368459746242
411941460927025518165357449640803707635921820480703170780049503051527919804684148029262
375914410938821175845422300258710509897434053801960809670600172944613159435393190215524
986391770306032510939596603820623564441379417884276524064389978721528985487390825419481
126849740034138058770293316002349752331437832506834683239029486296128917710704498770623
085845558300725385819369995627494731655208956179446276477688277511157614704990194822843
238680273470746738628147526753337201210107164668619616006337713521378211388572121554773
972169330440194716653483026828509805531003676041798533591083991213560094604211825482838
260805945116333374599933098540090956755312822504067646007341266354406248261660563576912
527998130161592594268648699500472477533277064999384644411392955896914652204299081224390
628600046503573812695217022563921682177336732020915867504500478898168703646096151846004
855741439881726097384472127464419545019833559323176882955783064580208982952763154907355
005465411111371947630436261732978590284651319161466655167663506411933457586671378181098
415646592962365769083627160722338147885085506388631774907872349191987679896746362545471
260440954168197799220710427640251548433196477390558104961000164705989515785294399191798
519780608939567864719736389009852422340005442237631164031518642793802179526568429978142
883832138959824393958877015579907839448920670917065520349676998605415590210576514771578
737867513712547304466033499423963787796271176836236896244596776440199697869302789570435
167443150980228986501287745403934073972426716050025587854988889409938420910016817387489
835845628935275170791170549000543621962170642940432780627987378667673022285201837041592
135242242759628375037299734959143125112923968531901297535611629672308486174323176389298
817540247039152280625468479103097962829343042133940228841293104759404346083566834323249

5613647542579862544554498896351671364445937935735007027731767348845174887099668250810432
8991556770998488740178829907494844233167883018060597371518899642950324199139880647782140
0200104111777478968839555108074211164804227507987793103151151108433831772887252558536680
0132319275233969078693015925808524191590883768045362608975083421104441911728556321464610
2372029116740614660357396431901332673511383180868544275306115682200642050776276243163090
0763395746802731529177990810125354943612591471988752386663993748525397978887930776928130
2087457027296121993908125462554625410900834305204038710838385722642176756302904930176920
9511528857908547895413427296790845515266249771074342215376569954921693119003567558879710
9695936089444056027985322829745923005729290225019186904996051519634665873845766286165580
3709262295490525587150988780191535985441716721357307006356974623664506999438595865304580
1695576681415188637516729582855194702568148675216121345853932624252174469525507007692700
0832805350536951638370248262056345211908643609455087863745556884165147477744660686802730
2090527456817539515713203754621706982626148475907106925055688718403630005305138681095750
3887480633463299948769553743481082454131098108373690707510452638146157137362035056374120
0611543698395643113673185097174963434058322907235834016182906087165871016013042159191710
3902364589705902734815681072289867195302579683764563989718769225175933155867973340773730
3043216755653910755754238916160702122717375920700563071368437478212135288989798668217760
2832572533345183230392747618654039192087965652589286081930704035739642080989321089432200
0851994167611717527353978882026618744562282022053152831427848303386860279965175179755810
9883331025625438230416018942339170386337835792103203778668218061890863264867419662008340
5377267213803004835349669601066224138642725292530005372917737061162348654387360333573150
1227139300475257703092179902691056833698681470070046680713190090063876906894203541867650
1074973393824758598638628919011849851605676272063568183641257922449828028996536186177760
6867781410290398847347627179023486383536719884181537764284357906836358790567005858121340
2784860173429927311498525829426381306584979321719602718883988881367210302529373730687280
4284252676818495032793295747070453752303208640889188204692025781143747067742104393146450
2786589638070408530744824078053943147968235436205409865242544613096095689729925320004690
3722765192645252347104386806162844150263272906125719272577676314022420185623514890590300
6131224408194389634712598106722230807362487403882344642804839175997118906930494868693110
8815733589463337313783064630758220603607272216512039408317360771272291089455309972771300
4131061415677302487550303033119745936782356969206929008684065517843550699097961301208590
1342082523653831972164002578438653155268518045595065276218297807292700162623807539845310
7874932145717674428422404980630487336345595571419216555990693160116854568047578569445820
5814652541056909421304151860787424364050507570011697845159928514363633141919856220392830
6363769807369642480231750529550137335897960359874442590363690717246275208403121238030890
2613188023207103014046119559039673478322147276210394285191545150346992392868606087894820
7204013155178548892421185875032601207662275879486610780619943166946023124667003664069450
6988033789419169279093056380077125569611138551110682307186263508781301659159665719956160
6392695240132819371226867383997102313023718606144240281675000278523701329742800573141230
3771076730526300291885432184952037726251725196654898586302752019858055528655670174124240
7813984411479456140361337235929100240288001695428252763700172605581740844534308051145690
0107092006853751074980056229956793937036094216558524012682608620139896639979818266838140
6426887629454986869869560183587213324687051751956171604085036070219265934907270251099470
5725221910842445595207830851434274897958314090611381368732186562478051997330989401100470
5707181898522937844381414345412750827098497919645792920802355036436353509601251717716800
5655499827883668720305795332335489227358143909560512229294254511059615665989980158806400
0542294318769492707622211102847618082615964466027043097290549291809577577590269624782430
4271968425210667370895391287926957103913170155241989566593796288840942869051952349190750
4939683374338510867886831129748407742561428880242025456470750857403395398767464470647240
1224405084157459877316927428065793845108082933471369745731717078012150465560773087987570
8705024420182513066251328457937966934267465917675447321287297992553229395654158286835860
2563929622701616958143610479646337016816903837002557364944013958190229025904302917933010
4131985196006053939593015118034855063021486381739005927865937796283974600501660024561920
4250559351652393899385785832922591711647601833158673958922691798677199264267707228044510
6574554501282171307485007367793445470248714883718847668824378185985568323003918292507220
2124723395438145081249592057273858216386717412145544407200077462566779999303388581438390
5224684186060994650551174949312624747545411490089929888024475117143941471715620484316610
6148349019046001190929625568516287760436851209221765372520306632610279260457121123463840
3089769100576076038205062694894518313368129757008494650362783044503424298855803363561980
4544505418523948398908251480866797159530872371873295246175261964989059205469690840452250
1974665477634065536705083961295269437798297198225507400872467579607375592988511922700400
1408992230997692507290824372529302536554584962933437019516944831600998169538217539750890
3931830881834902561945268977264011060413490804531314501071393769710546347665429387332780
8496507788911570443389987596862067766941170235325721969700633559801276796321735405701730
7387345600278846155575539188888190157937805544173152124711048525279597666087261897929910
4561575520979704048086756054469428312274540256332191157104455429631522523043608263630840
4221033568337100347482864619734312032202472443932293380297883927316609665734196648139710
1732905763186077594191418982874783291136684330253529252496479211019646490652178242802160
5844480119414837560870626468221705278855586660939730849211724889880705529500413886690760
3683994300818877934803775541180454695193374369307401500438691562902746936145881457045660
3729762794944061609311993111741804520449283545614699712876823501751445373892838376800720
0416806963953649250578693080932586432379583979185083667852687093943396098783481513142360
6452625415249275587799065325616332081271863536440504910181314864879728064836084966650240
8920735697536217190475720554079269663844354262130994352432518830971353082340787314154940
8096657487919620704298132176243781081796600003627805961686458583311024834331251029352660
3857765397147351005221181616567263513538413677555892138833561375461669015061175377649630
7297641275729796185460065305881543374664637502467661873682860135938997966104887037321290
9898807718930345582842479863258650832971647881368783693404315869944158405607310517008070

4090624232298342148364593754066689887990245296775038063088642462354000679205016940257668422673337762347007385086104194711069943580453445570489168572684254649000937125624761059366963889973127846586244379914413995156946681083926325223181911728640074558763410456288575125505680815252293927925978148617475452694776380447699595505060703040572307480234705742344672410312296534950506517011654313242852197592232503963491802465431634661291802225697714089021233932878860417394101352631152036911145392003875410900121740080037076406806582462780508755720410542581570457315479309584722082919046179453753955470558955349046238100816637319455023584810199222719291201617752024446686059009664014265572476843653318670352206558014589136521421488427956558662705933887491878639391231094985612126299412921957055098215904145913861125266562187945691785864140841846629181231277170185607642984140347592485979536413929570029539960004476524174118063609089107032993576012356745028948966724368311323482735638137007848180922045444878697136394407140683810853750237906792549199074354539111875441697879774458807061279467910261059726785068754968619102664289872660415540003559370620961468777480211955901297443475619036901503229876507442116594074948464211635836077421426796439831027568155546121057167915310722177793025457322874537292989194954644443021474349443399152619976461756050044687614144519713784817621178977241435546735467024250734783642218978843024064895284631438816343507529653333478207439874478442439483437862158000529594120169584499759566219531459463851706574494064454137708838532274753954622674721781641078597150626394829117437331691230877947558401624524346731607652792303833610840339322785923043706961423361851512020163003710647339236015940934876141393157801473755209993057143969093614178086869200812729995012689406338154594261804252487070552936951201209338500618267008961125574026292842893977981995365853346768901352513316544784862113897579395337638477043789600054374631526792261265540350013294479593320389950440369123410089565126418333271896351165130651176712022579372906101484235246743785474046961258221106399832604515812547546779293221601003968918351668673368303931362985329988728773133651400992007718115949585299647818660678895687304129796819333868640365429982502489760838987278799683224799486129062153811952781175065350076812703463806998532134523960704085038220311383731246735440854003549805598897628482182550298417519242138199529518355344031257843107666981982362890498595699397614720195040741534182884971636795557299811576928990294536517544152653286097253112470609774310064281021572319914392872471828409399674138326975913701920603934524244628182094162053668835891745861519946102892975861163326253936348579165089549747556108870430826578335339548720604361931381687421184175298180045024624013213889215413540631633023821615654645339912191886658802828303673099471289780816548788972062587681901474865764326474554358647966050544193140078330581576657539339176601496901725110135193032764125437908172466899998107870743298828606321054287292306846189654216257857556016125523403005801973294312555344214886520699809700430142983450790538092344582436794387491628148340152942878299190846211322390876235024637891877012267547630879194532414911410912534877347045290222856769737570216702723515203683228144986530301933624735824640022653017817862306131826734757455627191831385937038694232224060783415856173750148103203795100191326222685916207583999928414258360755023703675143734725006108456521231782069026932327170717018071750076671070243830088213495621142096666279257629475353656122311963034872979923961295610014411768430991443598065566847331837711114898251668605288243158296681577367429301750530961966351587675538641288368646401680329010209836414539013618201231060000077579660767714453237449036664538190330356596405377484863770568919521436067964328702651141982058714099621884954127590552896627174371637870154631182349958080516726387898690116970574332525233066881141023090637160965393382917542474057228131824852865622014084294835890305433176866938862725299013743012988285332041492596472487084288708681215621556590555201187669920995349288586951293492562058852997594598147350552550331514573498130249635666005715621294128820162299984814529158027274529433671346133989549925197261429546713411065508748687356789905052618418649467003626154556517859007474346830220335306212633198282054810388329437727848015435298542340690991512207114081916532842162868282636635610834323566621845834965842435114082527134184467043885460564089153311183112391186053987811676276761370365842848180731392005624904716232823299108066060664706264035428591900796973473018870585654299212914106508310502492118115256100637422924329390473242858590436910997808736870161271565760862527256304137946186853271590894846828027643878196883034690139863099353489043993961413138596462884385448545178614698948363593025023338886138660578825349382042975420534819660539437264347797913552987090440975167142565491734186574028331056378689930376561815662351820475549941943497152154891214252064707929043762128454036356540042644388587991544652650458844150220834818998082523762637834939192990877418633119502333553507950955019850442946460037920770326472316538371677976322551046150834470572786498431040691955211902908981909550664375400366598915712966903738260586886220943158567621416946277003446146558837096145418383678165326467179887047516242002769841056810499347974174742755566238344821872394925598724808179028869791780821130782667388227175699536815810068349441770141103531411070982796796045742155734677372931318357429009881725107070010270568161059092561977387252629549670643826974876315270996332315389780214195846283702358552280297758021406191250393900095497206896922935154870995796216818651729414300236510071410275558943126234999452621742518340731495182266541367071120543595047705298405516414408231604009414859322767183356104103195233484846079523646610699869631765885028193317090927754380975029691292829028718271868676560565601038836259747689933191815086846351020204443191415907723646830255391680889137742672332139947128849775979539773479229678936219309925112006630115661739065757368376763659371892114223684991202147557197223055741005725452572453855565056864605371122682267950192963103945844174558065212238893467377781174877110853585636510480400702351384140839434491495873363170337452474420769031046589402807065204049010261018063071440950890118365282910010426311261217230160730391427829982605482593748719707894559086334217056537691155991337169641525291258655071927483459371130756268223314419307505994006735363650372935500767980421512143519725325605622643945414113167472987783687140996758965630701996956082462801450491199124071218574805370693964038921234646978712255066960696516150681300606294107400804707570130916173507733475564877254736912252150981350331929933833438210788554383323618735

6907160805455809843670057435085075319940976596533687210434933222806188349920048278738 11
2527503049208888174846428319016539600630466502915620890532294731999667891042999177453 41
2876891030909976718814803093271626981236572060433159649649342053593097499554635341599 84
3823862049425069393201426663783689481202199761417986083058983829390067395145517573549 99
7170753892480315294109011851492614087594473721159393289263705069583299181008620840410 28
6489923463260345642370424824257063909129380801704659653630724144928183755330479486233 50
6336636375886561058906603353162911176597900236530121728575366201890101649815977724729 05
4022670786268338777301117009431851403577621945481676206437531968784148518425893448140 52
9583996384970253959762126359785945669110637060133227953341910530379540425978304137667 51
6749768787465299649178342336427000745475481949471359866806915666645853291490308232059 89
0281205878126815767430930721017613582263189992887973230932010149232126373267151796495 54
8968924776611843154567161757493938401616522907844089423150876687054665275793238805549 12
6661693777598953925561080406938118224800051350809372046686020039055009141165394481995 94
1732187190340971882559072629584283719770085817941850701023924759988289613573778756435 88
5496342561146780783340518710951312575999931805190921902266561569503928261947901601702 31
2243305754431260654464625008686622748043866744194420153942603811557827540000404020712 18
1901315746401093427335748336101469403685451256443203470752844838931754561764631572209 28
7810279912029021892238282470350546338309419447797469130882924572050292206249855523551 05
6541643045762988641768062017411348089234227288347454320119807266068956925882293270245 44
7753472176552839080617430025428281598287674713598313924867754531846844608380481508156 63
5419462565813296359550594829076330106815644966517880328377723442649573476209397575955 79
3063867100477743934006498340550723621811989484482127172777851389956849044762700861312 69
7815775722954647603359273556343018515256595829201382091502262008800317497285389982150 39
0653559331282828353022910424849910104096008081292273713451581458295962514381717546634 16
7630658001246761930273939749682827160964943723113557562907810196124240298111767982618 67
1404839546685238558072759405609329731240555537304479929325649327981148318332021373111 56
2324040925444193621712935732155195558269307036208693910309562625792884944657181752916 33
6105084401385563679207115718885798851349629804275613855008281669530390869896208029030 35
5380062556321403668177908180211648438779482785129378171151953129206025349939787617546 40
8001885491126509477377940338081385165821773654742122319328208269808040149053483955039 64
3892782029247237348271696437467813541562307929349307240336751699252188922405669614699 68
1473878851860314289763537976243242698191015966561818629392204880958915352336772256873 64
3846681697674177357449240277324427185217245824903794440288306738445640018536560746541 75
6053710825468179256936469644180207220767950152974675083841352311356162549952917635141 68
8384581887938427692077003633081667441340571715043076445505570862019621875090213884932 15
5884931463611851668192193333106871041572192225845136232219900267083802223487321575797 11
9113968268038488404028145926959232839596418866267961493051598203711862831094501976913 35
5793881589511415327250422495358886450529525188569766476775406419896461256327822597017 02
3337558520886222182600285359281446308610907067417161261230022530622943269439780326835 73
0088162486844963806183381290631720666982539273397400494758569587351393454769511348187 32
2641752631190577688592198023732093522988172495982321802051416465423317460267710479573 85
6951746726028070968152504373358982055047400803013301755815227169530967519720161200920 56
6230877542871069645863471374280667516783193735132565221483833173672308119816534223980 26
2474376558947675692163436877669156494799894490895333548260863798232539491667268994167 49
9834770469902146408996837582495058290814533142200622637026589808567589263050621772504 59
0274990999327919762378666652919186395558768793566387776474276695160789608393162352529 278
3250364155846780246181598805142926914406989652471948193396323136854634186509092841382 71
7252169538362006323009210996206249425060811181486751298160865486378491683891420244074 61
2537349911807444468004565780723476210113068446077979422132044175184816160101908431185 77
8373692302853393992756111160625500938380159311511113590785216256048538691432381224590 429
9772946964322273715189525802973366045356007075343804126696705867914369328092183041139 25
1793786077259043301053693860564531228257539431737223358522116815430443635849742772083 63
4227879617830153625028018585784844259719867132834248512769481082148289899874543172209 77
9240360832619873253615858956014193841365176258823231664971366187149882091304281551016 224
3911604524963384256540786205403968475984137295093158148777371240187179778380478989949 43
6542977673257015705381263785222746972467874241300793642328497818478384187095000920272 32
7654649817697185631159468301209971547273531735570252640974294862538148140078594209375 62
6383946867371632446500669475670531599472687811256170460064074445580742901999701262105 36
9442807140922166151821307934869897283700119529308107366672854879857871482235316879674 78
7357526619385423624000713911305675550338852901423771854164892294155671638458861411106 33
3383120411085276458284026102555584547225937596187234636430993963801234448925565298982 72
0299003678439151041868749582442954626212615552519867450944465290221964963295540000703 85
2105632196765824842265073312536205462602686845226609688804383804743726232331616660159 12
7936881995940502571999932917673393102710012595360946943796638596808326643193164965849 63
3977291944145183731667573688536451820230238148263770653011391128913691346249132791260 32
5353459199163234527758142476660279547954070433050957305857105121982912094431653359432 40
8468138074887538729708537544168280660988493506669963728979443165679268763670660592266 49
5503521043083435714033518387756174209630866622504815251137848588257002504051583861836 16
4507635919756645647121506198985206274610720778853409008974133378884529056064324678372 37
8442381162453960789731143333706096052597301996509374395456246686628124327527788178576 45
0978686549138923396145393872016099547277316887599733216771184419958948486142610389151 87
5753635313900844721671593136105659055230688227016390060464654237193404309850825077350 15
4851993174718357373044915069497572480079084269359829143830931389855487549423227449491 62
7921911744168176225851532923090627028862627327172713736747288536386212322152156598140 14
3461744208641223824870184217621137984800128918461150291340723510400996828163551558849 62
5934702284245296564463145812208779641481397405213189828785428426965782243489218621245 34
0214229187382487983128365520838811022350458713289651197529723939560011425296402196067 69
6795758147935979993141849812041111542822165132392559070987481632337101916113979198131 13

pi to two million places

343628756085191368235626377581119520159283010980245617011022257367211180753783939970561
978347858998010395864273294684313310109468796463429787772268219444805699924553685460167
533539940861811762259294277647153412400002769010274117619239922487172129321988282132145
815583308103276766568215762232861023578888664955757065519767142309504205762010616009270
364200234045097208356485771816682415119269448506815186293519056117128036592678213698507
508774982207319334861979007258977582664142806319759559863145370970654776647680072587850
063099401087894554709720638038948369303697002697458292541889356827697675923286776710169
803509732769360272883121039870404729507335731757223271391288682308969775881257914549379
342985359447300616559315055902450692901802949396531937276413075979435030186195182849400
682794556592773381062149016449883428475015304083572143224589265200027521484688613520268
320201235650451901911872323559167037232979034597875506946927721571147182379966807463907
229451229908534653261746797338216839409164611176212107568264733826146148780221208854610
941607249501461647180175574523157572564916552016964627970618347370461961194707193166112
753777075198944648689161124640677993406222196506865913705310362339187592819657416107839
829879513864770740645142244930292524076236866433594986611393309682004315015611717440284
570589293425048899723028035902438855797151315458832846190466314888130054461650106191633
925396324995668918855812076832961375023195402633096304694051247986849565872764528769709
915657361774733919053124868449462587226912095502070721707162187967918776070558048101519
229507432633782002791831155368264376788757991198116796887396317781452995442566577309275
671432912484702302278647522453354025140790949182546253529101439352289666482429845599342
557559177758575824235233897374748446334004725931357292624483984877701765481318806697061
502288930961065688203005792824776845650575916401812945586968803264581170511288126050608
222011029481378867623478883248252825039466886560479147243495936240857533931494128559506
580257628000890635688149232095192164003188097299919802546703286321856744877047667974008
052444941230192577061686079333377261667952372310225603045607349457206356031317070375962
721464901942508795425966836567349678477964017862775009417083925965179211371888502696058
085928715465641853891151124482809575136933274382978784025912924252701523801839428277642
307786599800714990011271862672569779749355858588276207844192101150122755574177976386270
585270654598893783329649518770115264414216448752086329424108541931861114096827628512901
437848023439124804724666828152772743115489646255670802912294104703665441279619458724516
005911000089599347648268573468246049695113680191131432130230724663416373425090192136199
058793486103011684913670312515932112231432891632315514863903892049073046753790603338481
462237904763363720223176833541143297333311418939424737931987551333659519236920605564181
536292745717249538204457474709743381119982520693905592573907077743606915448349546454394
555065175130465938835163084824746341099051994962367971035899271356201097850561624995231
890502155814684467724636155469783168324827569631357175583147078970917287783371804851989
545984382859669977686925059438280995311638096443817997760350113087073944851928516254905
890311087233296314882559207427401434402388147125455959070011936654709673891258727027356
852730777519396886203870643053734199636785940852157897724435609553832097371222349936362
340929899012531960339947214295787475147685543217321671274622768033233000563823270722754
529494217257519412510539167218643358594453492687692208238733143003926135466300572963673
315564909795980489924726693864564374620224968338000050557898268566769671142801876726217
786557664915770035246110093279468256367716153962437927712948381379723447290968522053123
318201578958447957484049665426250203956746823547335325896674648211886220773024416413964
005743958993445124070810023107666728342986005755493637474986490855563048804573498089745
784926319775144735958577735630772546785298233325468702009571974896190023249744687725350
453317273092423088905788306207272855498567467904848611804926653657381113003182547299877
874226322450523541721830132563417954396498386793829456411967522772180797079505564445083
804357892004410159910087105620865457535861988133754425521273028942416530758030330807963
603979606664242815793244860528724956074874090361813640624302017652262395586836828920962
749246091188942919022126987681974610643447889946335490493675843048514349980646095449673
288773001522154402956812345344894693259234979855786079219347904248738644206792284192513
273004963905388381579374791162995933734710442586573721918359134231189246812151005641755
378356731275277203394208457733093229393377414746291204143364237584532278001041819917548
416469079889666380349031404208579751276702343697309017820412023120163323706816098096193
762383531664281467808566072208949378140585082658256120641566908039133014503787374700861
916342078689658413132733633143633823725986924785670108198872061494310116009326430205345
194366867309888936587361852746283046848650865893166284417428158134399120583431847933143
825136125768653093775747737196608082413879789095686319242787821365341453517151201241563
214557726125717115423757523043561118558919663144423086936670511991381534032262106215943
974271207654652317651989664244752620471519096989174455118437433125604112060004810628341
774495189990610221341264453400653485761580063364046688273922022619214475711159414547570
565243538867082174995562889089016770823973102051948388718354983432880888607110361769033
231078137963055727389811291277038682799321659040531468963259863919429152076441218374053
356689581994265209152060722347870111507849459926379425827381004560937340373847005260432
400476515103397442629158978594216190276546124371007314153313395606701209923570558935555
864373324662393682733813766609885613860817560855257518818229823656205930398480268468925
648315727203819634275024490538138712272836538138174118903818629370668796552740183516011
067772154427487631866938516652689190966932412414576525175477138616790707468769050286308
364149318965595623542786245515875993374048688880593635094764046404226690237394346938131
980945539830596379527850040281880171518968731583106825414735475048320937668798878682016
249772079654622965998697092863444271187840432634846582416724416037900486290633522739626
326436896952186374585547737927708108320998006560378549786179281682380184437377739659675
829132206012553928697061311830757497364982101473139995137801195836304657845862188194414
978493700984027560068549802633573502769035013349159994106340615470787332906037072311612
403418705507883790008981769470259740812636634023320523475949462864772027962521136864483
778421277547089060751025530145464807254596276113743167930832271444495182515532025306889
930585319483180619097628180166772303612166067813695171764875979064176720782749004474566

711439030261848117098822188891664963938685131934691125989498624773419222103929318565373
462824471898957875867961128151109965937010037834603669908366054995393142099942349260195
172005600349859404096353991054727739469799041357870226320692025454984165726774279463961
2974391368521794850340618077285098742812276901141368164028467528875427664834717285451 23
8985915716062110621038611504145324978791301213806889378088785741135694023601086117274 74
9076863458679625973610019911702986963967536056330358598505902404485705231443082279855 85
8042106676069784668585613992053252487035554585370101077350088649519635411914539954755 70
6552627758435765745846125841563107474953677019045088460451789610356300964498325679803 05
2637110090348336832738009258557839639769788757455040977616660990279542918858930891795 09
8681231156305389211681423379581310829419130185387664983843266604890968801860786989969 46
3951952355412235444495109149042793265931689165384524647612102977323804949102697206319 73
1445768722301888645472310352033608803273451892514604283782723573738137990445139646145 87
7241578009918294527691148925644955445782081125854974629912272326464903975079642306002 61
9452609812386038941728525697644742742928937816441022597827808080938118848982866564507 50
4699035503682364045702878619264007846083106256689672105151078759980626053310210012235 80
8990878645255277392745448389494246097730976810328818104837755849053575436985478248720 97
9623960285646337036722444725602653139281324322602774944601780118006435380465617724182 88
4415379327973303165642625496445752478522515140333292180299436366892004502808793014583 62
6559274530369320858199236858240582449286797047004892934103674324968078716780528715571 52
6979538731761613935613259300309851874585260746042807145302953344424836352786309151458 11
1677033843498109034713808687989512890927047103603336771077814083484462468606147254988 36
5476714350789716501445276993860022792288731246161188868651454875712458758319486681960 71
3512978097342890093482996091613721718408056889128483203073780439902980119776744595260 87
2450178788482758220314160796646845338401373003633935223913261746422839714692012729341 08
0322401486297890741356362043551958301512406087823579905459937055983399639642725428884 44
3219350837396044412680349885498678014241201398479946947426725134575043284161152831893 43
3625782555576248604469892081082951581213080747248488317379714406575523709296204687229 22
9750573904325529358481198026632909340398949735809290273565026520968628278766926541661 87
7936514649996335189185919828112371770580851290889147989502296980227753697818326436580 30
6594608092400801768907232584144129719282424720380048659076248329843266612530268510334 99
0265765785799633057285781580448150653414729103401186510752757591462138595755579846533 17
4652421247970356296576484971699687784598414328136415062658803237353726720510020391872 08
6548894049163339238448057175638872509301231379069730344640283694891453218747968903689 08
0091910769648752267931968839730831363285516442848545438955119235162670552416610116191 93
9562321215404474588449317760332645864833269999532761311560819867303179877709688547320 69
7361110696873523008666132573282355451328892707358403815808959340954004776686337388220 98
9994775888725251392894710257611581130581373079256950891110603633747143004818074544707 12
3585696701876163044024859281029412448165330430019767812518488732429146546537476530840 78
5629909045373784763916341069713494831641494317725925735989877414104258525650646992257 53
3386670229379992990248338449796361658260177037602404285435251468929382667737884010050 40
7781498010651565529747650230253048184829022351663490708944987681119612151065080708845 28
9402983031913436947886071972309108790654545592904426962102412650896208002377548637921 90
0582213754179945136773755029344213947834384540882496830055969807227427147523461863844 96
3300372011058488444262822847183566951057836418070908711811976341346792304827999192942 33
4443433745265711851958275817242955697536554765355857153715878867315582373179120358194 33
6859024611689549358454843810218209805645667115227811115286631487979122410467150345412 26
3591740231050726757815651690794975354569546610234350635178926282007387671578418448532 33
1264748169641045009329526393910725514026380972345074214526634679124879921950424950639 83
5057563167002422209788884501423549332644770854207329564200647999045672828827308973634 24
0163559815127165585708760263193476035747971136732284425449546141241031441121365329590 73
1844662678111062740199008005740851336021711329191019148172385811035976323362159445906 86
0688581745827210663607832472110553988228531116230830772249320718760314283554972399990 95
4386747911904120640958163503069215365395255939829108364440577894768655906426959078555 88
9278014811531296052829739637830522396568797932913415829562315501056197500574788258435 06
8389480272013168005449241765541341553672296781867226297521935729676215974572929982784 57
6999581857012470410652755234709316760210887462018949830999056268035473239191743880352 85
2394519685590226299123405562366816842026140659460166142689816364896322670567145452761 55
2084031997755218112222830694640282754583090788009525538267001174808944008582442234484 09
0144393099507604587499296092919448678728424465529426035504015366304857278457506789033 42
0637548565180260586465453504923154636606721495597923199612838925660686245382196621379 09
5614180948196268023483737048644539098579604117131070124331472277069438162426160779174 73
5893060403221611731902362901081241463468239091014747845348126473693937803450866902011 78
5409226918907211390842736264008402256095279536622268780363107499295189633493724780784 24
6207738545646297469856781877946413327755959495919858741916684478301866887582598506335 80
9530473059262795878826628135669574185663518366296779633536616256900065883452847893261 23
0794253332120934313098617001394022451593986301533002758857448261555201163216558305401 10
9024799131744318609478889442471917656585739383361529389164631578777520692609899223778 42
7210762267547136296772652833826045803154881842918345790562055852313846748688098747574 18
2995143608952757114896969846591330551571824901726406346947383162248457029568821347832 18
6005021975794932795753714019469198710339192150346011748954935005764831184800383210704 55
1000319729946062669081703284652326380242224858174035422985108136931116248203781568402 67
0540265060927629574130039459067445279197996647299332750032387853445521820309695277254 11
8331319492983745429967434508349456920794558330895721941669742554468253386465610665141 61
9474027323306891805548871442397014824881413024333221160282289288031591327546040639314 47
5650451024707847827003100830623983379035059208538782367438487469071591071661383527097 55
8515843491321035012225586592857429605107691468925290223593827912711007177987873789790 20
6010405379271265542602785723538506654446667038666314508709916557153712069207844262872 58
0323455856531437854325903901816838600532754965792572606903799575988813473068888074744 75

471119322401485057132580064528148510649450621787971736653675812418556178181865092565915
089678588385373460069351497669402008541424119586157738199026180199575558106225483616404
106207630901758560745738847326450713338575654604173253371130280137894079385793643399533
627809231210839700233028364694233862029910331376974507251928193044810605361488981272372
755345364515179843869973757297234479617541226385432501523178083972557268785022568712168
135342539028347810191915420343242593460913311364251073193994337038432388613758805682715
616485493653802378509039348336224822731820740996845329664200662680372687820264867057829
851192845030221581505303564175689377542106313879430081690861925874286987546755092349447
396156707592501916836911293037748049551990970398233638263648913505843039964819572362115
740772576823369907023744639354292702053460448038310231444811459137953847492391572943127
807412044060803097587296697310718198441066933408260496984151577165929551948655816486757
794233936898827759003578942127616047047342116721879204887694233196384054499475111593254
796176239384985382401964770818520948648559242287732496687300301024626104466769085254560
976052527675426097221758042315321576735329075348153641113756403344937644773157010744177
623582318651700354720537594043178245991335983391817819096329898151391257543191389833725
440528150815457031022896802995772275083312116597704221998268965832469412245112832949898
729700252886927820088561271599497751356215241446212011857260152524697229091514046407531
921992817342803276185996457025802115601746209063589075186175753000042491967200921485676
401596976159704057369425903568270576217269808261258458059616706188663328402633856482864
261038593074900732614647683459041578797930499820505904321300238863933029787387361962755
332185958012662216325846640775464765793633808544964957096554919468161205115569439108079
681353938460597500671795063102561224795271996046570862538431272191945200310509319123694
795458561223792958467448322080383851926631532382845070622235541635575201604845859578772
796435664079289773151778925432460163741951653324854866212599811797246357628056178491675
839232577223668439231390463625063640230283172300137855383978440596035683326842799266546
160672110959606388342429504300233636510873394591424660824687867914224109725995512135343
919323395208857001590380446982663110695458536384642995474068040679609633400896596476123
350118154718221565202593800562590621502729012224260404147261584034278162605459873385724
563714777921604748274433441112967312376761198629786698630802417950047309905486807868392
900997525783605684232527345124938846464261174378459017654469540965159470872997650711276
266611757869601903289742862638348064602583518365195914728710254160903417107626677581504
651694625449579938289961786666098572735726068550669586903757230549955720906262171394703
968503952363129448708741524764225060652166970795528304128108686049739229536592431541725
993932707486079566741459387068412327500416690860257614875887987623593673475462930459683
470578182433923747399924993182137763038375299038515104921386268559486228289802909860409
840116650732228999532240562234847784269088277733330025209570716387891375728361113161508
282405867748061283780997658812245009750709847914399252401416224053285985178406026775338
192282678495278866724568933375945160731956862168560635120350246309928374185078076042047
738430532548387635924125575316364522931635117865222823289644483191026924741636339992805
353093665926179864424954948846174813289956813731394830959499581124735554883738711252190
337005154680402367783261544240656690622404363579199589191618731985304828006197422232098
836693647084018123214174762767322394733060040517262859354854118824613991225459936046289
697141341652030935025735593612611451396476466490210627544573749771447155080588826860313
596615759788880825134084175124084231421887595702639616666768421788501511665029595597861
841596054792017855481164654185831131412091278284596904448081819980631438903805220749709
996445926804268473445415578103344932059501563201963053976214180739908627084806430321780
024760901439771564231200722635434932737991573155185911006524727748519971019847797665568
944919671648616867079037571706648356280803259648676734045842056865018193702426952536481
695581459090936590780768681243863993425655353304650567850655486555712821183817396599836
335767803088710405729720638481947048233307935220906084398628018386708952979494559033980
397505682344935367837444146988538880452810081360625583295193112119347517762845206651922
225273866969263002566560571379874677472263219110395446393846121845585775692692112744696
565407157141418197949229614464039021523936521713197016823791016253905630796286769037036
699987201010551972758839049026661939716483481149742391173870422714295049803918440920353
535056464826470908831627172704391412142384229216009271291236006701029524904826889528581
360849432035380413569645159309243382773010698290740036378191072142201091906924187216027
755580459326490152150594142335313528627788269126850570087709441767082111178033916132310
778845476958923562860626769063681154808619448865059856349824207852248210273557171168286
043742793001905324214732698076035933663485592468191202394714809212332841860500625853791
085553950069351432572141821734241696582891687104773831105970509981349409693995533517245
346454725301945250021970423672563394199259591390300437260389765339723331018927319980366
786966546139807538001500749940589639656960444463410488891018772540175813648299270189893
187435147571951190164326245651420062310747654521955306220940907622656396770318223285959
360902816252306279768529106713880771274641902570560244612378307890264854860064900878587
772245828110243449387066147744535679373207145334080832496103686003164871160455763852964
000928899056590388078720153816868605784536026239537615843658764131568954201845166804253
112509061280575860518512612612637270196042362190166991290755289148425500020316639433680
320405711411604237579803585659563124749946956984951358676171716148613421187649219985740
236045258155314008760929919647446288133829073216882269131573555149018421727816778521303
836603763561290076854453156388109058718069408177819078957451433595938390351399592322158
154787478672620334227303779525481013433488701126988252706945729704429254738586395894031
625241817680103388373892597828563795133490722566439821360408060769512299054794220774558
697799330344439044199973576079750280203525292863656227166375301043481541454332706533916
509959467358711349333821669894856705573807822002173120939153007826526128684514242907265
999057095825486043035452329999651555781461768952857780536654894859263172911706262038297
980160841215931881886962902828715797871793688490402576691969478911970166423079447116300
479691660296336654116567141292665838641676558425834662650669041699510798199041563897096
165783606780345803858875490439836681107946259228094843989003473574576572212780205076636

1242474530272832779197536870946026921102641285052046477151131959097597847520095406773721
1741184399590135072729305996362535527518365125842604722808130501701636458298879529638
7352394422898404127423076692657538919129379270227358715890678007404859216483963090839
4039884009512873222983061853071494441245528974454318958561187400018095275209628513711
5684572621954387878592562737224001189285921097359177445309099137601857105133655268996
7982796691301664713645669707323708148465349387888981009948329822204546100201720436127
2031538116582991015118694304911447593744151199214827769884667239091980921508515824456
5200887185460698730137255363463299564455463872644523269557824381689553110896531340698
0412186385000690537990290112965560297304964605482196018497714995871601601863218792857
5514826098688914664178674355486398857672164079809930784644700415892529812120080775469
4448690228534770939508682132340733873815211764044476048345552930529995893020716502131
5576085163782416290715979408661795263226509087062500009785294598803412110825577607201
8877274907101283893632437344755327661099462982029247845727959599586690604600010182553
8674565658275750715858729686771675111487265212206589305147029816991145935759706240523
0427201676090897193644031047142701890112291972783281602113559756325548699992243388504
9858282629103818226122655992591597082919786233193166881062975254106841171086259870307
4086083889081617340183945217867616910217840007057215115331818346398109048550598419080
3798403931967652618254901449628263813136868301905372776290221408228253205031575814491
5214969340080059171842886747423281889208221098707510096094222717960611868752310865130
8794073032475585515857271295685166711502658575372152497020735665542874804818838168943
9294719392497078342691198162104713083095702791440448875294465222880916426932212089143
5326343470189414186549875808544389783754276111604821944985733991946416345105882759641
7497086868706533308767468551708636722004133550690898227033638087248324773623842767127
9827900047237503365691359648505894871179717735895375235172704532988089768706037171689
1311492611799076386817572277825030379948105309971413321067911176939082497857595017347
3274863356684315499742555434287981387288858840828665550630311692303426240065192461820
1294051384109438338309943373235611840834156109525474454266787441539452743808547815748
1805542923017257708254215514552311689815984157630331041024867859344513956337205688588
6905711908399969799521334483952287592283977865693169705450275524986037593813800658952
0565487153992594249997173171679762518307807180065173051529563382632793212718062695948
3416496910211160236464323589047724347766517250437018826211584376442266662047512782523
8870217898027267280277118227731521030358726420824528988433168315791666499256821611449
3424669515324639135704780768526898998489320525337210124674485594511686982516508631506
9023350608946133259647585740474307164510219889646668576509308330045681872757241680767
9216043554681009292559395648953329502797156721892027325708150627709071738710313238424
0099196766657262124887134992056972358693324073811177055732346090215798472234521382771
9580347220540258966441539053412137447005089033871538349390783651145807920272101479062
9899235730340324996976240607889732184310882449308266190802622049990112859749295562772
4646114689939294010594942097338175943828780250444099687534019038860177170124591227676
4675715236965521220057724245036539737696902851785827201084335983398454456817840625749
3191708774382410403186714142041720368114298950367725654607105012860183188433059026704
5914468999688471783347040829146812415474309300998153842585056327604739087689049279322
2409534991903416211544608213898364543627345255421371578915213206037656434612877334642
6527949078838192386444755770595009119444142257604748727334642460795924629817046415346
5133084095714764871552128608657847073529551768127582320953381637314429047417984026893
5113836233619036522193686948953929298610560654350579702715511242608643085927282057320
2452863375360040537159776632209913491222622982155246441341529456551577015191898692407
8625805270698483128954807962543531974171285842576611402820095349691802167787509064096
8938910484504664086575037294052210411280009547319660046500439442168485942209014452429
2225584905936638244027733283626450303053353993096987770343090712332576249414960161079
2873166852638845275126706030057076410952614298966488181203960638734084985550094173219
7966753274996201186977256904144690206247765655350656642375914691733307274891543405830
0707221480983167748193021736845370021228649151994480864391860713650673852662802901568
0073865848269567625421052986411659338692482538333787875213492229698895497703342047861
9807162101844151617722148191100437333711693674308773266973085990103719739779496102842
6186841275756991398649211424787814353108830701287103774766452424063043283708377315256
9447305575523791098461773531290644241044708945166904888069509146073182689449278788849
3950009950702290353435392133255894765250328071052854241242946878306631082799514039926
5381943346180295601431124328564846965317170173912538789746666097145394227727410114423
7958959878910545664104268147891162685728035996783039870978666344440474495023780537494
9169791738537209747073452739794605724927590824927825833506825690837808354569363668173
5915005489117122945893425019397020896398720423360813109299527818850402771287385744354
5819714490571304159192507557151185687124513861708461376219948138815550829388483756914
5902356397006311126012211800437038477707067215788291447250470385711950858809347550637
3829353177753808855894117419263293749543000850722248392228754394422626269598191189695
0525279099346294615360936163549447879175037771374159070623175967245583685325998421558
0316256244427655499025327501763136294312779601191643050971695953211299204527177449654
3602651536565828884193638793377535522100075233062499389170886030551349824542033286014
2251892444978971365438878760732583968694123983880820433462661172204379981330742351367
9908458648166662862011101678772854932272589731796290083893809378368862243092816036269
0732135646537153505048869164778779185842773790818861314754357370438396416108668454477
0518569883643545539414674488362256907582297656680517777748487429731521740717222408996
2027134463802893843314525169836380731179921467836533342174788805977284261602035843082
2445143907954874192059946703109376776927346001157263547865330378705088255862616916954
2736042627652426280382477111290356531092061375501041535163500294773602803042651560687
4503456586532737488831388713636012808339131268504056973058262367506066185397615704174
7427594094047776629585609930464445369693571312353313901844731016702853668941803502361
2619326787257731214972498245087303034204762585225658713844171927986481349699003368236

5925882060107514213059935405118239822139359584242128868015569496013163426588392290351 70
2963854931431202685751720924341677167595367385459589156304151543408885004160670533466 54
0828130700832897761488249696289240160421141615103617977793210297134762657997940215469 97
8038778479401598816085214213535304149794348531891952830115750065558526423618771644236 08
3078973458250523293243640264375924591800563290872305076297036484355086967621331082877 01
7716349377640015740770619290997731183271570537209205365402970776786726653277933085996 58
0010922173623890413584279519335088822145436519113434014342809428932147888483073872577 04
9742533246466094835351665781767074465378364802847096926101318645029312296165994373555 82
0292882023445072827713813735310873868156668468058416553952406315115736792154911263247 75
0126529807086865320787180461286792428807755570652927825941943007588427801209591016637 75
0414014314456516093920421399550727498787244571094135555045015722897094705928715478578 42
3754955553068827716278606818246476471999729800367332877207475226535910397549640886425 47
6524143137886190662271392033748303555382843444764459824363406532781363195780834389918 40
2072956331227424040632911369762268565400010623898604781668629777784031279489991707396 72
3067522162950965287896315037828467679161096745584887359347549108669196172246037943326 53
6717532667964275759439960499766806612040016533028504949972860704275425093720773858637 00
4154410260595244937075850036378413818607795676066806461723429500752397765145943294896 71
9001839450853507115250826284545345273757796912248333719234276158203080146280176756282 50
2407146025097160563093176724593076048688248790053847158052075074482030526496689819994 02
5157949423211590278410432542583533717778889923951746366574326137285181100529275734664 18
4818518156994434040881803768753519740663630478337284405581819299431249282069957456142 38
2665305098134466430096988357988999333648249647226676786609483821820069176447799515410 06
4831495092340231329855502076012976597654825143221427726580666575578245211435118357337 23
7730492437142595702958551587115638880779956709924041678394507854184745979089815804130 82
9675468152034660396987843938308183923747716278977138284443405134095116842773409217690 72
5247631089224446647891168794325132220073651534562311855013874215233544503089388539861 01
4611069074891095663596629481504685467562292894151089237294319319561491823182561129060 25
8364287817219492917534538823527993628588963636795180669943553947379606788386394329921 82
7855684125586779954979546133028786214646515713991659140284587348702705261069817062568 06
3586601281644979724666816416196987988677368474709635196349675767724117193889891161841 01
6757249065281230163968169315003082520148974579738890015268240237779830972400463108506 49
9741059120950157077584141489180528324673061835140951749137078851259753482476926500423 68
5462358436015970729618280443091680263700578592151137220121658572233654137444476594139 54
9643955313251616096580180992087790835257903757726750622322518588306075693231895633747 33
1807153668699884986386914370393791379690217509625869284257169129054200208155469777726 90
8846915524811744999365707260485043950809189432853044749398963006031851877274582105246 55
7741081225485874613984102455231546972860561599004919730133874530431602324644992651250 17
0425989598182556523887587795429109096721908835535777249868613661528049587621544173261 93
0411792496997142364803945636309930720835169224953962020083881511918372444522752636688 92
0992957667709566738351889661959616053753500860232336482872126727945718259271787177324 84
3095828273573981941079466564284153735209789228890021037317202758664290118383502401038 26
2269418554517462360055390533684447990378023149350091043351555983611865012604956960259 13
4576936587622265577063207518025910297097449295178885420211929716912663104142091401078 64
2523861320813476345472977531744085673221032254607393016774189560202729920316247975376 86
2780784597105213268572645373624225531925095186171876013451124337629905355546195917525 48
0430389044173761791316962743448785311550195253699780779344712638424889934358191719567 29
6122165205346223085534104546175351602853035317060140881213637333458637474490071286834 13
6213074664314991962645876623832479456855495264936646501737302526991190873488938380822 04
9031399905056275443988904463472232565858598791188668874080697010320268203991645965393 98
2763139767575867563066119124852892485699495545274758259583805926698056690931527911877 30
2059789706313787089338832207095089767335300855561529457849364066391245229601266959600 96
2624923623543500671735657509324621100978763878594000378111225548186286591913026546612 16
0749803532560234435636763124118158964697385615082951931800657154126306228994298631809 44
7088299969778236008373193275931245373029308031970839167921123822037759386912820570401 25
7724448615797828970323720982431419952191356933990701376486572379409018973102677314647 70
4915431246335314923116498284611912232044329798798865455048855028172241271964129621425 52
4408143383318483554853164891565559472845494159518329931788517978572333349821971674522 09
3982279470389483093870068880950095001760186510756826190182122598791106735765741023718 86
6272469935107575855314875850606889265301101838718983521117215435351275938264329690222 33
9493804597637855809073171634956875032708986570450916388205248079423522275356870825602 60
3182812925772700379918530126352379901061942595703521979786946037934090907681690393742 40
3294222448564906242924069041355557978176280993978467808750873788927217350541572168610 62
5243531297175287017002171389265268626196686871238222001808199810556711287217886692866 51
4086857307314203026029417580061475433969961445419578017347130745325184665024873918663 12
2659126015660236378626652016174160913156721283704302383185801183603595637611440219634 19
8851777358715380279808392396230695928944502881270771132659792205383647524900638650538 46
7352654712796013927057425335723992843925668211900759207612282727238608562138276534993 39
2172288313289294444554185591185233251318300351279137925207603882293290768317757284347 67
1056843279974476200796878545960902303706015474437242611467397637614102508918193427878 23
0418831155679580454124311921034413514902690955017999960420723226099427493619621259445 84
7015799426738408269168070342003733428173588242220480345776191805538441867061092786475 446
8711090427656490961336789669320618650990481167965986055034049205961741584031837366244 49
8788910350470616500092549942339456621656246048636275236757195846212709710103586783737 62
8594551903569480784184420461164274326860787480844054251871227322220026214347989542952 82
6749324038150098184804657007841619183863492574776926460569176755318367548227892033180 27
2665310931271164367933819622800959541330478706842758615783981596678635312212649107416 28
3403637584898673238782661376903593013623242961560311663457950334806960414680959998514 16
7331564166366106381359333702910558095186100259659693358538133702667209303857425742167 86

4290636425462777371245161425008963865385952758013222891865379060793284411531239571944333
3059711167462664902389432667179611307945823104411919007978388171522300067009440024006388
7592269436228510839890825228755833967504535432742412656453948010664801879337026477967866
4356301541959285025624393698024440700414898101285038476125576653219679989499751165166555
2568035436426047314998069415767114957179087061914479972522714795293279630735358761120755
8290680564592707528771761173126629456760366335713353483622574923160908849973395000823900
8022708340184421860239715622689469759011211164176352819617880020135508986069980355503955
5076960175235571734913676580503663480989176623747277794438892098679516938936159532508155
7610426498077187885831916660258754558558020712496020356901436024116067650944175116399555
2511760463554583545586837333273783538558665765175565323846658898639838629579349798658077
8391202883732216541805510150451839104689209505644292618713072718897597198522462131429511
9367378454725128419139172808431950208148721448681809122621419507962485836787251200849599
0524925366352557539713012825226663193126747271170874016874019818205300956901210799493699
3240463863410052260628335978477750993619741102160991533412417456322272914289845315460444
6793246316585082059900844430445292356835613059585727698497651003576373809488904378981499
7133894411215534047828538341025158627345220929813789580151252679590501688381107977937377
8523140754536423832677537259113972316858792687904119567525558325589432624746316959593233
7932807739721523413165102332695551004256713478165687894432590102030761299318706117131111
4724468473809876166132982158952373367323567167154561417555162388079712000152345311199455
4178338724608675919653156964945975434352624147173770273516235218690845881694332649543600
9630690327863678272173433787234212201277914530736152844291764593757545276655117361662499
3776258341232669470386618010995264161414469436866655412146535572510882313260444302059477
8295706469642084050663972063931736507351317300388044526225960365484863659694269514837211
9194053748866482246955138001416617534413109439766963048925949723208317530996261430784522
9933180165084520803236358264421016274310136612985587054229184691858535800025965709361066
8886707408368708490761257994716413099638885977280605725438367777594594948790395926558766
3192110322765058387304399870435415407721708153371228169725347084804606784815303398274255
3546067361331062553648078235171937319763829231976068793186399739397093292358465259252544
8784756066958736726025074964686593205347603330267022587893997155938467073463822210871655
4311598711083236359015069170462543351422056150023399316362122693161416451634672248700899
0857559784286720908705035059539110285561256499135175964493878468536271880164502150234655
6608575440062129328084245635478066713704565685192593359224746434449681389396409572880300
0774156674090238261424501639714899267150588062140473638877177652091612577430113475195455
2203749089616312210490435346518019000909935215196404714588647735602146512247941780590044
2351820766796507858002638510826166678848558950749158645126619309838305834826296907131700
6731071719728080776848480579658754441379929449107047471829728017872590170340333000923033
9253806104167548466988957313506803868617426399244369000507783234698240692427033122344255
3381686955456854241317436885767212185233924551451954366292887749040043945594665507408422
7025138815422580127799362494827612553010957594107015575245701740197885097699952595617200
8689434091597258963492835929491764187707351188755733523909753386139046117441975491085822
3789453594881734097886394506170092273558356900405856227237806491162179925752635876771733
3782545081928110810215902388311849529597030289045463552392776037669718046455009366389399
5271293756618364987721355743227583521179617697402893745677711710209601970860998889197488
2278183406451496324019134198595466675186304382178229805364768085964134878488276513359066
0703161660080738564502367408632651047956555552390569324136590617673191577439406138380088
1622869421478366058122796601378688103642239344928990155827756047672145288752696518498400
9757253114126969799968358313111636230166796763740682084735220764819875198822986305280188
2288737400846335839839835119179770057824481995867123744072885425542588206733038659351233
4884997970262313428093258461206187314490573933429526185556831586472346508233563111689900
3929807462373018589617512746201102413502530165047359873912996687731657887933146811462099
3006882223088720615833096858364793811717587232211880440756360456266510160834436772313411
4846187126693943558094472992220580954003417477116924948572249761131666871509490601304244
2787861170021809547239131794441841677024576301402297501831938044884481604777014040813744
1893954547596552735396146035191133930122313329913329256507989652308090598585740695106911
9203279334733972661495541705616616081542952178792420850189261589084277089990815410205166
3517964598157545596880694260180013035578554702715987600680733145181619407836722122385177
3317731783658569999622159384875883922489929110687127774169313747256956513343006306522733
7924551056266921880807811832586731369601120907714665794436628833001440815863098631726411
9063330440445080582281128154889161983229327318187999588512737838729385096990500690958155
0559968623777129423176197294040813941813602724051482082073795136314843354227939334277655
3686103525407725839996548618746068110854415993782634421838393844858485290353045697151099
9125044231382524925295400896002776538738663574156388274314110235995339222746614280895033
0235763018473899695696878579647614813385274626000172840076610353359970011074295886179099
3448568823861402212026281009878419477855669576706020829729728493217283594788507478562688
8354139604520503273390011597689975923488237896003704743509440126218101094815715133050055
2696734750515957930241419246771883779907897096741985620620603365776260662003934751458955
1212313889676805215858321485875621004939827433779291000856345960847872744334605561848211
6303387229105564330934672192647866563146898144200381872110934389738986303817212957909299
9112743076719777291578376748546289285218082039671437850793363736950077276642667558541999
9588569725888941718854732457041265453872792938355542304128413701321311631686167649503188
2078123352807867036057531692611111413192742877904464559074531083030459561199477889297644
9507064639716814514629453124294363555488686361248826561240064409327046272099193802499355
3896059744335128490659363042830104058174606123204394459678619943758818210788705529097244
0315088518044697672070493219326427327474947465735270631400846002270606955967696876993588
1612208087614289082438282250827162654510150765071122548776783127104898581258903148080111
5323482235485756250163613432445755489708401852562487855841161559062915706400186193826788
7191671454229784270131855635300759233917437150422523687143080286973897230659408747261555
8320028856056430543675473012697599239121409307820516347305683944357660805054953912914588

pi to two million places

6456418951984212667557032614040165264647181916825561780500043662639868126326051567703 73
7576322825572372240337326563439926101072777491347627176169780024202417454219633542368 34
2108998205098399693090096683351729155605180032912375027974138336346557054527834853747 61
5770729786028264231698956616215364150611916911869585076300887739841331443323522161086 48
0753762756279535732619276807505917919753822080505536085921437064255882040794994568053 20
9984521619529034874717291587880433789802719917575528412495164553897366903667977458510 60
4923731647493348513891310767047816893150928005652508074003360262558797465873389588650 45
5491059441871151078989476427723914132008590273134823051471147108625684424430556126457 31
3179759466908304134034969841895264128081303923795354233955145164368583297176727033879 96
6070502753808844666613249520847059562360351328876356928310404235142307368983022576189 39
8221045147116497508649579313310230516400702740418722925525312555050751962630901419042 88
1583786080187226885330334260324188741994743094870980900779689622152127820403638718498 69
1155689285267885143692081119896699199175939442227659858529607316110272771930392287503 73
5276746031364987286014859087801089027696481017841925051376836394432779858978349967075 10
6459160807997814986746212959429929495103455612395285189966483893151341663425625387229 04
2078570031005057667238782499422930433187629331075992163341034881853078945786213043844 69
9022597151095378424839640451956139526499409105419348027833381410505497503863252134416 65
2532578346241054313724205134365610070975026474975060187932989700495263510803848800122 17
0811344231159233018250655232900080532960950529674761782776499033804649275642460784400 62
2477510229840904464576088284008438396537572516330908693659501131668059031760653994130 46
7831860450062980954641179322276064394545097464031190004046101363637566525147633922056 40
8539488115656651564144995469997989615517085192389641735048616440196028228370231053195 49
4455715777919567588597530366609882202496886496240237376974149089622782626717164709054 54
5896434251358107215509931278700181094793378001942881169713845597813034375911744983050 39
7272005088385100204864527669175972913171516690414586990948299639788578847124158215181 91
5131193646592655022469952252065861237108617077752972480096285573811172643504397723991 59
1353337279437121493974863979262479245746309268806016126833281080137307606789638853035 24
7624795772962330363206079753131170032156315773961141417267096845791988967823317498025 00
9040927698398894869465840619708837118717162678950598037992818687908359676744386739411 08
9537196986182976738458954392441121744326072814344156833377839658009332416265311975886 38
6314575969916282260607790333739537806861549954334396662534041372393360583761357136631 48
4960228465675347318833395976940602425850333822826967160251488264837670313936615030465 56
9440359099795880865306477527207618121415394106818553151735880056317978950810699820712 32
6357146714531162758252500806200983530299859876534561587726559057767770523506800734721 16
0449806488391234431496994566296883810992317937596151519832650120598781441617932890058 63
3209563904496732740967756156579216951256957573430650111772794467699754846396367647409 51
9787469857405002566630430496930378286467308890740676217208162910009270841088021980366 46
9032006604898295265441973616508087089991399726806525622796418049464541564487670128005 83
9444300387771355707804656432991321870606322081514275062240463573963323033417129205307 95
5186159422189134174393464269860039630400508002042870415729593735309957716677615251624 52
4092900687648409445244870829637234794663408346767889242904515984050860716953416892727 68
5702610071896175287752516051145757923487582896882725005829643483257250194887049167329 32
7414414693861611008212073014894509422091561135319660150600951530042151879680050092567 02
7746169752691736633588478953104875162792541522915736474184880129370760082642006574185 02
4170407054317172875849453496755969256058106800091735532339246553976580561150626685714 58
8862313687782502707507793892985221289837675733944406123896019462548014968669537166756 92
0279450041465728456706853099431195910398723702492709010435325525980501157121226328470 74
5074590340596438552820872919118498395516680440487215475323391256199459302810597525033 41
6854656765184945399888524327757835037208347774127463818740545863929254471150555432279 98
1820868187969864701046561341186800539651313467232381249441404741459647131100723335520 28
1233852214050723861536931660894605930756758626679075053701611040703409092628267488775 02
8597523751023854526779620553060275277743678194057795338407659371912339041612299389658 77
5405042093316578069596534125892927690021058353472340561065145554293200917074504394701 99
5669836627702247022883042665069045512194383705684670838235730118163228543400184304949 79
8582621768415619917919935124153094915801007937477404039276368882693841499375805717517 13
4475581593103274577418757637118775713224689227224249337777113957558484346931432056193 52
2817599764563384463523667932694166365293829299473155416026227810434045971282582242670 19
7609692542335680025383489465631124951701468994004640091188074087077329964303094404358 18
3408414175801847899502624738956907547819266522199349674054172881911050332527900215854 59
4317573458873276958639774133830265422515748812728581889273284977696911085259385652172 39
1840425825823144703106476046481450202221028653330865637440225958804713218694501443668 17
5385307315892005190657548015823449154844317770817671252046119649394501886256763741453 76
8645600601421512612924246459673064284648851049761119188467204986645475999926801775664 33
7900340056847037544756287824934898408714007702362965041707501045403006968311235317293 40
3026317221317086176202960434454820954348676369524939415384766753809577355121463749908 05
0382956467090205230467398556435582380712001340517953581775926611792974601486921320263 06
0039317995648494444473161193725369420623475637450478267593973212703372805781343126621 11
7387222492806333639834519303680389259309688995209252035278571980992380576236122849606 52
3276970970764790391394231402133368690394808687684743308012268869946203470823716309724 70
4725028106033141647014474120542138240308128347191009308620269749091536072252751286597 74
9519507014468359570342661460253028153384827277644979007768898074458301080580762580282 65
2539846712184990443942589188021880629759956314281638485112094713499524227290966256213 10
5720027860252130271652435730201321995053171120419333856243218115968535314364280988660 10
9585436852601085293446378411885371826272165074541415429092412576342816834642480358183 33
9986607357730940949860657066158440701673506468455108344830410340714330688613506481612 31
3350084423362414174425220384716206858157780034407442240899773795250772272242522163252 70
7982864834239360319936007701578145485499732791495715242047771832598625574711721760004 75
9186861346576680074791388238888252605950369614563755509483645524943318493757958344708 256

5112029771305489059858893604904198436615568028063910169315079412398442614146314527207490634216746665622082073387405441027555277326425726662603254341416451806464620135910746369048140913948817349150190905887865886575608054633191153957951992269151017526113553536548044915380523524309209568391339236641860218486574056531386585891803888290100495441053950428190852170956897098522266049149770344256341520113354359560162080483345705678582847518288694326457284709415507066390042021847428414815072932114255525384876823984059640369467558558706507780528881184370543707668539402043667908794727576997674271743908241142251517023195532601492920532485233164430336141058134808121187844540168004984186371953775079122509816883959439843743490500927690743049721973216682690786818942988655432603934799602791528666316742452049935768321975982961409060960027750075411611823413659519543647682659743538725034335575607603094264593717684435256584561894133043626585493339644881366781418994035378125186361809861431093804686084238941765717091983753027352946053182177484980676410072780689353948778562080766076462886434975781384916754969480133361645855359234218744439048728220232748279000438097437222409055665798279254079201827191764073156868788990869823443332429846171348077941579260789437869378793218192762383208856219825626205370615705336509986660437335278004302978538582577725820814349306895099746902842123214805436252144109926584513208698356950379341219278770154837668462588485148035328210057622595312384395499845597718366177469694877133736029490243201400893695435246428467039062829612925333175609054580152415336269704633415609477147038783311408045532720266985169165505458088526234722700918568381152913251607557514732983104817451169011854473890500270796628135737332891521947351317150714211819622395064036066250793766101141090975719875195062790767199208415618383126232787053677949587209348085310337003707967698144333986531947300553955036137371690354044122744184825089725543411329141409921010502794060781343452745457489103978703959255767719730326629205850021841243056110147117266588588758581109801362242719178360150563008450612609580351146221868970392656006767752271278115086918240931263983238895143307309208061818367053327222199671221382439249413384503085537429139510060612140640152772992047216247469060199687536161293095944353196702138702622427471342529519834641150215258521719073343528760508969489859661873724647148445754004263211169130661080007590199276852772312294338453753734455619270731842077003525188819851000091058869702463614133946373640362916764850243139592226108914312458431080249185337663654737954281019800639467549165673882209372067955022453974952793604321876041673712529418689041678756050715819184628397499517761898470120141728492253165997642491396155346304559673700366982713924474617224053775638845250605738311722063309952246138275154384262484183054546161423976795807575792892955358231583791059988100013635583580253819304960874848406232442130196731285727975696380891589114151062682936713253390443220443935122562713425835719375807555547099528058650927489148560615014905867221863018145395527154337744115748430146045421047237106590937458945560185074706555125049620797634952629628585741910611986843374359599027675135124284216278738270407361790182264211619056545235598213465009844346847498934255194190612568539471215509383577839973398535979920804409140140130196920588481624165779207438506165929884295428759267653257646127865813653869302306432495148722915683419489339773257723831101860728599213827439955316707179778694631024120965629925156369070709797657462230224811183699337954003728904003552813838791669645146305017446678302677339156584851313373322134555912016942819944635586911990104670257248863039143131892690272342788076559567143550810085233287367618883314330625840285446138179091649151113898688610204244109694930399461881564346922592382271542373255618657637319111383935082984473757876309088819066974875034626160773620105614764919585888952617573029866000284492088330986935638966693573658315432198051146302918030353282391251226514579604511291362048141607426348368904872534774146049792896866547180431109637020693661800381564926460176322823546752577251006348108332460496397458189485625071592798514006882345658766432861696499447028761727860740162787937600313403053653720016336106731197354021657427455266137724641467970163423221094839273520539921618466844316578051485787487389956117356157422927107997893783041174634233172312368706298879939563796932343814093077303166738558758336589487921922929917029253219531310631375169956489555627920773263344399560699134012345566186600272386440912957768174567455620426969663979149248546317615555834788491120313298081093820200166708214261953839391351949433245587415772583694811634921404847335467048525276266990959144485709083160423866343355234799524193321743127082642757150153738114470464896147792343239188365976041732760691839647606786632267492919332316113187767391913271315644913056937058551533950582292262979366928009889012737440110729913075848391948372516387525152681209356155068966128156527850437743856737065968657129040745040213967864098050162871632426642267337613821529562365220402118843091692449620167039837724022900775191199017338872565494516702488423144667330169795649313895385861239811166830825720223337246985737787251767301674688527011564277582005939357098122586901258892772753477512459695452503898261166802128775738056368563156444219945818740281065680175318556456529558228861695528627420028196403061439105900321537979539693021851942326880794685279140725877194846904117641691522747211068243909658349368174072439125672601413920553750443877850971869061283089542144509045453485238152261213609136327962561871364143164942213935544220600533827345153079867230668801356293013165017655537704716300915062479212373919370575413872603778440942173025911250492386154938190070397322698270844593309383716314806112834117948631308461995943078968496411168708433793344057007952647802553242998827021223958907157262154491883119891179649225568725795187376437265210419618862359708076370825191601974822344820994433323650040151033080334509873420821279214120042048018050107979856172321647350440136981148554105881197260873953942549086287378390374680988320817221479683074313026938154363684537845761557520199479117741233670859357176920959284028880000629172087162737977415351295805029709944241913807628723085063578557022002901343270927772987375157616624748404739285155086316421530282083526501557563119590836823073403043927151810500275265003370868949842881323568496500724939884740105694734338056373384023824360725388733091113738840764500037734478470910018648045541171100256140540878369928869527413926193790851385229297895810628398059044992413872737639857194829128448347659740140414325981885024539106705917683246226948397361157181489852877065023792132178926953611637930446467710253991656844314443

2020298116859366166592552065597956848269167629977734031737378030874820811784876725458 68
3733424633452419941410780153692124159877833997466997395429518681820090649697610367821 52
8098952987606995710356909602837100859978989894633487209570802179233665748707146777367 36
0972463992215753219402369880425601500007584041175731957498416740333035260394809737885 65
3902886659099013584031964803732938986045942039868966162227548943655076000234581410708 96
1509394692645277394235114694011127719149136925437858539483527264857552160208720481249 16
9101852717995183756073284426677184767954071116218015353987918247749816185835646452280 30
8632399771384992835920927055530786651556452768959162186920347015645557340287669263841 32
5770360160596459690403225512310202955997098916418090712914574898624430297707113894146 28
3648484871576785677948781431251243293550247687268468946933878930968619372172452421687 39
2718595032641171831981357089270756010792991851456275028662192436979841781141404540319 60
9164247843921439027833868372641711554943963041940588806752381877429108551788002078811 76
7914778773113638802772856696940118272755475020009359274049483769196417417474376430993 58
8281149024158567172118654518195403274463085084900186721365271788742362751336854382584 35
7260976576339859353648790620203688881027015850058083028005290525005920409954009559933 82
8156055157808056927750376405932422853821694589247308372693348034515544089081803000097 13
5903138760369506462952558119323319104201739769685110785451020588411747429566742640862 76
2260667721607633383260924431171632661044281459586062506998686074775429041244464632860 78
4627920804163161718877365407205146931212143251692342945354332472824172960433247902906 35
4057442643551016576188688757552892490583073226688579406361072634713145128039096794093 68
4979336814875713298697813211924730599496014027781864831950609839994902924848226451969 07
6694936680918529246540254800772199089177291960931566054867802714844783766043906003147 14
6452296723373821018725039173155245869955423886136497929893405730911868982296875692219 87
8352678888481656201217993235571811933947899729411316250382377160747601225078909139136 007
3698161644961550771156247518486718642741875222098369926251107945876744271026054910183 77
1414939538460173089933449360476973301928779138027031350707321098188209904217747958244 67
5990200835711681784488304787391475344935193814011752088137059843984654955705501018947 46
3317851354478060500245322289869595610981274453720034050684331619519836303181703747898 62
6632707244065379439277750611738437739100370156045085424417182946223230987415992613792 31
0306397975196290621495493675149348295534255732673405736625456387820877247801701251910 80
6138076329419104837686206155039901710577549379437198149860227457432768735866547211937 24
7364836888473863695504723045861687577899262417991924288808686846632765621548945377584 64 26
9366017054904106967913965856504723142697926688920856929628784233165154008797079480444 60
6266020591420715726414911842725772544518517922522992289989255241793132955616293338761 04
1608491996663174423087794058873508394787307283099169773983497943684774634480157906910 42
0838749535492617591718541229359751899008922262172191445930678861976035622188127806388 13
2245635556068069415255297897496905210255587116668803946253165815590264701717990003990 24
8334019184741861117791425087957832200374313389994388787901749321783285709136225934523 98
7992119413582964860423871824067417098622812190185733368156857032359430919908413100327 42
4163085607040183967645421975659821525834228867964906175332880383375925188021355788935 11
9163336359125653076196344688595555679031931342675338330663164340607697250337461204565 78
5754274944560577318094734904463642115299853304353698384660984613766934308461700054908 36
1012391654874021552018074383204637467839049999351856781079228972587469036774957913385 35
0351336075507321325600291910315513107283177116250772551531257304962327220860225214648 40
2202842798593292836688799077140819239883785035622052945336156802820375313954033176431 58
9405632531123586823947289210428147770904529570491287593524120513586876032845142582734 69
4488545643193344560610392771958212921941310874666676659245749138539157944635523988690 67
6799695355659403400839266309489156370515518332939400050322641708848899460174496659076 68
6847228668101834234733580748167860825699263614579414341577379697276195362514816047504 64
4571393528525921757839527856441202769631859515199253706473854375073484198045852310995 24
5664026394525671148305359686783111380040816192340692340326098970410456803793483601054 49
2920692731323910928948941612252877254172588071576980029023276929926539672222312595423 78
9781879617297369231269862904132940693047792690340796085936969553082870633498853589806 31
7037906551023455503598110472263078432437887502680866528666197012358431075487193446996 13
5110246538230763263859894685743530656058275300173591298306951563954194337812121118044 07
0153615314384579873166672036184046659229140007286157047092318388883712590113775110154 67
2815683126173135845444955569340401606251591221445202630400731678624212347398415663606 15
7002295757951259606789094918071487219920115133860334201249033432690190315247111117705 37
6749037925070477108000780724379721999748678051270543865809773836608455714159223131125 07
6994385057495720844706161764642897016734248531307265311314113296922449973258986937933 62
0538231043046881167270296781793531514792526227715087317841347201175082919918140024385 16
5187293321922367139330218393718018284319234620686122487712481614464372623846672273871 69
2704343226676282133595720865759252993357046930167137706293723161840281281466540339392 99
7647994508355635293134044814997460821573143678277060209036200023656147948179616695338 98
2033036431519760589388240313104586462453903945756376887059279419144635131656563853034 13
8055116073830115234965016026289833117526654089514276454873943642900031046497563418646 70
6338393702640432719993444501765362119418398091405435431204661107102294914795728223004 88
8150989288005202982221956031371555319180723675808735157394941586386346592426492982712 46
8742873788710721177960934285282218507517612031632878650509567859784696943970668574363 94
4173341905638350440729438621752610060673994465714452565082185110811234734977092353306 07
3156034383366681280289728355294978346861207930653666229848868445897302368695057318071 70
9271048802093698860052594234685582421466323714836033470887905083467514163185997873182 13
7711360933796792267130480840063571274044761901699021280503624955146493756285725553472 20
1552916273600484207174507880795076184591360695778824861874794960279482146832553636964 80
5579475195804226596181757545537257578006385007882894288015404514728664507693636429261 65
6146409237609919442302530717760664764609748901300712832067009583444148679596428844265 65
3348062698500890014358597934320495470073979019762402183186450225187355768195384669571 30
7006288829263756160327876421596559312529249242221613295745040653822016726239861866164 858

1434298883151920315905796007336630592644780176824280579377519254991161387408165551961800
0806191448772948511662569977245150821371463281903651808773347375322139017956945546985580
946099509944819766231605827389739243085310494440787084726990640317835935290461384822406
0814013067392119403592658922911969016830434362279715304844598073159037135399085230555541
3585030330599426307530084474973121329228139004149372925731581039036715734392981372366609
6770703723997564711313836503606104054887160986190396184993610539032305603590895271652166
8824412076000293777205420054934505995932847657914451389480417920358508525299387445006490
0770503204262460392913479186052964181514073543083458770362016261363564753671667605974780
0085903142936133379293319175008485509604389718110409887752004921222370695558581249625713
3144760665069336788276888731032445424398289077351072677332667843932332201926219365644666
3220548552196362859886982124590594841453898355243171934249129900618146395923563828594066

1690297564763112622914667465366265823575832972216617099689212152632249540972530092760900

9689092981539854778104754603970480564097019438611243783216133017217352016953866778938050

1847299996113301753095363920200797294384334427132419629801071959511894080207607854224750

3470765972679570860437380133715614849302325710198132874943896149485487837329709427816333

6254436333082693990557968810494892037587507854537640505365757571955571940241392108268760

0908876905226368654030187082270188354758472399922894034079079513679637963507263442245540

1912809577059031587725558922077896911118686536928072383024038923627104980728883128755350

0717511334248096928769720158667518914217143425576390489594698580750060927981004390924940

5240817662509527417274156016145431594518522171855749552684752772716738913902901472023500

5989549615741731698904285539944028302783837762365010859093691920154027694868144088268390

2159342531859686296886770735568178360341970518419116791939661833729700019382298015027240

8308350801097177309111058708948946723054289271511824641544091066453295323935450519305270

8990931728114996492725306034702159871961345650020631071336136517861654416497035736820970

6258374943863039224877343571755966711011669312791920462088982875388710715363136881554660

7959015531454623653372771094192927484041691314541970017754019166194192795232601888253730

0337752336012196171161255538897835617670837738785260756313420586568197019298042050503450

0950135837302702058324470696463969223690638593311081122013967710006747724553614517398240

6578733436414551852124685804072880187810597648717763079789332546458009321561354841948420

1779175593309357859671946991950561912931679934411945972942434600010285797589940496943650

6661909757932956680542467013837449990948865697217529786249994440386325648873316760040760

7769071769110217133246166713693357767751796687482379811974164138932966510983219131889120

3061288321306175347306459432631236902194248765044265680064037237356520012431948237319110

1641586119940162345610705558803666260316221668784713489649977257144363570318750075329860

0063330828997051279145819777920607806942050549492682044404634256297157534094927435579720

5160157207979726607019969187009861222418896922478331223069883519302680013335158234809740

6911378369744867632078154664881263924082841865160249149549798644272098660518203571764560

8949935602043071048929581586506395191730563855938221751430131434807598669332650487479900

3805646835095656163984160231551047965988593168474522978453271156302256796382458057087380

3359848615942699923530317472056202203726152708976066846026088744711951867528525005617820

9728887196752466048954131079105130025731039262690888474237156518309916354673745723994380

3500092516539191810184237064182784463199649055288856993932528265626282428690760469591220

9338888263678905258896297992600436583512928591660168162711585038595099200450238288052570

8716079991485779511741071458789278928594455422975748926639149060612255901846742420489830

0696032609924160537319803993095803184587418756119655169410302588427461326771528604167560

2599716689009197446205707887606193824714430682007869995158218691523480946205994733672840

1618743801837624468431146227199718354041990857212670067522055708177908020764223877008300

2314596324376972484302281374748014994596782976524511116475628449488239112658022320723390

3967248735348379683470903721727807182199304579617087484668322607548311946463631629550460

1428918187034402551606610996043968227976008510510903936291091994211938826655136431104590

3773982337547022348970989382383349606222414458818157174486985800768017680983135448890800

7309801045982988406710128613818555977913112658579462797634402093254046425652321448785490

9837045212678648629596772359938670287589062826992794921488808890259752971774789067202990

3671236787634519865957080118747717986484510898251955339140452640402757528622015609839090

7436788392343368690293794902388062976992556920231250427705108943509783202370260907877220

1028883869173065202970742687059235430376889847491331150845727240892727685293202568303820

2902698549830926642798169621548264378964612836838042073209244634810624823762867848195810

0854733788917216503370931716230082783544609535500015708732585371829606975508170435834990

2204347823973727085823269636237017609767484500304109060401168774813123641572249254136730

5065969999743514683113000479043763389468381150750447983622497756891870633121598257365690

3430609810105810751283202846444444669225835877645410765424544616023777827884230143241480

3760775272286664586317686875728436203463872646470337810455803840869900184700132949067200

1629106732086656015272003570377008772363933708461915283204882311403505825354128671849760

9189874018370111971124740816615401897014577602302374503811231109971026466141404185726100

8956369605831244662510334421769869553123069183533005618851848711467565374712842537837530

6027002540478152676963868002814906708268471436627044938342088682979560559153091431959230

0537938970909123501683175231569389220817236677947717971362719241455588880860190380407490

4601510951815732199263163081536727786395339826503769350931962174930710360546827463851920

3840108589380482153796057037554136341945317910214402772440022859505105250788530056362540

8796203963051416750538902815483989382605618460596925430923050118448202444057353339948640

6232469861642715255204141839454588339064507002112028162690376437236786327092384808357040

9285566542256570041716643467942705669316946590355900250098110204615997069222696043039310

3415011238520208433078349276852128023229115251973791375174328405671719578654820096834830

9493549870633410451091155802399789655537286716719980635621882290898597135945659956583900

0719908411352900703212798486817378763769197501596503476210492679280029798828161442624700

4549931873587379611446122075041773768038423308899512489273821719117059951771345342994570

2972521521403834304607340293212929718359902336716755190204836798893485785428130740917110

2127491351889665038805950364886600193051774797377200603859479443146650104107719235172240

pi to two million places

4725816987613573155394736063610226081719549477487334657530769762382929970665938113035556
6698283830832766954761096688653181122041132550888898206279798064803011217227923341819560
7984129997207200884183938722113903472473108515327748366983782486796544853960546733245517
8282183737129068488432260975031906355301794126482389535114738786399505486022460003357769
1360318382569522393043941627677865022636715905426082179416260627865341378178728420381556
5930074463640673989667549268764939571849031321211365620239026152298406287309564812816993
0350186968503713109547232937674722247857296417081985894052169659810525337889233503198772
5849488937286407683296645338400341397336846566429981296207453255651661075482537381673996
9227716956365366825813785392959280486393462406800456128973677765649926444527556162223999
4317489110997868141300340878616096406890919344660106857817399649669192940715197977061996
7355632780830374867624281895329937993501743770330412678463941074239000407519860591815664
6594775860969999412559689622869881578902426577897779204528945726859788015120391878644997
1604999362226461249586877214347718177827215403315790874383331809450293538191572814541883
2682112481972325972143223494026289546957476210108078742584765714780160883340496255465006
3241474442371159678992370589827766568589771945792599269290408338932723324793025420327441
2082794436936354791397957972039636639578210495841631440396542409386047528332592434350007
8130659491794990770388167056714485647590147081936288460634387083087301399925372529836668
9104103313594394347713254525111255211109279872778595138270891636659095207417092752512999
0262045541030803481032270004620819931349774103396993570520081349069420803787220379046443
8289024990240122138046339798024218210683934685088493016791828966213042549333879965487443
8610932085149105332758773226902672412929662567364590687559148963120501742439452638939779
0244232370326490359685257821378096515045449893186606855909756632394422618222951965642215
7254180903209980449661381395312098534796711469934269493824514966747515985290047518052776
6122600719775722496708151505803914346182401252180929356344269769077593665452090820250660
2780467149336326778720581681597025048184007645428436549500337194233565531706110028442330
3004205305295293308676376028645654611038275374154748491009312872595644205697610999302008
8104035318669489262396099535657223357475743158786184590483196642956322272610527164819883
9854633794370836572964093948882039748696069433331514579163972074807234334427069762865668
5088947309498159719068311061200102867523095201110639978597041881942784387319179548374660
3671903556930383994015483738186262489261815817546723094662836205565121691747032788728445
7031534544485852236960697986889224934532820969343567616792601084723826258975990526379223
2591641503276362559460774627417043325449345744488947721687627182772707294796799294070337
2568210612950899937024621719988989446787668627345794135226403349817753083399390670296665
2133803469837272890532476006619392545820659289701415261295072742629227932465791704383771
6269320753650111960155752594694061918181848737771342683444424052930660057334458886905888
8039317748451177410239735877528228438595867202823987437435295921155624322438928963002772
9105687288726816161177303569527231697743695929142484462118989457501169031295742514828445
1744198713371748657674635397474576159541608781521949380382190631719785463648068772488661
8103918944897507305385580490920796321483089352318480379090666813452717823353224661252219
4992676529142759089092262117510817467050005956809313519528400804390075726178665775125774
5288433053553174841429175337424877509489933543735835955457882706037397391292269370301224
3965689712377394516731859670419317393074231020539449279372556695143497880554570330590883
4011304552420883774530182347148368540570383980300834901466615756278297245438449473653889
4739985342875432822747853813731163246992938367029583215296762931690157701637645970731115
4556827663490194825042032716235543761606289610103177892081300713313496833865486540725226
9994243818874654827286727227810546989992436385383891710091592717082304906672765961623778
1686404410858757454793666754385969867095547499965912023647186302513423428660312308328887
2542614846504913339140845571348974212133262795637514158859383437023288367636142721091009
1632643811993071180581370532052181871689034040842283314956139714100091017150993735501662
5049869802128077552418620045870896844438306344598949555144009876519220044348268870127001
9830394069224285391443443769252569060378563316363591695975636285511685563162452502775445
3761962942890436619458968023918515806791438606554576306665385508979167196722773975207663
5829190576647967180388564538818866035064543168335512453205238283278277210925962979474336
5082733419069484117477135866941623551515418976650273652437782927375010905109308258678998
4624186849472218794509286730565647293726572105569324973753017203003429846293990357604005
5326948019752126003066980384350399536224586496757173536448662296086766502211476197190666
8367000787861952572772496077495301582704019156303489627631553591293521702092981509995771
1897771246908544676144850835054133339784826134395314953719150241321492525701145762701003
2683591978855164103614737647596250976223028811188954047134804246179115385416323284005443
7108464269095036218683372877564455558445132126070936508889689962610406609727149015825992
6516847763506236573352767195383297892000906729856532354522274816541019480074074983918882
3027932639144942369573528899527704109953394755282701081943357336971843195081657518177331
3617892037222004623202250257101959975795240224447773214620837600834855387730627390209118
8683525101963976707841850327839303456164014187654569380004166672187859830183324978043006
6841370878997780609708751312245373317921047765321963226929206244411860240382393958269338
4876943864799215827507678016075065363560192478163288489506739317047508196646271951189668
7925950485598142253735349188202252232254527064031100450586493401962683243996750270941997
7257999962112631549818066293540715583610274971906518427406565937254512574742135652740661
2551420873683195358915340183005643761475560005901887559432489987342354418536298897724664
1429112985184953106090530703685290951747470466217592782127028442720276324221886500362882
9327344812778190824737197178733122826245293390331056612313694376721597019056278625102331
4965083850517849547462579286335484676475056193895604871272476315354271306025732461970773
0588914499576286610805194016087738399359475968793420630616497610162893847437876270839880
9365286890936241353974223097404401233773452835062258300768194953505737271247291463024229
3420118205594285875409672998247743329952532893889102882623850029186860662230600769541445
3440141543780274354652779811248059501088156886539095810551792517892616859476198901851228
8548533001971913658050934308651373391567144253106933455853593690680573111213522090148998
4322616396432630776114024959572757551801795894194013197745734228922330997391962454237881

5316373992053247664553480610143673068325795760516674364736620234621205488325796206777794
6589615346666284962251255998837366356154573809942398223413977857318118526694509219333340
0278395660522190434390795218769528629536258345114288337418801389766833483451992354372 75
9509972488475499853482128754160212142000716742527322818658471302437403801247212757715 51
7354380686932178170984693047721386934362393518517720943809190247679191235016341974983 00
1943492514392273283998952752845430980061397557007914170816782579339825803450530350435 59
9716301845528168292642279637951739982625697213931034888695236503388767235345917921388 31
1578797662440444585686266118761866077854423457825562175139151512175069970282671214823 53
7616753390299724794386940098439803372392608257591497122524969990916251682241883027706 48
3153811223687127561226085840232521772823899197546169668710046806668395139405468301470 66
3243728097173085261750040540584635799643871306025046653245098513711350478406669674081 20
6228084952470827367784896750668680665695204615935906403278260228102365520837977749099 98
8133930572493066866543878693836289431253517516138530476569608483426892163795317644541 89
1627305175221678972080411022372283886209656630432693750538126058074357155644252030153 60
6598273724463194200272636840007290391352321609780682089800250397115413563807418433383 84
3775594568899343275732876358995393433301321522590012083860051252010931866882673567260 49
9879953512265864607687845448841833834136625422196971463251892117282500974121983889437 99
6477424611182875649274008010958068107163190905554406637684199248303038244538612047639 18
0787774784095532936773126665062304634915420945503013186992838587040497769498762308681 60
1198225060378407782933137148693195576904124809600289428590147156300359952118751134960 92
8464643882776366816444290872354236562624184913110970711588110759956848824186276594293 11
5532643553365781078624936606809735256728328243880471495336516304463220419935623776365 92
3546949824861222404369330506445470866983819456132717316706287211292233088278228768566 11
2936704043109736681582156652530953192735760657553663381308114126150418274259197915846 86
0975661711155359265047245289013979730748365684566763766075030008038868274480925601952 50
1822877867751168351851390092399073510597032706696191540732872891168466075052090920060 71
4564638393565915655426687110625860799966340457758882769823034744991771278741658923779 79
6117044330665499088194970371992812185309204245501010187280970744329043394827028863200 72
9296823007160613009672672956269791838625419239274660390007121099734961053233558472567 59
4158335365038969578883612227716102099081807849942356230921096520200749681909702336820 46
4796210938523210558762150886567676848354321163469821573876550837832037338143199007420 26
3497842810116848957540102189754507098326654276214693339080399046753111520241502483200 56
6561580635979236181932300762882729466887837907885397404896953393147003113132449309322 77
0532613002885054342907789006340319100228559937419959054534198294838865764191083821664 29
9146114191054104072171837577155061513515392771400877552001228409781872581662708931273 96
4546477259808894904638744412033834039847054748264843421660150604097870387197604533296 81
5945603974179277038661169175403060560427594749273731758060875311296607988071723021883 09
1816310355469967899967681411322384058487215045110977754130927334808561399431389195645 91
7913746129612274326490289450582869601839766326676818486367297844296108245753273532378 55
8101279916957607536616328445715479677570022259203947901245647188595273323538013204986 70
7161550915878288956727461343961549590248105267578991639561562922800247341472909294565 42
4144238427975134894573060583395554662066270210014102767079458435211648908816862436979 65
6823419770822333130158028218768411671028519113734962555044156500322013218780720836321 56
7275832894119429300942017627734310749322216301696903711021196817814596112985080356782 47
1755722595233764640402399244994117133227064814092208903934067741659079335822479617612 71
9575790623216075333480442592524721663765328124917378791355454531828388653870756476397 34
0816244449879336143123185696534013864422093057439128727633815813871255067371224288300 99
8581863210156353494022783105631170331076712499040051320012934897027213099521549239150 78
5904214026893130046098656152361405253039272543131409786703723671598135087041444155684 74
0934242858068266918870587013314646320508191505624847600443520708075408782114949462115 09
2792335641676736833501642284278652933928327928453321515289204094301200081708618584107 50
4415762168102606083356828369738431971365108293621246800257976799115539990764840380499 28
1718037565345951838459509934009339260311050879753764133549052939570876599134289972977 01
8161429476080132837284371590590628796866400470614917846595143380897979017472288822130 53
1415145267504796951734362334726153303000930497426565394579474740788563667819470875812 03
4604862122119732683985031983988067512355607212314224839768206933579702545411426785688 28
6857621814616682467550295237752661408949262179410234215163541177570269072944307695709 08
9606441496588167174212166831811496370914477913934086779172036360471833759073820099690 94
5012308440297824319898307429912474750965950540243211346298334415639386846663845113041 71
6880468008283501799096954455742858132774403014003363683787123611627532239185200931510 86
9547854040603851429675337051449228581682317546757859933248970433194748116313656876224 42
0921196163984780749399063255065861047264994627857091184829307640052302395716940453022 97
7484337534496934791042788046497550916892848110273355938094404693489578483196619161915 68
7678664744209176769614601596143007118763711598184357094876341973991380856286178181951 68
3356605131978094532258542655251653405256418983604191809877564754700933354564638637458 81
8370719308992777474777651940071210016021292429042884377518855696937984137461959487866 40
4952885179702994403417092257126983643477923420004450897240195642776843574000469180688 57
8829638255568576955243348105923536963237766541213613659416589648936012690830391218907 96
6934638782699462568989432384269479001954491764907992596728333201502040550563958228832 29
8654215201273903857125511583389460147882679613070593684462714073176635850748773515360 87
8547105994508157375037687217575866897476371420450858593475520371592894144903845551888 24
7782248886056768179484488505424827117656020412756251081698730294789916929041778073208 20
2945391238728850578047115027943406781972067987066773468991759685701709642214988438621 31
7233301403648409062296336613973120512678548019751401068761497867822382951553014477543 88
0094209195811190845593172841912844754245930240434415604684960365223220703918197954739 02
3747928894306287558798955043463327292294265819818993849643390390174859190074549851943 24
3774688971516350617840447658172638369808979750933160686709027362906796736528277031546 32
0116423755537799298474640332327398535516109777781075212622968949860513517165602410287 10

pi to two million places

377241294082780755589199253507584971547702149091468336554322310865474877713862288757608
100792717858979025987591886351963060456666333631921740794453033459277301240490432328916
988631072549085903950130666659273011702603766298106832918880154007740068222930213859576
454235684172436497530339103424759546679776970800273759435800647152486835066819946207850
017810354281282583528653403952123279660353632408223180898254477105204750370425226479722
869915914522430007083320007429597732272579503765299376768720265918931466788798396187665
084089721207162147080505329655306838233758647809970173621775251826622594488975554791079
002943280737776954120378881938575336245355575553862151372157904856451955247827238390439
225555860854598378324204224899605866221584236888878281887503287720578409778708991012397
962235928130415428146206709469072943044276373570795194638240638535397538932514553204039
865813187666506718012855292092902813884649449913148962151109657353827367110519461256070
483211206288125968749690533254651660985515328470502072184489791513038599618270755253508
309417881533073713333483247287774790518106099400650621846957914160902586333765370269503
363251590124006107726551850408574372050402869641903450601543414825874821359489668051697
162204129218909013651942661633491015177709354187823411594434257301845846047967497734112
396736746076937584906354299939745453007174309149674014521858837580790841009395282512399
394188780009800085298325011797155246966298052393594260533425668348417106596468996024067
593181873007607716569646002749184784539283759773956101054362297228330796712427595819133
817907834096214082773026094598302411681339242540210247908295842719227209123104987777436
008228204047939823835763173244317194831569713300108285253401780917584652294174735919734
937221873348650377566376945451734805841274192964806237884746960032363296456071875008561
994006290196361814327696107910102484774499481774730369751899228135576935750144684547038
179643560419274820296641148426475538846160928431736673260951171414551466420877593721160
664005713187183194037824930356602642114555465509638156171759883263066063541502910121381
757407054603475894377765734334744331361457069549958552068159687192079552646702350322893
258869265521158374045717679269786983093658441605217539839796914164692130528871247348215
268404863354416036667164545205728789206539039689657009883303927822831246398832259368184
889730076202950191392217469639190081298224475781030178407124137118133347420691538063731
963420372270013531282315612820927307873336065731882243303526775361685144012848142160469
279280062614904237264752955289806723868980112463526170892236094195142983185054938776422
055983978423549606838430804446309187389828110323261749424902459296870542909598932771886
727881814902205185942496497864372201915084587252415137733086590163437389903690961839418
466048047641285774857339602432888485615948165391309950784326152761242743730419813219131
097138233235365219625656568413210099779346586712530980916312369454565524086709902579573
737869073570795762333041520457760151388345584741962374792667316394317081104616149280635
883891893012927765043664289222295248659619647425015639365130455542218413698115505602264
569224268844270921908249138797460468842621535222232159695297204600356284480180514309235
146490648315581470733739099094033063516284763645243070399022890691032269063357760364860
551940902782680315937808826592838678858928333981443121074324210574440779725530487580754
382718089738160582946051048302938386321120440632379853101812009680047840131210419317231
158801989412899950944905182352028550174784547276205986369707009621505367367800710401866
180814138596278076915303308597359792227429779680644323689308438222161613445029092444424
134286820459892391441005864948555982060284922716247787026995589742281427014367258362020
191046924111432481136567823885316616782305910130295772373949421822062853229253296628105
627894293746615051753207102325403956069542024998214315397713255432975868552527248013252
592049623639186428240229505652917173498207387727486453474499266638334680804728431021137
809271950366939837088898079287353281533984742606450074080844329450210486602352792585313
312965313224536877330895416706614836311068277941901052865443855254758821389430878387554
697438926764549662138072288425723934505208345421566445773935903272319758171765916091499
223005364777283812733416662253384147224262994241192462240097854472979829127844039263998
169698249831998810282024020189496060671263656610074693970892064689403357049238092701070
510535093856117942730216979882535416280152727203897968351604236902381883598872104029201
907105608751001679037111105179391713754662368383254144717859386530297056462626094815960
597311128207255718281113324607610421774775964548391117971361887347487868253984586689749
210617703503217366706570821698555986605315272702364292962106033276295129349217521429748
836174989730538729795231771376516956560760094102572096526413672280471890194845365773052
324718457685643453133418091260257540139039411638861092776307356147710429820371485588099
882807701182076886043581805556974289513493920850270360998529136566724200040934068156266
480004775036926701006718756549830267840394977902849840804112864904273731878732357492335
157779266546405875527019733174152553436934333587817743947667698653410903424182885600468
244271735589195629962507979082737456067765384902422624243413944410514766832862609811698
696299573291894803130938365787720305440635688808739217336431685630197495882407785656469
110058448506322174856027869870825449234350094624287811424792557092875881603713337049894
447935541357861767774259300350199487881893546457069989802388428594340235293957359036387
799554858018444355982316908424883535500656784002486228853977905919021010083266439140423
785834714153921125211966441271967985001403777378161139546945562639343963722916983970344
316180226380185350772747903835838776038137578386470121363152065502850438209268547160480
494467248770511152956399198461966070480199092254387591936048994177642432378782451275097
624575285549009194966343005632504215824785039438656573470302650698002722496538162275541
258123830022466850559270192809527663208055323913607488598549525769899507927614076643457
646429097181815270408434116864875195291524250698686972091219727276416639989480352938155
720610365285429980422793399098463092628786791888447458228183849154137902575761730557372
190989173358087060952118139139228370173304768818085099171090504440130249073627223745299
812479422165881185896380881296789286027173502479061364226666970965563306010179050522655
425750442591979884309629281030657817347163705682133316954675385257041275570407558625868
324966666399960771750714245424347638073499350925572652501928764086492761849717304589762
514871648859159895125537522822295535780927551577177349425449406463653286438873433421753
070279721078884393578408051947775698254173932129280352819043283042266081152776150337280

9322221614627280225517275940258914049567804027668536835600648375121195655603792908901 74
9705688289249537680005511704451470904027647709526612617891164352708494766333633037504 76
2668181349383168469838603671786180757090384099944166818885088571567337508594523816058 28
8505949719141157735194691638266329100093631969376262656339971438859083062405002210685 62
4839375451664538977952645501454384899174219683121931401372995118410097509794199237468 84
0954250132123647670753289568716490593510289684443170331513881484844754291156551495493 23
9860473470143099453095929657329964069617905657189155713959225222377996616334929240926 90
0212172351354308093758013216123137754512348729033561460914816427581049952002360871359 85
4964801275969333726114878220482714165428610972873853149810697327536968559132688931246 58
8639737786052796473145774438703755129143836310148088010255497501850563438374232649428 84
0380798143255578001639249929690852845898363913292517937062540150226681735964657598594 16
3322441515757367266219230425695566601862213826190880192581203723881546007760038590449 03
8861766421200157127662440876305292839786002937708353109004391865881200087431950741028 99
8065659379888701230973100697017955799260447386963611607865159837648650167943285933431 96
2812865551608452919059142677920493990411896788778786674303929975761676465546813424735 63
2581831341226626900374833698503187200117460321357251155487510227692201433444170414759 36
5038920954549976990021428049303569545892006008908812289808480549761464239812417065357 45
4304376304063168249955435897677978729067940476948112679095135574378201752416467534613 29
7580009812957790997270711903936380944971395629625872682383589471854165638473292427586 95
0984276737772192332175276539686606177282371596570821130355810363815182791325694461193 09
1588133531942734066542884074159970819492402918388497625748315339375254667365082843965 54
0113896630167639046301783547536947865239365547983209434681476337898406578472456142075 99
4498977412875174494581453465738378997960255958626928073950269927756990776084920221558 97
4346739263779175303666438705307148525194702209570892828636950493585584893209095586866 70
2334575543192545144251812880078499151418500330138018283582919167951015135063258915735 68
7948767419183823306159137685922349833225083199955278486255059084867455046951602420521 52
5235667638266643206244011346246184835538231919030789790489156492881675694951297606022 61
8515957509091442772246331356835722350994112552809162282424069411322325689953200039030 20
2196968217386130987969800721311638620390327297191995557705918946517770931086733435970 19
0867546375077749418164213662458712283625713096904252214240927414444286294676411115733 39
0260689928399591420734459182101298789382713047503738331667957872739006283488172123432 79
3167900780179277820576279424741963951177750945739527443895863353347367966070550527012 14
2544299479080491346447357381092368930786035662366461150744159419230799122635805375263 59
6249358838854994535786088823490647901666245572094788231038700999172108651627001741277 64
7893143063607031726839611668179673599740524372180120623481940467735151101150135753039 35
5361350398387654046363031729240400443934542228866504375595216796385599047414810736635 76
2226932664330642246622513619647559947939447516742830393884953878608666313656608760867 40
1686825424385950930983900950824741461151827971514792538782363231160391015979083705326 76
1335653050108925597369478256693522289815420801446620485501976532321021816166219530346 57
1841288016502644831777530378575721075721672703735192224031414870533281255602523790965 28
5517106044744791180673197071002068603349095233692899435173460169991071845070490952869 57
7741317941205305663931582598080094541594568474016733761993444644185952484585780679184 67
1826721457933027162864842332549702086814740691585705248301426251349130131793189738382 45
2549317175403435106305944151858517993322891846398568766128678082982141129066850039256 20
7747696005665324348516251485428270485614233397106339613269314052482118402280208764932 82
4600707929518671187707464164573645634222261847181242843835488266056541755904987953693 62
9595664972541183719335697984699399826697082328312099109341255994808198732203868645749 76
1500731501308035940504067340560123257097874696291882994649670095532299328883162376227 70
2344608416178629584181003305951772290600660895813030583131395588588048276225962517551 83
9426498063120045127181001922219497057669748844596592692997691620797266423414339698096 08
5014545299116867845278772258015085742859764318050407162254946512152689507976140983569 24
3094174676545181719566747204449842860326968037184108259382733558497438685580513595228 45
5287596361386027589819453100170760894427332474687242958911647821885362098458294683030 40
7511008305466766121316949446563866233697314905363048789078832874042072673383396925834 82
8135333246261196639727672957698744403654713600165916747714238618199164530627228981557 73
5662292266108971779771500834627946440936058431573206378347619517000016581060210092878 40
4435682065201945270285682643221760713581601752221973467223327780274398943597115597803 81
2762780652604670385750855602556081051666778183826369162112027595476357509273561033756 51
7976994657794959611449116213116791600460723425681348221091747041610254084248399240423 50
9621969126368120916434903792669349222546351017403410154654375283162070610539082122669 35
3971414467016387133919572456201351392040509186147323218293619518912364549208823904972 24
8828791425729913397782224781865210133914143716010778081000716129662093680467263377030 19
4059184785859655730889364507857936000878628686633807976368902807806198570100232244772 51
4039330322119572056967186423880232186334776112599435486499244747516611783603166952637 54
1604360063566323871059279357921756871199982281111089142464610154655356596212704209994 40
3197587351534398333195609894173093865474546514099938939769355392245864303588762315761 56
2586158746287281878133711235134587883785580421697764398525992789049962429065388962158 21
2221897887811625838296590736324849697798761200713068188337251900370377403487225042973 49
6099347660728486400016319929606695437071427118319214099244239469582566542054191451204 55
3614236488414816026053774866149864110283759516290999622091232981921678339239642561390 72
7753657077363937251198229736967869246891513716526486975965134435767122826858375314401 26
2804184046228693588973582463737878484474816412107338167587759228223079154982210804795 90
4348319338763433073399499251942433721363211915365810872552754939034970117953105652743 68
8889440854746664727270780909808046997309405202816129582446065629165507696802356145139 78
5999535614491856857949608990228154099318962273521074758244856724526195100482672561527 01
7230093439430290688053019264553530429770999782891887127297756731225080121569089892104 62
0610303265419535588308434633118423243266793502469894057473910493235573872862497868075 14
4049873814343511838525582585965084861976534830516557746535484925184706235846361128961 23

pi to two million places

1063475613487829196208080290215418895671695470578412636065128050970044865339456921267610
0746489021826015136002176404207593430425154960471216560638269072668810603281420474122000
8886673494151799724644431487852302819566773009243452527803547237133249811211144752719600
5172852639093240007410085041013534953404377507086826929095896450567776975505181699765470
7424907498713768970805422231030736998621442134449704808333886290361924620994170065952750
5234994560840888352771199512564117887487755426906953789016658371705059760688177908491160
1857021874607039200411210127386634258458451236459384881049049889171246857392696221806480
1134103477999102853345043363529355994699910797483860730938322241856554365049764945838570
0957547589241810795493776431689626073436458913015274465652114578614155770992104933437130
2765485502078197577561301902631627312439722426741466016940435362112664762066895571814710
4997912220390593604531712953095554012285510811212102657296975654697231782827535893252630
3833111069657658155611735948679547594241902955820124457509075373370460353622615497103950
4431129334077832390857018090461357962884855271634026996506312919099426969596622558949790
7562872128775719854241967347530158746347073305073784579614535454252064423087824084140800
7846279673687388788958520447092790810107121198779872783592423393355755619371324997959380
7609841477989821381826503432240130638574544885493589707048625142787242965210065664978700
7015683883864539959036508017111436410426627869672414199832654733411128487933093439580860
6789075434070202679378477434553526566512929013433723462897801078201155221887239373120160
0937097040686325536770926191900398941884114592615370116668658956367315052216330410779840
8867721149416004662110658605248504107954794838142514049963682497073968941442466671748730
6746316851704563577542423105150580986476464173910628745990473792488810727430814262952480
9093198549576289832417190896199838410018154664437808237533284022544160414895993190873000
8472472855748818376975818953479390880198458062894211236113335463254004046653571731667320
3865207845030801661957708090271324118916546201980708955460918723854372611387496629976660
5825231471481880032379500380667011859894219599482184048922604588548768880233759401421570
9154295287123568227193941659500062439118359342678164048354162384967924208087709963757330
8733972713857005101721639010281400619964029555510920514823727888939586225063582005762300
5586046067003335451488830880320699079612318449237973339003311505882948380206876710240910
6485958888394432465634466757463358495322414165188032825488208991824102940683182206511710
2210563581095080526760824300375506483382815036583507743940163080243748097984829566487270
7021341700558911396621800935230435395488825590266709000657118853359523301307008761904280
2865576790762169073394085607061588936193013487802595249967031591436244219977851983004370
9844414548655165908282041411393767318104087719596578936089006009298571878482454316218020
4538611683785858623451078284918216914586433097247677113149590308273461852478922159576360
1761941580341303131918416077594410121574394177453130001079505644225251093067375232236880
5432427912135305575926210031122118620382975027538178425417311733755171196816524092843670
6630140284331236103292345231891837022303444549741418863973986508372860686169436251316550
9806429404159609558281527947448940796006704015606635423305660888224109246225581873588040
2827934696642062741124597521522027613182620967360927370341863068042962094801519011215640
6323354619949604974083444354531421150067268251668779506672541821696486833308233724458540
9241292167319098180229151717027639539076596084501594587459760737822363718204911724903030
7217284752147681893828285852418664014070914099177883713956921617076979010060192305266290
7842195100992184012322062914006048217819490156194790264663131287502908324448371554662480
4940386152485353273663548381024000726950375367248133315649581319529832958248945699470150
4098386358837510527445038754237416350534548439059129801016033702406509141965440640893090
9287456350361224915848602375132873373130617776438334911416797427956353582643265650934380
9743876971254048909954397610309447012230170495967791613454602409207214946497104487480380
2550964755792337840850157046965409260993752929750657563244547864017818208689976035501200
4605289875237228409656415401043048746627607199592900351813908226465627800691749938188970
5755041245526281187695373575052501272801592229277826773719461371921651640018071110326060
6546576457256839904111265510326989847762004945257320763366118794668546807565557788161680
3365059804799752893885934462182727482769757353481167038511000881206002684512193161349870
2922797354351452516206626446755043513009958875998659114434873302760578327386061081960320
5678335362521398537214632713714373366852594289784092798477878664746644957883589668082040
5147767271907031262880771077178087594465928794910582812751317845119390219204362173999270
1041748174735530055149680713746137668261024582297889026864070620871691840805906684021070
1179755073163609712321042997983319216599464118767390473835823972710206691386867582223400
7683714025128786024336013754569561210121118668569527576098387624201318060097300151094870
7041864750146034719095600164459132581601108870034240951086006596682466186137003404540300
5650156032108961119564327940113323206244129685271897339002930438752682641325237281181870
3418377266831707982236681984925517311140484292263600497298366464717470320358925111903140
5526378203697482737644474657963470343627745210955607492099970685966310831881197052675400
7608418520990745264126830901434767342985506555550495836710687190384924438850171624189590
2415970643895517919761248010511832662108395180919193168647150076228549154633210029492310
3843480405709777013982744519348392219713060823701654633725680139350348101212266051149760
8782946285862320643474121126266335278215777473245248312413542866044140191905637144561933
6167340996637827149600978057687541695344544059947939618148914947728839344493862378544570
1071502666829049614037984996997990417731453498520525154680296797462648196687022861642310
2147762924124296583127636615013215995687556303213834957850419502366939282044083814911150
0686568428096030448529697382538002417267698709458055882387667508804699912381136464981780
0323271388627539998143064746124047754171778696333861396561788596483175635190236538942860
0987981324310596410750202204906039359870915954779421380986417913250939151430229182359190
4529115029434493134123621519818215831203202948600394027669012663220706625706516546005270
4752240101992387346902997506116316619281850976779226093100513544564936116459656602849110
2845526224785726874706115946260327304926038731909842727022302279363717567119268583647180
5781515513352034020094255732570241356749792983260661689237477237909231564483699398221900
6223959835129277487404921884865148618006764683647476634904354685227044181503294734668800
5980259970451471908672697542090927146372479398724290800146596030612664416602228604050720

1331815082946098885070980566227981549989279243140532399034861738080214933410263315505 11
0372233887514816204398816296144993711841473065439725635803732060537419376715721551625 20
2628791243475176956627450405183387160885984370464672049769257186263068174611117896507 12
7338941343126400421900228426884632219249602699285376809615893194890230425833901286085 29
0213585525352287293692772573112483980587096220893068466462636341647431789867000474061 56
6857637857107584274947429648579667625487697959410694492681165765692579106373912809174 33
2934276600823474451226468172447913454112832897445755146507869565899052824665349909387 15
1111696795153643282615119827898668897109310169591041784502488287352319238448624226362 97
3497576843928226311202000471321872017017837109757566855371393824758533811898056805524 56
1758925329012410446061386262529720144955188459248610761575320195237921069318224655177 28
5864226604778693839844215191391884947712040124803222614617113859479756568591174574724 44
1827556436675503186173710531284061591548821249719371730212674392008204111754985468363 98
4135677460883816733867417100470330451223726996329675335035568374255932788505284843970 99
5341517659032938402850692199642609108990684290371284529345494907934984037039501594365 63
4463139951298245813335313896483039546853777423867582387995950731279166633912135762293 00
8238137499508804241436770634118679457902022252519770235995448842923046328754923496722 08
7300742261853262394415846865082615318656057785769972925335180744046382897304361213124 87
4166495577320583031985264922840033821229619829400035721889609222760768921737321375612 81
7140127894768086018461734761335830477995555701108465899184652647160290624326823097213 79
1999950260020076779284281280102189046550368644940685163746947400979670522871746653346 53
4766832852993058391752919256842294613330335029926614749035530997059294430175663343834 32
2304415434370347646640492739502658109126488241517806384955847321292598486391433780405 35
6376028060686127816888921524835645436191658459051120653701944847924254620558790155833 33
5432558659101839153275563432530479137407024665888585517326415578510827116214091911502 01
8761617581702513117007944140863486314831905529615541131867774760309553999986080793843 74
9762273037037169749162292002183000135339181299918296040235929481622403757349964789815 65
2622682469224661822660033465631554440691909194461335922947647617530984014456964949854 17
8741721231360779557004623157574070164761282734096389797697740198761601941576015291091 09
2531227618383478679518524193706079791655907057515180512954283101853591731863652920291 30
8420578073926757411563134564106004814859065977277558923397304760176118610946668393831 7
0513676765980608645465327538444171933248210350034614387006425749998217425218212189204 246
9834596946017145410809671735478479028964900570936956585073602799626696168464311182371 98
1949954017555079372487801969337650451347412370232785817170527287540676808677865733191 9
1806541465070234693041076026043807649839841877624335388381358183139633229783045192735 42
0001244347701439141020280583513761524868346540092241485555890737202921946094967830938 04
7252254271539715644613932071756205105747947382556303404499840905518931225258151706446 91
3599454941078976605283937995021912602612047720514588368772774203939349274661742316966 12
7244888142028639142022781241935332974216945094679545206739569773828062800915341955720 92
9620870217816235957315398580494059913096459784367461688033276247131332187163639467180 93
6674734403357555262197725484420249996339317486166811685800241610193571815876391393715 91
5310763425043384067749110062398885979546113555397558396530539242511338515197295715072 56
7194931591445438867480941270092974257722117019767807775541153146444158888135464046100 34
3675463395438136573794175520229837046337812404527631159587874291541420416206624891261 62
3850077703928634847262333435064417462265548889643289608471692123310844633335053371471 73
3303319017211530774818159753187403206520654666303834024724044364192958515662077201957 31
9351948859162981533005510527995254001092334656798597064540514291065710504302628479375 93
5938805412008468120716572595896827794362993111922904628749893381516161241807541838876 59
7271700031938896653365652735965704729837055565100268027923851679503365206653091784354 68
0053222118117845683243680044326659025462859459910385475796520912343895032510595830284 51
4501945309892327719848928078784554674964362756461696626183648666203671557849813839868 25
2876195638573608520419332025764108658530820783460937354267441745879179816509776067480 34
2358794378816611199895956667944648382154771584452235751309630861325983230445664681920 97
2502934490357865898888405205528878640659389827094109866215273716175244926912647222856 26
7443179067066051325033145720678344046379514260173334959206264618127328737794001301535 71
5736623761268528321103911201661948115587795504039408651223604194975793571897974081155 63
7346720742742415773774044410912384855584519673648350488868309931389823442124854956202 33
9198190060354898518044067135803314087241326581558085633552923565055624340622173863587 10
0591091666902011060085192106206152172989898708361337945882584198929728135937846386408 146
2097321574887645854545290564856934762540928991066919225602464254529100149820094514753 88
6905850782157499016423832582661233303084236017133133019740264365928105169746006112974 13
4995436114150778901552316163627582116073452193855111272300896003399371087363400847445 85
8281169291728401471592927127197382253553963987645924738369326112803742994013875808176 17
5069360473808825916076549966028549415839794291304417893741349812851299433175675824407 67
3417695102424433117320805419812250315547648257870165086576670230978527121343260960828 28
0847222735516127980217943248638989060292519154480885294492491323857532148964128445650 58
3365153439839435370632236908681847489163408293026856719785796604160121522883409864495 03
0061363798475278879019534177434844467274800436348276180823399161008740511223516596774 45
1783081921670275125143549469966872668737082784919199168202590371147765982606846429716 82
8871519648270565793794566304849953425828271002028745751425313487886988808012391825275 47
4862935920257975002818219751009946291783785110346671609157650768942405764348823245410 62
5517145576852355740815455232660177674320947901564520546687532565105251479763201112266 02
6752483321395508993126832634930136851924284030942602387905323204787679384881578179912 07
5883998951182420162549243993752925029268338089129651247232814990269823023788861443531 89
8799215072001787269476593216105185240507686364891235175167193707377421624358856294062 35
9477043120052660625697625096842178121148882988002661604405922232933162417612290874337 90
2228780456170135772375061952160342686280629053786496887139338571256241696407932447583 13
6988591827299927578294929575130482504366602853237140205496447338073824577555825709270 07
5915358213622487873951980644753650922833873219737894509894881222431666501069873961667 29

pi to two million places

8992096442059681756961923183986619179340847425786815461594145893864060229613295003812003
3838976745020863385578266798810656903699908156782778516378293459936194336698065297922153
2215366628883994026803862183878413895499792007228937116950775706172400234487289868380888
8946932582186337823435691207402895687188566870960612738621934987326322409606596069917602
2005453603816589660214387177128755309837099020713308471230391797557448381005068328095118
8932927219123165494090664021456835987446321626557573979283730870286061293977236838581491
1993925841574254914633515482041412850525611641438473862157948502590959169401670192222715
5205159244638478673684403410197602254850585962037452021034019586721608171270192646070460
0795992871312107480351188250682335304496981265520956708088454194102253519913136835291159
9722281977965191751091412574906675271987992843990372741064588671698118350910120568349817
7677310095484699100664217037510129640279526680792620134649082657083731273088770349853816
6883018041591073508780278978144452516540770812748847379655033182989360261051517090092011
1022207100166947994860649884160992557782103329254222023824316169379455244507711661278195
5720289949059237178763071379162077328051900435950630278372052428607168631975683199449535
5965461758237933195492218355714063821726217011899062436301646834987994306732489748402990
0062675636866393449866870295746329559276358227417390476736646832709872540082565827407937
7301342125015787224693593360232027880213344754711692547244278823478857202047848109668249
9573259465069381839328994408562932954845234695473247072095768155000313628681830587359766
6552445629233370980039202446539708198088097516775900082945233933825387379751663668484819
9906191718923053029327528205228873899777985777574644330667384284668342381977782239415152
2436389874249517010656678530267666437085462406145075109823826508232921923116946593605521
1624370127430023949246000846441913713453739088171054329719962592178595836890022753546934
4419927056354499446464846356921147395454234209303509561912596294276033231402838156419581
1239921684357059261155218643672939081140498842995401350304582616685615119192470424806777
7874883871318987189673826192473974908922169656489981576717042890174496662059680086819909
9195664839871279960006066009833665085013172670506678073810536253324043615610980110847675
5549487749423658516371946527932849799057701845104909170153356861363244389487965903436759
9034956002166426558152444392827872271735945948307893872024847342032229005213360684605312
2929409749759889320134905015546639978806999170087371588917595677689472706181150301964728
8913257678481619091938845977305288817391379634191013912282881868957666915817506640190664
4257591185763887548293436299211719127105497738537301557783810188444186065783059241045272
2431966922764394688190230293660368932914352790067834549205228896117886051875408310491808
0891775926096257118288327086436346727818162747255714850253575093560194453370570429727931
1651832436970736387485609272821695707559352179828293176304039885438950572501794655341196
6643840611832817122580580931386536690163234155042355394880397710071250704105678774162025
5859908400717682098941448674622992276289202550858081621731508389755388405349427919056734
4487660483097079108665922529310175974785382557147194645296290878451956517409594789439633
3768568878413363340735390727437663722052802244591605345737540661837169580524217180032186
6022858375322597888350188042317887568940231975197437444613352597457897400554662442432497
7593440495376268236401505734726953980110100256582513119575389158493821251296799677253621
1276460763910672269184411105916667182317481206619477228053502579386189871073293114319621
1955835900757325454926544053448587623799704869639821290623046526023715456974639821850404
2406160647212247386285369145422758281589303257867192003815231307030123145001620383558559
9708463628872856866182958803815141259792427122280065807217537599560975302816813243190882
2675811213978645897799156707712334130061405072073787477542271334777231879138778060411602
2833892807322946298610943899480426877630904138282008249327643784456946668559691350972809
9296560296837842483190637664897589402297476523373707075902957322967641074447790285422057
7108331864160628348328403937671338148094153081003836346209867409231416257725926016424131
1076838385360967743938964538812198718470878357602846575850066264301318356377598343942325
5631956738892178647425115391464830610561761852266148498621173529930039419626797835115432
2471979210099902359950101850452263362136629541758790411552911630059509298870937205111995
5320951919761191111785656856314582374252733634874426217894373425559443882721099552583440
0480781533630123181725047611209858621391119503812768576842227060228808579228022789927013
5450268982898128670846948085869077368731048824135209253377992815178280722473043295605706
6622345618996569294079904301031800558838515495600371015362883393296308388611837572513442
2929623743625690286239908180896674784072154164821536466985251118356093769538838247792682
2054035562293103398234617217749928761143107112618698176716510102132817484322686049289996
6213744264891787470800521778991459793683257690825447049955736546073833295445503760545561
1693846299352695559825481436952274513515963501284438165762387821902834477841943484916755
4332208988657251072163801257455920500626138323533310017463563269678829979522312213359229
9559877714784256215281966009582480397960741880686481462221846735023846499620948290023725
1674715130176161834866485690095804452712924136107744855016454016588210094695318516709499
6532028368556343942755258676230939902626468802521002348398810810313959156722167521036400
3116827698020447084682751021600765291285961812328923919989837615465401528847640238956400
0800911170777168663258471518886521341810096309789246814277767444249098481946207209911866
0617837882725606027748920250775654960922415372184898191399949304356168693621596429177100
9525695097069900632310061085648555448231768169491892033538259383979779552017806002615040
5384660223480584280668080054097272424887098899181403017210837408519716844655506868668255
9576213176185347414437764098116897462020371211318615031805348163709928051005793939581833
9605381572799053173564620472564675646573375232496044286662754228334119477101158614822511
3290574723369645459350778630281397035026933558672025420653201911364568427852227113049933
0847401550832055345022071115182925037824584151595423857290920559093155527093715730436500
7139197066270720836605065359257807538799662427829626027190863678584203426179427292783877
2074227039258647899698885201728543732963891417985649549871231317421078111745847834714811
0113022006691881171390633223774639144278713501339101361465475823547313216387797859422922
5909287326603980617051451051893526138463564900758234863291520258655103110341381404910155
7351617886075764396461883410175654441487477439695871259318206839292181168088355971272266
5371195267463913547288909010252777742990668168005319870623247556963164794199431877289988

pi to two million places

9589073717447913822106916830314430210883271873693722363715982485112777376585439522066498763027698123461345369762104739754844398066496855150018242872413962930020784239040770171753087400621103886981275161331107671042265609534206562796480309195917122225686050866069749195871952851172018630145722311125595805803005959045178016209156003392715813561939615245058294094238306752231048760275683319313953706159405690082546716340768851873806202839376494141695224789764482741592439389073858703368311064748295916563631940759787977389368568253656567981194055582946890756209748639705158628061604955380199298790107269885264068694896100332327284203406188454212897943218868970727382736277850137270114963870884664078607942655552555485664825367262388553019570089909441814119681927822521474367302375772647906302136270092943519353752024482465044502817747353031305587840428905295653616695575746030446840020130258347285787860347964296622856338390938504059599633820520131345977594954959152824913675826557737086309853431031656991864774935235587698561676402569436794965082659516493618563906776134086344949610923085281595759473244692997843787506957796523406627034893439922003933142022159640475307987724854719289903190331136067538740992626587614582952454629627447825306071370861009325656109106290159370323457846510942541565848633675971714103682469068213645950225938235868980342052145842415621097712594198860517418460981052180233091349159305532364021180135399382739076096018812708570022161499420235228530119785151482540926695185656765341040444548548817766062915333423935588304097708038262943189443850770239040847501710409323686830979044978493238921559850021435870077134287158470069302471676631228530283911202961584833228762187312702544027509886977752418671972580398456783452867233726268194259137689223732798696369957194750828057492990801609296385648875774366813059933003010651656716864331160038178433180947698249242660826392564722108563028212285843591291142036032720601825232379462931092541025124541700516649174969785017658680012863254457378725293267127745162433436012734039785930222159961775173614866679396765633221951493429837679037493082517016185284933834424250418323030264100557818831854428936408740320396600889234387100234092685238849673228445668736570423431566989381131170854980556334241109039029402069878836686500964163691705281565858356477475504883191164843064598996632703398001097062714315487174370481121700620918608416245963219627581891687594715082363689276171748516314584518070543563797072327895745053875844710075645587473724567162607587558263162416383017589481237273465832842643349842119906790332769950618788667306344903282783764859091686806539844039317138256966595923657348223568760950460020695736736953943537344892878945414299449242296541992068707171799082027511232288302063740933243082840768023659962230745072395482913251001462315228385669963646481619930610803501193409855128773081595450154979098326100700084351632209140971316683905093077706782579384891592132099286598075166477627404210228706958031691019769066584929411630144904175524152840798558419204594222472240857954524899661499631295674599317874497834135019747600248558309355697881536973136321445251140829284181280450249199295984561729788649836529671737406535027575784653418707842130980573575850987089232183386027668096878674458767393704250610529455934480000337948441186910343848981992721780045698408825618002774024697155696345353705817713249665431707954879525776642112068569434074073694165211045301707775144950296642015085673418561330879369079908598888195417742618803144141748693529301286286876979634971644124217738001969097486279960894609364253067910417459357128319040298311315505930386112061927540034742996012976984567285680075747868256852655888055044650282472340621226723098765095246795551167576075518973671081866487339135554730387177148259924982090655636246368887442816354759738020092703372797235756205852019488731175736415208568879983962553950672045765637086678684961673992899051663954734806468841632161269623214043004303497879376589552559126127334944313749318755851522035048877154206128323215542501036958420117770605811310857406721768844739241215189067429767995284346046208510422989295590153886171777859625965902453747996448057375425903395571736901793975160019987583699094035346020060061145708129727286492441555885975024274901019975278569583453344943250025780204344440868289075077439617367055383761578786385387000953573335902594668119751237389838726653687955430018415044807205276494457025799468680349949294168674710474523631365047115269827810552059962650022454407342871399194980253833385058853936499417336369643189980369532114631746171702038807086634906478634042245846913559042424514028142597209433680394042646957622035197605253746691968646840574852273222141126346820073268099128359680433712489865124847133386599958155703624128431192371380520698554630522396286001693260924761875232125700995941645450175979130483158522690092440553118658153197845931405135496797501971591305636407967874274388697431812159633202424536950908108540107486745322336694887417447584560189776395844902174934597104770315419747217559031048955150713033750922642894743661501146171128540489836287823217755403355815130890086002311190892831719794615273396398147557956104816547218228209282412622440866173161182953146270119621366199594108793583564320932964189356289507521834160949562866054760820233943903693829441070690737842159371100843550809934951258480556142602794888117357782314109215630977556334356892809062401470430406809674541428500105312911014407193181060055619529375994409816126454374436773978928455823616806573056868188932905552483777388697883384821269003385525577296329485036257241617945606880625056743983435840688586527984713205682033268015840118612253710729945929721983140398249546430136414120937944647846329673080412403157916714681071721546579759508437906546392689441643670202671733332142868727930006752568089594724605400773921436623774703669370647989280683436306662357354918836306740896905345419625492185959482963529914425068124219578493976249362699766843201171783094789764534282159211054419253956738906802587429523470246253627205862445299161425787499201547834925160434238534924384341030380727377077657014743734358079845112149890213877261130749312518484971289917459095003932190625680772393254545546917673500211427825159153713922475215102619575181255589919223775605562518625777871520402423564300801544064737868647177454853375685133039577305055429841027452084882456380181174324411508866694172029225138714052593329218903930234849521783732353344653262693777473250410920550827627013601076268805734928341061501432117912584109328122674911529496919441405798335403820079492052726207312385833278588756477905672110416544166147076128836100062438413053105014008101079875557731522504246358724208134651707819681326647305265266870097539010235484005431910303058505073284856662

8923073610160979873010960457878627259697182149777459319121724208552038322309774373362 72
6009107917085415906540069540475769464526935273895890894656609355321694271294261140189 33
6817558212336609288093686836108412959313689766825463416180797337993114349199450619767 41
4038367395910433250837896097654632163442968447504718087753879660115946910584698569346 31
5467151967310540189434725327351012330555664924462530897985525988963444453829414488382 57
0967123605338919829313649034991332122842196608736575713694328636338387496569415447710 70
6138034367393954329945548894604432862117042281029757649010846130362580988520444565288 92
5386561835543746690507579482110698111609436227278171946884422390136435558225224330140 93
7115484011366213884108247980900542972492869087707863943383569758089134485548375376777 19
6589593658751543275029493401163628628419330604817109392399879190088420348729434372149 46
1239701703430337079169831624576605061362459054918358805245203073129842425880187096078 49
1816358337632141776552648466086267449477751311633746098532615677168216013147819044560 57
7089203080185215408812688822461085420684333127975848092195444438896671131444617893147 41
2936651281987902593291945652736873448363988933899843612116806898657975674885165488863 37
6900353752919877576930810573551739514437952727038004470490072857309252626316730990740 00
6849045997587871320935334814799807297803068592527492154032508062062967936802909636571 19
6554547498326575576094672472292408920613710562170097933992793206656709458921208390498 46
0475864801144552327813602814534457954387336599185402955060100187896258232064467145096 39
8089199675661465982370141287366468803859403266522240875088605288410671979991408544870 07
2930220172202603047863807108861726314153139237489947781917810407754525553693605459036 37
8161928639420022669648039758682263453758128535507962060646362022763410015625391199463 25
7867883608702524372526283033010210448943262262753220736676529109162819988871916167866 97
6986172106895009023640592917572185945847630689212470436502753632835065004334618318970 30
5082839153585060525172234422933189629437257761631522268739500581359156379950090704572 00
7260968988373875369824262186314951213988357167356388063005629032547514519966181726778 20
7896272799165637748029910225094724409419801468690188625285100423366590664307650167100 23
6787351898041750864760380556088271198478863911696605712575811161433220316239861953996 08
1064489115129903832189244967115181985798502770446978518416280732953187522170737542780 78
3674506060843688776930498302304143646891398370082693964060694562881641695291665443739 08
4757281969614641195715806636881312484878296005192365381669691441316437821280803774582 23
1914250665372731866015855576310005458819146086341512013406605686183053991092842220972 22
7727666420070998758231590762943912951563496720838098972472304203873278350860147411604 82
8520420074395960779396766574545741573438143762929561095311584820920001996834922274622 23
3203492297879772093259373458931853036322002185233292266043263277738699295254400374606 54
8044762949827240422914656005290459866149053053303411842327747134752876324177468600510 35
1968025950489335416177387650238931910840665213804667466529643577196052289272879058233 36
2671718010478707415097865327441555228155097901531432569914109132995250276991641218479 89
0353417341802888078594370474700374687166980729136987810519134823174319971817569732471 69
3411240421932383237581583407505032512102423272159997622559536081716396595545921520062 63
4938327793871745098766955342878789774664436385517065026504485771475878951663905261261 87
6717387540459387759244936972468702219846805151918260343414635153375173543983465840366 50
0780082513376195811125396059589417188047217874636046668507755956087664614979975907257 254
6693095018122660597563383204512084638644649947755674711048825818624518114821602417311 35
1155337639418016198608293258482850297215641249354354937014722183714909313527465161404 22
9884034725873584804710030497103678619969039664003189027014102187471439739304796671271 46
9674524858219615904435885750474711610835582761860115988679925232776700779134880627067 84
3058230376844322408372855777585082162732677755215753854931391889443311370971879765493 09
9063043708381210797273160414688741044273294030727774363782884423977594872346291732896 43
2364895042303033952547723285292251820978632941279227761069976483964461549803039103687 47
6367007442072013486804209783446567078085008512489260912708157237896817953897147166531 63
5417927413249374555104906770883053829104660986180133549273647157882317517022952925574 76
7294380718432352827893878730585071721698784087093608489127582967318450352330091008245 00
8946000168372896985534781567708986066370243799181871271374835942442963643209472757271 10
4180443346013051070221778247291988409544429155924729679314766416864679909456377046036 99
8870079612857346705087176357992672641907758648297958015096149717986439323118705902309 74
5168343571252335874425716502513078384396781248954102878996867215558351818219767292372 67
5088271913259289045705392169962315734135981016260634384197411159559871219485570791540 40
9912608431534494729361825971466635209403043017943612630797077809538770794982864536667 63
2635334220689453430306269748572960188084656490209749955256673401338028230788629806818 70
5205412520401199904304289199124015463064896075523480019294548752880557065055452354879 17
8955987440250913007416419180939972946822103570189811867672154904495028446259968376782 51
6887270792953349935513985117238049116955666124418804935182119542314545793295497329011 54
9763279700425725292885167605567069578881891666892649627826606842817888551356822010598 64
8644268973104042064203886202141593133433565060797764837286117478541013181953882096326 37
3978186750156702010351613827622355499078167081762580063265190907230973113261264539480 61
2746157639746972903819915758063074587538517334833468607608896462270121404016579559790 81
5513643179269714327811600039509295305301566405385440144681956741689143950050601298995 32
5206242564025699973540563356851170626312937882096557630557832675616162922170394518589 95
9392779546333735205016889846436488820731461399285601576461906088270052183882944905283 50
1856406504336417535013934985705100635004423277553280516632501555360016855860632161801 67
8822889859277739807082344301218764982988219507648793745273249775716467943768259786538 07
7709093158268698932185675410221811370662891505071916924055717515372979854679547716944 04
6087285834020071682558785035025806898027943094618467258018625571769991436669256776624 78
8825636710239050892975798248952227094184674434146644119119762486308867522691737956084 64
3416367635588308512954865371125074490373228827199257271519966000216669389560503827951 90
6623371071029645261125354820180816234059316123833832787215450909054427198032006442253 23
6001249893448436367593719142322778515962657684525307584855373583084165152477784998355 60
9967914529055371289933804173580332233138043482601916191028053475986623854151203895606 11

3270055649668928131675551297996763366535540470907939888668953068578101733026605688536895602118077216225892191992431173048925224971255312282117588225282650923582291225204138370082863879599408753314229202553788319259017888175978943077271131604891567850867837388122362887558552661186573367446022611736332880225562068614958467226605377935752555836093409891638296365998007307845001369945882120519742716629518893636617887245193879883914985074636701161462355909180891464878247698323775978709634559315457068005528207062946431076384817183641242844488322241634530641777649280600316789700191373441459952908181300827357120244541781460612372671440187538225274535151552424500790797536879918711567102845303187326563112819187358142050423077462972254036963523357483065620485610864093427383318334634327351271592642393900491279730288837834637442360464419658158438405319108290323239391500637052537770106594292190766393180810878765575969070784073173239732355063441358556746002928122819448262638691858136721604139546331879072561693081540996943760021476848236668959583386445843391397341957739544589473799649939650197318018758215438818580492440153865707188767878906058943439705724396806766233077750215424777082377904412690412076066171751458329065681240190888065205921445972236876022617302455546374056207480813993774670094125222153273448841706368152443582561186966261363834029286644970066037996750179376316763918069437678433862149089623561820105640614012377885098835667084731143716888913942684794853876466509841171954337028921584835072586019765152460415435606762746411781037958055110585275094244729655712945889544502046822567201062098620677182166748688559677813336730489413888303965661219189330587140477845513287672803014322099270529021061771392127737590526124680357832336122316724513143018328278987695299046550698848503343483983359922741648106833596317970504048017163575811696110898875265050394554508189045782033398880552733617405890660762567887602344899655818219507130679827984374701471305670367804002790888262609687517984330621616836497577339489618134418824168865022899678081627975625692274980874020887688103594369299203957265613050681288387614411959186240022364452480039479994244058253172682465135209489596658726936634884015099283759846464753424030561559065105449169124218607878117680038993076090690483506727951214030034045948829208453567271632950120071121468376544941407069259621433285678745746838018539304546243136985598732642070733736209526825330022465956542110218831635555322102232583541986992666491353192963187823490149158700147749921910894650128017261865842411995774784463808881792358967236549615822753535499698413029394932056796213757769896654209615618393575485106101873663140267325061995658143840958843545927104275247448553201536290028879027363715117097611575104474448575002325858148560788985128350955612212441353238781623331816561192925762099918516879242862423080170586007655853464509921222138629193200629166710405344413253099405031484201600328999231910328401724803603264117395877373643931580547563446116674421959053416946656368004997460891763261639369972680056711919008111646000296009990629766649508010850800515867038581971813055231732463017352876930349789853304607612670691519805210418792169386199911313682841025843874830863102275566524082814123688895195062447293752240366903159218186432340269192932376887151770807674952389899214924574176299185804334862960608936311062581001413806306012314943627930687332687687714744549611824196603717301173226721154889414471580764346364476457590703340858879379388551175194674235340453941225240645707121446590466567348092426153884175026364996764039796403950645360330584671065916086949364276706384187525076396896156031731192926863383238674463351811330830743913035433722790714103079528227165884110344434857882180908028208295544281220038019526602186359524610566606314957612702515823033352249470789046610507885186132600708705312521009241881403331109803432544721875356433943220404659540269285044885564614251052654795847216630572994563578827156077280214821750447870011247793657070567309892113872193059290580864978398631943466325792242834020275207962010766746046940717056095351333399376074927130117650602222407844780821493839639881200547789380056657803599043114871016372770352144947284480659802102462962863743293337500534210984024858609771594603462650708927576848032018361905498852289280953768213151504355825172037286016959609587642513950138220984049612226242281734043402808939972622577393303661029868199221337757916373560345378075501813255961569355010328329942384997475152433810011519501213178050138796465628491543324891943371832694701926816767596061591878897636526032086512658522642452411995901898187884508287694437676631849223848799241374007672294068073105280395354023660352099842055043059212038827556559304808391165973062450177252527879079885468508425855517383838338519943442889159122536441186696441712424001358879607219106134942308330297896634430882711074670005236299743261023180271422662226187505725439690773814742635221552448324008043756966990710294726405178030151618910268009263587698184180130345266471055199507316064326755040487453177281647974984931681635188811325461499640318314012084999975450565440566511483583871743810710444468199573634628689300271371764306960414783227327567890305080957691434783086703540161620281849114412320084399928213181848433228813425512488866865448527084230428400988313855490100379402648447623636375364651105500810294066099152481479263173087744064207095391999167556393167624758908322427072948295443281512295290316475060981071517094932166168130222008489990733519284840900143323698869379175997792387280564485863635604716964436602044525970486822151441978159123227475787721639865752760984089988937377750893740406584561154045344889826356794496286422470716132658749959558344008024494005256475311275829458245552383399388616021670954090395096229843697536057944381677828156635171719015608678501022971066937877214909889193684543867259730079749381103429453581189213029104857204996735622116733366350076432627405882954486157569614320478963305917252500296559546804876536262815497576765802778755902367873481625042453157027418353906499723714308623953642833379852588093656358087875872114167358200023785857917146541612112026034272557384221551801587463057767793952936789125329777005082262500819374164511416847373657252267778908745799682373452773273606299464292416996715508622928060803167877150190204161663020493507588377690661867461629647016770563441830896762661887440329705177881452434012512226794104121776172182888508159721642083847933339829566994034959379072820339780087960497013780702940145706183227529603485302837192226100593956449912415092778754613668231461289469981672588247118985447614664135974724040011672646403383299904015026352712185799138187518382154225304799215388902846165379294723637963334793127084664272273704354107685379121319034331192450

634520767333438091681290392671092987571771480282227330708585902259139290525897400243757
102169955326576133351851876638602761997003980523930527428933709016910236752074517701696
404723753863828765431904302903579819304468286320454301891421607505169966851233644518831
394315814046520685035597675284062096864840014632988026383254956272132582757344853558300
022255133185962288649772494481966641528190407028797109505677755838364707508929280129921
465508984652700726965716889740132432879571982172311902810990922494210691151942704477358
752026602177872997393804329178321634672128872843369790316934859245577217598633216922910
131299649345656945683126728480958429250935515615358682033736722013612851719579917906788
879489778741557950785828040051987951437931024097351375424452291066587300786546251418820
808073071926898391350492537754374420265701651485490390378491533578352391950918422941007
958179462613046216881844121746806220722871046251493876491783338925853594154399135800585
902429854085572504489429103113066841061052521529436405894282256195150902988534967011852
089646433204187932153336684750090937947458624405009441979525930580847057304417142280778
565703712794758093456290877047988346971693235516960591551290394654649194697695658010447
721221152971788542420630144935999036470488168696394545987395664956844680082797406485939
762888615420634495952047787647960222248140451871122057621282895120964242624397691077791
875989150916967488496901404178146248821899204721539789701004100445191637463548493777672
404896305617608574901906641992085649882441665925913641149797211057092004834635621911259
205315949520772857285350227717869113431709507474177404611259771054406639288875718393323
6000244502603875999517421359497976494040000414409398680931932864233231380731072605234702
226995502975336413333363768383076991222391147770558599778428742569645259730458979891618
440091187547381046980438055951700629630329433750112437691659207229530151254321394054433
778916278191406215516820884736345341979998879516117261028410632336985345662271408982502
069128670444116902582047965765068060833893544908621143873825659946434978803232721758292
694516998631267358751095484558784631407597172019624337085219967792883082041708362821886
710429402426005844004377358753310704188814221920924607149133502963690584664488320319474
101734611287867351794220941454660418534030155181556232143165747332666107989803109068170
082688732101936459561785851734505472858980078728721154172567402441979028843225315410192
140135091238671110323213731459405115614706721289593263819675803769072313032161582473040
701388589334636633597677154707019773249548814517149561588915972704031644349512185974704
146717150973113294738480850210707300489521237484215403899818595132249014418572919357094
375241592155456929631150144938470339489307624355383423543950785791770587588732868726361
377231317957631881191749399736458295599559616847144784415189854307741455943009162727770
640067845262221886063381067248472690244026426741339072193530058424406225946425394836856
547845053434905296743058974864956438929352506968728255730738865347979569737963739416312
512211357236612420140264683198752349137532591965158061938726661939160510493592652713216
922096224639699245339494168148769759450227569316017372978252259321139227972644699078707
972112927010072893164141328975540511298607130045424497219982559230173355939919666258862
848902801610297741472814721799607430468636839435837620966370592178003581516991294767315
483262434722529800380095958755554513635248529233660366613345215784920268506151949203452
902146178514203242331042284863520896879742184540038734941728320117627378226479639784677
713658735111930207072225600375074940781039463389519984544166314322973160808440498281354
303038336316353145405299148316425601251068208565690016030297291658467891832210586994891
004078010769247782572806721865866449357592377066019997260659525543327336425038947983366
014319930730848093451615088048076463666752908667169362062492873981488799043653338716396
911672736970273126537428408609734869729325527885419930190416842823213958579660248737540
654392608495318634134694686789235833606803394455761856487011325964277558202631925680997
158944893454073545166932384492149911855493382824457707668823052546979612822440415996689
237159295093923732119547894507408067744489003806244345752246115557238942268385930515277
549765454318083490238729198467486931626088717921512482924761589351414914158904235105073
534967969487491863344304793625203651055672156988823952034980523015312238521251326166449
473704612481860990143956546372710175562161122110472247926506088187921878564564770201918
708174098274263885178517823195293419048193157156404001782600804746415453642585796882213
147120219506870737039312153332239429647101433881763991811507421555422604821990245008205
203155158803107676568812198575038451204473602796923884894398504077669391919178038513117
904637264578728005664995015957625302767342474903557787303206946697620679371095314087874
660907190900547871502275738615622840311999793601481740181407268559346424708186513726761
279734277641240894070241225057591283320448767508382482335490062243196257292826480566009
677509285325730388834182425044101944383749082928907704415181513432790126318627093441028
058333197183938084511247877577905287996142480968537580976667637015694843487431747574899
146388916335043383627398851102955909972689955904715112917945559126983594293067385743048
698989855944326198964253434921711717619498688138115373601192528376348122187771094392593
220573709562698164645264593052541308176804768491799670945909756270994574641668731299851
777131558862076554331510263023608492235320184002464426949822200938856198141742352942110
120448887865176204772310072355773711756964540267737869878293238488465868548243072513224
599718195176378206516770173496390729119732315211045083889636900343634564977138841805680
298414053230978368787887332357458437167785962319311821299654426422746033116562189958073
857091407481709077707206012582553725598818255400017096790909741338551791505034624136279
629433752798039212161244942285734805540929961742218675526706638715401971649592580419828
457272339435872738491298062505229908230414417964201863239335975640856264721140987102756
842328471054420476927372279586934325516237287061306248948317683005950316273539272221555
960371912609270563209001688446422399745990762836038614515601146790867195227442253415373
563043636807658209294481681575624407583542094450414818369400724787199371608074714370480
527241227205762001482655673842585276152042257561677566344890835515904034755970552781149
851302508741216556160585427292302899331654735499079156121786647178134339282499415905014
092363201698408680599677236463118003230917231449065960183944335732467994721363667143093
322687259227699597866342198486047640383312151598246334815753891362137470506267760949391
565434449665030715756019052561493434123986500863349768772582014261603587642188657530917

4051824174917841215303222383004188066393854558891787620068788140487669276059762638850 84
1876717239068821513753446907420527968759386296574986544177629425187030091149613528443 89
2051450071551108730946649594990708997930523401295734938668817859272442308152159066064 99
6075502723760812723870585121372745528886177354454495938515895687751951802687798564825 20
2662409444861882867270542074750435367998458468021181612451191791640838822097788641827 56
8105850767756572864848283603702493287158198060435558799803757574763317200005449598498 72
5166885657063033528760680930815901814105937213785607881031512925317504110509609751654 25
3710308551748548992807927921650826702477524637499837850472341148722403887877968562165 89
1841573565939687030319350750298138289529968303573043060712075466299805847951077322904 19
1430681628702950900718814134214582841561163276458979779431852446703335722015183008067 73
0098434281459855594365738971990326286100716746911509026594642792375562493742351217445 08
0312134998741021050402625411576311412306403373840230248447393613277714317783264872278 72
0000313243799115845410732008325471765533577884197388111987830811612825334350013791097 32
6458045675356269284834551025317569761378314436825247785430693706314325509640762249427 09
6972762106167981630745864773136210291691319019350539173633877209593077288021138495225 30
8523356420091475821132150814163455937327663816462099641504181427926147848561122509697 44
1807399401218649576170877429853908394199011888587733637311313017101357779033475620443 95
2626076779765685385041517800286220260173983153578949045444271657055964920522231883544 74
2831119346960371194121860939647436968352163008411309212213761236193155509118775346445 60
4293737921516689620242547168037781827463859079682073564093429943342717920802887522112 54
3317901141491160047963896033187722047145519259305894869335049922335765207063933665786 10
8085920057759573577060563469345760388491080506695516093810694366212875882733161322864 83
1431471767211570461923561465003774053872176274111366017823585584517310029820778999364 68
1776875980577196904429326564149288895061617432739545348233166639979174849840274783540 53
5912002226094399053120707660196672743214667313250599196153749191206109264878195377790 61
4253518922346613960953196062526178425715869924378266091617174649716347204773896131486 71
9429482490291989419167583088892339731174155541726809475331027377979970981756504505473 60
2276786210697540450592614388377815161792537901060640229167380269625734343046453004211 04
2527662303055207247573930679272639371318872288012695855490424866322830702277401555280 34
2205573172609159292751328720443377723638154660224262722795524264047906912853466474395 67
0390153666448251186234027804025378088666113535664410691376972388236540537057203264851 33
0711800188621777680597953218065436753210222504280004399406185181288953614073372395066 31
1517070005713863153021329368553801848986969630285108930120217950647072487750320999483 67
5687172470029055814569840514467469450718871737636802873473556196853175307566120156930 57
0344309876149723068952866444156407483458808986525661664379720289586844220392181943171 51
2756411177614756371405936864000103588026389125969238170622763716762874806283816022759 41
0511462692288809129433027766495947249738447309337632746003710843590785997667180055868 70
2873018322966729256651195926100594158100365089290626039997891076469310195227174464519 94
4361699915556415641215108714382080886807522978508148022862341353184392056663971152460 89
0481318445192314929106328154027922489378228251545768271624596117639566886461742395371 58
6574462664399615547890516373252182578333253564458989290595192605865979867134482744782 62
6667898419196273605935202214966815704365569041670825752744588175728116095614818572243 69
5464750508302844307531707792355713293487611783908130291059918355226223746867115757059 37
7490937975793819524733163226623598269569980473433440261687965475130429346162426613460 74
7325269570311488146969164293369071948154548179082929107206942973187597197310154261993 35
6461532836182287015155903310706146530421700668825337970132344950607141683526860988131 22
7220540903094664606618585799991415397814484774156408225890354064490646351061543371940 04
0138616035071455973601427862345148657347962179784675702189899513333644381929190530085 77
3995045234934957189684612711376889575979332349533208953814539846770285124109139999624 09
4286153561549520156418899621259300512644209686597252899418435036681880480752910597233 60
0836548235701919868550926035004876573788295162923741832713236768658494640005967095067 78
3453610036744259491885819559592690251239311072595121211563382415896067374800718324687 78
4130780969382414829151895604275501754206517442088134014543607071355602676349959757596 00
4103616096121377362182022356398010145592493601568971489793336585499186349730410349500 79
0550971037329489219764058869953201896649335082043100488523059429848680178555656453897 15
2963868713982393892788628313053889870441633874853236655056254302382861317683147439934 46
1560931076538494758464893162151583588989339195673294433479039090964500201525452974223 60
9334873774857090601864807051699125755933251820304412057338911692494979374444181721018 00
4869527481582486075771221724138298525297670352688504213303463703205901112769270842312 24
7374039903446761895700102591785896614704561188690554318000135741145453848091623845601 93
9821457698015403674473093324214164727555219087739691741737350641459518460785124018137 74
5458837629851790660942517996950365872351329115406941185580040575610780435791910515438 95
2930178607056885781172174213915509532072119708984152215425316479193046160498117600999 40
4341319099118921551365122611550181311073519406748964186094028486928305502211992434386 63
0966122998376165898127473066900407133131532581930320281496745570289271198050230834294 90
7246105491095799550789366026346984656628188055490104387898957440931415296414337769026 05
0643640983268217633628709882627239743023005506385167528922648375095088613721983335346 06
9848906855685902444678886336439604378182364931607506979525366177704480786285210468209 32
6826682897220715910980081977801926495253830472463607958939173700369328966358022050659 80
2853387029968092228675427129133869994026633577360863754047202114992733399559638671394 14
1599506355503816227131799287614329892459586632102280507272017663282902813951362463925 98
7940841197742421478497488792853481329226175804296953605684964163335883612464776047176 63
0339853772671737323232435192979733423764606700725905697847782259010224718618495510870 04
1401552763492243058506497917469994122470166700310104432762653099301528420684246859523 59
1053096968105843118551037608085368103331709534908134883131172235937743874146218392650 17
1560903279402818993561269449639671433207829047319166667808518255197717288006277354539 91
5927890340107862889636611570807579263712515753212564345879767582279860562178539046344 38
7826022476983164473091167731376986543944139748134480038182981037549505885398354291463 22

753291226062391782931996213986918817711118424419627718789923057350447245775383119433851
793221285766035212168779011404776589767784356963513653291514930963803910475450116996758
780279997955389558550059045533297935637026407703334811205596791096608804654458199117569
673353817940202977420844671405547625380016657961957199126200807816682028859158624857236
155994016255477707914111600676407826080771078947343728991156761306850732249631591231634
197588462764728819202367627163751947669533254204908916102491648373349659172708001471152
710129089029612110404724620656228209632836266708889728464849194505485241475581339237736
269212766280901070396032994626272509471411769121429133539751301514317746716858402905968
622217080111036660714630206206422073967367402754445111531868035737119706126321435523468
555443824565325519496223092442226276161810763532712184867110387486332416707890468852233
292111150197900987237667401554791675344748589116281268686736042229943560768269783317351
7639411375681873118531093914733161347146429574802588866120984333362644789232779921718938
110490257508983329575231138511638411810192449913293008778472536273658801679273239119566
773773460292311671472527543877323954096440741744930881033569016899447326506293568124074
685916892546509211091423164339664349653553999052260304711487117501951086036214378879284
074498252703324251691779534323933805375341542863344920057275796819187421842721885884666
266028133915912265087032955629743121006084764623824061202097408858510971345024453456269
674845217493795199836601359599598044210553393057994635412156592603739545481307090026816
164735807530907005794698595121857669282043359313336580210439358016107908279426644878203
528015749847777187536663886871469284922335979702018592163752637064707239232807117749755
236536241706263154632700590266304024739804533530204093931304973971307917181514886323851
603514091871517272596320603977518189877379429833548962149298830651687972617323342951860
291979123542091466176185808120657850975540518126245478535871423498722824507628021855541
643937355728734131770795331826410695802318126782729262172479047867331323026028790147648
543358099932443723491884995859948625830676000122047336344668680030217744283089567321206
573109092985212685308293535203316260961238719270474910316941151648388474797456771234335
574429812684461432753371060377023811587306886288969394132363006060504289965200451060374
867696136491725117214171045397236983765748250928625319917610379605050700474527519870692
438307972081336510745808625339870452950365773947943751943255366001421055646414822436061
646770791716585117656108592356346094854976447796211655113187009699029140731514839039089
918159185783326502779539578418251970561524675181074563304570829594428891506667159297604
128033547451551004399493399113574003681082145201003716633376952121337532395906445150652
333790747504285781596952756961817847042381784203159924171121572817531382552899083172227
080319334018499746246615068641371786793594805932728519643357368802741431586900765208723
454663736398318691202096562075413488741155043517945705202192086628621570465012959513127
937440724676204192266556744533344472968171487354493873384801665428264237833848317565438
333617440873218792199714309719390756152899799193348168456648698943157601438028626335331
361857237931672366063675494380052529671399740350994071219337375857120455594960284445640
461306033622263621629341224576151165419387916848132809624695244456954621250879118935398
3221963789999498705755174877188610510452587091200155027181112140083303394599977286587045
234191667304068557004717286117263358849682710717450035389033631066658091122161122795352
0597356315423878627922117400279299276602723091008788896448671977510644852854236760680678
328702716021491220890738359867916779079846546847654432886332754592689976471361182191936
371970943091897609589330741950915357899815945626817403109118621361123870326632874592512
380172218592375964203971780119733013545486303115628764539733301035351993689089171658211
844720253940470931783306012396416727093121636937919332391842597730527614792293021230131
636529561376233305284546377449667838557241630555328610532755207843894044247233087001494
007564853949389708563666247235115549684263707422419853407218843317118086247851099998176
232258058120204907270236751559960385584667283973473259596127104496948996928070408723556
135501883486098273344942119279511596389142170133713625405959158400657637103362185943540
907214950797192642474168788661350962013130319398165644318423191036741420512556863328098
552077093239955742204583728924383094811084233008764153663084724168976375194193998480863
927695317901643727802977688806162490841933764103645096126040651273694733432136475166867
454187542353324904525140012619910255049422060899086534891218519778520803538297935164736
163639485284975628497148856270364254376152530348567914218138341546765630362935943271568
888511396453417550113555234226609517738178180389386443090830539927386531988392370825144
349766957951254066405582132495347608244642379595204674037169104022865060164401188212816
887278392342736929260620640964091959614590431451723416161791510706177671741511297009743
626357169179809791310760755444007274823165853639170769125919005551128507328081677051347
490741450119502481084276777357730810360845003755565026865827089490664096114629969042922
698380843496813891492479886224871671281240892627970065093741291428012018819220654215938
973633819322591270713038489421629319110049071492253628218620356176446854469959430764190
727133878182633847902690514134885240883415970409316671764584851653904600109634729323170
245268608078649180077024542605338592009166331507927787324832590160442171566874940579151
896771159131892750178044518249937438743299329143554374680946834026083464252681707351360
267844117117547680302578284327412712955509267108574023047469600264457118930180581121892
5757250024179106647302011294693754953338392710767838158555808875670613299964991589394990
408749778235503921051363016467163408622693653940345676951865277526856031286808815689169
916046013679356000288784865017387036118613661682337006376249017187035483916530088806575
237376799068155478888938646233804336788144738626369751444635331513645033652509877954130
93994146760112222850127827345575515961984487267288862169113912786444182650107159343331
816055288098093137576021954484236689181404876129698357403680117551891330057226994759192
287243969471072449770404732967513384853728989198514487912693399562727628630157178270573
552384501936652886942503015712886490989930558977451480649740071081376020676606100283353
983207243594567205949451216844025305614161150472376796871252693156319309816082329795042
589816674800878152648677364144935695842879538795111120900413882435069998882091565554032
892502288051416967879299266268622246705254906674953625013269700318245101140735192981527
091168287631615254533623132422680452228896149709173971135352554401236086188154541470853

pi to two million places

2046722994693907148818860332682826172282696478516984097556132809109049299420589020997586
8027011829714381130616650165606940509417447084136593172946036832314886783783401584666526
2779381103471856527342901126469689951352204381388359254084508757429340483048052570263
6746819999711139249943082380948147319257601152853824735720831491052716081699222814186753
2991179552447748792024698247835770179058176843376667776890217764906219369958965467659969
4287218010978136921367446220974783004092718190513763561232548612721452226168051802932
5681831093141396659245310344236884339706735287266383000454195146442303262301907189759856
124702358650054207598252489819907503165380324950260169372305831481731475243043594249891
4879189062802634091227267353344853777985327688970476167261585288351406035252708851992921
7133070578576387493937455594009676153752177828011626903772652898962034412615988106321
6825320644381640612917117212009556747383916722296235557461243901559905448832262644162568
7126870485003449211415757614315487883822624493825719072052822435654030668643394952786639
1978261966212889029317080915069335476093630695038779648380650097087712584207442114997
1698556158998974787651375057853627245365217806628977750732715703498547747167890295666395
8351111997725430882108300838719703001636037548232031811034519634199719570801626375425606
969661834362972690706622306143131863618116113316841849516129647994635408155166288645312
20105617962381014438462014132524685102641379341166216666044355543396726083900293342498
5605923047725430160485968987816153242523488947992749956804057508785961584656399688277050
58248080375262444099228426558107196531396214742222341535077003136186652290242424273397
5223220119730089596891049854054474276975638059626226908788476436765519375681951996304422
8090247196597798141122997611309966894840654703043061615428405289846055561052774316709
4547976542569994432561515127041177684024726299051846873938440317490922778671374650487756
5400352618233613582209691595165310030299470261213798326995515479430045282504041161789922
9947911176412173992693774165820202835024261155795357710192869502646054359241180066807
8233417498334223525119403957869035786809979573555664634818410923535663805321625058733961
2730165179209152696307741603539343614876508656958944166875931028197227084213006069890327
6812481364340882914506935350078426900283389692890036766306519621256911370825149526413
0730020572342600614347947841846620763374247401965234906393029662233773082064022870408809
5403944892602375593027578381867271119555903626438180369441026989560997022402685189290570
563411576345663453530917836449127065514652145274516095709269601981935148250423083093324
0208569382325737324655619783805079823678391489644132121190325383719305126121435120543
467213802491720844572406756078389118361442061721960932418878715390653119345624231430505
959758138968001459327268036990315314858981784218414086270354132340571406372423344162305
2011460053724335454408580478491527383560537008329841944194087857728942894298905564111848
9012798817424271309417325022464998977618499584448243196333877136064170050575881120626018
9035461258593451545618175684097314733842014951893758158996012087525756276033295003011
8318809564291086792993649140874263226672138684915224129903291462932026823734909566257903
206428045338516755725663359643282983690679715448949144144284457366131214716525772928322
838722519122781850333184575375231181388910468730112025332934330332281767444790920665632
501883887499178312452779568780325185708787710821321817542299137029990346340824319822001
818143016950158675647723184551735160193539741180681625549863346929742793638368312286209
0150084763296027154205540923472197748755577372771253584379299733675504135390096260754601
7704783200920900004370304772062396931123619969230694519212280751280626109033960808551
1993936257664560584547489298456610516437763230204762933488331366455334573480473571567444
997734717821981573926294356614853325635257380075373424585696273226443292539121854835008
4718726153761193599211755449468751722095340217149673323008543031277343008442170392235658
0523746997811952384744493338373857748511427462252203934675721232785066105269132797730
634628873726222419584671667202215168082910005267022364151265227407760046197949668504424
1492903303752615324755653009315314557741560785488843720415714060087651280761331140002151
7609289824898629450626479863972781208733447929847854531512329334051406847255746928486263
150354770925719144201422185887802572791283311779822123368077931168758654777139994623954
398600178217140445115877933764582521759199108819238300516633102828372361341272140722462
3795391293388364187931553299328948798748615386139152307468917410066261860777226791348
7136322147516568508441991780694861954601934089370819232141926382775337591945703264502363
043475687173452958399553670973947311374513943328197791122226939725459124938379823126607
0963822259670190083814532862904610606586856320978015085422334848110590617385229862052817
896049500732570427222020393613638247958310354325985507262140340985962778601721689559875
030328828176804094685209388640336365236494428576533381097953342025875230660994737779174
834099640562083733043167671087592982666684354670095997048589537484151152214502249945441
5283865780292853017658562910138814417266938379020705003419101213867913463546522874814
0715338202901919235146721268382751000173948051792235759103106294117826715838186378195464
884312297363020759072949613132264235510849102649984741887018127403987203067935831231548
287878038686720763454984951991134450991244247310505227252766832066034853805673485126369
319466529925162902626465894163413960915097218723640275500269701088386832494142125712048
869645658296361609865368598837883902802070607029639962089291692420117564629212717841443
866094448415307132753827418051247560470084561419607860495448592558130716152717681871096
104170286462445106386992799031329802393832292307860024611121256253749299206962360554973
977933709055091506159958074626476930706146547336572953880108465930773709264393270961733
589798755133298517353358057619820375607173964951210260568242153539432206578780654333681
668379183925431029629978625583138150842902346041464285063318207802667408575042965493539
544948651852756470881435132319597349789917141516937325688338933162833896451848870322639
893055689451839191243082932515654023675385004309455227522986219363499930799560689684466
187459894748823413664085188532193673114375894635657021422230371741481201272628291057331
857839227334795260680041312240444469069570034326579109561734228465513830287770817092800
437032752644557620090294898701726471822893276178823467995953896680114028668705263367060
063042612994608494995638275599060264777652197025375830641181461287543876098578289963422
105950225341504398260961876098352165231654331697721441251770038039021598137974891320292
927755438711703391163224807524657249729623124765093517943567483811431528641333029089122

pi to two million places

3777146612469044864551164926799346341556211882281756423024051694895444281683141404904380
057886059010737006718298499365040749470278557386272032710842602732695690064120155580946
913710129842552905449576450645756003740314945879082105473559113639906727806481459191706
433870697147736652477844338630255698388102589879309501971312840708918719696749394002657
194057221592958688345786698103181835949381027193116152515301740904031945172383224596330
526786264210007457363367972646143529714988846055291907822957213456926463834792175940578
051303673488795449473344645606796676912782679904942003628806990026035221665252664880972
246721212946167822822474271783410535858490938180843820769671226221556492524464101160066
383911818308730856354226721501721889134911144340742316720185801544096839417218455292470
306663317439699203209991372307939208706332681495027024183632373935575659483558643427585
271530364753467460118162312180861113799324835451482289863062536933279374737264046931267
375653401997300907614262122865011585689448208037142836120485831617475039077128760465033
612361352243121420491140962045858292255435749009027171143100562027796642732820368408835
142189973676612851541741701550559669295433553384988687023249020610644580716922863343391
855394434659741831033154532910259130360646226668797794557349045467488232753173759959372
322731037104452113311533828930424773972419572744011654184843155648940489213580557085576
275584955348891913856437916383424089396022097880195875047614164578733843443198087351575
166749682003791537961029734944321094760732700463633436612590711792603829657765048983399
682005284642342068544946993038712496466424858116044200046669339857416855517298369829263
584910447179338446832504338447175872526993668623375707985863799511764743787742210295932
621738817179921125649607665490503647530112846059719986422397278433919677740389582319175
573259941937900854928259806607678949854843333553305204429781468642262154639070566780479
389131776519220499357616638821963223572241387580488187287554778343055337141624291591814
407249101833736072586131305858393796369137316050463865378761619976568352789603916541221
197123163706463843508750588046575531967200804810632083118215379561380098353559526093637
000645317080644202888377266908268009424750615773653069536999464734442641799088072365856
916238996365175780762373186136628030006775952545698303593502093103401066548823876059063
096671525803190270180565107741796599641778895066406027884717068077927555703510222371473
067950065096075380534263982026154071272137856032274328861680241733894597905050321379748
466149030953017402300954957526179588969836097031429140848045838420177059333087278988292
106539860854978417702268001994317231256072796693509378461673808145347108132937634521964
744163193311786906499824823727616205615024443944723233791069608396885603267436594476132
436686239105834352637258702655272723546810973613675379988543402247829732195864738470798
498514172853867527792306584091743206050109910223892981893864572160416894923402085594048
059798887199075389944836245759181795872647854824368717842805118165701035999489616756458
144117743599941557415640541980940777060781817873278088392351665272981172947045182489488
694025397849704040125785017085252294800326448553982933954102504934105444614356130453712
369616822024270875468032257772246764538690691735846329099659789270857241360685294722841
899988811197694925777567347314920454188249935386075448538327349316024944583018400520110
597121122488189926014090339058430141050559807188444154763356093389295582703356383918920
724411566241363467937554167389089309186860803126378923091291660755009898084043087717386
876849306238533350915041060003830601639488536879210612389410574394034606240163718548425
217716754516397600255050226439611525994294308693498690746297837599701612950308438036606
600589226585293056378866958466784875726002532918393071854726101201435318123008262824539
075652638481662843067124140915353230173735777223170545453318573303986361162909280796514
000762580295868325211303562521349985400678329057981002626637678051720624754016353702521
682187355287204019963596188736069347306728409608128864989228165452185240832827912818493
863635272203008598275445998989995835111574368787888127048557173814857403078036294204859
420644334157901693839596815335852775087815743971924322779883170605463400533096961159954
373203941299551771974092493728193869104247191680745755805413172816833655379652759510402
582937660069379484763050243686693087498612911515579029908914751147143616550977810891689
159389043228611630980896160154365423970713173398762556138393349278906057471453816915692
648820151026214721832503409165624542935311732839683741755506978877246043985562610853373
740287709972880476114915778576510475290891138178065469222072171325415946797805559574054
495325587792843232475048202572961072119305427203445431119018432651599832951192425499568
866292451206155443548518778433760228457318552553020385780679964233347394328325507976814
317493529036535523570833622729540297603622459678702246796108729006536915811032977241127
196871846317120153108720228291216785136832868288998410063083059699732951240183437928080
758668778898496043772754027589522929366853922675139928237161596447373298270175090837568
027446669159111149779944667113569108892437919930942472130807308198426259314434796579085
670082562885883611446330706901910636068599518538704171062385680432411229940699769765217
189489934971880450386432175982864331340232731735034487552793783486413104199496595525770
669045671843550215620189672793973426821625686085922488113166647341429891013875712757048
305145943663609924661062720112440987239997104207565439150686310201357598460146730265119
903429865063967600069668795828289243397825905874856782626926330346837233212406615776029
515653722610682290383661336834150499989593428093201865424703607359076560816219599759343
820172461807695817834722127150399123937330805981643494623136717495999946304211763818147
830191021334473569265625880571016446879845566172037587428149099843393046523931220003564
424865028020022103872381508554360806108595381742785324654979231101510181266741384662946
267400340729092430677564917857934277516529509846000986282198651935014863141311338234081
864181019598887229593585603437224236723960151462789656548533533174007419842426013606673
552984075440257317771495402192754876256636632097951348389232398473093428272993909549226
286852582803762571040814041407092153377982471219334648252071838785374450072385259360567
659576220402194519247929124130758546485918127845559512533948537732743954653252016862250
537285001304537240004647444790745978251029444790475972689949375374692808933115543550514
205161236368344100734984299470708653487282616826119499544459888850359607914367111963913
932091120095413351288550899249339285379476656164159254527588534790680348593042101431778
577117245113741846243215533722405654121494232234673410032864092372275714731703809305846

6611410528665349292157043843719387582548918498944658974892112398043559253649190865890 66
9173908088675009132330542665482077157364025208162483016589873036086598083796157673641 17
7362734669761566653921348293456423991292807859979878815204292215190914169078549736187 25
1693099991322700067499723351557657951436647470237487661496440492613486082997609783626 04
9278231738894979224685209775995080498826972392495765987223064695118767799160549567269 96
9085152582265295227385885439302173427475574391874411376633994128594831234384848812794 57
6013671006676165974395896325545306708168429451211409122120091086669899891500102055692 48
4852372255421310716613919828276574298188291751833720841752386967682805910231519925312 80
1445377221647436825950860888636443467208040799574561042901019656088083930982871606160 49
1216360458690862228973756455741357430715910893672423316644773328296824188314921716494 97
2514011949369056709529615270432919617564101018514059608395422101125300432032447729045 09
5686728686928379789944533473254078320054283548804513088741363193695581682874607904656 69
4590040744288418738123256769967164692679986955886052080637298383211238624681202881704 34
8155814062949883062634593344998884038652605737422307385786640002377415312885909455125 75
3539336940869444293940752218284710010776658095127567020147754082982594365539007779061 80
3003710403019109218529328478241165518909228702901240416004521493170935775360816342085 65
5233201443845388958342206841713882399532270635638724261133021723608875319692482011790 65
2227508481546536063468432083525195108332160684317334365846805913015740878701822959876 58
2283002042528355669320450198198817145826117159840001172323248462236802337849839057195 84
2083394188330250004100260037883422114836730547496096779242970449998379947904405434971 08
9626589676691300284990960386063046240093337979092035755162551666400571121877172390300 15
0396095405184581699938643044980401039916128659347449558276066834824890933733862926698 96
4697053174156089229666242891438192737235672603030501103415970150390759411599156179251 16
5622892441767557720263927108978526059947131533570045904830124535860224570771605821233 21
8529587582203051937289020017734320619428734214752378860830070299796531556810301128799 25
8939183387796470067520270368877240584066436919090287438763388209714580101749510134645 84
0281278011316813989780650900740767464220963899804533262076514960825977452275842390413 45
0268461861681457953371759462268303063666145365992028030084325285149817881771272573867 53
5502851338367923056743243686962027275690495694721422424679884360411922631691556788276 48
4222391196274036714598974144543180061688629337635623975254816109201806289442065086508 86
5174388445174402936157089106653051819134408352417385390895294733126909002288147617359 24
0547275574100872211860248070655273478546467081003325288049487281884647664513871948464 70
0273983663967869110872249068944525449930136135982302100966496626582497907417933026044 79
6164678961217630473547094105905476778743627698111464819594676539533132602160451880568 52
0123818538359935250905586730481695893931266888871072451637580786918529804644375984939 01
4986408867291215615146935054468003927753771628002844461708872834613320160278469351417 10
3718981359285654440473889353364342252995356306714864357582266150708472242121395749058 81
2364726080795665391821078069759191962729961376825052719016801350182593650390431489237 42
2218299729435910504766510198435496771583639035605090274409454737620008662551895379897 39
8695524944209452836893729162255864458818572322005097340211242024270133813809750638707 36
8786224134616266076141865890367575672804946851392949246947497670448286278503799394283 27
8791220332971397543843644722779419524830053313308326594126816548143183672418519071645 37
1183945618853767186114634510098763556103968824034693227431638685638936692078262878666 46
3162305865623208034467033224148965844290862011917977518360789811784708762629615319400 34
7815464050634565985845395933678392047177816196115197815991533483239756111621222104528 96
8308713835459988065857801354859374904263956020172957868115494079887899895278594495312 91
2475824817137108859096914070619330361800303323891913216840242371178559414793817512261 53
7355292820484620119087835578241076795898728264838018836306051625874581323807170212707 03
1160159931956653211055908684463723011219393528829932843569560659719893014841894192469 65
1419504713100362091384684087143678688323812487187380582217796916687267705286949172329 69
2975741293715765031048614982649963942542515355227892655817659328122813519904998628338 91
7695098649870938852865224641624149800913360480941616720693342425001725333590241224520 69
6627428380607915709746101932343274422842790300921971678197965979059549127210553864472 40
7600831000588081819072447870343657454279475046660211686153282079367042283157677410978 70
6565288995809215120790091062488938646517486033663048808838585836547541959063529016960 79
5986671979195154726753999884776218884785107366055679223745515800961273463703729547099 64
1448954403570700509795971249707914989405750420165030739210083757394132816578085719802 85
1130424796134514504277763666054870573901649799638800674927633569990174214247086042763 36
8701538895425544860519661560114570743110126783606189763376540859558441673969989899171 46
8666484090241900149311117346206295823077870579486764638556758721099513646830997771609 45
4655720168122853937767374099428303951557494531692065820371452045827753578337982711615 75
3554754759598809028882500151326903062183753558815228008049946219926313951475900767150 44
4120102864042232346757214652225543337464540769554429637336518329440811481655311231788 85
6853449362565109233822325088751970064021785626240504392039311551274241981278661180457 52
0379031352263821500210772130502406624183002862477655911114130894749764170442763287774 73
6669514152962797298473622901963622315436319141841799696896280359277506155139875578536 72
6635378140081715318318793314798031663073541838249854774614347582122035033039491346267 25
0843739731331346134971878652215484795211328065589745100203248729792239256892753749027 04
2598668561490405375284504662614426425991229576994984595616886612934274152168604536950 85
8277105538068424706396860778132899310628551128799543943367009919152088861445556457442 27
9253309475103278639998286086647782469776693646959682093306304832523027286162884091854 02
1075857505952335491745175635055894316749129117086207384696004878978263910905625739599 48
7492495979011090191614659080207484962639358279265058364767670838301196885550505186157 98
0972185298078202754679077735284594885548642095718480957735502641837966220105606201767 24
1016475961823177144419881009610279477760819625624682085457499375939175577255504390164 42
7090909940333682021018189188094314468725119448839760726106428947637705083816809284728 70
4530854716107278666310122036688758290646249653294421504042608960795596028483768115833 10
6563981887154102229185636375546808614676806062617591254751326265896334165706265117268 17

0549301094065863026692204229894830237644323671421946364900320018910295105537319742039399
9303809428707866552914769288138564587814966477980823314826640216655484671340624018597144
1217542440978712771828778534134383738258099537774555668639030970006149284077302470091722
0199524620624539198585923713450742399985171822495428895132343503182334483788319295955533
4879010992117189922536529429365336258259539094655296358137449729369974651187533853748700
9421770808174570174225160409043884612157412445285020237983903699911389696734778804970422
0888639312843891579868614995357206376948214920930628125131228080499666262553224283839911
7352025667452522990840994632586468341113042084589880324142878241486082621057749475330033
7706021516825542168588825520528913719387724986908202539336294047304750500997044070946933
5891969900533478304463581196491048316381606807432397475188737745048539320801189209217622
0032541285900192128507800878010809612186997215672787878360378342850502233591047386100377
9003368195821534795312420332192379118697973810932850103678278806081274528288310391831577
0441480372715269111958913827350206266167881367389300589434987192627542175867837758616922
9115641969549780500502717441482142253177166456489762559437582676430949126055292857753555
6527311492956087911082159619401757024920446262077946805377695544163790238444286270760011
3358829372290589561353109924487923773970012638010390621036297100540090233266205286855122
9238893654007663966398392572450824496898926245992439438450876557890902818985683433451099
9619638409116437670596054841952535056878723052067916205797398008695875004656119615450400
1629767770096547018528433497764467949602803722422923120925815523806451507317582639784899
7165956107629449587379456851984590609369134973227024282382935848347620097279573203973900
8244049026524593739625729544911772960859885910979831916191923715574314777305613275865033
0186712879223339994092210626790946258508116928279669238172379855127629935238608831771466
8972857155910479691135290085367737548995512471868903957446600048479540144780288223208244
2567018451147545838594639977604121232182945120720779081768233151618455421959874974455899
0151199270623605189634073539348756409531560340706935958257608166769558749209824048066655
2756245519232200325145294175539646827190235219576379637897594175052158675420192853655966
6620145045633855278357125671205128227519609295390924578418855401271282860622031840304166
2523045733499198620336839418549191744312814170274683480533123636546408009522798679980811
6786143876702299176420421935770303028892406338233711883166689234725665314715426197369244
8818567744423683067628580803557766708524939432726759182713058028204744986831092388451833
9656929128269141358022850879202638155759445885778391763041538247774502736300479676856899
5905742907753282403183887419532847250575823699898418557360973793479262107564800127606600
0568485628952962703650334072849924570768216606004308519214499399461592740186790670249255
0918808425497538250000573641152765627882068316837456136116466105945296098764787617810657
3256994413006960404412757484805908154315252191475930781041409918043677066395667263443533
5624561935640751356546727167909411879488480868092309383287382250042851308615564673823133
4489119765303305903494885070904364085511708968159681885355596836776708287592695900122266
8130634079052180831417534288387501316529218766333357431437101016347796658883618869883499
9832898116400702596550204761199068845930641188013743352809379510983052205865755190255333
1324739355184215726088231016831817640972417966428217055392357331823599721055914675809900
9601571254536259146573583451080849769670807172669437183081249664139285293242058148352266
2744693758585695716870345022377577778359815858095456674554627297947307975693205085626400
5035283025567228667754448284086209981389793037092532642596403853499617054638608372303144
8948100137199901319701262801084969868327908059952507493775423599997878374465778371237533
5757852677710638143561367890794737247924261815992801915542363584092241092695194986758377
0140080691259459445559254670473203928004701116001559541702028795035929201770396860113455
6304449233397743545571654572714951735345290462215877669321039725654053386823913058096600
0212123248311387049072616349881777525063398803735283044137885040535291419573448626221488
0332949544888908545079248054268374194066918855516757216622110920899731852667678293526611
3290427617120717343300234525891953735574750926874315293634284496407717856739958438148944
2104628518103849643427735519244385401105600364091860906598300233736845914995840144499011
2936937258939528898348115564558106103079454580475664650489357678527521118340799459874422
0567550698461628297481692743403195744811212692198913243550319837054664442496096363374844
8655871436934240008426830694664726867821605430760555551571130105499636942141201452860566
7154928145035636088579354204498831255719599558707858336533055121083928499841126847905711
2972202465501387082052447492723419195036030393946034767708515347254338076913543021032333
1182709941052543731637179189960813384440367350920911106317337658740160001869730425298422
0244888527033172514692497475393198235035252261761609438480970535124570875131868926737000
5074427741207097904073463122620529000639189290990331964353335337782773033800956435520233
7281189341422221316384124662187626562926213165374744095230545401691905912102983325487388
4431499690081791766624445625710550014060366800133249580906410283487716419356471430409555
0576380385738522098716167959104852220807083954438166529535008774606811360273082488566266
1392866772837170303237832334671640641865105991624817670706345653146764074492658990400244
9294338203476654827303415226579042351013109297567830484813632976397632274673272990989399
2744963857214412908487064480704661630671832626973018955052261750367044376560638608975477
4809199526394403536046543998237735619024650269312958914022210598873408816322256258686177
4453244115894930355509977482474151507373411947573337355652762741851916424514620104987822
6294536895088446232117635800351184183592675986429712813985384144690969171907995167404677
8189358750274435035511807996777624987062847592913272616884488849391411287453205724835911
6236067416163446388013123511612633214157350735873789089864603319101047904951088053276244
3634380195320531364134355459364704198439973273192818302576008576753025832169671464683433
5884003914203707242670826017616253390298986715267562219813060352959484464640403968856000
6481766109033056079065038768968446731694854343891836893537933416898764610405060241093599
8555326643129997593902636796969251376936926159102285117216541488921572622359773667701466
3984585521047074763889722549022179557834738536300819334378988748156802769759994951212544
4157027137649027751507877994910956148957426227398794033221282575324578456195527149717733
7489231107175800448526108753397144093342364167000739474760533876638025429586355293691300
3476986899061333624590847353805299916952374065057657039349015473906556528925808194469622

pi to two million places

8401221392455111987603807405660994105231164360192568472205328510725875883708878783540 74
6177976015317304950058293812475052503021700236109573670907840223511622259723813014440 79
8479181330321054432443110271979100591373809348377586361399771943672006559245699381470 27
6191846661211804296171866528310695776609243771615983151245936172801039016365520466193 70
9252512536313965592011778214276801942804765865348581587470311992677097913350491107205 16
5243246231253831574487512954156035527502636796145461093444168535781027313899699954686 73
4351810344325525951693002330050592572799781590482022358905260479203482507504217173455 46
6425312528525947844068421333789797245599838145290249134127724342450971795118644615062 81
5283928650237212926436840813268942316931518881895385268180047631030778038992441215187 19
8582722254498999677875145879160239707666604413330594001772519682882761813890254014214 11
5403222188075221493138730394389483863488048858725731296054402267540119444344271239556 90
2379071500713861064198583888795688054863351728084446447248132772385956105213013091015 10
6264293276316247222859852571026371594629994013226550518804465739898003775442184629979 19
3304006051761618798602655576076891495679624033020920353382300641985751212805490876812 64
9998662968202160083935774256923900145096765518968300114985803950066246338616802255066 87
0759127483197530001455406015411980784113494829465680601707418219448269420291459193117 97
2944253523513933101121868831678002326637691951938989304956559636443049982636233222213 17
3795863375272901598026762438208490894364283124967962162500163529966893044625845804167 24
9071090414279544743227764055860644979937748159790612922202930619833526180042611796654 96
7580548290181068947372216120265716263043635302282129337245083943534370967854324380583 12
8895052786635657171288036985284548714950798562466555779317050790289986859714364593077 97
3507014015227544766934198239263898929453534319021801693875850287786879702046168231973 51
9928076697586512760646823839169601486711150096038459388200510615266462562727086373994 71
5025771810723089620436264155255715212790458354295905910120518508605199833582952027644 52
4251235773515361451329122133578341196718707585660635000297664587218996568468098354225 55
6797699786152958316206574320737210996844061946085275213997207746028379618294067782659 80
9969835866089743865103656242556250842354354563015121971111673283365507058324791727356 51
4276941098498635706661880252843453611977423490001604109135980658253251047772348737585 88
4666978580299992241973665041155119628160047321576507005166289845399637919414471961275 36
8496384841840783539219495160760752984767084138274604403017707579996667675686125361051 40
0391716817256780501389783718658379689761501720984806022721512087636711163552935619613 04
0209273964185286936047265139668755630400875358568683131412868609228255512422695956799 30
3502490113667709364034990374845874198910890189457051985781247844035757867139319708554 98
9380999706541057916902095988974987384472726137243518065616499316389735111970331039397 80
4092809479973433772650212249714340982378236519658928862792232779903941820506856983767 11
6623772151441202276633794917374373422831797279401193745390490579714461718360256732214 05
5219462186285914543896562340248945379811955964968707073021860780513193094678528484442 02
1284324993071541356442307938627252658520849226948439939485305352727068233586394860817 05
7740751697385212021062895941771607925413069913460814382463286693523126590734303668095 38
5956084501673932290965428288540978637787221825907274341956466116595934087134481202995 79
6040005764146848563841920840258326855212379549624891158626009400987654135858706192511 36
5371948148608571077083760209723746595531144030733949423484486152523722666253172090816 22
6940002112275918342552982816899719687610143851919012898807424280528355553325232571548 58
7561447752011946801115509440542965731019362159591875821948220747561530783348030085499 43
3087398213402707000311285887927967396273660920712697115138195377155464106337455584919 63
1691755255599189224089797833124273545384179602784595898647060095284180866467711094641 59
8431500095974947022075958749369609348923513155308795226753192288751932690569926959011 24
2800843780897580223727211128142771580921578619031438232822215843119736386797272276849 58
1363281470182765236169987038316548057146175201779780107490052009862225386713437898798 48
8104450302717738606821067182348566610328159840918571490847707412577372152962362895142 81
9394934492310024755294687688142282688263381992107050031482296907812706850236096795245 61
5631762537844390868388545471766250548688614538858940190650918410158852088336961298773 16
7275269197562386427609513694583842618521833895705186413263202522931391348380821287964 33
8813629845539042373128573855900623287197915909121817103349208828736572596006751034516 91
7303484066477312472853649869703225505802851210314581397165500540219991258467124752296 23
3047947620417483573396512933884625986054669020687274310798920093600862996258526495569 26
3422443490758898712075405572391778889837474006283110359893639753683143163791553508445 99
4998151790357191664595797264053563579622852232320999562529072956596862566647611821743 68
8936552658420973588048382356327038462917410427463403276404424721791963389233300452352 92
0028757301356303821117289713316943637220617508581520290849724636905566762892184262651 78
1976519538524643036425620321295916089585335981540706502452521657088226438896955320033 07
1238601771994269799874271196603530485283581844146085491345064443171308666656732474479 43
2280547339137606275818904283656586598954664885617098590233571893647110221915544801416 08
1086328652657202503734779658698028969756875969566551715729978174149125519450833779497 44
6698060268643181523422924316776326510155053746777092041564764624751249672301142884839 53
9706060560724653142273218895838355013985022479806163823494536612899404093559178092658 68
2360641984947863949273925514675962185564843402871639984241656407922432049921519352709 42
7492550972098376405547995076369623700896158531420828549780644065756946107401242830116 77
0917540880488062665750487449700100644817281703371857168769905205043126912675894235464 24
2632192681612607135255937799846876848766637463737084830913023031875975519252439917826 26
4027996616367094369112530086284350297886671483877355700954085095109254267237087162850 87
2049910014666606934352543968132422775052041208431177836208642591374140370189390584913 08
8530767180337775977981545040600450842816926539549424241734396825797942963322331213218 10
7801293219793602750388852631045872578879049933017249371699290336354529074965146405609 12
7548752881574874850466564788157133643242701577120650882647257091154529355415106455103 50
9471072017880017924641359721384299487104597755279842745069774265641488340683008092305 46
4626038948326722396040624646594574202522108428812826688967527898024346826557896265626 63
7974536891092934689209248410970235713162753302389020769867314325842760948183881245857 58

pi to two million places

734014097068671895618131122294700219784482415815445098198891529424289069346605626079565
664394339342799835882357116757511632925562149947095183522331245810913158655773591758470
369414848815038632640986605244160899442142372446731857685405261645960803590461604594774
723205628789108869923420195755264555491518628810370985562898184920845362817607417557822
466638790726046775548345174434818904968805637650325035359262713745149886240548295624736
933344962330902739113708105840712372655403944406661654666981279401611743803144254583611
313870018005106422250474212674953997075909715271031416553633477320054963549146671949925
664377111351964712248442133833448299147401916929709782733048209657023657441662164041385
975719940110755054813835675097536770208621480224769057593128457961205201060652722159599
745589563929274161013544777148660272222807878919031049330486064234888941265365091980468
047266792907970770229050202576713844368458156348290022266587224538924207586781264807425
984310372314483507393491632856870879893530898303553843839500797078015089931667721215355
785047682448066812060081609252490055065606882005394397529404297779977739095481218233882
815997862843934489379186155563845847942592989490543845037369764733322568524240216019590
447240163994449258915452220947381972857356081416294420120829692342136751905692105337359
237781600665668763641463665790435737824436651668510493519577657907919335490054245374836
258264105168437864450295918358983964405685936888263686371050276878385312777705665995603
001572358237147154721905671426014336879664684364914394999572722150580428992181009895079
934494421419902144689255392867691751039824675824637343805239735083028068718734460641876
299320312170407048304612916426319820862850786951729018997001434289682624417807819162644
171950445075766685632424567458175518591360435999585745988250355384667413282045253701080
509023511484048195305888064160408235523512941281404845448810370856942626000271639230100
860372142424842912649571956987321905242756339490162246454625934745670300567283750846724
982527983673495617708983651654782788942618610518327692085036036217800339152493371484445
501415791162505068909107138275802244650509860986883276779711832579391871621676788356224
198867538683931575658977686520163945282738867440651787566006894982173557480744432557759
069273637848180513357096270189715205890970811986200522767934991330405828456720358566472
410588945530875223646183843964025960112548528766088662848307636012870066672280267000361
494230261254165504582961699316384370582967507053229269109161196749361273729163120586368
847905250952735154380622219327300289599240793907443857366909352949258689440107342217802
437077152816124253190882271732715433823740147454362184333256802959940771301498332370451
096996545320274533507021770370706113819105163588030747477081980826531951055240343461890
804087728558871026191299109229595088225181920585188744198634862518824566545078032953634
810263484330308372418113655627830193910161835174032238450979146875216238771442392232324
557636410797470012553983247122601905304886666493332284863290538086835170964354044025686
679116884443421694025177269167200542365879524586487291931941963983705910834659556545737
455427472252563872049196484680456121634675557800183911458050720291040917746197882965048
612355528726975520004238203818320764255364096320893392454496759815230921518947305019785
351001529957353054281128364842365947435959609595698016202753023959419953344628208264979
236079421886804110602415874150857519458061568880834301854125378195459697414236778741870
667215842752231935277070188776280303233740286266042073050523785203542104257724425591404
270087490764352482693868107163766930730372723741717542458522477357582702959966498541023
115101387432370479915917879029944855016825586545153881254857425414604291201232285561488
9532717771724012262799441082686913997298743825568158111262387326261042642491914673031393
240789967378614329041420844114674153516742689733703219069028747706020088422190321251655
289117173856747773113661537343917519627079549216170937800543404578737896895794065840260
922267066639747064746191485114465244380741520552128686202006717232636847179672315491533
594924534289288748593176642699362096734074977450530705684314101332632877759213057623147
008743738450762603058757749787324207140665136179949569456010819284319373673284179893518
959542351970228934729769771049753586499567318504695098739662801395315247336067459574653
673422509659501098766962373414060683935034819838271821186844060176157656025470113953735
726783452694559607917709449717272346947334367803872237757479168689555204136215380142825
486737769528370384142793400440013056897835562279860713607059660446853343332247081996051
727615220108066710652379501907097493741821633529938651891700778898834763609236188052690
600640828079713434978942759592887202610785155541123973730460975923178834688072538351059
080021866044902897911896725936755213964729068579735073675718216974036670698961345787450
610971203501479565375164166153121647325441678927750724594362819238256233629810375653289
132823923222725067941709469131856699622676300847493329492377248202516805506663161990345
780050296216095097127831349754974849708075014692861779720539208773557354263446540469175
078783629783037122212634499537604587068254665673292775397228603678192473260758753750363
960555755152470447904689279007417440809916215353523795515931682503917084115733894721483
377058789369541534827316071703023202492909454665531205525012632253174142737294089358232
313041403596709104925718383520195357751110303018937387366729568834759002804669015028041
569220627941078976828096269661012137488119556311967779804681249306478374053162747606283
458684716459438243677534276082695576532344276053399712987080520898985614420659347221575
351357313531509643256863269976011816768872330904784738782610883002718765026082629252331
944769099404167002640065555971369918146697911356778065745105729647119638264380698960233
881135307214985853687086285871789268796297801608941639363012094163552302777342996301526
343577512504135198341187362058334547311853807457268433720690522085625501050610093794285
614047418455921179339272708597251152880056940280536141144924020924620879487418362248544
537016734793590201500699089471119500954771699606451569340980957608723061167985930544742
494585592763754265507909850782622745241428056419619579470161814101885939670292884088175
071326949126451479245871388347220957012545376287115461358447101311323201495490944640147
600030237632857171395365471490013555869633069258112640479200531728092117912870096788138
937329594906876916230917822286435333405933967916024289327484446631559457485611320451783
064649166224181324629576750918590298833323065514502362940434740549255611764221609388471
173418957407199850352736698693386698517025739380660230279106280853525493531661945852938
854013476198182979019270269975539762709721332077521428883136382794037795481043639684621

pi to two million places

6952494822984432296896920853355530853174095397100274487325283527573624794580127804455036
1060645585780357362625255636064773490568638324600588264572996728670647068819718804899591
8209538769867241261058123133718832815387305324063517160488373186348319448785524534021310
5960543269787362789902736235815268667728648413763217540668998973488261186018002936002236
2615884959038938183834781502164731089138369537380868316436990879808593012837352876220600
5362275872876794657916805763581432409253055023886548294925725127609771043084142413271492
2301455502491538011651570107259919660889103344587780201842019868725579834858927941157916
5489841807965598165292440028600089283308995984612515413473641247553705658072496073372896
8639565510344975858300171880139293408159346577407491687314019903828427712262333244605887
5673983859350076951311855631684573838655512292940803068422036256724591811386063504801552
2616706356496428673234596566937992435872932911668849839364206979703901915931945597036125
9262706370837171360797222924483897365994926322185943095293445517054009459274870328435199
3881402670859152894963595076363807323470534623093244150957569185048089195717391916531000
2415714293566869097067553848502610804074340645742634283252211020710345037453834072171927
2860930797090878640274037560341962032609518023331944660470439348005406358691029418314381
9807662636922920151962674547889054873008533422088159740328925356782478045723448555663884
2993651785938154287147347054077625040798071086832571272096595247028093129849059790306196
7508059944421798850698316109638043185757349320897027921443393913428290098389029276009981
0349716753400553502665754851358206981718943173652187372727038665243420592696839958587716
5807536293049174582102753301267023622273305213709274757549275403224866536323928428878807
1811943447754439431574633737742190514462638401483845223060132636502788451471704790583180
5834894085694942499441515548386334237720406996019335803137534497602844499515410901138156
0664132329431053540356633498250090053413621495974752980282398461972836700621058461397781
5827467657982601784729657646395894187742496331695884228391191590565640228193496801758163
8401394292081420882045469029946376520599881978317544801271199655622131732442716080219316
6446071845067024516046120117976382723921134833943879896290584017968636094325530065062888
9732392513616327023907523959826534893994680658059487686462751410940649931153419873217299
1431259100977718868694557124444935828613812597646375511342798457376202343562568983122530
4205902149070999674160321546707538816502965863991553151342905513331653248308850347019549
0556744841010321876589956758394558238286883108141862835473194755252747115754025434804661
7748038786859837871569491342785308294728873541204319023209059529543286106613269760762669
2661135211466252769841277340852419138268582809550758375787518294419538166996475933805810
9744197040887688325257374385618259110897501964314793725720708094058960553970982858004455
6307859986110829784519803998823094162509868026351628228075608165070483489644168361836594
6329769197332664504433270265296440733260835948712972056368036299922692205555021936131309
4392925616898258938095311543481328951849165428725462863519781030237300349037991793076886
1220453265131810138168987919567846678661443310580143812599127914158876671702906759990712
2292817274527854431917637751864885446905521418294754607553734560608556346420396176670957
5287454949012046603519646368735772929742823475054967865445927360627608989246978241890713
6691030092667819130305591951169569317831975407962403384211046644652404581868639326096463
5303347112921315436957144220672372701903216128316366068353535914027988526095314744197670
5764010907530604721386570676654997265613995625908185085303045559284076141246522180996543
5307163185074886478933137159804019105801024255417135661896120601106976120338712176953627
7481470240462879594479656892916665615162911773694618494617768316635945285117164140087961
0965586719421163816545789355943747416596019104026506996537608910884905409088076624223624
4525313285217856872117107507287258024370274958356646245635139772059647787347672137096977
8743722222285444150516258147590016001034987342164287379415209172808743852870687452996758
5063462361565683806568468586659128839299398749192804509759935762191530345339640241628163
7564573379859690128292741879627625038064030579982239393509589219952785102916463804783293
6209192780407715041873006891785738178253793126532269564298480575057043385936993433934560
4496232593724330435766714771166426205667161937773757701820493615978206551775960445574271
4015958506242086143202210279470032864409749853119493397220525601724380678398080630198033
0387140108237370202178023999196594847420810041624631399989387267812969837139653343697806
3946676428600241528248737856394129279353836077076083007500854688366849683447381380001948
0999463927939725480926175571232307228799147294996278293181211799953119139336829142170800
3917108392039962532465712426711480762187323888663027326326051022748558768582482996273785
6303750205262948831696089490129163137263488580498192487545535248873262394393675045016547
6893402068214585655661710507751194373805894134256960413139475819197066823026324234090244
5054095883410876889880583600190580085619949103323884501331396041945468828359061480627917
0274255056983630386819040807698607465088442367761705462260845439722219403554202652033104
4552690070468827745821456966366744699942884173471148070234795179430778361527574001167542
3810497822651699679327017909913253596925641310212031767596026367955108692598919360515293
1611639993790605221621826257344736334202630750572642522552550069625037213380532448446537
7149717057771217385550214036410911783897214179763547317648609973237020876571662328627364
0666665252244484423704743126027019405293253920388124561167839226071478001195718475045561
1615250384482208269186674529500194547955474267031195338846336750475341192430517289055830
6396064273009317899033971343931584059161008586393822138202271708192475778210015039163848
6660170908139091359533406190146269042409568052624010705647776618407365199659831201591486
1912479104532820849003769623579045204928147481448465817268742916011125678009811772691222
0907023785148661124371445919846685763094737512094243352004465053297391251670830182554297
1302260674660980052603919627557983895090692431900473756401836074549348591017944755771628
6315055488761028672918186758647664447865962782729403999320990533549691384757430422035802
6682005018352485651517034209943110726037475082164349588541432104573557419801182940065163
8459077831309473009929939541827181805997731995522537656235226168798804828472031495890625
6968944242784076171974785212111708675294460367053355570333619526994064352330819095743708
0465607841235006193415649510049733173836204004227344378915789653495861159191813589724449
56070703223227828015791485580886632670092404234203139271646896301370562218550399034622946791769783

pi to two million places

297015241275735845801308975952586398550215463677050900703329079755832652569733099251994
233523426743245262878343487803932099014786974131728112549644590427779736912668770753379
381052959792195600945196472452145664478112940888425973995022893156732048900359565314881
818133524884486986948006125364742775500050420404621034425555880866214425237932464586130
867091526143968878163367734969512409007739167264140941242164561853646208583805219480898
873774634285140004397923506702424670667069730792355322976556684570476290256322597831961
833397249154695255251435143479730725058985393503441430103727693308287010755526122132379
489153242485015478457009774568367395706462453231678342716059951325335038447646180452988
910089257614236456892093712162335777919010169852746507433133940386055838356051252991146
475251049740083923818804095246656978219067500774512734041374694859890324303842177375760
809258836398494285249166123974213403066399683994573153145921895618542973694163929127458
660214919325606290835478894538705890991028776842634490924275819538227715445507550828207
848836009388575040713380643144643510663577254200107215128214873801278325521947726666964
351963995138809766453243327235576842641531380978229804632131537746252354042906035801705
619274468848477598491780867455498329658519534616200108125422426766724465919850037372467
000945314183285138022443386726425015956759211414749624518491204720646756094059335988791
797905900136748853864506869620656149908349682257856811345564839572728890376856473485870
278747092443784170154003374569827482324692217767073805998507558616687137871868309680967
655837021661114077220785221064026826988873072896051684378352036740225003301294220879972
807335520332084982518794763661742905528375912856416497413728193651143325413215966544797
508896637844058341384482455733859312236750877569174580777066637614313837033604334586746
501571999949357092631498313539476010997460000053765814494573326745861063304932150297383
939354427370993864150604817313381076058253094394198785753236572332304795924957505802346
554857601740783004076489467682586409510388043743992695534150692609552099668496362196097
541990256689273001830398841155263548758410191862585651958949917543670137752629621235562
608907598144724510634531684374203926963412512254942553244660392238541492180254748828765
753640669566568544990494814915280503825352855646720044252289431463574597056054132111347
761834591435006804097516578939671178548516522920677835710752551395314528231222460774326
485780214710696712582490549532520029801187432980385609820847498781051624257385362174946
834560794401386804272597372177995976134849268369259049454472854534855547672855092626966
333515439531919926209022290117916503540532053541557323962865877885010012105094483643693
554565652987025104273373112703686432701853930462298639018498497790852704663810074106064
199346768348792149018237926231535776760045253983094883499531297315673811035767138095060
268988103260660443176435502995330743512178353637422630730894118909833411963860551522024
968443939425305314272823845197724460166040399583547279137257994876905630996389204706293
760505652182054213457997735548935383680336528207608125148943624690017425014836948910324
978639419114600540230621618556990861024673154864609880202781073473737770491883439815296
069606171715477091558045391472137944886307027510625910079537399951394861744903022978921
815233955882101576496402094591940900160611658741111198073433510338190204012029929324031
443298771647219418758925676345923219909229015572393372888607136940617761645093456715228
049982747509278302635993198163991685562604495679935194832038447039650098998538011265861
333379477552130328891619787933732768447413283162153821350502232982479519855842530644062
732756704048542157953717334383013657922716976581525422725319026917215835634790655014339
322211845457743373186545271855941021062294882520434730878381222298316872357783733734451
559948232923392695789294479982420094939266427073943232747132009716034757074410728503079
630390757270283805020959161755070005016627792429318124505238672399685851931779590388404
675565133255758214943153517959498790175939954018699586162066971538933325756001809207795
785012715852501324370267983815321683151058802737455893982251650796556806740552933580416
458692852316528925062441034507254751746696957664759912065568479831778838624441646954191
913411663831359509426984078970554946580729459831838654621775210631145510433633536617757
305037406342920891277502236209418309382037942049430183256488151462174731495312272496651
940332666946548174825341725229124749705114616160054281880540354198947234572559079014198
518298148145990271405143266071026229634919481259342145337898724583097604861888669585130
390729173392077225689609836520912793114821797475064465597613403875759224022396034735708
491833781121498925829532335444378531833361017437217127075561619143805327266024494866847
503438075251839224592395717830234714190223503503807941127277487289504002308047407224556
001247620731599212504578861913386499033912685147243309106085594366056248486764753301033
636598577489631546413465281763741261581100781736701249544796541160122490093946034993309
994023927494381350400802944899168793936676108675503546923865708941470394616477845589307
011895036016968430078158866532913562055689169725157812754395541621519521053001313362218
719381457197443546784636798831874133330551292210015747304782227473543623380608200983366
453685586892563315746920243151683123406729773141705079853683002114655362735781206338697
324687906485958758167228676488743765465044904838056518022973211179540565437944615042021
563406645608062174908344929657888829537930350472608621682320154984933606588503695960666
607613634125466771426576696693825333353829686677801855476375874137561497408516836293864
444543728252821275965733418806978040386375624321359353338276914364031872205194448826998
998678043790503172651953332428847616800651629802990750232478834434240656782881288607696
374064960256307156567675052037053083913616628392164184509828446743051228444239833464130
153027392175042011926614572750828139037673363576392544605297160176537577389894691397706
717184212302966812872049713645325528993086401233052322605408161138788925319487861723276
315274802047699701084142223997929110102895691963294280332770835352538903514318588286319
374292654222129276285260042254539799810059553667839998923929180915038777361400928153081
940786066474842382259576352851784924147444843684342520668215659669197697288057366703621
563551249944361226689330580394804546277509788735672716123758257138391394108724319551951
605337651555589253551826979714756066893173531053191521229835917302302675839274164931422
439389744310087449119622244807371585249475527582841371687338856845139719357174351376651
048952390290170630927122646840669278348399886285959423967936456417355871840199018757254
634606762070626278962322034805371836351794691087763851991107837936690226478961426528198

9949894409583646926514431656582124717920789203351405936678284940177989652979815054754358
8031775856225586906101230234010936129553535763584329974630792884081670233667889701281458
9588474281994287498661437711859701046078611831222856749469131900448256430282620072487877
5253943979012522035179906701088644441733329427038504777629594843814990998912625348880023
2470112685386605943766222703650826722123222351572550376503854095253101757586873383531199
9693602416600396509280727438071375470484844568868489196872123109819909873074795320541401
1039597361969675230524421640561900526742759939791372686862785354305517912575047157660181
4923718223934849391969295394499883801965726625036517574940331196959794117212576237138316
6511479626055791784402781881225538340896398270497890725743044303341179108109059950541714
2207737379747750303681258203099584411948679998579401117139332423952627036199270637724171
0339781113332382715682773420147905472616543401932267144218019610533705026333742731904551
1879487134852498626668222211111318914455762210422898347390349981259778110809013186425670
8891036742980304913636514213249820339892382623311457754007631638251268666848503158411111
1411558864267907213460409221705074598247207024552431403520119565313924583310091425363495
5878979074393713659709552725556666007024202839100745946246631307454467511305993777512048
1192805564972915123491505532781022986430608405376440989174443107699871727600381516344282
6065218306921009179927941709319362942477458068173353359701759893286031485320156876979152
6521241896700403095917277570816030186945971407998244363332875431922140735996952626830383
5659584520895655497781013274539434086438652695414066042052501513083780866457299749256902
3761709300281616225031870030327371448605125164072390070088238239068114739958043555903515
2412218232982743667778903853858623918147815883532581378931913051647239015905150073292581
2746204289143926675349521549429240927856129414258693729514689314270544422700093241610939
3444781762618884916441692560813590775794473862060159240401206337499895425291163409524488
5314231824745868679213510269662875170521064607847049746661545188141510351673498157830884
8050629025247278491328835858571968770316330097553904904884566448974688882484225042274103
7690691547824158619851842195790945913926945539349707417082601299136137293319908996124475
6112702770438892717017234886176319636850246720826698760848197526515117846839743308317262
0487854030332942786443609114897628797413020336753926893185945801618329179440100958324984
0587506458366412769529285982657703330062345826549555323166532305637373512195284921489632
9294238105955982270927599730329947375056874498728129347026066244776158346661704916269758
7179758724292914187975074878217153341997452680557322560031417046342203189757820773023701
3862469785041650979758445271645852204355139759287529508954652280662696943449901488020048
1811864203977422040427026695544603299092549594352025279648873458005434584684945297535393
1583792911305703761773663375795239771087393379547332118548790619268542240083953610368774
8991522104521065020038005180834770931615105441297268508996642282464489776423231947675601
2438097694631001688776057256798936928008650248744676082454595750132838100012297473056532
9991376611276067858345129580303840502530416631173982221137922074743939663003407049607644
3288219873398773338028597936098215354655910072431709715706097105968688690664790679515081
1011519705136357516361120759637386375738584999837864530579030443943012950410974378372740
7321582107022267675703966141860877443969096247776142982681257255182073821062914291792897
8257797002330749298885823539931935639425268061948642067083324510468170670405542134187651
1641921977618868029589218724367391297929670217002604707954075998806965382947068262947503
0799209178054501087213183671030214034123993988674101404729131764442090801109198035244325
1133365358285271828432623672503700241162594822559744880831760673701548562891186554465505
4563052137790450573281851201979665409430270049744632546121224141128329366437940321984471
9276656109271707635594012205535902446730707378406810891604840226316131653788226130623473
6493228155229195223409251839601714955706253393019190674597189900776543585394537305708588
7673393775225561887590862665572607148136026843048094633781089487053325334693152865228189
4850150399380033663878973389341128843457735233219997625754387619478290608410492317086975
9272668565017717845357601444068717168690095280680343189335630427097277870650886333372977
0310015932022324797004101814946764613369688447909453611490174462989749323003758025319178
6955052416250065528426114373076226542082682213459467537033662184218165664434775720963000
1368851514596794720360121393994632611454144468275264866158675668160232392170474512570137
4865906164308600588556879208478336062463041958464174708303366065330053420620428368631888
8832426681603517521400746402690075876108947946351508449596170005018277089668242632747559
2939134464826875600962762224250729627218837874695585396808698731268923748126981350128705
2594852987093853722712995055630159371662858218865916052074038057260330451319721679291471
7186756305293557276882390292226621973058048731140117530813389217011861801173372514366353
0875657342089417048072193359887479736426861987941028542129529410436548061666466560953505
6812679860837723426852206158774774504544085972412357281362939502487721322918144746035283
2409050040101773662698641702181670351818974670512042796054366627945217494148564864034236
3759590549060136096930916841062916367946892321891209127401951706180387212284307087960779
3113725441306054172989950548378827727046663864191137798869740632776799960810327556562871
0906770161485811851671525572067310925943266024855593887184124304225616774615108389728838
4242258914850508472905176061899777583007065652520847408208842173373910767398114880773334
2201658891141005158542238840636728656725089712885038452940316291883714453787866130540005
9170501111547185436355831033207211981334858631153423601920293718304352616908440195500481
1450658937687152384712418582070564413530487420515614512086668096886553647307014517115557
8399645835678003490949039327445144117799163796310338254450612672966298207689283473834821
7563761713839607939250893828751938890837424828395334225664013005888776657917835977404067
7075017710871939114578468254250012110401156781381295662572551091764603659677805799613085
6282001150125167924849476048498842037333939543468668590235457523095407381530411675919199
6403395008232212122031948582192443475529337530173935181814669205300676835498326089762679
3603011784585548752703073220035324122391029670664816719559254532213478249740250027027484
0599816837621411818338760834879258098138151661604146420807520205374549580130513552975381
7831795560660953452750289995284305259958631497799031259959268523867599757644136022576042
7065119871932352626491081301935915996762477542005468332913608093323184530910326642695704
2736368861586841986355898696621612636942346269870652951646036590889776309436953289214977

1802597126310791986342343368337985427815924375361059523136767058842514726866925962255562
3338825444915338951480780353160096236272362629321093838121344292596168977607129610965532
8581263856352840691870726909508719908487588059768061543438498330786223732995053859544465
2872580091738482163952476186691942844092020325285863597342736521084177924065337699487009
1904006536284002657991027808588169428112419867803212676611908212068769439025689831855502
9507358136283325881329349787561199657064770323354601359330157371869985275952788401552231
3044666467007044570167374730294477842583797133957981023419274301141663356310472002062230
3466720043473633620918607406379387410837798372659102262266281668368174614508881058679940
2091696237070267227078556702466966215235923248906556541142321653001230668315813709501117
5164974741770771047867114423127030825964970280652309726522595350260937603204641810401128
2570297152909966301797287496716078187386434604365603126009130801990559134496982430598810
4481214223239198832330517487617761060380242248693360782698348799053187120193656571881113
7988285470003661803376461641808005621106566571354457380355721627020669870665961163026669
2813351285972342273474043550450301843661570597586025918972971762937820758515214366300558
4413754353015287363863937559202494019912296147831205339020402152495716237517713920740448
1216320695616878374406751427661181935704092622544281257244674793566990224000161627935669
9977379362229328899510966718814725472447442332450832816113588506261778134752263774166889
3067961890698173854262071168345202086617225540215131520301426153635292417624887240193884
3284703145335685532116346032411910969004980661636537048300440101181291865610897469806695
5769191358515559383379158906798187857369687334916653170293483274482623496789367134407226
7722684084039078504473370916901619483417492844768576605582389499762665707260959172810226
1123708822042404896741798176105971216524341898769732540518357639068967524643049459814403
9960198336818628217058607337201469935697285000231851541357699941300289798454638720609925
9916567500425745577213855821606623818818432608559086488423057729024583754833271946592221
8906082255771930310244284508802380411824587454595405991187938986652434677760671624111186
1001010409073491303067136969073215943484812974545321465061611701587079237826776752243666
3566391919604273126428902144018734759285470742567034499691767523359847813886556570589999
8331860123461503647593817115860385697047892499359044412797604188982091303483302149530667
8261969030240608250991840949624111714750193668254719668444733985153038785005110697902000
6497353545593857570788333676889246110793462714441980272903059667094654269466800036557224
8250538537265700346455298437548560576643654846895919870325550908590937420980486241309992
6743237286356117618097163687365885258754292899868407066740913105233311691390175906117779
0840555041040973012675087716768600432640471731730874894457953682806516682884188097638876
8775167722540150700393369379883713582313675501585287524033755398687094783975615794630885
2599146212072385609522922010241942264365500964372816212365929564921203419310855804805579
2520705609707331100361087257336551244363974172687751532206542256393391424987919220299924
3040151353261830425163982175998799403271770630652960196594660331660919322521397851742202
7560453282255800911070557016085197780657101431630121188091255806498603094804914667610077
2074550263450695615815328307044196446264440197895304167380833292384654515372513331684403
3152563896589471100879881182953236468004613394537670149121770428219428285050662218846630
5088040978357011532665491625524952638509627579674947577160349234735962801761556267439906
2330347004453811940952869755093858067970266114568484484630899149431515248817802441697440
9989060390843944185025304735724693730561618537934068829460142540221142337313972408574449
4586237761932255185568911036463768407051600360614064357081184779777040599564120010460014
1300089078808937759529574504765503905318359914685455407570254194453588172823021718408873
4015606594770669821729663865913212771651925166629124216002608240455641818282420715559930
4141284792123814859514483996567150576460271361040734548881707227180083296814012039662223
7524091762593050119645743753899286461890218741075016941551730748796555794341221340186619
4571111414514347679110508738584379543228146239251032152517806102200298506117151152292446
8440623367092708922645324107817862349300203648615299307013684697050663695876213038719918
4967362628612877313530772131662208880690117845223199493653227244317747904408052101242668
2827577605768520721103235336355173322284658538557907259299765849786886901193452927464422
7525558504878241741596277439272695086300373919723243280964233209858067469745511572367003
8795593244575845312260023956451863424137001098615026519509650127633382819673659764355994
0000898370507219226837562534245796421520814153814035283423274574528218865399845402774441
2225452294226541450102462883885257064639199041273858637472377197018166150155930655258804
0244909298244953501327780953256426342626343279899516866922969197876946230763821342879918
5340166382658613898105566506740106342082853947818002045364226967901634799177968142697701
9624314183701793323920820696391145686534357517893702466426959525960065432609160427090032
7733412487937622089785456943184741943451062039746655514720561706101946122268668159690448
3839429043099283167735414442937221925763921777923422377339714848818191016998201827862211
3181284536326398438621565509677653198854178318558674158239004305345117073737286728018882
3354578530695996778806741796430097938413254054481238083190305865408515322785423135428837
4235375872168823632205507065270649525135696367521210463206418432239356377952099896545547
2001696835107430770193988419787070947694987486420085547074572670602171274095669266543314
3833776902401314278999775677872417957135722306063230523856351476313255726446759773377769
8628417509403393812169214267356386462354340206694306350751394027442889765823700307564493
0106573451869821269475788504119196136104609343164407415337439577243588525650231378094475
4319577305451878520720986386116227304334532958754748551273383279722191850350478718978884
2492871068961821089941715867761208380151618886514540012858597164630136449651605514938822
7928344490837674328198921142910431061110868692165527920798201649329374581342150765511887
8489276738254879351178323516112085517841083762117528150078518547781860767942864148432333
2331910972668500329341289140458582044315576107787857625774238384899315723739598381106007
8124677863585568996527368845240942528704596435920760579346684324145325596356248748821117
9120818397034784641754929142248619881698358368191292319024071712957000768746334505854460
5361897413652911487487882667012766317504121007203157810889243766403677309117553090409928
1711961362105392341387334225937678768526257135122434134824492437863318852768275374310049
0354455242143571369313856402904000656993625368177991878558718447307705825297435057486664

1270846544260384727445331835298489368389889780828901862307448404708449085284940303943422
9554638527408547590152688160003747725228117811611574237139819507406741753431841463505444
3424383574519248611580880039606683439401908987378192440400219832284520765154792113692555
0747874165836602422185462194643077515218613558151988754046355176140959054373857005250133
5809390485967216202160698441615690787716840681274143766140911181396310855991516614810799
5445153249599213668254996327123735625684147454143123120604881956549350282357979943767777
5718356659006902041799336909172824104906231327124424941601428516096280779063603833584144
1992709154227684258177499221993057258032595123771201299442484070122796867944473418067922
5826579349589147678883789154409326316567006088947310932561056198803086054865208774564544
5247485353154050111681242847495790437215941553943670290712577865368975422894964011618500
2443713140843463035435806477212711435043136254843866092436155003196510855005090715833700
4150255688910245840041819335202521244984571676792556822180232663962315709645890796869944
9413734326211814954738978562488217201055827898453333313654848945088878567519040445959266
5319521581192448981135290221345503835659827193694931765657818676004700773169181645503411
9476677438018159033309930256659566680601025092653011515016062246861638971930902143214177
0400219145772685840786520972216809806340340974308690729882230975195198310029861267048900
8177882768775796117833022329018460019071608747684231522405077436077933069971420826618844
0794046925806327424727504156574254671472330996260395728653859055288800591122257746897622
1928238904065552137197494522321276833845155145768466112776688207883475885860064886573577
6316527556999411910834661445618670681274946339286427366151115589122408685970789270075033
9421987583653597343004106955922676103553305993105917631279351162971407960650125329361944
0136650630794157005021118804358149453379132085873254843046365963575445347023689576407544
7704415571493176867930227775311988218483730191178628930694011158935995127482991205306811
2955192982493827214779554101294437734517564251171652621616699918583564687464935792409777
2455133280236293667570990769785251422664119119358543450137409607937600508655283422806577
2647802204206130795169963953324761662289154668469409691451525256341542697672709654289233
6696840319253509490673845090202972431809123127018255195138346002937882176207767661470177
9105698709532770660030841618105920658748656005148395218482499254325485613250858519743644
1707255380319640692807156322132217334545187661557552639031833939656842002946070111912900
0374032234397670106259189549411081661777562192162312113709031397199510930006010300715333
3345290159788935879859588478418005253796008798347110926578754895419075386106622018995977
8954059343787394706384236767750218336928872866052783454452202405805711362960075974665199
7771111263096946950342384651053143366709511076068629058782907882208713346420036439036633
3298098850085133781496316618857710806631255804432420769864751221658236162083468148414088
3152527539605062706669652630190259384874403412660635747377192472521516522939406695524533
4224812999183265659442375912055988200472864204206707422287275907409713982157237963214544
9916729673080286486448406832219033684902698992971020092041186578702517859183575055270333
4768877563858546273960750997614740057219289818296672620120311458260815855382692225101388
2561102542944306796243640089978100544006780213427670764254992593671010228467486622594111
7529405166755581862694175059399711537675399665098330161573697092270055730969595564927233
8518257578755341788875263978855596462447492326407482162854233803633594937489952680601677
5414219788350902373105768750328191825905212533184931830610070220267858052785156302432411
9555393856571133062252246234147971144793257899373626522022284579923034601177104157440944
1246819729500291141398676155035259981948073523915452858111822298181564944792756355332233
5062904962376021888577294081893514505639433389353397765545634086655528265813600081646333
3452544948450595556670516310770887250520662602256057362921445167472113886016916293005422
0512420641555527240194558735950975262553372909389990752880852423425526878962326127453555
7578081715309102459635355394814762771969164198426111827588625890580436615487562068163744
3408004760016441984380293734146828677717616041428541260642564158093743961591292427136311
3673177612445899338591777306113329884665524574828349252311945870699891238106008382022700
2659856257633391903367294000126429069966838423817943612746986659318961904532222329360311
6632960149436871828093274304915238932256255191114730075314812880720642684222698113618555
2015447980876010728760945026924945406148795259778098636006966910137781412349002068691988
3892641537672255176874951520401488303120411302022780645964589480587137799676887349391877
0379821439167206459086965189722569169850999031020309665736330187721579107878142644572566
1741158418587769963563529157661151716117556191214637721380113652263627127850352145333066
5842404271934657080561805642862699883193272737246850808063725345867743143564713432991366
1084587114567017680602415639874524038583363793983563393494459503071442769421392165933511
5161002806826001862171080275899091634546246022698586717454231097729730601950455750212177
8667292686336651258071050500174721336920726980719277906330129190334291822013011713020688
6124637361536176394805561122679035959983458207368030321560508366401669154640260872605066
6519849075749621296331192046086424701059966275204067990852111188739395262221717183945677
4358436625025940756778745520688317702101822639289293331994618955621433939875377741823344
9077638559935400871935321557810634349264692166801973069587793471722254480791181111963922
6392764800123517725719274788305783971136690606455254331919032228919360954971843249100900
4540662729238502740563383854859018826814943584364584398026261111617176584281609795765255
0678761179511632399263517003261953415092975005534048866409658351916597949193508634882199
2615981448436924375455651122041948055268230753324769740884727787435452359272108804052622
5835368689199319844609897160844674263673591680055284786340626912717217317175606167717144
6347556161980788439031135847771642605104747457663614383208549936721974573997978665227755
0353980618914088883859093213743752720336230257877956104729285638608513091577846496000877
3633392031048977816919904548372115769326147221016937339567708656913761110869153278354055
5689486050710822954248091805580809585284066735287814803865380214664675714389654758086044
3451295539355130958693211086299311106058399394249657601065749524026449463655244424073055
9903652849089666480404679455176056890276317171918768727725748903365671778563823216530566
9212911505032641281573270750113835519789309408910748803426109088274141371194091230943711
6786961363607247765710462348615040606854704645771878916603821401434750973053691031108444
0796955045623775381198275521595213650187756339707354395801204719660195128825105450331733

051621634109051819226052554631226435532259295747572882001626270808236042444590358136199
045960449167540375537272061819889995147771614932760797999354053231793037435275849954268
171872137473000259335615361921111292616389162184695669562033564970596509332371688551878
420333041807503066505560625174160505262331664091925223825887095521890288129575052171165
597917130825340460843307977465476881666919681447689732848391792177677642710321517452447
950735880763289054167316031811926102417003817756596118256541525180967987602616343017278
370327961732925081347047654857656059010727673521385459827689298733297583299446853656599
192720272312841968966632593594772667223500113719502646730844926286095985262052240952182
235920031470698269779266722504236559291924920543503544423909408805762010504653092697731
349410857279976380113049279739865584198988765833201593343961046875079635201784729873173
044084272669658406096180546546563193025050149598884019250855963188092332801473038791291
957958251012920437765334741089180758072471272402761666296862622222316607044875229214715
071461596077335123827216691545527291307887613670400334477105207700594289972711773659192
429913212080970648963125588439119442642483455502072746152267209925644565283467989949066
034174136847615577313440734698003798042141226713203724653210732173573760491932062755646
766549039130286899780152779120272475105329245955273986420662452957180086809157335553970
196512931004832314704135049429359651182657240498204431397567031470537098506131461559915
459679080382063327112705397643894613068335246691567644480584790531856264957839354546836
297097508864072557823669299065081270643207674253904368857138109407458556596741811348102
672978012876597705816267284575615932812660645753286983567541694373351718691544942980395
328095625329671764742784192171054963853414283219862148618952679148304004542302437244624
942769588188513047879480515092402218472726874326028962048598568318074375214862909921339
139921806953807443634711062302102023990801664283121979039310889890286812774491398778160
123696349353837904833738861649739886424085640388600121735603712566302824193284413937152
603550255365006846913799512533015708816931761980595766096113281453624863814374708137609
973392793407138710218356044379465595763210859726380561131738627799618662828106580006360
605669651605002754632000642838339900470686106215897801359187080238837576895579111713270
218719138612446092285094662178800912356646714252845813168832334386170263454531063573268
617141732682521099271958324907832321928980485122982303379378592697226931958950333541260
677636845192027990112943513290085258960490613481768461244818586345267324944123950337024
289552856674557637765483130391544907241744699033348963010326960851264053688782221214621
521942382509078188940264353675156891044885683292128360258157480099845832054876530840824
256135089359722960318588388580383581856644121538670876760124831010463084474432188014479
093367474688447856414817445909124539810323008859206371015635875651643095961127396401586
176978131408795073714288317760436689819626413475006971940565195045505167742397940196899
862527890085237445807328867069741773963572545060985423456789852042507322860709420263948
418286692566061665444072045775683939323122654078124783166688018030184225045200535518686
848505584530852538497542612057943053533080747750826496088529445572785003441395589379335
051184030225298706162915059642259996005856489572336531798691366969944278665789125256846
260414798110865685571739072133078933108525903333113718587728703492627902715673291686627
166749081993182583166832828500157570780161193169221931514754937755159804654092839991094
937420103717085608605881785449005704104136040435137642468998152680609255401123465329504
349180537477356166667104692983095678920316482063931739221248405512034780632111316813373
224063216455415588237846091942738088502838312362266549744300558099898299574258432353764
298631465635055283560477090757473282636770439630979234656297949344566413960851464371303
932136774212470904452721540687921542630642597260292018994655298115142612604900763867141
730235727276839041559723450666693866460588292012471144178831782234821533891876058363276
181819433227695553112581904847517462905620013468996440716198232054347184611020511315550
930226825107491990149608178562605108859036584745150376384915134000329516399106219240557
283008103521761979616832213981692408357639556211657126112192905087163255258558649616638
254193591482181876195923292056995506376458182685575221152787011802994335467415362762077
497854054133303136335423682410108464763749063885279841490076464697648540094796358954975
461448137636970591635699836811987525054793069353207570766780148447701424716241908166822
490074207111864881547728917186535967765395799335033427282146054169649600984706979585592
643042870363664713071314782330611576419913222420646099898830762685836055527409904784676
107604241784215062851755735299964786255295428367429870664579433758010140740211618614484
329765744263428528704778556308309631435278783041945019702946575777732816746858087453931
603937253315899280579434631408735860861778826334927746151184911655130681846713677348823
341085136403947939208876886336339461382358344794081569610914293877347138934237736191096
460564244474779082076049660271356168954106444832136598082938909729618912118342914906163
896386106937520895346883983344467189821243478072387407457697554507436846747135024858818
399665568196344528811941833172636825050611864900394125520574571203603557802514190435267
183721921384829905803224695842432315898443251039654435350535432292167470407786146848597
625574461535118800314305699549278471674544972697612839332518381972223283607075227812928
130106569412629487306342688373381817421706086475482763942423914027532180429519034116351
704698074233515560578575624509992532017874996366404734770389855873065076038709977318431
281098978988208543559550943253902371895216820233442455725753078792633985509016455942373
396625223351648750589556942172972448959988250892321120347958941546546030378786175915716
613988693268737496847305496532937821475648105793808285300532447080506569294223400109593
482946145390788906616264021501307353300331920745637263770770999399922886212243248802062
634850888530360107234368901360642758142528398785949179979611219637975765192452186709608
809213711197750008781593043072934488393095757415924137528597779729189345385050803831986
774590025186579172370808574164297153807884060713068680361982419715774763895072534684045
691927595319372237022290155800656076047385473599044779967487499697694271376686955331951
253377640985870966838632639261649456086841403745684207194059507017430354691821509004664
939985517413893851975731215682616228622318810967297476060130283311937161140874727067625
585677751199566674861519649129701933180849941096181392964927893609021253544332737506426
062429941203273625582441749834509473094534366159072841631936830757197980682315357371555

pi to two million places

7181612215678793642501388711702327555577930226678580319993081083057630765233205074001 39
3909580790163771762925928376487479017727412567819055556218050487674699114083997791937 65
4232062337471732470336976335792589151526031561403332127284919441843715069655208754245 05
9895678796130331164628399634646042209010610577945815130927562832084531584652001027797 23
5612923012605863536011649209902587455521403969791153402241589813245057229232332131045 39
2469983661127509161341877972271870257744736124388514083465932031248728192989830756892 39
9493819598321885103228754680176531265348959379946441039256716271373765670976190037824 17
7952641403477592636637423901248853818378590472910469319889215610792471433275600029017 18
6285425139634004575898174665552114607079932283996193649114731212691945301467755623261 88
2254742174925939881766761512788878297907755343511217083990188345223321969034746579455 0
9447689207579306911954669008259571716759263187037589166388932888941845380700295318809 84
8102997242480925591550439382798086201698320576363102564886837333675568675751472233937 86
3999270437059938350588590556051836547971443132215953828793278294890967017460767160108 03
7280410348481522656564863648428245411048994130152708936078824806301180992691498515905 41
0255747870215367884778425517346863577233343834064175890144702018804220184060001250792 87
3534368375279693390587583068899000986676275660627135899542946930880709000058236957793 02
1108867524302635878255120563115819373842312741680228519325223995892981844804455795800 75
3843563554453054923787125405638142343649250391762278734712421891968913826306731028522 00
9679086457117716531202213996237431344790855045029007333499184252611096876138081628885 80
1127188341154174941363451388680889567822705813438505214804376938599252578850067847093 39
9732629602951076683755523081557048444813758213270991881704028121932722609107605374893 56
7056325226489607815143554977384885571753483164646216519304208951078062789549490832143 78
9737583340053558953425457371180856230185443410941852895513095759677695824021819543455 44
5476524714266831898042366841672405864288065331048692161272550541092444862842360466663 96
6876147105844227261274811965288811515010480668263347806452906218068056551796298381892 87
1506926703423108833602202568832303517003217643956163405051042571225816099315475191043 62
6338758855092048451506724838755449447928555786325977466367276246557645720319428334042 10
3036158833933844201467867233277984922209152052247727510877973604183486583336431402763 96
9391070723796763877080890754603191915200788416149577637354864681844110384241658239699 36
8228164036303887816483214327153419525007551571752929863369367387669458325165191195465 78
3936460223407898671238770886661104613626409576399196556202962257718226596229118176771 73
9432738524884463626361256484220692998405099676566648012269769258631488877954900601384 94
2462986568705249080143013828833620818097308846341463202287814457468375675939892992920 90
4464825996561395926284077104199213925418262167430242564805735355031812607552950681624 62
6571955565519346874759336717787958808496385713012993667026002902050282544011011431866 84
7185798966085185495002791174005549660444491775609721231613832797595286816514606061650 70
4910013295507759074803446445817922573888494673311069873740862773007551043643143223828 40
2864289328146284941294516466682725709916396716860304098896168069906151005010157355140 93
9459901853359350306116480868589949545536606822008646665467731919478415409872106957103 21
0711443502372237590898186666870002784068161374415068300474704099988121015108507287685 91
4153301632210585397382620030377981993254477255381930506775275294360694554874566703892 92
8357599870603110753542296686789766352654948475149787023745325355515173837887571126709 04
4885899490953019297247245015141602225347270911954692283452897738670849450731650048552 46
3079998593758709276604725212645303134109242273645706176663764018635010131098558835785 72
2345662789021745479660642561550687104323243317465089827413759393279514596662462631060 69
5623382417335314957896678394140232357276877148742499852158783565171772625661272236516 28
5216287859651365442974751712256087432584779557584810016953652394345942507621428498531 84
4629621706395990477654710754714511824504465055839588367152006868604747683409009363862 74
4133893515465048157994441498003952636712805496773779901381076747632564945824463028590 73
9407007186227896154902883272081773653153812106351373863454338856975538711323607892409 95
2743598767851293074233631417475457610754371017370518718029117596564006858306889791904 40
0320542338097796934370412775504842123495715391056192839838075047507552554858915512975 46
3425428051095809803904104562357300278373896590245082101715834045199062439300561020243 18
3301065722461358859726421462332629935897949997998495497456330726186027040255198976954 32
8697568114102694346410103040569145649549302325195907391424103852774180281241157484916 82
5961480110088633340600972125340137378807147108482643552978717524130359717451374126395 24
4665242250182381153256186443503028193687010515709590163506258885608650419110926648805 28
3600197148485235034894368224801635175602794139878180975894463004014506535380369455779 26
3708249993054025452210069419526613918364306194048777119275938581850798470680776630099 90
7568296889455463304582170040817840947808940834029721137139622882961616518961851649416 13
6812717359499449290179615187331515292647675294606074097945941738119841721595103442863 69
5869237796757507184025299910766398837664823207338998624373533147175425516695999357842 94
6222267548956769987388220242216321467884439617525111080710777248806218867936510606900 02
7175998874184650753130175036043120382780790880828116951856129138528441792145628605961 43
1353725785397142326297509841625285671159802649215352667458803873918522600841027544562 17
0980641615814433117525769185303106714094377655211810055973596666208753308872247916949 86
5246485523551219596632236549209533208057033511662731410248240307470712760066379169780 27
5237448478923007439482047204360195822769007904309654658811948643937207699778546846542 64
5047253566591660361688710908417973658791335665229114280432442681375020653571221973637 79
6122865657686886710735514325729122697598417745955219980002460011047067788729849107363 46
5360707033770156698594353310579828190379562164226428784659171455865359157196016347543 98
1343776443753318165489698397800119782763230773284443241450932082125217156412560540057 36
6149951624320346826076787219396436171374198224725794421479759128386518786054802194677 26
4926078141701617648704799472184964104895684069512896288654567042232789397033309317982 48
3926570753537607649085267763734377127179813927869671132570696222468195339362590127422 45
3053814413659304024468942940082227463307395618481273362849103278568503529113857682364 02
5710722088804476696300283570300099249317638529503040891974222916479855621675706765271 56
6277929393073197164350161379526078950593975829357323522152170001083511847520516552612 25

pi to two million places

239868171374162475066366990644839218127067500391619779986318211199510227152971788053425
440240321868144584123751856983721699137131344773738909928824119538001706269856834125048
865930432090039408943452432203058358047052999655904391642973642698721404293129049015523
275958507902692457665096415007774684878886170454869215528316049558407528070920635151616
321500593655176396049900364796461220865103092840415758549979551371187993386237846448651
382925278342265933077777740990195162092569319948623908262037817707627039551938251780587
117737380744242739433265921281363945383004553024179415631399346101913636651017140197155
995708024238421303262215084377606114926885700045747197662478919600303333725528659943094
199504506862852949106694833172184221157550391183899655884047861198855920911227201441015
204849221790571671805141371206884120463932961125485107911976689009881928379448154669960
355234923886259845362296456715940103452596065982286081842345856104717930854972475406784
284588843704639739065773262867403683941139595862633458467045288738639699574753615306929
447850883148449012225003386776121522866927603666764860568529417885090796359363550515585
914503371560774038450172429909909215409716738554049379303835475199754438483482505717880
819340093560352443327169697271839249352823675641571876861752595644831500401552963859998
025886519830126752446868856337502969660706422589532283451727293386862816326267680638377
473627517594594272166394703003976640189437210974729981851186907356240362374610461958085
377896644842539613121384640425464040441656551124879345387303664724746330332972622659009
748985016809265452536662571488220342821394913550028691091819428597728472435263727367816
629784338042832095279214573572380011707187593489878936108178108210550127122508877928686
600748991502286297027061839006303832673313385695545704483590553500607397050837156990675
847554202174995088683426647711569000877853601592482872150280097010703583054044855973882
276809065169305050078027523216252935291247475240774191030189363033712341211850839508658
846464502285412734117145346290090744153005025854827586445529961278337782026239817117908
152983077095812534533178977550733398239202563370742361013090679121025999030109029255848
597442405949765949630327899159635309568832894477885552180609165010911782683220674524633
216259847011110981416593062754977981733940705730642637473214924215086423336407206723048
974642504555441273562690590346067393545480925566444546349939477361093625832630910123467
269497392216287863936535112797072643772839452659443278347478579339402241513735483031764
972305326810556887912293885879178646094703172691555603802280773868901029919273359757023
781925391263659016706270860597884665085997973359348926470382865650861766953741043686806
608325493204653176787330980319610512402799309530864701866886033471558324158960416705354
475526090702787686609137595871544631630345437726061950903265001028302045761300468293648
059977619500346167638970750626882356067252948912202197292015160258683939306779838319012
046320885844777216789272669134017242280348558947361341223543578529835077850058122734147
005729124934847454189375769264019741048188010668638154063182522305782153384397722864233
181458641702277018027065675896757119554399459608242654703576099658266546963955745024446
403497040494339924668068603015644136801446892205200655153732393549852661337753241779327
791089999312424666263179797388688249690424684731930347930361333305790389216167524210538
309678212370574599651080911423894194283512207445528922777586136633496462872490611052391
922839947670084309133128854318953174497340631465690791289577985975314863683592475181478
646967086071255214110164716774322421637017073816414567873160130208213002403524216883886
643133730689677644196348009042199682789555382474557684850693682538124490751997253798308
152882516467456119880867522189843254351380069196716387142987462009572720677026654131433
285590275966051728074121324068050566299124453439220599112037656661417981958014287203238
829143682312224217993058006678197871646605106489623116882957382975717474588343731614066
131066156298985430281480459252977597926702560732498815877081783489414466170479551409834
544744021994565591418199513785986221562779134746218830994906403758570598705985163540532
273090076245417562751118099150469036857167733019227043639386483639279458733663224796864
251054674142236187095873299701323198920809703453633286479007451601086416390906198384571
553719617881126200131786568115212466004274541507768484061222793439927595038338401460692
759306501914427489784699699002780546688163762823447563927009629634961066679637742329728
912673421989381443660955889155459213329551855751338311422133071999267567656920417708629
764687550926010026345942121704638383710422040665616549277666744833681288312291792299507
680647602518626385424657849190232130939803335250429112978110538821606534459525827018464
970206032281518676298047592294645238702570573607946728955985925098999458704986204825877
016673578894902893301711139452970644916125960683006452113902846696639754817408215506377
418553553103594659814405944128757463385787354672152978511984114028846495231421116922552
803249379215827474437568927791079859082417138005142170795550403168977800868214183182820
315155010558112889184759955556650984183483583522655330960222983154663487857108596983249
928858487953394794023246587023245115274630857661323731094587544027120537198663449628531
442274586379734586172629582481971345131689666375854676246999562852002254489935311318370
040462629647296927727558931352493418702951741145678638804417640819461987796109316460281
814911975537932808430738463637721219967812060875682995003776835323903116366789941635652
056998758108430345243488631060330124532924969252532635705884486313779352239666851416156
618088822717076533627675876051231959738469937303222655355020439912318068940458270029983
057378405882337495842095017692633449969163274140254190763850106432525466883251789907187
347195860482187437836834287900655771963945772704710472250049357716319040772786887627698
0188712551264270443156890707366069219609019969747555262980396098316589837892224853638775
819415973180685732532925040440220343695819525598443539719928510265556238353507846663937
109771280744253504980457025213786921002442190679829114226772094935128270783758720298900
774719247041858982800247626657659261072348440562912696647725622446382312426343294594051
883237901665236187780583510372240359388023779863501745078438966193784371237751213443763
089182604618085382631652494557960037789093831527560635137382694657985185558222708241955
776561860106652161800093661292104083839138089477949036598892118516351182583416899076706
218369852265532192759420076015795296298293296986682691633370064562901068301033977622048
122398380665897878730707374891237646697914888389240624334837479708369348019631614942460
919903541808080984407649245156331813487948222402195047315993663159964882972616455407024

0362275462816330123302289859221830504049652006508779753525534509935645463076830044196004868499061833883641465734644676805947286339198696294769952224527296021653312118403543664758270058498711984141934909616341449430710732743149287178508554296015855355077949175363671629912432716370466491458338515539003303527433063289058639902531570976795307459977366485243453062933103932703915456057281693337251044713033853527074313225296036408756169744405456625200784215318802447158257671794444367223753577401623489472674009435098253067856321275609574196136543383872014652461468129843773498065867578796512330904080136301555308230248027931289591430430922777834473548152770971305835154473677120161192471696127809488511158602722648831297588738794177877924955687036461840173227611252601841145318185966215523443251806829302752253779144710090663034934540008137405368565409185768693607060773238811316741482014741034739077501785716359913012120738945366626193077228429294733441503594299015508122521888713945211947135995824936285013846534029190449588498685858028611045426887489260831909084390097603187395356053436027816831907656532798265894054352763021472794153930625893891607737489303184176926965632105185056688883037210197442561862339380455355904366473767712688932348057510204088622462151919692642348300547945701769453382424835020200226613589543881374506553933591322527185242640599351354813554689023800967994305748637026118049119837088883914827043623845823060657379535353910377338476972734006285156617536525371210399768433116290910059871971103025737319922307166815615723121147447923486050071999756704710496711437236158370175559821799530202388599860982509522294248367605256432760058787859353816309239737103541557078653345118066145870344649559161097303152504497895399984274830795906818826468047392875647788447061216445185180801691762247294330407259394253118979076922324427222949501281612758575988902723981820144796798385380437090931902891324494165572341988130576703758641613899443424224264877186367014317916369893573091349057673956733004520583936873057111366349826274397571395861466797138842049294995397897376193681744559272493976827535354797512182452925732421748277136518854807902338328808232916881708836732386671438828566289438857481295461702302880729999182094607745714896762466932494004492282151151518682905057420082102743609547819595108571470386562400591566709322350305889255209318786528083271709442372499917681379798289688126565037777362206224758568689109175947628654344603579137233545291658529270398311638983639230492055376322299664349274922629687399983442769336339582436344429736195035295669721798652087376173073092627289413855769581948055220598159027387493217120507387548879424079507150538587422763794716008794386922506136955056996829395855657831574582102200957984252870859817709918178651670972656719630409069311599512906462085542013765257872177953142808888328632624301972998545180561805845169788781595470329752739627292335223419242778056537016277281612121799781199577889454892644239046762394058108773947403031409654311227761095303849883042525521281829794765437951876900780586935016500583211507587639270393059453751138495194208116300820913369348061702025525756622974208136310409439567027642093133197101662736131616985546989553603828904952500825686988929060551027889485860876495069443683122089935294956757519765888173857109910788934352309818132724170904905781065417998428398772658024424873108121837959363271161996186527914945081055173621499774083701085666951345165077157885953809728566352157330358140637258504402699398579048079816014033374714627431167128705986400159324370721623817656046802601351509323325236021322277033660992284441277243533665992684896584936317302066274513169449964052766606666943454842231448121466228599436571584217476215335595882168791599126155228836357216673879779361707418545862548096035838844454635153788219846869552151111521708681359241122887416506060853382881131530858453812043306790789117519700466707651200284231314358559895875119918171279882051211596721401410771575775308036882422893400965517697439246008514917471688836705998719789164725451275108713122420100053506168725387718958627727765584316990896936349615590067745440758260161723788594586989339237695216214128523116086891744231497774816405552723319823457120639346736618729238488351079513814927695809122790328318737633691528256830868790178191479488906017791880596388088162091224742316596139081848455427267965907699014762341443128259070578598506122334054909018915004067266090813761435318607263184915173052065150803692715285315138907694600923724529188772724564401249923490922693966521872025235939583746178792684963972492872572564267278327486989882374297714253425413542346022102160319700742340293826008618935566573590972263958362079249034452659533423355923559999539512820322804564557617566121781948898829681434712928319133792865711552163826025516809048850014020215505047701798712862721848083976634119573580142343979595637004282589044254336178377526527418820221060132291386365058199526099629760013173133217556313223921480920726614137384964442594402902213691132054431740592192507481699975948527988798017757138377744044895492788306747687528821517117486168099208220322848904529371279041929109400489148675171610342681441000975601293512358085532565185642520745776658114101167645364940167938799340112840486517547402478796431629087834813793705739942327554087122968721122928991364354127997921261440862831654873840860335736704648629100656936525045380946205149035826753556821245889631607404636203262746135181523331191785708475816497769058746369180682858050841140068366679130627765229011978331217321386522565500410662350658073685810392833017400658796268320733457039769812645454919314526207494620980018428053091983664262073217177854144409829229369102811575198444608690050417525548336307667015929739560209334978479673112433540791640492766403315359020874460768426690703079181220486526160791971612994550824837877392796477282686265809288844482278937515134736620403392721960895814863162814180176953292049292587133057162678029814668101814699794466109569438359959799761003428890218850979846953149743955023988556675887522317208575985848466534079268357516523763130387622402497177891167497302544236998154074973462447493933606008475004761886043353183416927678421310097938129933312771875697814578829702158416621300048810196099837187650559414836449137519305393657026428052685360549753653029833627231897063639805603542171775345299402573092997057418579977702272830636415457502108386936267639091494150896453423370514649686778038016564478359407789116607589268563719989851078229016656121638329043277306644652907484779409251877096761792582464727062583115199247718068930934267082190061636572975989030274357119061299368220860502190924762615170222525421594431015343242284316500505359242094855756251036476895296248344779230505813010848214700899275682606965513633420

228554108815058783720613763484138398853403307178971948415076957748166191667027038723373
169056471544049716965517413667622661957671586649790203882441061828395871088242653730413
344141574077547314806877384789897084363123179598789828622886765731775530964219212946972
838337281112926976764764618230918079170782570206800500199864048582284786227427470652888
793508067783371921938177701291304857778163257255414866199462844789872352820129946752925
557439010560056080380226580612655299536227486584739699852722520515431747381738654724185
952981352953565505939994516823941956535993704312865296291306850612614664361677043552084
271429136070289829943141857972430787922641785165187498850857284384846060485909618772884
260851945775900036061031383545950873340972260366075620517834847395396518668602870003583
179418968081431381104317818128435894073624773517133347374317032542787428958173021066248
014722124996047399795180035121006868049666405577127837202228557455789319445205490461596
722016021793182261814163033427265844580014092433765931409378753156802624673254221210087
571381680871798509831307879655098273155225927980952727732868230640459200085966199249947
552498655432347843095978585989604655403299011899247038587916811840099873881943865164871
542537109858047558392294142244847616624689489615520913620289752363275074378353037386275
270862741125268440733053708845149591274746417767863977636695063435452607396911398474158
926480816047488882921806335801552580226762835098993069213269296290936028325939091376465
952034074994835308906252889710182428327239995420981860640520349849438807197022517879577
316171865851253723047622937213192801957429279480941115919166529621860483697273478199048
566748426245497904439968570279524141789130117635098665897024654384474973848114570930768
707586016434011216902674015510564018796840180377540253587337957329124853489974864873713
015615793697037491268862990381565062687554337109725873792388573365711988146488219605280
510064279643933202247938942094342471004793850429504837657938680933281903481322129989923
243685708257137115992938121452740403047964575365989537537653470914678112471195547529245
122999934849431610985591533522652356250815416605057917236602163073086482642165486369856
587260988090207522362242852130019753966700225137893048310376682312042374636504437204751
969566352449382724291554708261465304208430858554826358902519324832899539063550831478533
050080569778375485652616759741435352777571849368475925777145444324365809477317334415287
190059938950329322493928986880718899245359633492318339488316846337216256124869298854732
330487286194198737665360880265127830767731811039348457665389607012508988950080180934349
70914456709013723374955670331234968084745680501295459297764861903569746564154197214598
536781121383785970531491640310243774213998434005172008262868695677445055329032295929605
996598455524534755463147651661971493498306878551361826037166364682184210114922271969706
643943242761043271470132497038101081852028306877507082370288295386367744237976980595884
8320526129446965988947602706022957624666726264324333463863804263691255245722723694490326
104143552956269121789235801370040784797378192592200693038507027901397410780142134272044
277793572063176165280890032106678951544338478937126954628030864752199725145040346434429
897089211999027743222261271547423494447371428455309818741654118942170425230467119617820
832886246690885555834793867149842566786626840675930421161734712744513689712096595078795
548425136488285084732451822449948179188901489064314125065880033855172004245987375403362
884830102751507637721433595339104689224804741208517092093924828161214011488393208499846
594652487335761578197751895296792085690669771122854095831094564568809446452219142954936
135522276856511653029116623419131484238653561064359855153148330084635740043331191493078
845779359452218911703802327145702097132711662001521484371619589276718060187995612882736
776918304959397860829055733590402387512204854549395927460255324742958967089065050382464
349611132485824630664665180060092432503430628369220335287282519102783001114914671841588
625346834592186938592822403565317892573375500688110713617727210427504809288860665008035
739042294500823384694222780197739948566445358200091640884907374929481635238791231019992
769300316002213908714682700354025708145919911179279180012935351378034301158179777678266
052258257731655291760198109923844196766468236814572617736278462757219231597052177716070
3619237590129803224074187431121387673993967380280780212799124656767778910056722158641162
854880931616902162631007245800727635817104840257105728332161195060104356809425037070310
621639869948764968310405147356224229261239587285932254616928121658183337378911535578170
128139870381947648895191403928578566676282345902196113475363177764021834594447162138239
138395513475544161480202179395389665805070671606076264201577703338779276512173529609876
508309675773155358989430902189303817595844880230220993272161420581018044684396972279699
951522415583617844823191251019650503914644676931970969283903039221716236542277249310539
19415964426272887874025688351780240907800270585444451083451781560768829999535035655671
471327637064387167147401108905115052624880281852750812885813477235142159032429072301228
545963910744430138685709384655279356626515039931813671546093989127988821136152845835332
663219466135001374079077222640672545433813149140560242962853077317830077662551900018906
747546153088546777337860541501485799009165116162759160832990839245090949796196210689093
044466241468371012131349629064662827249810435884031733745859729955399295512720320931966
671178806030692052603905671368900476439555798566203632177362955769858392517227439655452
068550800381769831325992823110284822629933768193950424819403122540758315875842310273562
271402368087029124188731295130415999506756878275655549018543128687574479657865178843346
355016098944330708315653209214034295633699677855735611002787222231718469057225575716518
2250539067268636062062638519753594623800568067069261526893838883600265084969779580222025
854602099614411832832893376667875096411624199856191438710951803204471673007620605469957
789248098953795961663759803988228729296581815755505791542807729520863108174926214335865
355600171398264958324378306707374783920199010109098489629075873541447850410053356916689
712008272789728280533936649264302610307941485967769204481673809232729463067975500609691
850926866159518284649751117742096782791321615197684414870579348722797996840085844329404
080032639688895312130912882546814077543735818598184894450507307314777877339607336152370
940914071103599096532992224592632095246991956327727184347678245923665200440348015576349
029603758583786608847827751773726928642492695521543720581757865162178469250694927952569
709840361412664771891068805945133322773408525257552604557037085857710736351264866454132
886145916847219138619390452655635442730678738701052736298991593460976715693464654858202

pi to two million places

061663046096768347010384936659969965031803730591102220313071565930752982844373060219323
953698062338111481761422965184199183810191335133709709193398578658013007546890781793054
616779825031385907036695190661052706885263667964182844578076573018943716162809843369627
751696251357301752320855626039116957144089743490386501438355408263880393185502902894490
287555444141916759947710384499750097758806727780314871527220038429863934778206073816242
863381246151805359944833731503747923438842540365053192065040661944071655780210320107818
835399781251505230864601961586105044156384382215636139147921600417212857423605106431547
496876642782991204258741329366659116729649333822598931458469630242764815946177635020431
329893609111151188594916989225106264850240445377352335843455423877063386896020698943250
731042194214479856185542681705411406765140086488452163385736345240434411355736137869753
973818223577635732990175783690719991139484911853269802367386071899201701133621421265261
872851528824766362395790482723349458393787715037892712822907447099096326404224859708479
492382014687521100291150671978423780987713136757403269436182095589775679297354585570002
928173494003687111479107485679107511384590334188971314193435370020492370708405273792391
000174846763603292397642540995309225249841210767002871033393780985254844211989124171260
039925980614212015375586270805247747695541175206907304291479523932277418998607434291335
645419155702594787292704449635360238012840854268642754584341668622013040934100959022130
092244912873452449734540787774040600390073123239609834103648245450750639695632400518210
302946917820093539781944332478548094285779790024719353343660993831766625898872279601753
215378064366963593972951125481324760598394140508445260239677876902566830580074746811627
168231794357048726613492799726868877385188093643625465511957733556936470294166086810188
496976820678326458864395364934026340702803123079091446605778758436416153795075656780240
604127470039201064843148216207137179356336221374974692950466881155464146530509660791073
414587061024231325255463701056547777301103185050086614598390686290134775939192579122679
075500100395428492660257271430972094148516591555280370178875136744927769166745736801310
204391363224491837591768341432361277206750559519720156566795122899736576894903437383078
381815070703448777731675174618859776761134901608379410976443643621828030537695319227478
556556206088713314981214764323966769480921538731479440804009774304955543554291395212079
930714097228766924973609678516867212844718828345136835236226452189862327508183911046257
234203875447716848760828782158008249181819040423771894063954300711819853865447826369631
742563521096659040452235275961390770573305910986801847006150779840922154407492831409466
940151519203030265345583071202668370507265261429822310358048000851264950247307711985581
597469397520346068241818466257406448256030514167886934511061863492653225671898605063354
431200295653155469761421318364142494492682957573716907784444755274742005597432603466697
631757495922146723305483427533342090216919200280867133818151765134672724678560180872806
985793062516391815900734019178126337720629437217584348434738567376454354936764452021055
196413204190903475109089559010733926850768992818767336590183285996077004219611615319892
308668841564928732239391933687758631957538828286880243222928574906283469880031800978211
278346447369943068589274459351460331554052161323918049602851109420092006810860967113716
990163112495777335628881475776751337975001021482842434971314482215556944824425129651731
005282933000759688032026971471430557997272702402384556693200309336153400134638822651285
849604943374136513620175909550076887257237908399677877104327832361821807928864451298570
472294591887639017741016215492529443404466334553796035815259084482716504714210649494339
056402967120395788936520858735686384856403823401247326470157568285626763148949096486370
878037736537056228632422749199532968099537601176573560706688962334065473634628753017400
138289335696564141976244273880653200559357195259929416191825008855942980493562562347576
387675325026176810510294378948280038181909401633385362184824477857521114190200115722154
891014126574623409409760478059024692915223769207021504308659984976907725846512005462308
926730442308399098528361798087488216517501394768839219596197383521902537829218427400267
994986835504779867088743493487788426349113577417379378963803266412166087864015102669353
853929117586753982070454540099155480994055594771630227826822881996691199207772768569004
256445342466159455191246984940917971312767081216355879737550546953960201992891686869013
167822532699828311881176520517131424030310361907703448600685769071256829912292289041651
418435412215189270926191182789072038647298902831244963716722622607947058907730151579237
066505377355736785429749614128323981072567805447784550333802582299900614329006567458113
138370620933451382728850951005894284364009347525826153481549343334392320278401906340610
398822072400859571454785636314427709357818646870031383115060320239013819367383658231106
319423511375220125930221646506772395367401083460724736060018732335957259674257792577875
347464847438558336672633057975180627948174154553576632455621990677154365428762769211865
397630554166772737118471255681953742987890612319779768238132231036099871107441163778245
341555930694652870645143714518255481709386745886225997193200220125679477080345755340685
593956471166039497550936890740914114081986740342973089775931954775849403202835925283686
826560128537934159295817603428346619281327360098610986205325102030365438914766958145612
587052851476579044833126326126127547596737061392946434465502398563499069044171681297793
156045424478118151684323997331024904086120232035262599455138091994394242264420730669942
230207875737211345397840918929527188365151157728433565072833963266727277365527127783510
960648834916111659729647707283397881762071064236965358576000178613345742474716467547956
055760431599211599617105310000789312031291659211814559773632128131008845996859651547991
596733482069257258694465104840901607254303794793743992494817035408492948271673935387930
918914910594027388257563442674029204361297196900355320278074601299701516776456616608662
065098314906518399081340365789233077742757928881785397478425345621594851689925631541665
380460861937576610115928271916265498459928441380917909345028944622269874606236912491022
267635698462472987555305122807730603827231476727333923697147318274650823570649364585741
650652752803365758850018679830223336153074856777018962152237772807866297099713846012456
411393353523114439413631136412506303170575463867164921917781818872780499865142962369203
289752132044164594823111651568070936018232574719615128398118245002857359411366899363002
865197086434968015837737896070734050171102922190994235678759852991064219794252888071598
268543897365101995698553713435634484449994806623871155983268000913912923052134531871093

3406054852880528023480091981952064083701465298058753408827757824197190966845092572735422
2915973394209667721622675032838016836701268327087718197447113416474042333643324178455755
1429177479553973263018478223787511033467842821574771997718724907884569287855914978807211
5276216777487839372594667697466283646364324636870088891169603064008013321225544033259900
5510759601301217827216735090870830367828069535618798243249758543158370245511038222955266
9283087687897247615175634507109322670487519583543893267268520625106523486429896817851288
0346257186736577340840303172544172060562613289306744282270727818969513042713693386732533
0835043523622680664723620365359551522004610868541956015009851807742491375204290318837399
0768814511063319459431968016034238314294701177907339951880518632249363303737428987666399
5573004912311542547445635751228217960314931921755362155975173541771496270578666609166044
0129420641332056019879886479191028082572373344097806550004676084659077213277620485344288
8730641075216358008586042833817269955428440736006869980254634660144658650587017966628011
1375048195127464755134410941297544400517363615053233195614027633605043714849935500979333
9657556934728358078932701517616687071678505215157824221181977230596102718022113760557755
6508224510733940623404403522459232467132972924729590596607658491677679758175455500629288
9709458071169846248333631850495054136032446037122523002840366443270502682929121298013255
8704855451849166937175611706931575295002611807299804573870328033446422407624565753091066
1177342684363974742384207824933961324358878511433221439680160135550669500491595007129522
4097148418320524054713082666000289406977374674792357783279567686618221831164536152981044
3236739652109901286166797655130483366709558198145730160957319527868897617949738664731122
5813687331982562667229155559341762250892788702151915522317201257618794622116819631251122
6015128786659330839972870209589444425483857813403467475295709847983486791009529393177066
7779595282935586933838928401788831265834018207413113059012317875493132600386474392184200
2431142502717934449849109538269804121337806395730645594985291207547945343378821860367811
6869162464018601760989508005757040007613617583726869278889401279370879168731321041767499
2748146988683866933742989452900095322620623439557090365005347022096836720699559919206966
2310396834888855246790686617100250522735631931622916480153383634406669330611916306523322
2209677175050766526506928945267104663596291397570417951995407638592691435297273926525855
8349226001281210594341546090490896668448353479476740184091693440821121437575991844978933
5887981199988378104395450582209265478153024192434287549661529872752062512421645652414433
3769982120575773976979255452979275548311022227497804116723290650593640701326838116937111
7861709365336114661791958895816380085858227273146398812507431885930628540477252197325522
0103315021122179520602902261508148427987160887715683048894626628110875165719315803483799
1380475609981192779258193816013463820567615791947688858111977946573731371412896940425855
8642608461356780363809590304941436012976511086995327556982120044158668042851351485946288
0094502680282789078366103133193981570053235773709828419626117546312776400179255663368866
8410146625787264487540176093362492442105565222772843270865638145258939894095286299449122
6389851502639968925280789714373052711703482878632549311351468566743745502638079331233899
0987581997446383546982888408867958361478572004108178195267273630793701973161108035513300
7543426319758326623125873440471849275378555505659744347641467171749341623139966841084077
2883691239929036817624897833842304924086916049333585720849523415746389995902555929846488
4168530920932239450099842609353650712961038293914178384059673601317690142192294468824777
0343309980382663330910761357953567031620320156529942633836389917031552967101157171386600
5381110987649809142253221211885497858375452673543597066911907582773427165708311055881577
8202657719055716555229917225925199417118207127471635220224166932627204294585007602675155
5722364500253579838951440306567087689640251984362047673669476247213222076489353185298133
4830566157261004674616090217820368092789803584434628331615296800144294383434813251494444
7836469971094300627871751901583054975268249706681967084240041532931426707243910919269011
6602701227549972625900669589063641020196803150521418400014612813151431704629289361260455
1399255636599087035111822170092048786834138155836860001022139317162938175161481422045433
9766128328489548792451489203915346873564127793802822945618826470036726138100729222789244
3834143289121957615846825889532244491262372104309959217485693600393072875597764359278255
7219585173289203289373649695654810015209406770758433196108019191526970359449728205424099
9217747948855101892579673832190733955438482190213913546103957287564147009796204801679033
1823271129616918837055865294232028211967775211030427824285323075765857525501803872772311
1951540730843007100420922318378952510609510146744802307784143329653969117485526374925011
8136516701659813169341969516820963864880921930027649289087855906578223221964400442443844
5503444471502714727566153260498437996491046373298777309466552416081745527731188401660811
6921694723561545945639842474351310735621151612254696731133754720321066639057395945458355
7889811186178324261238258387629483664197843693455941473887456954555106879533894113498799
8990599543328825651993184489233911989102422072732503304780443482482631679538318812345366
4320343976941067634205074027762099250441628598738008542063309404490840874629501180438766
8199563680977412512147825641458109605463945483538312120741133967096575761170921701569199
4046703005278833850119606881285436914259105126429379763596380442614466936903741447024099
8707900609286591456627747370705418470014126066636777256970661853998281839518386935147377
0302170829143866834918955739571146882248943834768015147317842088430908014658858769187711
2165641204838346950799106078824258535962968911647995139741902725378504631972160757532999
2476702013619487991073630985615243470277483287408707462139789750922989475628757311491233
1819192777775556843021379643624950093850855751306244946161283158063084985942058804578700
4078785515289719360456282644804687204566172880931465167028752283204657047496845802111988
1776297074091159420791493023781858413123207558236599809265253524234756986784289710312199
3378059604896982190750251304431226998315156450957979143815249557321035853507966260643344
4733981207748446986453521343942991395784011881459016726094688838855923487055032101404677
2953629067700802205950873829180852655117756318673595463745915845363391393801788926019122
6055103459350757967533983990829864822423213165178011470479533400954086803420795695416411
3111313550566611792435082263851486519280295837109462616696734295288820969592714431338276
1321007621497855663453697598846376520810640392838659143588297104990724241461189967117100
3467866306169690503731543153385041441444438577519281212082102519526090256985712782919388

1403014206202610683350398499458356221159155783560795763515299610604051548273336617677738
340649685644925242332798264200981421894329173443574786681031630253844207655672219010059
2807309090078580449191845003829560291915383448907964461571120018222173427521505846427775
1374514677885234986818302513692382696295177021772692528700944035940577934112915061449 47
5258823639041486185734141833290220129687836159176523990639890991571159559355676459033 07
6816414419845129941860910715526173219470295359256820831108954817488732891594760874594 81
7098888310190065088204222186688229106327370360854471074710804892547330897052307538333 38
4151108973076302032735905789880482107806930863805361885579645761057623018844633724469 61
1597969699932476635084046605920344321714425712773923034115042505708630495725119876770 19
3866370920645610897938154224397738336028622288236479758761741488277700962489510182397 71
4125693022200631359251922488938193736190554035615770084653297526555503049938525731285 40
3824614959879327056558531969324783521876282827748273693948065961564891925347719840553 87
4804427669092596855956020727898300261155209277910428226631010454188865270353166747292 90
7085642544172223309105163560882086109904820627869701036948655891645399219118345402280 33
1042182748410590567150586091536669232563402988633518464432952007768643087419675280937 64
3264326672223112216556883679826461784773799624030049157803060204202615615506942744707 75
1415699612697991889112020018520257531724731667342763016248578103288925588957286530967 03
2275530583655334625146189699577629252114532121101739056503886949727673609383610760161 60
0140322746096019444999594979560584505532465867268155161409925883559627385207077637152 97
6810910648347980716144087277808789726535124904062481660472197728303827874893466324517 28
6119254819946483638414932048962869598306068733566963690589365710036075371832280900465 00
3332549451895453049904322120188516364217673353380493698898351079242738823401116782204 36
9273267897176289052900722793913491300733621360460004716889166283482916451595575378539 27
0614526965115028268661827758959898122661607881751724267012794003829724580908863980581 20
0684691912473047962433350282049942906298689379734617033549819695545939434973471194757 37
9757075609834103180496921651060009701890479059375973618113106497769154523557929449927 85
0465326579260730807671345765865029635193009323862107225338768892375712210477835546187 50
3180074726798105326172855576086282078706151858108607723302186836796716173007939855760 03
8996928378655583021998119832697537073423903902551709537909981777451373543552410778301 13
7715106908354935724496520163885696393828331150882836185011203512192635905462087400913 42
0032908867884573358610322762654808555613827089977500553907338886543813407467982328583 62
2007743023980791242463509493662347447637899454502049465865105148224418487140692473393 50
4624165585734481743992819138230905721084882235926205995688996691483769193218487552882 33
1045688426971807906897206412011258948396415953155910514040006134863689803300410077913 71
8736229031809775080956859244043436542169929572517284190594443274133733954296303223783 07
3191515300708542264445331886577340752773309300815301756510387598561132664967427779404 12
5526454470266616909283631440646666445902424986897093042461375318074737655449457221896 25
5431147366833366548611261949045173029357556767198284129574480585791471023826373544252 70
7415472746759206753801335471842434058662892762390587031503691069578792215619529661735 94
2596696087819400570044851976149808011313579686529299257307266080513163138433750866201 82
6705026616170517489695807454139305366056428978373583107420825718157625642257394885586 54
1942455333270836021679960477083195772348994613032988672401373815618146417419933582888 12
5851851003573452380690084108851670384822597573358453759824332273371242637479071738265 87
2435383217590548414039161349076412225854213741249284502344357782756293092404829217883 25
9182158683915637578199336078132879823246441195326615476687337967421828214959111641482 20
3206918073268749665647622017097835879227910058900894528715294738250934507139575906656 36
9185873619981668003303308835942704547691295638675516641353685466468387281941524493238 03
2135422770561273637768633001546553312177703061099449178162875941496620821749616891458 14
0940100518841373683179775869161117376998183347548193343643897230466831619894091319344 37
6712833005846268148660546073888294840353873975531902316135889033380330667378964181470 19
2799027197674610146874678869227203102321637603678745198691400258791080025354503894916 51
1851771879389960831446516405330895280875936215283363953482102878891934889816847121236 44
1834836446929587826327112171675866817153075038601574967774510518809579132627923144462 34
4961654210483434273916666511035444005134621391553046515006260739649020444665202414534 64
6435187022672381915311989880688501310767966553500410383254183728688876509101986164700 41
8730683014524064726127579026058693559451741036556030457056138221153316701990330358072 39
0729231061218346168799095108155640903626023777930601251270365781828636981457266704987 54
8870075077545896346760198384475584940469478412057822410457990133039112200637541437280 09
9548379449079438695655528395435630072145167325684883669565640421615071142293602449007 83
6280825724600793049452868598881978278076451731617513416052208717279647128293473188181 48
5575352150425326558054469241310282081249746390190130106208424047402697337939310848383 37
8581825080267219049125423944940865558698557089928353255747252044089318335726941045794 63
8663495824240152999669616295760947058092692691833853641895791798437251934688851501035 19
3131233006973567883201641609609375599781905134278846895044381392358949001307570675107 38
2178849969474636104245658180542313581365134325040581167835026770301018417413988676146 34
7671414178542565749132051248201226906896357960270163212891659104007383230877986462249 90
2610391651924041798479296129014418848985782125235002195382505140434477396055467593383 17
4140940994102654097605237006698261434020749502113687327164411246734564863786437193093 22
9333928546973857484350573781499362266905313731758394375432403764197910050284485247747 58
6528521594286946386079297224622805428644210294825978155123651488392854201050433849070 13
3767904495820654612373141701726919984050286910053277743495746375278631317740030946526 19
6169365526691378634132857811686346986445968824637437536206203243297676495139497815447 74
9011706860734137987524744966582977900728098950003532950081035461558161613653054067782 62
9403695568104819395225624204232904313035343812731884229211218497338488080293489185929 60
6219918089852558758895045564074917504054481701335071389702036643511703034177845566286 17
1572618308757598640529788258389693746161834363756511220684927438610688544093117677876 60
4395786884846480346071543195389209187630145428628519212293199392095296030135583554952 91
9585652426092253008596756367177901700844571391144313843100591576816750945810971046966 00

pi to two million places

092625026990015096991869959466590247284103548378823322216720817938074201190433940477425
920741931877374968046193309055548928929770822295763200540037271738836494743120898543392
350210927915802957370122260708274608641174781364120788430294826006552814565032677027357
222474210832190140844993769716739896642085714845597399302025520137522791961684768818728
745815443237906112429822004833415967961258481205249381201923489000991212920168330895984
108718855848677116453653218245365546996318020032677679992468368277154296439045093330173
330201621781928636615717115994056708951389834533996186393970638594528210694906634075061
042709130986167300015909615767302501668288426152732375026632084486392149052178070542762
641598751393940233088093413195044269065352613033984962843984142833279616914133550276722
016089926390267425875620953919360120504649286172177776065596275374216822214679095880056
741434725158601217856786612839687866654440983738846615459461313896139435544845543403887
339929852878363427481025375249206115264583062354172885502320286227824820725848444055087
551327061675270150982618563660013232534817716580290368181800999510900807191530263851058
480925832469454898972399654641232164009214268191900709396222118646033241214911916683625
944407696235588414392780575364392053070584657431629103517855282588177964202147774815683
823425565001999070753042954765390211156096939228041016247490735331709945916374292041545
888825153339539458061047439332662015578320737680562791245123903987255079370676631510831
805307195747939237025002546536397109416979212944308885232042642105170758802232700591173
698649081479993613474727760359052269770401216857661874999802206760931748305734484130860
138844731486970500646809194025697107683631555324255168354528006544892938735735676494860
168285975410956522720384873512124450311531627712171058369656500961872931803369917020010
352646853492507432847597453672047420570359451337511043839592098190427021307771306448768
789544430143070667201582190702800349586351509928245767041034955974656959433710946041168
074029836103035892402375353327906084326431386805943029160548874815762032286270092767688
495736387544338460873690962837101147652836969678836106497175524992320540980982995467425
791773337383322086774867955370269450421951155972443507423408091337738806591337525835458
003783563356142958996222086388689503006323799465918286945177527550524141149302644977925
146317119513290169466788685738714307615115060231558641916899576429811727905964272072944
981502475187258378612109015187434019881006991818994823833638377679739239371758027302035
314343810327700447415853220274776590939283625419883553120717605484234353392636007477422
111061447562265816389058030615122945790205781255482017320716626636056866082223851625813
504837094721175316343702397576598089060830312969737599634710545777869091675426959634625
115527674538843114668858873050400734446043264800500295051494086449828312927743991599540
068782360695029824012812837075073679223892835953394483457975805977428442212859654761078
573322955724409943083536749439233828848662203154265534449456042400957370053193811155583
027717533284257974453242605347223240517097653988473357855706242020075429409966387470386
054390385814845871487804606083865089453322289030505677297582953736175107965791881244183
290792732785164859667841672476465482633761893388230500296856063172150643520671310679318
116663216306271781347806972050406296173286995006355833913682910205784703573531824084753
682056449486529971734105035355734619350359250373305946704092946884688293700272687389888
500939502866553466106117725815072042695831982702412951176529798387426625355572852775548
023218639119849038441045154404186627727437787826483852943914570910332252329465880267711
928339771679388196723121278115173792078368490115200282373759593656173505895324967911388
162142759582102190309109349165551014527749896717584164167932121597508498691588548922085
616117832286595755906462546473816499766521939503929506276876641351969037146198899116876
531046803438246604630793901970568187998493815306484251358313885935892320896594200836592
423084627586015272979689454759600701019360519138736967058659017380108090554371675063659
543517465355230083569535848062358313786250990073398329546239584735330530344554189935384
180294397330349204665142435194859864134893436102080910297208881888776972772892840431743
383182176156684154993366331264404902404101529718726530125352514780497544647265911394423
091253289207909721358309817779828178149441225688698204166764613953412976526402629672122
539180030166281242295198050064750161462851249543965705117789948627961394484093361754571
581899755603703080278537943032505013639504733197373662941121564817007508560444806315137
063552584291766186712229301990646303053060461487004505358926862382503170497298987957729
334235461988049633391377727018145500914967033623084124385409983843705812481020257565506
479983551304123987965489691888509216160317664927602140803755057076389894605923933023407
565505713052913670577253166875124908920627388438457578135027547891135670714351527289621
724023236911075659954418027188436980738649851301547954814726934882726750189151082358637
398170132799092918179506454698925152564367720486441769353820704407216185336871092957019
213698206551769205679903021627162051936528039387593925174101543365098303255246030127780
871882556831611332118009671295095040609324962552761006924175489740127732702522171390539
159501556610364978777696810105776455952170042828369560295935006704456505158319330730176
565696852999526101660391068654998927129732994023656477573303600630814799137457407399241
132359171231362573147321157416109819932467981175650051598274043546299608667739171770734
822104143513462023454956722593190917867911538700690403075685062935057087822947167091562
009083097045556888624047504799651681090422230823888932091467491286383460594082803337865
913190839401068297948614759098801655365202987250982404071967376656623490667681246689477
946371129675755192321066434374087000103699485459852810352154720460368442291051996701478
786783092671391425223481619622903832668392866058225121737480311378697356044740440702595
281983952320379832900446659804139384125474288052111427038713915639198984117716083093382
952729758762224144768885209886646123735001084211985268628863430500061416333460071298272
047432612534719521191181261259302621870992544122272708181294606068788878773286196175337
066278803708702225816653309932801611476704801300656985277195872945727068477480553038181
034705425033002101575513297886481119112961690725241550028773214890481294508686553047865
675878399093809083316058800040584722985626630205885390311937773434813130492463664681597
558360869635025378607307802830800278221805811888970197815268729172999732456356696472338
798964264318906236646739877834932824871376125310611133063009979427044221966611241037772
994635881040697659542601520633737419951055580393999940135632763889366911620579513145257

271126624363342299010025468909376504783904229550509425987150075796909539050867255172618
486366427715536048125574443141981545770075853524151532648888046140069038186870720758768

4834461736641910401115956854757901755784765940276183863145589646608901681420812247265 11
3423925585736543927826097356536922420129440266182188695417224700695935045490970942051 86
6900676168946533708136822316677836807377873275676668075315835911196418537531734418070 06
0550680199649197319099828618224839488590349172345080528799981704137031400123844033496 91
0522611555342743208101358743829168093106061059588161797729074795656471736366154357046 99
2313131956326419097134300593404733115190424818895784177094814433312924282325872555783 97
3268512526626365950188507431682437732163097915953712374211251951463783130399122116398 46
4789766993004536603554381415286507593066485804864123277408394140241530282842631498219 65
5395586802776878246625296789588752024658439309958178182893651022233992055827013593368 02
0611139172256812090940839439018291718864252208537288902534587984370643374398714998623 94
0855239841755430722958212250983060263444905477443214846747126751348939588454455063646 28
9295654563467254595774993462964166224601084767557580496486063866689473455746069336831 56
0328584611198027225581890951099253035562042574740676224381272044012941347230349847000 76
1470100072992271364361007821814937460740116269325419029717801598448714483195193041664 71
9244015280904631025375310675896593891220782018065538359031070418738563998267374399328 93
3597905761630324356147740399926034896137978956329753516249656427646373944540621923760 80
7859136632883820029293822100822179561607772367087211562877966016751562931714759559285 39
1553976857916094413898512417683395515720224925168127519884740939870441386167954886122 42
4460328735453262455044546797757621800616930522188764810137964000175908312452973044557 06
6388946490906765307135848030806347248223959872687459229142740496037228172454030647994 29
0032199388850948376362208230751120289942376059851574583118428317651234768761198275769 32
4160803852917336395930221924517215687445544974491036192639516963747164017346118551266 12
1695791513923120153163803092774895576531749664404767892752160481733963310085239025163 96
6337662607934291192041122506537261778755964790013745977986232335396887794576871779979 60
2180955291581233188538399831361406465334609535689488244814833769870785377182447866619 80
5009085093790502186353282074876580876663407038002677001395876355661758588746095615150 78
7864437489425955564534874904997528831858905803155265961539875598519945074213669718231 98
2227990703837165644907526537767841613020446976753181025729049635981024296343340758693 05
3532503882829700806448263077860150670426402101179283980237279169153858599333994609020 56
9094291330978981580399673169137471191708544415020621161284579285972471535127354908388 01
8086019371763187374720982788683559611262321758751407360371158761338943742390747060631 02
0607395545992532308224043850645093736642168740471190722349628125891457626931417858308 53
4578680935825759677515811020142556710191650134286606889146178986591338191689625487610 05
2717553792469779495983001515882095606435944795687265247147830002105080076591905069136 87
1066027573936819623874636960365356180448792336711963610964435903677803869756987000661 28
6009089727637841706810110238783275095560498634378250407293987110124190626811847588717 78
2195756687163910962856131148513102629880184454839835633463698589616016388324516438337 73
9925175615382141925383092847275741793531579740607724898543424384382453558773880643684 36
4349620462208568177084296982344020471191361813682215677312033531619454105452989155496 84
4754005958920530570270812362643770963471195042811363387804742941401066271871766461434 00
9954206703556133362732534488319425223526646097517871676225283332817134241991129225961 65
4139185467587469789606291045609607608698578516400507883618933128926303596731357988113 22
8154701349703213729105691188635274337987464000987190849198108177584248963771314900610 87
1025507363644799082308632208516929107811058224580700980093793034594248034906955497399 93
7641471432924851815748529385711422921498090317786280905847193805156438967231509137891 53
0348471925999580280050799979897682631641806709990832996209575799355791674069664451150 54
9259994834149897951354376443615073878495970237311993240526764724517398259336487915521 61
6198665538804116505671637373740872470375437939036617695192957295460613177618460647502 93
1801875655555033904882550146711297524929604250044693657854970788048904377347860999426 12
1212246912003465728337060868407699173246106978949106646543195068889585206971914268441 46
7130256796909917908161843343953821576359323173590060476018374028271387912311930665217 35
6382552974441855461583761873215681575535943102566846559238320247588359255951566301448 33
8733468028772541151723651806439184542603290062282780037938508686861884600590837105869 16
4944434545650668683317144654694489423491958520481878178581864186652628245115786007000 34
6777967811528108873473070340738091621206231037907068582652638805098960059457151190977 98
7815461985683940224542743719769048248289768868040262432578504078460446484041845351005 17
0187869236198444496122152062906692049137516087297732254199910428180972262403198572191 47
3967501393802967133905374141891311826148783437969387834707191356691280563360167565306 83
1716340211946181202188998478234958154166682407577319765731900329568741580149128223293 60
9928807855328924205287605062011826251197066362795080659757421326269879340926926428884 69
5565957505962325130984536031532306445792555666055739549734918975207884764401952215093 37
7418261105878027766306575323339240415793901200053330746689770490718955358184510673399 86
7614686421760473134074559626643769676892192679088676378709757458562077951038556837810 73
7304860143483249014829257669750989327827179755087313598536537306866016968165132389832 15
7282266174767709782269046759050494659137413863367386395493417601910809384333524884677 09
6548681040709864057500834779190326711629914415764558863978276000403848287428424438614 93
5465158867157312719217620481738577578321941015646812333082742889505193499577296569017 80
3125827128598708198180200186426405294086375940413881854720874887907438357580102215705 37
5121078705356029312117031516024309631319026540753547826878251230520973920926587461039 09
2753300386855055820939189946694708727128946595370113772878351554308450766041871254355 10
0678627041374829158339574500601953400360215643624388818318923400263909244859268150550 33
4726194238998749766926960001209655375782390621057005919784207705057310478568380069315 47
9085201755944173155973470918365810434270916340942240606825712555855011729631042009856 67
0150738541387628608348441090397994900631076865259626287746969497707163288826878287650 05
7276006039939527486837006811516841954237948114406304811626746914327413021918202763722 48
8888668107509182776598933938856876638249141831178001954440842196270702506708338523563 65
1420162928555559700486728259023209876665133554382247699378109797376722404539201626500 62
5597190560021991174937026626367175777192266507145965072522863464184343231710755108558 04

pi to two million places

666217200930802825159436275318302935346278844133152752351594876559887995046170338856095
054884305286737286173769996197529663476443792334066140274083722996951510312621131088948
995460588905459764060744920276424449342417666313459705955848355249221445267177909721367
610794378016308180679199951121423842304886578459372274506778692420555460958393855128817
276236761159778217722196789875001301358757350873928790139894027498981343072068437980754
584511692459969952372233191619371045058243833104059024950010904180340197077884592138733
469393704088242267353424084030024067032946226993995473026833974047436154180927495284009
114363678909453527042118687516903973635367951735480552415434804208440853721478355844254
783013658611062402218146142924178574076799508197070611795756127606788754413161027026343
821720009740730158344216978998138293949416967316934122811432163603049679274613247557915
557969402137401412101494166963982452659604063492820645350464763675265001412607488918771
742814744030991232704805398789204395876358143955816503170625854907078558585247659698094
437772943085908679342481083500866990359292098395114572163532597777036011129571411236488
016786586259699346054394980334033708571506370972146606360353792241828319825770046092359
480007603647620982048017015967988161174930685991710171368583257142063944662881146946123
488022028274665524407979362085232558179666018227016554929919141776451647431129448442602
487781478409672150910808239857149545255879757274662510277729255190487290751796470524928
818033615454052644380943245238825091022409370569515607954465230238674297272721111523377
627644440776562544438331250172414578536718858657399631224181327651514676701406850787332
925883997907216233695381942832166042030104179994619404025963881925685065359815852518209
832502937546885565055533977562652445147847298695570796867605584726387835882446894472600
635742779711502318405949059747631007798430459506571752665210241246578268880009943747311
412875854650420236279463485829338049547758470233337787503043049092959451644581399712860
323341348581526161789503031017648512903182781551072044616008934673100966253658237673368
000816336515813168780819116985299641307152298089926851411881878401835775431710370497487
607761587561419320077687451406856009348023898390894845660946915572963531308441873754716
066400696433210327170835276144803053913497222203519476885434709404097050376633765763549
501615092242429857701617895118827724370753979020310031399436317825013009178738423672220
723593036078578096900768339959803816693350007481986001412908850692889696635149670912210
555284013929884518337747244442804550130685078335842132240909023183112937309893263224509
968084903600924287929173084369688408666907818840812509051401974124025843605812537519492
798710532516981340641771380661380117193494478611233079937205839995199869238264477952843
846598190424271918660981408426499157746058205263484658647446181594245306607911404908596
600872160507442825865427311540597867705964990595159312949083760920247678150808879147302
364187228922196644043390395395999381489642669022394088598287035971042776153065200186457
889045850067849172792388542153028924745696893457404202830329265521776037609522652754348
686338717261615421275480184163072227968865970299063635792994778873978197634395837425395
020588545304498756288732523815433715433283372579001894720828519820697346145488541118622
190112069898499585929071219059387499150066228027081394790538092328039957683585680348289
853702628882548998543588469514688053188175440985077562809243821181791362720550609427419
404356531708785619547111006489092905050442576861317703067553925332322054338859533362080
300806211134084566507369509673647710526308149037568836023555707825510124247058618487671
937304302464034509994963373705507621763253948180925704599463379323379637367036393030264
619036219486165542950215451427369006302908360164689918584264311936458743638163947906271
225307500609318770694729948526502778488467870332300739109482109419448219563706466818339
754840137177488484584509024651959458307752108442458327710439385345290027326669556817399
204147698680146028165307210992285249035578237503360757698120301173632024289765228329259
111555675497741153798752139325047753087007747424387247210216427850628387163803261943237
613560986936969237450051080535913307759833147723018751752934075582215826597470889069839
584815965211756789245816503566688335450379788406236977268040720112155415819678417500591
907469346485459657820948043704575829006909626242630413028207404839224464228505254583156
348487040732015949938241050244004109677570683989228912567495690033024761099880584753666
056745904171737415972506569201195823321489155840908233061654823793435902032728934659612
878684552969174113406322661407381216586023372345917390591570158073007174200231143248974
502010431861172865466955768348903433131847862732929803160322407309556113290476746009671
778198731316600479919561011843289348693551484344966509587370851794700194457039656716073
371752480592606726448161001872755180711312575273270478548035377198071239226320322969200
091235473531544733933837805299484477600858247588640654354884252744910374108677024902432
377615594800097569216580901686365383509260928868235259127505342745624494215729250729539
149467890887964172376971836175561556454145381931878387995929677541104602034801724033140
587641225808722220436063140490615798832761911472898002566335265033391746122368431299301
829037454796161352320408146176362410862469054994992389033616031457712780475870724998140
946663406198123086396602450554056833110527919491172277212606136441696919556437921236411
830788679866587562507445265868556818414699018451613848894967740330295507740705262079572
520205221242507031661529560743563531432750029418678546782720093516743220489401651368783
038220535168959537497586933080707271892356443567583317004611078381819522558143702121067
524633687832076745111963374795519128939138625977645164918624733716977720941085152139953
362567560560367886882586549586122093268291393150617024195410858774463906079666948859700
420345740546119420970785407422782843557953905665929420080807821081815349747884417949658
386185049950095229679950404191888109867644522504129401746103335549859076924175925701506
687062770409014751834151854582782442168383831825250869734398854883503199405955070543475
996412272843544640125849467943458695569459470986821921765116348782457813722422400920193
442703479821250481492858526922346714373200796670761522973844520662160578325726478661429
241744561667057331450700503406501738304247105358027932984401820354994633424085445732702
649924708635480216702435233490530527193036061995995957784662929164172760713595038243914
527851218457131160073407299161287438673026144795490979101613892245091907480499692768591
200228344312533648540100634349730472257208362753759543811112370571968613385115333839868
255350699757820042714949293594817995662225406269004232947138869471243516182102123324131

2175324769955172281103517897914013775353872927136861756383904970704540031221188046089652
2664788254916547992239216280590815315434711303750019830615506118039221856111264089351563
3472858890709228829401965202816446938956039641301910180863710033748824148646803294287684
7199142774070687477771578048176483529100515731810567892413144189438923226870676065147765
2069304311477230585595833906011398960534134638293025735832426555850244487652621226022936
9806525617779015915319880522269284602641457917879166247341657781602893004480838478661997
8425364434996381239301510586814186476557294835273736739024582043993407411953749481776578
0798567247857780306764917364134232251095500942703279053088817682691756430588162793746399
2179174044405413620861870268084594549848708337214862378121458549307257974154998462056410
7806792044907873099138117705588011913968782988659360702702369540819707150391514498554121
9139876995668178565653356269775207581581915957594630377347682658907313370161613190786712
0975725736105318096342886946338802837717248173530548252334139735963409674628602247488432
9177192984491380750359881395212572264970194847696023273524686439310186333881511393424732
8342634852603580672702296266435067170602680439132157139295382045310997757200088105934362
1560411595244992647865696702821632287125563534227319915050469885110105646664413505873739
0967245651031846285840394481901190154803444128591023778749921744969801462586895202813244
4634959338549767227392999829783075972482124443051880383191320263440926230520998975595066
8097138712490657730862921976625526732118282575933988010177866059011780597703052691886244
3323476234643963842478995779284764297580504566491690499419133605759235967803795624402155
8630522493298365271663880977286603738615950822241989763323176370731466519631335067079966
3027725798772253912421386245786787491654752658168161509129210223185717753164760720227977
6301815449738418361600735519759653870859597539590324629060290937754117655106334683075733
6506519078052511695382735583096803570748391089547813437905924143976521768111551100397799
5879671961319947277641521300356248243923892783765968718454846023218647150446999501753577
6752314906605306682110776357840535191789673317798563145213083633024620616225569604185077
0793256045965318415217468542640773974771820409952007614523003226089817648345367734021888
6739800837187063779300267505872399812156438378599494809482230564845221567052450763368955
6507224041384633230185776861420411556192394692248506925669903645011509767867339478893266
0665981565549422588593647563374780209553811731798916880473537822637346881236592720415755
8104413653798731384414075493605731045076505545228165165982564417121112247780084067304822
8816684387561665085826330172049791776229417835827552020592092937516710946235627131246222
0389989622932783108773372057205150306860187125571549727918786119067043356905074515462366
1659999562764826311876699131931112937822089575664055375371491843353565604779696007537566
1501011430354654476211431354725399153085049185088354217992109892346710107755459069156822
1692199579116398965135514853248833397502949714249832981642297974253309867278233985057077
0758277982677551738563143426783240267161105188948183850479437307268848802619378130740622
0844978452864135585840822679853485364731467927780384483162714594966648772462781803260733
8128086485392528572458915596811799992021401404806534279637744518720564336933263318818033
7365471145628325645065165631004619032934683177059245167642418636918184002812889932253400
4589106906625601045273698654464728254540193510233325684437033179785560193790614518470888
1373028138778121537980258934511739125492565343190433077425510129672823507193474837243088
0326866635855402888379535412359263333476896265974817935655420599962184903106756740308411
6726633462472740594281121206777219769316301378752459376607344204172908334121908798727255
6630634286858056935904792790376402796627442671248282056581024160868135549421640662351166
3846046622098173009213501362255251587918230562104541860996727484765445606502552317744255
8013786108298813279690500054858930190265429265485530813503536842527647951957856974949477
0510097160909266088264378507055073138222227186202488136108001182207374190121475591202422
6558065800022649306896139575034856622456345535408525481248447025626234817909735144052811
7544928668356365776918512647344172368247637070389739122484199771269733429266119537109588
8449991637898331251775986874272908150093361669624155350761829392917338072044522241864944
2284616781777404614208068859723926347221833376822462834604028634437112321798021770416355
8201167961521780436740911891062798154038452740975948872605787896115074630255240414813666
5013783330027942409972524008009302531018027863190105582816615741112171835699318671448444
7953622512704805234125459039991730123399929226739649895524772163846220617387644131984155
9953289044040516401998378978632395769945639259570842479740336030688858518156618607940755
6214332002372243954586021020178886162541342837073610809493330524284225998038151921469999
8403034757589022148112433377773461758492754308368104834972583640976744421773102853247688
3909562945053205194186136989035212501636503389264285242693116202992910360286494131506088
5658384759581779783237523287406455554374786868363961650079322158066792734274024741515600
0558381541358232215469184449737940564500947914881100557502030550412272448071151222955955
0652506078275887230268021205141800667074318165559735842043613422585492800940300440941655
7580398194328888161750761664263588292490067567004346674018005185183588127437866968943699
8654366884340235884494313891188048259623134591586754860004620433234247487444744940147822
6903902126838762649638679047471277539553546992377398867171948424635863575200940534238133
3999061197160622045129033547966300315945425136980881108910940111346738243809894095215366
6069635786103631632133762017381708480037324650323700508465850556005966930999989485233911
7617456472810062374682911002147070625324454703378241619126339419601160043019251072609222
6446747909443800901991671735348479286068490645672600475377508901255124698458573073301844
5723300128958259870731979210216087469105448430417136011151113563182814146668906831280444
5530915469000672904210020707015831025511965229325730141008203589279147285553245541393466
5882941768130796573587952317369148028627683665756559049049463000059344824015415234927111
9647369381683565972558141889527327889539234321168642997959216959651307428176473473623544
7721511885471866997853577071143190518860792684809998777639508857400899068570649641462288
5875235872799744938756645456716002690539470863161551392428857578129367993983517483552611
0339059454312539035128957421538677744496384640067040841529088634544750981207150645725811
1710354230516535384505045183860148772247628499682862252940282770912459500906128938367777
3252337567393259452790113502871085104777534101102622578236726148521106175979056145566822
5872871021414973236220883709109761908832942956750012227659111951478475721319487291624255

078423506332851676045179588018786648916185269345068459432095411174723326037841819358203
924185037051649878132410393074884039242227022483384371853634360533337730229456750412093
677067977898934544273498213214499883836566737095121881322913047643889806705017384308113
577608625390525557603506207041315889077820011366844999223480070483070333956556971704669
442487886324684091276018170206137952151258456641179593343544063103351859134495723259367
564547684030071400138562081660791630040076256156687327818177651784329374433645317000760
410954918456787005943845553488168176844331172230538090916375706962894197967514327228764
263318915508604238217543028807883655870850654509686925191057357914563685038546460265341
575077791018985552612916200253181951955849415325131849783949859989935390082675806066356
331381324010734516965264832534436493810037208041661769168716164442757461147372149218554
603431318184517241168483211866319624233286552158885443373892366247795295195447972293558
192195331227651139457648639605222764304035727787288596717363399200622646692735769214239
533729173843739241613504972253581065941791921917893335540837048849955909871226940571826
957658602361729883504210482776895475868777353615346904919758986076064702565475048605352
100549139104889825661809073779837337249511518627073736136689983232857487408209001521945
533334111084427388832474524568180541180992795812416190053788809226041184924763559953638
524020070910844905404246622870489353326682699462912149648780108791483407296653013223653
303011900623594623082303632781949766423405898317402254309657311701805471115702135628252
893301120590307818548335135357903182409031875051960535600947318850066069174378606704656
147034034795182738005282758965504200276402167341920519576567123117366419753170002776711
012495366732167143189746155484193230168383961159857734756261336509556783655357391120778
557456830420097966862002589428682250045915418242358931541436850217093941803132907088995
685318068774530212723575643929353536045784458023981010535281311548371894601915368974886
549785287161521853637100964893902232064437962855745655496208729841832171344203830752840
642005094377746129125250896064309259812860246489884224923348607475996437257410065832349
231576843180785246022543334950547482773472633558673970258767509349767215616266314404375
954073356487037713464962833419307129474269294538385465970021636295074800165138939798027
269455868539114956147862078812593769859980757337173500029854815280256020331879335354130
587389744122891152117726424420019422383166059453832227603036249616046662599576196473409
198892058089590274584619985219209326110678430105305453201191464754276208440176443855365
479813725251165864461897763451968408223738988904430611966567039435747680764685975305895
222540830331553247102949087686461550482503110152283610074358169144014642496472806882049
524014945618142084856317625512103875314200949124693044273430394880730722910939709119400
750934960189628850476142457455050981380656524653014399102308153136654846947334726387651
600611386544467157075075019877624437545401326899411111900335571338786948713226543658912
798973034600831795409659949595237765288686972547460484795404832683268076290899130225082
806773258271979576991006257854368244789166006345239536579747671215601618331871350133870
039947182164893927289367569588375075496942039022503799157375790697572327107289991440316
405179842606853794299292568060776899641014279129647997500238172046624871017583368889294
622791734279284187442938250670866290137177495575490804459104798534069551812210551511815
993097027729825986454276221255515875621181555369892547740789170321515895651306236623879
336851924914068232622300207569101278368008574012100992189783357245578078619709747175612
208785809423847824272548766886524351335859742855265691486191960382372887371178004311300
605863539885834075967983776985321827193778942494423007922656559131681716936778030978977
849309735389044362225566665111193503928973874855793904695374291040765977395542685426324
111711605570765040583149272709340178078673269748722468490630622575348908197632131017104
327106996441890129583382390450067131999029453105792046862460004219053395424026828406913
204143206286576406980330059554759026005330982642700680232788752557743634631512469224722
383415960101530083400221445366197343588801958421964441595460320139834418802053994141684
273160118718063524913447434122596520304531988917162324751254021995648261436349731268393
445204985675141923020511435397849916647543092160048159052079386679535788578519665555517
825783626344977536107878544906245716101799710495510795259618522918146189614223590121972
625963475983385123638480644315752213379854915026210912917384405208115476840340530169253
859283586368652119037511365797795636221826675121314173559182238880730103921167369888817
216111364104900615805272514965548320976615256455568927168183421428909638367291775811520
830527097472090443212607220008539805320479769925646083385926230608457699732983872482143
045848883121649071798597340741313508727922912531669579665337585274527363355953399694801
950287814100335429116047957239858793903398833370668784330427841378700950342948267423984
903016394592319348472907914407908204841026050809969170069056490311531518372319464566283
163930945396706226977746993894269275239768212433725645262552038911960396654403023444433
845803767904537999281668999130903029824057317612560282756325391675188598180745834483242
464269847397553329648179810230459689359136891531948348697651597970783225844399834051323
916793251780996083955384983850599346207555909166382314160918530008909070294833398527449
914497460276381194162569034694702300763013084439615650041674923334970924824919653130485
398722230286635246487397332425030523706532380767257599918695219809284930097130748594420
554008854482665741166285338551522726759256027019560985161450067183200035819619106234962
836431271502540610155653846637472147371674226829240883657232246836703965447005936702081
785727395304415327584744890952239885149658410801670378571435832952294608173252838249079
623844911397665196507550220564054079450698687486793140544630090924368338293153693890752
147810811047528964160791513317086429597200956544319434729966713151722532397439543036198
335125875130912141038455761687362730143313895753237179171430267042602004454665337561719
295723990616348042202021852723870439562918415675823993049750394724013691151748064294164
890188119691486469109625101422630207598832589733139982435367486199359218040712775496249
989422066126308296128606468573382391797249324813417916087682719409556399951479082831894
852293715425560900178436910995450146540633686862395199705651684519232418860612853266893
475215403260546255132622039596923221658547838793082451521968188150647315933007842335830
960273414246859474108886913720908551809824834004205124880182611184552624855956592743221
345717106772577063555970232404109134206517642257569146270612969193035145694397264387816

1938881184711606466982799656086980162947583865107027931544760147804245158912325307887291504641587327521023091986248153528599014916180637588873437369584127780738104950991935137568005864184619951835406298458306129652984965952795355481727699725931949699396069829506627369687764188065994590530316580292904135804112903676935766362467899970613379713245757859335963404137909390888178336680969502455889501038139571465855359340653450278980552824362055227647540308578901977654914701637993956757679563875294519163701876016423776556900786466964202749764623821901754713553243982004570497689596062151557728504156957257837971178845064309325215157061984811353382152205115331723276191198643895652023063404734883222840220868619523676037436065834351215011838544398711563758953671014411460871398785572384739365710026932283972675322348308955195987766895690410259687336571891442779738553725380838902727615738585031502869606700378593723306142212947914948908928409634272352299535110905787213892350983938098408609928505876716315329457022663284750863299097919473124720502013216278396715409976134964101394283177438084559681604451156891513916236232836937740926630204367047162301804355757466457103106277100725700483872159000215555696565132540558211223380777394928469566859532517425946831791752636908892966355847630585074861993882678207389773959267642292387562942687732270566672591527113317390478875058147307720184417889623179257272982848506818006012163424706334560551500428273133486858190131169234070546061605372807490114722238902623836879909167668792931703323625773272354921456093930552056659290786290185897925212669555969073458642517432460304121538376308039861457777135753880511802118433979380338303897482312733260584287424018389386200482042581772128532709393766364822600963677807180492901956724479205191200157767288377976298328837590377765612158793698108964426496642742956449444424437318668073086169166572408368569628124230295045135476971925988733250635604268934284105592686219286517913415853547975140113910454930691794912261136697196288158070475430033976161355522205894517346948953523173291494108905960033629475442845881359479134241719274098372701815585115651057219309434917756681010988334073499054635222005162554761785258448559938916620556276758248415593032663811226945287873997844826705842772307087083409740119950138067872376814275836124221057825554079037634313712003690367812134045731627738650734848428060349247146728329384528536263482982448091335418609590343809282418962632968925556729359140882845846795865921104573133763566044294990268099043409538493305691851123345065920192196818565468360243016393171623701091610501127548381053040082205793177066064022445106067925876751078209822826462977467825171732380495539910754943941597631214204382284864528819188457174453602486842895688833777450011323927190850053065432740917128086981380030173023266553322037207069507936505258926350318572702581723310456191025656280491868097710006894171127688939974058984092570211687135814945626263789824985209661327978391196531832048441535301608337064995931183344072671903512344298475109264921074490746830497148566750472575037859782182680788712731425385622908571276306465644530733294303935873363769654054225169297208129718344184880198945872212835296504296027466871564631372364742512649236779631431994145827983999443378350696501507755354709452081493436717703224113226461402018429889395331149912334193662379020322647862640756062216817289912997398297073887574255719997083932308575787031464934733603375123385816101978346216298086529623712676987489490898173234200336137929486647251847885704638864463065707428393804192645479177960974929186331038786226817663790791244250898907025441492737246986362784840881046223220895267317559916472568378944302547461614947586921367763516401739713761544886713823265700752642791611859117891515598137020584447571060466297886108664532370840320852970544040931170706478053907040033430996561148623906950986443300830309398092366662510361263377617770553644586125191136968087935594643136189132097460664557354624495154461199627357636141663083751498965046860596210791474151377762433736938467610682882561718494062377420884453511039345186440563381145980544389098882730236057582587526810954554022740271786734073951532110726545325286695715309376570548884489540012507075355169202268037891338227559186049625015968745046922388164226758238199150716707933202557453329709100112646345215980918427443086236276795013569188238584321328494659250409049603203537897122666395443390287899823255960484693289642286268554458617382057900171152816292464779439611307755831585512272594298411425721639307185741934899302535050665564131532299438205704523599770080982965961756639912826446818922136116786472674730121151973848034667218551400222673351454244731282647140364127942985462550751264370710834440366887490129169586983678673323750602157191233275728315977372599450919499091353989119981846130220152245460399013472809031593029067079242545393470259511030893934888914861257342968929352764911879096133490149584345017369077474279785514022720592739446483283043301289538750580939726250457292166965908273659679263650123715093211034305038484321111741340330685128753966028799671875167589747968607114323715949176382027505353487947809809595465229786720134820926145430927940895065928922423913570777042334600940446989945605532795246990123331165548479184850718957882142430904414147707636175063517982786476484844063990912102279749936424149016727820134811561792922568053808342448248593401110076917417543760239544056278077310502362111747216912387547161813099783513701463792306504930402528116701287227089338041645594001194593859414542874671139351738948843845559443245702960862532463269841287945310817035724725353504973464966202505450587126408920791156351533691632890924524446999459290423921742341639832349855717325035468589989735368354673586241351668452909236047710316502924594117949090902988646972561980937337672792648508868481744513213171378627122806219348059967684541656987148919379873867866607808181672247149109856048054371312777433368154384338297712009001312647725584012876633558067748345768810872935035921968212006976331979536243508901188769794202492173271961405338038004084858271910447875357270403745207075094276509956071237096472375062966313402930027715511481175493445048820698778644102776953982080477744995224483435321769795263064973385778333674034127318816947050285716949737410767777509439777859056401253187360058192544913614439779844246627064295749295833720989203077565678526860537478987882380254838456905920560864889244545999409813289109880923585602156622417377941048951509233700155446149212217949295706795439880862314399844962864841680058820244754255330859616638695754246924665668446935341032513144983221961510403037972122768577698065466106585403781197847278102566762071611536749463762704507040511706994320351889415532687438039318271282752335367912422

7712466332035535962372425487649340777885201909645741528261030067834946063654206413645244
8147103915839616993692455121455273641761508290268123731628998756290537195992437919347699
0214183219546776036245757210228358495298448024244638440071675316002352057333125169803944
0472322398823072457259578619254652751772592073272181755914688555936560083314877331607144
4430434875813118955173412422416995216760934089448153155605323136554407176663299690251666
9175010921827434883670812469993230779028605323678828735037789410483543854387902050632111
1874923062069520738868084294719939430272246307668346852483461754474533920484880481820888
8606883968935512395664159354392414725603140669195728642524704143808215698484074606262122
2178174048975514934849388024267870598008091546326046700134696841032823182449160930385700
9446031532167343856655091997393275256557331661296708349475169378174963531855055018161477
7897748667974112538496897483598051590890866138925005084672614873889919573890626317047377
0476209862604899920363902208634414011833773231398008051024110620197823842221653413294744
4495358046385130953270748921950708134131193905556362766406031527821171991249228651191922
5609435668272859163936096960701968963468696170527997135088516218460995793734778749603644
7904131378954973747119951810743984612899500120650348067330579895239527371663111173640044
8370192522791943152999196686552569019728568666112715072418180834818778556393349068888899
5745416674245032309759590847754843918119080949390200362681004205303891234436953950405799
2348400663128295228663860905869734844590917924116627730159410489821147056701065023574677
3426780148268883811919809638600386998335889596024926707620885142371840374777386559506333
1748847145257772073476445864737897195816827264095255133159659246845445801097930126414399
5676032762936312601698376045706027828139752660797623375303640346239906017919585632671955
8032424261005169473354029425051981489100593502184603081212278874202951827561828778036633
8962642004195982127404280416279722823286531612716764473708806377159518150846956152258700
3584196945501195023499667402770810027124057472948681220280126083429477503765580557807833
7555701611149473616759206134223757433697199575798810429283853378741905713286108357702644
3756829305748162633889969023351137871929261637950117232660444383317822546233371202391233
2055488993881247533701332862114672294207729329323830576445577530636970098234629231997000
5108864755707017175590992697449358277744923985062822509163695662012081445574216165447788
4077647507215861139522786174074846685500048491183095332236183514585538744644305576461288
5228690020625520855516885215413226281833643901136834907350346256321834315142150109054677
4031977356430634321824519726894097949625787151236524228970744345717533945075786104726388
9673051967766265906354928929365943847821888168136863016246692340172166661618732069016155
0381196599364717894138225625445736050485786950453256964908737819761671038928011411655266
8273608402758747391874975906003455368967169107023174682350173089635298686946319826953366
5612936821558043520411263534705397533676188442752197027727673154475008726192242315923955
5863834032381082304435149206642941982961446624108993858742443077443208223958380492852022
6508215598752872371657051097921098867914571821338333584679052932446919964404538011861411
8407664423881133929611923953274196328608267238487830290791763654357503723401965213298899
4873037364239757338754668260353231035813673361323888751753170650998611072661968845780922
6334312196574235115952811082902792453590309016244184966400748253413288334577202457787588
0678519210094075172814044242122341941901514115922979920083221764545596080969682952768344
5766386735760741270405782607954705869742063707587039720723103497441683903113467514469966
9161006605953456117906193203147464740897886878302934292506021465982568854963537607357100
5173632037176664576414338591026534494437601613720574177980885696455404038877222799434211
8036520188893744073314476573450932453901834400144480389525925392618618481464157774195000
5957848566932596113880476541667436901608551837429068981208761122941275893593211633899444
4629367194883430155364025837851474514294594806672151280575235576269918621184696727000122
7721132105499578090655576310056121401566005447093528734042723289361435054574282696609400
7234476825928834638000750511202245933885230046626753646960890129571584927793113124875811
7345176883183909359251004672612517059600958339761928777921284405568995784341829850791511
0478138967565494202133885912857544741863841094508394635970231935642536427462361903374077
6115298682589977995057614951071626470754115073605021068787510110924315229761177581785955
7660225026405537841640261357158000103823740319082490662550530521205734784460041910908133
3158382165114465961166716196325814706118276218787945439257031051445129791298293102118477
1967751262819372132662528241299680001230684244152793171435859889174143969724728506330399
6857464786636725991813912574601682235484131079072063181438820607730854321587372468670411
5372304352342184871333426264623285838842914910945372810929802550747724131360902381226488
6410551513709162143890939844923172893238197365434439772676899218514550973829332267882955
6480698504362470319196483755833914453686385342553017640653862820676772168538556265083611
5590273725751417710255991182198784480167990745683079096851038057985082251994010616164177
5417538998913469209184679190928059817646234812445550680489944940134515186944672194912399
8364575302157959007806876689034366817627941874857280021834498569873133544254293377642099
0600065336600517219960033894757919930484531635814104864189226934325903666972646410427233
0103882999765634667379805455747223839321398889623223319815589112452537941099975192317533
5098418570969788026415335211670582505887171241196516081601907337925608276674626887061322
3182554937510544162622476963506747274300177942122398421210770350278396203036430372113277
8943591141024880091996592500760965533699735345712420481148581670254163927080443230211011
2368616034967285002892386147998214860681020946698016737229587083956186070302831500689944
9767683040668704575065293856564826562129412231861037201683773607119471394636513870690511
7591983876860934840227114213347700378782925320093588126380212087713696606565524523346633
9220593299377100399565453676530649072456539323712823590065116166827762268826164790215700
7809419981899191041272323364610474234572071149545921414722175121818936704126048811514277
2236327430643629342606828928920251720306733123017632108723037776532644050635350237366600
9809597645925787485849504223607890894057442331548961784429593388138848633698600724277555
2855265942576240993309261629265329484620859542325814931361405131637903179959675168496455
3008203589541335684402322377273187614600330387973951534125819443334720105962975046393955
9413463228367693679603514084056972720708411221298452339029297857708388235996518814959777
3704431529262421605813960114170605468457752110165033898461047023274954009638277518259088

pi to two million places

1064340240174641580171833063991221309281520378058906870119750291242053501923819828375722
2519986638273554885528541092957843426734845356387395190931003694331900560759337542632599
9334737631609941264115586694111196759873099060972024755971819308243348219870667317545022
2216346615225386739446789688590309022082714211489745235774715545290523493190282178868211
9240761759499761131246659708022371836204354321557914964127786630978800803837138894209133
0719168444394252939207115285358054305076081924444525776989381547733718653369189114005595
0526461164127523568320326028238299213832903275787807811154210826344336342941254695488833
1436994261830531497741160168081944212222015699619208342286471177300468644051402717552722
4640446568267671729349422325290693444737258830373554456546957629606758397595409824786766
1994480933417057844124912064220893830747408885990667421330435168078099640027714908899199
2683002790565219265311083116106695163712013628794650320151480277968507337353162737468777
4169343008438910652109949081900554669544000199098918955667979760290404469716686812566922
9203148000329732865798653588976243396052719517567861982302843444815115085977981001428133
8242229832962255206952448941751814472024665147061801970516553173443124564324206999865466
4291823590140804636949067663452689945320618663487557511517014970976564571283736813195299
9622506181990259498776315013794184612401351330093362387053965653535479893198123321411988
2161549791261969187883391566951064638150365647937915888638264648502310118446073879199644
2971683002324635700562235499030151784748570824578233432149993333578173522088753855765388
8806034944646022413198353420272425810978037270601693126644612228382980158542316860383400
9414721191944066400620261529963698102373569865421979819453962713429151668366631478176911
6440979157736189936732084127701424844518721288226802065071786137297767049917704161670333
4228925768536130099874493513675949011369756162992944305780746531739679021919603472620199
3690851460733289950446797755916772545224683309418971583790681803972695974872104844155977
1359024602179515083934062401359457672560917471712010467739530075913659973536994141061911
7213126947773809109216880328882701049374868530280210363563159197990461385540422162753244
8452397878644057807116294496712708691054505914068932329854691691290471796011687543635822
5721147970056428988781773400342073849531462571788788080085836323568737603972941336591355
7180330637540168015375439411151945039474025062365476641661657903689786479286355291796055
2203106065811045046992410573907893430672989738531551305400054280850683585998598680808511
5397270518279766043375804691638495689554212839432531338072261818584614685316615413439222
0358101885858332283668865954038370046067294963198085372041109694174004482059771538001433
4970092475005920047941764990257279605782169644690809298703768186647442914593632961425422
0122241558528176841790627781703375876389494452121064828851510403656618445753695388719188
3993600110698702738613112536834417650437417935370003917046708865000858156845394789767899
4525528462269887210321667943144877599678574364631999991905007904466971280532223748470199
5839035098822049479463819140999969432703020600958072478670619668070693104064030645176466
2099596943946305242073777986342437751691091401565088431980980698110700851320455086246399
5957581740102095241884495797736997032158719436844272051127903981551433370973822918960711
0094493529209826188419750225818256889164409216185927077578397093704713636746065233409000
7740397034740698910081519755465832654916619541885676749504303887503132725640168187272199
8880128463845835282224992783784683653494254789010724707319297710453922235071915141924122
1130716868836434188577717889216143425502386624889856943094494232355267644517133158340999
3249379030859777726044372464096509180804172067951608181178885043731365712809986349194044
7277284822054200817211038393377273444625325418180780732625006908787657389874281666909411
5291329162325219353681589675446283717695080756852172903630146572635391854989917040946955
8091096525285271303514467277171062438212145321661690773660281055756602436061565209253177
1682471353684078120004340875642005916183222474434704825900895013252946068470593899392777
4378731913219704493064999365759787248112573940185833765328329336869869465460707820949033
6934842104661019852508435301256691975650847957040527541829421337437460657759563023862966
4984864061537038334354873277221205290601932187825936392179582557332671340825858009317700
1096504492396840529204230276654886186977950732622895522686559513751296467836456268163911
9690871748525058156475707832371814862741167378399975649094277827797725661883051620270688
6153151392692503823083561542222028961713790920620869245537890752788071699779617578037377
3226514514148386362413356977276015369288753705718223999756436465467838127514772668381655
5971661888008046876549117520064839117226208562722157581781384906628093517358218584097499
8803200479063958513269144187898895775429780226839673974782220150071793447510455198229755
7570639287176031428527746066876919759147497890436986357572924726203422512356685042712311
7811765866004017829986736365652448932293221489324560827412323686884532981649822601018988
4115535207782088141849003914532683018786919475850477818466259852238219443187160232185966
8022896375637728786584403972428978433864414464738758473465425833325831923504952543822911
1540647047515244365643882104218794268222660452928792728047549789482823000174761944457533
5051150800705957135970535123304405733709577642458818077679174692382058930824815641534922
8873144557550970755121978150175832649590237344092847872373652700064850704015969115731044
6394470104057822799031679838276533440777606765349506626939526630743111721791799309957188
4593793304528989679436012767541843570592613179896374433065500669820214155480508203934399
2226093596079565717745976166622466824201521168660398697914784891632193306919898685760777
7301876184891696509383662229661771768832462063906452888506165989375554517906162045705944
0354231462082872716823474162961885609422893764832967920351419460777728275686696890166166
1552198675650164809301350509768799567849120584286934104268146266118169196312192260773699
5533455735656557662685018600784322333752629259095712716213408793808571472874157647214399
2656366580077222496070781560667780207352717045016279775271206310258602126773159018728588
1448207617090730772315983272364698411364925329048588241787922365109878104218686212449633
3860836584188014185114588648402763330011586915227636661487900053435927708039262321544499
9292687753145412195749162785234578137987284767858479429735287117292468763647401492302088
7373045324850906622816706109087950300101727998084786147968741221686183674823363250273288
4275717884488522542316644818713318019790758979167570412586263988325077227430246950865411
3295592893380033862162488691416451157370884821832246426144844581143464740017060078983900
0992206426869991660893737055334766898788727930333114690475998571572953357646603540201555

pi to two million places

9660686866961427060772670189470946912387133421166405866129345109397267472193914365865 09
9243892140541305484670475839283914341638005939753099627833333836713518827409186051143 37
5029081515267530653717184904062206377660339971453317343175744075793222287930233658583 59
6388198251944327957124933734275285333563511696418197538273358721278998447590791139692 96
5922516412945444920335087669192617793934709464579231569363142171478374305697707204662 94
3194622050193137448308467113270666772921174603516320631906475190653497256714174357346 02
3937938447983998915092402803048663812005298340991184904192230628254206902866414036770 45
0322377131890375411752523539994246625563318656324843212023978407362647132422481438485 65
4919106636040278397671136534397767327465993022433651856186364922133214259621123917768 06
1412617543336223657525038240248790001170522298791079198403064287800994939994455294184 75
1587312419070525806007063669394988606628272410138615534525861211600049697448251612995 63
1068697764668492275140836134162973695387932044617636583073096796666606964117497695239 77
1648695835449124116435295054730515377777115106457138908751053628780515067054030910931 43
4748983314004631423994849764221389801457991095399470339468653282367951147012087259207 10
5232093056609069596433321256111080982238419634174939043974027376789048147886509549026 74
8342562707643102692794116681660467902717486163835392703699516947106745446544623211294 57
0472881157715453721838617323675231473945100468202060801459272225584948795651466667633 15
3997279944158624558456310634023834325999831023178268011840099564133220666377797906553 79
6960488110691971887613794154382431521380589931300510216407225370358824723748792099648 95
9705582242626444117612510549148219684869324749621292136358699159105676898806487204287 54
4415049758459884214454736725156862910812869712723439973260272868183623458363129283903 74
4815355973640681255815034364659834802769347103175671475988076683349678976222541258770 13
8737144400422518409600699741585504983699213959432926613343799976759243868935836026374 72
1155789769933007548047711869275420627592195301590709275903981638679448476493852272976 01
8742577512513871889059481032099882135820816750401096609670903398212090804989646481643 26
4930345144220100171089551812647257644334814451922937132030786520941883492609252027962 65
7365702379971271658498395386284890982014651634039851475488513817382111389591247455265 47
0576965786483374231769282232214215797988359857779783156701429212992283073506656062161 26
5342427416596771378277747631499009849255409904492072472722970448938160125726434444976 37
2173393323704617252659454345417082381782182896745522732049224481287316390393203682826 16
8797936193990551963834547860197722252057057694203107494418561877563194278658881993843 70
6755630919008511267200008962423522057482701152939845732039149088495457230208941621473 57
8743651738728622601455837224102863802832374567614222451647089221624237660953225833709 58
7834917497248869467311401231441853753268504982032461218380928675266286983542976594189 98
9787303823900225363868259298632998828599251424820432263843778151861153229403374717064 62
7923355283767600299973897971034579324486291767623812530960122224662468179833878622311 30
3026811895269360316385427801142153588374677160279727785347895605944364287234333463969 553
7367504570693088028993732538583841043816963388573081370521647539331420243884369471642 04
3512842808411417933452057877908382719443606288049383176061329488405557985263626341299 66
4982243528326045280518784357435305920661233361024958330561897693184255581465469049315 30
1881763016432290541436786179288759817112318007431793897366190098281387986373913357573 15
9283022479030577713986420891294838498442234528067347013652613288604128518860636888402 07
2158805189811298988172130696556773535337869549493436482037561326555156009168006278381 71
0683746868130654981148380126205637127592091721790823010465197362985514337773652433794 84
0714095210046546746819908732448254133045369773560003549772198395677419520526215544930 05
5931666645670059679375591318461409879044997262434396560548289620240292253557743681224 83
6673125931932362757785519062829209187532527365677029792871452000296751965787979234241 12
6185981056307889692145533712817831744438482938308872735983651679360727956791959871192 45
5806017356875989712335502079588981431855113678418189515346304374084160197870535568424 96
8657971113562682271307504446983504737448695295323369357237208058928078700239850822569 45
7536683644002331179791847934923634920163643445551778011213945768541310763082571876023 95
4955989252438894812708255368651332894203934267133121703227954958393841045794637854843 17
0397025037233979777067599972812388647944617971470703148263075342019020083400692204557 05
7967314516252343665399876647254460842892664226018551786150099018932473405504710635668 45
7118579611199403260201885840876046262154779745465664569858225017624832708675211254128 35
0702479517121000793395438800188758143572656684499958346940532114881283762316979276825 43
8488581349337751356200695531070260426632313002938500155864172249719925639998573799493 93
7498968811753353657889085913807068515075519810397674693434326864328859768130756175340 96
6325553276726887111712532493202428123975748854095596250720507346595876517620575951464 90
2183107333628321243400439903824530732065696524660090293962236811136599717552051456317 69
8448408693659956732630336866236616613397232176656783712498567968698267821249667847745 37
5060635657039033621289106330486526421206603919618961142038221972944372626618334945229 52
2757679061714747032712903856906532551447488348146461076515968643109828903344372479712 69
3725568813584041597085245387748299540700501649427928366069004431024436544407854548314 36
8979082464065339328031537234757219213649494330128912631314700594948789614664893647198 61
5050083740958181883426788002161033447116057525590572262286737216239445020487636980417 78
0416347933767955082875655071681838131148015146978475205769894233644091196029173738978 84
6784263708481642421932883504963299974417503760576043826893821068907278766738151204535 81
0300383380530178592159621194721727436379000326289232289694737659457741079263444834265 69
6709833686253393927917746815461934782216883558105101822598433059309370808080159250595 80
8915728534365981980295885339640128273778964333186291398369200395989469992012912402384 61
5376357276421161103889001708477269694585426373479413835099806709698720617120984844801 29
5541349522810771222886472736864512858542196415443969269484866065116274088872728730931 97
9126735085099812806700308648945100550602678985578244902104621057826184995279660435186 03
0319598462508074622984781667417542119642994688983291072623747679411159366816736288432 02
9462487961283349052866981356295248861155351909838673562181745936295322120872378206125 04
2856183473214315410002994613900282193854940005767542202875030822132934854149476976650 78
6047450004223170224741941219267483533238672288724740982931147696391227782288671572911 94

pi to two million places

7994235995199298833361933464902471124014668054181468727050365840808586413358866592685 71
0480863100894295759044995728232920005530952080546352969594240858994789323811151217133425
2261118875921047013754081787926726274192928790974770284786993316327987609728570169479 25
0892306074898016317807250836797938391537727942351454295608841057131697598158149777910 18
9524731404827332545354517162780204238104235431228600118446139946980656755511604466958 86
5716277111481939440230611290700997677563424596014544226208594974572937909444310019785 93
8111360362575267143782153912246888624716728784619741325322973754295262780038830929240 20
0573953917340977418587971789021334593819280093318515575972126525523686571661910931835 79
1209948094257836523562116442469649064567619565850058761818485218348989373070525457709 03
8878335836979496453435071016342250058237282318515516068809521451163728844345582194591 38
0985339450220902211741201005175695110745171076910113608777935733128835523705759761744 14
2075409315594799906633634896922659243539123114373290877581168261537621217089086789384 02
7768584981043764636915274476679119586186728031444517610702615154002970448177190450261 49
9520472235183873777685689033038969803349582149759065659588036245337496645675873245990 20
0157923923760087553428901845745681339691239527713053010083274063307446070742224685930 66
1761229947104162206239258439882148108195895849159067122131994077968464911477675015558 52
9902305831414657541809637138328646448978049343987456828181194848842232906209745250490 05
4898580627412060076974126504194412945205245851930051883955441860398702655158251737205 45
2513693157134207615270642573282564981554241006017920578959952881347135991584345879730 11
6099507607074496768735820717881434914528406321400684399407145349049926352572351363400 07
9978643874009751045233222746294206610031373035614072215741300719808133839133902707656 43
4365416747267256131564337329684199569180939735108473008783284038293001699781892687778 11
4642923913093850888792052758510019848634100226518574165161453283729284678643381260818 4
5447690245791453859437358504737882785329741785142051219080017495055596709184193924106 57
6337226230328142830478659199083675341740842357045133302909890805801440678984409461220 77
1611879132857823223445965173626414724977519276617041315447300118073854619801542201046 87
7949235174664724894240491492293995514276002668443025491914970377506057565863117629221 71
5652657381210615827369744019617747949818111317420098934972582823478910698252640446718 62
0583370522900118466559965010126687481607246624483290221393775423512461216298812525431 03
1696556012951495964396678939767791231900316180328724466132091093092820531971247470103 02
6779102437194816887548430100334327392686693719402155497830228375413712688114123086097 04
1106915904408206743652743193141879617836074047192782744479774981681339352782881648668 54
8961180017720604014051485926600892453891203563580838222662654126769564918686960996767 62
3418856458245042225584301668839846448756252569528339404373154296623073976349616798741 67
5124148069883819322631039790184108859561612013407291658212367390137376369408354819335 91
1427575625656130706256635577473792043674677232098431995593996469333954455925349468050 42
5587331508321137934885471900150069508117850486772872937746927547845298969762155262744 32
5933574137910900173377515149487145509448149618156357466408614582214002917714481336552 33
8226597470500672382802199745214360666317046301674393048577392016223725147497304850761 94
4058231375931877793339846416042504016257726661152998152215900085296248573793933458792 91
9022916842323713469251102370935595402517424997033606891155045177202512023928482016912 48
7694342561869369281908273455349229722334258687322852775774525195165465237355351434958 88
1372691484239811412188164603192824295361254054397375180993200851046451515946342267018 98
8806404827019276757781234599180969593905228404653675682314908567939862253708140535151 157
0488090754834732613216699321367410369834272517748361423899042239929785235620398098899 36
0413696242548928042103480814800411716967412313724288085755709160928926003064527907307 98
9403877032180200147539187900862402549623194828865923280390791315379261302720575499615 15
0234740753602471466085321468757991324383485732027795730831011831266075447195774979838 46
0312388899796069784623039139802618480579920197313639207245947808093448656764259486546 32
2730780785920620304478081540212943506618341672381646501539015395641930880994478097749 43
9974365787567157765918599060458426826866856435818621007818318594223586984189296711074 36
5660926136526914586981261588831348909435032428417528618127060065014243420047352991338 21
9437740670726391771057158482883334796396052874130049098728544065047345110772820113424 71
1385843493874278109771809152090284204395157703112835554993555486261258762373841251644 12
1269849557132043871040370587754890308992835591120633069360365022032195783243574262710 10
5816188997431117922529660806683293749780721923113745661648339242760496611381290434066 25
7068298720562180810295859833902010400924351622346762576541520266189134776680050533740 23
2665204803104185685685378247794722888320291363778990295576192209371767917639438522160 94
0136981231966145570508807955047014015903770556677356800652026238485598637675937597188 83
2904783479676989977268246478945717982494983396669074365275954185908525069521933811226 09
1175555255552238761706741252697828836888775477480758659377062537311303120328300213363 616
4030162976118153063015153177666045655903689869210583409195285523286133413532888223717 46
6371883632296230981964361661707496788666208431257368173127328801646730976372123724356 55
2978536219368311496742530530734680817620249424171551299285581955414820627494228565207 60
1119015577872907311074929758983710617067070473893610704774183715627452057536217394153 08
7738159446436976707369411434718837938709959701672890639076162714354069635515656908176 05
4089913934154727490575059255684683459003667734036486116297300061109377762490610649801 17
1254138049573491550464808411456248822581137908861303711551126544698104587986799019717 40
3343176500797093834400984963904931154294795425305533503446426218036632885002250513416 68
9509991277612453780822362828932273475408599036880048383611587284561336661615825371246 81
4543387858723047390157907514963164098204859655911728526719707626246213764464332404728 83
8855753452634565449797011825727359202844975293563251208510126061700444012020215380505 85
0584516562890494772220182417069122413080851429147987928248480927812595732692807878179 18
3997436299946140486731605488466790068865330146567998661708220056800010544921103890377 83
1253674663164612924619173510967493158742807049606015200447119372359068261115043387564 25
9030177467913920997464922973321408271030068758715919168374692922961675617399879936483 51
1221166309310006731609422989089484949081811907939367907613334931787627877125124997313 76
2457588568634937095603126830333767885267014148444525662539976069941904378641572978275 27

1875041331436555997608337482045642199423046997630234779601577101998747369482671686111 23
2707205272055880526295876149629086939255629267415046973221504488624436000271640267553 62
3452614829411728682320101752361772469091155170325040190888226760984837279876080307102 68
3447679799512845295897694290944130778187649647159415644763771529512665905295245386917 67
7045553716122689703529007260280535346512362936337733834858680578127099883807825752128 19
6009294267381324799738560175665834814540448639224578250000300838183500602929502949057 90
4038087271070317061683495969852132590990867424038017131693562481251013175479236963471 89
5741990125065383842151629777300300044776109535988508722977268409326212475244093801323 77
4818757995515122255219049149375411810954325930411075546047774708196110365680878966695 17
4787110138403672219938908626123306519255078024962725120424862367086335577968713393906 21
2649863307349666650541662218553151352451533730761780916053915188048628421544993197804 27
8452160336012076779670571519979345069461164538978596122134763535117416081505717970610 46
4026406795512793439559195969205878566187404508477997109779881700518712224401856899170 11
8992157746986316872709344905186832264140145933409574561145280674398776019687405271306 09
0685859906298779818221800781448224056322463782462456355658852132274772568310714623728 85
2201302127097069491956769787285403739172053263023608007318179372835724406630199803098 75
3229857770090342711930487308562643693091724750110030884101909932018516238142756840959 69
5529562603926010781188530199070254920799925574350605499319346394775588986743810314610 34
2145758767905920325427320513323762096547975452949940666432277430875039725979100930909 12
5406153845396548154180727559040728947060567322221359528995348916225213479600300345341 19
1409446799712335027828430810684059269027673576449185401569811480356621006696775062107 32
1557314848844944172521122797076970310539511483611671064489751305094618259845719164525 53
1662609381590438362515602998770375181898980978631641373676625427507433454648688588714 99
3351072311896487510963094890195619271883367231558327823936650581801088442396032965267 67
4918384087500762026459655866726275770965220690470045225480594228206841651711214797544 75
2211258119930200367767982135842826676785449979332118832261714287867656127667266884209 08
8474363260627594762359528239391093634592855902550606082499931128499555557970868619470 91
5492071937804414874699811820862229210588658192837564751363450500967108302776519081965 10
7522992651085894558343145676920118255624951668233142184625078815531763135921760161744 18
4331582920718062065909528698210692646891917069067700637888525504304846914387513999035 65
9090292201182226852785891671842742185203963941924919871354701332561317931778346577906 54
5181755006745681339064182694462294403153502662074652047177271235461185231094096854427 23
8341610903247486918882068369438904918616851806323473687768766702625413229285577016987 98
5847665513220136436103449985282913914068753067964838336956493239906370624424199273256 62
9293457796930776042528662287454933624218795615859941117825310978502776537343978781793 82
5409180820372453356522051584482477138797436812782548698286356817725987070942010352826 30
1815315403687328042532951240932744537654156198382106971170849207925971209063701767084 06
8143961659786177389010474832488164831304862112527325046778583501517393194722001999693 12
7101864156095058795975437844794458096292070981858405755607785043400930889540261933855 43
5630639436727527666702930379900508627094644116680940278600609394396058143043487479290 25
8363080949972832191261341475268747427217219010477993518905388086381809506951421004613 70
6674799919213942241349173773417856315042825830396395148652943430047962898841585842887 19
6977035235945105835063761151418901239589776941555720206834902881326302422549833914316 41
5645751747509245513636428454265278441313104552960256249851954604527031351451353972542 14
8171314790899295237417091651628863243017198900373680267707980494451875555548748312979 51
4915202362107830343216509140732551915507456372014954587143074400116360841149663906234 58
0038885520891208767506517405773311913372117070225456388831173909653395767685950753489 65
9557235940492520504105757074662656892592512400926256174959899745515148625249947541378 17
0518160474697965334139129417122856024784686769394902161425683121250982541046819606418 79
2621938034243129985534976927139416821021963857567947136122949869347113633420010459306 98
4268814809227248912872386169761736297477419813586778052518018628098205747388657458451 17
5595087705197526548596250803270125865478928725555655446692959614274177617017619740296 45
3928508496038126661812505779843066329906604346048758285282164051694679155898481928493 30
6303487856886440025448466422217251499218260351727180097048051239352372135693012599273 86
1963394376957595850269057282975883929482099026207867547312302362743542594106921938856 26
4822437875827019532211602437993665894346774746026442885547464770815554853106702455628 19
8215048428953349307529913617250295542871647714185741141751347731533398128387556812799 38
6780596502037163120782201608027721266190978683826330132137721476632928435912028020394 59
0843572067459139344682480995499474299180695303656894728401419261618228328657716074537 52
7413433008051795658119769613479079637419120050649606696462557905868405800012732043517 86
5286253981494901033115124328318661799156898804760850919479155409084335619585333176392 54
8561450955797715236698545493460286376119254389480109795410818964807550513250401345885 43
3770071590151749415932495304558904244672628745004055325742710029463234336378435965384 67
3088834319825365080315213368368782897977755682656476885502586410302168387558835733148 78
9492392545233927782833070177981389615196302531086160276812079093943074039991544616109 69
2549511136063285761108895940200992586769051262016001392186086519663071505852979663077 79
4421599380819462119847909959301185167774901946436787749892019656235956721284698667584 61
5592412314969975655745285206374429753700033134867525106172809635044661233917421689460 79
0862035791767665431484463586182367447851803685693021201109693898452594558332461326639 04
4216477454399872990537332439828870431988597911967349148559773977490511921696688536539 57
2406399560657170539305293156744164078106179189000926613314085891084206605523242584322 75
1723625299232220412767884434272668290763959982948886999402214490072426500919113267876 86
8941086237736179726612655321107119223534686870222314886743719705567909895793330649727 05
4861789444019072947434983344895581690728397073947893086527151109067978216142874581241 82
6285896383992512805437054182553704776655991536708537995355010910737402719150704798086 30
3723577839759475955055407528641314642819284902671760342727181552815846188815806265693 69
7294791818761506940260292918542005844340416104898248993676978357564432316754823339687 14
2741601379182234530126730596080643562458014525034279518081949622264616630024851771380 74

3617141445055899904229034274344606886031686747182958999005116617103996718680915622355361
1443917739391869069502552313043907243598517625676364322154128851855894909832105889349 9
5137010562767500384766852473503203351672316438309610953701566752815577389249277992012 68
1740775379008716225790774821019051126267126767881233486684293608882558282736161613033 89
4574962936454591330502271903174055156359406794094633333862607210066841469713482579785 2
0434157777859406881567445707018053119118473608097381557755775001569568523183939423447 58
6430283117293467132091761673757514655474753197085894493826778830808070073881754117820 08
4608326590991949325188495413389818361032713500332681873104638640933892293376138055364 61
0649589779979943893009574307223611004413588731498659972707865897075742529479141827511 85
9562834519333127843799709606067026048977881066076951974091386028860077988995846923062 29
8959231218558040737052538173148554655654020096572523593162637058144612741522433140249 78
7736425565052311711573477622564504053830683813117067818449360967641972075915168585809 60
1803200123798860395784159316561818282836292722987252927839284293001120168467427302824 11
6466869361052044639882320760408507796310117641598953599551481605288675479506968279917 23
4448221915360211464856923814874753555882189646017719217943759571229723641536950928136 39
4747497031421872783749454095315851896510808361774845335167725557623274729065848207881 57
1667300981525195801703597033940051089592024944649179974038964726752892940931087156771 49
7782656385580232962522431318292134064984266472295796421596622908272000444371606753134 46
1715287652396265353457938029773373563250757720189889718633419961430796432663125660266 89
0734428821800286073725897755096472690905378127507765048140473129006137599959267316263 29
3355621203849981416438322122193118474693654324269408476013463649534072334080110365183 18
2428908815711478389816805488417751499334044572496961003289604734182132815841652775972 04
5183981914602239864505356709356613465129491545823319756179475862639824755598467290292 07
1883909523593133527913462735879157133003327019793024894505670911965470286547700487227 40
6886970096378842690695725709404634427273165391020537778702065465434425211161693985932 07
8997912480102356799209531931196027425759331158295524372651725230777828707952304851198 00
6529345313339630995580311237670129394306938085198957931181373791603725915607015187660 18
8421459845439869088443169611311940111635310854018325704903547168016756854382914935332 34
2508967022085383042189349547089808944824564965225673030383063525716589320780435767072 16
5125563208518294112276602609171693880706756255600844134581073139531690185109110374495 79
5259807390956192591993685141558730445876918208436636368281770588257191740979695822606 02
2391801425238183995874348622311665769987549490828130266911270068501427907662527547723 76
1040686725997363906224273587647247785398458846694975735287639633815194704113817045604 45
0817563727069717123161676245512618501940097133057140212886685408365730536113794139753 85
3974116640261636270830388595306931250047984884257911828193768631488648466683481249630 03
0865384136308545606332831232049910048355930174284206951516034734239278590103452498223 33
4582741942228263730133802885034411222042305932785631479402899356755661411848027599503 75
4159927647824244166216034799692971667750848783749658110284142314466497195312332028411 17
5208018321246037984713346810914700281627355957201926271110318405647415152012607788065 82
8807364833700173930874767677705127386842871566397326837663846274140294201221379796393 67
2192134997069858793188200831176082620818556104350233484416789137238260700531086361843 54
9173190164236010687792782320084297855615614419962845925343604806635838245370590769256 88
0020115801685466802991960915951443193919524138262855533621674333259322770432312012386 18
6849395294140424727608976556337862681005260896353982936212658220369250668415427522308 64
8509294399281819318480320633104533113985609207697511706167825493617577002592545343932 13
6665142574129014918175262603796667501495426595437425497900048387795582840766505835534 20
7689006360236055234854579435421610650470608429433001076307674010542035286733307925839 13
8958024802412777710424151094465573989442704194993864885852139414430587001939258141198 94
5992211445971714826362891815470500901525858057086765552274588272380578467572762429550 54
0037361881551826030365268138424350721533474606611856427259699879701819551029393851130 75
4619408954987736835566347047172314797934113119446724140559198745962092813180454424090 89
0637195079261324229021163069090882596017289827583424817476279448352544469289720791180 80
7595293810921275022430159359147753268687437921505866551532647113730594258480203976322 99
4350937469450590346724994366877168150523649019070662033538693718407287975600603939001 97
0899695400390159199487233318179060086553483772260125868526718945247186354272017648464 05
6303913392875733329887378100761080790764299591671514908073136747987370196087379514138 68
2745129354167570128872183095135840057252251412949279554729227540788994411536810179364 50
1775815394730933046303807294688704428368311871924168920018833467530213694422030584216 91
1870075820774087111885749461581299403673955301897758801825927272291718788769954908855 74
2551097723145521760146517054269555831146358256814262863951488130074792069822268092269 69
1089469701865684490277048312993918954564987819444928533764466049909884480772319842772 79
4084097493048073657757883435652983675712844768744354409400642785267356111054980473601 81
6271319929822508142785704234579807756914765012844632023955300988294699463054881292057 79
1939922088724826598914102904562402456911251248137484178566555204328005073695621031892 01
2739415706782329198437736474606802980749150750226938532442291933896864310146644303758 98
5478213636144284846886692213752356690855092520049860919999669625192055357456114255592 65
0628254964996403158023067895254539157471824493964460256279429406388802386986293531926 86
1753077880058292709045096717562410120063261918322176918512510615424238683844712541667 95
6637417437355923142162538725376248635842521976504289309938191312458364849600880777733 14
4859391314699288099502597887837660226106237120651109436758445761972880376479100104447 67
3261015740617128547624825396095087796201265680395660108687830516790508695899592578569 81
4036998878651306866685594971190993299775193264261379992487086442584415568065666122863 76
9258378714304846349504179527995128491771923861807028047450525955842120451376843584385 92
6461005889172409780114054978843142865569175289319844710762301342299695112060106076769 91
7506216106240400178671700009507478727448659605557945584619545175811499130650488461106 75
5521206987897137860139053873879137657848154552436773263776395009079610197149561555143 80
0736563527576783285549268154304229849516385502001789484592278412061354586630790723104 12
9729522457082177199639970924726896334133344630539724509247869781670854138268345919934 39

pi to two million places

5639131776393985750004131568157233569464254560041461724540165129823174269292769564245680
03968590328411905166583072684922956735723480252026194711545113088075062159159922338748647403460140316854029366391560077486718453943929914867821248591831779197451494197619210622919957866099908160780623608879891232492554346858518882677295464592488616495628044618999222033928333318247595106172356887083181795231548353567771780143500631085240771682178360975194487062090103221397760320763207173040106947301931874677112048687216563663160381284728422789767663366733261751766200733147277806150896827098435352283123932749740169167905799018239444414308238304725097164318575308424550354827278558111634249228714335105193181541130015463900371730034874428849432704390358585271371624440744027977856947416236616395816146542644745210617356141885070188083963599116906588032666469748706961625879083165739355452537405265411924997654796982770257783420890711317175762054160081036837209396353662382856117576657532701966244427354981833975178743729071686868174725937737092154530369971501271618804303525564919334504387665184692040713326690800725039646339937959661320336822380379379482457257356491607985354796832876789135611271422107189268570045358107625205467959691068565288327163411729159139657389839123734278826449239634473308602800898394709658859865262353571326806566541058726638391429982485564272705787896935213230993809528875709860599889011666809477959409588687052645462107879758391482081568050191763637449261869331652599754673004695930453624269408402150539467714838640220148864666944989888098005463828311963423915759889411451342285189589296402822767959031376927324302935356732564293555195488244366639716377680183010899362147956788817017698383243166614339066340503632978937924876126649733656509266806313040694108390369135912529357309576970547107916824395410887324028434516221272845943764906068485388747899266501056748922155205301516741731755044920406934137418402790645744560898065772660742737902007409125675041087374447188026934496325186022152542573120402665825241909888982584772721524033411864680890121499367088434955081043869291104445896311211214233686222692706639046587419124921584735500469509687552483014800476011104163142690083865862307810339609240679808604642346745776020116787778488229347429275269899720322047406182761368235728284319603458082652779631109758400558604348209215876507267228165255941104077936270984827241020665312997988594920122178935946313948652428026753607116490914512172859372477611141875957420128061162038461433480204483256724170343417970265828015947791827205556799838334497670513839776059612093709003432201278059587131194058033456896420278738869873150553794514459426184962873741860046641099227906293258157867071533508591753001953398711115326580042936351143373001857681497173865295973441809952420394045660641965761337138425551149639835399814867765770082437398631983287791406380933502309794391516114603013011895585897179460628175352035274839344830393118258538199968003058733928109573862088877202317843534746361685259630306488432025192214358728952425458009681938424333872451802519345253548877280300599704621241961961847009617577045439262033473589000699733764853384836290225192611995941165470081830463312658496808704905118040416871720273173320026713140145983958668202300803103066439740162311819384303531739832606318503624487693786693301090431946988758390120980315395234441540510449330799408230848036112702803403301238137623334805142304337697608803949450646553172411527783104626069766686800044134508958134044430105830498814169143931549820246384864259541250697853874227503193068002381007047216787601666416175167614304377196881898326182399888628961037288260247593452266860720154095929598105777829143145458973944675507148277395288837739052675872927206854838676731077380837974043452321270633557801093265237759983743920817789474094508634587377907762690621253033018471659943671397109753634193560839316955831885241641530605664298971720432485909296891168512647523547451236503602196115343079756035144780424289045276310064825051136075161157632217655947015313077312663097352643396799044907060866686939546743018105433523529718905529331084611934690542045288993327873942758030130113959948185399415015077334654431820488362509814519760944343266901688313655511247858529315300712393167093989011786900158473858736738170093070900326208370261315352734974327735342600178895652789712683124985514121388525069269433118177363287417181685176859874114093952677123785259699468002934430264438296054559998531595003261220307742107749280242230832331336891787642011181198129749298856029233009679882137319159007159710642163476480783494106907508818629578465390714862236506849034511482382236217897640643126920600866514110928290677952130080586392880854339856994422989721298373703372664044098759187298470029488498038468350292838500323293605304599877488193371687813474892900087399839707662356652596456239263173858761046505827550280682785044914410868346037356888055502193330879293458150935598987408982162204300967724505319329456203917408601276495076452116545057317176619459050941680224544719673922659429517810243184528203169499820648309394672403513801630708799554180772129171551756952409719197068866046201236591088245299265320315104922449497459492042081604903182645908081477268305570265310954583147443951258580855214903895379463647026686074086055288664181427861974135788206887932437303884651834547566564803527538729838811414483084748403077792660320027802883964568990801406255839258963562267340738625981089880778642793927635260239321600268397585443329625129255586296608526858341367202564944528122892110679232029874590318092523312097192397463223122980926456553827466881462403889415471397944477286416279747492219806384872509817339275312305630057322078185090875724557590061076161901562185638824633602279325784168451086359777253212226868593555951089943375986626240071490966972519360294371013999249522919188689134011339848619592983162291736960611697848752615483783936829712645405746060948822948194910309088259839699221643722894174374873806607355871696820724997624039632939479060514786917117654518503106457479026611882906793919558274058746292220749519754037558343727509839431581177527746305912946517277675424989681771055547591413396305990386804215449458101599598783347817923425669562360844835083417539771889282500253167343477677911247473460889065134796480386729039118506046672373275688936858000525122203964410461172767442461620568304568751873016672488884547086280384376567577917562906551908501947424390346048555542321350824898448736291810381082517385266248680991716035544803393330518461373716727356208698653162506124254153931589753458483341431904471015595562150698897416508373317450827688734283984242838945449820675784639861502984799024652283293022772596502112225050605399364999735641071012435256246313048118379587373971572411354479635301123574841 8

819846622673982580101073021546970365634341374304385205631465465934587834098148824426260
725103752734807235931713030041852648242906269950927724328545553326394242598363601644072
268585303101124412207832604290553076231843181039361770556782723809489821543861507867911
300631888983547046105199535664897227110398872107179331915026203204642758981024800654823
477355971942080226364580972779339119568830539811018990636112771526680464605866863209443
377045337878781312461586626742196668811909159238499908802230811969885812779434330865519
862873526450672717676914844056416952659398412227783436079898274886145284972420043719222
572708246864025408195131656038506029892743190675750662678554114093789176003845271670725
724433711121956842139515924293207653625240846082576305446795087470360642441582865891161
334645673729623793321677718359363982822373844398502141412291818354570071743913973086365
756377219487247131461861736652435794480634002179034950926757446822396487186462956076826
60390437622849821539550344999928039478538072755537242491438955392962552102377386185575
165203059602712208123142398289663607680890789449672318266513803655583979955340226463068
841070922493878096400169932570190125041353581476966752262086515194874118233626844276790
426076820613284334675304213835409844021325917396613781035640553660416815761690889128612
686885230231449600293228738508877115152757919761842866757879367657918687628792415750935
270209804935461116056098674847712063858219250468923543061142008298190843994590616895892
675017901989986820402727445815423327219050782239765925247281840774636860053473932292674
650963230646597238476002018994806281449207777767966416696352792127337791364231258991414
150849256104059719286058417676865030673808548675888222251576353994369118410071654248920
9287677712393310928858880828385622199509613320749940802378041252225123140879328581480479
976974447885027546041863489792041224909016008967400247890292470489672610106370913528496
647605094161322512165932923355526518660407029793803217625007563208381547684132422713874
998277503186179181679924252869833673257045957334824280884879509367096738903904601170850
681391876921619805398790915506034540730567996897226017068832765792170117283110843615725
326861058855003514167733937305831580748140091412522579299107230446499256917749886642232
252897560372487307087151046910200443873680174292493253268623425936971486753129097433311
759294743697770997577705976888330829110497720426767484703941325776362067079523700732143
950702184758784200969616188998382969389149444261551908928765410898686641167449867019470
490854546066194210131343878013138144958270111789981322050685916697174350175170756752456
877482337468946820975665207596016651008160357063832368406299597447448178896032819359964
8635185557583321481883442791408099390108635409892526628305752584764014065327452652664569
451979984748087483835038849965495890373237305305742260724357636711556667470619509257936
108936977358969194176554616806150269557592980703034328771740839003754638113576271916734
764833541144730043903130966504260189473422035187544597751723192303210929475296952053297
259172752411475262875180979347006555624666573848600309344981413039827890930404121418870
384480132888501567995505340965316137962615700924126565127670585868170278117123888248553
247619760662943080915587361948780426377525416777921595678285815380770683969530843879556
551601994120237661565317927451772058563931036040617002995078064984481929134312961648435
627281743209085090681993924814273953258989593916764061320296985206130869605294925464474
430981108949116310534193941040133930339597260092780479478357880158610912682148814692732
520062500510240593094946335405107935383005119411647424974722970237457780314998437209308
827225994308097558810188812124903022956921431982352470040786195618109079948063044688046
643050785151957718114693178067316857825909003988393439315570637322961205963577025304449
053104390285331463574244033698432715890646087840463869608314142275399075474929646214597
531984506841621398372522701659830013068516339879888501810532912052352463992475019433035
467140204525151026556951459702436908778991497415469186574918155742115747556332507367763
083801374641763599801200498391505367772490542173012293905947891887831129463952335717442
121208145596032005852753164615589596996668677279455607335642274428182930344144140575051
419165979154452232356903353363381175439893415234450395317577583382522187699969139626197
061035066057298295844763225062231916537193380903722627996100184031865849597297931787596
2142003970307689850279930425951986408380222627857730930820867372994507892971040383174913
493570873389033992669952432310362621303347885337254063573004944154890861548559506534816
304445540920699829427698900606426247512219101057135087752748595262009130001494551918169
926084252394257562574029142778156606596727661881228018320713164615666059100805971131783
624765221569759688411622948468500704086399021624908921114235797544936806468241188725055
6303888211065457264777980414995982243403828824342150688933573974266244304153268914723203
542185722691489769067885233194035672793324375015930409034571679363134194488615530416448
351268334147594838543835439609401451134412271413164849993291999913445741275535170052521
721604291749628784350156011069647018694481852111792695988890515973387039322515312632302
252254631926972312234355934168695005597741504295263395765854286150123343302985413754694
239305812013838595682488823753844776199839449001084652795456476077394840547065619919152
705030401230013871858530559731632603699803847963516003383824122963883261756698818256920
224012434611412275238752350052616692977116672073375583703882471598179636934757561716567
127075135417931434469675124866776999212388499710061752179848521653992274621373173059051
545834892680585103445364788865242097160666890889360820435935126628493154819144242909854
375815974746992201994355772222646853141226043612149851709798714050413420944522365986946
656697106614168985041052797430649573799225922838313798968176078270959140288126987938073
048718091802262889541234176376570305014067815760978111668044573076713730072299887878283
814409199894286430912156267211715875094295803123901096134787441790566463753264816769313
419730263938236304019932960099410427531209605242102429145654260478322427168111020945416
771470643286481248005958249169321921753786159933186730209121220639961781250477605574906
053078411688462972831922286622046973761572376271716792298360380292695362514242654555714
475989966654422052791542937818532454786691631616991236270672649451216753106883013130425
123214591254094398415795511911739861793763023121438270924534044281732169710952139327405
044597073628442081058597859107711752826604920519017725747174615928892924070730106591290
967482892707878727687629347559365788373406472568074071996053239354074973674242597165774
669411198875291207041885627925049491521271571186884514473311360287699619224671876349127

pi to two million places

1750210824606887654972706544123009627512581313871470461121794493808919555132248020032 91
5299161390040256516278445986803232170196332082816751484244179225812353732670611855594 29
3533692353612262470380444985273792546414611333659665961825908213530616654746104531095 78
1626324491189484052027646258745133348356776557707285451575882273790912270447300335470 09
0059744271958784654448928585291711898800251571032228072299565450345630382376388366062 65
9938594990202352265643954503439676024265114050871870743691502123950672832414335643262 85
5446148317364133208508038400469468676190195137852554420637720268272827545874963777832 05
2937939171336885960724115192690096385331512298722441275916081415575409261057110808485 14
8428294591554087782017151541587329421027792281584847276211281700721519688963678721928 29
8391393799970678786728522126965700463755578820211544403176122658626773424354088386879 73
2037865776531494229446991985036602468904554427371843249638582774712612609688082098796 22
1387231596135417986878376826034699826957895693418364711695051837764245076272376598189 06
0592277500165294095582129278402249791191000267594378078807999147255552010593187991423 20
3018343968542247214862490604132971982263600477783857589042756799436814436804895612141 82
8718694917603674795822012614665682282187222445815146197977333347701234632325885305139 31
8857688542620529028095198118377467920447082048936247460039159829855523588875565366638 20
0597139580534849112254508674801026640230100781828919080737314346650096723324928484015 21
0998863190607640659770390092331119861307737815931306618686076534506835017181128392216 66
2653641925760974205149442342804634933785190690094483256662295745517213838042570308340 92
3493994274376401929604889910980520402443328730947449786434034989767689518365061650199 10
1339451737903065880038552343001673465156535085302715349667842254603176061864038355370 39
1553137585514863807706891584910628293318952915015790755942576688270941277251550486236 50
0781302739711823405293294160941814642432796191429256354433073812389126237535649358868 77
3557473162244091438191064036410097205673533958631613263527267915222481970288151828601 47
4807955314488606105507843254514908567829871334045176477671034888634886808426207376413 02
7081097844540698761014282144600382984008179385602176031825997319734171444149535189268 131
0297227746523755301388516964290682730204302611678178143887114382131044506614477067839 31
3659983186548933967537446288004870514521471549116268006883102330573612331041991367092 90
7982653068644649794065055802196653176971943811263038136588009563527052879015497567993 32
3664545080391524767598345531023526059738200526782086774206894585013515900742448724890 15
0569833804324544862058288345525196288462037686263307753316090589657710994948671540172 95
9649603419008917253340530394453853259097354074998554063207030469796750072146749667938 37
3660769421462586459415096274756057736737670009711769887012145178902797897957691472192 57
5517012380996780346062607599619391712410489068620239709414360387531739799883185239835 63
9754347539004094203409180618709690004048528124332184277077710022620196801163685925035 99
3617713398335576695268762863886877738012669790542188764448502526083018121018257791573 46
5553383616907085859134814753810617737901744601369367976880667826205099376258750787125 41
0243341157896709049548503283148726551059792163497903715777325136992490013983461267177 85
5653125770772528712089030119671227700961873741072453614395683073205840776189690457652 79
4113313197946579259779532693669509010600301594769035116396269885338668721028316958706 32
7584873016596465807036913729839296393910977273740192999964863472112227259963880097686 74
7530993132628051648558579702029056778255253648262719008763183443263936396105814048748 07
3248987080652979210778844299043317210896244710358338779515211816982432903633226653281 06
0208785330959375494286331540031618842492764849607951607737073578787248934038854774338 78
8988567652296271875498753605368600874546443461315264267687809831548968544915988705241 14
8845548169889641537408620273969151335484336162930917353911076623626829091798926406661 85
5424619966744700526171972246929972807178672185009569363502759541747219330511265643373 95
8604765709420264469483643602311600403862033207708715555252065839485356386945866645399 785
2379646762484866497528561840523744148199927505104568239795257761428947633152198012953 80
7319817237695720431830651983757004701227675342054169347005515409234239116259353918003 12
9019247070400696244767634366427797597916450217311792873902483381727677665550033018780 30
7553994855286636947921548081263391761484558148047658946030105615216447093891834098739 90
9838534861923892482813797911947155755557729320833545522757740215604692526993805016543 524
4638489355574544913929682656171324529446378718863786625006109553936647731670867773128 06
0431839290195211590038400172006436095725901073824946135324197031420860823112447152103 85
0445990585962383992586913373745239362534355678557273453909775062580802763223732803767 84
3760848513794393925013661638658930858261985460920670927316527089044552124488499408277 19
7786799758218121744325559277617890055048595386054986540380192645556004865931111352186 17
6073768408573374298952644731245194595600488134119929837065001883055442014151236905496 33
3701761309802903944630900308126766143790722239918227817571175675529730689592526585223 13
5997472715513991387956094764102647321183532831214570196109880110540122287539494776364 85
9888753245054690378068840515230158401444336222665717585446590736001157009363219102587 95
2655950978552656107863937681656585680702250977016791018026793828710867335979323089705 93
5129421284020545948465660463478900426107042837705649157661777868688867600367501801718 58
1381401116996579206228847234858142950296931441935107245710932081716904087431524151288 77
9772207943969437146335100113398681279079047942318184538750060041856436311090976195278 92
5224119249463713407801364009948811250119525459453546071966533403059279775590289772497 25
4123645116096901085907047453924116905886146771422950233519609521243114349970965298495 63
8285527541437437728341261026193264629860548392708270821076523666026524528412987817869 01
1123185412186563030394147359042309082900849192198185153562860154721738926058322726660 79
6822952927119077056302142495826137982427735157923503799075290182866718571614166476611 53
4818047442884938520908917730510623949975011524777499541693467044974366449849759903113 58
7425843548717000394600121442060607316712691522607881065541638388125566944233676618129 93
2246387977835445982599340636291610827725897053503536790868543776847303406467084217796 22
7742690568284744031452394283978616447651889991312252319111924211771997476105532402873 35
8362831218106540339212119987047954117230156031158259392100137572272056916529440852979 37
0522692955832280174029763687909750297928700920091416567086525777004702469113156558178 70
0123516730181700222009578509786605509290472644786035725873804474501168601861503973253 29

416474992658478808417773509486357430392228800328141065154749433608794290548027405659717
557867981049783497660735287812389290219911479522363221589292747058930364613617347015065
457378894845145803420516616356688086116320673506143161368129022380754193348725454031940
482350776926849804956572310719321572107714543144167287813767226772926713215980470474983
808068062761130939280900459776797285512363894640133268496484386673233642989414005820877
389380550380750077690320776615783449048603268726419455760771616561275110011422490697370
837324694141111609748421585056990743420619715916544719180004055961248140290642996622849
288751056517932258214873934637435656184528578321534302155942519795790351202929175339188
756151986413402546560441715240028839372014257394072412103854423193767550360984399556348
767691227738152045062634020568063048215990269640260130381224731114738752378077940354728
477520578014118581876076566930138771312048451265047525288909185681854919928164307169719
490313932473580306799678101308606252382537545073319120035172551501406227280213105790773
768361275515593916620996853184541088796365358321231959430249011977141762448663373751149
970509727600079251238798177730400182170327604562994401039017479267996672401406990830476
534926552711821558512776072758980174330126619069018210566926467047091733345221501559360
361505161150047883436352682888406644350997951080208344220860257828437081761693101848171
566035786879295813860746968039339873154585695939805105280125841927566802449601840649665
611263515091112690877265873759360567561868955432679288058356809717048474541084977580078
9092835072408625112804001243866302609294617062066963658231200919111923705122329289450347
395664568066960210673258928108408454748445831699368794262853094775161511486758976038663
990603339824720322811470053902425785735038402361628265851337877345124236791274462295467
981047465353847450116465388968649405963509944383999120932825602925698729666544857094858
290451526833937526676362912159069556167286023438650977739116715383220155217856746760972
471999226587686811861459360383703794894240079275380529969814723802348816779799831854974
696356343236496285455030132440104347640417267660941690737623154324148392167869420910904
464494658665884485876678161886559487138956186146792545654787306994453405970243952081949
188141026489732019021747881070995073637254043918184367526547172262073424556773156875309
059560002382677730895231887602213759274986645347544296557744794151430319775426328499384
309253848575605971465690991958432021851027957105928408441121315708057973771708770600209
298359063291504904456641361363954079404283351903300161677692270860418857440373007378046
911322554226613231641480562523483040847182976384234509249675316035013078790587218642169
970407365751677625953976488240920447117806548862535950676262169418665518892880081828064
462265061704270564370787782342539384875480914973784941965494523053646848842925752350614
108558608764325773727595709554567555536223462090793524421722241336867213731054231955555
231822318098358584557424652718714001001884016208387698359959278603562919931687765493043
616571590712117503914815886511732298861814603312792681322407970031340838894228110033400
076965374465490229166996224894149603666925045040083252520310745954091293261349570002584
189397829670164389985795443990828571637571075803636012698591514203803887413903983055111
502868460044318166939716079863173030619550444318687071924725756315917432269921023506268
446552376007451544828784791663968440254821313385669666907861865501182243485600181771893
163960301011673489128925071282228368958033017138661053308045793377508018508347121429684
897812533544626215196141779014823806650673865677612137898530376888100098642119699976204
864120988603423894660916520414691265313583000777955117134180555766278433884444425839898
220362354241272075415512266487829308346231229331192038288682446535266557042316272195328
284292343660837000612513174008155769196267263713136729172382698032144999564732349893740
810691838938562044679856411748767079719767669758882328945716964748544180350631727302541
510464775223673543694170255018723377721375380276274756397484034850253358455222252847272
385457772948158426315279786967377543007108819804993332280458786295411999085775739661996
673664679480839245844040192078753262803047335286798557779891261426257709848982087553886
984822816790891098417041598204942278568675402967285589021704705815293437052070309005375
615321243016415886700423956692354177601462649452738056680899956318567090370492068451187
994500703196272742432616771802889234311579152061004802439677482735056175887870731464565
621397685686161874954249799306530943181732701644978074309953212076166513875935092328313
348288412929989467717857767659423754535402545572500310081397818390905250759755180084134
984366943907583413488245175598484150058338212023822260990356237995508875617009900781254
127016837208615267418593095139067385952106703458911556956837724958190903569201941319743
235208620389061494959625271981071078578610277993722913856604344015393973253024884405929
140496361209631392891810154023596586183066963100389925755681250110348812046241217524896
695251272593903711537771629895616141313724508193122494358762086373886846063984307002444
443865310404385555576081735107842150394478356247159725298467298515942638064347650315706
436091201867204822194720699789919077139241580915477283393030903269314974442626137748976
233306431412253876951372534978731886743359917765730069440463704358417410404064223013158
993542712430926008986025525977937163149989001290367830005979969884694255847085969724281
425921815582862044399721641837498498878961636208366591059238671041585754122799710155989
705129530947901200207167639910770265149865140665808602569987251443845084162940264807030
608051395289327526420560739995602283580146510427486596304263302266830546987698062340176
295626643302254711490507431144035727116693888143117549022639859545874784082800179662047
276210525548650418072281670921448243296895143711820520808066551920331510491307471702838
309607063780245589589536192847375423746230589868423346351379255173628997095630590896839
558665627550696914936809434283113556473333868542887599992077716739065238947911867696797
165664344905240471004242908704803351836110837132027932504938475510692518645400452549822
638788977416839895743426284241122495981809154390814372576288614344940832828323633533323
904012240103431451446642259262335060063605340859020624276570325112578559912951305725778
711846063258988735169802447432954165847446469244089159625656213117466818005116896592219
609792952808200754159324222444972623680457872546928367238146520113282645779337165481410
066844860148676028466463364060697677919989866059738470430135124520889642880264978884159
296434262115926318065390830973044585424083491137938929537584542209824406669940941652125
911974428694838639921660196662809760051443054787930384441029221686259795969292530514148

8925560994266916499579147865182716254393864909306330785845253670080056602278985683224381376904787028897488805381479710649667287408174452185561108234232121660116766817702244889565881027419279514920407136808382895639205808871148770647693126569490884162347708525867072163081152179973933133912019643450413848929650219805846195034464066927594573718749163152754051874550559899690083598259113221536518871098125279127036521084742098013464612969773847378052092334289272273510067930381719217605875959268320740488240614590823891381254897164790377548438060282704456355412202521473719168984672307419246104685333631069912781850795212693905259508252793574647368742992034388462486667915242187693870496780662348342080419682659843183957027823886614298198314636100913090316176734229988391047956795774106489746319620421866242113348619571858687229843663942805059467961977633871089544582949398350764829883307801078791518390947593838570357259063302562061871946408236125119594750504420610865283139912980327613199255780867770758418832525460768616440310836116460654332086865256872079008947180943561124714803326479683334770924513434018502250888604843691793525397039533683382657421862449131314634096327719274788407295763315195845035543468112904617696494535968857837171300906028970724982845351220091840852850030260141254141857569731740636185012567063871261339073662776917863683840585331390992268171594831737761310452966454917290909105075971740601645160329892640592336769569263161243956923903670624909427594987100410846744536057716071304226617752609443802370486024475034712047482392370559144072373040672882951378156918866031687790756938433017225236054633529125018462021204395330701300108215926127084324099614954018646167218872397996888954365580955063598689961121537322766025132571538457106611685966007828015542788019906876938680209800193976164437422357377267054677282737530079223182619331428457496304014764001557426650225044878909797968400034205756561675603290064811804247237887515243228706786089086227404204834484182464103594673674971499469500655077261245721827690739320525403447990422750121557570703802545554418726180389975843035048850492621022945582610518356316736986595370052922074944756148685795187624126174555233796027560076261917007326122526192295637722642284966771794441620889954047558570961604165466677227869727806059121630202220234087268564274645519192872086795259376899537363036825320909010547662778379010726227587994067495966517231730782088562315849883215875951542288217583124001156835369863512715972628367229419541971945417417836550976381157341396370412394771609744642489123973382323321485326183547126867338174417677671981460236895866648321959652898645788202015156469097013590918275069515984703362887458919058661323784979088163100400948637742937542991355375393725239286092750865470975836441515311022425162233146119506768844967484604043428829453055686305570514127207344142118714451918086522868575133552455109070855763167275427461833339046048480092213031590206131486823315519820207470736593987175458313011626455163294059073062118141287197073705352093387722653720837708440675991595809184601285591235640874173776244028779173350776933193040020420768405975299589606325663962793852248583323885565042915575029065841582905652934830694998823869798436215834561750429824172215385582757625361458540276647670775195628981468296582288619122152907978599097685214404600333826771510288154774377393256418485453894466500008324974836555450059443633898881908170146814768707743494374253486253246403814417417418092157204547359637740501562367430255860639144339093299018826467119419930915299131414492887413569174413316393830020489279332112850585390940226623402121248467564672088244221628565976802086504853741723196378488185202530891428521605982127985412880097976972736742615493978382236304146731248599904145402057239185786636356432534006799587549020489270085719875953145143670052851638584960497980773549307581192609389068871715794513071787733298720389360687131868853208315692259130041285498639088223398374913630243199396343246236338707060287700602497144896452968073706924662351913159923096887598781566391857878068580182102157128255872930970312523698658251866400457044433616961409321209453807962425360426648068134940659263137593636882429252228100702873014663030718958313503729158209744459154820109628096470729327258022569705285245376863097915579749425754149851275224732169146464022887776279282297618920133578872986689471594312678840707209458222499145740830533787545705162879414942839926762680668294082022797924074813649001337462910302509848689872978607500077208938914294000353852003618434577259853755070921606669961689343364363411342160114848458304145267118059472028984540258733674713840781068423758912496533753251421479735412218775465693523839798156085753627887724863278079288184039950465350118947953180428370395636773382136560495516635874543653693868222898985323251755762668188191417309786110661671485502684733709293965982213282850896693406082396877373173187562190372049704981744058334179100860829701126209685234561761924830008844698561105615984621757992030892634664801506899492065617686265961547848052025701844284712767586884807420148459342675081476794965520632168987013926936973315467477978006643750060872398386910536878933715290021536533109570082548531091833957640146144249887650177162475083905892536254109421001843841530848869583122493608350386242798589624649360601983750113191602178089351798880713547204798844651664951792248131500529882881789475554545892994253962998257448622606146696255755148130518827997335067267202019773653091944901360246985258361789174909579719920327045921803315271692092211854110240364827323956296024926997386479577202606276426659834575960999016809653619526417509988984410906696117928420704766639376030240124431728159961609901552657208999809865435587260187976926461613776736371763110433606586937154075972341494175649437870284574935324137573987599184926108151270519090033971259743051430570297523951776099583708208716198851099447445306142656395637961828534214888678191931648715097320778590509098824919727945988520622126543827404897029466054737550366236762508243277810275775468828530786088843113276039499642345224417506029306326409996819911655702804142307784132238126953198850883949876834734718298134553816901566080257191530432292891711291189934986495429507742755737527443830965983732604958568562172278013578261235637615530952274620193820821938619974538931407657003872841268419411103456736523449579910174591210255157070627515841269018145418659193332080137519221494391311390322164307810398345490984097789912671027305749583358884110460762474028662813671706308591895511945386434989943065729606727446887524552019759310660042571795411224735030235786929471911361641408769474652846690413351973714468286371876204136070787809149677854038222379031325933707416713376670300074174297889810942077917603705239432210746250

536561210893912063748003777852982810399961982366744888654858962175855103379602689038290
443620117930662614103968573857757329466187920416717398710326247875543226747079766098661
328829907681159356563964633273204741739784265369199190646712778194691142976960118511471
239916246438581746961620165320380690418516622834351648532800087527289125465685629701201
490423918431288488389663226033266490504625424514215347584116081980944438373488836186965
881706169576409427254428138926942347770515747622285386000212947858872097964162848784823
402978623645157658953918944950241291951379627932033037232242176326155841484954516904150
481594884532645326553653281584852361499117017491452565020673881465567418367766013622884
181398544575838175149216798753623630224986650429880939102699718848802597021704485903020
255281376143367273000239314876299433294208753934637510479521663759369485928545296692388
206119609240108921940593703503900889489567554362043651223813504612232372382866856077012
147081993400632500504015944476137330293113072648737428865343019135662635609855517648157
420585324060915889455368833801259706382927178595879134872875146570278175062411446232761
653556428807613855546695349648767049259826390047902398515700112973510422734980397930594
893217731377835894642742149082818960757531663530474132916766726095819881366751938560800
708344597969448541187459637879570518688388410024575300697902067470221895246733062965330
342647425250643770366535796179927235140168088257230933194932235877697472080831632355032
305372876567519744991243310734787102283251543468653144891570018572714713571395281870574
403999035495292821984194312175107601239606851671559800959248113233096097277527290408919
323045138192140595486909598636156394978700520999400453737343156735490812703704510200581
394407500195798398153412379254954412683033510100983294068693167631486522184212629750354
696318803837229693273196369743552064935541432839605757761496876843255071820259896690012
147396634121190161145159208872722508230507932949494079983035117439139983561499528317698
741053009191646371157608033511517823544752890951669668677205601074949728003859896889690
175959651632083599249657918715316651765549612330531322300490379802892581565885346905592
047320220448582157391494245442560076874875963196383371954892180348425782151467432798324
626713901150955473503845122706744168158795314683679900968850352672289130370503480372551
647510513950112069430326188844836012931414513267422774490205843566135720229760256924370
8324414736426416862463644112821486668230832788188494762575231475975914601911830264703265
364351048628723308503013729871313054656816719716890905587991979239236783607661069139729
150170285367070569174329791673497006282391209582141932733273361327046384978971169436863
631742875989441795970005150265948402811553367844892890987111274270014556977186037683662
578910129563249421739324232168356599413714970634463931210308426517347819460763480768667
087802256060833998196852703913918886249039676884620136475618424530109411795428278950660
962786715984160333835560517106636856399954515920294897909002016370361939133668720115887
227353601700656960651464471120340094309551938779368928571657979945702661759806750658584
312007097858804395220899315610725582190373890438252120855893405296268706039463083862964
357048224731511988823264965123392600187450587221558420062895641955723634632331736133539
635988828154163245820868282051771812674230009967145556939963211009792089279391206198925
227872517706263567390676915075036279049590617019026398853848242279793595660084466646208
117859652920023970017257321295381222958572475480299622804524803151473644476070058401470
111615758330158798602567053286960250376685894688955023257396063113350790062326814658960
061059183444981762694684360620874637260023428731048777509719973729761222048448836153296
814104084191527315302909848106065910524127947206721677857638503413335352953768616347649
393982206113351167018085933748920098562602163753645763780873079879441630313293316895436
085481882718279108027649824309170051899256377720671901282595644396814549721198775362462
428587348366727741388547074278988113931036710074026045809509374008931914550371596430022
079870990203065777948063315078278940388496027687867501150352659995965391190385099575413
618646077638276100001304674420375584841941075474313628321747499691401725627458019098307
308145989564263710707449293185933599496778454639752986682330066260519705295839655493478
793407689272982386509972833688730942514477529332553010929388909505962166350064585168264
565447845180421527102664145472557243576467412591500350973442004140137796425302765595800
327792109172338906938023083570007457202077507051320768536767272669298257278806558077391
819386526298795127174848702311339078814879707566297739766886745595026821138724850673265
308769626032637977883817039169225967070289286107234695501370892227372415747352146311860
616833296428089427109325645217857757270417410463370921490477550989291583656236696358717
002254740672067093192911298504984090118749708255655051363127676195118237393101788940519
552680738315732701998473140823010125986798640249312652998698370675174619761290586419715
101691903811305389962591544058977983188111848636733518061096425229334202379950798747141
179724133194970352120390227690761693998661539247323087743933199384739346554960246748514
595610893832143463804437642967447806044542861177235622881235455493854936747031830938630
059948081007775522664098580690978871378998953777268055931062615837683270227356952548162
868397754059831321881072095957348925727333607100177373732221572506245554076368125818209
9456433315664416882111240820296324249271540402575558438562079972481372543147979164509345
891752823037794163366234595408270359566741034205830065125966067610571237001968882883335
782652565408580060812476178346121343470426599076471885291788165598760336660595722972060
467641459384056050985003866257202499991628345347601214095757561311377201422225379992258
192630947647380824921053667270233756952747512467574585656475437187370293203808164716184
881757492845639807036059299858141720705239145558163312398386041648372208155081340838489
890108366584225927027318649588730149154750457798656015794960986925857407280916123745761
202808037828969730368470554068069623149508576585780179457693095467489289107241726997087
828710790670221214053753348991612709489247267572344210192995716317251677202681645119541
166379138729020266989452202267016878474342065359295371655476871469752207308654175307695
502877479965694422846207438005525935426234468118729661835447521732595472403599115499033
810989292538332301108126011194582913247891826145659005709191480498898106410926531986415
491543906371153226168013913692077903113441179044933330532860222488114011183121690798555
552765277012968970799415293440757744481215215833914807326588549027739291894694430223060
336436124352701765988697804072997295130502643362922997685978827880727384238664341055561

pi to two million places

8673233405243304421265422580922026128079377868316809369096503930718518869575540535551924
474010188943418513036377102371916542473855697723434544164292942818412135505616030340415
197305170567200415254841430480032671364932915510047527288635743156272212124828080394804
923399414833168816848856114246398562772124899801046925246509842976800281988738917967161
599756299490810318094577654993003353372766028986041305137208816749655798938564649312549
305261794154357932439967264638223127993043825027457046156612228021160225297111191472604
790679114334610532916291139941814270585956941209348705329859577332450152356744973314305
978034707598118104996264834658498942908313702882424640767048658194554011222616640067209
749347735432236607890982793629180178659752966916049133180586454418649019700531272110948
672892161104504999285033923823873427727498171068639662551650545007777297633522839190936
341662686730283859776285488119063765714156748522900276413007228464624776604071999694613
008712309698169611776402458802404012277418711903828145364087877021405063492045688038663
506375827652148287252840916675958836895125614744011243810584148169296968942308269772458
826627968797006541223749113713683518273816786549949908789667069442237087859447037228743
161509109596631446466201164070945609747984902336368998006439917728458070166806855050901
362872368692311375217531959132603685491079270852283226079413342844843869645538234043507
367517552792845759993897084065342413403330059654225845197743549315652525394904076905497
553575974283270739261405029025531357059032776602472873309375923278180121805754870154845
632321548818821748220175933259548451238642854569852437940055643359240775739243854042343
670253663096934297865742526702990857323452167561710123261449059510089220234401191017621
933825534855414517694923404167850494813081108037421525083298166315034405878206840508107
976351147787203140491779334056996245570005100693036175538679862105922113749601741647859
581459938472721179339102336288987467100288436021052170030892146382834642017249337751459
461845171864693842448751077048033105857362742225568981078359505681344078125767276467759
112661617171405478682692221335071159277571118093621635640070319886152791549715815318620
255816103602208833140622334375402127944239891115148261761456759828494129184350381437440
406672132027369655909465688771867557646895937998403428439243642831384999361064558995730
821752409575198040756322481727837193912031933520095796798854086628858417158654024753194
619847844915636237942278008482519000342182972500368077973282965382738268270964577883062
907325769696592585022478834799000892773753323379739199484980079456201626834352930708261
139150815401573863587479108768681933062234033983618409712895981877562602320509009083023
276126078423856047623994116114319918232855723035025632996230987986981372198738395349508
800511056502551983076631001619269793411377996508716998429506073438599287426692821325806
226561042709298555323533190484297057597918028685911242678133278920001306637330850163745
789963926841843244879175040596605681343473002115212736490381460146601752804951232356523
323008822931562318907472490518120317167734021314247820386106747549184568130175936336823
259562941510850663733241345865262633044277892032996301166478431055788674965560193634540
666107980434433101760405369658787820701177835366459579770650091105134671806196456268045
892855313927988654488718292361747895538135482243227886132438874537573063925535056108022
035560618016580113110466655516833625015396973279483339203199122342273318161913617394790
971735209725444187397616167995449157406664226388362108258654824667000046603045031769080
346096803893749421289724462507372378801185645049543717747930590105413309154632762282060
176177726835719949947210008588867775294851461832830022160531607450286227139406225752271
536153117948325395295552742443222018538412130706088399437290030324169622364650062679206
998011745074151388569172249881228694521886700365546148784018091905968464151740997383864
403572306976416525888924650133328010595770874378130562626917433897663374685287599025182
271698006671366646708361635272676281649125400544732434236003266930910899393768313144242
197382299322978899509262145427603142396254271230092705131256849634464869986491336884943
639089723126426985662762511195953309015856024338982430606220942126898955033347468679306
819675596149852797331688931123863928159704872146311784898941116772994112153556595564602
254923636029767497358206868245320519582507730379363207582534854244796943684323958300229
664803194516068861469246983068428373136376435257635957350148686021954365414064201558168
884496826588112346578630251592084439687290937301123018832008241085506946274755850721881
988176641625353285219936784509184182428655803081514319124717065152153768560754252963760
776766790355340420681950971436961093717825001040346833890047236835797188718171866901601
416549973181593902753404297029478246885715803080708271595273198839570827244320729419541
477376160496369054079066248569819214941742723336708611922624396286500454742100377481547
483680701662129709692506903003943401395314388437285136146794894202202440990456468490128
084942230059632100179967311469204503257922153562548371953645571305199770032984963931022
431021180511979176412110170676120373178870189983365634887019288887530215727875646836504
202114486565320041827358101052483179031323330999480349305949980910359087330683426208375
330508051534731571048365541101858028938714764639728679708648175511967444202706231028853
344602840098064273631390115188257354508704401278803203088045899940139285783137032014631
169008275408416316137011378228175894830757274318712663013695540649705124014514793856161
607269587551618897759220093725494342263026666980219056424941329713649058451446070935457
705336461927026351091734441874165996058050721762322370670527694064598828269651738596744
318504935088709778844538955509746429914941415741063308802091460270719921816981821848327
678140565798031917860610615469014750739181327737402257584650193362927178684413597620946
794413405609798352525324749547488537619645521287202144933395185716779718324013615168083
948120338647768328115354036761880487119525767514184498817143299528951896302121924404678
378057413972164676150352183581754195988015587818389877976475267585748818331264167501749
810815143956777800430391656241421262183012497813989961269749311366312453454316541231370
246544678922138633261672210957095956775874781360274393282525867880943190187831786885162
815506666593489901119484204008654954306003803917832796210697202261917310139471184135238
262439808203804472560429396888860804766181485442792435502261760433830379987532493633844
337963930510645729014703350484425150829410029885717602209436030402689545253902264156554
722967444069816099570489706185759483066124539686153354855144597331415848823064508729062
747461691612693811227240324559977885023567656421042563528224329378227841337453906326841

724811294747773030374330284546143127261463231054005353426874295074516071168211325894504
737799852590628088970600841994325724353565991595671844195326672363569542603751340160430
291739059688164139004648874461987704713895350172088782930056934588329435186236441050520
472850795809082277452628379309621245247595045516612361438921321303632724583994778552010
358178117546910511108562484397813271226425738183065824881492113345160118758678218223689
876871510855234308438356729146224210574386173325666174196099936951870348029905664774658
947108378273848943931133757530927421892233623870241278137474699439306717379258961666208
232681848594299372006722042496665728301766117251770474243838873560953063077199230280701
657315710923722634020253364566142117210250498340082245863258341695067797490762157337322
084574771516728370216515257272690632420731081126675511116947543113934249625225886563205
594333643320042519380168662236961207972080580740842366932359222684518055532782070923498
976196712171677979162425780181446520515623399935707138000186543251139952406157139627860
042748761022274703605630086648981855636851288380978403164513015656934545393867343428477
068513872235729288140058166779277438303949914245719052875735721815448154682097769230039
701931588853344831834588709784166083322294817764296269862440018956263080105855695187343
003230459129313590389824489278817580776525982661720431484267406228422069358052956902096
633837016605316358514824802924808353213333866336552655846206542647208138004228866116415
553859799434407771856815108755523126448545964525682108101363543381796045718686396281912
497281114155466294954354934914680495744830271969878908369322495792479826449028468729907
283707520127828417701296038008043731360423501447543490823255460170757247290533176026007
168578534904476930884455602408001824755381632930999047059579577019350151577466033079725
700373762325590290641339382501275562649100510616602673582260016757926804227712563042088
963082394958703325881815901148357758853131200675974897321484873801355473336822289584377
169733243596867061474174214824198491715314444568644103073777827889633659175734785985830
896048923788864995744858433237190790563871810177897048941413043979539619304755136966890
917381633740202841805918926362634932358957961233483397612835753302169311053950963729495
957963237906257024403791754295515233462608114735451617283191517000014474899195218776476
917796619525687211950072287996953030485369675763637399079443068124854411943468769764884
058346616632214254392467783295252912629907876264227606330177169498919015833545190931969
710555142543804634544788337295850795248169098208668872949001376274936027610045903568066
045868909058431258692363897381507878483913244730620982187600698931976648193331380052981
213588470461356389427927750975444271361557414241267121343614431457693594318507539740155
634504300721477338036015915644231515266680816325650072654491737090742236099788501752860
483087906977643378024401438369175765823499600233283998024261432303051647115052461338692
295596198359018526989237206162766722239182968662229969012836837210704579973823259553391
994273015201586329031086623608902874311511451889364441672750585782363920478507306914908
015077421383261345609768350470320648086900699266616858755686557297130854886848503055927
422266456489552175234253394411505534725112646874301122506920502936588393469377742735748
818930547630545685454418508778334402716495296764357654554237608683138550128461793410564
132515360011011623132486698903783227670525440629689664390164466242697687371277902594848
672305041883512738771499958322927201056908176477203657453830626321212878234468285712677
402161319942376745969026344380275818965473334216076195899910713593304608006360114340785
902223013814431675618941610083866684613067297879981602699198450160903644503543006885749
180789117498869185855912088927342856240750598637464206718268858410438991569960048600167
802922715540433478128168320697425259008189993852431294865490970147090348901658048570591
023809840230292095484580270313579975506082780061492870342061018970592855562503690666603
519135105061273768315360391598296544096119848260379215369282152317537330450074594560265
041649063874521712882164977966308965320357792770559757387890470897607282272861556133365
541746541733325319504614471596125613447283876079497920347277890260645594036838939830512
804310689788320922552674030990727666325563098291433773362487639009803746799583633167074
909560157767806252841815856088281251037304566516783812329202644333428924799895376370028
470882057317814802861566947067176088301797629574365703863766543046743001822177521707979
011378132020917101225944593947603263352210034409607288891180582302988000139210576990206
104119422259014412676393754756585523326364613691909630559005238784905808585472189314329
512565509891118545790904743098672451982156484092911235670470447158165412242094848952602
395442514968969986468927865337205038152583553395259594977924503735072336029292568166440
317544720933502346993573428041628057585312310210403470741055394794662342247219765418606
633075609383928430245561876321522936953094967261453911550427797429139462348116389099455
719879805891080940522085663477364025185331579289422226129519671014878689533990411656262
361099307862774431260985791519865108219376344239750924444318888327216723144731398927565
393874690848378599983903019284755990844803139757226257799580238371821182610778254632805
016627569293348171019992136884912579741783441246362148079932630317896300773162287411001
414744682958710964094360165114643704044042410280223715486434163126869710296745723799303
787091192660226885434933928846977833183377275536002360452697553331207405219443841871317
065004062303142578963962462550805346231951577272160682467971323982350711214667903334253
166105009268282697832051707970606477094753322191130031634327466649212585334869826284993
336490749757627177624238424575293760363736930665698580813242457184231916598443524836383
914723677091125428797492840916714841056061734952555303808025489089798998339974419688164
435811599635419445468244300187739344851567862993714124642025542762803464528194319699851
567759428015890117242991147492324496362966536112256462081778084193736717417486930319865
392314510420342087874944638342007423694674167379137718541729302110672529004985058939500
157644652578426785050041281505141101676109024179981163673651173343630520421718834697334
409607550618922832080020682321882552093166826822405525067853326727393544627953544503469
346708868997829659766680155868555439280549691600153824644154729634264346247442338095616
697486432678510537404871530783285248609214792938904006191940775854125772182974031111166
845751943377154666059314963226057639579506672379248045566267449590160990445717007476853
018258359111847027181738399220659114697386904071066889191440478183940152554984309841740
429629710037588902667980037542352007789188689320658825673592163374899321907433307763236

8063067660313305809936503927849966330583441058395584235470613925830296256059372949699I7
7865370073582595672916556375753338237946506670281331633677877857765403844507622370281 56
0026646876008330083477694106522423132196281717170517457473158031468377944229920625843 91
4516306421928076234721173733905412416898122211627447460479742515349515614487846451905 98
0558527492770651104696899879122692095436329187565984975816004981782653422680617557997 78
8709829039794776921498971388294412989921906111009727334464422545870811256275160768956 04
1869526381765629725352532046547322321411273494490007070709995343823704310299549318990 72
9038940393600846457924780974770689377792360934824065290804468238738675355043214380336 07
2212290228590015216874744186876802189742358281150586845563682198271868382822712022953 55
9526409811922062625661817453862043789764774736740130114106191782005477288908821348907 13
6348279155939588984311141468449293179595876873104619946621724299405533634560437264288 31
4927165898433580145809725762776274800868304148766120407151948274950978760628543052606 09
1740304158561947710424021563176127625037890826423309483212927690310813309633553644609 98
1134686582629193850306336466883852804520439120650966599343761209721035661834491420632 97
5093591105533838199628637611122145983100288638777522538898710716930772948289682838818 00
8789055758962870282203495555921919642820702660166974704507588241236492467930828360440 47
7992025567808100970827073963925898237239903150169736878073855053728066007033935902989 40
5855071015558203951929452755763416105602426845117319470938660056472209964497000579711 62
8683878848068645754521298876212011332307835994455796242751619061661208659222884439993 71
3691613377477647967546404475222365198791890666869461536340304978504270563269736889717 67
8169635891214794508856032492035910500052702912506617445710902488231852308738473375760 96
2962300323163193177051241885518874008914452275838720303933463413703988374750037000413 03
0323746168105282278963024065719542847439561363059606636727661637076243861517931768563 78
2181670726366526372749429668826496421566909876575563953898151737153919037387366178108 51
1437354917182691516646587954314366916574559914509134698300554346973946815184912852674 36
8216332335002400415347563366555436747853435663843492308266410118637946368242741476321 33
0855586089115991598069299333358737999211709877610100086351226966108582377960077298252 99
1642245245377579078637677001762042169801201206477788168122853229701582651537591047279 82
0858102544716438909214598859930587389980618144437124679609278960169959305673114778765 16
4546041828068081373056411394770033955105929671179213573561422133497701035557890393120 36
7950779726785520691309610133057995736556598984214080110895777496352490305867214403321 22
6633945777603366707398986637565550566837590195411170400240417317419194394595874100949 12
7918935832205096075154456583107646322020741556730073492983922375886024363302193587602 70
2353834850809030972404723067483383427565419390175663305343590572925362906666292442988 49
1888974930180463368274060418797264455287909103745901527962998641911206412766341665467 26
0103101439464825427319408260519252525514707780650538193970341483331883886244544685875 02
8351367931069259265542345543140287727009644625932593620484150997618613559311505374332 05
9576171817539372376834948965880556197103384605676517806077143781175119854958150265954 04
7897160571142748167043354307983627680857375203598045470718987444627551568650652899034 42
7052422663290067377061407384533145546369620398514634845405036245205155448090190444992 77
9197019463405204021649225879685477096465918960525508020437435632499552788756401854856 88
0904661726647219303539802576895540756070945089216392751733624848727460634457746061430 01
5047148066380230750784671866268665480628749533438126508529861793606273054116087751320 65
0670760095159696922847060025105918981686056367472284834036623131284002783496502642074 29
4648098719581798919823422678010991367265869779572715637083820603846750851250378536868 90
5000292924591381449676103680156504940064501720281008440749824884336338377389451464234 99
8074348754194957065340487599306296865740177229892207041712379380368689301950471408053 16
3679244276114987282434977973451896553103515163001928410739851952468072113643847868746 46
8783890546685870146820040424766266911978505277023299502552375920218239761314030131820 78
1109424341894035028781848335699875419549142490764928185012769692584507982309924563974 00
7490022041400408149827161772232785971823207444360963872983969617695232630442266714035 64
4801369501115127361959523389542466979960902971117729497985260181209594985534793603995 87
0226035386844460060036455878372081734484817289849623343316168523289628174768566947774 4
7841176667785550405402032877231291296120071756182185749660721286011550094324832066067 17
0837338439244717497650650653958694806527498510964350441313666311716149092748008947971 55
1280380890602661233877021024608712602369236278204272063214885695583869244556192954397 98
2824586312915597016112139337667041604781465453739848204793206230545889183023663591107 08
6132361362025300605816801169567677917218658389861106410284466693539512890787367358837 88
0907293933556892422980921899937585652519996830353957204303815423832321558948134879730 33
4986225669486094404523710963112042947370999832775237590287943510148079602410853182822 72
7206520917140527163995570068548889930713270444263811713616137842164926708454473601533 71
5351403225226693993156469392548886092771946950951027205658982085161114169375891858497 767
3395810739552055411361167474588940005579265012516802256429132904645122461016116377589 58
6003551963308462636809484298685578212459237860644910862614952838673857484606243294769 75
5650601734470232245169325819594204801744386182779009979102699794363773861023435990197 29
0266572176934905312421082014790107623442211764049817109404564671479648318500364607065 24
3737991495754907464379355660866873504706880288802273716915066796162272636779142044531 13
8129657889219079893067928079519634385544299305266711818883169318151533000819834137036 33
3781165136447844956432026304478905688325391327568550064988743645297240190393219423619 46
8581860924819914471822312563503610709908143816820094888145747918340074208805492824346 48
2202018090066089312184871142202389059614812957078986870673096328350772800358466472526 33
3260925467988315628147216653104290355106381867239459262224918971857466227564481923207 912
0334629404623716455086820747807921200224810448240894376772964570375052148843238334778 44
2365762659625774764216633838696810515560040511665892942265564305067431960663214938326 85
2324112410309829261839656672653757096227270816376522222612795703789130273622424542802 35
7769496697979160378060206119982965376686414860573724378570868569536717430528588672794 15
0486971707923068099196355970906670860867044395694929890349846096408355212432082158722 09
8569426394420680861525393226385146418582255088724356407610075557986030430061902939977 50

pi to two million places

3652924496276806301619395868967484282296966828716007054121885360112526214726199193457 01
1249413604007313710356012354491414456113262081479722024456292008554596767107745262944 64
0369871156673705233042276920606892199721228765159234917270882505266047849195844456959 14
9545052034397221960570224287059634771994874534230358497345554348357522590050013267928 95
2996834528300292496787153256950895599473560087716147867358907024223034525677092641861 30
6526224026655930237961089086540285236354325701095038570359225167241337030774595282986 81
9959453161810997086177904319521494540869980019284417834044175757475498417234322439525 42
1877214719142877946227042455110384041022996321821486724478514170043239076582050744174 33
3913902549372260051917805528866835927835968613048448054765207011038705456790506671332 68
6146917932097415430437005900575140245303811350859762320498835003752190107317242391320 61
2084534163642542090199083978951697491908019825707546953871760435550310293545603188254 24
1234153802602357998486127102726194594617302501506272017827183475079537546142823224020 48
0035347713087358554386583808986557902901720899215236260285441345666442100755713174115 36
1881364359683391783668530395519789793183344026657639581776384415020474638391694349403 75
4718896359566538837878175800852943310515647806316755976396846284548685176130784923521 92
8758438571742115372742662475494600749829464689113132026049824266660617823621675211624 46
7655718520142770726517054576989526891586548638365326577466124782875861639551056907937 57
9047468002414808461730023500651159437599504722627953558309565685290552412433260659123 96
5566162975159924678569767286095783870829726888186247764282460889091310277168092334173 21
8186389413029912801785768100411845331705487807299353922416470846592767695960086133846 72
5891792224684019652360040451922668805022913825254274427296686686844346999364644953850 30
2964431418168330679566065919224226864233725897029749867798223798697035091567288250118 97
2850543669699911822927643201346614262096674034698918812334912849878609458508308541160 26
7918749563839972658793854049375889376265175214313400915629767663484039340597113895896 11
1779237271973120087119028387238504209951365020895996502386923874311230076493895458704 53
7403913276742754658569143249223442487954978543194217729551032904366695097122905794105 42
2297138498333184568167800471652743034462150937374568908834277265499795897400277158697 28
1756716217747160623935625132132183377063081236199066762068912611033583997668527940228 00
6933262304276436320080460279454398218569268957608829810520347041628906151426955663260 24
1235845684774022161533175079490450618994277541648766812830608767680342801370615151759 36
5420040664344334264969630135976957110953844130547513738065151146567042728087879211161 88
3867025832614835203359524897998800598432685214221661012138642344447403554153618125451 23
2725417903752432846526134001564034078029508317387068633858218904309437268251977835434 14
0441463122505028998509562964721998989599570276017161080026390045428066701936887008687 37
6096355170539610915864885224752259827608509868639319086571570514647943064980794740035 99
3567831575316978640164121027673592744840821479316168720146307368326807553235550723378 45
1966583395961990088575605320464904398718780003877653449848279127297597926785893238835 19
7843597745199337906050047835747413011343857767012774677655849643653499207291127505244 78
0679900247638081336083769159343073447099908724612469511584999056938945819299683152569 34
3465601650490736543456168992441785418309627877468642564733090801771234738274986537042 76
8589418353870662496945358975129763649216620225195597820488201122024520631801743185314 05
3571493988583535207253360592816564736622280375125433181190686358045011060008848948725 19
9038611570686001291733949132919990989699652040286479747867114112641497352355060282805 63
8673189135206201822654474237530759147972778345485196580957117670372489113221874184884 20
5375456643048605484171435420981220790343901468246430804763093999988094364036237904956 66
9083066589344116671233071908768869022289082455094393576537008813954468575958221592423 77
1337592890494566991341957756948720993455234879742554959734725668245703275248563420580 30
0519066333222614729844663566441394741536163903196368909701892934218412965470642120441 57
9830928211275512323094119704687036675585862914686658160114690758181453238862599468813 06
1100858655243133038885236178754162451931854885430520476587697178187186060324509226705 02
3562330752865400634453747477398103301753010639285876706600683662423915934474792389358 68
4474668813439768412895220852628080152478115809087282720910682707148879262220018872071 33
5834731160493875032024713505110973415142718851074543943863599825168938672783207606565 84
3299641926687729117880553063987707589284665465017568713885097853054377475608159818905 38
1285975975872902764191458768046930395798940250280628449443997726788334603498714471591 11
8640875088480118250466062694928132550585446066807219280767226318541970608997272793288 35
6308679665821656376621268700062086139857953368721832976694352586585218102197112799988 52
3286014542438864921063962550811291452452556165662517453982107916583229482739198907226 63
8085938777066692017984543733825758335616544148227413493668899462756623005823486416374 8
9129257434542653687209861633471954725605376490261109861602206371517203173107223352777 90
2356230116122158872190683184769671215454477771891976364767563598836504853738681471118 97
1719168094424532016562646584241532046095219750813874943627273911539703898102205020721 13
7964289816784565719998320957828675482461992552416895064756244587568047657982710420649 18
2800496114178996326534938369709961865633562525191279527466116452893243942502605394991 29
7678948685545138364629861387059300848639514482754280925425826592818116234558573696435 12
1661127570786587873891004750949502074143285562475323406880694603658663822415920594770 11
6272455350770188320341275061651643409892427663619490660405147590189818284704145406798 63
3774958971593303210863967948024522360219047075408495243809362207420421644466590941861 15
1612542143978530353147142204276050991099958425827065342806551699533764135405683985905 40
3911441118107931792354792978041774003118902834014205913178601761814010139067839451607 97
3409333145292883356826676108730896556866634147469799092012916395177656476041001359230 84
6407781118577017147547093052279610780512922592213796676795242302547778686909869863959 32
8301752671349257394247665703796711996549057910346451721056316111628475761605028027414 734
4205937987385709393659976458406436347102341051386856098154364194620709348749556571515 25
8764739863874662911205252188619786447193534012187708564430494976466327211173129663074 01
4742885338807139554839666151164358327963270714970366472045548239075241704873711090203 56
8887842627143284596829485901099120563280067781728195594808359588637380060203312581204 04
7411619510055707663751391852063501937756689087912226255291511541716115943112322974848 07

pi to two million places

9460008427361783694526536568638908922379219148889415134127643812867253543943836658764 98
9351173459546582210462070156946249629829686586946032971642961058302736294664078669548 04
7160687674012930983882701126941745767284965389499677700135497071758739769885357016099 68
8719243980125753778863004529960603894245253298127688379753282943478474717273694838978 44
3031544557233271925819567795752017684059491039451024870405297775461324139574737024958 06
9521906119355246561300760608170366063986369091968779673269874206984102154041358774349 94
4248453987366289979370493199201243310871771588908331459369497513444150308969888483367 04
8895335173438366613750498866788501897876248896041891956562134925599579469582151271873 60
2229443457603123411616222118783062630160025615130600896179588091337785993827573891166 17
4922000480194665397167253199996787824646520704202552895806129348792215707032350510788 73
3810356132469058912542801673145222610510975268844581739585286749700873281737026503911 41
8808636257860068677095357094541862720371829592329394290073204568908538467692807676706 07
8362731454661357464511861419269968400363552457866825286055236087839263372094955686573 05
8885355242731718303555576619560181320696970523147285916036574152819379480390780570177 66
2496776770830545339964410411490744480918935442371587513163712069092849718818652651795 69
6250608768086379929927514381917551314128569838049905838324664071843533573031977719234 02
1733551252475402895946451934978941878211165244207196479633540280842768258745505143075 40
9128869703382757982917252701250517485725186567605296857571710439851775863465492736334 97
8374578557751759823118887963560642928826106915221138773046907291178439687957179969190 87
6747446268939010438389308935906815055062518042834249848285474537390911965500578748523 61
9442673653389203529338179687943458851309654480825952165334885331425859066052701002099 81
3049869066106986772341451502161803321817251780261638035508006115148516501870348138649 66
2322983550474748140322134137111339042015650468629586325733803273188936337821297631232 40
8165946581252935606983259454820972718723855864012085426527936578437086121319008494446 31
6692276804749918097816344705150811603055468454777271216035378405507104475174017789158 50
1440333997119754393337190219371611995932172165043485986068599075814866338515935225398 28
3863657870494267924624822064659076371160799116669227004834577383495050767456171329677 41
5741936054230465108639047074309668302135313137383915401860054831581999478536840792438 86
3038942448666729162465474925205755454998328995505246549396464110672470246181615198685 29
5740835066271448492976276103276615079546740980485564804519926700067541051461682403713 36
6468707391805762245964512112517214928446078387295065526556240782332021634284227449734 06
6706057854813128416170006827941652408616053049764141535310762475851611431896326641215 41
7647883980573951816708275462640191334272469434064546092333107500561374885845601058707 64
2805510491568322992177718771114228821399142351389997804809074196675042934268844058769 30
9739698849212657406507384241325904995133280748918486312513660051689098675766189306418 60
1450768899664364598478242476379218946820529332861816072858625490193606396597217345974 24
6685348919077150969594715984149588357499407943076371650758713084303506596563276142496 89
1702704725118289315301163513640123233194847773556432256180885507268633615667027747665 13
1684473045892330468450358273413921552056559555054402700770957118912353700328939384719 07
1301230061494850009773304190162974081294121310074365216967694737119446504302675287456 52
6427117576772985100227318153147912186678231763814300133213572333164083676304828635024 62
2929510394464197290117772306739817085494282131685487863584423157565151264246197138635 69
8642648625570072883284977465256569893588812206561543274352350542264304763070175962039 98
9445693754077604277206428566156427496406231822580913760461333554092581093486568339592 59
5977232926299283940707102662050696026191475071454672953791936057733951421585773670258 12
5547777805618183848629235782904863041169675947595731092163648745128732060989775157444 36
0628320107517560505898637739464267237256633132915147636294596529573750455603308507170 21
8653864979703998269119572356546639664878836705324991760547624637368838647432393680307 11
2408140098413134187190232461398167284743485685843324906763226807882989699853205500159 54
5642584642953957832967136840868605551376134831638199710691041124877229308565596116289 25
1943272895673579308548825346218805978303619673143472244736074019048825436693838594986 43
8967158443464978194846087046684244058850581329094867528103508941891768736013456703984 02
5409539983912926163964011975850402830598747519890383637633604795218009056399940711149 01
6348811608797558286887112371978411312774047497747160299039098629320119434195326130687 55
8687004791150584278118927360189884553794280093835596003055564249057493383113158126893 58
0917182167939384608913110009137821696025696194391563857097721085294853408402225263153 80
7576481435211101786060796620215364574519804287924921559609178514675534716283451480976 98
4781066794695258043862789330728544307551867467403178992059623253069684985062225645379 80
1933240714701053883124572159752937376641446000243334746342161564710908006844857331009 96
9795675086132295621539489661290707608705945773211096139595720683769817333217722722945 23
0201171495827462461423008792456391539977587988893436051928092805342757289772298270410 82
5265060322474743465539712444582110094180852882766428721553037320730305369998331909546 44
6013490580027726873360909675283536989869152991883777944708673208051172958828522346346 61
3521178777082202451523309869472048802369839037472245069452923338166809156021611945380 71
2720856939944271125470567703391660685041883392112071020579718959740295371913788508239 54
1844729720512583014502236399845888188480795321109613607068185762270372222467367863298 88
2619535077488482192580101028167734878013651737085975159686350692761148681867915433063 31
0728405287263296042746501674512098057246159835312497965083360962898465179248508654985 01
7657926748368303140237251160519048200492375688697048132356701387122140348525975716982 31
4939210350445381249343854754042877162926078715965971000324253508180067985047846704341 07
1380230002355842292762441880983121027849845688281905621486116883829213807695590238595 98
8882718427141298315357465948383508863675568725072370057882080806516430851305753907148 08
1251981972754406042860721307233112513435473512282372636902059233468408952903944181747 22
2180808802420374542922819505156020903257066566675135522211083893234433013109226745609 08
9648928471600900168434256860302406518387125364447910844822863457234286300447771535676 79
2005126801583206845145484620537724083170230751185826958827819683280629318141579629230 29
1459009238921910502400330171598536021068981657309597054708754304915350757283471021367 36
5907829889889859188258506383870582323610401342736802145772797812650351447130076685198 89

8899488125113240712311225423655374109305713776764663088980819746673440194853967427464 64
8583663038582926833163247475856073411295604423504815136650364369824130652116487321361 72
7815379001118438764792147306024859969662645976895999616672114157678742716314574342191 96
6125438248680637950656707837930566454874971508171264992274587707051401784032520258374 30
6723979387519407383478213542492644804594169360123778415575683802193224515671981468313 19
3517176876767542936808742932303253319005776115303778488076717915873718241896292349633 57
7308871330142065824064720184368666477110195082256544185041410485950923315164175668353 35
9035680597774882628603921630511428440147711227677359780055699685645156981544404973095 65
6182081410723455507709488931646638715879782699096720104770048662041577436418510088977 51
2294013386215725597966730300333356089650457129948604508801030069957081183274122014028 84
6731876121147215076082864500804933506058341244005349924100791897897156942929060194596 16
6639535739005740828189795865610630214250940578026003878066747025727850423102679255268 66
4164783757239192657408335310109199460586891262969289722799695616183796586253951457765 65
1389915605969119257014267769719536221269214924793301322253417967338045456906900224486 31
4543699280508959028798536399116496104569341635237381672700798248649593725432944147501 74
2253488007460536685274107358142140481381926095774857539832396432202767689829529932229 91
0059111394809030625786719468522075436723314415632747458413678296951820909075753738081 52
2550681581634512968584305547003405730937195651153334118822579118429127897624427424866 88
2894608417337727628589144343810796632418581374338465591336125065621440863118565507482 03
6595406311994514473552366495307424361262208221587779918117318239987916772686246981001 33
0899344134103255651244750981673830578388002793178236626680057338662830452691501217498 81
1101066266316999074089546788872498376069642675607327655020295306432024565848803574340 91
9018457969670918958359215806431892115502078222540589621238504836333668640526509539885 39
6018687480254395990775708635647118107817320835882410789339848496095045157862369069379 09
1029833658612505780061282042832989870130361742550814179298851020439277603910088639919 70
8838942244343747624450649943253350075532150135691645447402450634142172092637049441628 54
4935517496673951868233770388227017428395306486621402208480235694617451935044520055084 29
8960638549788480292304687650931824304560690662412014323288603587029046926956993772836 99
5725611799981416803288920099720767647275016278368676505409280614368255993407537108512 49
5335052183812219579418437687503437478613569746659621989929833979417746291399911050338 56
6080297250556471584663407144188104066816046594308008476074233269689870583340799591401 17
3561401519392927093827126873574129118647878817379335307972392211331923087224051198776 75
7392537261658645494269708352486288423730510869231211870617197851882091508255323763828 74
5420058366872399087894313476817121025215776821785403498261233445733212541743174510218 53
5496555794341519240414293857507345018732263112696526148087256926960357068121518907376 58
5098910753621028837150272211457153420899618755816531597537322869658706613441043924561 44
3118789525746217250888142210732883214864383691577885022055530437938704387517855349000 33
3364632825063606684345881652288584391434107449088740095011349925750083793557515655530 08
7911804733639802807273511498996544983963827932308692981654726199994170358824640368229 60
3474754109759259242750221302723318111936925255836794834445355812231850163728754602455 91
4289361798984100341014967367348692796553529916764681396203188564826286366520511948155 19
8319369243601872033962586434774202443854834329270704086360237177118957877448782109083 10
3386033757165356383769148111551775783186058652713314834704707101474914407295267173437 31
8892380411957913758311717249547015105322902122778856375914412683590991477357721896785 66
8394789665546193776843719924854092034774794359735964590751000134136846327053994236633 18
2757326027905931328900717524232979988202872981545698068575260272085345081518955899801 53
7985910833026265950340924822081696983094509905172222977479309296013722311620706318059 09
8098652204984126751609689936908715312615410639114099719516652752102655137462106723975 45
3675152772266427332500755330381155779249017108197999874003433206894954302021233491655 97
5534138475433458908459721056512617527790387371558021681020134897623004393933012361956 06
7405655072576447626418641038275225027502488152042120144894873427315767243977751651960 48
7715913993375733052156316005828183275242512498428929504569726627231794999822413892045 81
1392228819709672866813419965854508925414788186612143164646115500308942954164150080443 51
2671000771719106608163491160104940952885915071027319245300921033761574785281901117852 53
1960939737707588345355016210512061861846016959484792422655828320538917434141914253387 34
9283626252747562106500268920930549803819135050751583315362729526902335454906730690707 94
6290270980430696775777232406825890828825087192289923845356124960533609093119877149768 53
4918771058781525115763217826880038364325320845549496605992316134490296229826383665569 56
2390291636399240531845738319733605689630290156720278747647422923743120838753647797980 21
1706640414003623960342990547821460015964384903914086217248900119363936205294911649019 01
7893767741770081841573646424822803477659833995594199947423513808772027356499653746518 05
0592837624590615012486354904059035044355177396875212569939225314054666167532600959644 01
5081447134085005994133586824015679689391652329956902561795703187363460025492750797585 72
4069987753769610695549112455929371681615609269974014748594909719575283932785029551163 73
1283631609414002641277143134385809364107291831885578184391193113021591853505785782678 13
7409515139882872093298720891708796558403379958180364109130176058364283768346750269744 79
1732736392549826609286541266862508695840576838267736355761018840756880462931100001763 13
8204199133842294642175476228763268895103256501257806609110154182617597117863011849312 04
1707888481703544459114209246267425738549267599297215826630124896079160280796283008237 69
8498425662446273054786522712475385565870151842593678666072623746429361493122303169711 49
9382137162377734101893468412498583333897120574609876533495437095445810473411945604716 67
6371488771940018585038016653918844353076806785521002842208061018153259866723259360862 91
1942795285067162032159562926234421085030375100458021678205859431213178498651650183889 71
4292810204672766483396583361220050412269954094524631090541486701656213111418814642514 00
0685943421416229501372411950289404024171529620214821316942687345506174832577489901727 06
7868022135805201631717965970625499148910104082259013880692712259987661906499755397253 79
2420978780527903122745353722125723678198168627754360065993160516085471941176933600655 30
4932700154996409412462940276077861462842633799787082124130967482772952565904033339871 30

1663605205152943872270926530023128571127739343919972429465616031492288646383048211826400
063463224452327691249279637091900448243861527211738977530537784151082701879026418975985

3921758509491453718196397670500624601091814374473901128622026150385171539507502465700300
5427489989986632106446923470524485266358419869215485060736373635970103700820045499135530
0234227683615660923718376237661323071794294035051620589837880575761639563189651328333790
9108098628743715032183337937481271006919655660981024550343720431975792380163893383788490
7280096200674952639522924552530003915132865200430494808400441676166357256110336090890350
7870266660457411866011239153570922307993506275634358669412124140690200009842972885997105
5262046007359616597124624587280019214922823745704770055555223507901007425779641745666420
1813760767728021416763431176633476092805124487517986175014697651109231401998502475210550
5143218320324683862236386980214973252917268081826916022640145198571598986306364769384670
3462437353362122945267400297366321892755776313361433052212983732893624168380976631121780
4390646592131852765757542452149706408572612694855555151694419169634563494851895549099860
1854707492170919397215014181583554915177803493082841627853032003711388766971247016453050
2962514625354808846593732198132460404700507874553157774453012722167870969923965194481000
2766772452209826138895761193998258213505953906108253438307384613766434248142635030356350
3280882554565828695145688348344389864084706996158155355657093597852981815412281024931160
2677166475542324543736455209716055534344760049579913059586066547480919699758523696391050
4306053299698617774555702748403160511015000927509830541778820701652887805648588835393000
9663784141645146430654160945346492749719443088430190083243450010711224850958462363311550
9565403500102538107765487088903824482105960217567469170761266444651549744820261769359920
1590010846789424863080931109744521680530599687470719493341482127894602404789986782270860
0395217898191152575910924409490610821941572188245138787525105378779992750563508926517780
2275699283904072625943009940683658411506015542740854969171093225533351399055771430122450
6225359209089289954878901688738794414981766740919472190858532075881342435241468440946940
2873442303467981486932066690547644081879075643327809690737431561065400350336256344104550
4056116584256661946413156995920281660954660404043307842960942336428006764252024688437900
3702664327383752673444566774360447591172345544312137512998716053494178990624074624805910
7515454214501015259350185794979677091494277629876776187535917697665693353620075264400550
8272233782574612419177510398969828500878046291161247029359770415712278682084150599899000
3507649056813531544960105058378252154072295847814120922607738836858332569915595827625740
3699100895388635258579158154629740992889295565810872246879891444339328678006124344797040
9409511498725372216207121828147224720560791632035468453582705399969956173342655643802130
9344413441781432747347458571334336964504566883634524530004149418719459219484336580518380
8472852895732812318742723572081746464906156807088101174985927243048800304705360659826360
2448722181336220584427696347899255367395689161759560083127512536373114862891828853199640
7740513065287390111602795147240376520467967527572094612954884943477068473583886574086290
8796841386854129038445673491791947243547373577962973109501363685634098652485123074424920
5486001220483463776965774869823809907533872473810161633104522018148205171024549027040320
1103931752914123585900613219958071152352465327308245187455190176321232392660919553605330
7933941481022223920430585099111278122083417299854285324888997493781377149242350646220600
4970162272257555927202823045246168144832779459577268405740735243872632512177386078725220
3739508292068258624582958286602422370091641145322515623508672835702592065793793580717230
1657825310264002591649486575851561663240059633743322405836357689136380996606783331520550
3819429817545186286239527622246426426761932200572609407080074537976602069321506604806540
7092367002545547809374312963446995119364570227770428949198747314651599917434466366431150
2703209631814474396846660701703865683892904137261451792052830377980015967332211262139210
1109269437324265147451532902567797789786279168007426457001776585877276441102488175387090
5502866366285702631304326182167559973407551630874118481015994778546171992758064208007640
5039393096765010791279645756176777172313252870877118564365479815572132267131671034283330
7502648092327092794693618754237171323743211882013104013257754206155347774738095493302950
6052863142992396761724259517652596037853677004865430205037905560798497617984976754852220
9656890866878862727228816635729965074212178167101318446489839792745841046091740195535940
5218077264960164856589633127651777256587055663862467702202973067161904964760175553191740
3069427296380243999015786258795403804292501039962284236096579961742566414610828782467130
0865386722519747486475160037191223164329373613721418190451445398538129030657845525777680
1906421502086639354601117976077314460169807722149462018215475418346815698669749522143740
8395574360990937730628569538919333632277884147783147016358795275705807090366941321434640
6721421620873597452407196344419139871572266495623520912489482903291737923475390647902060
4059689445858181883163725056971508249547210011430612911212796492461448149452673204611800
9483177953710626138560659281107487802047445339466878405823283084368252349828787878723100
4380127584326528875932890559713908709489343291878936358752309503197241494331147802681510
4328645709073399000646460304594591932175490547053317780363061108023837904132610217065420
3977768286739489552343409114590980321800742174826630803310943203510224497654058743052680
7218717511941262182496407053305819760578030758724139895291925934464883286421322089511970
2857058085172912608097008970995887380884261281947562746814442658908170193970695114201150
8900969329212635055878691414422507111789576412606821203649731588155850616088636087609140
5410649358397382046054976991260664869741727460455620073077997025851000427207211190438320
2659934125585254218741031547322259605089037109905124346378284634845172903396094077545960
6357132459100961756763029023480504605636796417009621159392723484723154635934305942047870
8391654455907223041721402185485729296347299206902594182993433563283374619158931099366650
1490270631455346378192132910495364916167968633686620280920590928177169636016432038717630
6110189152647442437196853402434141040966015175233063101005427660172999592407986052657980
1817035438372244189708677491155714139590504587071241318866159966720310611786784238817540
2900081730679000770076910331594768471274805145745012416960942122068563871719203334477010
4312778358428303612558669722799227600553023795754766876108169953269680779921056242198710
3776427483934593400770564856824734564374292896286732430071327569721178286342863180307840
6080543896703122565204815726550590173590318065618665763697901334100388048100426918537380
8288393764140930372369738882559938015301602765093895222562548658203110829663871847889790
1925727351648652994821211128582237785207552579470411165167098301067388636802065032351220

1422566629020838518497397969219459740399672259293570084223216544494338101694044001464211
7270436903725308849772967616083826795289327244373323332460130592584153258528018542292699
7550504851231651320117445020063422897574979074186647609056661313606066440044968969697211
2349015160103161691153912636141154812019175440384297310648061380667240679453841987763099
0520418186455398856001605808208530346674199922554838143891426886039051307311340207810322
2056539906675658811834697124731849574210492598204123300698106660267266711696986925236999
6725712083960758106599060488550652559453746533370816312519731527315099997994543164744800
3653052545522585838843703653135848914496887615578098654561687941532633343292732632005488
0732988292692523956547035274150938764856434363473639056006758434295779253318528562866300
8026045643506051869466396882768368303391132105827131669010536965894843353034394424917499
9437032897561269118496786675396745756918673960184265955632586698247362632808326013041622
2007837763962863535663722658247741018746398460913781797021993624669433787561051423091600
3496737482753218542447344046986722814462197334456906559322604513631677572276976175152955
1313685227957989500063086648583338358879170854878326140081434318685536432540900592565777
5696365794515185243227890604661143245633459438333635555708743837768195479358597783206077
2790896851046376629561175361129876481100581050628573833021491418969175732364079223940033
8761018773068820804520987063069665011545608978239558975650821895298053734294241239025166
1320915068352562360643229146350152926876256308839547958673606574040467597603190075156022
6286994491769911033254919036462191565489658363878526176254475910304043111983313367489633
7354305033470837868025988572971423570280320308651757777636626671414358456262372031729299
1827115658477385191493002681144009361798336375981553692131194616541151993329792365836355
4700798531153404437551172697653606369634960745709208910153884541024742189227367753929011
5254514876178265326992234093458834443606532356957483067475863959823738062392209697911300
3750092472776748120961480014836982845866882539992254174831248779909843213888058066575288
0496316938509281661241484204855904365375260557067208866019837147246948855906002346295666
8074869276448530988945777025255927655136843481797837680312522006201165499621938376824633
8114283942757573358775707616656887162930170018370150183710395854417950429051793028426299
8424052204885287687017992868810223410809450030874343238609852022593583939779916995141488
0658359577079153360421522477570541381583059805172426820302391389961878553314937010018177
3759943534051799482846602724271389232420663582213735640255563916124604359314445647716177
7866866219075699419354521024055957858709626247578237422408723073946637423015115630807966
2125507698184341575221302013166985311539109517086258592117075346833924898471142457857977
0929970095577749053693445893021962862232942707370913938488090580763601623358850707116500
7059220903422702768367440471779518562158464237853809484391702154030916967118014483350411
5350152398578786248240022863505124509787530133362927489565881013559983576664132584186766
8819941826558770925409282025365756890916920648701645652031351912023088447792142018232388
3481258192523062906405913889072670048017050238028310754987743595158610958478458204717233
2047821419359564352300682408293316769495863783243373629213297448576488445727694223181466
2916904038740784342539986547618600703163787244496872408393259030407246300688392527437833
4320999780031007394854325596199591305431833800668084217007647172117108083147313042201011
7879602727695518486931563000038647395859647512708180620202911281013372150482991415981466
5296592513913475870637105740984625317445175515929826926313019563163278023172545116337522
8049349478213609651767830549119371709810834938941184111728452961564504339530580776362311
0393940021474620213824867065613243912892408055084364894505142562230869827011391775839522
0659478505746967206455657639948900823506165158225565200555796691516136075405962475926533
0219315383957703770918635363344128644336041922563641143092881385230927858644517927850388
6204597542368888150965913430139335350472349651327867496158750834395837196203126669839788
7897116632211286759597529257421038357996509963160278725067226121470749202825494016182722
7916804905828715215622649998606310950999171987955150379163544412158335771216658935147066
3557670774177455965046723469454445248778458271464984039905889737207190336078242914235433
9200614369604317659974379932847178302267780133277821692301466942679873820594282661912933
5915550241534659859716266438807886448062050279682904455324822018847278226014288024076611
4485697222371000494062350674316122741711142021592211570174269652208795884730735033377788
6843630900466032245551832580310406736353432215313256371953335646315412864393601892189733
5346291253323333891122035793313455212024376390413029200165188889232334403531933367434155
2687761776111174702266205540216938347703591864563743276148501609736069422327089308607222
5875039556213370209528527353177977315724476882687105018858384863039847390144310214636166
4483879364810692156136003580721472419463903076299621691210963202978789883584021172152811
5430062371612253285159802246433991570443722024644840539304518739429753938795881825815777
8171893391203385805833591447523824338459959936562998642178735541514331313377733241353433
9458095459787784886429886558690531155543771295866335954955920668668577694628692434141977
4371848682290803852501750620796430006123119146765019569729553334365643626938747146871711
4052440922880724819709970010298537210906476202665190151089449192159380568240426536818611
9821561269994572756486799464215799287829129083245770290909394312383347883415365341888400
8352149658397378248332578501210947675253707258092841993390656771104604298884468811126933
5015712428077052996412608756182246277559805968030382519629276633870099886530770620542966
5884164696345591633462835642275673914571781582360311976130648510010157504833030154770300
2491248535410412867236140306031098349430425067285756089151409672762435903364564307491544
2474254059150845773248437634216808202852072595557958941675923129498538571039827315751466
9592829773366492648873982104721127762546030868318566297528787714424385018991417928230911
6621968644740854954015857375029628552674287372648052720244116343425544291700571631803922
4242276210931840189579542060483531936162253735381429531430785253308147812045578155571522
3993259521478795622178433397101762017345750034375046620133042414462151114869950426685555
2171910185262342140921878596694363227555129650993284624716694708031896836340029746673700
1764473427552726213171859683344751368681813276114950659644366300319556094875629789553222
5745714956690143728768706478163327939054276144749325362371974192670057766902686266425622
6088989982012057934926592812602550854364029026921121509072591968524973435083718315581033
4376731936819376536706717274969210859183308824058934625476797100626860295158177455830111

5793702824731391147386204076255741738014903506856061259796793063294355409429380041815640
0070761300759267933112611042843377477066695066937178313155677553536161451776980714360
6628226967203588742614294540517539002302946956999832630045124204704504552732838619833178
1520003824904847794211792678398582650955002373020763311245160099150755007601964345824
83
7757030436165153733364345907750639479875369456718733037166639725355008343159151892198
47
9029219700501294339949570671150484079399450120256568179765596114481120929933105911225
03
4785992142835622524850388002898353419049400711771931121130717962409362878013075878453
66
8699614135486271003600750793954306732703736174672646717212329456789507275669597745628
08
5120230477070505117208513422935342795649753485784433574206002892952641607319014347585
11
4120356910175199319424272210273389129535795331724390419592833741522546851074427748314
87
8926369160857807930305359014060162192682530265309807459500678630325596620504874093175
34
8477543016697409925823301806975932328567426213596776085475257805605396880003244309000
42
5155019060379449983293845229926936289598267835862428095229881133017270023200724208815
11
3725732110353685582212702478465355993074747067827654259179918439115411502208293307749
25
8950294842996022677709323981603331918211680594978658091077234268648570726238517624869
16
6305687440613510670928647331627464422512380368768470351891205132023999158743610439622
71
2383225781529440390203536020933739191149988188052506873337210927007623310735888527641
44
8900288612476854229864742077875874539558424177101421271522039801426990158806628570988
05
7170669390661906903839737440153107932039771274766966005235054333738192298976714214491
91
7439716489263052927490811588855806958731606178494591691335020180992855067928872598465
88
9679953606692207619854107456862896676884377735027571541775693759308175237274795315394
11
2679619035049651624575897223932625391686215610458922611228999449544553186315026497329
88
8967944207682156105940728402460139245550763109634853444541619512655555400156267289301
34
1725474261091261081520946866256290361685737304611262080764959322815470544830377503511
34
7698848019025948303315818620060887078018481708426164768340482473633512851170910962210
57
3190950226868606609098318347407350429928393742861568633323604382087094260763964893300
28
3409105366920213102221196968365867274560153129776517151183649532304670225441932352928
67
3901643587304707448124431252703415797224272067089797211746325634858996907176554986493
46
1326903812010512798654252214573097813228208365707607380050367561315036555164267868437
24
1690655562517279834906175869631762303229669645966076985711179553344107082615503943408
27
0743961988793513627237439087907454814083186216651674174971766864210875274837423023477
12
7202867189691470943067636953255285550797981563279957363872370541335219111881881915997
81
4561230853704836058489914723122040539677271963041708808021723976125906337394764734226
20
7643086750556046072053889046910366775709707908970521372322113646476134023183061260694
03
9120279897948400098116406928795025114744201715500978925343001603389974140059260682072
62
0973881060936296099225855166066184587101173050040982911351844536392604813813715340011
57
2727658830498040391504087812182836779274443908740579091424309049869725489556999997685
50
7352784286592955514937906074037563049095191410462279513883576477403636200554970290375
82
1104406427100628690218584250834731140095073133517443229570708229140961225824352884680
79
4167664374900949458402528654264995475064667640705423991351515763519164491832958715725
74
2961113935304572970690064957736803201865756169123508829609324078992648763326933398452
31
3827809515419818884812817812769800119885474225627951799264719498940838620269735522855
90
0812216819507947001795903182303581131249192467513045138450693245275941257007577552171
45
1485933803958451293973232424551802100530793930763083998392820178674982785068950203266
19
5676769500779384775965703924828360713261571106503833306005323026595298124413242506694
66
3413487940194706217040719279473403434273841495003519732752537421076728289327851838324
05
2520194678361955375725340872783918003665228848815705159181907633594202112305890828972
07
3649519071962567085941148216750097042339022053206793440915575334308141290468561623591
72
2913257660037333932938359319174500047887388160711155511768400904292133740181138339902
48
3727227642211555644048315196414145164945394267001792130786890395462459624347210633413
83
4898355230683742111302110613713319104428516809395680363250711801912877643304921639820
88
9856125213676243198409043794080140231662231519663325253678500339080592121728688442712
45
2260036211086903065933084044669202741471547990398338408669480638943853330399196845282
55
2015387985185362001805785817912788288849159605926426985895403398924055242096594304622
20
7690681586048525395674325452567215745891739703895861513595631415528593359011295369399
38
9790211247060577990053030786153643063858287045195581991491302540836331315922277874452
87
0479872706854600402690758258359419858425114751102469755114751007026709772806380223167
49
8867056327398845374941200988482645607721637433072298252394181880971884703831130820692
25
3425116060945431924127151876036263826822198412445823118397506500062983127036488112388
80
9896167873025398330684133314990086539924601304280591880801090027701242360017617236538
54
1063336817683048710763762827850153935247139447715908542757419881155228134030846946711
10
4382711454886911980629107370639527903874124421992618869091781113272707226803963690552
78
1945592606425103189497960982406969118071394794941134340452728781721862161136431228178
17
1146410379803360720672872600758349868920915798822892822003980061851300391813097438523
12
8871705155509070155355758024252982672121232946464635444723031758464252660977434331482
13
4868381914113894841408442214050709517290652128725280114870243171831378136855890491225
53
1180544881187948874007008108290933298879686131982009074716356982926060903225091652610
50
8630008723158807754338242205943930741000872074024442925164544670649816081495817041806
471
8601396504088229227114133210486811274287680719140167212758527410090431072291990255822
13
7306705212869155010148432476979496405746184737937712517731212218787401773979631020754
89
5691209977470121282453609842930997794022826552086370592157153448765452782624148778796
05
1114177064220781428558458464655696331401231798503468112303175590392432175482302565023
04
5367559275320455591565463158747938385487460642216409946847170264563279556260628602122
18
9029090273325340317485876850803098841808548093180592355545668304233540732125473786639
48
1686758072475941575687679923528559710962477537357849783538134085224397004260758556857
72
4843365191950314601437403654106320675575516470758666509394697333026083845242426205822
47
5100345561748435724874299880636778110417982526631726505240766047199076217610421454592
78
3013137194502490810591813772080860161705126985472490919624104497092565255178667366010
16

pi to two million places

845945664452059094551179241832779872846506395574870652733027193203505389063047543617864
388839929228895179939828726590454311416259810242014694535109537683447326179822728009500
931967455351323373348932762797414824498286334119070741396339042300567130833374969755707
114366686571962587401232606674557948093635363811792271942909369156300218972765091813164
970143772648616781228324340281150053301111736313855858785372685847040310954106565567162
722983161297620739006117448879482884850881925054030897003425289698979449635932553749925
801254540679479443276702672902121099548437697928298894341933563828775765484398068724045
212662961014583481924230663397834737516774912497027128441222369721542076535032999372365
168820876857049583002004007802778047019845102154543407272316663289014803610900743046044
196904984594628606304205404302702958743005633858786114400956088549020840458994447946566
008868247453788927580349324432039968752789158241629799950877198859687493845164247462110
101850222605585129730069675704818658480621342984344612686258743745206630741603373617697
338711052498336964375233831119199687223586400453245701760579376370958606242078744701128
564722962134482036724256829521468447493369284277955831282114014096597674040832901159438
314885600381728770042042647731890685545118846421868858750727274929104631789558264769835
592157224620408173122652906394382116852398416539484347780618834348949041736822743617543
086311338496165055973475496727326928619136308086876548060998434481604408154456017587584
413144775012051029417173912399038688384684836804647271806504048282428680122370761992811
488384318396099856384579516937999145424465986926614582772641371711901375713705892019929
287843450162319928142565932571697909063669759796655877965616839836409216964477733355655
413787811937736646228341968390014553372078483211164730725968135546954559885771877116755
030366001543544295483308259552771078000183657980963551924472126223191092348515234300014
997996630084499049587581779153306500197542990218830253151365016933641225099050478985724
176728382656410166465436371745690080291165693365534335353633131973760292715710934038706
121245986223933472195350213219228284375009213932717023077028718097460501372211342861045
585248929230757198025476058788335132601092696781694885621452113034190719561237118114519
633340298718566606841301175467544197897573193642847029648514895951212123452315992751892
6822806122989438149838225514964023054335991619388992446515917946364746666431252876786884
567170755414680990022981023755787793928992819985146657754927510063163308515165806111459
114867818370866102312651638689323522458605852349154271670528518449450282723132941531210
755024054638638993183453146901101996816041634558651769359993096035866123936418651933371
763977400754389038908199381820186981364220768777215062319486496999431820218765735259367
635537424122963136429383517351601248194266346415732434699072412002043291221691385170084
812745846958293047076256867344741950630549379168872540759734198178419403255554592988041
336369701385940688723159215713335465021960109920039586501072419534111442691701924902011
188038056358079210411416959527132249847476034577772620037273724335946986570794980057136
122821078004376589939385040404024574416426182205451274782965032547067452190452859953672
135187886026976628190901030955826814940486561255983869937514106613746281611682830234710
019965842555213318092568641374707341765619996619825206646998741719885774662270157275066
644402533615266149780683961931427687882271150002365270949948374738126066377184496451281
240971792490727594297652349918599000569748238974923673930751965047631259719181675408900
949013885459825721268195729599155366298055637407597488657025467719284111115396835466043
186998016116962493020242312261423687768267981280059864992597558607707775778721261194088
754549083355843928735863585584051377750097366295520129463445917038426415138810861593925
498475833631481438453403038969144309048241062267622185499715879306209694684154957250034
982836974288221051457524569069370033620894901588628815141224968698074723369975215024412
523803762360095642077689593959898600459698882778880623362792728253228791167688568401582
931751052293598848828454954647503996051036919576202147791890557332819138936641665401902
916569529830266817126108492237804803329876065946542599158897100456319570804957828627577
678628718218871930716575143774961676879867584658308807462574806215096419704050456085195
001753003576234839992304448887059027285908621191921976853161891209026112570660861540583
849391910726489838663114553993553054204085566427659878799456783858427182764830936135454
428443431814923612006793870511065356139675655712925906505984595311742498224933753652011
961969500145256152490461281058823049767639819518533852676301840492853731435853702573012
428414129681670192160182449001368533105569778147845703757231001527352389262891350171851
892689413006418377119910974979273610887529890880071868878749967866890733761003872850700
751774675051042272579895881838021823024959677054839460489968102757947399116200090609771
836973571640694633612326700209689661244855159298532959985826808698024009042071915519420
889741565350773879986022243934286200077246586910083322117977358908242019193895761422906
296151771107746622562136965723780263402411245643911610146128947202730735460950973093079
816669868895807891160239468502952905043920070580669564899513050323906154391199493739162
684124714837080430528146880534563634358830085465726219640728675466347396636129365912950
955568409518625332243851011025297318357726697521856586749171527247326374104502791549405
038092357549023443622477211505817268883502812203295215613553778736803725568801657142367
564502176877956274078087974190704019650941602815129337701310710077937179893285160363486
004028679892866531063902351479858762634067646411731591486929503029693730701939546036882
169214939519590444297718885683358567331743055811730672868076628434631990297168505855546
238936005877615834333546215166727780188455795269438915279878596764528146861386400904825
812769193637157901341338941271576010865992063084789640425698818791997279501834307081124
525274455875112485829573333954765081846885504040304439114024687472324094902784543602621
112119191102792481335728865287975414249788288327715444051743411313190667370930033486172
485839375389241810991247234063480362748563854528664464314532935883568249748333900642146
851101504350965947692110045276533703779864818927523273284495825491012428704544871692555
294467504554096176787361480787833355873432546356261172461708495169209528441698523798464
731479818579722309016467394945757307787896609689525643016682550516664545359719036797327
863541770069101685407747763801455311172809346640255872586470245057499320189390563014262
5216405286174380723991216522768198019663457770728382762225796006969716188404107911707424
149814254179744171039262455830886876621082298827288076455704524246291773102838519081538

445746043919582417547275591834352078566185914284802697622870947127318435170098200850619
604093584275612710244689044132242022010411038891605099421128254329541446645523196076935
065598515848516542119538622251715324306932619148476977050577927223968177323477708684233
836549348039303733185280884078145239655041024116175100186103827117951780343143218224259
334113668307387293866003523682100540511430243709461306644877907356899579034975525133549
858010035887269659298754463456065445224217227819916757152679628669223557442065368344751
203122548883063104281214743025665061524959240970647686237364242656522163883822344851873
651242849012608877468855707785392873707919955324864698338975431337112778297610932590598
050028051127153986301449563489155594710527533475698138562077454681407922809821592710013
700576393023685119954597622663641153408669327318863925177081842657712168820300646618031
354879130062820473115888591700974977509287836340225674774206433035431501123302839310001
364674871814383988561949321519764519315928568595492067662999743522978490202033428072156
029594540434089598763426857954431988226280744666307345859215589786921785359681682628282
154012873242389484496275223003281191942112598313168698731457432581967906771012482249441
985295414807259290277841042673190250176450955502925721924409140756622170056512993409828
490545491800175531180467195634764927665396597456220282304641304991576547873800872472298
883854918211168710311976440264376185466616630356837886906621895882269067443732389792955
466715552613343011399355747716748196077862077134482176290499981501595717537803508506789
804729069189851537212328510118279734197994943397299856419340505150138753742295780826322
322974410921453218483913662401724059023577471600973036753603664684873051145414925103403
990446321266031588601173532046424819212090597429991882046615775246050367052381573392114
245292783575557240409903083415913489143353288341230022793954672923136585181298550262655
799545665662621918430888864596519736002033376293984410736519329252652080175574935862949
061553306134074962706654195016463304987916916187237732115941602183940952326063212239305
272525943493310416675715603196759338774055217517444213109775658159014065578280158493514
690926179834445836791961620634521923222215457215891955722093523495150525582819882661712
470573890982385534968144642573379534323010027380082606408894199045167479012833533999557
143941358379417047704133791339174880272806525812833235986112341055383656355064907585537
979869571883040263245949007896218521242861123959998727779045501964050837987259433460174
649391596315999317561136845485835115621541457095670126354328866460329007994726132053109
240488134373712336241500913266103384809169877778624430619041950035836727162904728038785
603548847381719747728653743626327352875601367671273767090097867455223270214768799870432
061037500755947479730111339018082774921726394173657212785575380058107721740810969809236
933358905505268729509527750110393814581035253191478365364293976950158100883522260072603
766551970847334971628401539331618159549713402729795071949581985577933421672659015378655
242029700770622619242690025582947222206024231470552681160653604454931946232311283145857
402378864204215693162944993068152632336392496611173449712223183149666232310085736220223
844227973459822017407927193372474850163741467228567734596325942938093660657375097997312
080816381542995809702229327977184622452136149230318444270164174420086403597748718050953
360318448624085399718222835953941390169820350292263615557605072193509203506514704444362
137776860684240130693777683346357826584801090810991029103877787635261944382784078883991
027210058671436785271122749007284462837430657853066472593486408734907732017129192604063
761757438560395367348873530642316391085464251479903049228853222186708802415782552360909
811920743357874538379346431991121319472235879802524943452850674501323234578533075376448
412329779770060728553744386042545294814174719070699463586379345268648120797385006793234
226653753661448273671068511827548717501291281824112622445736228118618701117502942592496
368335611218007664333087366896090878789995756921875676598603578091723940199102914977136
213726271835660703772466147034469964363547551544801671405079627946048828467912595001653
169948001843572642044011890837226325391791843690034703224007901621667216020217615299029
902192087746140429639710982678527880886037990290770807152482335706559565918127243206802
634942079662207361649786611408409820926631301500372392638249218319436762323586321058610
430754506710405783517462110348683942449601111862941890154802640160068570209159806245149
159190410899727729716078732435069537914162033467446028947431796492024430868427425255529
223432958068087951486723042540124782417346846137938099073067850583299083898778741067775
530986656867018156618201891493622445628383679721907476061759878502824296404038552210867
880252912209268578463841054889005966208643029619340656511612634026070853241887369430106
857964253081043619619141041899850703478016095324821018425359679328695054683856632196247
233466493080316046360098569917082246349259069801290728462727962501930962494048043118556
802598862927399357750283099622184612153295510836426991304222667048144139298049268773460
146839156145547227820009273889134687275727146942749040655262566617991298090963507999212
134990223212253880655422959386963398189575464006721971986435852343861305251744288159082
161422316003111643235104058466010658705679284067494045315540824838120845457741738759608
832060717780726103506053587647583776348205689872476561534465892129080778699239974896875
767905362803144293486793137252996965001080673751838134180873374745465828225601593500083
514218756293771444859408140980713428313808947662519406627834314154609290426223431512684
052936176475353979175632645909526240772492712942950573803658796143607147241672945983796
728801117614877552210112184478094762397992662015807140467208162850823665017876870205536
534840345588544245056744782555628951187093582645109004755647153531732030465507146568264
744713911814330304699892001929419901848566917122558044639267541581686974948394244728736
768100444114889292221177536649523085170222535478751683498615161461175485467800485208842
319861324812790936263381603809381213094136167349869416284895280374419983325089328032966
307826837847944011336173664111985320725893567163628748458302406655374886630005814693906
061433469655628664844800152265261664059584894483922816629867438119996986020846070731070
562542738286436810593004494516932855622883158229354293450654959515562627270277324997903
171845645622771989606932922553490245573730527154449983267067450118796610751414392572778
990497854226892455943241812780979115893335333267318234612851845863042392385531058041909
330168725151050292546579019362113238827293437194907793934373159860286492252061666369452
979382269678826233009327918638919635379727471753440644479181341820520051981383889763113

8888349293288645090156499607726753137813583936824606897418699355378680743300352837112 47
6540814224466339529512462243852315724026900748802270539732900627573956553662651539941 70
6148609705972400369647435813492531564391649576548190806118142556749060960039152487404 30
8301805689878450467185846104190056272120344194439844478446019217072217880458821266455 32
0000706273825873588848691471125394102864141527055900714606934690086450966535722677156 12
6969971737745163127543938764531055269931825668861237011624834126223646905643763764649 61
1308585941347301627184196049200239155832873368258767744093359597116969388688449900483 67
7327561010859739647481064767071311220477531398118700074795894569897106661066304628317 27
6531697690661332008363480602762005060524829302699418947488458957655040469817236104959 43
6383886602965387888163256499873356150161019825882845456790972141338288970689888621706 80
1136575920434669098557117219394059825059447689950178399926378990532664856458016388251 19
1774792469933802349094548869601441909205645041902685182793927516423323769404774793233 95
9433873153813698616609721653160778558773479909850797744763998432600789991990385709743 16
1922757540636020543821165260717190969240615039792631519627450427161494891219990200111 19
7278296078443636023044560653274420263355612790785645014925453344107400376438269613286 70
3695146683640639960603135000583391898181014373431465585878069820179928441846700243176 21
4432402784733327467750552104661324818804413386520353167371484401853020228387119355564 21
9620974539108533928287703287827406891339027128686299921896339074390967219539823245796 24
5943777344896754084508992493048683359506628601537829700445343301745818271817144944993 10
8420340464336513364097930778095696078463471262368404844203340748700797318240855042351 51
7201255085995896876330451778387229267302536494072167626031415994877776787732496989229 41
7419019305037384049268019991177138249464756364417868166408288810420652030825527261455 85
1713231349872460896849273108101831945274277311072359242910223862886900367092082239440 45
8814304388484373118991691975170151175622264558438225138710301564252881546116451332488 87
4296598108615156901970104495039760369460908061367087346641113372005004995213651470062 05
7499434720715508750614038928712755369436405939421382234006092003047988333997814471220 35
9459448143732840559471648387521739892509654450229736080717408193417879613511117821579 82
4297145121641437975530494060698200535830364078484879509125039686920849221305598168938 28
6521700261071133778698460878352456355835784061581979387158942418282743186585655891722 59
9439412235309989679878259171461422929767654926165818557384230805978396431912053732627 96
3132611858815083042846748370731554010547368113668604456569775179195591943715026889722 46
1875283109885010535469314741479177083787814969382973715533383425194408009370717727933 61
3369249311387737778782825558129259674560504935382885885347665378202163882867585560478 64
1943175796520778438415139245717113068814684572041637454197031419011240068750913561465 16
4491006465464776308006061478096313220402040793094680297458021651889906138043884854965 52
2648202588245537897237691175385511184512079830503991735636787414664620095975058944174 67
3648331845946653220330086952550078397572667024547409294689995253498997106327948492995 97
0688168096012640338885908601888153501444577416358848233718232927723665702358395723078 19
1282659687546480266250716169392808811553972469889217387161154792630683834529596315243 77
5470547036047869332387735115426292306618733480514306208576085652529071529300678588501 65
0726291145449377489687133593277692960282531251710614145019365146512581460535543244715 92
7807831903673930659448202860112546792577810207415545107404341942875349357133454301619 84
3368858620248517917808956522337148348152285843407842920462785113548400281552060892069 92
1208910450276041441943402460787964956638904742557046678013729620736177446068774881337 69
2777635770943178225216508225609015662254160919278497199885180548636900818061706108260 45
2140370233507188522759327567935906007958558123820614998970932176885553556211114806599 16
2428877248465714918058421204677883078687708993955208379956153609329706728316471576594 37
5735242889250252791776966416921152247669808916847387795477064790783100172400326457722 60
7144874725415932850667527005249742497721257594840066181707182178678758665379846717233 93
2323634579422517317872693239126737469745757376405273639805843606839114587774732997489 99
2325401348656409690079043740106341951060866901353813305049176352472014677475888228351 64
3123485278986862896328896405323846659190699659645778638508257960550663006283866308348 02
6904498821546272844121089001659394631256098868113161664244047072112594561158387599815 43
3548399685558951666208189454348093970667204950629536148447693371809429599909120628931 08
9623257560564407455070617934782845215024876823927028696033555916531504300860754967284 33
1131777731426368225176325627879404286008760467750978291010525684621873851419726923792 90
0743971335499017703285671945285180379033529197777414186966601269267386824255050297640 96
5672574425470865258868629139455661437246379859258722092118614938644165718301024586701 79
5185321801389937120037862422413935140198209821188228879536531289420139838180749079054 81
7674916578522783030839969709665736008621218477069460985271999072332011014721596259398 87
4451748193788593686802210311505564634639648829929115275205112003956026291170927782609 06
4706864779219414575830642612260943987080190913497400880097271131478409200285092823579 05
6039902771210718442261779800026204708526993187340678406051113187237468370980543434995 80
1743652371293874857279989047844131453528690900335953532326539343646189252978767400672 15
0923226239805480889768814975547239333705957003410057743184019079575432643981401930692 50
2685201834001039292540333791048012696456434958997132753546654302422123843107170778975 62
9489113300884957857840176917489677430490757726418398562084312410677188673096109322886 71
1621147794660721047464917085292437197833573597369985778340382114102833311259006482020 33
7561533020806675454396461710608332443442421300780168338113344873329838791588117100930 25
9338288197097236237805141607724147411697042491219569478873058196793399415623587588866 01
9744825594562512603988866200125734341215806131944962325550849965597569196261705030902 17
3327350094820219017628949236816936568835594205681312775393631774479002000398586270614 90
6260718267451651879271670558437740007911930538587498298537717643896030777211433347476 41
1929621129397678289080230443271245899374215897135730651322650912754818092628405668055 26
9484945660813778849831177341704676378681282475237937482314290171087085032361103275190 51
7016479028602026220203813399171270394808804515509693931452414785687225637607636797656 65
0356574753911725319444657594338497653853554857742277612337661348533270999506201776497 60
9164790710299200444444044788388580810499658011497941420335106906463982277642117919876 37

pi to two million places

2956624966402635941595320497360171495718576968415820905206482304892318436454433212717751533499238135246317498227286074715484226309236540503925171578443622662623467967246539635148073950275690496365308561694048855455084900042575211830697684105661871566079201865839848052110179489304161528971681775867499805570405111998417369401780967744918427964971557129237593639105778309231577821814727029829794851946583163123839701039857865600457129967576856381404539441447039018042096807454739239182804066634075451249894497343177602401694500220896223132742719227858752077177626370116817683875321991012306861957996425130382208801290905897905103465236025274668813758664883539503600084342609382482067316140749190905802540577113742470147261082429598461711317476967156948091591678961008327110050457023462097213259712295213630545100958403141880150451138936471831079189123679420365190868561410756182253703887880962834560351676580632840317762080831348810308796165533229915120367196675948319516425522073675911779997618491433336201232251184820518016214162554625959212049280380924642847363061942919621590922979555791221389169270553804556640926853892699568644682854681906632459830973466597620807835083131173895599997290455863914238795219617362871170410977842712503667156583711997681438755077059927986388255657234909913437963860916281897689042193865149156436715237614467955720212894541716665767937037696186885956344503357141517298285966662735416120223502944430216757107555979321461274945055867238975568687999065986799348580659053647915342174130132966161117184794654042552657151135836574681857198427044364586356832376724639673244782271147519089217369144785415791200307293437243565123431350607119443393887489991170757102291725100167217213737031101930425505189636313398265494150732014644892655618576412320531424852427795340908050032781316421299499453588934178509899721017346534058836992895588657144143654836564275321610752981109855469388057353257093927229475174525650389496927917422089676884324180838226255531634446493661637329520380888891294406522425906290789928061523077174926903611325788296723919302602003642292805285207201390080277647219325767129610149483996503621950261487992635479628876065330067999352041575446040723345628212943382666040183684331274646774545885389524289778973169541617678804225730256748358907859703820456003736142665913431216054941264054640827516815080868735352543328005058489612593609778839536262721038218169112426996984563649433295336338348616573521997940181787820550247850249382105243298688543927554366956987412125494704605354915997857561241122775687243370618889057030214701587138345505039114404114379228971312098997797638795986525813277767074501859873887678424174700567970250439253579088496534034858332077702618132851562804018456063534227076879265178650498310158965251482133902843710869050843937655311200427932204918510470747217979010188655824392951211163339045418150649615165633131434934820413035197004106535628303340583739221919787176446947285198634334005154739552192975507771272380393233828024458823090317262487150051545742664436981615408414595549073075285604590037029454586086253851293598021957152369515303710092432060719343782344298668337178137302630595649998238644704045971516052270441757361143251443663267704331776392642389619019558207390681043841403089455275101204915628467467091132655131704840638235333675961676590995384580866567443336045441074292859872432696931505738359487241944285466943165809427694988591713234265973744324425161276207268053657299130970842466751201121077055123539219483158798244822254227385993706832712418496474411658384760020031283158840801340380832973760058871436947963952794624032346437087213253110581413711279280396297310650417676795395001193885483475618475814294433837220071832808366537483482965487063850470331217074419923974526805661593649168499944865138529738606972660537455279474329841708780828417735968586933380840428178454769197446359441939512815886273196443098490891906009028265815564603895792322389746991416973931186314183810347849750418842887383018212129492789544190824765916504462025068826643223314033488577285566034939170634311521844585634058634014774074762021138425310852817844510588011967503511599329513924554428319823079491317449132668435960490458622358954987282655418954079799702904217260237148642139079166628181055946272402941799554667256634468063744927328579226928377633469892257786452364383660283349800039138972224296666629606871804540407033571927197106938463415819523302270925752603523218905749718074034586335788680462994341952001771181280368106012237728731605187845657107090742610787487901417670940658182337680614045609836668800226257641107991712682717032457241362306204365733902632980711166126206676163597717888114917826027288982321231004349985135435085253938840459134292633185235311222491065155231153580880701792350798740677677777607495494180331435495073461563037109183150159548656547848984715522311942333367469704480266675043623458979112264636312467621911590169794282817552376315015516476227242183709182662777279651447813024029529870473597330699109928008748077079913830255314311437997717616716164687770942708140146542378048755570022774608634599169780961707906477907716040263708338325013833795429031673379385607668476251992389223141984923386394353096369692814879841674623454315610414397333347176373032920063272132940072651397329466801014953179871634182345361437022132475042279442386829083030871105194044514119008304463728392351434411442381567174608715352660212558640807279805611933890665348428254027677989412646081036242885659333380525174034519800859297153039553799781036778995491298825240716937355762987055030916135966221170328221494085983518157573958420644178334528646802975564113838489331293660128555926095621695213634176609938160022719526537163958475344128173443090518476322532673877892610796445423952396360179211103115354644586033484857438259274718746574815901001849690442154592592628647927781883067732140210426528621719135928724432168712518707013984428686357337383395950853016742169100693187857721659499799650001080175391416970841959243022704375945806888504993322313046072735297414342216686679212228191985085164012507841142689889225704917294011595706908619005473505001730656533284048385114278071943675150318626623066935275794402084474656870757615051109230151549417151326961809746489557071879739323355744573924363337211608839537156786811386956522568737479480517262056222408780218631305907292390491797576370549426369785973376080037593900289693258213920406419896770592273451696366772429687033143154874366714153264083376998167386678704484414984176157266531029373604332473619387096132785634281668386982285618843964764504410522457950788761332486772431131511343409248192935391838244205382936688074062799782610547599713556793305941297156693981492525086206524003943715016854657885027236841603499158635222643677186516886071535025951084182

2985160545927219151371868414771455360151763566805684029843597853104555184761852904063678
5629680716037777424856965569994783232411507142708848372995265563156843336744251338679852
97736433420287326092941785838280994051591320793497084484094992427134171638220897100165
6088517481731595423822916481006424210572380559322712397426120135081564770143187729123306
89251660580158868271901975541539346631136148953000877306253128002529388080939088493277
09192785708898221812528596900743786766992403957671987826257628649595810785976211005377
8114967276371207129372367528162705868869275397964041650983959398304572811915803408423444
5360616771948864789903330722270315497713886362937895217932544023846580217472930432224997
97811556825647504256511574399105272580960216688621221326088470203689769403870369361992
51477135168784722088129654330914185276878893288601857369847552142389604630259021675840
1568800562087337922855242082127450646450447726482523693383697651616866390212062687614019
5494903755944290705864024242357306381243468658000953729793794809994564359934040708367888
62037238281976794327603092986241911323925277889224028720694323427404057561774335866062
83903928007141488711539942291214012766490328888447794131516521822816476992215487430268
50243496379051614397957656533598900865259478831800334076517540428164327750050190905980900
03657510382545576040846641328014490304538705141991309219082878796761051770973224588677
23964116645344650689410231411645614716332313953640376318870096864122812627248640079749
62056135417906055301468091792793092198192935428222387799408425511679223335752693663896137
23527639893858827096178892506711014623299278058153660121035463507658001793987421783693
86823253486164200649032922945653593183569805089827875754080092747425440438275406413654
97208698287079951635794651545623763038710324774203730811314787702654307393596626450422601
90491902527893847125038552641971617315674337969187947282302863983093854075676194232545
7466328037455060478888073677974346388660291881218606096932742338144937982128119252162876
268705918907045148466035534278910281165593812845120991245250752717747611424570301231104
22178662850261512291074872456906790854174425925553698141351011144936801929168926708598
96200284173429512598680322930477041422292675128171031126914792757368374376861804039504
6400794922400032653133040676688090305489881890079642039502799741818886745561944004976873
97118893409912548160700653831278880615045251926736580634463799989244409078629286034564
9172321843080925655584674371544926354401408614145029205834403372452089129452559969357751
77720815411526731647546939949525049821738784612363827546122229255470102371106660059563
6583524463037204171311939692965406468265920586272019133169626786040832691604079208415075
94603938551141337633518124749410770795510934496658103506117891219588786785014241730039
29355027421609852834503708187056394648026056602909507361523224432994071120424061899351488
05518135607374376781144006229753317241253533288479577304201015440217012918667874564905
95519702992837556102077481176030925498899943817438474164266618911689062616265695913205
25407744393079562614513182211622415796165791445272459047175088202476593051428346768811468
63475847964339274526858263062286153993330588176535762719396495750520322014276341206692
31618119227096366400675652500664604395501663148098212131159674612422709897145610886780
6040032255608864471805689954732581806798778434693082345955083479494832229243839722759192
6997491453129639220958632063128704091272626714974397163013999744705602690808459577410689
23535931504113716893121388439925023752795478336623747894748412845231888917962394241402
64264627650602483426116878770824086679855680905601089069848195120741125550201529489776
44324909306790843005973384032355759094280443116678527081353023623467000175562991045496840
72884528548320322287991200321396795413314842796273038873776364368806066802541568869650
946900632290323818467433952815642558548236294719622771638441867253272439633242749068640
0384244297951216231418245557474640192687916377314974360341264953037223460959472087557111
74564646166507355795031988875562695369495849285019206071959465525129548381821643330611
74900382720557789321265228782240776969228584824885291075320165578816677039197186160727
93398277629583084130628982320702473115834124354392135455897523060225812410250528208775147
6482405067365722054936121414382855178940191838088364926906703318727745513450373478747
01749267881886892873335750159135347795631122223074256386880281998271239916028281737850913
31295544729617381900517950119556888239207587485904401790348110983789003543167433526464
93060760472479173795424415796596785473157896419713753955930595202992966113331787357294
88250739782426790739656663289611674010659624887512865182932051997547426894815392216857879
07538139717262008323340986469852894199800705588680739933278897310226665836945696683342
2605087563181472140446824269812515750967362494584484344628839811344858678450551423318444
94360577129356999958682103478960075883114807340498113300944918080172945033844006099304
60567804869213669182131255697044231973573406086745833823270043389829686408975255495223
62705791904284973772403628978310510442409889326018610412179204187514939789993145461358954
16404940731507458386845961084152488299692167395571581450668231419750667488592811012928
53065487540810685627315858892116170643171945137876506590201556480722726856943327649855
92225167086414749449618548521610140395529976483109342199951853547030142966883713170776641
79163133631469731789702877183443775025191747804708611725105926877206927940633467372539
35054456862692047059810814080917152893109512992805739975608670032557452296348473235174
95019461465692392428361519086379680078950669037948290380425832293159822905803503626628197
584392317844052679445117972727451755777920924639393483796395058354700409501412192196233
268921883475652386931480470877824122122533511921625391524862347061778401863041613390552
205767528417286071516576815808759513334484725976756676428054788911519123182256311257907
8192817548949150859604332728464243723487517578360863514909554237381689777194417486083
2962174031364326072340879832446492543974157646107605081064704360442878712892960359248798
80718965384301408524375417772607797996688858776255476220329773391028828068738555778165
24976156772836864900963022914307537418334473104292418268448846098396741207526522449776741
55271380577692937441971109163455868354770772450697068212965433343656939612375625699736
04759187387058134001000381942655688179410201502828764711003955407073063803186891898595
0911009920447544917895574533763686286980724707530870041476800247384302176856302059915814
957907617893931945448883608487579263926961547374513219764993170271505961619828901155883
68715298941662925955084754511291052004264495687060269323475456543746333035921406864515
840494475658732479825125550521149322020215318605305107332356920514492018396595468399872 9

0386605826317761985208253534738455483049370860650677200745928418279158285853660824174761227804953369454675938349876945619727732727223650315037381286468997802767205555241776834457918323757703898936410378555144548534953288610478055131635249049543358789835863755743721822220810380455418950726477913597186447752838195042838346449198020275176761042580037053169202239594495616362479794587243080542152433556256674345156831319687267861065419751387107184375980639123137049127376579025812944754285723656284936601073427480232712504306803845577491911074701033096554017762128654954392360456769809455636525150643123629839338151215237554117053941443229281253319160223851209450755353214186482967443654411151540596356378510697739923266691871804642394309440278034050027108101204225319013850022187844028650201793079660492805782562002662311428500179054822220497475491648442197879999765626271532996283683295855621271085112330597696025897251293141369754876480979768305210819516461203635906024054253325927403845671618486861086021787645878242208268786917150037650210372132293428729069331957350809939708772692644830010890327822820630428617683351489938321449237269446719365384227817755935502879739271275860224516133892752679863279939606047475384739892952808447086153929134487373440819867838728853220757691381945878526809738431036330880558471056401507951433741828982979714072554236134173540052269891836642892189927063129909721741573432933608764772488177864022395315474075405691872673869890515480899092630262178984657115871195395741923892613891766650385205279211278953075938735248770427011907070265278740500725112106861200959596158699782535958453342782922079791529619166279270074247010078942272565945994958454905187882966321299934145115011027709763868033490157568576787075523323972080854672129657156220510965846432214904453804900474372651712484296088906524834189988226046129361238583890219534857306791137837246559711324072551280818047899109220164661233253348024800841330779122475036318260349769357766076503639668044017023343296848839782774048194773082647664734600248332095730863740315170274016532091118457655492927891934410755676897776621474014108112020534438293623542185736443269883948879531898194586506764414768804895656208576803875682212562931592383899274488424833740996697153962103071635860263235964540326763447906840960698352003609058728758054046369419714746201281218664057490329671310732266451809419284458149381827436320071570165669283935299600852175367767505167960118283704141071842796523898018550856558432770947775670238086696825773939736482511074508919711478904901610897393557090701088687684154992631573040497786984457512142983664322877466820855436055315565015003398283330950400279462540668551585559806635657642173873209795794045714186781993565694016152185955498484703058077580829765214375574930543503373934071902710782930273435285547499500603442080021210362421544782700231736060082390742626711613215060996574665424014984716562463978622891460853283462062555533604660738950416340705421459233653814012758608826237868805925528176406413685912794707166688919454105300243118472181538533806429419480484439726748395184990705056256525177028504284578522669383680705288577565497487648659861933480707252827535460101275597320443287991996634147586051964371478465834365895153957808505833228022648104977023726235647403177677285679224785881185938284587695652570622656630988885606674440597747890614414919888914214169062103497773502978930693234280591334220600205797053790380119696568294206356052721428510118787882009046778649971890589306918025154076119868952101308262156900154359378908361028775536334946581003660846541882632213524627443416849268828263518295365312646609033634420355307782563890040781647238679175892499617085011648643322297126535217247664672936950771263782410756072075812567254497299479332854797367409515269027513174658066326447367265935136498179387722170840152747061739087360893917173407079667407713179950516521233688437262857203675876621933321103673150893254862826200885525560366889159658053024905165916765078596707653732391835255491525963311873440361449156272888927297913751988672577143334889313203957005648831775187778891229489721796989865109010645378357371168615923566840760972892584595485286656770897283608632841611128238658487737106715327754140853066446627958897209487197977163063764924087921035024040700357586666821879068812302564858656342390820407402668574620471559003770598565792273084752183658195585524670640169298384680683769917787179487632100899578464521066045559101976938616845228000685964719028363678232701250910540617737256959569035324673285213224257298951759308557877395749351940367583160291852071298361708548741070034240398162010945087072442299952719172248917435724752097942140419988468605962000633789556058807710445213922931731932879901061450251947744914623606548628278263006273971001844426999508800793857337231623573836393862715510911761339171661863100693800958608029189722940619205207602876769622153246572447199745518387027716386401809818198479872796339702880609030951284992860892975827279426961143995553788875228621491366586471179112213763528023225952083007085064839787626586295147363953012542903610033091724718390026914847205806775163552565257183122228528866397937439719821032314912428062467735665560981898043597862381191113744079240491882099473277367808374708134030012055808352850485309063663942894861324643070658997598326727990916160571682213364443405327487814115816026705657104974381247149215853659041290006489577868108637672639858764931648099393525613829377481114026756791794826665350957515819574333765178821241052945826033568746419349855367424574085867099700264424393768231665574539364176173608853611041837216997088160639600083130968194137167640498174865960809388971090579775597053148165335851930845513529650811904832418931414183449355344638634046032245005132201212665743350209266621132097146021735483833301311244584908226997640410980423840439521776256151257540947884196269444239604474880388488151006851557753986949014747534774309767065251999921166225843544546684247952141526742317095770270336704646726039351180733874645910425597330132661553726298616736664221727725758308464265006881887844280177301607361031470970455202578288017161872198108584446480660215456726684854227263280443525496823005168654059255453503659055491018977607889299000859007062612959705165382920058061837767972923575409598037531231487274127533849076396959671854061157426720614624851371297889868874759019325054328234914725802764766706942344670339043985808479878346570813808082548827620819875925564779837685469470302326282443115908705013729641664455804515279829628652867428262493608723726360216464512273589889892444789250938142594170345759329531244458358173081417452878237374736836868339689799065053376586836523203160887478191714255420428490291565886951287399097747600509265720329588896322608031193400324014175659323

8806756834493734235603931635303423550909965864307177222998715254698929471011194605597 99
5509541270368804217572633582249621085712815474958425506336161591375333015235409213178 76
5433665403013060850426334090926907011102744403006018830369125260399365285512982627038 03
2577274147166519440084659867122064966414850337001929400954155377823338651983901365807 91
2434908452458795598336127608861302257488538574980068550750242978949336950815862126229 25
8134163643212473678792586329715210427768057195329210657200965787094561194278394118865 20
4775140918764000270662577402676251800853876038085229589063889296606434511518200491029 85
8139178932201913930891647807019258538940670574127422932163433995193161903392442890067 38
9045564816906542255962712272486304951956599190376718654891755122249924569370982365522 91
8266003945918162315985930518238078450012219949427982620664197302217589470255948318300 29
1190788558320345486139235106638476293872653318981033248875170842355801439465317022040 21
8662233486707423489880763602726090133770698322506871501679419308069328769060110871283 00
6735385783682963108848527449789480544917781730515386148336425569698505693547831919835 05
4468324383925618517827581524704820561163402620949229716670445015330797065566675718076 80
2802919610629777124286034412708981319147200862389998899276783086977186667487601213139 78
4708746553701239405543447812408322685311968581780133865172515988578464222603451540409 31
1584590899768956723661569019876209287665011122104358389005795130918216805045753018785 75
2017845820731113899866119936217309467484686735262729752182863976019166225799490647689 93
2318606340051240401880668550698893828471534886148639408199837393597136076649714507065 12
8983955238423788980728003425281220311449866170614255157046595234975670875470183492339 45
8729153882181290964437664952492704070162522629981258230261552153859158556376965536993 92
5026256832708799990897716292382588596071038353461879989697045287227044380462706633865 29
0846562775461478447906478045742913065745729731944345399543618355810391215549958734867 88
3967308103303687005351174429705448480025581348705947480093486242150388489404457625979 74
2258470268858541825023124937629084798261975996209006215026395519650522732977018887277 36
9777891096700002405018392934959284315043798194710149058111014421304342230737675568351 94
7792895041871249981413202498484208238257857888717347904385348733136179264431490523244 37
5389469597307771900752012210971784809370842608190687094276977199466977383080566980220 96
9982683849558631509095463975668788448367168276827609578215960608131249996552244534587 00
1032732118955643909058458390877097383882832004245058325285001997228264773187021828904 49
8087104861663866017492466316668234890780713808311662103932476597250368129061824257751 28
4298323180247630804923437667965185137048079326315308884825627830174213300385840904035 84
8760809163099788016427216165046843713567588692647245005888263278559180841992586630515 14
9057450407400340395352230955856981232215403045333769214268516114273538693324877920490 24
4685411822500795319690530543251594793887673311159854948703058831457945856656566334164 42
3537906872845008234376174900373597494004939092738390330583603299148057321913450187626 96
4952183531345908653082013920053125144957182587423071503707499150571034488047928119697 18
4390092964702740588925546976613765843860876273651049702938877873144260568486206518248 81
3308352993665922051762273317243782751139474169859627543672662953734271707020256737093 52
1019303447291257646330175952106577282682145850114347354400707450024615172294949841342 93
0904724640207729176111271516911405440806230087424707002052330485332849115486909081325 95
5696576068724403472459067542351553838103971468885996779887846560233401215245706630275 07
4342810593762834596937500858930103253502697981173285411047042665094327676600716037777 89
2972767502362886526923425014315711378901889652580238305381451665837042141263068109301 91
5530491796740776674631529609230564464811798486504516996150312000649315061156375768784 63
7267374414301822016039458303284651254316748244656914412517224965465078082203244771051 40
2698181374749609895745475059493560000302969709783833852607333500084516397400402993155 26
2498652096064014041850890983742171389647786159556640038291131987005521938715258969487 23
3760030526887055333052216955380486688301879757636762744180482898439230846488851910970 14
0091429999259961739309689825963090410275407657323844504747500314558521996622462552374 31
8312264080405731278494947653917159260287799884328368627738736358262625870513409620252 00
0863000083384293997320694917783137638107195172711324375489390391829261994858606802281 12
5309093583534915444172387414290765319965825887641288152565568257944637651404212725679 51
6601910277885997198341722805573843062947945938020166749145358766270671083640457325590 49
5781811210845134413927327392052303638641474591622857286480564562462015273533792638345 85
1241397168677874686710013699075589123142638422472875958663065322203767004882839100775 39
6001595018682935644261497505407297796579626247015020919418167453876159827631527928058 49
2314270742376889476681390944978055333208425492535784430942535280193818542286308120758 55
5507840039639161878936340596969830247022093863675759213743336164581157151783432908624 07
0347425085374663439000399430304378631330502492749385599157679058350415291652518414051 98
0153648731305198018778668330408981183733458971600796199419200672236672446227088456947 19
1266729883218097345290466568775321890979438661589601762819619116442991368957886658352 39
6596651277470991167150741195804511321397782066188766577408596851652895113122531991762 39
9895622887559543643883756931055680141855995021628477498370158749352136421750811050303 03
3262700266001354191957304861164283108023787482134910125300028230241939568139093112889 29
5901000967692546885390109331442889820680770917366239969585440049738583119181470722754 72
1682056332630471692816483492228772779162091722150672747374158336125613851183563730343 67
2368969816816788338705695694790020390336702965651493966613346166102177693484182562183 95
4862413209143243567329543789784444402899212172526505488757208619694659734120271830282 10
8217150144323595507442627504624823012073241238346894456769524941861898211062182620989 31
6449391745909999513313037945425850951480845634370739020302736599306111076097231726082 98
8456783679633599864121028790478348896593295453379419946023102511852246745982239174946 25
7310656751009988513093817352309164030139849172299596368244840771232548530521784190001 72
1331652959247479056093472090142167883822667681034021863192698365168780054950928877576 97
5568356972712304598379419625941762793943986763904895534635462625581346944289986007629 80
0536683459065232226287867394572931189312121839004210785046954384089376130404331939678 7 57
0819736232553475026664916954835683559598287736433710784330369395164567705505105816790 45
0043733910624001989917718953806961019070542757314041472935804403558958815854152799237 16

8925245853780632737764188122567407700455106647252547622629412271098257210109173733493411
6401760679890644718681753183215399546215136010548394594413162994262841204072651673250001
5381268790532554780375450366433096743572151777450637347869911099335456770723120260748401
0180153504775159699400331096877348227957047098036915687514748073866389004556935741184711
8511263087029100299883356900324152245539515111492642039051442230894882332156274223939681
6326719228001855339194931193763114862027099155653271102451872308580178214890917618312781
5067055679526651234933174750058592379707266499359969739270335550475652075599755818816451
0883030164414312901470583683065583214408655959391889449544314850383704130673600882795211
5230338602485798781428426426385990271278699245555771795216498663421050214214582044677981
8493148338416460193177083462920725689592236529241672162253457467966772664995480743958391
3766614565130409677432761707168113070972414060851389330530806929895925547953195790210661
3925519874545059032352020851499318965806307754464296930203821473111606175789361189858701
0138198237590049038675290604029680504605540969596247427802888013896519836184301219053471
7088668608473304069747649154265877829334802293008839542482580567055935841873317093167201
5935845389882078019749377414558162087636249381007324142446515870222464018827216923274971
7008172158558260987509983496504622157252825209181931838004031016280733752210041952311591
1013879048638309613310310660806227098183175555209346737367782776053565354079174746271191
3922061539866328737204488670178680167367670667089976490322504845219613085215790005886981
1226210156538307525986660520333557947981569940951727877002229117514325593376637383036381
6066847137205255410766970202869956221368108960466547814636334762050925769560583546868021
5848749579604146434343877976843817973061799285798650186752695020887428975310455911414901
9585342958744360517274512826387553362771141332840242441312250274972830079527131307267551
5189497709932476410173992061765906996188060060365437915447913185891119092686063946503271
8020651897732938072779634489584521001736075195922307394386933478303225153125795532472211
0802569957110866901873298964703104385370318681167060294131885524835815382913916563417941
1656472927062452774013395967976845292311129479049603578523190609921955881715627872752601
0488423861839738469238687206553146480435301794148721289729504526434722587834777687387551
9516770017244243814981549410768628303190109915533557755091592503844006060790530238457751
4536934838573819575331750260235321313769564814447134680692145549944363126074162588230131
9242277640951581468280847172906563148959261889229911041648788195789851290317738929695511
5557569908988468506357270966422592843678922021083726284730665943871193072936031590447841
6552956201893104133498064140120373025845266853253479009612771764399700935739107117213651
0051736387834853110349187551328683711505158276088992938077491787349692514701971086015631
2296899213162016772967785057270057517926450639756343789825938029900703764426494456693721
0461987933482094779831379744896820679111614212210926203404421071061376189706417689666211
0970741033530287159839774769228839187754251675406421684230294065505898148129252382840391
7436766255991622408016779301704915727755605486874360614380153358395103098773572649981261
4249080799019463622685681344417325732311801749894132291244800986005655311405049818352971
3195704994754818112629798643153957652745964937580774475610441431541888284933173286820621
3394105925181089580844263073084164605055216562429372322421462147864739314242764294539401
6796980401878067926724444618297525485163686646171018865281507083333516728503532659007039
3523001853061859641700108464643667159072543277227184942607695539064233862162506028383241
6893776410665118636180286316275075304469887336486432169660066459873593093261111126824381
1654614936674324303049258951238073482109979456800372472508772281090360714832735969281781
5436909753769595857830686301458607808079459224610890356991867607870080162324092796572331
8580867657217600369171326692971760286251688648520464344314302196944098149072664591421951
4958850358783655526158480220174285167540676478009408510395325876741063467648708629814731
1891383586624087753623784755926343438422725083879256909208981692050355439817192015942521
8759366656891215626086190419684645146588853725015315248719040151086173231843570946411951
0232524617928658825558031544035987247147289014905644028687915415666242875572913276181991
3236435901178278658999316819212512845999343383218055082648329851394761054393548743878431
1151750387474635844032466904357140622714956133356464169048436303544507695662342881060781
1925079917378679280655199050401430922612688674823134949985132228315826777477480800263761
5831166660382951018639935741810776153876387634067228500788443905041764784522009549779201
1705879491174175187657788858078145493377341594894061253055130364832802908480652723863041
9166386279092820203308678050461143435598422669539386318372272100955659046921407450421641
8033790923928056152683758002867892715868508618416627965395052533188540442027826967531671
9038573835641939923961980439087436577734896139290010648862696820273783451635192544295011
2949264954851991022744443414020569598036342023604621042011232181972026736073998212442181
7510930194561833261260188474778671827603721417803930772115436519750679177670395232933681
2611618720376198611085417917776843352207654692114158754331786506426007820305710654705471
9827395494217107283320892323011174915177998539136209454840672281998395072580880072164851
4672947402079200672413152900125835299896277085975602457261104770743416791467840151941451
2434428829046270438026099600463502597553623692071117603428889420017209692698466910977371
9667220824142944360005689665581315268707122302224488201575958060981626176984299882725421
8373186895913054709197884103637344797527394215632212498000415180863114293975885911931171
4159759037519150686678886097905531023271789560097277945231294446061851200527884923122301
3955502095210626645713184582498496981623417089753251857168813181502520158749816176220811
7981847494650760808359093389450679077410950684513128117192481918421960265489663087603241
5918872454975125642981582286252894172048604722893499258958414842829114277570248879757391
1478365590125453837541215121669261459303318714702987477830141469075368059613545794425891
7675459551238761470656774294817234346874862989197869903054546886506205571084677213711731
1563241583890764889958625554566860309094613421435187815125778976960777832436103063256091
3430071062881933010523168555236679194739705193471012602393776008193500668748983905538761
6295516056848003479090335424216373172257390162667752577863642151079834795634069637093001
9350174940996012097705328046394818310929163306329294756558539311206050498714492396025301
3460966674656676370945123983630899971997618569469144869792462261915848016995988985250711
3133563187780382168545150804193604689945465359827353040286058733328196648216999406265031

4788912599032871410742033343367828909848825614005588003388694707055480064090012882619393
6399119994539578874874860720772968587695579800469430971035106792662532037825222628409017
6984583262504911593424076679413544658224361019864840916868031629253631541839680817403166
6987242898678270516782031893230837094975719330572794322277977169525996847261062654818455
3987600598691059302671365884248423240434573140739516773972185326255882566339301430909644
0777640447653923174738670102608820642514826419210132744227190416855787021182699431096099
5216615468531060958955429765757225355385427507097678721382098617235079720999092714031000
0498075439008528233678795406338716712488436719273884571467477173672028597020463245063233
7233651844640013092133626952446236244989193473112482834846099695925944325460382353694699
5286480854995335950700556725014785482101450092347712889016871450821458582767151558421266
1235071335031856990883629073071494625130140679406844739184056270827244738047635562111544
1531274844948392103692055465501008667883222647616108991784860140808464463185498681157144
4287424325147264636530792011439414937255819075523273923758269926689227525903181143690333
0258958028587072890269438867215496697058948731613594868522997160182845931930256556660311
8916742348108580141592483997869316738502241970789364684119999148630409253332805757043111
3449258729343286044212547875551111461986096249878611879771736355172435181952960029052511
938781295151564800604627647060125261906220840958643469204307837930352491523471388319682
313630859254580459914456526459914458330885520806827917443559908537454504754680938983923
657663300631572454848433106045903017255360509361895725292915577685984723192428926867732
532716284167162052551215469973689648669302877915688569903703153942798001249290830172096
707035828268631535375245053983497147755428940045388074866758523154252707573755464857646
268001594028465017955786671228233781548808063580997299468151248186928918914935611178402
795629803459576395430477752698203674944510418694546478067769575842945169265143470432420
135479026303012709620456120375724168132214753121283871722695068361380377071697202169014
272550751177950972851902897929351768067517517072774206615874503254807122875763082600284
940084601060640185724342573867837105060358795150970083174258724773745359624369857717058
395759941938621973480930188115802236428357651247840813569700083734201254556759037815966
5823679283838120150473462764774540512095199118606494336525819761507798840786250494977783
9843608364869488920374440113598582208376889217692949592857357820355711950404789187977466
4938180131277454241874425462577415563865547184893109536350000290756427663737110273955822
1930417257450501050400238853860622118940267032536723079797736855434051211085213945045933
6707192514538613844172914006394188096431155929600927559741198414581102631854578400828744
4455647598352313800900985010306859700472185371201643181818527982818705389831053593144266
5678734499046645625854571554906239226419431540368527648232620779299898780620190494340766
4957501029240444842884066573362475176904174687572268057124011729002797085991342086914822
2586567726516807440443313373480431759225664381524645600387956455555443558118081042231444
3609465617776273394680245996202835625443590447241715804872188937839698211626014531804322
8460658418816833724005368885952793487511212334386870273021705241716650044891722302092499
9163523459119627892629265674839793608134095791301376642343338311438931075370351284696566
6021234234104615081448710731504299989347351022309841385321190094069139413559670849018755
6599425158538797102638021948103082380137506752626103988923015286433360194883426917214966
1569915421436265806676638064229745083599567925696523123890823652440300497773804237894900
9622134996775605408355122570747616087152365032443582943842843780650616169530094653870566
7657179335213706114945770284345838380274122889677067062093318796567935081097728239030355
1009732928581719427684192865985998505302820097232844000895047096712396079210946965738822
8505878054302639831164450914214002179002901811941058704856842266522480156683190887942400
5028290117511561975920878025722103685608699647084836275678007262992287900965339394630288
2730994499001812459756570678343500812944199384550481546660556876843316446218321199582211
5141406992175403126032638047484358399541706332267026842802864338736552678232961478941711
1827776522578006668269868299420224899817335605909156203049163868075320885440643776085966
5567870188843300921686803391721303349414154393334646626689601068507267029794370205908155
6036332868014855816123021368414511560678513944216081618457883300332644089751685948956777
9639092122178514674948968672353357101778868753807795421153614761225588851677236098050711
3659882528931352715995323857736121617828773496853082174735810260977644289207457623421688
1356243471770691398624792000311167284982490554082924091184246130770758601101570092967855
9605391072614193266931001373956494937260691286402151362850383563335465044349328217151222
9153045420744184677475520968103045655329122222362884535278212253047589908059416662942488
1531242481905527629531133132445690938544188467088514834020824497626537303341426775469877
7380112566270176192574225840599307094609559227110867404628106933320346232040948748259022
7073253663783576586296309856991338013432121348554018165704089753977199165487036720235266
4651734751816622880955408317895976515213619091144794455747106141725737017066401926063833
4432275445466186569492407689196265137830970567870715874658698159950529958493540039976400
8117913397645896030456742471700425645993375755946112529798516907893356963415795340094977
2287980760877096202686162643344929093762273144937207286490406000505783203315955824671733
3299857651485707639126693531077936639901294928566312306271094581890916730827382393093544
9297585964982628576431569466652161359025550962495799377269975011465627523471472575208655
1081920264463350394076125601337260814560955849693746734212237272157940614834247805166755
9759176796337335180071153779922036247229620274947805823599690150638957122100883227167166
3350473571087955872984696886268690238397559542398334562008941741188091048880466555794666
1318021251077741735489926900689019239142802574240868467022907085991124848830089895413111
3695329733713708490275428612232145322589641284729816386483565940775698703139071717117944
3280450475689251460002550128025556566241585557332873130927725267682294709677871995461853
556883702987273207707776404584906666089305829789874037762280496980541023990019691254429881
3132958235618503754446644285058264643540587051491380923041825784411594526766293539781099
5024368075125260626048859978085292322849335646506691902243199423890562961214346687665700
5276013596980755295722857329913985460860892469064738792496680345932917081885695007242133
8278953210159354862529491796408112931388855455692219081827159027280749385357884008411655
9513952137919960481520701230255365044488032241428494817843338763289706660822546934336999

pi to two million places

595569535411040235243330045581422275947222465611703205381046216288157794027499331020662
377412179943203648246597794482704778879897410732264848104103189000186648288692055525045
398948074030514991893991626454329664338446088308266696118681490062920382060349715354532
335434763834704761166226162069874329752340118586866047810074168658686566106928956970846
459191284268804079206944137698711041697008454822908235044453468298586976140392211366888
731351427994678147307835320013497195855103055294292490033841740248581797203330816154405
454944075597553124544067636053868402888707060339670536161863571190194110066330022732743
306600869238014558086418964221595786760775362222868634600089391151079608846132843705572
277402001863434335506051913415976000247202479858176669332506679740504144302461642837738
938101260706396676051995721659080949333771108387347499942363454681811957985374728724057
906576009206984926473393679938047661300093234174970961403256271409663721426987059194293
769684569204340023706278579920730074364907834378520698146597259147443179192769430196437
417374459152795200744765722125163192205917445974253362871871484878403076088087793021904
901402738297051818380204252379807593507534069732316739589700705429580103970052356399588
263099303707763707915241104195108372537043733933524242871834477185981694624151690867394
607370279810544939781310136132461406156416441503293845591001285206279996137932723534245
069852352864657266844033433748002436099677888057705304005135692505832347361643477431263
530041505841044680051754552555009921703377215892948907837386058969178170121851331538414
251611807334980999958467524866610907679294601939712431334382567816267322308326950197592
244243548997339904745967695901365049763598304165974197675972935775698022460465054651678
927132066161847677155904718245285857744692625530193768645228099859833989091453185765249
979534420059387152605214922177703311878964024248914454420243950436642964733010670087339
708467311350710175408014140271689933348965652389778980870956604556490184249307705738554
941607320485176522816733469694948408260591791636353623995705691053597307270564725839745
486864744447424188699068122486998809259038408267134335572065968364488675236146828125212
894148698518883471487888932165768615641252385823963725857960597235816706182730225860697
720634973600626337802675294976129435993344170077998977887717332532207579806692118472818
976099087743144138391306159955411024397285672351309901026045092063740885832953416236122
510710102891115151667106614793617770293885185560346216942384478525210632605500686143919
085110973007891528954603532945999970157505803600363184184228493666443219858974483241262
221018275155790819538529998017246148771884690017019993967342953214542776711386418329376
258885097049778756458444232914941363168568538972505610905506642681310791910069029732259
474764754009162510725228177642247124310812756973688293395482457379438234868510089377101
642469847908660611816806546082722655407736699453609300042700575759083064557480909585763
211801848260494003621160140722267278428176713648184198492393873924659059361763131032383
260478210738611891705556687850542644615907966303911149648829606899587630647512684318838
212335886332535848414461080521131856930047427380476778753283575922887321829912225491585
632679636092810024431149117008436786434916994921335986690946807925776305636253482936225
092907827231767644977721656837504091553180954333619126497023562550416280850869589588046
680311335488182017739396210701050800664718344380946625129088419382525061512701820026982
542359884635960321015803838553790524111375424030661255925060751879617293494713257188069
610429396093284214525976146253408205788772913562426612229210077439776043836266567193756
680221988504564391557698321634743350965808064914618652177597182723773324471796768363405
125747109791146883940897787119310694710551758036590675710428295209085054130293396515905
981976103669519092343447892550773002738859379526082711685373651264725500791290828889142
712140454672438561147116011251901501288054899621595069916514503782323780694925087014288
953211065958887963996255745506699250832619095448624646414034066213125908274020725935211
594260615874544775039424844090679882739125670210502355540373326845072710517655458245508
070285593733450115156227501831247892447173165611634816453796476588952205823367277559797
330611192895877492499712056153811765488004596415305163101058679705795954934356492009004
041776401672062355563067504297348359580221480108770136457948992568294745367659086568356
679343525261365201167233941464392264911451625049539256147929742331608203537465311290068
884373491321218628754421155926958846820025714774822823668149639541106848404851515519672
758777664497231860184456589824848340782197060021951480792703695594316774629620173938778
833182276953606488869308318465231782591880756435264040555881002340078083273427738385322
878739323613311553656013351253116906601310484586150877661072352962672171464760219141601
627817602688623698298341705563482893664839479756363025407181860920708311918547950892412
163882670207305973081479568786780560972482259146689539209155338415178446637186568716609
061095529164161288312651574499930129634703083612974318530606463838256694854997323438318
468643976978860021384533270297421084804724223430887349395146861737426627520356879717621
069750723227523066981713322344527760895016769171762591891516410070715888406300789472952
296953966784819920532984699867072184756746193384801710883268117224550098809443771750040
803703108608114328269201873006504080895030745047055991395385118114844815123020350731454
056536360880159612806492174647895385532750986588897060972475255579027400918571283469985
561413954006478343839670589349260170125868795581195231094101737971958262888948464982307
453184044443254208368722823954428569515078428650686907366594637927018362362870890054132
221796517963136557413022389338201606561554555341702800452046229363106085312008529199180
060541008441463779799807174969038846912953133647996137885823117301528750823818150905767
715047423937183445564647210723832730120990985401002606186489692732623201482701202646322
396532805494802202480981979873806566726865061691587855674635382314888810941536540438376
249696700084733719257565984982078080833853904661204430813291360875947936454340388828836
960228075831174003569782917275906812994402892670694884474336513779440395622390385727239
715313546160349759239365608122680060083315264842048593226686266510604025671618926184239
842871019921100600339830549577753812025597931386447647711955327208592349853199307071249
679924768612153794275810544325322671049211151270370703328827094507337945337191983352466
764025179516715104238899147428936920057916641640695698216709410738080949117725764716114
079705828948149520682341768254576988708460495775999704901719640386764193852351673205447
455421098697590263452079794578449369710946692566326810736720109063287115591197082125337

0248865957959171921840352044766565419569302775196777215580517627330302254327661379286913
7311133812617644844997052060573326071172388837818772654150187152482766538466452556243567
7747389642462920726451207826956590050060801825274702594027786208790865577159642865987817
0167685110384620108140423584020885368290714161192482676766203619891236792175452166411919
2275975764667747700556846774098888971992678426391277644785896001739426452730715847498347
2444160370460323898456830130332239785395564045248969182656389495950687446459638015126762
4109924782517505182906172299453997986483887818726618744205865502312859330976973436955998
7586094094424926157681601625000420757773747367950093529619634733140101387729922953608014
7289849986018176672144762625942051111154282651262113832581882376446946194940939088289034
5661704514123969053765531379792252265048990264182142409759742639939644332188244438158349
9893723984258159847648499780547809825879491226403501544792308777327828355661814157910241
3000565249785800507567152647348326762199022503232945872547121952114339536802886495780085
1321198191141221597501827124221544418694025744182648242573898113196626771251807533253541
2610382794626449157176872996599599143257558518081475242964459072317980953972162295208723
5120675205393492251441890404572759628316836640856972782107327343390285709936766006955544
1033552840317350147362837665891691471659625872033515250241382383785952864992308625963242
6395320881572188423096414469215539717733666031629067366470709545442543745692089194157547
8006732948166019584246116077823253609333937803039370174630917666409903712857743695254014
6404369252116459939839915327516336628382265216193676581706289346480667895878781031684431
5147295588652072915196326957533420045673271518858835358948814111408895890253473111244529
2160703369946857917294495280967030626957070966005861697214217704874042587362074923044277
5893697184623384344104115157815044166881247646282471396898362643275067568568961963154954
6768102529094129981176135730734614577561464606470019996992310712278164828011602127235838
4984537141593917080820473263093577571614862490340773159622948425161071263787800583427912
1177570261242639020714433218042045982742805547955186965918957990647864414064986230180945
2184956901999344825401613262201045529961487146073217825242700336295504215085435072550130
3166947862058425931967963045805527879079032250411980477439074782200821572004040562698760
2446970657649092379202879396847414221978032706748711268316397304049371733015117331139719
4034263424120239254007836673449834587577561683650958505706073875457789059763460275851923
4946232695602863324532308615853763131001411751573240500198707969130102584867417376163839
0895130552847848129373447394396159037055256692412669355574457441355914710825915758838637
9343711174819828871942179027576457968997581906065244762240490561301324216046432184031251
8671554887872794721614703575751132204846233766746385886350939685006656646664105826317592
7651327383566303241479809871473743633060297403749115356577465742784568092349865132904298
6729034089517568027626162846856528879636228629928070347878914589021936292039980123927476
0368110508083242821529788702967715575515749127810660723358362303826709179401397403651072
1716587327336066413750626913314792756275777387788218390671376887858575239936314582082458
8710098199284986742197620935628860275602702052627241441017943626280106999924526815748776
8663991990851055702969878815256606029151491254222618578894941193215375901679875594627263
4757184512401952376952047086934795292753519191631534403822741629169251177423200292592813
8855912932294350315393576417607605638253125085480300448351023570487075721301993490218411
8032926247158511824547932032660326273544763614416310800949565484140516046666494525367203
1107448889582108929892614775380108146909053603929986193274667615913597676178340851053142
0856150987900508491686137229440657125999588572686828925263848862562278623377469710100909
2631430951744067789683326479115364774254329642039128484702107972352053210484977198754125
6334925538437000543515598811748144103240505370751477803024716187380547672373982587299961
3933257226452940204813320159611637276232756270442448609728635229762865534399925785281900
4043022140758694792310231845243192192419812503645878707022367716306298076607211862719869
7400622974155557412749233470191813788360520383896414334092368344718335646579477951536200
2888783452614905419818111829677107326312830959879971804452538187371710923557111092900752
5376858005926216425325124238918200329094994827343432361662523076039208269424755256867319
3820998143369395885976127101755638821842968669877771094469438907836743674315705070590193
4541371013606131326427057398068253508898865279075448704087616450790176659060994832462927
8665789007154154687306793252045026379919982497240180129450859144127825595002360886903676
4870245394239783819037326879511585158958598918403036947131188016871361695090628234805078
8609022752707861257547178335990208947540335742499584015667765510535679648289169556332894
1472473278437234589922209691769128383634442596317011408921902933345806670482889728451285
6970530366782601914316388876576375195193155688335816517369405642078404296468509606676623
7773568326655369855016771888163100201701227467134166956119598326466764015915300936745093
6961539937905534873972741412414350081146024212657115113957740653498474699063356899065666
0294951077376295951831625077719899422238952483112984975553401822128865200358669329781562
5314132398852780951810799321475238990895115081558435530745693353702617916119406917794916
0724806386901716341278371439907657912166172253948925830501238124205329105977790961615722
3470578653583855947899448582802213383102004500499242941126521625548726593268978968316639
0445350292792437266828800507068605130882453106948715545236426593641893882609369718897197
0442654848059466034447700464531508781581431554378329529733370931398563421867298446664623
3769908491538357439657115859834362514850599161099481348437446981776461002389265506280917
5552654434250557394107665491277360025092721569430150308198932709038280298946124448403599
5389701366178928607505953871082562212538457384891884653208921958661042158357863317941144
3818769426035270788468368560830684345094187641767118369328810966802873386167248862312164
4695293365487303126790111685924233133956946079497765485030306593339375033914581294284212
3557638753239495533184049913207004779975921406904826602503793818943738571240164996274990
2307313859136201716777878394021699374638891221735880639379769947716193625354353779037944
7298833546919904012350880687099748669643005272434866153169376293089570103419090956002750
4467150841016464323283773110372249432334276623187035710497280015319072020777315107541597
1044185661667965842590199304685918250085564082968020188096057241867995771344732622158185
4841049533628408731847181604240967268493174608477747066275292044345184693277487141949840
3925918690488

9613220749970865577444061218832941182152244344634731044804774309932826220218806268239035
5074678251487127433276643194513852794274960859740902643367480265299710643099922724043536
8638268474557177675899110050649467963426214929433575677364018889799735537686669539864920
6330941898157623614676931131202895777823039136984227562866040623408057469270265851198
7864983708490842884142219042498126231533938153942907300639565870699208746150916347468901
6327155121101172731503097487192713833941747220138183248459131690694932367116000394869541
9205352567907793710545684913603593718139949046129004027580869756367830961890407162699
7450473373462249567156151635939677972270206418692114983837727724809533594567279000653431
7093118291046310882321877411594869475855828934206128980075734658941853154201763699710
9799208349538013299655292157020546910821567985276459967356370057911067744921383319655154
8181307555323199153007137038569894628129896340082227103038470756339311684785135696687509
9783140727068940100313641568124513494771130810672693095746324627172396898915667046238
8628504100149392562947277911347005995135571505661799099523213913302821958570419569315216
8287266540173050072176132564571976709854467411924565325313413764877427986246054482612350
2669817949013881454990653365982875263821211488285171204299139156804617774304853261376
3644803399454422616188722767120385717855970984313631348335683791498507003108021219922355
9797395988486572874586268088557533329707955154717779454920357430872148027512660933113474
1470626143969615945928084828620051551296970402893493080880488618222816606368889147812
5600076782447609223263290069739039975143825843313965108190632187693917770035503059949273
7964093804373676851438185579881618604946979992411425942376690287949544130499849202620360
3141574145724366309438107587177721789249981795184180424610082584987861806528590725495305
7918864028330981137706452693164099876629712801790401115284335459728102783670272318221748
9045875771682624729213293866520394171695335781074913749855030834834127999586726639678309
3692377805950159501841993402048755941703964069167036937220952903594474503498181016
9789482945555968577963926739802999664100321832576553677441255888664204886990529350906826
8412107541440951108147352234228967740970289502229921065414040015343907464067202650309738
5680594674973189499483268285331531077464906380623929787453358556311252880296634688155
0138073603863782167131468093581124368944979109060035130743599690512377992622638597059117
1313170501205016724563857564401240028283208408048038305870435655916940552982996137102850
21433636136340713409489877297113487128085866281022813218266229806861252287392366411064
3498971783962411954444955363509708069326962602007875188033477601284449545082039401794855
701948050308532936358928223326355567990364181596367964281719007376978772438794723956286
9868306348645259920399113762560938452611832715217627340180756952796221230566603977679223
3333660540421017261419967440328009090790086195715679896357806920014617097156079860275
4835681312374170868753206964301461942695542245901529660866325450864681011799827650507638
9795737238610784380149779943163213215752937601935199858961042730874796515238022056961125
4728365216804527400102165422695558306107623411578608431226207524454263709260127493491
1410853566087750121073151937717513695080728545541442657440866477746371881112887576635923
4163537004868252025662165878353992522765088347059788443496925089392516197615281384222625
98695471277389491096443440643792519119312372611855186476045552679663600901520740845987
3291576964057387510090378952901259518151361351341863176729738905137636300393396197812802
2839958009519199775775557383768685962887622530583967840016473170603710554580059312063600
59632580058449210970971458495254446533311954128735180831861486532903780643558088202104
6826390045657393631028424741860626363319002725708949084742828735175022628656000109840211
0464526346649701372273793547132950125925812326759355354763473539742905641358711892249
7319155455011349434732130776956848747168555875754612771616441741634162352236691374266841
6715193674967181895967763474239765617328394027071087764125167991338666512936731732626842
6547161719455285780753762471339574708152745125356262708436393940378885828754858444264
1679038836159657998183149838628214098040548981635562901759525737257167985443099596088909
4345924813875012650638741559904626463289307006655383264471988819043905005864755107485579
1900440382496504633879012591124536490567028654144036581524305080108213469659647510504
1802097536887366502239230886360386047709099691698576949702310675300030664741137275965199
9703702346468720732490769301926760124886671312254882126355725815243076555733926484605473
2910112673143256705136383910581690809491435733658508741049739865156712877791545289545
7228134494482153405878507967596365574396320726326645558741806162815278149473634497370899
9088624172304580321315533705186501799762010155737276254437715541382342990591142278905842
5446497735063344659801594979304507314280134095533675855768034961218338530478544934191
6381697803035957400143408360180459275128445139866481454541006731883657330998103752245432
0325398859947957813842634499661738605028784861666585119782621953230816497741287010602591
5646664020265558628176952024825976241835614023651807061454736506889341428701859908654
2140221609627342507824299489079416519225482358641197996142961294874026402454623683755259
5672423669456126903353994545998695461773493921720274766512361614524234767541187002827155
5676865936924245838771150193741113473166678060725178981843526565195226369453919810771618
7211702617643698610461770978673179725507152066918928242934513345336303983846682732099
6504659230680296733212455710759981257023641161270232224100953092876759331618423266524475
6033990946868199215440411394735275446185311290469156214793807381229310989037544471977157
2769285912389817928455784592973063439406837070328723569247704523544551417187483362789406
9833201440422697773363191081346513427915571487858889256202850080764975021060587970524
9043554159079712534446539977271089746175281534582910434429015361579496712153566067384883
41953541145341477671527787089064199683384395864703528041349430242872829826039910739269
0304835044190741136553870743917363162824245332735667746657488620191516003575724529925041
1115136481130665116836579507576959972228563382611184712113377119285926895197933127400932
650099304559491777473100550105925416425112023602056674205146165726440152445427506469100
7401881035020741675951541836915907248051475469442196986012039446204057348671807215030356
3256051948115257568830709697035475641028775782646658870800622202851195195344679870203
6908432851220159085057693256538504309986420836735123398276489398543611745230685391360742
6458374752332287707718931638407122850671293364440954862742284915005526754634554985328221
41117539052442789094717447302962603423830009655262939195508688712657195877567791513713

pi to two million places

407552495543120335761071410738809012916615876163173627234429851169362186279696645763166
358213979658529111297953102970285308931151776393216958740256769510312702463403859230016
697499217848410062401133801745181903156895098336236532613442608961020746278790202881771
604765596177082739927789407503099731827361425908571103601437958924025167165160414929507
779645846398606012409685134052478835177944633317820275170601506768278701944696339009517
123990184495425119658007534589173202979706611768681310398215413649552146138314796406675
583343869856283047719766843949598046852596665028920264419224636297694112075092893564803
088024579964861240684974731680542450819674712512187484627759418710717872920972153613010
562824948154294531173463931522571119716281027173775677810100863978753074846976238419507
766782082005412154030477691484971521253647351548064284661166885323559158395350788398635
883708672776785787338361804299973463988679037088986767922257369551703578673837525083880
758716013250125964133511997697969471053665484507058505461205714620180599812390901847361
674927468755263436949625172902561809250673725553441593884563315135487482260131931909518
922289497137795462348311961756612800580661824128007583654420236261322988868546979175713
097790653032523126401966016443229485471328842768162940456856817780294190912983684687667
969158806581382774896589984265945616190972180141366167294051781791852306269095065615218
093424473741979652964141541185008792044076974035292463720014075313721281932797852150794
724056561544278875399110955149923993052947302247394613821021378147557899258844167941677
464766832983817006884221954861972352932305331559252942287112503845374137489045506604165
819507329431461676641120340459299198179929214872889930212493106883973794884743140898322
940121491224903308684473870921041039863163417689578490322834927849475923242983358730878
034126177806179875907332965150543691389195794557114280745330059173532903622150012697686
511364799803212732038196468224472262314459659674846703423116617547730295931812340331802
213103935146773820396098275701168542153656865104658301666156563492814998015130267335950
630947781338517837089233936140552663724132438657527813875543701604045347328877494747811
190549466438486526723821104380542257385041906812226325106443032675559384000742503165196
5584543847916802428906823616545509055566079244291861423507661979239692542888334581770558
345555226733606427245074889226027744975775864393063789828683142247756716055916486940788
187937134235657901539219044425391573795240056447393649869357602495617157076420438546397
050634088528339147945015189340431893283747341432615490168442385592934428828667912619401
414877927444857418816490177342371150314367940636660534416170645229731436664182359781667
660989392850900073856061926402031802569783298917126865831979685240145046231371138739593
162429811024585432062918343449431371654262356280242738812533575283601608772348354720508
163152472809072629551030192457794844552585867080981835259592742202538308633477250792140
150817251771753344557497262213961915849811350101688464734830960445104599807742199255847
251966571098958779807895982853885979778356882225100447509980822851032920078067199400112
036141036752265643186784066645114068070972454088875520190734699337905125201848409078989
208527224500292469017912642705475377185619336829782917027379032190713375248641602223551
676894718585610230291566858293965718334412046615384134874569081545445947471702495483919
960184534663901341124139359367781207707359518410679882297660390699181352856410130222462
927582806485904576061842527073321519585411581100544526075930982979258158591232578605040
815764089251571454658506243174800528388692722107840840946385634569149443997881845060127
738114122367077593012676423473058058158893180248578514741138060484751355562309585290738
953395655993902978153766052171933015936182916279269606678403534583809063810770483742180
066514889890828366067388590935990860932722067650705911897077477759073832782276883577575
240513841725676523115039654957445674203546395519796986761525234235362221688963174124485
789314760164356674132072654911781952632151974251764033389849184480189530793514718671414
201869258276998566800568801846729771148864942940626345968352063206671578965491725951526
782165239256646290002825466504719310299693521944828653119848703488121261010082091920111
846555574240533557254876177196868205025050109811705990063264893929868828736939116042139
015317494188012938664916377463734283706375498198107472869169636050989477618594743719022
165013608317758996479735218733068274013111754303421108315296948014558767002893601978565
082601400482648132890272244395652037933247005609055051828651267353697273154195039159042
813600042319784475683838590395380909871739068874665115258807487975031226017699926885307
329422904935373963560198980477902100890102527561540819622031311166797134446669885067845
408481089037477574604033920215510136423109065447645989127243659797638321174384721916080
824554599728608818290662916866936332331011532483453389034087843338848902977897340663837
993830099240740211662378179073318172741933469869204925516020594375490394702861729395752
586935641967753306134175103084342363826378314013702135956058721080435988233807669117028
301708244792302954210185625696756934303270130017625356727742686964541323668125250741296
062745460991626200613297343177815098737218808551483660852304386454546693824970411689650
388154595456082431182207356339575427726896464381793030450537803987346359255849586381610
191545071386286769346990656357414284296990544242350671553554376956691024605187180103988
265659962594489019349191614346175604921260363013461505727478426511791322278561834618617
025145849237651793741124994856217543328111016800620180281156359820085108324503154698131
263901481021156873016681336656493693842644655752069571952108346175003698820819286861581
176429073437865612608188650465457857719024632537896563019298204182940291454182508124290
969252349054941929885697528466974346616328463165028336776944492025151165611199994999002
502567807504533594979589996844108302729229602535700272665060135100820264684739299868469
845178897709478394427028504099154669438286079459206356824251298642449562445788026709040
546249836033490673529058699870411086748114373484432899561697662587853795605933466863975
784661742470306441516553963571334290041245434785664643445250541229474079917018647484537
2079637371812420555756048982693592271619866481976086012189269700455156288261715531457378
404916162606096477524384724394767711721083711245206217639959017897685110958661640783001
331817106296199296884729971453922896422848490207115918759560178440813072649489913901918
302115995196859566964401907080045167802479074278580807265649709830211243386403592864649
230846047499334936313324711198961161271364571711894520382017131321653294842599897788070
237775178765859529386404277406856620457032831056268035338259778191081466610777758806372

8182215469645442069166277003169826705314915358865682569712910117329971773811820069859 36

771604936140078171391313330626569300512089902248181614501983873838100809270269311382169
749857067288901540622030661117810514766374444680903416958002728057584263422983334795531
430660456098789521165128884821633006660723752917439384373173535401122635346542269274954
188621933198298581884538108634344250673571120291158171973384990454998559885167605134573
673113303025279036785863615449709867659352359343625950476338803100324533950172464340962
630989994436708257836142249331280819779011061336218766663977003396106465475829050520482
164194725001101156991575869751406574011777318115799765579798863404366817406827121817810
222794316128956584798258104959715948703200278655363654717640650377976657597219670404202
872449843850302229407494544733373339067212045655547084540996572901817580376628758886363
038367084818240958231274221067917286313565572114475871393091848759331019500723791038700
456299694343208468143216532622740577854667638090712224117092761955318562881567751939266
614221094838739402593700263417538370779644830183667567516551800779285194455321362022508
269061663782944649029969599800105550336332707997648382028474465432477006461972580767055
955636075024177397895591548860447292873473395406325652094056574151387695561691435397925
704053831555105755248629792651044693643745463718271097372744133348678132508253968565211
406742392796174868231282283749008544188478202590957512527025510136638939263276269343847
948096478816523915397420081109633824941843000343460150948610955068114404202894561882140
658915161218566101308195936453679546695677847426426402078804755025996652866292696350014
485359422548811363845315718767751761027894894725121339214460360847250837375287211126932
508741626623863921573601421161604720230159582943775231181009825281065376630961616332799
641733155733928536535732445246523308038776381698542615049662467943534122639544984560795
270961548543849457928511060605660077299825143862068444120576139814429051855792712987952
326271794684049069973872734597736263156000312090717377675815793654622763032718267906660
445404205327992870844776516930461060932323502112427956008770311687828655663291464427409
474881820586302690154169749240267239004496108307924694288692263797914209618677048876312
895632307937346972722355559674697017162631629054708944900948101121652479229554374477495
108994565531136519417152390947697736525050760686740663388577123134626262471933926536126
728092149061241966252564050707148776455000998359199597854670827801479162135086654645017
438903547711087681600560072635357857464177175479319220248084507024511463370610205589490
571050685932816669435038052983863579629746587571113601222327277229585791111060537328360
322993104441616655922766604591955293927377662106553722776963550397424730442157698849878
721013463583953952113945170123970954889239226189792446899101425095385381713169492911219
490079647949624105447263302695715022770515676569782973741245502258997530631228721115827
528295532599488625134815903797869671520404769152214950932001945165375497016718933252744
217359796141057761150624010007941966088605105839402670014726604021679495426914234612050
961610427757841618404616762539638354207873247330368374048594859557812449919760333218199
028158967902209531637654007536649011048981997801919981675854482828062237024634774369792
129254105090907184809950260693678194102304199350360752775786048301214623116005787048724
095410834630543285611949799197576078427673921569948346512569103321535311838849510187796
857709538759674127741853164907800694834139196102848149405722901619284974533093670788022
519867939314263204362875667555756202945990222695026736962552222646216257588371809187018
651782872279944569177235152354357280734880287809479776731257485268702725526344268923139
887446423193440684548175417238521819670358586570473059335962439335876850729223038574474
810934103363039977563610615506880817095551882451383534920007422997784689578794593027308
434418622670241009943690347356727867521231124668813650255749657465501492154338471141312
696108571971528982200132119190550498769109798931855468637742127321021510549128093318934
024728858494396447018349905793271660098171210699590035029546986583405770183666711064378
659278891613794180037558959352607782634411766382440810899202661136635956370100141764779
349425272819115455438092032371936889314447154735039460063599284142920896793933792610416
480026236760600695409046971772552424087454363822256107994190881656653134818488003007389
790998533031701885901976443706101966895035878079198050574385881739559096854206164229361
008998954015721207200449711496686974839830451989077778769469993788721865505972646468887
544292467426864661747075328044618386097772774518404277154285611540249786303291153071857
939343610639277745523607920712322128104159345106916049416357443706919485893352044626797
134248328401626978883199646957650780075851385894616071520695088634749401214113716855979
374398718985279621994923112423930410497015455236392812413812015653371998368166935394211
964665105973291522806471759362013932705286043480401821640459334329150332854934037914523
248559762574655884428845374349845948426503639100221714595429203997862457566078300279594
340792026543075295773445464761724176119593609671251733741289393750420943489105690378163
250810065474447712067436032174734372839449377691244362941040552366093574579950634293035
364624718124157699807336294395174786882933800209533811199575644096131703043286404168665
739560644224254359277013716659436903241159036231682017483959857954911457266352435488601
970669384760118844311633231643365570831539809169908601970461506908058302796363360566342
863807534770280280691738546452038440697330741465181352401913028809730628246581107375114
683244464923694364297187071052580228726726234351972911479690340999422330687785375406633
609814833474046865007924938785005664332910940734084286068877114982033970480144936294558
624194910542846244612038848927053922477814287457205250181643726331862220252631409233710
227166185465406450307963045940932787002146578443252499590517326607696871901517804434028
315602187905514402668000208659204801352969337679326492370900109681969126363116102474386
957397725649096321015723749703578690134422464814165473661930904468181785203972325602038
454608087149461993817382166396769510151056885631565841571907340746301561594343245432752
177728255461716999857180209593164635733854802411521576805994225639307044234098000073654
616129120578512093224186074814346715844884547115765300726024355563764495906317576858982
590271529886523756393580926914623976032142610635540206247060927705428436023488850755542
613849862632928558965138784177279535348606614535439248712456771766570659136587062655794
8909779720340816494448979768023978147693894474361487022790790959610987547277912638382185
582165631763001863053287682450705268510806661349868983587704851041059697927287890833051
672759996677025985412085727510530725059390354343207942726528442462498810576358576806663

2159100221046508230315072837910935201793491199303034954273737779976449364857184705189766111964756911954721476898967283119221492976900129632037725543677593416241731405747773132125802931067069048495425182045090860090001693701570596608016227824503790433564537126956911591770813134234087443802593879140580524947066855967113153330371582548122443574123169048143204205406665836807315932962488032084751653271641711604683880703112777411351240956177265285212224631196837703623175431335780921956877899477254157199929999627669332679718443602891298853106486240440785429867217437778353365741865352709862317637265696978146652957473571265321354543400604578663289540595046109057009926802375998325116904036401461243632387925048144006492647117366590599706873673029530122107258273165354661574585131041508108307456237001999915384688312567860635421793641123516272788497592545956078727304561243256273530036479812228434541373743649333290640620409925914416893878712724982844550242233952643189369729008361719987345918315518604025397550184748255121157096748287577091008027447970487299563141446836926691356747155457149396882621402386584906131244234036114522052502709662854949280994585663598421377967702470963994756297476777864972365041918062219431449465012390719724826122222955524590806012726420441626763577382901083056263395528466670039862101252202952571914468051034173210131784409484903938259766273322810408200116368052307261430433378773999621874480417628637442954896017964929630283313467762864675332415445087623440251916899871888288520355132152522369466233481724679647423808290697472510088320478858212926949889039740584799751490285577181562767815095651859332288685492449855396427764223504913186381107569819839676064949277242341661474126957004508948829018572451030222481439955350804981205208561953081920623046345648841447538266375340825980451024845570666920982013873051655052884146399029905355066938661282588312596367921286084420191847022172524295417799650650985654795211867209756713082894826676752230650347593861687652075486208915085601737623510105711557320805694807518367745800249779448065234636501796692954423838153671715163651666685774034276762111341526797562115705652828182701688824449227035157805548688001169043400422280441720670181374415584692562677287452824909125391522114316892410065622850105691087029955603796140393199834727341125510199187572245943713470002755101341991242535834909941290760272524539975765601859129298335098762188748982964351274791707594565844387856151571533758798905194342932115612991307284038167457528459891954444696300243399774391183044829730519061087154035449100346776160269693769180158965465817339817121947451005381710821920445648494933222137532802542515797668131592283770308124617825236411432030489734965392073618561164093870435278265867397898420193718312266403006605738776489749279874540037877516373428221134657104427948634441636166823052151177538902501481756673856665429006169628456220664626184627397360251041282295064310536681635128921097293517719055027396713388042919048638538161193241010581644940662368625605653448765166759224600807995260890616899520849611851415236603729342127247774509640547550183308406741204637109505526291210560278647587036769740188589205957244407328080082225059367036000784996150848928033576037655581586233445303727926940678073048308848266804812564411153304318877683036326099564154162780940815723438529077247915489016751784985148338146200500844850973318212885656027229472421842049200143381524889374964354233078039064390805317649070030142305796702821346151080004331704838763224882991185981075182772008986689047934185959569775348593482908269232631660084450960352930352767623775583290128767407799765040877493175891747680670172271869656755531502493650807019001240626734863438738876655116997282357664540972959116059445155993813715379112795277674058970451495964503998940968314663950309790304037963740217821327095619324633236781914287420143040619710812430328619252751917221478566287582538348299163806006154067203418532901881120495145828037104012861213805549632706630268442748986486109804589458938912881203383895938527703575131962663994379189365063793353106118188290709438193305196878322730857891246453155916595091098510265473783936453086236207387735051443003201106530764469847398891830237102939440749990655339824805515167028782406637671084663179464657926549453457729990877443807283255991258981780746767328967315379496200229212320334252677057587538607450911410845758096318689566480528691409240658624108513442175444997088725237301205042962948680168192323399025527536289508655543489529844719762317245296775258216859730902496275277239260946789649937770444288632795183931615169176116700119378403507102961055233216951230111531091409258917955876894970487067485657495589470293348908715735360709164425176795429332412087671218925575279852318673150943095642489139083213123549486277000918963061549723587830727838254503014179047042672268539323847048064351597480062828262342750002502943087209354673045454160536205830769587278494802307586346473461630825338332605863415563240244757799187673255448310514971627608357764319956184210416618335347621901829746460192612268626506830183239691097338775762555971753408811685892229763095644934687546757062640644826247051812764401357337266837470155961567798071074743661424195960153909114856679416823831701587175952322468081407889376973452471044613502201506906749435761347925935187169027026562413564192708252490771671303229296557695630321873131619657701086942150204880412991087396373478489445560951007848737203430947639716046221494230097817943391742261422932435772224339873478022398867778754634006083245051390773436915952377426165072330496315091171966378331381496127378805439940932243682843065812305465340864811307301931656003252364868989747924057184968617537385893988762974113507337018797206157241316926818588609290472968809056872909512668206445118163579624390384861166619551871199596240775654517202264166735379443403531750510204633661552084290254428184107726750665578485824588749474535165994457959440851384864617651774891720240444255461552626139190068344596842800606133887723371077757766638818309686518621257549409546267050923301664435969028314935517338272291027965981737216795951291837374996243617181684175605172537719223930617526771573832289654655829613957411129254182447280344093932697365203066598116879614951406132269726768197228016429960578132048344157660750777005819459485848561126051271457273864226862321084959181673825234726142293968183555771656150708597874237015361057399920039161591223734384994099920500039663266230725514186436643184392154638056647867510360826462790739143576795195651747082958190406842904831503119271173188972544003895483694189710581250251474461253170381294995601227939497850801357446191274844936584893735139914261514574512744458239041509328832465037545875619845820700521450866637623054277080355113802212775672933

pi to two million places

5900209215945552898823374800005281116830714366948236970696839317364567145827458843901336
6453128599471799169953112184622764224303970144342647801934767325608067810539740430660795
5046039208793290743900328508201648464183643383786731003838824888766604415034184089663213
3031289755346605516152689621639790098017495214164174516703823564826501746988637931842163
3201311039943650692014326905376470313001121291114134337713510688328839178900593074523114
4421319150153393921242814896884118367050218766170820262487070850102891232701427815143758
8215036487676252854651759274886088704595981064479115038761605705741229132691803542651322
2382273651928616368116633318902010201684556237525108375540915524200573111329005618184831
1331579245448498098206892333045653367440728317189946886566784077241368229712320726103152
2495444192433131850484983510930223823768000572536763289854967762230195266965747049182249
9623879705549555816577600057215475597937827580905524746065617586849383460375218030172323
1768295559605276864900199695499627819383135664188567820675345770068894815985660938777651
1155511156382958687394916015846316338166273158878120922878684950254664823080233418782367
7012199953807396570676611721778056054500921929002135959683502372558856236604053793250939
9191653932193880853715300241500743556604111350173286887594241171720251304336646753351390
0682683782670806754548454590521969087821109275425552355050940360579870558066222587544128
8132125989081365168890769804358980963764669343518963376257455436012876485807220524281700
0836631073289417797253615085262606881282459220592104349395922973721825823485552303866224
4167453318956705968867845725268545432559201150587964463203688629017104606467906673681595
5974244441249505672328879253769452383796377232039915439982829194741752199379619323859992
2317722518605855811045963846052387225734917620264692348275913116537787657659687806335427
7153915842606380846546483377559373321689460563208287836758803561808639140990091440401033
3652415498501248727958699404201548057844782389571236903110719107313186545459737581015992
2487955194588598943591932619653808858045593580777511133823872529085901688476049048521480
0108683885695421495158543137291826495570310248111330347987612172934041398965580538088876
6470613178270404216779132713266187978697550769233261403967427611188957902347414342992509
9291502372568445650534253790127575309941210272820206789682778581793596859815581864678489
9313555076776817537538694580762250244409728955157129381330289772892470477197023846630112
2765183719824284676429112725353830224358840274488085554297049644464584208338768889252306
6729270608974219839076023416744361182471316237498950595944648519741547965367740555111399
9340394473005178531724682080304952241042025489328774098503667096215171282868631240513301
1948837929407815310374007826036448783863586817025481734492215789984337882789014125386915
5166776972569861323628531775229882408339953897932923485147488630519973256215142763856912
2164946098713906585963881765959588958030405316019940571775907746358420773092036591418935
5329330396379227202726363271805535849741172017714275866559065655940211176280275341285547
7265895263581092694554814375886723197039927737350154651326323572585677812289913721503626
6685131251975095344649646814215446415675683304662596200922056090050918726851609947661498
8140269258559418796528018772346003598275404955437789875108018565309363644871681693770318
8569097519609232188154501567307857527511169481701995941689066976701742517363996842709492
2170169724105283908963122493971461977828624967264848511222045446543178004848245453408111
1894151610958601033144560369341568888559900081525707005969478215851819047502000448939160
0140279963571814421478661361550460289040028340990824077127771174738991915732399870412301
1001858939216645302176518884612894064972760001512369150337150144967492115944551673233226
6676135313710985196374841989590801221082640206840497019736048617170144887189073127477443
3916933141742169257444157968843429819945817914629907797765653159553467374496423755670218
8339698198827816577100585802068218893056097096959571082704534615951707085811427783076514
4657918912696697462203073482956272240257263470313872145051570024734421349151473405228591
1366598067142930123098285145298594999209038353622700324882548281103967466118502969148022
2206682192777624868551436071600240689654102278855436551038358004147514164578257887667190
0309112930708520034733597828643434964596141171649223461015569879576734648927572601186310
0176640613544108511668111938298286542597098819586576093837037522242422248178681585775054
4109149312257826689542360460866950818386756938295937774154945828481772331415723484590055
5068706554483420531689023073770479560623538824289096971811638481179086975435054688782246
6610995702330467035439748861554249959265680271859824156370868687259416891469136071004339
9428961092876329223669633479770896852772627615453656185157277055217197240445283654674634
4149727158653918805349385204798255193594086337896274910680791972085528113202505894953757
7776317286154466772737027162818318592610642778874387630049339891947575832148201894215471
1955331985771921609343992088721353383525872518643987852052135202024926648331478229257660
0460904494539912176975288423731546117271983530929893010627226440428308481167376063487397
7507461787806073095084033874187359814571770442317327895735807934749414089489704082671528
8535704403696758596638310034204237585268099372656782550641700288980868412860889006668244
4359713305387076770695508991428764635674899776777877488157816586131530762780045583257703
3327111808058555255107310340892698157957322666959060071035669487617411261938926644564541
1409820862406208779259140247521752358856021622000555786237767595257059968080991532474833
3951213339987158791113606266976410247524693085069546232213400435986176596230074205029981
1515143712973794970792299481663249517791402107006780554249263878410950056283270688186001
1367556099146107291634293265987651029363190387106902154329687201696889297938960853119232
2636184402420205778343902351714177794617885210758329050498362187787099965422021624957086
6750998756674791417232448734525155227248672492244805628977469719183784734865529626208632
2341014270040150422351741920633762686069388022006030211796711279581181011975705131141775
5980358129594870552749327636723954909339013286627890614578945176634371932449002367355382
2301964319281649990717566828217841421331765145474054520521608909878877292761400523046527
7901524542732880408570774714423147995177477513338217066589166491895677250979580157474880
0632214022709492662153054526043397023613093424296080384926358173136459983754729129079425
5686669885922637931132522382176421424139120083506384215146929141571464688929021851245994
4758652115118147225504060631649994621343317908161691119022584033030019303111797550526498
8524455558419314199541827013320746257079231559607519808151962827724871696568419739091181
5069965346557894497146281540495930888111971930216233328931951650267548937336657425926288

pi to two million places

1727166377708443108179521312925390567733968885139350064072051986906807707548347009077135768850426330398087847486874132827395855752916628067491356089857941782851001682950272083328351631316732007725539184971874655152826615104123222158678515033027726133555102013539373831766222043997119648909948290836912152876995084830038627216619176012636995469549797219413032222930214659471856205700661114319210618616065185628887134573983159160337347372579703944160683345429677581590120332165905280056645214665552736596525937557729807223726618679388893782102168790063883958615516282359475597869083848800710682745827171711139369100291341833966941539269050194599464353184685336643879949975427023727561212961917019365261241077732391146668639475759253835530646997218377908292637775231627730477727492710457150076124687806774968917652290677839703458572720615252961455116186683059472478216317978673770653162269941968122135370420555579281126365539320846961270873597275966510145901606149664035849681532424169481773470011562085111339499806279756230654620918376862697578461558329149413812786151874840934287567011560698348016951403184777602913185118816643442970649934317158150062661335412741780643248613484446428346117466655427026837242838069960010409346352593954224191253990764867787860278917140947780604001342573136473485101945749721669259004620459333747273101285590767828902540494621236036770412440268561016498257047096143811654076296700025458316621595590894850402648607025269859862678429468886975483989762821293970939090178330870798403792285267377224226177548445196332601513221105994625221378452094172231816133932979228470083825385285087680518645995215815094382696508452819565539899390063811669387069935625410029752474223092901201614728604568128908209433357403166512233800638597982546530828208023672133228677983227307895827656051974833253888885273815297147862848863584948744294475593379345974702922123901773629862945360224389594165025144729550330926644882522924039405917095990674756016580470018560878884124770440827008932186711885302599545773575187394817064445804728744655033261272503677382923214534700111061339162608983094458400060840404630255972818589010729846891305591443721607039801733276054359763178927288871715738065681316482985341523991714679239365559478238336589465604414938270660093557759616936618933849201667456034936526568546716742076153597439196803767766954181593899476273878232069624530082985914202537724967773503928158660894078178409897534883194684036200219184965294893637517487859481203008986666233853046385169055853635082515683618983203170925451513388444135929744017848221555530012809355983058361210425037258224489842805113468074547977090431253759362773234485978781879083819471362117825104101204443500274081257499749958578533794472603204563717827083072122852489368910636707256578632862540408648874675138833743724548365764687998303368955520277115089457586725071172270100604467233011608752349964320445829768896674307842685233949000238722316128028617330956174903008594804087421486724961875449041843941568067629713443259277681566964812310368577492885179964295312270085941072197587748529710980409264173869907843541905779172097119372699736658142032684660951228450423491731166853272426512907916747152056076948239313756770605909789727068237916343479023904995104086643409720267585126121830092160439669799625759290520121288949682043983290977005506032286371043844764639349952899173710456473360611059632729652316870901970992989883570094051477765519164241584078379753844040071029384829183994829172723807781571251772961452847276228561226239970159158606311290718499215772290322036279028364635505109704597491980061593756414242848158650892638415404045405505465471836635020404712363250756662885005566980043950063199545223125635311550236954478365248308063548965409276716165492205109750098002150029861245987513532464255562388235063117654593879416352719744210440293015501133381004282169121771415573778427448363834686611842537733141031579638412484173654192050183961242679941005967133120800194998437520342373537669283876368847724471790318849981574986611875844736753940536020562780598930676486102939402810856573387257802578106702560784267297383169827645602432266966546944550425825706876317786037736542159294514617922432077588336722729870514748126720304758790238216352009472602593813109627761781993243906892391491958611216512019531557178741197833124784548667595114991785637053311332564160257237057180475158193423991254296632124295395296415943366154675112988328215836403401385885188461511275491588930519287827261389564384963556526491028607436054765982101475881425329117108743695733015672908033737379226500913297735678407108564853547151195312742684529047334214991264610534134525285406962367677540202499229905255278357225546747814147145628106044258915021960298139146233161281038957562040267354049519108955259093553015190734839597314556249774950814858169751926757045249209640165389233112537609458475073680428436662494365929708604445291024622233800942394781594957119517939919907816057152464832328223575305493769397413848074441956881779336568142516690481254937518542783984328285204705594609771849619069503214152122104265168157381740447281867260269893952513299218162857610553340698071901202474300780967672259518438479687028281109606806099391657200382887994499430970644039334943237465041896217902408483772944777573459628765562601065539858791092244925975837818217404768564045548560228970202154828677341275512551579397114275120921571613226881795352033295070686259654020346589635062157100685381557088185652212316332398396500209725862904358373084804444469111178651229418119041494739717038255333836954864813844712737246114242684415166619443130681869801830715093364223569955785533100726262959467697979706858343888773593829365022762391411998794053575790151189125259386554431154115632910902642595124991363117570750976891932565033796382245185986624364677463584796449490954498424838396761285575047683726163404732444862612564105571193044714161158540080327862136049720032306096076216952632955895887090838942570612940909704471486349456978033699369977888465186707793610936623532960558574647300270720913195197728724930477146721839198532673306686332681706217374240756776788350207760786830320303717584935202927399404491511302421187590597932385268040426565767849660832449553735378738544769057159598051720485070464541553176047409771631878957394959528075996361751545695282124850492874738439721192159540953138321313221056397329935460905209878988234589106086607177584133907628768839812951410642902749752884276655768358021621091563946464968587191511177440417187343841108864363863158826903814384079711771339492910627442856479284764344960781053725593512680418446657669141962542274958776861662886143338359748624863255810897928696225524406440386038841014123368286932428506749561030035988773011343870154784319589781842330267917284121184808851463548407110907

104689641571588925676701571817988920660253991102227313626057975047619547314778836080639
841226025471626690315410833196795247825615532124763931203139414474930951910109266920302
220261723893894134303708749687012024895796874275907478759843266473637413381597311024024
626143305906559259383045049319543182532396963146718428865621081758689088643731922906357
205167244049890735909416197344483559674747388844608888073743562261771386390844931593071
875055339525168624615123773573078409710270714905231559505944014609270759073568008901927
670535389267351142741085424219691792059927028861138117027598563442327826206085645276626
522933988106049440056212212317955117037579117718516884219376448430531406733107637041212
661149827572450276968303190238224752238178619224189704443271898118551024856774337349491
806426789998959074054542008685719761991887817079271743645953952307862975141122147196592
022481382744080374121633072742460133093285885867278867598491687742470995643741394577236
750173254415687274281487988725484217286502105848713274809949374570119653032926485044793
421098607534080490312058791367790144645105663748530810457778317373032062635689542128854
918014928080145819389810415993388828693751481202598505298268249568226988617884904256503
939377162292448403891205509819601571511450123888807743386968279338790505456418806537847
204736413069732957168229391773960801027758241466531846376772253643976510735823424371909
200561593546250700868790068829617879725971632106435068705261242715536558713131642731378
544298653059981107095870392727812438538211919756202440295031397275850409314075615742962
539687375364344474251687611028054968203112477503720136285666664179369825094200533727764
414446278599021150332500311983187585396015257608059906268100562759107530557679665667052
664400452847767185943858573605542596080534461587687190394440035054585274839716628975104
440263615495948191476886465240547677767712307161114597426053051512738162071806029531325
160160571941111974219124175744562803771690056704282000793219501887527262947318923835667
335709545918405304576692854946627236967514969507294357499933420807622649258313051026915
877752532089744406466564840500108587697900183588988551711153148089496025372750607236474
225390048432898925109352733694650385223528130536014823621090742006594904045665989370705
314619321185429040766757719333975508487358495413856188659819617529147838755277521521584
701479061467608560682601472999431274215696871497918509556491922388987591205823089286478
647735053499146834057098097449526182668412317054628938463662265848525723986763458479563
877098682696846127438345132134323577909543362057514647633082226901694628222852735812526
628352607825678338045342617461301773907840044830322610412551590523539438096806881745104
386836869416499950014346771110006443460572095085804481150185770224060714630472862011454
590029331276782788867743700560631541942792355331913903688525036044043795990897425487436
053008629242709526155371231567483472016544330574060799988122171971963627292333243296627
077103815094363579971642219126049283795054750249506885457744705491270716875450729982789
363911322909514041506296611713971333286115458999726974236090408144933530449236785409344
131664071598865939583558642780209143866682521720712647345242608722566586152739147314518
339671958260274830225192898987472356321915391821594418019675543957424288309159945223265
542118753715309027369348893581674908887602756434318624419609277572974145052596614998722
931982290751361827419679484952976395333947664857431138058788027686997476243032886719624
861149914998708631142563170819643182257273949060075811516446197268338390826443084299422
518387677613995617977694781335391269236147721560699897064619246170025335043989851688070
085315659771541209660465970537384189651070553531275498500809742689509908584410107984915
370653771973599201784340660359524095219573897158008031265343637913878686908736843668349
6555652686590662444016891209919814463291615093934426011321220077014635032805185028013714
699463008176703193495402316894554451461470991839362736702748084177601653022612926295788
805000870129333507422720075250089328170466230263289604228540526355822973127407002241551
272976713188237535882155294085636776257896979555841994370492354319525448282708365426115
299736789039673868470753338091751675806532998497890853733679626476672768912955660976040
715386612201451779546353336594226491671431257955127037864873918269103503961830780607466
647691528731898759619852855498400460128540559513620513632350067824567516884982768342598
061839716665868498679892950895046538708732188820932228997341780456779128055874520654936
496177338904383039379775015147028927017228638719650933241264178898064527938584157927721
930882552001149291961053537716832480656576637908569445944298481762296018828863506997923
603228161194568856989553151300140732136185556555171444804337251520153878220666025550088
086381348717936052246443651709493840292259264770612347091052993164735178447544110105531
003158415419667332855450056321912056442613487478423057150659185771471244161957385671712
403873974892972272530709582167745503947640950920431813415602533002078449176675287544890
355535061220111273828587892035914220957294790639256843928502781304493448562276409364946
533890891071721435661567994915072476810031150830902441967881509263133464016633135692096
677664305106239931941303758159011628139639503454440291178958632717879322691975719763647
059915428606049851652654657972408144946552364348446491569048252282636966360876612599239
731337867657017513834245782028575209250091577535109016849818312366626051086127645355906
707477125708496149187776941612164995511552126120084147330738703421785067597645661446781
149805486158317171233959782299447872592582948305831273933432429667419473280889364708120
943262874541848420123196614958510156318344960578692838039816680257229196154306048979290
720813933586221202277207297502911628684514285476765662462998627777474453093506065438090
248484850191980999522932019150725570008333357139092834588578471567397893608850509564919
384575063710875752528897927984655567661376016920285628121233368299699798872933841915280
734807998449277274124254845706067464352566593268719743884071792061812097586831189807883
219022066118569823473805368295314476088200254028957614743235959283288934992309567197289
692679558166037224485367705857882804738191653500666102630505793775601198279993022442248
2918703340882554966281854508240723198004714239335155050151990187828327445842193598148316
523486944651211486741866533290172149700358679397838947239686948008821357159150528369290
157334843072139804141589531684188061904896651792591918437267336533157752971368376010947
235097766003312871094686842522293477110011852216497352233448272098478127546956708602237
794171271843888755952613667942300622502711642130068002711165711882133599635885113767744
417456716290124173332490952376568994891056613451142116064231156912195859595065814731217

426711113791512845439867255387502929822184897297878391529388422812424937382900311564824
108442910532414047793095368887224484774757832061798165403109841868688325798047063534537
058115249276736138962018255937978495553245515718772745315704575908585341919926601991008
457845647114248141646364378749928946060713194886677680515046051078676679684905255265800
945584926340002634655905477752729848169514639511682273440666072992287690106348934478555
197485935269427606797645514825642483022850628522851726402286655819570113379342875147275
320952281791913856132631959663297240225915082948569424421412472736995889038384207897273
641194902737373311862367409166229943051586400175052139447329059776214485300221840628133
210370533051273156071079997272338543268973107040821861658961179727878261578776453022526
718871336742156498863869154552898957299117086255706750371395022743562692344508862033321
330071857956512268058516963110766666944732128158034775044295389680097551613420814508704
503710330920780603366327977177824935278228207381571403266073134335258096981631451228539
276694133495869239397121710886582191513534435661773474036400833004440458251682236147687
961430092464041553918446263210358217806175146058832482762296856328481587800468254283593
662867486032980549220555363656302197158637322149578254264725803012858008838368152640484
752327099764450774259874937596947502651267360910598169127137223742859790397780989470757
2950027424002422339141159562933490984134356650505200171116708308053637370424263488870937
370428618174818471225596276944363381855871317607562800396919607871122502455043157230042
467851607322355644545258232062683625143773364053042142970255570818865087665776606480559
357792198920823057524398112173730807641801952457832841584404278046383916670744168989227
312493873291677277852129969979871180592895382065663232198235854255472934502124348680791
619632520993074899206020387066294338756865608355577736544629265543171950295738981219543
053648195316162685149989504624299237040848854739843537997564762833609841445113989141780
466171443644963130535234511922196423813443589934628418979764819641434014023703411790280
762882881162540721441102741608591340696340243707994968484041874462922847083924832522399
963780278686580215563015008058204405031120785865915308372888591176033651690401721081028
491146145409127913141813711954911100840460792239621937705056348125926412624494696309145
647287775535646774946093348295780906580806623525218081041057614855435370855071993846313
356012558401480328620584798803736652103988977177578128872025487586919986839877130081025
161290011137183886875507814034090436954389673872218972621734716288740261563621060499775
307700810756549214121692699052520117941464278439511985252362115689467559170171992709892
649781679599790485001133355779322365744266186946695616115003828505646823418493983081310
168267350508550862589894878218557084248530605579492832433442766792018306129191437199225
490706476646082912100476211649107836505977685745844925406847173453673436364679442385841
154212004832724175341949849746426053974123622708644720408112956859125912071349662847644
460851822505980244157644305009756368012832446111746023755486332425436793472656200901932
056709498609930523140229796645318792688234217646295704423309300509163231452719262960218
659964395606657259763761158368282017559983383553032681100281125430227180108600193980546
103311371822579060828730650181329602926684716520661884868636284416942619230184140592881
295018250306054325194319513671104542001711259251794167625136989958535035278392977191357
123563485516982469108481660508584936089488280547833794826496273113725150703584593109236
186722822940671749528973794109466038748176587080846225128761218422446392539186249996707
649219205537260588243501647534263928752042929322521888473655059466200652919737474368078
127460834853464622140831671303410945004354208727395735173813398262145026816892351913396
435433316434826216563670510603063354197273525731658678481424302461811994787098559085299
947124629883468114776006576213852162349267511238439550340255663687439467120600741213527
696558944526080214534922755569039064281533640906271879772653687900611949347950111162496
832265802194622879821545184355892274609199869648477921622044158663043067372953496037943
948629812135908008235507492569070422504011493870324447455477978994767840015298486238507
974026463918346445707057928343818399738834621476348439763708290173074500134366263129593
544075171940206426024108395238107546508289984938187678553335638484780397963204978617287
246771145767637694689978488251372578698672563252310349487061115327252812756236442729274
990855944044946035533553858245085978741330436758888623711382861516968692317626652368962
355227118337427107740390734753354574370371664039997900733300523951530502154309053278451
494248367040631187369959975502245042264035084996947254242990990467452244606510315794726
081659700197108484293305028403179675045299063006913303623162561831814189431562125478780
022693861436411187630662804262404570243372642303672015196311656226869287419038421695573
490122580143642617240582213053845262954865693241574559957276553216058197725970258517851
091247124526794155823574706758290305691789578991178660346005854519602618054533208903190
780960107557160898692974039965565954844763530986837560770815236100837037986265706665707
898506571054107285667797145072174181352316675100430705215441528491866385301255727899040
966880021534806812970349730537612125880194499582490590567537530518911500250728303689464
695695345960629377068830366192351641550823773084617577715211558713417763742746188548329
658337121900504414631069105433091946444446632199887907037566084719136436505171168042382
047107324677438677252082253667951695738229082516061360444358021179360396336732802516348
252903977129198235337358364622912921578876950504072178036969961830516953225269195227161
904389530573179073362141363068374365245917276720231062862359380908494279901053937795027
704462783413768779696700350368094721575335435929767707835078731525467939282029402835750
834805498846216532219462803735360866476596460223686898244430529302031584138191294785134
479107623075454708385793316996148523949205417545849428986386812158468929965862315452760
422812662176472595779919083482628270441871971841183319704497876776563524412355483297372
089028316754784868136491338259641103318940773528595759636090325145677020492870620512362
733181541798128663351260077053174315407450359236385187654832746952275997580180762266226
650494683867583207838590299339051546829706235307312840275994297914350220696058989218888
155050024359062843774196290997254602920515018975543483467367646736215459971695371554699
535463450583974285774574223700661787223815472654498950145997055293199335034901622932656
568665999241640280801426028306076017448874878894631921644044477382653277249001242598609
176912441950964867876490868183809733832500674776633311324570782869504879071054560423357

4963126396470670214820405595959620140157093820331101076232619781904292169866314981121627
2768774028186299930601355545241190193221498458433010968635061218093864243567581899008451
1477051236193040039601809228285059798852414796797452627639981181977561011591350930780844
6033069121228622071902770204593165350991773444857129498598131242326715177525385209073
0177425776477694584075491351422901480883368159824655014447026494905373236711323348871075
774854312504147199579446808690666290751189176779781150058282437238823065815703467844731
8535682805739518189866611878504147047058884929347854030706652886263993219161183027234481
2993993604711404709017434939360269604620792096047685677175110981234672065029992442647381
2995596224732223134154778082069134905153227866018096281969491536792549636947593243089717
9869510779553840985588397399594658776336556564165895159080463746547016958145986255112624
3543681305801488587378250230682382193319863216313649118445966627150402816462723843524045
7657719097991417776673265046352782135572266218334686664404385907861465646794672429294569
0669007024634879171889960358254102648690882625406765423828980644160167624864333950868270
9702552689258437243343483638787113398475059434686174575681380165666934176904441641623593
5954163882459384260889673237369245721070884419763174469581846652324454774161602977180060
8384085813036046269948977131821949384911271721101910309907903082073798684916654940128704
8762971565003750032829130662372042691268168364544966688821596208557862450300558144808494
3810093991186071912044123376869507296128070831635294511736001272439099144672619201285565
3841979311135572466666227376640336137378362310906472439949267203395460099655420133007374
7539360236131529913180260467334764060983455282903694153801373487184381551673676730306510
4180731646382664172296257383891567723049314454530704241280808779640228080070053024979392
6961946421950572256752030496598173141602448616887723961056060389463301554347191524171443
3172307031525714505685290789583211954643162647950819230419231721396534888635607655626995
9623681053451891074127629756183833657715057315166528965547822854784797391811266492705849
6787960289852164501284033573552781260836593113932082856827196331781718915076360705532089
4469174659708682192717520165740676767352210549545486965855962709700377468507280110894092
0463402783150588629553325625120858952613783484155519477968952435415994962436279256904545
6860321868130605850090586899292358153689394439114270681380767752084830084030226749338769
9648738948633819271854795486140361351567437924141441807230917566341306083390646446773104
5371772244604046317761588018754564732643450822672444809503223641407493261415823032533460
5612328514221630945848919550396583258524374367352028417004107611979585328274770025973411
1571908446889321021025696070396529422071459119602618719587532331985236781640987570989636
4820778032580875913613064906185499962402436578070397852177249489727318701472242173953776
2303948241587884051227122006868779583165688013183110767029127373314224360882335229499676
9329046897686346950298563325127513514895963272874346727636262291226191342597089724022794
7704307792449874812886588505220397258738173679330634010059799790588440704861653763557345
5029500264176645479768413173368953846100894223542675857049777469000292277361983721489072
9486781994629751760265320580806466862080402522475456550633169486892169969201948511031987
3550833921399590839914919149796750801149021424193596846646823659263740085576346807930711
8870860752766746714666656537936261799087140775329865860645954615580867105809288815594004
1679639124502167098464807441128261556615169481967364860654635780082366785714564773364540
7266154887458917670557337154860284750328831832013771331036253464832297905721554143529144
9331809446683374882184676702297756121271714377677296388744664843268534525003619942088232
3477983892396817444392822658315338504605746819339659513949412207947159635461899824866193
2487297189737765064536718207849303260515112131029931321984701449470588327698925358715642
6731546473276343158242854293476183402999872006968183297957013622141523103658285932039300
9079549651699151694357894196009783759873571070721500806777614833174962003996259573804490
6255303664353446375387404741418059446612582834427819158597370661257610642881541722951790
1581910468811817687396736292802026564745077389393288963466752825049624853719210261230061
4439254902617772848360843871507713176322670182635093643926532110432624617388295232236451
7235769488481331162944759747631036267645212674062951973783101125751356581230654579663046
3707078643808129411622890758604261833929656066567814585149459019520999794688276940807340
7201091891362935991778087895771898275466833052733805741116437934792696605697017939612455
2021876607004407087177624539593353885136524575425499802201575672819079891973943195016475
9892981146484369876469306873280815930947510531322621557401710254540172541071498451078403
2484501840787756786230982050740796048310060099112149917153845651568393692629486174825156
1809584620638024325406394483473248403744646675727080594500599689016486118040029789517430
8401280415381589791268419869493694076285704600493881723961915142910152080558501034125290
0687194864407736998818274203662116845992379448686943459592832910221633401463275729904668
7805920540812932149983731750029229610028470389673026029169916124787853019958440841470820
1344975316440700284499140237862713816609706489148191363432068317843256675199366275527900
9739236122417485997272416804712399266515881759346484711783095336348618254412371186554431
1715259983199284679613255746103741088297766872517730611666982833547267021423206645878982
1324476392412793287491249072592245789916409614625633047042676550913711685479931925563424
7916773336999745855769973151641535545899708519537127636250404003325704054582227116217060
0234538895165580509881462760352558518485082439588682597331011245847639975221897368755295
5670502440780415182211288252021280407352736318884823756725875599772679743189065476538464
0449155146096273131309514630077589257052099238357185422944580775332928443738234130075675
2243532926188431935997958259825140153159563044913999916774900478842211798612901551770811
7103373400764409291831094511792272995539215351492898411620008985477463414885797842344169
7067758272588053008636377120702729320365052646342791826099160856254155223054987450965307
5395633506461368101982203474962229285008861697772797962549658135239729724036735895410314
7641372440160965605642570930346079904973957233459256289574649278899090121289302759264623
4780959956060442676039078058526497035308302656692073027517368388208951737650786376298996
7985249945098584722576995571028137421247644846239771326219210502946662580996042931071270
8560553806844495480894966497204232067851818710378638583964189641995223572217180715732296
8292530135635195039796706263930868236623481465374362820553025883984712607383465611429321
47027603131489227

125616064722024226133605738817093276196485192501250825267923053064252218706495452689935
046030903351051885438437590414528281079547531437369149752640555707898926216287182626758

280048110345181379760353089754618642540423741109057163866363952830697096549504388639489
092265330983668390221712286170318704150170408581327150830021530138782021947725111388283
433025790403643365389424340094036556798932608187354541779616338641572570773675428489064
9721985928333280165260070549372928538352260128793809911092110163333232814421675129569467
328127665319810351720726242426870114697878327393643392756456477532551580751314612547861
152068913521837203027436217068568145694582153752866891798204022345324641125473297325171
116832048687086693589575098103282465113816506108870958870521962691807397039529045615713
585099447231847584374159537634209368009830752236397328637588752713545975447724260231887
962877638230654213380863796225631603046054233672588687072684276388049898318551890830672
412506283174151115825426234733146591604602398992697599642656920780249965676671214510708
930525515885124271282352187839893047523316765733154178342468663493721318017123531219901
140912637864920298624853889848844790288003877055459331047320991907490787626945795496196
286629101253342445896842655448291237139174287735182709426773351588183822398594240622198
110192146550961155322856585959642074615954184080348741624229656734920096872981606727546
516981526472240580023652551141497293431278893114540413932619066523484107642685454789460
284128295577840619838420531812335046661955184944171774434561522050161111967466602724084
519939567635190921123382970568321567853626774213388141485358752083342084444492499075265
474361118912348413874572421020316426175410823842338798611824365365863221191972080402260
229679702329564632391941406240098943249159346280764180402734694102121002100038853585070
425558011478145004232747312996563083940531862415745973550236724626609607603621928339799
532628415127585265996070670086997967145770814154909707545493635547251650419191981755645
103135320870066344880229745178612437801264625250269703717494883942997924082422828884219
571472384199759306019356744020049329249227390601629590390627350419362289386985587396438
474804140501329274210878758372371017862345090143362153921285189278600247906904167232550
679540095334113432355109261085822697366236774713773475186265743284360754556712986881351
131950419147179035563367792699438194043867180426810250574884118901922908891579185294235
653940674399642667018676625170148783127026118667283700281977750215858507187403989345700
139340457914113038894724433627826822086031951658644513479154297221074701573699555742381
376505994286891417412928305074006816575605488268860785410252984387274572141381200013715
448051063762334575396960481236001821179361586443380129259266059091945922209716276448376
811694253809423295180296147204862343042812488557921128765332661861575588623758914976635
9762626713179992879855782279533707776744671389089788854126711610112477956955001893571644
511840044577805108234218049933900742534402771812005699624301521655667605025338235914856
490569448111763417690613942079098248472407973252849005479302477122694387162530574720739
843156945453828516231556051361247930254494253286660746275457889448319872275700565248787
905582144791866966451761443514120960493856298541264102307852161157289202149210708500794
754908002951166044588805896279685671032484245667571509508738352033525660217780495948403
915555302546514654866091308705238859631576260935158582606848173384648024366025101947978
459720062552594800543785033878640075983455968386084481397923877498037624741958926542580
736896210097055812978071089379002482533861894017357747837869026233527705520115956800562
662307031041972029747315914741263949395688402982176400215511107707933358030743433320971
956547621273806261219556650697800271309133777623453325092406825550599066490778751753866
228503839850126593208723527059140915410440925082615878321765871085234857479670871609726
657855818042083184177594826789624695246105789577776916577351824088540666921567783688733
667765464379111528095688221559794726980766969326419490886816355988924853662946529777578
217787259112280999347614330368284009606059485584383181585628458018960811114901565516729
114190753115370055506568379897594350887270824531433704009228218349204285376397456114911
436188605057225799808863966773008259004639917175556434964924499553256534350427293192847
402561603618614118786437228772181567098647066414276290424787381728381010606536627823593
628732401563714852487717165725030483587850950867259587848617759771007643829134764486960
274269682957023792564632842409256513922627921519426886978662229776723278286232342844045
424294846402053391766241774210462386243525357124558279948408821351492219627767513898399
234552509630833296797257100746474561577711561951383855468380171673929590348268564858855
550301700345613127381144551620481429929578030537220586718630322647863463127246233122798
329629080695171224760184500408213175473331680673322004044775523699056028458188908477278
635437874826664430099706735157138404624610474020146352886363664543505608717849905327637
192574606724423517435028172147345798240684081850689491532553320317616579234939005085101
254864620207888986996230288130542424230457942726301274554491629099205677008170406779037
792257749345819058148862688883721513957054710900028839160040873311174698766734346901574
228839870057765718746619997630402868171820434414543086063671428337717717861214365279564
512662920527875035692788734770916031960016089458149508076443468326817611556387717017069
611735779401272923399738377506673501566495744557602546562135607273956945673682196253870
047672690407481192776534284776315796486882508315861961491805694818348758664924326276833
255766614687449054024866182177359728980193391310486293444573461902254737211650877691406
083378720566433354107677367386269435875247792631999103429316639902060797788321643539837
653642225973510476126655057755196615133814722785683561020471896982941295367776482823433
058257465820111403474139476383338288139913401475790898350420125447852898292762965360400
804833062325779554115998681290634903050202695735813896460437838765308554977882299340427
400977258630351996526788717910614882029306029704873614509577656885577017205817099116892
420040413956656846696073167885563729634528986393613751530810126066106921542040482355330
671717524759007499462069713501987117182283776480664183962876031896207550616364113281900
979000417973770467169147912096509337534451561362455512853816947158860289925396954090673
828205994407096406623007013258521528736165970227093680236761125540594189599970701765307
983039762137086016877603529670714611988487691534259038351543015968389222929325558282232
140626366733240798799278601811530440294114321580335458400397847113101565225739983207117
4278350959999604658979471882093454532480357285956192818162532173821557134503087516652589
674490327745448706192109565777430681864520914739791855504372992928341281161890215717102
355807381018643830542649806930037957979565139287925716006008898243874290090075977912049

3270756544737899881706492099745906860554101160949763061684915868830773020518377644321111
1225077225096208168767864780069218312496076631113740378697668831242328928006244191764530
0109692585015637001935001609692575254098146585034452216304537574457775078223368192847784
8913893302795801634662959442922911476922947717613676627936308981060162044726989607195941
4989461522192875858817136272438373049544773875943317482876962005024545073025578080093
8739768010584624279759233361163314839824082255539410993331794343866384193072825100899574
2820245067402472714962831292594582387529382216103258051718600261196486299894947183244145
2754889340471259303318874883228947169843359464368810260938209931597423626598698399700
8158734796917613435793790947658045866677570212254525709901879945031976781187442296803127
2400116661763784695335543202836438602897464420207109155234024788164125966721348185183417
5257875196232075767338616792191940037060021569009318871129855539691471706342554008240
7612725242630056119699720743195258356995584513207381788158904296557760841438240940161197
7236645616504760396839698674223448551155672202658351458503592003855354129950786037261423
0485657862103182251019321933461223211946413612002881215977363325627922700208000662493
2893476904231421385788666031606211765437585371346946933020945368144704415938689812201126
4300755694206177402745594959573332010949474138442959660961122441924534670808854655776440
6176274545719140990651525039600060111873730412184289331617030902501555825425118996388
5255125949216723300112229691336914449767033221831135498319638998446725794514055066077351
7306128901950894349345062266429772322281927848150128261789044480122432045781739547504194
6322549290061415756758829029645681041166828458009298190285497230522681273842052805080
6003378441921234824806599619836002532060720206352220207458883794320578606225962945844436
9614008713334554017620777018936353409967324456260774798347395281094674011526883036196519
7066511276315046224875229603006482990427959135785498246838992880327367567191769876072
5654138663788914289629183490890056102047254028998968417409420783400589635669223666144936
4683897129467965372145909331205662320016816796775223444924404107747450651175646538281883
6096297983814918036892696819535636695565675934135007800620937061643821982849291860509
0901053491752101374761720258561993376242055543650765653765409808656270283434978476820021
1135537253542412715312165481232616936308040089111352072961414933973817965122523221287079
1695605191063654012698767414958089400937969603963715569057239407800498703632315756928
0964564641114192885472977064751322904183532448498659877783387570466015427967455187279558
1789746732823143022592793596197496455868020088701823485545589626644107871420842767813584
8623193674573762870025499292719669348572254726764937209569905959300232205836502730241
4833614184496472076869108267698013172731177268295469838223370591022133584965128047518452
3821152816141360760356606666311968546440038790889362521539298997966022070036298101726014
3550604066829414100109847117273602936947359841945360144415210653594478502897896615749
7502900982989189806750039061835005304695900350012049745864292183817821660552628290534696
7781285393337293335335627966916520020303385566541041032559344997425010346515149317875230
2217995884209799474625735594046915907916293562027788740536606555720665628743016534405
6695894233966220986186362151818366266207384995764893929169972256061546841945800867922019
8783254364724398999326492320615017853734761111514929201067492991171340464779680517469850
0276761468444915506249098781935876745900515230298972171113760564955664324990007871231900
7616259493415304383132131889024336889847667733324496322202102578488422895330821074811
9098739323803775160860459965685116807314237730983177143263691075229024752484980662217064
1522487382526195891797391741182899508950365696348821608433289901235121254671733858472534
7252444349653766122467740217292634562269263912186332242775650607723450075355142993553
0677645391136959253099128863339086151350903108960993995687995183547544177519527622938720
7308566470061242365636498770263681185568655076835342424242908757343654292680862050020371
3110791576601972094670607436003163718792509118247371611528852097251770830856627588777
2814691994941056833011254379600982330667419795412005531863591373938858104550490546489356
1314346917979971959084632590663776585579000340227876994380921270828019263255664354880871
6670581965292555103320961310701100833037915068429714907454585167081203052972260480083
1294697550647980353447200708005536474068146740200521592344732821790224130879644569490464
6221862140950785151737359216078488066010294334021006468802491893157400329152738575427866
9739360881707850477382916979765814791572050262221469847866531207593066013531427715619249
4412527223385973779847659319201028806368161542997066065467378751192873748579917717125
2584416607292129370717956907786490484192367256420254816506097986933944658238128632731470
7742191329850053439823660821453838685030278355120428443931929872750833270387634056667154
0239953894198586960689334286551935258811342411145859960219032005076961600778259963336
5581871490968655483513494630038688139943037802414422058988678146492384170128428736139098
2033572065680575012600188947698807186479713593347462837398824109939487593302898158769727
4634146917462813365042570029818776406740858921022851633761695850356304419197838556647
5205111978033797576914317890315052588470656702639144450138416749246825967180697610347479
1072466688724031680185809651717440390009717996977781040833494620211735514755697909270102
2403895707442863595942351075904088828510738648216431985807949514105022486228384139368
8370744584795518815539795346860631568903360317935757995258399174217690513437630784915950
2832464335141091764072725561467023714950921015848925656479905273076075873082324104460432
3880761960336822980674168155654247206279023165111986735838091314856182490284447587967
7772704924636434659545674164314720400925557649972001159797133724120155123650948201016074
3851729744409908589193118420364047039217693446772262540537711045121463221233747762909022
7372940858945407977163088578753670159932471180781815966942330657254454458681354016822
7152162196728530843041352591505106301174181712511917279506892318499229312715605745300871
5035211563137383226297731808481753298397484028618690765311881078168742092330633485588684
4304364142876507053643064507289470485182661860837087866217115627810226432428465390898
6485559001402426778025056128642357392220194474966969631607457360806363767952786533858897
8654620422784624557185618808306518391896706370159498793456023605819436052547048028725598
0827265222778700021971969603590998659649784177081212734146355768836008589440190250279
3332698383192931730575865294454212494471159356606799837759257438088868397528932929618985
965348977387754180445629978086982819917509637949679399283575856066240975048959751541530

4650419239260641451073950155568810807505511389575488597915018598376119471679086464218 70
0156882997462168840723586092314154418712000235307132104804941612969671806406534569262 48
8318121289486414623835237263714163905438437795622729569619465518505676250474592135024 63
7390711317364944814028702743812838801434959321569989442202356240107813472944880780926 28
3573914716650507620953154244249370676993966393958140038356968746289722104539153635871 46
5598859290932477996008810287757589829615045606168441814186445979869306587687748288674 07
4216086008883427670396775265843933546698458723530635453526252305870185787909861094493 83
8181002960982744431069721174140235672364952651954454462759169997787332024695894757226 01
9296541482637652010170260110777927465533898395248682278166786482030272992607891607283 70
6367225391172430382409538972221092080231274299866783837891943863047647854299835914323 07
8705288799820144572547030342022206059054289507787219816751837709459927404897538681310 43
5518908037597343384806485717119114054478626660349085716872882606256248466705407364021 07
1512439671971138468554137576880992902898887506571583554486393023364108528154081687243 70
8723496892951587550929581633576439422093662415564660661233250255266809185696742947451 11
5636597676146144084663536365521707077353807223305999465475308329952657647405830291884 38
0210786882081676274879636031442623040964378712506608046717057284683139283064568380267 36
4382858110051208243752683727565482603942207978438064310507688098066139907378702198845 72
0041373231318123580812284488796174936830604752436694333560703340437021689456215327692 26
5642141337924193939685931899862160063929882187290107579985040960630828731591546929364 15
7717236511817911099565168849942668500836196599010115886726754496802230386768611630762 38
4232304396873310601726195264167343021374402435338473848234184122876254840383600422046 04
3663010396693426363453335931103648563632571773781258074722765115851596550815971653489 79
3491862588328415461863207745348124492643304446594733732249502243143961233438505154080 00
6745129571188929009303230285954083751575065208307112457821522610853950893389764964234 56
7128278293253006708080395196803763798807734005253357056341659568697864598511633404097 52
9402674081546104118206539153234229490470669841251858360645030903477786741428618638166 66
2353463401272070041420132737661790508227440840076284986518585186456202641481064258416 27
2962732066138739855611611733101052097800742069586666342112949058527216198489999305571 61
9102708254761706783154497160651232777247384106882188616239792997772252865933559445636 69
1199191713135940453270285344376468109261681354047990087841787530272919118193007899033 02
2626306246513884020010329895637478588222296897446676261964416397372057235988573760419 52
6143235845542605714076748104291191813058524425946045776063728157341485487715247741007 33
0184117628178831407551231310509753703820815713451487337944635449886600613156555933251 76
8620143063771919290804442260537537562508324791812072410634288709870444666104320091855 46
5455112431399084257721976064306474260962572367950547187344740818985633199309289486748 00
4976809314164130817613647638662035527809790246930174781241260080910342808929783526495 05
0147181689824164408302697926171343533172914107222322548518146465731212494381529975925 11
7180120425864820643966972475836981915466640376990288658372165612138933978178633613909 52
7072319360457953093858465818902303887197664824693330465190345302954250631011191095404 66
3931061394503230906021984273163974022220245113347559542194335805925533968052917118119 06
3961438265986448343577898370978814281316590749542239998140930795109953757918271919161 71
4627538203813617893847909043225855552725666733452181459106971099070157136723366038675 25
5506282715190532826245509499373612592022984634908314872119933694994108492530274846471 71
6678246143766171499432116259344782503266914617176671984093614647060835288347825482348 90
9322418076755475155501210975497020935432219472739538195388602772511727139532924830503 69
7414219306745331995669724270557705947652044820885444564287592740909692989666580263507 68
8049237418162344125209874492006397872881798905564462824098112185996330786927815087876 23
1110919011931736690244096389040306328931287294915418136821992164600706351814030873560 92
3513430459770206822378918968077196860161413742797823620409392408498689355791182795575 16
1771418004980635895576226384815113933856541275229311396169034114823385261229113039864 31
1980381169286367655314661824115406364930352205727672220178787707092795900153042176147 20
0295377585866758865357646809504936267207172325337622063161741819062930106522029861538 32
4511289450334806753098511973310735521532157769243474775501882340973014530728921529813 64
5427199985900370480781548516051084831457753662321879864513179429040319341125584609020 68
8762017033389711530708132728138616179073409226446443259584213401013081856095028990240 72
8657327509804576201503525323980957329008782910238085969250352947838676987979342664039 08
0880658340343346263586029854136334265725052791075107323133810671291618223022033745479 14
4041615831603945656325628748507091571346065319052111970799104754140111541082996647854 85
2901137707777216418907235455801677579706380025983026122487455741593936021696850385580 61
6059065014853828239351321785552864037105541434010410327328147357951271945189322294816 90
7040708131464401815210166175807951977104443098460126801497488322897460319558220847050 21
1169717137964309667376449311013992010646146219319860257511718902724494064758835247675 20
5560952063494051732527266260292314237002070520703272734602329654336338589814109728168 47
3775527587615844198161862262514159310826346043156411156250846966691961677162015233014 94
0467355388626111255850646885894269678170123941040863474123151478021972282168089222298 93
7308074168567731396798800452045563885662699488462075115791112880847342819195236146694 48
7076546188155697240599421573778388868520436783524895861211503165301139309308946636115 88
1570782780169670656510450238772125182295659514980293585257683255749954139169893819916 5
1377981616424611634643043017841013535708404345850744012388007941787612072597755632384 65
8070007174244720197408431363501496803818548078774682276410774691665213514511669894092 25
4555725163815805454407943770799924628378346001053621407739272387177568831644886172076 86
1824493405707524617414090407769581465124527032598098192974147985454722385939901583979 2
4195171209424403734799244796339234890533739667700491851538129733255143812224593160291 41
9902866173496435439292778861330247860161916172326659772013411775966410308090591702824 45
3308248140758232888141005389570537478011140257190908818562593810851267015218472074456 51
5847174735165490542537147259510094694947511823951134313060065662597940813936980809065 82
4101152775395466634755708457566258789649560812931484052176637560426638881831184014148 18
2263133312354460049537204056639838160733322957004230514457860183128690926529676938061 31

pi to two million places

5193205980314920139309456993156085283269505268882798555026164631022649121812093415660042
3318418122927753678881050877423532522551249592100637837771075230923805245375605513882 66
2261646011738111877990938625653868760938608513038781286267892945562209973676551975876 56
5764336551829716649359637567214476044911134930980810316872054729234454684998918919651 15
9637657662434185849837314945829073552362683003905450789109864531576128217556653664171 94
5766309754056633786536958653114905842430798262139182449449689318085835846229952765912 27
4509912633440355587891607916745887175875342951008779033129702947237755547217420722027 90
5841881380622930923341189551703487361552571130234287560942228646233014874742165914414 39
3471517571414651177211517253346558771486785844310114720285278746469408395857204616039 51
8943183021452786039015734359285968508506026795458183721493426138847083019096513244345 47
8708759951540671476483828639196354127839036489459790734366541983242984850446644904985 19
1981702525041496201272396139986920613481396860782044583589681355355923416820462840670 36
0015359683549579786465703856943134963340082922017830211714657178203434651435377135696 57
8358525500791505358327219633980501826127819161157938339696846137733981237025498772710 44
0070416712524199288098181352342684998075764080113203559668985547294321307241401823164 07
8214558734929830872763727953297475249859328373236103761137839458174460863868211418076 94
5520498505840717552376271303121233005602619219661088335134450172590232447037310784931 06
3661624305614177235608919075779596241446085882693456324453206681255193495692372994800 852
8786250881174870968437449942253088362812622361498363257786573530914660039936911693443 13
2848036746965286429075005513148447625496238303159969782189069935919675325804759796406 39
6575717653114888085349240942794608367860540291227478274501448372841523661083178109619 76
6189240292975579085810905581770050754194908177874320945987945362058511429891424790011 61
1120470624323885919649593717133987620472592365923643008829511525450206918055446082702 29
2776965913599879804341519462994333045520867775571283983907912518310111088360744191996 32
8250802011681708031283082641515638300208917045272375956661785301812602020400793292190 75
9062475412417330476764972336969814590995049365339721834161738146760003286586043473905 06
5149533075629371936886859582797564294240260452775622665476550272302212281549460463202 43
7829094681922599082040785293466043760702203962742780879372743429205306565333422348056 35
6912911152305396790551243165863658227925173786938700354062711310633982565006248314400 26
8237509509955251944986588246555408356418884829725881126238105449401901221011293560669 11
6057494622731092119136905653284541287621160135340397831352437183925861436429980512368 42
1582515270186441008754484889820880548405493951090921811405095899894818896529333925423 57
6963115191017513000521222727464782175211305349141263091415481062938757076567372897247 46
8313321850649869362652765826373123960229707594879302822280769647712486184280605076780 03
5570895916198041784347186253142986052207654480261954205167143955962435916946031030188 25
2614681712871922817151698476209982250392498125372116295124684405275033775757162599544 70
7056288068570547708424720662689657867701946164445254379189209778944000633632974034229 49
8059214365185523167830554490434635837445299804901379085409388291832009213588885440657 09
0835595647053476427101064473890730911090337128133567183794808517131930613671125475311 18
6760804805419727159286597928465148800498102298261104278077874100039001536306127867185 06
4593407335576219695740174363895318444094104637979910568040734742213650819456582638471 70
2967448926318244348320848361946655413765253816409948917261929451465492883057604164672 33
1085507267588432514081443833695933668848133389250775822206486950220386277357372271696 76
7968180899313366648775664042738072995079134665196038607149877996353598164835466638714 46
5383541274436613755166930264389632785285748960243711718271830926969613362289231382005 13
1112423231905428654081706467242922987899817033188859851315872969714408519760842522837 44
2621391230704737249490914196860543155000271034055564264806009919359585878278260919290 02
6993915455552179880772875628041188187828747110413378339666175059240245932956226324886 20
6880576661759597630201942207946660405844810429819926475478582327954544245159174006433 87
6608930231957726062961837798498587271559040784241324139200519730307429991546764582355 63
8380410139452815177823438655839570372319298828846201688700444172924570084450991321986 88
1948881079106439645054193634794106609805671055621570408146231562932784417539439164207 32
1951005250625828608165685310186994814358312403955911005374764114111486836637842436468 01
5182121770015292632041629983229229050537282405615290885096566439555325391641181349075 511
0718326611153534066871200634511884149040946480613450730557450221504085146672302551687 51
1402417519361553161566770808069304424822525428431367554837782381866890133177443875775 56
8326171185433615889807241636918893591395011809286640856368447518927119602117326757563 47
3790641848188853522073389794830700387813305651532866484187736963872917603081568907589 760
9991501521611385406820869067452344252319154951131209842432913734806557690218177793184 36
5557334164148450884014165169896300659891564660662947443347903713333799655588288562542 384
5945227111084705747678099169210213402527678437440093026378855751501979534201425569229 53
0228143323085587786673918434760876530612762486975586160531181926168442789338060554751 53
2848394320375091404817496529347447588830572875798056739411009374944946501934088458137 13
8379720018191515280934357082085350696477584928513263232856818191670338046865030260453 40
4882790708226677055956587285615989703003859286663738292924787412578496206170861151284 70
6051271231382215519535402117573946872590506768764240929579871627300739957495269777332 85
8355044988917419865709837159794706282047925318908866804380746566348548118085793045975 68
8383760760345557771265820309495984000637232007988133958614347841561478811040581477128 96
1836945767010471606665503455407434680575553587277968309473336304386520846729665043171 506
4667330723047932135980445076493799452422399633469053791592935451584013506732255591874 33
3710104937574985497149779829645131586899045579278194429627047165221755861500520033073 73
9611653434851630666435153950400618018026583182760755014620791063741382890751808192065 43
7770924462827162808639610329932435176594936748582831324731567156948223663370522288303 48
1305229524079949517151466042018547955649083117632357777564224819318326046963523619321 74
1217679541792082580004409661205558899284446275987422787719049018728920476941379526892 92
1295132283588880913284867187502794259733541903816276833085681535580700873228699314290 71
7438353241749775484012833567088647875134503557818579713561839547570470821706236889699 37
8195021955165917761551636177014329285196973969146867189179848417931339610649783108310 38

pi to two million places

146695403464203839361418250871279727286800630793169525769221343775280974538621719729743
746562018652776997749065854721495227182277833735597298960582872526500056285203957410860
939611954926801747873049007825251874569853475729323706044760942637852356810816990852332
664634240824832529858666183093496929825407229902479175080600657536382782678683934534124
038694054868859378542882911230898169313921072028458484415579636634013166050962968943942
514108334138841556734006020347702469481326437878698029378723112930488282972515072383181
312481949757097519629975737946123919675752519916649691951293335068597271000318820379882
996401478668778436322329712624042396361088070576381306484018994811806502065533633794649
610731003408978019471810864486092887499024411307303098078524313570457392148227632809380
243440996359837265618337774147436331156733710277358342041903488460172842329278868391880
928118476792922454257683413150134660074585447054496861693850200752702279701921159712758
08890040821871106337106188231093271705430602689336802671307248676277986222461782535747
438873286079630222925097939466714300885741903821843526108065126523951575072700445385440
001280149485562660425505179546467182622396903449879154092382980859589715700978865201310
554829477770405686561408546889619363041462939802963281363686598467939206102866136053094
779859784528192073491956173430212407713937202077430440230322379592019109409061039256944
754401813208276880099180668245433535669164585225857287527404977934011749890140224266558
984973836632121328615811693731790829287606218773799304218390689236595880319110491350380
817702015825008337944244766880462202558972593848059455297675687074656978750737887456859
332784961728948060924629870554336428488931303931668018597616742639275388797879008488555
474508234643641765948586260827947404835019979874102381685739644507377881433663199973721
611784377981486670922404192613518256912499784307833843336036781277538375840462457354359
280057121393600767391368947760360468921382616507106343716676042172983942862581473262010
241371867811288229439558632130383880214217402565025691273087261437150824671053057325951
504705047942006444356504490333726441704870653611563810282274338978321361710873526668310
844440893697507435791397214359022378304106607045352699109426745604039470468179481479931
475929890964660417522978365744577623366624196507449722241972016952613214124967755159910
101093084168653528878477184825089184696415540803716122103104810457598596863423173044573
361778381660389545032990025512176896241351625761428778745774430861034688745969565555979
117779817112540920998753872966668150608006663513791502641219479764865201654716257185247
310803652549777716211824816868088607980869504210201956090011033591919401893598031318665
802857425972697295495854584957327986657124309906421930815448479841602557488992816159893
039216031949213985292433183180719844180368301297494850077670264765730271628977047181656
042966205512739091585254354393576890621508209652416441129575275962527881121822076057309
867428719287783120138143779853846683153654469925528434752897621863400953779907576069853
845161839054158273680767800363185811617773525395983492056719679589107009725000272100712
464501568679793299769282342148042436925307107955226391386005612143042691197632949570109
073597981550965867148751416639161215768976599730149380552803854602899629676748013889021
066840865856267090237304700913886124710614279486722018446633052465396255413391367214105
921529021590728843408116358760114282859398993272905402305752956161228164040034201717968
893877032816889611661313348568612213709761735668552873795049078269118972689640533504280
324075276683728140291470015310249078955452802420633185920128947636842533510851855950457
623286570643659621222154368167340661209694809199454367418585645154928896222886234256209
946092986695775103764733789543727749977109854155435814404712661430570860991901614538766
763600432559807303403977601305253537553191487933834678893737986171190434968437925156640
718698697923709006767503773489510392802146599415591046047182336866004999323532470594805
767997628723417941111097048889759693711069440943913420072929639092797693000966410771935
882193583018262906629080908475887359945257699708924314372988820491590985026543094150187
647948599141745763987700321798307898418761322072054121977819794145985309326768701384208
047514789216191433975358985443558662562179256137537595075323197313412221784791405951423
270790060969801220845121412873436073429567484581257564870111862611886747832409206495499
453511978073370983367275421691152842160312236537131014181196861391137630829646523245208
201656125750341916283457829945364259131418873374025593801027847825122022947722164962274
465744601486786413607939447788346490625532439252840714967430705192036611084175945524956
467637634656791979465787050059606142110050444305332784607679242238844255238855009519831
873613931191323022152464305414480081975578210085620707480035493290797500628161749357483
172401538282400818293675879202471058650428855952805344648449650474955401405849283633361
020579463790504915538771831205502968884047752673323630093745368471014844795614263391649
090670852485116538460720890002360634415851584381096763121064017653344951993484666599537
365563443137181063188398673730922774852342535604365363776957658644540131329564305576234
308762909793138294981134807185988056095006655760957758822508966604056125437199582538199
281727158009338960198047465044202743963100533404358539938874690754736933705032914155147
421763008086305022480963794691588080487153850649418842716240384864459649961026056618082
191565452229895341226441022520833905121009667886849198655692647023127110384945326517591
546224709042542334823478153627482344132511620736561487867764554985686701170314189201632
813031578247030071922739017441218502495910092032706984416166882445312626392390770609186
466049733461200014568810896236871297974704311634616687177208321104334448037841094671927
003925642957131640642781149639403774288433054565625086242917930756690806312684726388926
319664590878806537468538253681846469181887492240120186539069316933508335631741332213881
986071115403387253771477999286146903622949005112015639030634781774644245278654455868506
827667897490872717903501430926207739382120321955900787717260630326862586918338382561252
350989385034059625788762497521780028341996393065260212022838353179166088281734438031353
118234601525728302019490414465671389348698831192929245041265426633287228272943469543341
006166888539036079960782828232639453101853255156709278505161579356461326084214660850722
233342842750916176912038028815505626140431087476305929152038512141159536364008973376961
737552171981370574195952511540391011363215302902532681068748747046905501767593814331473
524885552332844457629209273454621344422823726342573370307726344793085074186737561216913
362258968541554387452867250121039861075149980739200130324150084668381015889723121719690

7808135170241526112353510199074163636932809780208930650030724506906281989207352463347411
5455976977556153468405530149782209387050136163214463506244633941567729411517305524405988
3105453015438590009201801000969919243987250657018412953056928532540504427390481555096722
7530945982647599190305322245158088792567066807555744163980844954785452804726675726001333
1098737831136313058048558388413429755264500353660696627583226066397568809132209043010055
6826685486966497864430092914198749301734661070488077370783615215442454833414866671324911
1512135715180526794308106611146815545078748813219198110804914007144203582425301062683911
3816090446130333957429245929667527753440324421477779212740686818015797876513084618216611
9234731007962412381305822477439323701061401074029375397631372192983732739973124106345211
6462420537632100349139903391099115555367662675587915441048480209924696823919055551645911
1766619682614836461023841932856911150341819784996748359924795588337192756711485308686622
0433186223453208268952529597451281084908052402578098145297108667022058105726998801130577
1883901503623299611746246778926796919045071352416329610410032692473804997109456903040488
1549037048176849984768976878833539621093319114868067246927279509014389462142951055679722
9437792447820808655733255430046281137794148076308795024428214588686151369437912745225888
6186106165591400854410767860711814762201119584185086614161968280422500081607299958384622
4406051034359146307297254814914336004579411874526397197108227863224563119808924728583788
0214545771898655519194751948676041939715822013611772649752302934326195278327977027822944
0840616957723627847858488676007331748266404246175833290916856367219731091530011969529855
5118464174446631537550488272985692006772001510241922993628441630737344696583439801776133
2966146764145801322203393410039503879900964508682395664793061036502620466065862293626655
4922436446601190868345238918721109649957256456430221204417437942552584465955640987420400
3849236861028095428912640667166241202863338081879702531556009577558590890090590759583811
3827872621037888118266388867649707447076711594786507325084034068354900166588251796408011
8109544589346100062346971503244368154755056235270030673821441434229767807852038074796444
1095667291144702034354321591344274619431690935995065739921703778892507963778890722428411
2194254674573152097136328982051916385748632650178868161531583150481889876789355468535911
9064692462454990038261222351894239755935126948581813690814878676169345592071574025082744
4722919016232015316965925964375975264386225681169486649070455103736131612024771127524288
5174200875487490127959564767322268499643172635974952631161164135491679461630219914050700
9647436502830843725380000399076933776486959396311686558646357635940277856913226849513444
8726522218097823434680963330322648195460890091398677364626520117469019994649748166866299
8629286980525811044761373413445452164532061915200728872476253907338289970558444956694199
7516750097715126303808798446624666986420463600283708796813603369733482124336947504162822
2767612802533064719318808977672577110461948603265818828725062101893822062851319155672599
4306057251647363597167596545784581967915766494193960341355380669570181656539774391675588
2675723875888098446686637372496572738942116315988248675264705342591453342349797717030011
0656955723970907144836074546662614197277287666791292457960835088053032301177818649270177
1114633433904337818010529040193231121779319589232425126148547743466181792599722339560088
7996549496273484955479950175364021388591018690770019729249166995545259640505918532265022
8025740090992171264418903542377458615287853080103153995682839943068032632475283657373688
3529166127214698682889047176801040011515229896104037351332374511596828435679052916010433
6357551716267148660909414060068575386769712609349020718833483814311091457887021153500222
6383444282702684740002344777034181272687118225786430965007751806768314196014801248618399
5131629348565454669820075540031033037786603468909934356702321077887259841627762254471222
6527425427781663680312920045431626595274684958852160664242477514296044621366504975890688
7961967116967416335929851286833989958546246256171136481800126463530762811119677796896822
5660586752579594008188422952394852312568323570795079525339853461762600854158416351462811
6169623632378869328071312899603567021293330026020395399578850400701543976938757656322777
4679321851846607947079470339271631529682145628932406799171513811070349562584790137738899
9891688396136493104072673258958365156078285340161203007234121963697403333523673612119588
0665586103864638509992894398157409759971253123669631042323829342474061407219393180789733
7061614024109346968925205325062959152507006057138993016014160619116463902979754757279399
5102963369403865186084957988613637999459660437560514552604404386272000601693499569752111
9760456517809813533962644005523528287787529391044667862766351841653470004527426436719299
2624153914783294124003604727946670402150429948054463412673738797618628865378176992845299
5755057971111489762650882765392996673796622903161964355747294290286633985927374384876277
1952446706745816989237319960527222598229158561116262260124487382807051031268205526608133
6237324799714683685779986632738275861283247985956408432810073048246786745421064280100511
0516572112666701824969076382100667359761150166744198893398715750416727397956093925693999
9670082547800163099494639125074354003940863370981513584406725749639529114823918029194588
6801462935880865624117189402234051424869056379491090401676530088813394816617704767211100
2650337731794971898060266741420491517547840549069171569765923145580742228357176699317688
2666488282508089676139672962624789725592833774833391882249642160534573090701924628489955
1350280933325447827663741528945712908940564898509330506752491872440492147901375628014077
0454798596121563753658453708755973309500887831105137646099832530755661944676937202806722
1389487002852119453346336567520234226088620526470563704450155596974502239255950269342844
8580205220049829520102400733370796302381541140681307399248197640479462615023572159444033
4198424589221531374465031177719131841044824927077214680126300898405608007395562450858466
0125034027658885072068284699704086102317498428000553530304606310233637660376605408045811
5262934220835634713096511519565070324919720126997008211775412749958792381680773284443044
4240501832210804009076680122688282718253514179643379527740445632501434118238218304233822
8127467665249077131792907724614365478770646116108480595010790688797932778402789418156755
4011955699942394004390817694239924922083247477615954891430957624985310515964428754438877
8573785446806228854065515941802105546524778896644115827333646878522284357820332029795022
6178794977987872210938425859706275573831617658368501124014760508457834907349775833781933
5921200532939819361805232670391845096200310001589013896954493294418902939228355526686277
7555348131737768192631657956978082896748258021362573215021130399165745132014454849595011

9997691523217446655979633388048872057723479102110927391042968533745443526336472267414545
5883526514769587503856529012306049017991176918354169851005325031665474459163093906691 61
3066930949216133592976063693351880617551197829914955262404607288855376811239479452628 81
9960557782157683737748580882786062283003500290826724430308055185149246252696068647749 18
3631247389554753171764393899007630415873706265202481793603601908442745804710080278566 15
1277626452841279675683829474380654559875603905476889855109354718519828203054167200743 37
0492574238651501306420422920309355190247972690203310025732114617157566470791436508623 53
1142302794551475821237240600213796775337275649210262206942791871324767093409787858206 89
8986678201671136031979971401163819621755916530813492286526917656840582207795469425645 52
4450479005820691372099648731888346169371006787201948259305556083685061388884389030727 85
4075603988240289880833772556627420149221366824005446608959389878196480405224303997288 14
5061014051641106273689203421644828519880802384542361830973551922989519217773569515979 78
4578285614618903013077007478637583863132850976946473168783892460233151457183465856336 36
7763463948828105926045895180965611622972920936984886457169841613353387284051322169087 59
8645276622779044890463110210359546751933446871826367611781879535800722710559617466061 63
9945284571343193854922653776175082045238292342523711782575173771363540383281099554271 38
7286634517513339516689497293782415603373897291154737285035884509249123780962304183946 10
3042067610095618839160114577326255235527661322866787228484946134172086609473005234639 44
1068961376037782335223754715441926354929480275588065857330324398064923100360904510559 38
7887717572693668933652811945735426880884055000671847150407373095768297105370395787713 22
1327962726362262192094768057794794301876488783086179545267998202946213503850137993552 48
5383553281668177676933009838751411162174544578854846885105708552973794548841030094600 67
4841948831443525451481269097290337144174119916665944555248310828159771686724911610913 27
4736949857625758322787747456278637791812727096663392355639677612772112303358386262858 66
0294410319041265468932918605719216731691865623544106431943966805391169121078919283893 54
8845555739237349929936769822262009036264856076680308600314702960807946482878914508106 51
6839325448572249366146981589472410948716151242724418222420270200646609898632588179270 13
5120296470563742504940475908065273950956473906022595446292568587710035708483046167402 83
1022596002753301213908789622342731570670312557227314869511270168222387006657385849551 11
0500085593030172105420530145604754977368436139100405902766062266496003493209795921911 85
9167779671796232217909156320784585845010465252244346986882961439343185109413281906120 99
9560151328658969789508923340499431746265400033547532892826689768860122748671611984304 24
5000830816433393737100939108940321761933419255584401160389625510624870620933716469887 21
8929144926652937320110724023230537921597577735297179012246461015289067763317584277403 60
2015483864008372092036683113407449717095022860588943215754703721414747330761380518541 24
8964854839944322085221982975433059038755250175504100275950989554220660587164843271390 74
8691739183058509355345051351681167132243707427035147200877953594377785460575011325995 77
0854999900827464448230505663659971485375968822166025244695835769751065205163184215908 25
6014125273058189296177995815963899575118203890699605303902540208224216098277107695798 99
9993416043791931237564371371661460947959526163439923133866163614357436355964139279359 17
6978287801105894068344638124296132359670491835771063739651280082108273749250844342118 23
8489193775615296004362861354767345038881387937612495074580273542607604303200108106316 84
7718138916251855385603844282439476775971301862840611639246310790331383664998122459194 24
7197013833799345189313794984584919331280484568894099833767135979732015220503281622375 13
2401406082459001025945582522149739725323990283514025021651831346564996120454161491734 73
8555122324793665190473287164874551526726654809694360260663937008030495748079910887650 22
0934704019010992900564729739044468809091683511449170152348798064246314496364931739268 95
9339800716791937776570211080248358351163674299848140381887675344130054921474222866031 21
6601393399142621844954208441146830865409999074702482737641149750770808743076230767619 12
0283618507254246530091749672040362539114528694627626781541311901693669713145192241899 15
4764961610091107435802682685486649836349761322028560331497929557317175123324141543876 36
6388844734957300316344769050451349617360166564956697851347928891223276132683466199182 85
5063175715561990649516122918073256223064514199884470996239228557933728599894142843760 42
6512046627980209753379739661788612309548681848677278790367810659990882812164773359896 96
7875878964901729443484214079439097617786914566259044915924641840188104711891940150114 74
4988324345151810936301593398751430535559115485213753142474725465293083820735202625056 95
5804821814750779939112759178200116263840629646417064688227808321794156837314862933892 84
3005123604999172225581915109999207422418305257430347417587926192790775367752832184132 4
9587206235569543246712084778756889070651650482646091952186486089855710850101540182875 14
2121379517633061403367896832748520010781099730087750840535181778499287435736218878580 61
0279129746576793324580777389724601569440209645094998085534436602909645864182314090534 85
8551876468073673991736707315229778095264460808272598655595500603869569613973914203880 55
9396064717699420119264864750192629213284959784964822565454726785873496762789210593295 62
0493667713558353046131630914445135682018697687663845113108013328882415166224892071594 11
8552247112847582525042840926423311923434670450636935592063463772454568353645115742561 30
4072437835961365968964177539915110705148283924279317368320046205528820906167485690057 36
5223666289540239832687184703064484342563901473141774981500450871141308689864686362913 10
2700378759433108174156080772983785337550340436586046489429769010369830425504511766933 60
5724090651993020708324237640815827404137363721093856537099718825304755645108019727108 08
4518496474200699243290500627731736066297362123784562179436137986902690909280691510650 37
6501667699302225748492450417548104769459259430389258435896613918178157011553874418001 01
6715709712740307201772364451508364908861204297030072230625047891755943492392059747669 66
1917409394008528141595762458590115783818798913710747875403399929818915259438896377084 22
1417840249852092206312372091245640675126725357795673975259762528619650066488912545289 66
7290522239512565433972987415508009116372871746980666943083066695433060990117331114466 58
6096472936840095216474846337938716225790676146805013603287283105362661387469426354349 68
9878499363176857236893724370874261322057844426981366111743441779737268381310269371334 25
7982857164161006976244432364660858302053605552010966939192332835455532876726475659023 02

pi to two million places

124186099499183243468566140060521348252695625592274793894562464884104064085796356808901
014404902668244628185979910557364909742778680159614104031522554462642892577298319493982
160209410751063129942957296120670252137821730493226482195642517678469151148366298004619
873649626742651153898671651714460734903308202466439286091441195908342652344071304883212
912686021825755574031092138319317611626809160536199090507972154931653101200067119858297
653712611506190870381443116303572808809386065093163730928842790270701808133185500120048
201761192199458729328807319179284483956511982696429851431262853556012584092134288280278
071559040207982799056745491530122248743243587575647964829500570100711651170016589529780
424508420708582834716638442175086747663143936204991247390183904495115313343555529941923
622074513000029156296401423440352947442371470956738778877529736493289870139162752556450
624180390414043492008976032886743907393739525964570319437175653883649239782669162940829
690334226677087643177064566752969854243512604375202400953430799331922165364645373537622
540545773060396641258453293944927795585630183751492733886782835283550664920237243103671
376792701778161791190104399228617472071132142079226261259370701072933771592684627991603
52525959474665898797878416773030920622295857553901283928225527743891816223384044476236795
647647167068843728615727342948490901152573265538524131455357137418504229615691428593526
286309264852984355749049824831850010239665201996291736831596119358909082932506037320976
9102339381335087754007569386714574333399690674765704190276013562914303499966823590403861
577246281781765632173265571476863432938710769896595541102095587133173894785700972129807
873835803417771686726713367882670534151401129885859648924557363046082221886965787823862
150018047482876827789359295551439523809503461080069713638630730196935392195596720125100
649091593371330873695100117094728048678839818497893221596052174114434159108058730299947
364993896663217989471699888443857309446840464891670428745069383314532135171490633987309
421707923799437793112196726116429125886941689530302086803093551376812586812805729575630
571525237769724368673360429466379193992330729622572976572347570012278829580590155536075
380194001938627417201555550726594956175560541018143230895583683094294166244993884494926
331183842672559812134057296271029876555420875245611282213540627883247627201692083223486
285608928275990183719797682060951226276032170973041069044394857718973206106406885547165
988227728677929955539653306415714524325200411135084177703062144916421194167609435534067
252057767997777914658959164675939411912409130801606655568739702603704906557112740343690
384908595002794131653390218932970513283716738205402531709110392738410887224896208452822
318480799139140127008397930179576329207696456138979400245171300653280369757213973184374
750477637032720677342116510650464607839899857290584303620370322031159876765070704683107
7730146628418533910018254117957291273198232024582493339397225075675471331316846636822692
870802333293181608194808022913141768513856111068401144680671105607251773055121759199588
263369340638359285898627719696660268134789385209747297819345990981760059041781100847526
481228686182712962458103984464672084707155916806879738490789114168575484928321131716716
700026495550006428359648340559972390707996956596720622399697367935652340618958133442425
016190166468908331306313266576481014244171368710328538337848542993038145644211416830102
359435763716381684150631226012538404971477593875931544567931216948642335481994002682537
643514309072780217660200801025130429538149127474634876854752116325498512827111470684299
920113780035582595479723192598406516977072833753642189258557651812352719727336502871306
385479524712482789334647309644892062030841553802237590646873078519320888600726402112743
527201509940253468384566090451645926595610991266750955654794095480579897984405564762339
80160326376768077031507515774147170888218582311712977733876520055572215527526256812969
927537766903496023887530276803749058583963585280359215153052271642573275665874890419592
113998913243257903271349403503419757511345458779230397863336487439756006646518610316550
839198983923275809833042183555861670093614247257642932197820741520997843628489052853066
398060904465559938261039668117525494766649399886553025746715193428020387906048717942361
614928274923317810309005803867263418163171482929637150394599191951968752378352709529910
298970604540764698286449713380969820927609039645038823785159470819759420289466070585389
699170540325909756413658904250979213336374183770892073206090421827949125505867777995983
070864571937704771334414478159672524055230559878461289878783771191879646733258975074089
156069953581579102607803955360673774869926097973745893913915563818183154812180025137018
313398129045308113822144731832241974047585165715957291989172401821483184010262188475942
270358708739331687899833221883280765996135629028978204589548384947255982506581891896832
246140966883002107122022453355063572303651131391780244928313107668813200167323953184059
452332385523122183695084987643071455409728562468325343318303062762135644410633658408349
373034928786196631495837141886148397759020976886928270479844792035659926636374099719436
3678140232416189165306189197352224492943578444645846512724481247689407302853410781661070
998741414457795165883181575965853790195404568710603536784613766141960811197346375771619
790755643418799857782035555419408952688447685859577722384260454087426473669840377502923
5000506491862585856332504122933059449336692280591481179789740092897557526725860814377404
688246590652844542012221893908820420359654027253774782347068463726416908569209915755745
356060078256656341530182996574977092303231146000325376551529857793010263919767315760051
673473200020378977167581322737920408143951497619415153387051992263732505295637192396198
756358323748028464990440004551303598494677912164458699219118463005575136858957938345969
744687634023015089440591498463367847550581606385218935539854243460352919156977105412506
589650994290209826311959131545781256305603685953378977220059274225679795847259769894971
121468021051886656307885485858438731072044200267657813916936435113004289852998786802226
431122418214848218142741458527410713762557423889679279270713613394362923680320392744371
431890673156105063080621318057662508275732364531432417640091389979849814538346845761233
726619027117434751901005891430655015074774649845138864327455784630292575510078766104208
709696817516988456667026869906608292608306091690757252848502870487943073723391204521644
597350336270622409422638554152314708534946282202321637450129652082365210750864686682095
745090452935190639992782169471304277907156170614647085318455410962531409137205013335741
550034355324469736993675109108592043865715464435960377773962674053283461426330248845113
439456951178499870060594012815058368642357330543262210226759959334331728695106501990551

8542827478425775777000170041176122346315200201465254925613793455043577522525924153634867
39254842719341965831579500716671133518863466032116465604181370981822164761743127527378
4804808843993292119400067652791843525999316399277912075624045719679651393771773282628906
880669785215051839061695450432141904991128630337564605776464996985746322402159256962055
993989279395429340572227640030217827122396360174034874642716029299099764656713410895391
892472850006495806808061390442771921017873214952713319143341795701687436251904322089471
214848072621450313460419243241698814921370856033147758102191395260015641560069799289441
504698665650800732132276803639129458071632067599441617833637325774878927512918757357625
670183262430475556951948952230148631324482959331849924181597516052825684494610359114821
167035381240051604647710508456609650343848615353787433863785744140905043867711196976975
021483253555805445367513752296248203623486275849947043273781458030118463581894144559399
840789697941248451832323233095936026221346917189708081156409426794379380741869244176515
531297582745791218650376018649770531492871832908985300466276423537585138620882946443394
863008754668920486694512076888037768611458112237568432771052213965283233041251681795562
278001521397605783630122694559280150049103360554539407203144085737106821863439087354069
162665962433149768044558829034735259758723337168977017321300429424115018440482701196770
127626998351208141163571596039959697698992247408714384201414303233439663815793294346504
006513291783614281590895455287917436714647734103329549611158548384634682173252581181410
326971429456388495425916279966888225087247047780037499165375025970997194933888980705737
486425776333793778890359860561374533154994128606986902368551804589489862388002629996691
531966256304155544405375449648273995992768282186020642884824485954730853491394259423351
704445748983019860386407470376608188967110487159544121706955073203545470002953715135227
484508553391491299754049143823897108007714480890310939978127229533689280616652302776693
625424978721511005998791479675050064769677266779863118717812028213305866327841932398413
095515923706808980854376763172771080397189595906675722537009398722586777354596402111542
837316205424591270053551881411986859174792368667060266081350276075043598679963191191868
401897400325712019037805227216924713222670941125306332086164702938262839942933394263329
407619892818166577633593499669588234062191149573591340247781366375922122641794101268264
887623690928567205194970877255314241720993042518805530335618928267117233985596522794084
092490753691725056566710588284372425082659146648671026839023022165606931142011980841557
786890869202062885831659065443076584654417515628745397741206451685631666083160013370323
430778803927394772156078388540165539811205093367557846784148136252383190193424533860976
858019459402548132252796267505698431862402887532270574598616511232460388262716477712823
321382549337276238268983775996617055577303744044727164404698458388165923142290117438361
450954575781216744631693715970915050807882246467033952659292157103395825096432825109603
002147056238688571407505915496933287597568582413705901243575419542845621593871754375964
695627683611178818124431321880498887248464380365137905020028823898961445890369527849336
763652121173489762416930067899346609873625697119039946559688148881571976961351791059311
424339281865649699711877076809457657521051697379580605045458218923402242902568717502729
787283772560095339199105656144927812891331844300061051994805529802410681455793233968875
108970730710555363951357609346307586131683206002167657469925967404148720762882947754282
426865955160956170530669196283646990739060418665000093373229542917035265374845467906476
335029897009195243451376114298804583817484120953582836486569557845960489845282611863335
932600720258750037449344626602428947301333171701958043192209855890518250858239232871062
148280498169382884555782475752087750532004744155511484334990373715566897979018408688114
335124807175088347897622330828977999737899478283988003871128461075459597346561246713404
298201456943831806140056969194233875260906990729411355231688625930449904356296052773013
431932765918738210997474252756036242716822787385617759996319906876388841749185301675455
425035790440155680053540035143722135755232777676528075608414783738951438670981725030416
776224590711138807956006247282739747111172287457718065836735664139570060591367912718373
482515950829417929199527155408594170037247242347665014260771887119524342115419323196267
784018507471775494291794208080897853197262516047477867675392461331428849628821951292468
515900523682797374862444174695498673114630861624671570381846209100344365884612713362894
814494534901632231851022058188517967042269271800677308502969095357934910588882194648863
489067603577816631108610450008872606092366007497342309412872557703185692239965688979258
089589969713518866607371099724461315137658186906155672258031806894130292554385568659274
595795152423953373114279095693551930399960619249952383695160812490533007913212839247612
673013502846075860191008849120607172535795849231604916181382380644233858647915800400296
572299906463188602755695789886029114266122121335316902801632195570825524667186177288411
623720858203949944796961473056938573874026790720786673718610522230688226891267585151041
171017171637020492122894679270366694385055266456745884772822568706895506057428128149706
981991506320060571917829017645870029712959556627688680348894690207524832071472003684325
310145619510421158924520470751687934791013573974654067687498356984715749904374091339436
896176536669571719571905736418013749363664239839321531420895111053125945281429994009128
575429605769501527223260368402053074798235244921218826727389024126603683743593529141399
927650948652389771356052480135653888712923794250979752156421339843200744548601475759005
122514604620733041861899905063075499603487063318997329013773608317872957905155518547597
993432763046140360397721788790936720929546161894543455488077295780219828311883483548225
958134987175368617096495221040748015766378148707963874729539435113033129651567985389339
848553435570481281013893168155727321337867799757526194075532796876561950837164466442832
874490505995644258306142394476466103674887246661879766073466531605211399419510290034513
547278106834781335630214343918747232702419308119788933344939594916949045948313579956680
756955891704115334556895703939370391167283781613920222978199743124481788200809573642199
078764072642209420363159948204205421732828179928258562489125818312219141964016084197909
518657600894983288706503671297945733185203472101057031691629050630807107790947353089699
895674795660050076404379556210074887698640958924028618152859468553105420597231693812006
172801708855033788355648303852646340503229114618781939962181608965429780390022600548706
767154653959293382158601364220289231942084795134193506112876221859162866250865348824940

pi to two million places

9164782387659773854737706352352917001202760097498437403614888649601877279213287473739338
801329184181512424083045968322348191759775675444826641461169553643862986924999555678881
781788590497518935282495073040408514168008547470172155363985163682730594333334778484212
782870045834874460190597263159136039664331231907968084178156540599785706295342018544960
089458825480544668504193709666880952856666229072554013586434203905029102822081824590270
882803936648558355195874168508909339622334029567025938810033395738917346767264565038948
352102741428775028225575743099651548314593599868746006556505074999162322294531798762344
665442896711146938054647477739869908369254044519102884030156287601145636633742044694495
975632097714419110200922054000729683352473068244092301985323439692882458387235862199355
051775645061721512726485182293065860085241331926364165312452229320724286163465271282931
818101357790345452933094351516772403189582581825164965893765793066554632604439292715100
186232415362855836396952770206442628713069207212068571048889237494258619583540719219510
366933868682672272754194854363132887117071620342408571437556374773801255507488893034717
153785282593516154317512578304692474657024232101138127061528320835009842584108204527759
520604922865060137569870457713089140558554488348491049813073647344076348000099995693613
703302244812156443188765962915807601882531866004063889851566878566231802893818084831690
183852637486600387656558820811174687868983287753738413169802725898042157708919172034044
772374266633646232634173107966348638625042759953456017431944184345068412389564694646356
268500207569947370892145258802980089656922538404025493075731849517106246274272667655967
920103034878559729811323792437718528790668600350265758141103039198495717927624999136839
350403752488789143330108831229777730595415365846731465075991801626197183171154247804165
088442718659884696727505264239593251323044466316903180911259656331247455596762603907673
163764769554638031675640353467250640647684802093103038625606004799062896458228508285509
065025263402812834868648256451906909929129246475686717325762815121935627556760832237783
153542454086497686343964797636346975755086634879651664768600266367733280359392526690801
692777541289577373966084856585372957543876220533818013804019508755344710195407213309471
149182109872575662946223894714824975348622954245784394927277094001242766391280986464826
468110113477221509735110977170699605279996225805036225029946357959385142790812014840561
689496147333265032820838744068792316934646139066522120723199071475518258901164418973778
973712659161753271536367463381971340414626192234955290399046258130500919714697956736396
329436664824235936747520679905361690064530140744366704401817005414274770232780471864768
894642327141592633719028826226382555589259877371547334958125879027685029652496588916611
147619318581499424571407242100369797906472939019734687315360236462791513489666954593153
373452617115003215390776039250836429501719081841561335108685268975552472663999414919339
969821985333519667234061414392914551667838267314238048167999102677509376754069292390467
048068463930505959542942660649787768399376104932931947204345368921225517411882728360337
026303306241513804665424614208313462850236068293260901588223365969861627828299275959744
174063644711269296618554522334406686375770816009804010867509043176044948386734976929056
571673912632854130456748149385573268518881607175762554905130059041334986190581202035245
466417417317659760860936793979686012949495080355210120015176955572364472021341436973046
910034921942361389999213522234553847613136444889059042031442366911612982006391119865095
396349317352423124828821922158359304522824692829946587012065124328746106167253939148208
561402508626648699749099755214457570541851539109513740530750687433963474754969067126093
306716515780184980732895058461829328339008468072595712111256509327716488556626681816391
516663541125136169805400257955397886778602144381973114296725514494152882229127425040417
337716442759080083427606768919709985362827863088733995569139467580103006234626675271632
757058221212930375799645883514402738413235073238245349207705795028202281071823685952885
827971575607082914921426907146414254499144574509123356690067526861535140684415530142034
300913729567933614613701636542395075372610356514783774846603485595063636120373158227039
602812434232349845319942317801757445785109292233440953182011794538990629711197047931585
130119220451987388821087390330868437567484440432278247732916434575484578984473852960349
223856196668568292142116114878540829005090344166732832861054480755887520029047051513543
311884064249727077624100602783291645071624712694263508684943902769921677767674830605272
882780231574328533154776862856199388923481900421218368368559032054215031500854425637584
512786486037901362590566602087284329328213889865003014597689388886846325607294688896717
353692104844470688038004518990872112330724704172452949649929438039805630858656140021896
797466041063088115195158573792637232787757940595232793543558946700144115320743349078743
781103082387690481919292036195888936549367732716470431026603237502726870087738917440828
436259826718567598442234407795469089359688869708766997468496327165017189713268349939924
184455975603409622902171373236700358819561251326159322921595723572755776432158794426587
309763283226528468643935971468512441909686926385051246838590362587665588982552372458470
027098276914629074966328286001775638667771797935728382920600048351695628117626540437776
916782930831262544938376907635863828302172678008360102550178235131483125102108917280789
458322609921677449569781301776884628772264692929346041879649195268858955018945231980856
859593112936537174967593925706218798534066425966133054945244063716339259806407503899595
485942380723447877411060385727137876413342301404501027403968873311822107484995521274957
932923859160327749809792249045775877131612975369419405899970830541968466281013649728816
404657007888468767087313699177605266712799376204880422246890222980027317410018032754903
602199359031742920607512837570981563936611721341186967821438436193000039271539630152103
625484982033577433369442035728392989926422771876271371620757964757898192751425143789838
7912564296235252555757816018978872852005341482243558084376261407444296996187744232690308
489462693983381921125354850088379640201624880707895810545639132331355794044787358679340
032840299510022293146125013131589765943648876639355862007366877674417295696863011615477
354783371548667679265179612855938297183513494593284896177197215039403217542052405610465
480060055746739078366488981506861789573165740994403859010950465154856969324947072785418
588800871250815438323522146059027191601494681221022971235764090610951590068344958273462
236772554645068302829497480512407918346713208148951821921810371257694532597995807476120
785294930963873880347919929393091236084544289939286563626836308859152796014194084100148

8124545609415778093940056150605786667532914282599248360561069665798000554482263859333829
9779608324511821963711870374149941360898424538451325427922383934649481345726786489972657
7489497993617178915701478870813611769312056919834758249855155388329366103876011389073050
0105421222056514107854736675514254726243825697061739714362525773554367287123475331510802
2853859668085163363844753857609129743604605775386386831690899937349968147265876729618535
5258227276257751609793320744722949640375233341091138527090124746867500331722873770535691
1952559422251104321869703295392508936647141552900014601497212216864271679423544577343329
9736444890690709168590213558539114753151385318581298252614035620379451763920751431862299
9181996686534811709446063160707235283855661185484791086145373522477750933883357760923970
0286447575649538761523899477163032494280364247948400886336968300872413537539674510770189
9392641317608141932715239984273386342741815151353832939813614858398780080383674269320247
7723764313937058027589539108791351745609365883944499710805996247607160169235777674441053
3320769563383670675912410648503550643000602374718676010189903412271487422490816341740253
3046864982010491381951411521468867368800727033339832462135929500388059846790122288469314
4752376301708517785731081733610362210676403353145987448907075516405386720622296175081982
2211897843832868843770192718310235205214072669676080711698539295785465421121366764052259
9622348828208119421364114176714753184524265254425153200115941080588613274226836692286219
9493683267537010608442914038448423127054478407741214463787531400981272155802408742867127
7830259762461393732764605854505143625751349208175861719469304837438996177820356312167841
1090631056231399071586430760267996813855429921285264165648924446517509607108682899245550
0960629621868075459424458255139936811676058977602566697801872052328333755539998193329633
0326587690461018903976676855782533287949137916397080025280400499161939350603222898649694
0428771850457846887028465655897556788355519897743958748820658936305527693233896305203782
2418223104601282602696192356618421323164734519378758565383656167360338457488782631653757
7062848685221158657519030142241180444886078077302131670161381906954642738350984821432322
2894789488930272390792615261223190440033935417142485922733222024631545116805601266690302
2336779969175247391601847324227367317291951866887342445245554258492965363469454884479616
6875554085755985007723145102767890940655156477013925696674045237321608169187193245521847
7810530414603426501371391684433223140086711402174359509606045251720918686954025581696303
3263793142101899206838309686933943811988616290108875418470766184218643102680311242296633
1399747084959229430137531525797183525757988422141793400625502911603948587316686904867982
2685636632891907995257186893081746745152148451711452296597571160739867422888459760096384
4203715393094883972243488102777533811612346977266783472555940903457776607086515476840770
0916202794290410532398211932856572298181816313287774904840970963513602874327624548879835
5966424161258068208234298491250219654789529996678840672361313707000868557507146557298381
1864048558316810405786785509059460613453311225813566566147894135070265867484703099473274
4497666815821988773025866450872370989985239609307808811211481521175948160816675087442680
0891715188825640331601449193274718489627988311203519563614282668135084267162090966457732
2728371533131677930958263296089646769631562349174906961053921883735626397886846804588501
1896228140989326957947849238879490647121998191265057823774850839758771779416162091725927
7631170823916258826176106893886467303664233225289905963581823977504356427493853650897196
6991309971872504489598183501812445169844648357964089470838155314637861948854278198323947
7666484690383569231644852890309127156133866577780278666711494744015268249723953961202254
4275946429892714094931271003926969172360491660550046134920601927392945914719374725459826
1646185889915235768029898998279839892189925273486053528099349101667852632464310128734460
2578324346342892716901384290449670830023226560187891414605863584672456236844617753891093
9914413917159043214669387939675006334403546747274544526143540911560500969923740209343932
6274823513755633392558821848356985699935836555260256310860420503583227535891598263260180
3707595501967841331814583689952852181576555020241337548712147691636372960357848804311440
8298563606805940834015350180185156466921556339317852260437841549994341033237676588466262
8470428491326222529745206818440979093619842308120228738345257420203825202331913631820675
1290572572623740332803599916573451960758130134875020077796438184235698879251326477844910
2513387381327198050458648544089635654624599558099970799613413269704300505562190864973028
5435379818751635164764269244933212643345739345276102382727615859651750406958097128990380
3564806369433101777607132061325499062454790756556574718497787528326893200886164086231515
9869847062154390172256891740833804537451410151181410378257293029728754292833436674500764
2984150000230147578754035717914734835025497532676463272733994948779683496654906440302121
5684540313117419234931217171222252534067904917559361062001980677030408375938695794119219
4916719262400867922796639783740273072840134443920062055657241523259769896834029632998766
0624656865671519378753756719044493010428113673248320323481883930078233958835978078760205
1821570371233534052474256396458937184548791179973338264254441945351131049990564828650457
3427758293464293387094146308079292803297574759569275557494390278760464438997006823625467
5604187449738823255677064826532016092637715853935755783060088533058811454746285297334012
8037835578013503710349003546123527941112988536299930314671145703966311649925802432851816
9037117290476250704404272983622423247014681647026601156353510725783422668835359441857617
1223469281139646616867815369655820819571785276562388016280767563683393422236074354565518
2714779683762276490072229415482109148629557539723763844421944407892638956768881351440622
3618220290212616570568588483592415582352389676846205805989752187961642530425640952166253
4008676041630208696304561001029154814436103096503583802256629952699509188415563964301663
0471794446700699058137500293757360128213152980398261936278498454620795576632629063174151
9624243308330663952763551882740462094834585946331571373455422689759772496715162794005652
3753504868013433251752212228804631864331738619043488897864372419631311380290153629690054
4485502682940906515670096205947131662070448236954388993659304221416321568794559710895727
0800790067767861395679887333634126133726350493676319544259483374202411812806128940239300
1658987480104609478473520292931207428811150686157256513783815208908004411934818920158555
3546496192867442195990309844318889737867643456559514729703484612834891624205705281692322
7356124982074615422006498884595484620142184639569455152465712378186114360636312015652484
9220272964030209209296042260726933326403201598504770092750282321360661128235693646835703

5231044097827052382362736086839535176995905119444155901615154537147012420907519407880103
2729497767734369002840812247815603227939237151128971087916700610380579524462364364319653
2554363928733020249395267903408881928035859597199453265085156970144553306481248952744104
8962951714276509511136735347147834099082279463348359910823730980207958992836288376905335
8469486744109904902243929130209090468582251048584849789253848820742794104612124568455456
6282299002694841622393071707984073679277359347800371282416083019767644458887452273351287
6651166291898903367627923944264877036471758707631118268082322528074232106108303493213748
7300925480055572613644074408959169584346278057277294389737117249408632651841495847226089
5758092763133262482626449108744000057152463791377020054820589652316857946053821364157940
0779058443419917759293572664727706958931238722423144096683062930286625839745249077043421
7504030720066161616459608991156442375623407851492899280732951917444260650135600104038832
8294967371974738252694628176742690322001570447632796261821581998577386410119171129073923
9753559051065353983539143283487963773932847281444770656732016392335057403752217368135784
7971056052702312098710707857545625700534429756071878757143002743517790832231604042954435
8486967497739171443557252568612477174195236578332902472803521433298631084997014456402146
4880954366526631935505218604405421521017256075160489221257691785296475113848932856418251
6859533209004700157332466884883626921764209084783401642664082951446224564820407060277036
5442312631711609267427724840568826405330997710104304331246253040794678986902877845498285
8473174071157203331179733499710272750671539596838958883477799079128319097839782978133314
5801946986403915821858963047998302053861683160715175460724024433504476842043627714713881
2368861324136491561768017402663074524370459130614105242802300740476724756006711933158211
3525624275293211982304277666998608300593335360387266223831613528211876421192340903412691
5058150847521447302899105329851707249468713898921764663590183544393073415948694561628888
6935556937488195764952477386460696017202712505202376556368407284533349628399569587665353
0816471998013607992801665213636158201782769333614139067701703382822221060126314136528855
3862465918184763226601045812495779841545157735113160048384621753474465241162484104539924
5292940405707597937130941686342792150548561109571922683715871175803591551851524773793477
4088459202757794362672709288107774679814635853352329076325533030445432923441879050372379
0543695759574755641778560467638474445443941527467392929996118928913524471802740698411355
3254145937079674989588311521192288737021078498539973160722562420416048139290636858097205
3603683178632553897219910320741119773023207784728746057531453612900332180277092726924548
2445461776208019097249250742140169323962364665740686024280267631880695455950816985514109
5896339212779098245742798395431304044270234508555913744943715670427355919689143258727856
3586503277752204528054964102711505271525098737281632492074556689405327824068464289068255
1693471652157138262818183091986616986411261248587180894334599514940261982856769714846710
5927395462847840572803699431212769253533525125596420396492968351022093587485193242484292
7211695935887697021064060428920314298361924013393119860603465395769117367554168270959907
7977588016796073946932994226920452783681607222877105956869858288557968412888242435966779
2603308987325764147012615940878392817898803023486374890330898605682126178127777357883124
8068322777081629654054658595757247100708409436565707960004775714054590890025750113410135
2095203050777279183946821240414432331002653830541710294405495628849511662805080282442446
7113795966602601427311542713880924716385650813614153513496815922257603258747440872272416
5863404190937791520404381911759960028474434921337369831641793042606802669509249392301980
8938858211257183326730162974479542140521065899034281558553691167272464757281064383422889
5436837236363133406690292717036282460127354284860103611996973372265676874402979798794652
1648850188205043291977488194884251890724987133093623673968559617162268693317236307593864
1069806855655262499517096410955918902884283816886636866125293151456174670878892887456072
1566756350961041492977203301775177691659934607964937542655405876864729169401888509630463
9254115967122389585522332083607681270634329237900237297508985349664554470430655498794242
6548847691924782439382639596424262646214980874985633124275005013758947590435118678350125
7583501199129487036050538205810573287774053863578946166889462723712827729839741722448416
8594057652992972406109081892353751223155877053561763181474001103002892870762626330927567
5812654814496148406893735556943510119002555826539405781555741755142391023346715855004402
7246623795365347410525131077518578048925039622696493973366519453130958658780381275746298
2981573836589681206539606419204048737439162605583711110301590915242519539659110872018167
0186485347012922133890604686621660686255025632111205524239924297426087369784207421806716
0694312349695484511651395548574670675079765918582149549963759435426913717657572124155785
6937559692819015732101641396054781663831582581127427109662744206348130493726598721695216
1643309484370716805000255185310270727991550888591205780538290296867836134794468839013222
7735338592872763739806489119847525329116106835815260125648037337390518901687827950098790
1891791443819929116641124327600877203476671991747747496382404503744972168995436956828543
3078265082206280310589994830634518810797753038069022384455079651904014236609371797122044
9893897286687443900950377353042250380603929296738431069278884116232700740731042857112697
1661307366656918904358135081148389751783093894946367515244714184398149891958101430513789
3423852387726005976498676279397553611303116765612614776937605296916598102558860931178061
5128421747068759567461959806762550279482635107467621815642117424174396612879177205732543
8796802285123727936334190153512853082976270099161279433361257706862825317299738511140954
0245767023872686907803429400048908497172970522721761596147970413197190238763076072246173
3471676080176131970064611274324420535021772305340483992019572666686555192649149160521708
6072018956123548964369885246797235105585247402584337176871372606277613569650497349581121
3906641454690534450656228985932869216957237216530146504986467959055199024405440673992403
4653942822219115260869884110702216145470135803171503972075930411235796901609319302121226
5415301542929581732099171783455836182805257085611505902525040441563955846688057531960023
8445536205210577307221321764723215256792454576821990295998771852204591704107941169583764
0712670160087353916285596515525919253680862059910662399509273006322511704138733383340134
1327591453176377413177319104489864959509882017599026839280389559537956720429798727929354
4961213426862872910229158102409954284125025700376199683936901626063688578156755053965382
9024783670716489919774379940526334298129015488901498093020657920111449659297913965551360

838092996504112697637904238812162722416393105282849984492554876008734381617102750350672
712148186304749391334940249990032821723681344771479572697608026982539835498205968437570
410195234393039912128345390047096496960170796679344121074529447968212094921091037047905
502315605183930731093513065227886923603277280718636851321303895009640203946581117622092
922224709299981261291677688060136723412136493288361702474104213913248801669410548936725
820127563072521793081036867538004016622799907422346498224501430588817150314565198335964
014807435106815230086520296659293328872127435085999722879767060599935319193199364060455
882270681105973354235566201260622931420808304974591837405741384921819435605022255750005
386056891116221406125952645774853876900826609604179389076021011986920939456570997835180
810788016170758337170660718420841736288951802607905312341679136352478627805555238028928
400935149572162926042397902538597305331369185604228112878180663529336026735294851123986
918520882220190403746531103942060428837017682735797906444394576198377328267041952990518
425989545328603822462073570504572804536627004582077336658095228068280117408262555111641
268527314984086429861714437827187492541579993098455498329868175091145728687185414736040
149889461746621995388358237539634917123485244482959700849534821798330966984545909274163
702712775099516881363224476072068133375641501607486936936891363279977986510951180042350
374753577244155652631115139835147336127767654968437082275295494731817789727211598161267
717452932337792235887176697429727338450791130618179845552210869375734130386212028566601
308211563856033551833626167865641708075799861734295848764201663848065573234614528993680
624017604405766811587390248544816382428856616053784627783021357630287968241504655464526
013533189758041401772308344400952508483697004162675941650694716672110820720154343325295
533779558347913700425385363481459240766795595385765522717548126810151593568554259737062
732988465818960099640126770152248766317054623446583940409338972321390357043329155124145
556484560271277020132872553064746060266755638123723144174611180597838613508438023970846
173213169403692778140096971775243802630206203665165969277644497592084709909600859500655
399487474143093428689611527559404663415463287534651963023164028444498658772811837561860
456761038752954943078051821795807937193890840811121909534762040375557279459955497433540
236269258069096418521696629868913354140996601576492529315180881451133188747028871611042
638803060205937903760426271043276224708366970785770839865445507284201461754790301370499
825921403893033274053735354400972855338025408342545976495948811298709397280024176249899
789927226337976126757570956972015101360109847109876763908021719974123350513410821493259
783376950586110868199656692082096243926687979227448012428831525871917201226630588335756
599900816972642452154477331460065169003844648026759218430871053006521141731526330399857
208744577851551958207467238141402466097276591358719835498004841793220307455330575861840
223844553169796500419117179918352965402807223411358752836676748916493653958506857074453
742463087744866655968196533605378245532073553530970919691752561852192471851274878137849
796909151074922146267793710878316645317577256943302772528061468988769060862599987308200
565912254478986362681859251939409189141885833735429582514492096848147377722990288665384
060809400323930960456048836296434358336896504016725840020964633959013350962652397944890
195777229982329879419657939812629153000245205266767596427226787008500522657826496421597
190409356040459528235566861377244867562539645511252020812026080775508088407553805356624
486189929342907289367863826770677613979379610074738702182397240941795982636596919000601
935180019127453315224480042326428029235894690849907644351935011457777560178970897018855
337641900637311978249508075184206728986619606133104675594429199668456735304057579493151
190346133907821340225296968773992636269364475040378565058779974208313733452753529774851
666073807345154719469167950206001568448596341796600354436841187223185526791356434196652
966156381775321742445752973511440972735194825897563717794816907869853882109829436852320
083151510912555666909523211804263886541474740970616880407368824211529313692077160305085
777262147240236694736990028686090896161526073138309923436812079791723940678254147552345
523063754515232446930291396026849356405396380611122913487792767442341018490026132534812
277106120663925628929963427410712658828735773742149626131525656575668656452970578286390
826708940929684097845302355090969053557986580252405529417550001729120223589651064340979
352596241811113302230857295282863314558934437075661360927912720843503451487865560242921
689050418510247692543341269783697911585850625268122643802154921808223182152732839452618
063840129149059039072795603494739136566860924840249618791782597266579093955998715715253
546432152467119177729058469389893615661046980193355524281132041280301981390871744322528
527370912096038678858380168211199621585541846298508046966905549747056764277361198125349
607702231042141551481174660677041909822021338077562719882541551656166467817455711291140
709686374838747378496209085541003345265676805390786590207539588121367060333952858329881
407788605723726485445449413079667837952442724242938069342169642795653060568806382427235
462309725097028774124004938774816065937678511960996354008270806042115844661313090524086
355845882645659304898217955429425154467142669147279997258353634042142777025632808660807
877240787102652287393511920768213836093516530000113138647999492097816135395276741675216
927730927241213816425137944692317437712290728330734700095962438677395395665229599597764
318519988226902267819334588595712163489133715637318571349096852511658018976246003632697
847617787772501981064939331988052601854903284615589772233753277997445478304737317556854
895966401547798109409051762685740879547613280474869331149977328159877061650158764110299
746607195362569813115278815798182200661436804035395098446584205794258365829317996710143
225499908736820403473328132067262982747168917937131900559030419311115927359246901788962
771145400151361106329004551160343812390599330859858510961650007350446893393479746401867
882096936447868410830228449454314606506478731401218979663531397266033274817293032005954
085381944128675294440166549243625515563754155191335434279855646253752184508559015427830
178228697599968364107930973205862867572363338566529951358115854142892931541556835353066
567210720017983658118901225950018370015171854790274701174340968995573722828140317707185
105672060132368902296475953069989987534399661722854307786765593864021488850136917725671
101323835458248692843434366699500102511723913073649841356805147882811529128762288814656
106313787142852299896671395853791692080019236916384032548740541082370391337955813811868
054200636184857816141134664824368761305672828006746017126610138095691287128886532897078

555316954292311117341082068661228699361085620580586989621817126898614321163366890747141
998985284387180782120223003321229708282529690758942480469683986533322469537853103217200
853441593531243339756324347937651562659667374185316195133588214505610144082828278277148
340873404475204432111727910490373597998271693770190895061681044222030174922167582344126
582984085327290976256061318941780521642753247477127479711521642170127055033392420791334
441337979389859114041990726948311555873809408415185817043352088539396397315629735985850
561330879305794974470192231400089381806691367974243275386352657226700322375836575883624
051260726836145995254296288364378008827120339272018561140239901277505999341130028970933
580938164659106278112711981949022173475256944414316655939047045979227975907271354278393
645905555234846687709750822545603493244240336233698891533316368042516922101189209385791
921325005087672751300806706518010107802323962415009017921884078881954292746887820995338
685101389736986293633042392199354038037663041826126129142281457248831104911168235024582
158032058121689079135991938684530125471407604425831293834428245025587188063393406279055
439626594302385454421449567056448726043404358342908004458220436563753878946920010865185
109557086702663661019419964596196647195427037047087817376652629362747578974493503235225
678207374792702112486927649550340999092203440242363207273829697876937329645553501063479
606077860673108152489706749462894904429657629093842892194009361039657652058082112980120
383925580117361211374157019473723187420712048802806711095438110040745223339072997006748
445026502114060465857534082801942459495944027176989792365055416926459616637078531594493
991730462498640110284979056233704805915160050301377318092898242318904195949021083030600
162057116586683106214396616973982767523812602750869528549393095166667458867606761696472
6449558789614718935034054624003357498305334901951452632650710332270155581857047238482681
841857148654097981133406160527886853963384755787946811791373587126144957974147119404583
232807720380768354730160685196046657264763846264507579733377491956324978847176533104308
490726606807722049812410497459790390156604952752962731948347391783623202666790454064704
570841996491170167127955015349765632727062756704910573547295438008846556273821442246310
088450441291061309115636282080716104308673312753424953046018347930298056145131840186674
605499710183285562903407692331379900465517955197214375609590246123959357655063165168645
627744356168171993757232476266348592014705232381942469780854248368031011548566603573986
776300958442609542766528841019475951424527323466694741033752071953134159323532566434295
961452178482154740441442251414882153490420257206176421525626107100119359415363209841118
294892941250127048312546619296435799983823032451759328510368291779834765236510432790993
781032874420309615491093967361158120096586359775123534417106855017595453788214207966110
624822549766012690810825914670057370786595751594589056658543563019535522283187262717196
280429368991652457977072675398701086660300007682580542917138617303645544539525190921858
042425527978448797525845745504212544928257953985898493330579041809480316609309755604426
741072440305553314801234517404074308728042762354485612231362359091899241584959824509524
456607765405413765501312105360696753635531157351504325611791771187379416251552713494773
194525183670833120382510882560369580810715139264721755281752771334811931768378637918832
342914103341253678815607092030061153301024918208978925155731485697806861920984868859001
077256729366066998221583556917021626997820622171863274856991815903466494617610106761633
863529840085017197181078763288098303519361687604033760605167051268788416980629394720373
559131288131722360686334818972292099266629320496122179940483328104947100976538861081271
259669947346197911338387645687660029195583100779556554560422186379977133606429225542457
460094417401854052060596269444937855719740252568753082078913859008203434028172180809847
866955089253117643480956557055912039672916827351598504375462149223847450208224297613364
319091930424526178838407989287243791405924127349950903987203453960193158869564822678914
518480988356975801171542125152000972150629153123229096078101562637516560107227408246986
779992772933345769218521549105339619687423793100437761222167664087658528562961021450108
940162945962306209247129300221126858150344948795185073184690095191904233806724383922812
499066254404252378961238158039863059900884032522995801724462303197379872729614225756198
434057030392492417170852956614659176779929806492205218755828191105121005256197118980944
398236223423904054495693022646761562164819511993979707697464589027435584279523196459334
577999650914693190414367756812007885139030069759229462857332355849574717204632610989374
177755660982802728586342070683739335949298405639577734905488111049576528116020912098154
152129457147440195936776195817365554431517537985787664958476619872392051604991580113291
643917251598822529884499806529861174185392503643718951023960398712409599961012503586665
47477190162600875538870195720627297979314721651209577756468129465842388271891995551465575
557237852971745459779577210772640461802852796523892695856408883751240110230143670868278
010679685938885198517143205881803861318677783842525302969127366608151541456369500262592
124246265165306004908241716933531310253221035518997270465409417706818941708500141688238
149191897194021445968368761313871873122501115749770833491604672977008533532516626471677
004688790931179879646841435615042359084762622786260209348128079832129542074703627771434
619515041034709719065187386442528635196459761505186902260769921530822542892917328763765
489750960458726447860533447743550278715605662923057981738938339180876461183754210797668
882451028984519936976998545832340896227742858674744256410001349390499959058059405358714
176097048685549057689334815806356530441056010527278593892073323434732155060080993370807
521164316558981738965593331824909209497815749495265524447331465478634612390403433939315
624626271337591099729946297477181134353273105477453307793398278844306465338548474263117
780901685730466527769709719682320280854032288989830700830414676382607850387848766026550
936327919241938863998540914154200715957157625562333403790710840064576178552138957832300
703106132983208970940263418342887265137822560904817842109406012976000712680618871447961
045152393466165602550195892969282244812359837215919583296633991399863555474009990876596
388382788256968466514103709166800736237178199750435007173748595378619762063988291832577
075549280835158726296897618263258346594391558415465614854675820017018323496174599134546
731853262110989493691942791414284352834860526009802926216937304205431416984545394333893
912120820547753328076203433631944246468257041091521868320085957621586914288481575203648
173557536687439638249227205333715019707541587478873646401654505082093072908305729545925

pi to two million places

6496501969930648624878582931699595910539847745462945671961192208138764958154690180761O2
9180589502469928843325498847466950228824997723338387013613653795992377399199675260664 29
9729253975984768301973408667500689931464279410680339373920257160088300840708271870023 72
6046740775187799569425315423904090330982613168033543742985353048560255615331261694506 85
3735232468593155762803438547483394365692355719988742339344575560290507712879625080430 27
1826784868661020237384630096461186280463094990375797250230706412845022851714784383964 58
7086008263986077891663250075490144698716229540425781174522548798319479232957741838702 17
6320495961886907875212066021387716400007890210940474081648086140288536055566344317001 677
5157914177038373235120356774293832301830340563847484857231102836315517646363933859845 19
5631225744592093119305755186608366187366095312353574985721682905223707671675373378795 81
9759996116911389337864294160037875866927210124454830856703571515288535476748772285135 02
4456742862097299939424361355657676093163707958937429922172048914723734263211113336259 03
4406363132533847070601933121477818439567903015290079416378716956257276647507850572024 61
1705901705436760447151032783338859212465203361400985338023257501572604401456625464704 43
0328837988104252034538571906599324216186344257323566375303414303633597984252616863182 53
9729942177495314800377881186411409744346139838403059593686930370818549620049963273454 97
4448311673182683549930816030486571825244591566610907320047487565878684545072973557167 71
2847078274982534187704254232174903730711870059127116777305028358493777433678367390407 00
9310732527546767695942504064452005690567507190326265575600239334945547166247111326857 33
8176289697568712292271979255266757581217940464789435431202837597775118586424986854753 11
1061520695473531472190687259375791138235537165419692617202124902517268644291197483936 14
2859167558448310538400185342425606308344563799565435651776456276423522279536315455968 13
0107521742292945570546659172664096839998305126797633151321797859299141860157580746430 24
0550031290388282141327742186676899206356752020992949652098611614516267896282079950947 54
9376575838066055670554135008373893848990691545464032437781317189214145646063102565126 52
4371573536405551524185024041418141723340406885119742997791226829353384097072894020788 54
0657624095329723662074634471291697995044259732586149292228146154292592649855342474370 32
2733082465756966010129046440059570995202129011931985518413206438882393101019360484110 70
7304976818139140658209466880417988047703732434838490577255450332102920223407254951567 43
3328153612985247296833444157120202362805779249909040668784088517029723713314196740825 72
7899937943195449653921161692866912901329086425480183185144220904048089854425997971459 62
9674099488096211568141941984618743903424451817517085638627322090487466168948095169155 49
5610656836403214798265812741392707478470766577310148743193768926593362597495069955165 91
9828606001833782895724086594402608432453528438871394898542809533855888661944641332839 44
5349883224822826996985134080958769629309447301891067786272865360953503389623320694496 64
3352874760374819303920721679553775896178757126222704410345597351407864145503745978445 72
1066552509127344303865276832148313474569218178814748190703042373350767564897973029998 90
8210684017771866834605943882697468607499620348684669334913552744226589190404807616616 50
6108924016791940475511342406990881715319183560602225087535610077289771473298908810052 41
2039941589110773091201975755785998784548564658832442010228775353571213135123878884729 63
0924281338470366532228846697219111951637884373025422212591879484519088236381067852480 206
0559960439789752360474068326360168903766136692582841654734377099660979789147379637441 36
1373503135015249909080054538998012133566146084629417176799117439094236663248014496019 36
8682577288957235631488844879780245757838249919944527029429214694103510227828009638133 797
0197156162962756964951386431185172164751294596545133346844115322134952727238478594048 97
0123652867278893089990843543507261812289420078388677052146536217097153126827069795602 12
7636111473044184618027654012035575852554455195872467670122331764024308766263327756582 20
3921908041519563265476283971852223344696280340881941060029518158834568248372344469045 37
6842170950585416455563893357521827978689094677270961567703160363561757576611151109825 30
4022442537454651024985803248668577863168162054943125973929342162065228034890659150623 14
3513190088571502393150382163400312154707064382062874845164456262129440343024175810917 50
0563624466448689487910772038948176124413330365529163656223874505519344170123997088486 13
9708768182256625905847045802773554385608308799918245983219464932267608055661050545681 28
3118906956298863315552060060303852864430578147529884300241853550476850610580752056508 75
8363784578296805785679635813622503976797491381833790731824137151661520831634381466987 61
9713092778993713035175617598325225366844307736065863546458076510414092875283079673182 73
8673465284911940254871928719440497238997288719894131076487089978367788512562441821757 86
3093377836250489738183652623396062933532101068322177324072232170521767265991626838059 03
6644043621633685091074830646142709158255014744076870569836305675499128283137937652441 80
2356144598984866850823753009701576710810755998821225007413574673831645580405671489751 73
8548093115181533694190872957405145613038244024218874840583902024819182347588200359924 71
4558260870626820195336294712989713782874314826495275493033057653503532462456365476958 68
5214749563895239703931439002200233015554536901854556155829599311327648675877889452163 72
4140991267206412628450362395141436121963020660629346961764582572752282676092420400171 39
7271713261815103271936559355419240856741287096311050111428296858950724033275748783577 18
1952359613497741373085845917456955080156503782872744031079553181798131820692150204908 95
3575441610471857857691622518045017327834366332288035344054900938554970094588242086030 06
5430601111564575539772534206996290786191261001457228994810806704869727349929615474162 87
2519209999015304768302659733044669701530211947925927996587774899303990435170860737865 23
8248582631592320868918624582993517870949412172769522143346162963893030529702690419496 26
4386513725966810686529319158307567776690927532228885283517232703702074261225617914191 61
0785635991369284076561024478027585981460878663935675423062634202893344322744808135456 03
9021347181031267811516887101284516166951897851027965161536009305379938098995402205459 71
9151126832781261820423113811386147740175139727654451800189604767460206459733513292393 16
6429091929756633682182683886945877414230951574006266220325794216616725681311703809669 22
5009185685272192425722164326808870595016123420199798189646632816683198353375699843078 46
9777456461092884458804202849985424404303421988412517585792940654401176186348204924998 18
6021719494971603845325162171844503434700618955393082288653884397411732090639701915710 93

7028698666867730595953319254928368090465438604755253032859369302384983641373011226 98827
9292028592573770720833329955609683853426634448915209982466590308421482849336653409 44477
8032245907984828974420276388578859908477426881078957848457696405304109455985947409 64790
9804533111316422361774895708611563332500518100971722217401823975318212258047515451 80069
2274812445765071366705297449914061024529481190712669656937782953971017225365400443 93292
8175685367680577328636358035383209960639932705452098788331834353749998865601663482 68640
9004979330474963737660993472318302221752743936251966725104890329897781869133786281 90031
4587225456022146683134723573572271091836086236272682406301564703736941353542094803 16819
3670134985201255413208337146223498823265325545760016489791074420357399460109008107 99537
0435418003841296226556451083310266257160544489548392418127475122150112981398156258 30908
0205286377398184093358687421486759127970387463223050725122010144817216524127574980 05538
1435680885381713462732141660104875140352452583392308176191777104256720239694745196 47458
3874663759903813520070962410158081173190413913068501033439659225758224066437635018 50585
9253018105617148326472697122676393472590115838698418194708734426229420211964817192 00732
6325855220036944791649835515526047233595889313622913142064621494891776007512373126 85253
3308880032753207188304170925416795374647814877336406323803648119898384426574210248 86516
5495264783466910385114290902948418378943978215944573998056961269034262787528412859 89025
7741754195439106530557206119267303961121072555387930507203921250894301475794061853 98003
6818383711391497455446333574858977957496733209951942618317947380159065807027718315 59968
9432102511921796649784390894146005246516065960023044855323342187006568100098587576 51698
7186064807043416639335098745297425686095866665929914257108934534682464035156707078 38472
4085420789174501105698147242439555144648778060930884207167696573057973147760316396 70971
5871218411115857872429427231853841119275692388150261653218685774863931967808862048 92321
6695342151156314713840656353119370429454032020124142969160732725816747823331278925 97275
3055561351115375801105956751558122816558222346416127744558382976861445480818092371 31425
9244518628497969307820393797782729530391199314109632093764673009245073633605810885 67026
7463355274197409317111261638583761341536420710425507095805785084299330713852171911 41182
8878966333872928472357066473397080358812450848117988928797417551364178172510911181 70931
4194802624233389975743858592356910342816847204155102323971716917786483682751886875 60435
1390094223124877048279014764032069888010176054792874954022233018565624421108453556 16977
9517712976810450742036258096907696218080016680876091158016729159631748671749778274 85793
0154367727730618880773443714074487737720323231568524093334700627369562999268748831 59732
9342013263476479121588488626520802716258965081000981327943145689665693993601378125 23147
3097404901010357392951602742946406999536923335433293116678299106577555392450996175 17209
4467493740324398725986317928328758818805526051710593572076696825641657850092964482 16670
3685343011968489639016240695044818201450914228466415661581874406951812182778419500 21803
8005152742222546249979180348957046858602721756855132093049909136568860888814970313 95103
6579673572339887756044293362434406037036043495093338281559081298782553238735885481 42095
0946433236912739674506249957162209274156061918156916372244711131851802899985715705 06343
7344672669641926190559777576960632069884661949060954006314736068951411232827581821 91108
1815618736691728506367129357264089317842501394239362009645925180543265678161349921 16909
9650060440697838160279296153917330917613215717646752371308067612097645967524901925 57822
2446506220015559764248756052694874938503790071175907846930725472242619645309612520 06215
5223637785889036618592241107466392496888601677049950757512872228014428970756642715 73437
9315757820323012182896883177277249525743537390108141074591422256466559807283157557 84264
4807112245420707714474432241219348914840780852493457237582202860869308891104386605 48219
5900950852158215214882196952279728240369334844636478593242929231190732356035887910 61803
2893505799219964300517300878033349766839572804352988827792898742934745741989761397 15135
8427410447633628502051914225152013211791766617839763056984381547886286532973419510 44440
8379521947541352915679255778846526736946140171835623825011985052619430265305135676 94119
5141726652285513603288174424709942136480431652517861780145559952580780975442289351 07911
7640807025756470305512626302569218017053836992313811934067425551460487644295816981 47861
4202940205462024790951929017786288058555750924075365692211789527077132433159593453 37434
8371923817553912291006109507143767639666430870146137694067497131820884153258480464 98243
5188093541859998598306817193697342051639360982343004161599773836252492904088376212 96185
6233567022742717151602158355238619782354882350374921297544217105825909273186342563 81141
1106337015947872936495004049671124794715117319607737991914462295244352059501681421 37795
3741913103002874631882396013932831499363381666226547121326694685979002409476680921 65433
1073672943413705549625192457128920316054931370187773943521841206005310826138254824 38612
6703705127180147953784400806064324196064107590077022630860741640711735463745275403 86126
8650418729075341794190198931773706194924764062914888325197650783623807223951814431 44554
7650957925253218654028131581803020951363547798878076137078863992004404688366173579 50589
7286799561264720505289094096252745524768989926264694474365358793130294660529351775 12753
3746274803804509970538849514340109534934339080089948757259785079230222677950508206 55860
5759174510263746586108037218061087800664127963674045493583223424778263085544852498 51125
6735531496031165868260986806064708316387893805150210361580132569655365104488014259 45732
6999993160144681796025254412124680919570057631989803073333706938928428247902602083 24906
5417995839423883309122289164620783511716792988609666296474830828213604848607908030 65468
5063303164898958521532406363723631635091870144517423043223420108343784747479634341 32153
6821125905984725175163816215868709165013322925557131242031831548752518933652551535 98455
1411146628636234637535701149855387138618167121920749965899821493891403841703972042 97413
5683215032037609373316436217358117491962468688344217932027349173921319898328187225 86616
5486384169759808705160878149700420411165969793857061299998630422811232103336065219 82324
7619994914027814089097800021026849407464331123898735540514751584854434530772688745 52258
1518366341216250371821372972150667781511896861501337153263983454693979668257676360 05159
2180182844016485882825538880603893202790250170344527694425055533592995287946608410 05853
6464630137102826829186774295556278765401926671432033822534762603899970024390641936 33066
8995791594986962094358549582670784689694534155849162996997523046431553000706405278 44927

882866332114476885622952337887353456345773015271100492773767979751523186701257344645562
129187461876777324280527503556664086098156327443592257558025175542650028119254140984749
865727008423692231927182204347846370994345053830795904299453379040345136045751755696950
366682059570975551937947112436558750675415202762649274535809141351231398801191358777999
622833895098225478483194907154159320940200015684416050060047060040379956222791838283858
219337585044489717069932313789495049090264548610291294273486837910320533051200759913422
018339736226719000853080297003529171407462183783087760072985073784853129478908788551955
424169116076551061766580632896108604440745755187057317515882295099959811917496031559192
800883093684195787075407292619367974453254349875843236419394790598321598748072550555489
144047009345738317035724883344280647176059668856105413810339080959565237561360314934436
760856201557422273477301104104218176419661106214381269178439990916290759983133930612661
731075739043262792614023705656080201778259056233420543527976894261094458742084823443959
132720213740809633371023174892627983806133112373999502820449908165073881401078245710451
436891136441834460485834959443611887296706663417144467373175987237689953470138124665549
933218128982867601467158953218320383947390822053232966393210556459067376846636897980657
693054151639022696146190112258887861596282575801039504841831091951868963359532873055996
670419459507902505764231091248590068785481149615885010994959139556172973641973333219407
481035718319969143485252284338898618138076976700627423542260207776475617984939213141767
741008983185724327975347345957752688966004847215276420707814446490121535232063815283171
518607906641062462672441019071016575765177770398395626346128558541442046032030291362131
778658714010711912615580116084814368151484832136631419538080449847144560017123238323935
928736754527944792436025371183492923394709928418475998713690776372591525107285709031828
542680712557326041006012855943361399080958074367248466950851558749291497849996400168115
523305049827804455868263637562900033077962742176714223411672399772678595232222848828900
585343118835380536215316033780573229681694095264057627398648980673463507724195259241039
702483180263805755038351883221815177996985817457168682367159248033973531723106854521214
5502739646100263587447106245099589107592818279796151114520893424150695854780803109319634
949049145456937048557329809743804209334557218462088499677378576413264984148490427154729
350782846034972438551981779965424869060575526708635424554577898435475770171337283714838
689550040629973540838925569600382608509311657609880590042863793827532652058908435401909
614023436571787678043853708011219843394114641879826277604246121811757672409880597335412
461778643936474626426293264504794695529629354044397412141008356389804768613781722299447
049710599378993421793544944080483739371992812701214430362065398514589227325377779375517
570567111471949844008834976880609608777373475175576183205598353966894381602999540205051
081038994698584765508643694989931923784018892352850739520728054861324663943124708252464
737128529142368775064562703426173554415283192504741927705961825143736902747201238405246
268426181708426339462864731610225030405542821951331576772861121496397082486850340927798
971483897816839923561728009850713095282783604994359393926047264383745788007848480392344
600467394441093480544010425527826467780856005111842376882105813572946319368531461407866
053529092593935294719430624544126091186249443529113841195454403927335186591803722388415
025675182599361739104937387964558145527839432125506522417663401473871841376369054553288
563551885936238394745605988781885125179416784493482905343435200425441676073056260246009
453223763559401430803624712979634788295158652361720411349865564739631762595198059021212
868391462356523278262117105724915105232188056103685326999090193252001957144996302781018
256206677359235702893099835679356572369826224229302408438533615614076151243341199666595
791507048313026502059468987149185865747488303455316364028485114459988603773970797345417
220021705539152769345022599923745393583830214085098647378432840735188184362201715880546
102455214726516809275272498628747649274129662470209506268066545778689039520913260849578
556237896182343713918848413035283727477040966631911314086050874357461800597471912223314
786422262045470015968100738955180836981024500508765677487796955558999668663372271727013
354880596045907632942524819688288716023314047088864344865984699970535130609545057905720
057373591364309561099154234310952479898091010437712700778113196649628374830158054562162
932827902247020241645474707948501307401425822621052914824162626174725023020620120701439
826944916390166379957063852220176163098302842882108832988412252613661975128402796494126
883843609744252282511293587823972471261207369565303895720235336817553686446242434850614
938456143227550041950090948927991456404830241599362950858890849663318820800778256407877
870378176376708510737878381444099079593484720226187352964802563645801533023864166742224
302323374456092155341394384175107164014409139221211458172760255214509223614339993605448
917278730208801524259801253997782919278979371274147016427922607627512132141067386210325
336582229342826492037106508353632531829277869472921664440464151661965593189468907365848
579472829889384391450620141591713075165378669781350454560889281302986709490821795288068
048724311990231795353385405164913804693365118562776559620909664118370096009458889022869
791836735063287954140559256833451523131488588123328504224155805561772994543082591739608
398487552143212477661020272058380670130750597970390177573763250347606663477918089156496
0273163712034888989138064186050273450052587457391162173058997459400351484500172969356955
0200360458115575556480328991509318149986728725244792827785111379060732328537014350651051
911367767081805124494601785107204386369879770436382254062477331569243624332184791582404
718989862641215617912389124640683758612656279936428675635091389718496815780970768466425
335838570275284285060783966936776503597298668761962662116174695900740584956990664937223
878799501043862719711452367277949153429688539305448820110933885271355530377559660712759
336262180026934242949932072011999030696190274616992716558047684155466216302230841985328
601508083938313075947514229358028467425773317402732749936074334972886576994667901847745
850100516211063267780061374891714232050053607748499547751849903692305018587282263233770
132379223724455714184683101787435755153407790243482965852301615672721197665405536761412
432295085602142160568747253457075109278926032531650682776471314526569779876602780269815
414416663726185497064284965610404411495396186344043935313167861080168497877869137971639
260506254159446662684831110686244571718936880446373066216438328876448758440687638959212
102448148183911308923468825929483897399880054031486477238414436024923802813708383561561

575800998008588662740989299130163805745485260071564132640717166677074492132987488798509
867554568107175945159275002255024783896377250042572089844196772793344835301942704905231
813680838522582146728389454412352348558931691202105985364471133185696424629517822711899
403181081665333188142522098640098764301514287798188755973503863081654614155977982265598
457880985130479969349452113622612524403173400325892471755166098999223778866989103343721
063702778171815378730493276098189975501444551242422601997092977037974759305440582626494
261074019953244143320578441611340570731380676755335948208021285442863011752459113836413
458042686622510280394246898370201936014370732785901720113186061365742577380986029050069
397540738243558625164018449738966719710949410387838480554284582993969337286937579510726
815209370676561428388917596465763986104502938890634274433926345174775988151836678011421
818826624156305259002350612521258619227059640264349433820359792206496815829982446828089
895505987599301655912670304547922979831290644083274956385556228237848309417227175831138
534349584019424204301449615894094207112065067572496152947124866662663330492386751234745
179047968491966374156767957158300845791669507622043268763634849267887713575530065838422
965576309847089257891761240513683023723700019473570910752049654842689448987120024416915
066338537982263673719757010701781221627988243835402744230824965839405812363817284715562
056138314464288615576672179234135493211578659479776159144171220934109573814584708114890
129801565935173686598396799033241517159409130058199354365940641971632179891050968288533
372554967792993047189922780982593173758295377137257650466915557915490497655540217087838
456861915721069997581354220319993522766700448886294386374832914250540697281991221585941
906623235191542743785240251551978514850130127192334910759340747639990513499880656393517
597814609200989146777751857204850463615153052585459237499054367065962123840739811299460
471774357014763384940731088776976478293182447924940089560053669530793810813496774088531
405551168052817539769263579510941674408905355490400745269923377496624425591312601276341
379538236031828491420338020350348656464053492466020117513410276864866717345861669816953
657078995343420921315015651724486214223673134557657835983828769468619476787272262870742
513923293926458117089204091393012650847161699845625698445051439268717452790845458916948
462637253163391815110422315860623967598983465380243594264186704607926257910199222898349
027368039754643262532351651713912146503498775672981538311320199904701301282698490878822
617143362245100712222436632783082862131869696775289965991360661241473502119176358730470
978817643077380939843801782853901376708928097757427642075081093863749043068126369793423
268781338167676046621193824421008388142213891726411473905324697058440339389151816629813
703695863983137917580028433516217275134979155793260314225974426267535836619950285827588
821537032890537591279536067406089306823316166736834906697838329791612649305894234522643
197103837543310050329698856620089728422905770289910907749236244080255554780882456029544
577109580919540656267201308136585929974358470086764703273604900665027270593020196405172
239532653852829096497007229337466191910476110154327509396463926585719399247959635654428
578921519443115422258308561896675670282691404773960485799598172504784091150620122448161
018767090486355572414920395051576033429371571603509628865698822601203868560562898151316
818474991580231956172809342934207212069974528968901534658856829484245106116998681494534
434172544196501007184596512509899610577929749517829823374384049458533168772283334721148
926515642683489462541026078959252353334750892005341227752053284652406110069001466051589
209748586872228602612439990883061669760697049663999645652346773870715530069871610110988
761084784796843395329489079350985526418316021862142648774331175926273918713775032210456
482689170831322210084244365249768976777749659868424990176196172471636025783622079442711
584134416170028677516754017155368023153957927161211363860783739133937652326530026694284
924643866516764020902880908930962476133812843100611603549841555863107894466953227348805
839672314260654734158953894743751069049530983797259000435889833556612978799716185154677
733885663842623796499463237437902058390417511705246491245558457817229663466108634189026
977666347465551682238892765877940808440378337088763426733804780414113041792328962344191
773456920609581976056236571347199705398730893864951247986495107516047114740501334238215
048255646410252480994139475709704248473936637451562836874968966248383457465157978592821
951803729579350133378113802382946116662950278418989550042205060783561594722035331726446
113678251533000589222948665269527269361128882402169471044295882568662004922440907408251
640862615492949143860824564580883180388310089095348787577726275581772047667010018188287
024334380608700543692494017366628227128653714923694684975721496888366654567103453122842
178982487642376909885754617464889068846985019473567484796233188885412396309524965156382
117364473933368425110188293416617772408387877821888458542492785908205356873872394663013
474133117210261783679249243526323026105648614536940836280313489615417111499888319863949
017193937248525415728242017512785939412762067236990125210645008355090427535406478651322
039944044701313293651109970090114679128375929372283797756836244991123631808064029865042
065400911248832772879741268544815496355997830396192670641720518489772292686912477657417
936637717063341999588087755319732231883473652546154400356757683827126433212097695659554
084173806285695858433903835310211014708680822074573592509271591007318787697414394814775
335914082398179669141426081670130310367944541190572417043145962074261560886687342902605
344221002106811454183593432774343577601392979161997257450059803246169274267122049190730
588244217538832540838767283260474524176157048242973955612949049687376718356688612606978
637106844582306392507181040297030734338706201102340453151847877572356855323108201297602
557783216149313330337127805382131265499563429007892181399760494886390646056133226581074
450317126166454488123429666410478317671210792348081101999930448597839641098802481653159
772098309795970785235164256201769840616676674441246849643429175759571470466517498585065
221483407876155618638607828658357483780257973035634477560016885233499513577779919627665
793677578876454492389674110227795464740812985971475791583775697645111904682162749397616
615459184182768961802818507475838775881576935021445084832669240456430579040270739263110
134272224123985312053303890172334642446423607155892845134965407190388484829158102179085
902271242441877064818408637055497592024784899302337707594981287002531592110383427771229
440051786786389093902485924862268450650166368696011071683408876038811999563903968752583
714840170874288927187786938264886102173679485722997992596862873848589987789215722796099

pi to two million places

3875677082989337042645000282274468938923473495763795758911474237548257997569047456729l9
4929290965753170472908199855645482970497308210990388218724479955962603210289592319488887
9839938772572620473902204732705636834300008483214999822045756598127678752571770874600l3
3911062958659137122650182129704248338039167633602331473257310341941804360346071185655l6
6819517075056548918698572298134943169106528514925051894179553136161916121027566918743 62
389582044110875525123789196041267197339153108815232229707880880066670054997212860277193
8071648944795312097509852926506656733698449607329032759927968196466817829716043259950 37
8839782903059950844585077842733552744385435014535847559288402361317392231713631711444 65
8367856821555716427547514540924539039521159419416355390081213541376589400155222279264 23
5782256291787585398337250487949570562284550473281609921090228882120867123999805999283 69
7032062834829608015579417157498938656000114306720827205122639125758077266285143905673 68
5266971204440112969379334976789073100663339301529731417094938352425186523340500770978 75
1025654925465618599279777331799659057521132758737308896510687552901171083967549707878438
0432459022260009741341019431733964990195404686528561755436200386067253542395618667450 69
7298109406264232646818410793927721397346474649444691602976632284884905381044506342003 74
4145957752386380021674999107967853559300136823536277684298577946624489031961672328560 41
3553959308193646668220539368694686080366160055771470617479486996758324456663286825781 16
9353502397792983919203706018942939613077987328801513673634553784406375156188108992536 09
1587542286461404861953578573170640890666481664163506322626677472414526567822749005857 75
9325986027660137908634423549148668443029652362044541555783230497240936202054391675751 58
8683070418853201384907159002841032465887775245547197619997650490912779818406937862613 99
2037803857426531671237208882849919298799341405880707373220222757811404597273932230132 37
3583095576982483957088960471639642379292073456346433105875861005115127796149538913061 08
5102993538004723994966897538085142890683122438496321319867937569323020598851740643043 54
6447095940008229576203499772748015702088643251254439741259938119111481882184099289895 03
3362207585919420764525101988528467212484809218057445337201748883783380049931461707573 41
2655901055978880956115336660685417602657424182307006382991613805051808177051985908158 24
0453248546987289721383818482786721255000618579273499061202487196281635471923321075465 13
7869504069361910613884947454129832673118655027995633785475993105477446701778258860527 05
6228183594632623937776593293140672398842260623578605846220433007352944123730132855009 54
0281079520183644212233255474791774200486130898692957696521425793144925827490374370368 12
1085352998009223400928349206821440785090449227604964764089406556772657077204852085580 24
9616043080234652894093978415233455356083672844497215429589672923571294129190422912786 59
5643136342094398185003231494443767966463815965948983216256768042419328930423907485822 92
6916802103019401347077722022597522155379490155837477228183401868960795710217448983157 86
7724805335490140180491532911083898586259381909304780976092115411536251923414820437958 34
6641577370418147513811966422605847293018526583235604263061225886218757653797072733028 02
9963182160683737077813845186571768313143192703794555098741521776325948157810791604164 76
8454484296689867792008829442636315446102009742854327593146103955399245828460118978785 39
3561168237850108949676655745254599084251670705790979528146570127666407471935046095596 06
5876207099737655221124143470465052761354730600626722019761329449771463593215267065470 79
8848862212528287277919963339416143700682231170612953738212316372551249920830739587630 11
6521215135088526068136270012813632987630670291703528387731426335510680027174109669124 32
3095944815652003917082267540882083699549120735001677799760626432067353280056442071413 79
0502017365481520162640721955613829285347378112653804589101205143762840074567078989329 51
9090781767392758779867680743150941928249516733231321920870070916300384891351035720266 32
2284863316476234216967317100954366386739882111334411907964652848932507831615021574931 17
3717385063319154130551904058645303413506478204900445636288275949454202116456574989828 45
7841520613197302165934553094091431813646542711006161564077356545348806355512199443149 12
6786694161420828650811864998217520698087998978071536040793919941864332174745408671926 96
9538344802695144822514871351568658335511018199851360759135509734437835761570316571870 66
0034602415986878254773465839758486366829397491534376346441607960310362387969469367692 49
2438501137477035486895960933324509051449732881172885953525430967523271650338433602974 91
1724801316539544446598069727666894671022019363026479512915108306241884624039223952138 63
5265976058178673137376478777187063799982388430931826198397864002783941647378826789741 57
8123702916120930347018514030572751246253890362279714508016774830642631509233291505358 58
3203879836455806005240558003366027538446930209692774693791458304515171238207455175302 87
0947807129828361591292415643557893120316837251868253611122181916398478706550677849320 97
9824751790543082369203397937399844577460095437989938931592475505155228533148676684589 19
8941044379946613546357461474964275891212498628394245968407811816302285053947835309741 08
1530298543082813920146820341094263797669884730033235568846149286909154969428247312200 96
7078864832133359503068933753277352194578031878306825523971634696663082560452580657677 43
7325812444944552745033121283598884082839017502166942836965008698347041427765019478833 01
7456731520839075858418810627803174443576795781739882569963127396230693053646889987476 88
3251846489335318991185389655726289643484017464044974488043322089032226867592125626870 86
0798999330134272252793621468664872675984964568978357885364871744314421217875013160952 17
5473991467447274125177739083374405497385520190933566263065120671445629803441466153197 64
6777985662018225023154781806555038047888435649651267965685079878183657795927209930971 76
9538571515163470437321502991768569768937172453554421726716771868226181095913518878887 68
4706748914315199821989660422022630952314982290480034810317343298316066034335611872726 04
8282349270337127776368567157782315122367770032912594827582121891390352588483630730491 69
4631257314352267814458404815675916397074200619144902826381422790591399460155784257220 35
5741808083888070190699920653333914186404144434129901494900696830074738797639465173024 04
4482220619707774917379527760673015209197601000762311773909288735707403285306096805723 03
2415077455850431496028421636098075345826748204650071705745688545618735223968612895728 81
1298902963056797431963398482413122250777183495197999736364575973352696912673023920141 13
1151889204635754362130730838438458367831793636081897110106802853270249352807554035200 89
3234270496200661208447058132532793428876657010526316781719662023501513467632418845603 35

pi to two million places

3842796726077928707641530104435458424069533633408410124151106747950489801182569708770825735353828797174778895993248479057598224130350496108140176338492023097724520311227932237077500957575377846961909837574474326314455721757275782268577142145170066093494445212341405808339548689285058828932603622704944866409564967833146538415742700525995412426597861588035550513192505768677516872468599907398663464141555518741741484434932519238087440824918969196256527507365998029521869367642595550224550173237375672270115508339410666363855742428126380775659207372209718367058231753321790172064983012778172503492837200490722473728880476244805897006879345694591278719194533544929075237785147771635859420024631337584181354428843182810535760823984737614887160422707322447119943634165598776813926409103579076031346536203632071353311644437202153144906322889823115402471531359222495138072558982024738128035308214408350321848510587546749414067663363909905352481942133190469217333475022818623824216086127117647048511780875643067893341158800699563754394176179748171100506042835886983097795177358312147000443540795476491794024175242739894149491595062981068234428677132488480299163869313842897591632220472339001790853579997765034417988094865182843032416242624255441854998835804546684051475154407267821246379002930081732058412963997169689784387307420725164651897000920707343851181899796194540024951989395920476822234218271473299489498936417411237813918201849224295344474741229726352379624202676382049366140897033430158972264332452714010076233827478963043688243014556139809556349427659698202395358060146555568158576658047061309130546535444463931567800682195775284246856395350005987008538181266667442425969956882331913813843262199487656355826650282337358744111763284214542749251037456497778240541893308829119101764248006220817953805325659558860083822439214622775271705616486186214301918417439917452006531912604634927822651374210378027477918424776516752126833116137901601201834535700304597165113280581825148059253171008481965267638622036097416502133953075056723298679575064206258837531383658508683974443370476033762132008310944427697470434268403675764347132981648586253885122348822818694704183531557681586335872310531417111261713206483837108500493473481539843512261839013575291962656800572268165518372611299005736759000282207714469611367410371806433474186772131013081910938219263348938073660021633971325826588231197594883228401223073125001665188221944701930987232024461376836681562332983621267880388005287235284658172777242762368405160665335870603135986939905729469461714301412290832555927735219633256751958951469540405853694926445200915944547997092875782883145859100762866311796200375383255841128583531026237549409239818305472059471936318853151441273588793607253117103467522873394204104183092437746863792838069725293594728176123833110577645088498194279349762288755422716192046947388833948349335235637986963225368833664437545539577048312984827493284284965746862776007391016484498835769090766945457765955905105906300847722202286916808200368713568051717502929933779569273026658098504185689920358230173978159603538308401397581726252478012902901315228114983748196911507042836061119402248673926020054034847337161657951163627138301557436444910668733416852373370421748512928864199705747795531369544066610306846160391135776537961342535533487333884212244367085823892272312214200045560007369741301832549448185949240179967688397361098812050955097643287236151057568948677648220145778604998321863088377846903063948152858040600539404565274436458710946161680363551478235230610524232129335345233559099004124561311687690488770915684688769611743196583310839166436691191783105487696755133066406248600310001389475792659492517326395664162380417409599838410215004234960482511484968105884947127079954435859443327801036065577473983182854054166828930332459302441243525997317736331936359114109453986979174263845741058770003961719970787802110909592249116545573869181070548088593481233478802919019123851302772591687024933668891360863987631690032747946799718246235694854320007891866561250895346760420402860570009602108740899127326744681954285909446109325095089064469308589226197240162360279764600454498886784337339772576128431594746333651987474523900082160046384719350660902696308996421907088513222096809681833709027691016133910254125747477313729041052027075949934731640268673751333745521995505501374571527413020027105147580260484331959446413430189335080005753551470786684541293254586723296802087861651846931691882920162100205409911530192428740463980690781433749443293103748402261598823432676586379506365600053958172319214223925771969910611901814807421271828153375454776592389230413317664792204947750178744738569661392630426994839907826966682371449053100047167048626863583690429383011313487669092036096492967302628739205852454319540455136385064564517742965294065827745016568399508391759601478883964584362254496371966949179160578008843178213482234855451243041843413426943014386164815240170423617059310565298258377322195974665147256294673119601724369322751641175748055792519803374994170016403737762219195622207153054535042322453268831139942317701076599895731910619741334847167281264847660392804705111061101359515922690105565274818915915641844241942831660267040741656059237096686848971548794703137410432667035565171849358355786659433253947271919502312748773827209644859365321842382761549870997780451762357911418668928011371184297552928778666084110724095746535648842542565297562467212140819229517224685415105522794750678519508557014445159999594482390630027978442128236831296188837142293526672535075542996266499026205449518235878857286732580058360724663707616627035526613344469939485265601031662507195983044807828663783004562140296186317692347803811421226411894627920365707902463717888967240093056405345175373885020366354411987874363247780571989262471323698724106763754934457066852943290726123454134362017705363653800762000388793056024020360773605254578377440439885525999874243205650185434878936504363242257112808935162720249815932059621631128522298962990460481798927857718806042450330520547819209228218939090798969963966262625596420827039378309811382638956046576510331101418903985340119377497841619592385064030600281039566055408182243059292157592895716239375844960352167325916217304381591102267539733884480155729493625207319632363043815604556085062683296125168453750999345831939833678965750497865047761114108313172658845543580671507921748427946153918576743914656871820595750128517218114828207160067494590924245971356194373592882831334808904987555929386525497930206134479669687633203628383835221196550935284840608555977260307191569753635687171663624919696494513171261941242430543748725812737064537451880846789532633857963727048087385835045237330186383505183508557209029206807767298072450759516440105007703437877813993357137758437329963152557302903703130

pi to two million places

3797420852855242193476918462644545843061018557971537251806547150108731295040211063658 79
2996796268932932035735182287178272998050485234581137438620764719199602890484586746790 40
7727610697952439309530403823061746269939618861054718828188953047476795543463399472585 59
9209272403023665849961347890580200774668969869978799212231094977828989088375670330006 04
3829806230402125234430816167648725551205002670053287657850592449027213795781681643619 98
5788775559041425117221143184309354777380102896115465288490300362993809554553390160884 51
3076598754328609068101706009110124521384812498606366066078021592701569636403034676361 20
0582132151549142981666194227790714898788776751811200518513372705035160829553014862017 04
7029285225773877228841729711932555080192021059897653182987690900217519910443599377018 71
4330807339252415576048039542616199960585080997834685265963020238315880127692709098724 34
8736989833709990956697827335261322207005583179092251488917225565181492057322061984903 37
0915476271288761575989443310360621471244749201939029262734263025485917867272648591223 98
5016660886370377393306930086453210779500565391774919443046366296488855138998647720511 37
2906935224200484397323018648089409733248149526371287940815475297214881719549046388132 94
8067660507935314292607120402095410032939755870150357579336839887903800057657242000095 21
6482324786967090329368082836725138877359787324495984873998412585770606063198260924114 29
5415607361653536086019614296972183565460509495369610919646493568444233751766099669015 60
8282904744791135522624021295017040950144634544410435666000209310796755414531719380336 60
5225356584382381811729712857035112420563153704161165467233725279131447084239724912163 81
1069407869321365854834838643742028239235451934561280830070157105011334821629319529209 64
8784736948637563779207061220453716494123418270094741405340081725933040610413468527853 41
0055534748424545483183372267754234300081933370658256528768132825345123155284102623666 27
5163120232153456811997343021724987223019954704038463656154197397704912959150790525499 38
8150565562303127576823436236857066961735070334531040320635970953969846259147541599329 04
7191761053980756421773550906547753529832116340359206664551209866480302569559876590369 05
7669082668741717941832900057757962352280157303530402819337597379969013283912009153527 71
9381067498705284998135701506651962935252149679328453394980530612668226082708132526692 70
1402355565404266243102411226213826138953574884597362016766482283494042008278222588657 65
3531228768251518693191284816837925143258106082528403137845904030976206632070008732392 02
2961769628294104839180080641263445960067547282117710410030913046346205746769257154359 25
5079914272701019814065110980005894915470982930270551502338438147786932693434191830784 40
7476131236663562579308298684649290051410483714711426736150762498783465012295705008981 57
5503190870692363618182666525885011701502511805379523769340572437920975502315168589866 08
5965207500581645437226504806589138384294021315258584091126640104237059694905296557713 57
3525334243380928390293106706157073935445161011592307466486096846410267527738629875516 84
4249242705349330115728782201330909611183627991378733314136103031810336860969461897650 13
7154567292316623282971848623545556946499654426531474105873957529426492553394033985793 44
8532962657165145474175903442624999990653792603364209456638062470928851495493586233112 85
0461703178084164423372120605484981947889801042555115712677794750910503694100261103223 94
4934756474910910259840119020282341289037417082564975912567535387230218950755773519435 82
8449250088476095370683126648823994006036011838101885670728774734166722760274073718308 78
3968573845111834460523662056088731928036541682558203151254426757408895086866965351412 65
4671453212042685346162428617599059946481580678523830669146626390505655612938924556359 88
4849008703288437144264379852258047624342607766602306840227143181005742837582013899065 27
8878726475395804564347666961249695426202011599408572130116366732859815767273284488968 89
2093226505035509166689186406173349493589207709227066420117670450947216330374702621603 39
4973102216038783298849262538351095936856848304815076707582456266211139782659568665944 87
5575408668035718703608500116515815866798278843730425798387620559057038074744082740919 86
4898266686765354493950301339282294876294109252694551204596486262472008814250877924320 75
2628542206430332782945161651446489191702721507012318330011195902101088584201257635829 08
2565678949466300158874871598610076931288651902275410920584539632839385766209375631120 23
2172146392033233952859030748587120385098289878551185200867033694027681950845149494711 16
7278050557603839268590366026914238499894456693328889854132207723503564022465738692710 26
0990531353933398630398826975580842087122717085540171910107981096155027547956470352183 88
6663897947060811108216832227266255088715885332911357951121944343211799252031673594788 68
8265539200030167527848992652487907179424919864945797931728374377582930257074185838674 16
6824348244860083769833311786940646704189545101137141124406825258242969304734768074530 48
6152177455229080977970049634585150627515373659137556786342124092992573752157062640211 06
4332578032958784301352556115029139427148951218584662018693325217073216642728808163116 08
7306889785706222096843895014902671736594865064275688143516268018075146337863351987077 56
5265714379112685286901142254230832415906784556830127537223218092482698283106560855330 53
5123016525305645535131339627619003699933615639734490285053996302406234603290975438675 69
5811594455172427727163151283857987575802669075350200853777859621909186226692400558661 00
7921604446911262716247860332756618680476890390701267014905320976146665045921272615910 34
6168912579038293853300236053719782863960350960248467613926147704396013528164164049095 99
2484480240507032176278475453612289861285592198026548128859242694821874229757641839032 96
6958840166903878362224117596226129733436721949699535607310457037301638636582801256222 84
4362072181420368075024802473517327837680841240201409329926074042449309826855138104303 19
3183294082782393056989110262169922658561062159086149626830623441742526097167260755270 90
1388816920081481741431586435210236104795328314743502273820855936224521555377057000182 46
3952758348949699391251890567431987913356298614943677025193970187287651344725508664199 70
7537920932984815928333453387178899259819431513157314622809741061010485563909233324492 95
1446958676393621084368096757772416886383880507480711529579458200295650336606389979450 92
8339454872699591332832355257665210645458567187039904634473832379109274313754447675688 27
6488936925804975077388811932210865347214772918706870577567634801097672091385488086338 61
0740409394561892651206463566811149148609393142908882109395066985909764455695904910979 33
0783045591104243713388249172111302012651142203631792259882052976063763780959248411189 91
2292034019611488386360915265276733173211665951546614089592203118843927131533876807166 19

3060135944322640468637094929966440127236079378826262657190548551783322162013252747052231
1694328804663018430578162600964839886226771083978924295624633817043923377760198527207764
8462262533233373430783028851968605834325303258274636809053932803306951779050665820572176
2877537467104355186546717434014882882802052577658517525256412319809791779746518397529924
2058388225771423154047439163672622379525907033234536649429287442132225782179152950129722
6820627218468413894641809178149449839067347588296389150634673434178961075502697066573752
8173982095596717934744374886807102970781604072672867948509531230550469095741468086335723
8784504674228087300293682473109657138946744243757183865281481171196261293525042189436539
7402203227485564700347493791610567631223363350957336738051491082271914028019109495573894
2208314576681548988563676765410364282244071995185433850096677829682729580923119278136459
8400291077596559120433539095205777503509453859067051195893729843290677971375953494186024
1435983514988012901330669214823060153533770155326042889313328257626301701436101225238580
4299700289408941260947277036679479243859988631261715152274311322879620279216082197258802
9270309951912378731547142107179082840593674872623845052052125829818504696762427782896351
8101325688987770640879493843284982615756567676019502344223828028440806692520384774281646
8829135244661635457705378611588076233157002616459492187967092894670139977636866885251493
9718731546196056068610273989284264742677243354907208864811354782085533755317464550922198
1448599558368314209501186104750193852421085110521098946025564753833194846509693957037309
9307279397434379429820354420411133887987321324779480291673565705109344820872810859347473
1117429091638552316437637240993753804993858453481741394100264736854547165064087377689367
6477644763793693392801385492090357754816162664035092857947194427758468986375801936792677
0883331246156087755541006185964046268663499748526784417068905640153228240868787308934542
9069717769208743376456642327212081631924113818861326159890294262215859652063028856171418
7539574574843148372029709998714698475654899380500326153582130698595864962387827792401746
6870488443441313826022634462391844967181614879197130332610684361725854176338751329989884
5755082379200205347311734933178845429573446425183054378842807159389355121150679161108640
7580026524197288802878024560869246528852855068198506912707540133345544808533726757522127
8592346312597120442883534842348532807759489292384420418152871029442051966182221711917739
2730737152755611186890744780699284070928687532256825201152213171374632613165824310858274
4439741403033677164399946874131825942090322722635382886254879961371941202999555376311291
0934561553810441475766805063187223549644379327196913483803625166943261549910683181802966
1346891067019718571985783431015248872195655100071876513842517126929648349573872946541559
1108574288706368389926253517691198656386699829794534626915291572619622085961941554910963
8543940456505001652066802074721664807408022178992842088424498097628388077993292837683076
8015015203039442614332728774267317780318415641316196051636347088122870168860262468577387
3475612238938438583755684010511732028350296933331532278695631348618411512918849635789501
4979855095631466736268943004252069383376917379613505413320373604265897132431104233587289
2334600219374283808984890138735465068851108393856571266417221545546626095694091896208361
2846019677983222087324963495405903810631058538849244911711522252544689315328557184622125
2384915281638714041390799436861894802599865920627436770142891606780775490562590714321386
5660484100731759853032480959674479488856096965914302006148498632187241157177688153614710
0922659927260754112157396004167581681836338556357793866523907236469209220857546662161216
5387447678033341103069819866204708926273879615293221191945848728619199783026462817816471
6304559179136073822246237278281269207624309423826870607509612480764049428354331943252284
4122444366838103904545921130005267579622579923639497045088615273121374448236939618353433
7336763508642909387990860045572237146388410588433192484106172339571569852244903280570052
2585454925464030798946658522485962774073673956442956886682016305155624950564528333846078
2512372212617190379937179656191519895273397085675687383040254515398314790111866360282860
1197567684109017546986718149603369876694208776540205530796656305713506929956308145075095
9748532354967094265415072247007462004616097545205263520596586681002652800346980649073480
0040872678865627926000202975825476942980971145532905189169428972156347640512910444293969
2761502432334918864270003790657563680332925341990479637089857293742344928324785824689915
7360099148567229097473200044314865011564962375660585390914975252650478238020973347051390
0321764444803672533326812057845956359545129460959114288987884215966636735706294656333963
4135198997659489830115087386037447127512463174540165412350963856316125935275196628394098
7547985867819874953675750894251521280838946979856291794469814623055360213696140339468425
6923208585672901978782434529046668798862289878284362877756327087425235365829436398977188
5412363253940785905882328346527535926613415133557165261991849436431522587852938298720977
2033347701678623555256436526655889784358668130226511684597418511207017140634693905053888
0057470836259659665833692869061839309770037562544724590020327366908175115252919513881241
1510418626724075762550868871305838493595559876108268615283044570043462915992517987052462
4731213222816141475672491886530843499724747469373354628797461201801575666652923784381160
2002485053802633303804390915726701844812517476691587086240955779066344926264601782177088
6795072177820498751969755796333367010091046319051757130698429383443406355426301450690543
8530729241587293408877983028902425292030987848039912382349690662285112454392977727887042
2620734497792274946884633980837430799111334821710692908976864583042872219970835165933653
7937551726698140501940496100572894024231785364204109633772422963688721144664078669934807
0488885086081887884153003710410932009177901080924295471760175864583866456591063894655989
0643774755780596965781000443705315990218959451441048920053957671557932398202554402371168
5148302538954165182064430426790097843088410482349876705376545186904845948764122910521326
3417174035440773014109105784076952147149634085333835068085428343638590991522953416962005
9121268312489798634721955482737376536898893789277337767964041935918094737829513973471660
4224988901186860760542831129834762431582712287125561889125605016836609101956800302218574
5599239309421900439861325072580365486771799973536389655301468271933949976361413292594728
7342260729410356602474573581361745589820858011616883688811836465457794101442646991956963
4828806447782365632580007048689657617985953332642933588161315350433356287218277309315369
4638134546867694189147298885325871522877983073140920058508505538212660794421723977674191
58858306142

pi to two million places

7322242609909044178189341642399485265935262637143283940521326332649778235848467938897601783451752853422488739306707208737963948740066001322413870377929403108214645279232578363153117187638660320018476102982251064785010536763200006250344067377956675427572153373742793470483381623943654198419905163461415311390777932877311807488127809497703217991505641975741636447915060620282481735402229677664197579416840619321138023497260031803928664497897443271432940504815340809579309660288526368011661065803977522762595604887590851821489354633096517122289842092278556406917124423457847800481127003200332993989119978440169662622304084682925305507605919297296873286371112596640911542318793059977776872740945536494242904650829978162737373840733233103891774964253647513081708672599979551679137891706792721007770500730499342918771985104964554425900643432106471700817660012691242819330193438549641804431102300643321186004492896907495467410917210991469359307027911990500888091078972169581156129849419070422059951312052500589525853710471475805315969082138253970534861993410188512374456008770003373141214734542829532453161479032109846076279645570061075761443694186504002836235806914978400465944621879366050507040964827584079530246759697359235415681025933317663122893069837381030541368612140999138911632116738402921470698421303955357982024827716124674220724529703509616063365736495306840529876959516526251349468304028836106332138366344704963462658860837601133028221868637609234879072127162515432615731531815237411387293922589673998634973746225722397603671543741973491746450479177861211020360803586755676267982967567828505574462082153739952113208834976836873013100924730265713187784430115535891767401552877320825841630325715059973193174805760352871385833021905227890127706381789380012142591233128538624976578129851497641254313294430322894183344365329441356669143970462487530735600526738517307320364479880082407449361500630596378815471959046693039121951598085584676431129150199629488445550903348253320530822830242061582866714942901658292663183572908816975462141006234057986429703668443412721254963212523995911669933447753018176684412478541856467575700462698525258623899499595430282241500229060808859835735630440182705550742081009855976056827829615549500776955399762766086256776015679528471368178061773501767512150404163490851005419144802247561039226087886883850560311298186551796776652943749177470465248982771901790309396881057343973435829897243005067146375982340074288278145255793819306137266416296291895784363505685437767317368833749699093897038658009927368814596109748868203402510916207808114841369521752864862491708488090759891110550588778521397266871808399436479808368642631145769946289677801619378298775805449252044662315364123396247060347958837321721864432560460598799841177299955978343444307077070424882059450482887720434101123065979497020667694841895596970926762852194488377064969864676035687612348681362753167471691888425212504914713227706690436864180970490154997211944101386278479335317654652667843760953192962408185626137374551637696363115122005786241104688734055789959629429795367986019300460591886819167596264676111778863211561613446021361277707715020673200492773706426170936658712220434357269355578310643447090405535046661269188517273486403467091768069577559375275293812814321439802497577552452693525271087796878327704266191855045118479236999095617461744277678250083759388063913316945020071800612081855158763316238554512822176413277954874702086530448917953535397117930214441842104524686505350592752178858571143404087439908790704315431344254295299782320212465723437447597355878142933223463461484418940957737921614080167926061465306046875991756714367452743068690733898407407970315229855971236986408617669614588245183444652956865975648515973844457099755032189339654345799966950518039086254422330299965680538112653735449990650671301591086299187872121944756891045209911692080908170652966881053911419857170102543990076889539689297478456808133271944500712498912499361186779270216378231820781374943321356360355539897235378402130557858078670374959006987294541625017760912781944292522386955155889131231046410014926429474523017164552841905023283564079230883297791125087522291862369867470975138578013178952358633043602699760488629918548119811221744140730732636972627718570014607834154879827688744833042783538106182378363043553585880423609475111273779531377774888109725091109914497987338602455671848376075161755816934393319875793932666549198226884909989247267725030338718693384188737841251525691512011853298471727431758973779137832107438189109710153504751592856921971893431270251346266506661704569416840846263307062934807653631673119508340048887456769467208667657551154077074991709577923280893998022622210759528616983096900622171108494343358819691849561482786450717802993227111545357076538249267755613844846385813453947846797525653022976944146949095005365669751944297761104947685004109830192271968804377809182543228560532741555978519816957765587975935069212994947473015883592927095792568764526173829256869453102391121381730898655946528719802085744577851670784866660154730182534979515302125741342705220347733213416694056655049655238505558965345767488263653289728331453184242644051061542928864723345945961988640421728633639863469542327332450766839140579270788889288391450472927074445376283399646932594461383206011527175770319941455622237184996688653240308837457284079212228785121983857199185828934365254876610167987732295359609509813993223872262150097742930093725823994695955548533837738378211041453004756744908721457867119127194871674773923477872284099578211057000651503230635885221244416105132507990100859718439950728585950451653073115948839758195003487066310282412398312277474483051995791383813984177317167163352293206078867112484816204734084551472479023442232881405284562345829194803191153589245605175793036697120079074785175514543711950143989110919685335883447110419450346079109898357351419256732952582096169805479468966501923377387560036745936639290425234874099712396371293425639145864168988070216269973570719893910033081762701963125881759978302829393652136366260399984180116997429416058871530230083864279024544271468610722570056772967172628417846139027470514094282415067197920976369951084465700739427343076267398114758784239900081941817741566196177909794017450417475558632519466450381094782557198508164394394119884645979993967382889703792270172967860310610741322511579139158108840310525893809469106839223673041607131996222987955169340720057807516169892116876033004858375570244777559445904958247962734449014545357052614634700864790378953577323477706922069221192909142766915512755344679492320602185206717377498562604103627319458968422030005697106044713943695176488570415663168278742682798987290319556531666113302121497487090551370598450041420066248820538203289670007292921096953571362838325583683281119080870

5897155480415675906931001731973575498918072796415671142941914051780798702963939829993 84
8689681907863942573652644861112834397295813350205356225544058264076567954901455612532 77
4246860394778971779310158104152400487670094474444974142951874462441956169093667774874 03
0155810835461514194573707933958152146944447505813441604082429053584086900443817863312 63
0335051848730835933047002702075731156046928460037144921150674152647320978310727102339 22
9729784518992259193294137032720874955165072767979612376487589504562097257947774708865 38
2791967928042385654317904001660815149918289019996532230501419603928303322069356261076 54
9879582775741924827856716834920535003947622450405823124400600487240353467109484214009 58
9983671715335169038460759062476726286882657003794226432572977766349177344227444673899 14
5010392468584474765275055873059117242026642001027991303604765715247913313568243814180 66
9380158029155031273735465510569709084276282634373742307848403417528446964781847141725 11
9604843632784188371479408492751307456013520383154516234607568616937962627894043896146 83
8831861762886711790482382318173089293271217339159854126624999679970322512928359364671 89
6757232959666707886043419367532523347371172624537644547278693741273844619286486584037 83
6813181336766314662291688023853760415657708716180575806431283802594261253377443878020 26
3314160563270899552806397113465161214196882473556230863640674847144315562776699111248 31
1087880587088270290668078088828579994325415862772664493899808567387498842989197707544 81
2079432413103384217991425062441272357392904679816019195572486805841628912396671445853 10
0305209562173525220599077724604369226400987769042765290188961319847173186646111772392 55
4865429923121928842853615931543121608874014173299965151314647436479526669512772308118 08
1930010248444172955807796340030828322925767594674761316455495200743053609209708333594 13
9490331080587278690415410388914865729346163930477026500288250227947363323176183602143 07
7600632787944144786778308281798110764536159987603650643053299912893326613555500871146 75
7885171242428189030950580663572772935229800450576355252022042599155805910473389756960 17
2466314188053565861814682910603252384694954248249722198861208560630348166512964852930 92
5643643710408156726032238136683302174020634568723709217458002884131140335919191606591 99
5909048092232714383850525122775535734830857388616262587827921478354627932065163057205 19
2967404891027304431472663566681154691315079658902294996386775964685500839396851408128 29
3888041171345381444152656979419963730198592755533747652068192684704790043316534180074 84
5808598025569148150215509687572804563306236202852109432396107723530994544914196182847 67
6463142968350572764875689383098978505029358557186241167739888223730070374938065913449 44
2670185581137209141638809463257214429524081966173101887451026805729023317069023401522 64
0868700852273998953278752572512053907738332187441989522086800832670350798629501100533 77
9416623310485901155699563295362865479373483414072283361142642791840306459504440007789 28
3004179897359723474630587477017639408962494405176897948163537294548660753022699570774 24
7850610784036729681203539480603437444421207778613226184171948735402076489770419669490 44
2252398989203050301179472356648534912143894683199310958560341471462774713675190814338 04
1573290917490783563878132072929940420454831326062183895978793787305723418899067184009 03
4817917247583236334412076565238395669721950669554973512289908409524732103416028645435 36
6863999542126541698803459442804011716205609847937735388519951265156104934099374501836 10
6878174153929178882027230511765194272143004764252940618420835699937712851495082430008 33
2806023118790709144278383365447660824104739665015120662599781549579195601467000118865 16
0870401459395669330446887486702455385681525433509517542202807017713886203098675690995 25
5951226757022712031367681888505481373113629553379654187867586739386128245873387852174 06
6511531328887120848727797131025035805578775938042521627102343103984024531270260765293 11
0897928303600081479170855457539722066529209261365089539378038947227049496847967634942 68
9306941956369288147931111530323603108955835077123507894542283712323527887834055938205 48
2450395869515810086164353340253044581446864917256838364269662604222714102749302249761 79
5872211343202063087379616750652644211588664395537051963889144973110269594992729388806 99
6546658267423990970859618059444420495367947100467423530841992952062068905417432810452 26
9510412419429294645130196754189403765724447098503734283190250877093148213630114085004 46
7664672019580807225465446087658378169686028785219155883708464618608831378133821940107 51
4449242709844161351784923496303579952194152428895139053970141865470944467852548007561 89
3112861506071780848785150080384122573760521541043149972590732291690924930228198534147 21
3125684189639295535984182950597203994913821359793170215783801881212608779276436638222 48
1243797566227759908762343920924037962275371753965418838046850841071434123282146704941 77
1761810439002950258915144434914287363840594667900065667542124362606305904938430368820 80
1558377323658289208541680834170295438746247890457291423811989796906931389409172594587 15
7511443840716966911593967870564014271604397214545099756662506000471710407367258200317 206
4142442700914754011572814980542383939910403670022259248499013789795428672971387153724 58
1719611171363158611435984506140489263300645647095408341952524421146349787060519639647 64
6471772785886951058525952850381229641481231143642372488286662427086342837131764861826 22
9826290240693003398833281557794968053115624831272273488612730242318141235889923659047 84
3272034439046966372192121383668660395540444429260425495687189708083965046851027195846 75
7714179697653487387399484858956571641058239461384863974949816454143455166851231249548 71
4100728960311963625327229268831975963403611972429080627329409481052512954922094142658 47
9716945225867645490627729960094345614104379275480443687145570147194739843096063062566 67
5264151434936483471865050145339013791878520405001812401952737146689689848996741695788 74
5273102503741555611641755311119360014118917292666116183424759112051034092022685866947 05
2478760303792722241905761173444416150632279077598601836054694835999040512283096951235 28
5999369053717277134689465930552146194190725224831261292133457185427154796510876623650 78
4996004770356401832620982876360169052212623885521669344150246848605220612939596421271 55
3133786830544100014154299832028797282796926695976585698224321952385910723671178581225 03
2891671668592659899158758504787052653286940550860074253679153928320138517555791104241 78
2429068513966783543859831845290017440932203658423823314467973246832747896125244266785 54
9482774605754624285743143484469072297429721090500282316768528100883765950106231769626 95
7456217371885405422321916691428247802457912119111176785332172463650498542406798687658 88
0638810516078771365317056113903584825207663251046531048040853624737039307803068646968 21

9074431749664543945998755044116190595451927703516203818443427501729261747244186300161 74
2915840702720505542248555756938163151812543662769188116183019774956650194148260601468 72
1184143495145976882295415977788165316341527149960315979955977612929958255809747196175 95
8467462956264590306279372658029451830105963277440832479817362119198293858751679664428 47
8893614673433285525471034990936093344925212957853916483663051193257059367592816469795 13
9640546925299223018687912185660895837592205630331592307739900785403829445072026380901 73
8525244927244949595413256214952763623826236825606827677154884580621002115586769740449 90
3847798532137209091684163749026684733763320365714330093247637282525895838143350997020 46
4991578764863286689509876045344595348670774708006388720994406661248459581183168487865 09
2789053424593827643188015701050198219273396460487843729608146126626907560851796502584 96
8227066889333759357347433152015603710335795157982687000225678967212261969237173869605 45
4027199383525908436484716601555430362499355057295928818235440479301238803694560012249 46
4378307003393654467593273595604455322177673892242424803629211028557936095174191213210 58
8311431382948216764362952324204586956530948271547087645695542765414564061669566974257 71
9873499126567004057058566860990919661211748092344703295670646385414908145546657874227 35
7623385664628126376102463388519855947162171200769016678189061028255045378308352349536 61
9656649651443519716432674596276510152586880017816449838235716772079031429448467492292 33
1326691495584640573339568166314141546454358949474471345066927203917832444669424927456 57
7029189696459758816296286726974583691677662501184266843275371493750112860867083414606 29
2571579409179980300237573863799786078705100588931346631134368553055407892425419408502 81
8177990506050450356085935575322214888972288473746287977732841232151798094368711228970 34
0635053617626902083070653453873461688649827715635267979181182302902968276768043566318 29
4297211694764461750793172083736192880032513571878629875947688188165134029729036980824 64
4151356316991531412515900958229632101399776732331146612779421118607626713669239653391 18
9333610905867346787031492582344604770277492336398786391724147825649622073999307632046 21
4874394057970861044188580816056270462844993637323187864882118174094726804403818743528 31
2838937056536547562982403107801589206209310100501497894303981704523381554927416777536 27
0752842585792734053309266398260150388109914722190863624010706891582365805136766065506 00
7027370129704125413168361638375116919160133177521739282450608134078776189659903287957 98
7859852259573493820592261434140218904128248191180770564951990798593695110636467081547 41
0743249636101715836376206450524950726500391887743141286223707020802874570485219567931 47
9469714340000051131416934832652397201375770093327498260386370189448925796257278954800 60
2386466348417563506336447590755727584734565465092207722063151132444733687570923538552 11
5436062053662494566971949137157053122682831338857457330326062093536903612245893239692 94
3009486350231193641837211387977224415194866591642268676450669958662806097207793966789 48
0167778291525458480329861562603105284260002792569752762972775856575366305824823175831 14
1394127670637845448930853141486785894743331632697072447100541044219500074610635287890 18
0734851231262047962825722802367680245251449412397929205781195737713624014790712541186 09
3879049857427291667205644966884225638105371387141398081509924535137790879602817226132 99
3712498369682688318693697318449520416716396161710056682789942882018918043235862828212 88
2707463105855802333303747429441645643316178154556550819486105003618615487729322318309 81
8271898456586411468624991950945346742807058914583227882052407693598036230232220614707 50
4813528283355115037858092745845484740371239834398538587335702954552734912403414014244 76
0488454636779607325459544112353783735801301564871770796992653586591313134680348483527 78
0670199085041918524255265561470931668743856190894442851357704630856875794183278597594 69
3679578723109957386588345424525625187029557462586471906604023909170687032350658424438 88
7711935117312843620877489036899801632509791921098542210251814031278023140830380978403 57
1575037673358440019348486846210006764428921641460785639563521530697754685497698920860 14
7918995952079159273318700635320953953035654027768349038006818496546502741664666481990 99
9352741923518493428242209504260937818387897033350718054326819592906576354646706867139 10
9029051309507446384464513162884338650337903490886458161923362891766462726775461943515 46
6725859690418848325477758265402701719332210380036899731634015626174433076506158631100 91
4865090608023442523768765666066183414036863710228578912640993304443871792778346657082 05
3654404185944494668203121846013841777413591126048418720338294092977251966330922570310 85
4463766082011825515041137972489098044805186432878536970945710497005692335478544376568 22
7763372664130832723383398425466615619344009155724424983025872984113235401101151259945 75
7916357184616296802832643402687531124677926494990337472835676560446044771814999659973 42
1468569340543850132651966734542731938571170593336025446423752550830382729760277661512 41
0259231426629406126975047071555335122089312016860719044761765613532836273405252194822 36
2653570937991080485336862259975657201112743545776897652361416842673113974656383361057 82
5715405129696773338435138241699857078902485467824545813562566249477146232723794517691 84
6463920562581448092871353273482925121694521975763439322975154640683506980916957435057 39
4605536325740315230131906156558210384629339959424566448563246403866621673923853346164 07
7048879393898963863526122710490964406185380268342340885415288227137454371458533199277 18
0963060010455759466317036880686590159310953636513663342489585154302548728987564203346 02
7362209303848780499579271554907973506170722388550784700523044558831301086104292728086 31
5239038013431171158256061944929372245775784175677121523618005716242239261312607423053 06
1613072601932820897267824083839185058653670756574172862980815398236010687985118994332 70
5714564377168101573942093294361013846796453117194132179767557231395546509675510128681 87
6191266748545187386839327058906561238642920309203645274715468863419234797157173815748 96
8524376079213177910070960449756317656651753143642416207371662186122154369308027468478 21
1268800511269418457450704809754609313195226077490962962645100062812501640641792099768 49
4461095611528684978068089674316137773620974335145369183822466710732668580365959214789 40
5542955557989240679768277812984278070642181723631654370167610712587616282977619609603 60
8786409667007575569911644514280493529840648920261813771930350226711850477685840767303 89
4717011067126964655298487461649150265966913981215823003147595811840920041405818855653 71
2484678569355322812255893504387336010443855228609257090027496718184724407256382216076 75
0925168769855616568239255327823415670031005526139493574865966157974846769073534868114 01

5260456148978065039522462169146705885939088873886775082212432592465935385982512276540787
7331498512905189885988630213235487044218386746340412124528285964781647750041438962614688
2616994135754980246992903607395080491178292225254197661545411221495540943118401786473774
6302222399485910314648843386999470861649919878104604090490642947211729599296927759310438
8597274412186940487288621255126276294397102939389770618396918162885160402027025983430417
9912793881450712826289470054498813371997559749617046148962363168762664154407667409134733
1503854071967117493367945387220998022549451701814221075419797168220132751362724972828606
8630791418126011767767567896960883552307305722489743949195127127639201849238891435679008
0957161388959310343326919543348580985265970561684172841247152430509273527232823787069911
3761274830413928916383191547062385080116016237396162228222558629418438050458300064827903
6127746769792346248094317338682370334896655636627141028775274415264385162559734034472077
1813992813548737578633997247924182056613417933194373641246509643489180051690285945593262
3773993076404209101569769719477758555252035509510360766539399273820974424122542372683083
2959807237183389104767133966381711108507809741596451233059643437737291820009526361017008
9278077585593371580314989398638945974049609733774402047238594664223763981754049951938188
4259589640106340101012433663888363878972661204487914677575459454576754474125462146256111
2692353494153191755212873625928441426416786799258170590178574277229679001686344872637488
7664231012704993659623593828037020114193545433642847936272862107981596924273845601967222
6341409343837693178649987868711677205254579513542002998537163890426391103754674209309500
2129957795925905833180607195989519859903015297880501161742424785508534728495483609136311
8343843566126705200563790372924765163740870489287029735374314924500029424847771142760699
5776022648037942673892105225198679398514192856975742000368326393819912285700677317312877
7402121980631741544661857891294348629055122562600193047435467147192335961123979807946066
8475409474531177487368390451085668492595663005072930595999944429993623120750461149304566
2364745081633315097580563420976311774823771418984517640960193596319563592667416286505855
8058390007911231141658696000733516306936540375941952032769521339835513466845889619613099
7103905672574683609010228095435925887329121224398745469756805347734170627432186949665866
8955360820337670766005220734225696159975350266953509870048022982929432232455926380797999
9937788907569256848240090828829879160567587149272888141615163531219190362385224909297055
4257727128221666739657355463700516375200164629782660928269506574042260976034672182535555
3820558881128094655810146185042169517073432271731605026720621515220383761551061097354655
5570657182929320687042116643471764906380702720727162036949003060359278772243903349126300
3020859641072602032060076251079267033076891459128272337429427727146212857057419565426633
6011746761143806100066367152544523763013270910501471362070242788292633305143461336248233
4046809874108329181932030870230036118521054944334974578448209122744976632177554605566744
3530607631239546730989366422111809604938592129537989785173827685760161581737521387106077
3819959304808317391794761281038133041841159009204049933151083957248221190818187454046299
6538500308944583518179515756322924614265048889571226700482027902207678212953340547987211
9445122429186061115135727987182241198810453679610238070036858921672124265867759451485766
9985608103165118569478847998200532321187174301674110258903151964420874876595296923586133
8882255099383271315947669341229488486389402734995136159020870542315260210742635903729455
2443389644778824646243049811386866786252192733152787349322560419195552330143581730884311
1424325624096725869621559718470447416098070958302276073268755494359336717828187716294500
4494898232005749236771414491873639292245739417768282478844799366063668799545928870771933
6523911637805459452673221623620873541605652779999523075901484927963630488976390391739511
6710739980376922417568769889315154183722852489395868960509235409176095585875601003018866
2809013512027861159997634412147540331954332175312121998375251031998276034424943768351422
7648825541167896137884596796227874350136719546214803774366120712286228069017375590469044
1629281424245814305880409269912277003573538012185972650950194933842338376297387708112433
7900266428277133475320484810921289241663638721406452619762292721731659036033498561062366
3258842491310789834666809865958488683578702019682781505088019423874552611670600411075555
4512902684441451747884031503399877331848723636070811891625334025584527217113757629758277
0238113039718782735116142256472134847220194177536663134669241078507650840027274015030177
3098231791798143053790773452553243691531430702054537004934471199826541320169275024322544
1098774012973923905995750994439232684056245139961559845171596363515327363310835355010077
1991314295321801537385026210740661564210996355491720862242994951601596844368564935191933
5431914594878464580790721399620076702131393305209659748708814304254646709730365343394544
6065097500754844492152755293021159744266765060374740192757684993143860628290633156633266
3459280232348471626463263969086947963053977295016630065573149634402647217123156604896844
3027597790636990950348955984371831086201155907259654131527048340385990094148181549627644
0780383631493893079585197526280041153275334516811576551123629441324485764363039058963199
0067342219004737933993856590273961525988294379583493289749356253775250667288405664089444
0169771981657822935147742670811741133781567691502169228585986910143541374578743112801300
5835836482841557101206919268456569185817393193503135722147491704545148275926993207941700
5292084001384265006733575372097674425104842027872745143645749873557007501870915585489511
5999781969154363069020551139208192864973139783481509976227725539614197271821603650259499
4743152832051959974340950362576348091108791362161968267244401883740467268496548838237866
2344896072147060650425788453289722473185521259839761157215758141297959295754087763789277
1452389847229305429278768956650667415454954001282943424845027625436342625711191069917555
3255690722583241329714026330799605450308577430411568012412596612714956872125520591880499
0210261118695320412297260400661242994846501668058480664341415312830915565486024549876700
8571170380357954668311879143548240352180364258693420153723955874215850587960597223512099
7540762113491964109212857590078600717778679119438934562658263289116870900711695314068733
9010613163423253125052153025318092371760074607680616982176930264065635532534831104323300
8802688050276844443023582660787648672330755844879096259890839748282791141060392540535999
9583304223597464214908843021168865833158595044601081154568128157667000569293569671858966
0386729556436597151499881466821623410215467801753304215414870104712833649477618977008566
1107872999003454404614063423324588907689945484817282966644901075245349248904734621228344

1436964119331566967257779669250765961091643837476961615104849910496499462890181142392908837650267635087106452541306004019396801704186401245866566115309060474061876229441312194296383591744405290062389613764513002596514730570384793817098826799806723500508753470480911304951593257493300195947152297600102453228942091952705948732424547013840540422657368405271380781504392974797144901583794310678407294894925176876255487922809768291875748310799308084454190556635089847106253820307249009268484059018151845353061450037505903385757218294411429842052074263519853454777975586230536069508656525294975808078784288397414035548412766737635812346018484967994211658319976853571633063510327143301699168858129181218436862964477707665474640274638293982321203477533733013054353967399911769744997827935360569436551439370206430511779195297639502188470422499390805274171077678773067776573369798685718640377953395049319145159120672984091317452695049044127334684786198079691914979100303683919114177578837679710662759932365028624409288181307808992686489905214952259776351709411214762244232355320502174296089312645085972474657568763029133106156333705142618167178930722984028507306608849969955129361024561282062107476154728989201419061873990131284361903777172764632291361147609740812996405180381749834735046216929876151389137406446509978519751124157131845685093661657423242159392407149394845318990514646915713289260544958542126252478479370767197084316094579581948825604919027223068038570233388150776608154414422259266410763588152811399453592877777153275139126469291593764323636069996685748206005881920157313250498971727168701549327694216042403792519383700843917364781860429722227671000810500002197743546070412440256489276196324324699012693085223852811971378916828416269527883083063093092373984083773148518010369839900244651678363943680751176661529637465795771799431168749650773126383953675600590326156070118550739849963818126741526667303681959592825405405310133411375727545437529900703239029063239557305064474948576351632766738262487353850609805905172010721823193660920089599129054282866259013272222106600671042493681934726172848993034814642449765869550085209390502480318033795437181144050152108008067068960925064673320464683642763064393717719517959472794419300312877011717078237090766463656263444978260480681570390323584704717192974045193868831176564883963847010336820715610562903569384509157463051246570341367481730781006943633525789525302745516232139478563958084167844613631870593874799986327120770201347604684177651430849644195409162305975664988022273280824178968983210444561923691504651056688742474292601857381506047699933807729385248420277356159702007345429173735356635984923327360907890597742059841610713931478514205199749223248157267838195927536356820140908400783566507358973146976829832001999979545181384341583912115097258122279212951895352934168263232367832743404701053204051596461901110820005090324054008990473644308588698721874451403685648773260525180707598313498841365112399071762600632811787936487968529681590491445450519046639950640584675448352182128085714665453518369024210369789223219131134506399699308305003344393602979652816554377808786710183531557369037390556990086124347820881658038478692407199359563964640999775954663474310697203478909927691285974253329677044568054306745648700054365232625776269833764469799068852977873536086089507872629483192280284489265493083095969322242673999743575110505839885113395580729756064424321438127712942851188028506351577994175467578407647329350128253540024854448165186489529450920500208900621317630296030251515361385485349955674780452752460885907366593709543563982308428353169367209787803519417213427542442173815375959691623290731818640149125409650252136912518728024845426950073644822965857816854569956491411714521519766094719965329127556836415096975865871787953136845343332611134596102573967280983333230411523939991028893079114907780219537168751698869689124719411289960218672081207711538448226670200466135986304225060939016625789398921931790140703368170697948538373186314630568239552815985880189963439001721888850849763127814729223016670379292802265779529219131731497424216860273778178125388697219563807738372429596824429825745932926155233786909965927770274963354356366062253278932272556838981028710812660041818251940496470865730917051192431856144045277965180963737487826664793474314000766773686156165526219768443049141307496344820996043136764079851050067159765593351501366541483080145557764653819491538498725215539555069799702622766930966673996416854659382644783939545828729376848858783218473556179492950258676015470809831834420683763536132095894925485313694220643516230910084044045522947306794884895559159707061069356915607854958543449765064303976545064135401630921788341076157898842029134877434766803937683427959591469891866274711933490999057186734867864086824899569824961966413421231869311168884411234934419810018788304440538800560818933299045505688504051555135441073314544799230941048594945417808170785874046331677974131553145499082758198528373011767443837893475491238523904517520543663887937477353896916463250440970424000825792358665165399617494628933891683547697168433320854612645478850760926623918909673062500949266404340900582428565391936013930675106589294421575556037626843546735354767174053277288610188428987479278802634314279883474807245581938896335503799108842322004280337456493317732546574068587696933630190298573482789400562581401237092493337500835192300253680196453535971901156957654161053501189239661138763303073239469589590353776028524379087572686940091075920722004471138628059895837806064103462692924772766337493118196522820753112729368588140610515441431256007286030016526828313868425652776653812002980321175305419256753461568719168259300414286052678912308178048811923758851382573607492629109370720053037983540358567467386176902244897005246006557000556557851301350491023221340042913069044121338352892913715651509430865780278241574632378191724693666831002443052405031483839733891601167409491255440801530104398868086202775269643066460835478506743714368408494613000068934186405970103525666461476451906525163949327102614592462010193016791092554018943645133640774912361591483606883925421933622119050851765785744896142143853296994699481661332190142117414268443992022397370156326767855645341051101093151410058267060749660369101038078015545483945191362360490410900402298484517769873577844949583140979639913285885278600447113915939278151617491061837574690499060524513835425176578397999135140169504365836194013746858274984873726905607458185225249485785427820669170250208283080626123045907014681023219495698745697368937599398473971376375512158043063074671193084135429992152263731085125144195799715026852681264911768354596565808942828225096447439690770868364813632750941981474217138564101454854256742639885123542998287701758676628872230981828154393533810943434

pi to two million places

5415165232856293487924921419275836589395196460151096829826838678534814595833948458465 23
6727263571023907090939655443607431076191158878811482700905778163885257198293969090169 47
1156497411211651250492653327192803772124115191648778344329492803274379788201047052675 01
5591242855507177406441963455613244426806774845389693361139431235490744348955393644313 83
2508445449719502567227490309520126258031394033681165052879105124244061373475788100864 41
9267437893982874974594376888548038605775461022660244787448983137263675258448661557860 60
0790360363978050213015565251228941286063507581512509310424558587544783965761206154259 20
2599922801614675949634330589762989652447607087828592420897083400230222801905072713105 76
5915666805079628671384567411220464444629488545517117668703979776179260053619373270085 25
2046769273707781281168373048072631156538276780753748860553208044575955559506226249492 571
6861855267869255116522174629555898955964097685893799356268814551155294690424609273232 950
6900414585943542867234324816849477645993883626845884622696734854733483157752697996072 55
1300444851545385758107375404196543847170848149708712914833818838923269627604491585368 15
1060274968714486495468220845928982861464663180716414016882905009326922010717705877625 46
6318283827702491317819987063047025394731618773119516006819212924819527514268051845428 65
7764684149030895234524157594993659615814333569138735323300223578304429321841535784102 55
5616021272646358429778219246139780525902332735499266692315439116480398208548990037501 86
2236489099498853797811945854349607247685358356887079360263710709512017901666963781504 32
0576862528994947357026899692398987955206441919854778414164706306805374742843257990810 59
7447209901310030533158059077565394399887426940201551550329625455652069905549348354811 77
4475544653840076154932967233768526491316482329835988736717388177332941796953870739371 73
7306930090575036318544955673264845889320609797982372945556158381235237214249007905197 60
2895442129211573342483439105295213293892936962708027015606486458674193531135044771647 78
7854763998653797259885669405655433972171693642484243841340783204270235481851736388328 23
2449666930691204860695308407579935450648735262492120187689414695939712873000466536451 65
8811110503327115750350052645538059162083946606488062552142145305375757649543047213159 25
2715758969908127750309712525802217680828392000051899961337375543683275556837591671400 32
7792513087473826397272471992157573634400307951910389276437434640429631210809531131335 26
6684538784331154680845390074886839099292762604686170725995214245065655389020731188664 83
5864376419276837278038340673415328525670993452855327784925620117759556026278343044675 59
9304247743773556090116464242598263199178678340224740329949850107591605201453255416435 46
4962622146160537098284796292082132745694538459367151397372206886798964285105476011538 46
4572875383684638165448140746461841027331726733968070174814541059415841080478783640813 17
4188152157245550405377211834777742753961856689081465109047211220910718701452179993470 62
5636367262916987458912328482890245196669636086113612808347120604706082824150697223641 30
6584718307595260678375509163791368786563092052849277519994387995966808827622582840125 11
3699579317264608655614175993335332864178264240411585747785641717971492747866020442799 48
4170508957811141030283923071905554967906252223969575123208215978674203684205865473893 331
9194051204417799825453703393836374317621521390080154742304873811178293287663032697107 71
0116513627091110241499872384336742844600940034635521328198463316257672682799558316995 13
9254114453459071328768525818241423575851932773114091089395562671436797960140280525023 03
1035124155427977408627943675299823980003095190997969048414623795183781105663053837144 88
6359385214946227660394955804466371101281428041664997435163808494382079773484601118670 14
2105857641501723463107598377799655113900319357230064926304861965491272698617042911022 46
8693617941439369420809992801752053135384130934330904233926053028121032097360671534997 03
9366256440132767163644580384340917888392124775047616024024469917784312116977355303553 26
0232925077969530656045095852489149017011644853120183070336808894156476547445983687006 02
1773015927771827151658025176248695011602377413584815491554420922222761972694123851491 30
1906627616668287445493017611083087756529662651772856870057794020793374853582555293547 68
3233547265595386274432922317779673143862465831176485639309698073189994111808747647284 45
7694329084752864963463319228939100984505906320185054421867255902158213310231590282700 40
7716362403180647378087852963683354807256142640224345452946293130881212433614301733289 81
0461774000029776094151535632614059175864259890707110384020800955273660554066063956247 60
1520043738794386243113895854422040321034675842872407562769841645592309206923671389801 84
5185471492910039533965337173012518696230536680484044595523196014210203469332924780574 34
9662570008947877288833064912659858710447182051864723086150529511083971353745252816597 56
0883240257322119958804131531634354034089542754601416249039288480699004676557194377455 51
1770708768178125216192230013999625680889112692814078610196870683090249182505668926870 19
3052797652571545877454636229911190755149278999473316983304397540831291280601236756247 16
2129162300256955058420205431557323060720360457607943619931973053157930586289169970182 46
3139939134008890403754899785543730769487348733809645510798596997504001127050091621159 47
6846249535839818886764657508908108153879409963656502115884676948148379382095208607512 76
5761108195591141968911098920865185521742027584750078329223145339407144934372933160619 25
8391298383018999152160834307763115489501553383845841595769622186366763925821394044335 04
7304604386698716631561289811935912714986708883899074660889121331682170288181814525384 52
0512821955281359603946186450031642793395559804393681497949395934181103165846879754198 10
6772759613487466191352117959788672883688023533042072064887135733332242951338781535407 81
0832881578887819169592263441957515469323420078135376184670667740804348023223855945745 71
1226730870254199926206363090818973711266317414192517305522311129831035323871775751506 48
9436741591551552151765562490277584798878845964305837403719258896849872495231269001958 97
4781305670660529407020616722052848389836703621193663704990489338029061549767453517414 93
7034524815095477014473919266619022806233567898940292154118381592553286753394693555554 04
5304591545740975082986135162096087140228259210242042336701851484103808784466582280295 50
1881660664156416262427391898055262442515993258980836872532983984166416007024902755759 84
4598665381154345808659115352145298045152721712430683925770382669956288490385922511609 33
9816485264080589581162873451781761184154532533945546737951564434632430508404942278788 53
1601096187704971790366548425050853156055796317670967622013488127846289540777491411695 50
7855008953736355495863185206541099630440965536094173537854720541384802896696856379509 75

pi to two million places

7375746881826367286447090238101659785189333949041999131201330000110674154079280048176645
1164275491851546977686559371489909293578514686727144935314151075267143570853389530348948
0003199361020364935019103603623638655939932278773490462364015165943289775370442995076099
9867118490642370041973651154705637597140735841391680372466511238168861392554000356571405
...

7358733814592917814930891079323794354460590463713708049853419337308119087463913112174496698264426536881909688332218916186685249630601308814313585792645087835634394497799788473788321344541063480692782795382100081116970475615331560021800281141031457661949328822932925556008377353943254479040670715531027023533126834649100868208442859242749892823941709498706633168780642641792771996087320333843771544009033231631775475772276437823906342429810738428498928542343176580767975398637251179039686007592373821533305790195525337569480231133753504263074772962202219159397772163688107072503458497743646551366040616866324689680686954495301837008120127007949077049925138491813724698319819302319249025690065695842811700540131918837723276031056066940552351832118716799100237691529889471645418257450984559345397599352494265648672994048860612969866821378313272517018425553727199730440552264207663379700487090041693579594007528696101112078042206093471800272603159673793593631193362705303146338185485618299943320847649334166281309065307363100671462288246863490740238539752365419320378749706636853049213423432517803803927393429655077908841025291753728600593844984281110997399686666211102005809268886689194311299266560892710972258852242037259115949585064647167096648010458381439278314298440502548444275276781730215743187276154803074818125674948834872111545155184313290245976785271229109914883569307440591706284694284533484290023546990654175086608283480534118537921473348645293788701640413565868213737768225449391658671775957380011687333894253591832069817371237971101445869350202387156561777467732514930215230264197894198641373094183059466578075217540790569777169827995059037533391839047942868949839697833388698967042964263801026373156197816870214404905832123336216958228700462043727930746455442404957538443608870075477341148659711035300752652129556748237420213810312901304676980887188121892286427402677066420025719873138344901262488709876832695428030285560538728697695160794431590878379022474356687869661760593015745354389223481219767014296075856158629299327008547678784241121258567579968408262535653209343388335473100156440762991521962645719106290325082227735368001302224149311112601053364982617398575367229059437820267496572646803470531988567768058284963708296540638767012372918643862549949437995800259856909451912860148301602951496645249846823004315771651820581680704646569735280469140757917615574942973222229804469868281108486902886508762344856320802787971882104099861256325755204930277873820529580375126788357539919135525785183036679586432792674576927139642033592058152421376215071698931340065944687000497349270311303328921917717336626816193964914882073257210864915918873306983960953488506566197360608264245099701679866272727415978302640195992867475345335698039500567655147384450777872979121784707690944762454645137373954530579094390343486267563889834650131921830477499593479704240175684394935223587753622718232997059986206456425628942902953644296345007202727588402702965620309897872140003842860707133374334062476016987704978868313201966238224316487325661254231003104385252875143135292701454197116594283406472060082427603989274624237380375898225500423520788384580584539874899999571683315332276702308565932894639765164344726314146537516763353972629719426827210778316812251006246650770586351080965114688975989864855514512865098100132560409462090389226761937486301837320276823597692101549933023247082429808377858479325122329134645859202225857437828652069531842191916746119726847538897713983905168586900595932747185808582953191889007581391072187615853876738276116366198100819611834380027225565832657477071996805621684933573655268177133842785645806651254423292673125195898836357454161343602084831630177820373226849236731481697817373864394200850538435358470665019842863368948526249843112745535820231079638602943182989061972294015233086198253815754506366471638703933851577619910072270881167342056398187661786579604961431709890950307878007494009758707906045852219344791931668911298591237734330072241519705472314436365171931560070553112574842097390679175209625410178373724470749931578154343135460373647908497691497391657705382865297296104265139429243594645420437610515351840037944998722654377338704884767933232756440453460739208247578023380851068563123976779207690835981381734060995171347489836953576161083250619999251453692208812425284618755986372902473465127854221277135818686914107687274493146789370076162224923773586469928637454162434314167901937718637677890413327067629002270403029919194021461159168639789233079076066299257964651115290337955080109542420785306877229759567684005312515990274709404262247739967234905658022946919332056900784250462573298680492159951900547857818990185103957629389805279351218914050810806098878627738908584923536263291298846860097966921611967370215456886317795498039249753737795131515097259701283927276658303954972829471637246262005111583829738046358168266192415120019777145737075094004160096170422109871327975993685858643476010270734233221512629338539943526750240931648525891208854237662617952657779085964008646001185720367276418115080947090196135877506513966873254503560694218756221606323328985885104913492241566654019449247865924029010948126476097825148812830905498194732699941110835167571101941330161641826554697154403640929299397693826508241355329830567695554843791362278738720177892637526068453923025274385615714408142920880490533497300377186165198492988425211274334089795775601186927722713652654425921200904540224463048401352172803617935255566191201927344847552170787094442022759106331572878693216105588506139235463827579526156513691986751961840788933506600915043548686146130197119666997265388173313873189688064650821945371056931327297491836131426424900957734477682095925819320689554913764276778440288884103184762398499468925435924502664127567651486872639433871979418853015806153278379266798170298512462327538903070804978690767041325656401162001300628075402999487326220896554241284545834017456598161512521837795972130457105364533121705900427003649301775098353999123536743274624912291180644519723707548835335734994187398410280097230089456539873169368131113322808720548587390320938353307795005284860341691745902387701023093203870372647468444051827821891124710488380047770214923616533577143218671387874817452169566055904373578645225670546538819105363801268777007970588380730745586899276475924307746007141214951294578895545245564104972795015106449450425930510516007414956386768991619452775571178026101770652653391034503693128477517999501530257402316766987046774838049889762348688327377476026779802217636984608669923206997338000523858122227879088986317328091551047049574491975518316510110376551064319091810839867007662345620441101512822287138361370299729352116633841847972603020065600469682614192264263804619401913634958207805468156068864071587117530660998344352504560241309932472 8

287081949355732561892665537392295754858040793159132253007406136467211318686384311836496
952632223615180874000673446550598362956665005570340094877109988761058515978618924278462
834248493978871187322041228995944900537890955393174160178606730164371639081672127097298
013285024070657499482705514292244052395469476454102941299708203131060415164045148076062
340931842108417714219928065889958160297658440927847469177243758175141970649036200751101
258841549217856807640611981667047077081949746323453383436611612027283356815525944369086
125589267064753651618111339496179104006544910097045607477926117931857828930864512657739
235900220846954611779005211666565782216001768346976735256003358726409081852720885284146
172547554790354957189484287341178103985147614775711682443704798040986358239409668040363
363136482238699618268544818054131856676013510236364050238441662364751466653513262750858
633083338063986518190521950431033587336544362582778266450324741247716499060042602316866
055852285455003363034515163848411416905250264130288481536314592128561975038801355171465
723363519456878836968300205925880979997711023594029734564608285817969596664266817529079
097711085838670474457967474160498417364758410376841156021552283911503164896466494951978
087693992921150160832295428986210760699250500625803168098616692532231817146890450982734
302777071968815066385649469485808494196397013478367707912912938083807204834919452352820
659411641979147397542216599823664239171071062086867083903225843345498569304287065813714
928651417706028858017756079026344990213812928838868244662035362909705524841155479699714
545684262869149610838420079224414932925366338268760247472404209240640882302480719988468
137777449605971928365397542820235539344226895481049807703580862782970524979008958875537
849835696032176361884904661867426680797379879162747291396040138385225347043964738496676
451660218578778012341099654117180378212339852333990294283499025133705660997441149547868
548134835941305567131669130794755790618749872078315653237992863154152121966211769614737
304313297431931336581384853063400532576176088133948861447172473980276715286001255340807
215001260306631665881687277080271026810801232394661259633297872492513047778395733263470
327510762749339543038528524924766999043753255668298654270422159293178528700374432730840
695243429734929208567592935555240231759459414569192963087638880644882662150646162489020
566651211894236147483575393226981066623511493256237968808609403220342665088165711020765
168232888861730257789979474089611125709050068852796972360296511777211528431184014123566
423155671538346815965467787371920853152823579697842631700702041178767112768924724566318
145654374780676267677174546845875998730690561023891919115670682505109064458156977272254
112005948038840324662207843058627278907857057608858239772388034580578245566661129138240
401676699548487692648672415950064948328670588146893153434726703943059737578001650509122
613812562687923600687835470364489403773177785196858460433663956787509600229071150758421
182973976670922632068432817060011192527915975281512486595714802942771155140958129750865
419027006287682492931491748918582171718444364269175389197054389010919707403652419488143
360129006627612973013854909256385593929839196377663759645526351408553053890511349413836
327554454781574649400530051102631164781953207429030343840063084325798585987814923050493
895258669617408485513516009187251492571076667718861235084158972239519970742083537683150
076497114012050308853300847020877435623413050803981148023545271945671353708967408026751
635429358087610196797560191808133202572244644566302790039640833355253228266209223110227
094876467610631844354550764908387783641700333746118499359042589488315191663634529830219
460406549610658846178898124537127283629802736159198788566016122803738237622014016369573
495092194011841913892505776921552676681108707059030434444595406351079009377153653334169
344079361036496585498919187595393195534851017278809934366789708375929080196970085138605
303951543307514437275130108981935379001756668285741372780938785718840248612921014759664
031989660278837310142677760556172680613118405775264323605971677786072149779782358114110
732710126236724718229465202268319701040895878004814201078577119458013193847000149336929
635535041200331052496251931174142669540689590148330093043970604932332331977004222839602
171673018008239727918802413884712385294727925817084132122499203643499989237019066108568
837481905311131195426386753841234585062590566449567901222271749082019044520332821282132
210403995503858488248491982994922466252388497460099637803541268706167345721792434404766
866870125749949458481625122657979099077121112358636444070001547135295995451118733278240
893682779558799265589778843417082583830188658936716745203578737137626842208115191946778
244140584627072024821985033188916237957705766962112984341256102299327141232024091116949
363712310206244201543747149536881943971176273591913099602444068293165176801119671348882
268046433096792671536193635509647599961519710251042395340465251786888223433637196657176
246451403082712456783826533824064617180419150628360487839654988758483464485178364436755
532995083431979672576037548743317580393586461179890633966266728543339424072993713763854
8480353463044584394085066791466666779296531280690236070081415096413878579038392972596314
920194033240276205585145166769757675983084993899601721337826079996771630815289479479611
576457547812757792686892459777664903113083738993131831759251233558694837296864983338165
936560319657002317745822090249188486904704408322689135150730136231731466987750163714451
754929787322409485056977098161873516478671878383185615549037561005088258818995159916337
697179494358961543285586239555075850073720132579044738388311752722792809800855480564530
811262649122857539346406336985960992878781181023908433301485174803626701343066583914468
080144284930383588174068633372250765500427844346364900422231684721079927302301566747010
442956281163958578272760748475466783157827551032369460308488321383370487798254489594338
787467101670787876441635640584312882773289328078677831009431570273297722314668141509305
888209633215013313861851932675712345447480549341618523052932187285720207205312868023145
4438810416966337722627629364962797778220632427659854050609104410898057275252201004557991
063084683095655229880005342519263895674452740072958101259778021289516714770966550278053
476907123389241109074163394065827471585947837010098661575678230561860902906832183799824
441034226402449885000914687759782135188753930180060296699676445996204596723932714724039
764390573385830712399720370683992205810203872755883405468321178022027794530099777284997
457180381112594640639330976113111142784766717776809711070805899693067623376856235307934
364894455624200765074903287814776390215992089856322122762754273251762493453849167959324
380786255303488342570028901610514846651987944062944265440838284018348673068205149837556

729717151228447118472677533712707669138140248901943678948895519105883068951194850124684
749235096458692249550739729939846526213677708967112813097878771869604606865551496532899
974175317550675968018768771083722051788378806678316567267386555149637107771980911744468
538967683609222540874951615471472142642884144654943381399878179335636771314840408447072
099876827183046888969856887331985381177442544328694697283332636117486986166341641586393
312823377473300357306717977984752600661579219646144005711630397244365336180317455938746
135513551743096576075364028802996640418738157913913482278453640477387705288755458906053
214391787838127707128324660310205367810664249171568993304500066432672325411879761802095
549425502216511521795388946463780346517710447373460670019914814099828496235422762719433
251403439080864959443189588565978159001253483267725478007291615474336858444209994767444
837703573050136729227752622884179623800584253698596564431556245870419193820468665990076
841630044357229776986204740300197897367429986326347814613851660492045824123828791695380
815859581363567292080210589893716325774317287115734103317945490957794705247496424293009
279141304973042859527025403218534196039952143451790512112867051215441204402199315885779
859270196394603317000937343832782131179805830952631349641563875469099467363726269215509
951419974059651283950091385763856893338725815581373241700160441331341039551662200397863
667187389486424706105166170391523709139890010648711920320144384343730717845180562512804
263657817965215930362096306193495295809821296944066806570046232067433817989270849622324
579963810599886318610135228783405225100783726800614449840719147630353659836453486595218
363016291330051830834199170792450371378743975121477203024459381007354237616967986079420
201862892986321728648756684663866817129465265739309477125424804811328414562681258231394
457759994927260789480607894852456499041249419270406671655339608486002943927763105322018
658530860877914486735946408046760954496269120916904634949759952741479644987706191486043
678375339855723147015517203836028346678972645674270545763362091512672964053147825157989
351888830860600053285167174029909496690350877141545480927193660002249285629753332118759
232562216542186347981526453640124400786571752789915309022812393784523330548826200527414
893742375621129536637995897668174037229550353429814255617246768130073900852010344996085
508548570405421345962666567434589124704761561445148802725512748511656106683424410576766
020868238377572993854002371684660324341541642531424845329415644227976594220821620531914
573826781542506925725775898386627738129128766637717808126505223102856490281012845141836
774628069163610797593228804238449051005368099436362376287976543271318447242551523087501
851263130196713685143194580031723858562526407809329010907442552102354250484005963854566
799218998489835163485625866746877842650704995329010304645192929568154960753512971830997
050073705213282480724035120320210341587628411906380143615454363823338538816218091577094
575068228400223247598846529712308848157579451350853419766546866582596782083013973414906
027854376547270044586472904336611234187870939226334930914798369656694586009274383890165
340028608505955532330218458955751322997171454036525493790867947153104602967454012464955
563081890659049539399029815468265505050680622408958283686864444830445503050403314085454
646009651241401499238880995510641261153855516542758426833092762865624588717138544471461
400605324481172135711762414761866208527275919156124566986020097192299075558618307411860
551900895948068046797280441634194396181701995890839213620955681296679225050467132926411
099599714347253213685275318877639176450018134256242538596252589966358503201823205339124
090369748375950081352601927716733677993028611591005639598595964149939913558676505387335
563649395305618363542792347602960620887191772052917385058837755063202627402422408176559
971691266525023647103804868517109052602064269033794994591216468165789155151824346189852
140722634040960524404661558340652790085457507787492265304284347200062852149411001637802
318690032496881673462798575519093320473114985470265974055828053016062524058494423589170
757053814073402757856528948046563395535697436454627371977447741957358803581276643687906
485870312539062481052927917942687542133303376623828047662744138860921805898777799880997
310749058846073300241359100245303759605669693318875048237534557550134586998096891169220
643385116588526051346261209863474494912782726154514774476250729944966412298639874239459
957921065835761702465280718659344876095262840792037461084847990937863326107963353666491
365845252308900142223153054498122715846067688927112771150988711249995637364426624168763
343936339226202843927754785715799762828754793014264271981872325110571262479985951206446
898460815540938533886961965614908441473721960901296657528464549334031392823712515821783
661765719305856048382059527621369256653993619520612056962063365427245773781429001509716
615156833091949169275587943867647461610595598972541807955512408424682588528784212928634
655874553104225279704238777218317443096106166253440336966615747533778334075653935370418
581234874201606356949436788880121098734053539979550105897058519532310837719249211435588
288856296786682309851105895556925849481866062514469198048307405917355493392555313417998
232306591571418430466647095895238997143210405028248819241467595374759526375005843981607
073531860080073745392602287926776590023976814721896390583209494503460698914437305103874
231346741770792718311222362852443845791381408889277509386916192520842321167785074170005
000142414276330508644058952129246223163687906304151698761150856393227536103409464891112
424074458535758458923923713822181018679020632367976699973136858406942374755043166091704
237183394814351620084134453186452802647046749996110169887395874854168353902088717917248
617754767066775419186261659100047003379916705768667438669607052856296372631068102882482
977340522485596252014506327880090880212520428186050836111781094559367791322688268238320
254570041507432572607533891059079029062295312512264475056676312689118774383520639251044
351868372564953142014110868999000442881517928669270779219908706893329331552216843336119
968522467121925763428860818082150029393453388689560512513690499836837005456530614115157
652316498380084974172340184029538911732558523651940093320840120017361942699415191820851
137481796183799123374319034170575565875241965060838362373611617952116308250732546161963
644663942194451964004148725625161320554140202320850618698649483285589316773922996295036
800337740687666562681901591295207879156962334111477752563320337123486135962286652463991
418986621445499407840676938370527929620527899958436355614034768058314205418836902171384
832890937848446077730582223587059006821412579409174734490624231902439804369257510426814
093231823428184892657722902270114752621575104423916322091946827699491162639621249911347

pi to two million places

7439923659457948110285492906514619403209562890738419084956470377436072710401452363578 47
0919596304724424503417449794325684927059630569998152213405150381866557384303010198578 91
7828626183629543775068228859621522258690224276653625122532153919793345690085028386265 77
1312526100726742634157933090934961736341039978150030563705302370480102525440919773900 93
1379956653942090134755135528549784964828550496262532546329879408412516413257899478063 54
3008676853042271738501835409268104479215980206182385053050309053710232921890256861486 68
1756392896314368540771546281932414987080953152743898968252048927101970475843135733810 05
7447678538901193272777078248695709850517175481806501949026878078524479201429570893873 00
3842001825817935065020536738800093034489193955840679706198623039302019416289372694415 87
2101581712903757610281499068150631907091728225564831858192329148040142693056234209149 69
8364488996049681241335762708251261564392496181459742562551980565886017413195498929992 13
1394907673305765097331137427006237880417665599077422311796651762091704173716814909197 14
0775239558424317414364068331295786289443732637055435386221631772052621026640654286324 44
7848089396656273040657143397425587683291777727978063437900801895566357927526969309776 96
3460900196963145415575286142458219055080598423572525027198685954869483533261986573591 32
6416433360893955758259210311275276001686229936344117525718152893855888404807520328235 53
5287924597138650972317378013416291962987691302901442371008297552715355229004043569663 74
5380899989009363518938348585910856371841353090097945069729369111528060607097995183565 33
0940317475174934165708073605256554547546353703848703382702728527634166365083303318606 56
0595122636718875050490147465541668272942028312157019880664722134726652639148185724611 67
5311680779236747429694541175322106812209074657683861775442636196270607260936080622125 67
4202731027537387248367045900747365708981697928162073451677980798359957549952293884686 12
8293446133453982203408058449493321251355485561968405023656321823509206876978267600781 28
0775663990523438005235076356533838556336419695472701035107599272466068642743096509456 11
9744427299606417376470516834302740126272755403299778625000891872338391353611031170931 27
6599519659670059995524994680303403948217253857754469587029827804350287154561025368193 19
9585413597758181883726310730235268775869153632132142944176785038969217215766454688889 51
4399027980435436648150380891625912551776997909488496845822854290479383806320901393803 66
7398525098466087188284619523424555435734875302437429402619761518570847166647115324764 59
7389463779460792109451882897404370184506145934096003395296373470804740502022673359598 96
2434320353543410167670464769587194524190995959598923094966975569057690046013298139996 90
1101017889790665928244338934858946100103334293046971490131430895518764631492532633003 78
3470245713932004057415255075296427021420859721408239432102287684243422540345984683625 02
3637886626211421402602163422894459277040316530671499010672735502515433903465687310274 31
8180144950398536559017079446874333180825020386242912523397871696959620501610210392845 88
2825803911447734846169137385301002470361036115208281466205829674973231060128814226120 19
9382101374198278129311952910851506069186857879078661119228156064212434331775694794907 59
6228792191997887270319017372647630245819570249923680472524223541240376859270130812480 87
3695412900272216187487794791292706639289918868722701374027498338781235767266149708987 06
9885615934628914215809333393014634219430168992274322031768109633832976460429447424013 64
4185658762505469470580949746689333548025153956804626295105698730536165688125827768103 58
9605107105730638468486214784930402582184769121213816412996352045747029750953594050966 80
4678782755099548352221010349266052868875517608878108599137823790420623038906864789799 35
6912974910552594729364081021826135082240847363783635428025469707429912394984902424175 01
7206230496405736647937249176693339799110698643659795859007344031643484156819925961972 86
3913529788980570676687770447509143877129362705948585406150905862564572949785272593613 37
0273224505661072429852780110275946644040479367468092203010223691341353891422028825688 15
6566330006679469548377992950767262449367122046333725055725101837990712007946801518922 18
7248892528284176603213113015797717473366711835410739095502687077007071226080484538485 38
1931143431359974069933226798639950496493005931886294319308383723711820980952366492680 79
7619405052282395266166129472234878176200739704824007855106641247238936966231148448457 77
6739912024790338499065375528517879624884424569377082538599193745044510378900609469679 21
4313674732937598036569191442563779446938494332734689766742365967553288970182967687416 22
7316517970621648901369736800488374565963582872958943314965156655748617167829908813013 40
8700920228060722241512269425417496266460511383333198070269082309454205447872315715643 63
3770366442923659925239046715181546987104661375348710484416783018864802944823090403676 78
5025453134162545128712812727504630565820944112045565033428183798544672537876687057117 00
3208693615256633911679130266334723703717911362791286600828651441236399830071385758401 32
4357317776172791790166598528516936636929303368856394564619273216078406591613028695964 86
9924383338681202462487259432351724748515943997255440021778542616944561930736713553361 49
1420651879196641622668400880116096432294720341913203655317936335605517575345256105436 70
7249957548161917686733399428114344255644065301106739032495455999731119471346278011185 35
5452843524164064997806380527794768264085959899276140699448732025674524425911063854259 92
0522317129817157751295373004608463582732540390740788467216960620279155881445581178646 66
4957846647101853737481841001099662174117815866986303381084567146652200594358416250621 83
1254337824835706576491031211417444148093541636821914929834490601656919688556727270465 35
1786202030691018474347311913689112622876592990451848711597780517833855656585356984929 65
9833025991708503056129774280195021411828163173048495091312148892136865344247529284919 36
9772270002569121239219903107879934082955578409674725996082981810818916164674314541160 53
3488224917178671454313978010486232492101361763829987785310804900783816903440114649721 55
0336655944392888306173467704663961861583328565953972081581674205305111665566511304834 18
3303455305269621552838421132350670046141530524363408430651451439219402794986660381147 81
3154530506209834302487996031615903278851787161174863874206707696916166987810073825220 10
6058797701025770218125592141842859962200170704110235262472515396645896092074496069221 48
9837179811735563431274300870460459188654184283552857580272672888203820307425645512155 74
0872584144597961964876398400086337195677805810023791698841901742179319096209300190543 71
3995223910141438230581390798928437066691192815050805538443370203079949591058375418476 34
1424545722795298596656734874632739592597534906926573573537470098393119979690254291381 73

4627123524726697665879443981963845035994884209234902474677859216325172710690136795798819722224766108200382799983338718484691648461571093847237335147851832968305335902887848964730000878709667397898041061109907081270637533497952257059625205007679081355848072161756877266207555778850148390346942662063311026243737111709147838471982374341587955933349495727373839971475052196958092952539590908062681936321780960983857643610644074617580197822923581768438857054553338115693780959458314977589861846272860796470721433552053775123798454966942408002662173137300031497426519939188411561200965303118394789638656356542819839161140224219584072329074112929656641790634561324622475896410515348782844298486924406611967477410092012480106477491511954001997337917182359012239971209283100801218559519263821068487497935900062557471708321852980492971619973918211461616747162333207897273010040899221935736910463740415875760741173155409914221982038178061847796911612912729972416155084621090757222548522467694959677561712015061262731297909053755044424918119431878848736930635769734626803390216545766566217209209815082346720392556513103955805942545004639853610006327843972429873571110628206975116796926017902906670528656040507136874617276154128847757611734487324738236816823713472654029099151221997889171774182050799733147859447793696389968129775363747072456936057313256813417637537251515566839118560435589243780379060948067654653729176690229745391635727464916041178659233425987035276626327469723964773281761050142098958406202296184053754556313813302109130360478048242658484985563927150807471240341717379364259856241142136648496388799450972075533215396968364344065905045636642877978441232778654190471469257006753165928049209239873991209188554539246462825748218001511064055547987308483224711787140888713222342219469098458634784489239585316078136786123111861312239071863800353603590976206054173116864942538818678452443711012885098064054009488015453243159828764581060637989623355462216983820473748814017465295990528993125717072207676443518057468476334834808927123748535936351659398584383281838449746826091084078956975574330485995531088061936814579206840411418592300229495935031251191318717986480444780901969584805303914463919566738909407647275793467032982956752213747026814250547620655627247339713985091971339329332494364657379857495783036520740987926103871296192095105153389055724398577058945494168183022654385314034902690393950273412869345777386423153458702942608002321020203310481501024231242789667194960716457953549696257597921124961107699521429810479338179359748751255386058369914870188279746231363524467561009315955998508312402589470596279499401494775038067730743069476977694269739165568706084526063022974227615868449595901569149551656148596447530223787171646016659630960746880219874453648915209708906889001843810757632943692647011662541289634123724367581741825560951065323959721897764429445108913448091126000857732165522835801180610352840099339242540944097400667646531397938805541310859696403810802096852952179325440802869503229643923191578039397099790729565677883278936834190889535816126908065260170729052360657696681262163571606783205386754516126090297841114898392793736796172901426977091437277136052919785702585331973985885846423045855075839960514250764775752681727074693874955104247621812104394300354189023469459818095488209583059848917049997346518092955608312170961968280456467146610033509967206709759151660289656426661480409206512009358456018114512493073482040014158338982349582333075615371719506414615295200255739188920500953855557637135624613243209286637276991306194453492361343527362373287890186489417136871372812760624012021169728137116450152976821838201904065555203350442816943747155836426381805724874857858139633947169767201999735788882226702917632822016560780658924835460787071304426690592255547261569195411492575353275343956468873105233627886451353355845085554878483763117545793245651234338151226390492504469993509493785403948499283001833165024762519767294198472861595475709428579515521656096882021547152369299384429292813950955204906752688988515657102864567901537430920706584024376915280452727169725049523291948157469032429198279355201672226863089559524087385149509249283156392119862490734532916879970712606392931549184467023101628220521918500217812779714826401536682804738460726342473074908995904409980446063913129519751583984611225347149613076600915467522793603766123338057239997313197506549365456366755012035053611010497680263930967943207619017867309377221850847730626239813111117797877226612005465827244507948537810770148915392547354559794192026468046460693818500230741866994278630303685256174166299122368383184396657533164335505525979166690879913800786961855019556226065178127130751323831450249763459011891191992886556901515329727673998393367435832710707542775594995962025544288577428234586076929990486678228566412372091876693371559530512537058784640329587102579431756271151954761311018115017614273503158541979352448960955046834369091533979624657458981321217413399764562156238184427283676477971352387910713152125850793116345179598145756897699187957865267413394827933716847187438310917460841261146664654551607225610540354184585983740257529475914058869112693518769557466894185327058900489543589626049203378668460413403226608563344899530386217443011051686667615290776257435833691796713519195884306867369314726965662901751641842197026834839279822898782743865250058892197036680513723031816978131135387160397733696823406399783071124766089390994470983949380591669440406867815434897479682183542493910479109179774290101387642045435866346944586882051984913794589150396758731978970429158663631757903320866486399607245780870161039442812516019707692355709698066988375111376389428665692726021189237671768880624122230775154741412773244704108360708954176291760067365874537026845480587204558858440110470001396777430349055694227566777389614789337781669999918518455609406014996391438831321675816368904744038969955570033156082296760704554269069644780137556567330465762292123379965287693966889517906040308768038757634280676260250765097452111093949881701053896483467029635342653806557533970599694283025268688833610388077684335185841400142982286145404567225931272007685512571084785746390485725775432632098892635968784441430814135696650424024128208279510866914292245978716316841055831789328555605013362485926979628100474247057443120931289518580268575564785219822477250503973349885805159691481173613733947472017436595805645428548233480113613431337803278319477696460340492674347192038234504175373291253872888912706200062748411631591954646303815713189742448135360410202374517238833882982917328891762404371386848695759067402561168009182466564638125518976321665610064795226688176562168450549099411403659180400913068885823810805523623558387963386182088044394973732068499086147002476098952472733308237029416866892448750567

6926858560705687606220170301691948249233246719829085608947585809763279601056707655012600
0289152061135676072682801421930646125970086812294771275415962684883047949126993897160580
6520754558909591564544913846595978696909930223062845701148995888809674631045569787047733
0478768731150657865875027394716432681898401989927968181760575646309452529547910187228988
6237012907068698268780874749343699571517202939321559112657050580844898466160628311257833
5041964175352500724695480087908929181678628071216712158862499013832369605881619132308255
6267414583594112235490493815037160622808462949701468125146379952862299262808901881826244
7017621867844395619410360678600408044390047518419852094783996941779191641137814871462244
0213399108291633105989627567278036318210156197218372618618419336629554294082997847788977
8918064149827462653527539232547847311397315998894017571948703843069269301388044282521299
0701291923118930802666806486698699184931926426139487374966422153474541222438257782589344
3715286422874966436669911312730757188639317051901363874806278031538304362349888091146422
4576634259762813700988728238473397274302025103114296643876618437336330351924375055098777
8833221004895279937262245889172376576239669921168374007756131154887058940195321615068466
9241977667903694263808643186114462982814281680657154877922972240103464815312210427170888
4892821972137239954120885318100303423351718452134576270175688212497198963201121848455900
0128266864731117988049927894757842383705717385123054677697765440390659919114391468432677
5026678790645706282781173009317644741373945502811103939521374615614525244968755491428000
9102259673280472613164293403113614571518694083125751540238528674308438103715429738072566
6024497105934098698746812099074301880504751653063152803566451236525738955199452193184233
9411964869023175381837611315007742252723286295180272538870829169222063886645004561011233
8156417605272542714450093184605067915085668026525241216647167107248067986795955785842300
4819546977727003246000361982295642259642731958778469418450692530608675980255288356249755
0565302503615766997898081755734440543448922077410231856883059836697808978685392503181699
6689720438895892211615024215843107881126701645165245609950802961681378123713996359000366
1195473327859007366144179973297697690142791852091656118139724903078096017838458270415499
7920978581372420346779817944314027364073040468838884617861262719816117367316977212062211
0097715501329932918712977515865288403441882079158145298668029452075816645292510406593733
9087476490178297212541360755399046515329278874472517958000082941987412475015686690200177
6963874752486175744400162312052840141855407504325631678199308364091299955711943111683500
0434626822768975609667874040686807595178887024301312197691297774416592309108964734030533
5365578999222654584496484236668733222794300708552871866584254801079569699349279417501233
3253781751964577929479323736250497767018261168133747767034023208341194048745560665849188
1993167055907222908405651722369085496955771869849429254887774566385044313765173936311122
1470197541440932279131541544044254794996029349571079561664440286195454421388657781902700
4140339471037183448764583270367269987926324261615908696513236122783747601025354842943555
6602474803851571399110787647114373122865235024626150082784839176285769662380410731834088
8518222294298076936602015498053910891787874107954281195040737652614696600476695366359299
1864564611498754520903976894880683981538322206672404127804372879157787780492572247703122
4653689477505383387222137615803415546384285347552629614555399886913814109999620419260233
7236739975135910602664687332049526498020634839712185031634357288771245311140044698971922
4844365752726853386536173704519326961578365844131906325633532386080565950122524111246322
9518185841742240637369277859878726737126900140698919740010691467431409447308641873312577
6112307656617440582089611336772309107368395700281827294715033163453611497841208943504899
5821928170054440150147186843691069448953420553915940880683997106388288678377693653406600
5841072846977076686680720984191073183333215476848276919628555221761189235408133267315911
7411862032928384628064172108062005736399530355619952531331748238740501075567898962208233
9185324628600954368585596920949917622955596822590728519304275099817755916818026422992500
6415513064915878698062548410406075796238999287534315933677143747842344251130296892511800
6697371893579330399215941416628579248060020089218888789052658042688713234986221532053122
4358445789496423174140880161120677798780862158315208901572764656326222644937626760997638
2678451545422380751312441635507071907986162753694998824305732638773370112793850644304244
0611727224139754182211363845116014588661695995606098205174747906312268016190314629438500
6814123746260455085176866637489223065875127697098577101884313356726680203938937097182255
4751661330701055906958537697453392427179075813892986194640033104708909057635118237509733
1583620326086601125832587774130952359228310026785157382410342980603730029319476588933488
0313668336788340794152273583964038333084384958032007972221608943181735820728222865982988
7104876666377596826307778088575637902203939204568479571357058049965515926617999549991611
2383678252435561196635308400684571863616051856593939284490835234602214682054393317202288
1131130961314108138748795991893332112431733158276496294235157230594396210488583347923388
6390458796875072596933816246759642894360961754852892929589769880968606901249471152674311
2307616072877207136639426595364846718307083263574116221915572152296673520712132254063733
0015678466904619538338317173153786769178016365150600456437785407523667593366757308737566
3424389110868783185295400911081703080701890804722100369320395834044342162131886055370122
7820774999959424139492224534374396587589970216008402497407235490340869639584551938467922
0804328543812384805357305088951426976372470981000974301307047381139421481258976972073699
2615050745537369351168354194550111842112932673530395039536570029696789525764050203488077
4558389464249774099584124383964604716275690652313064545167433092710138765801110866587366
3184267216183157421269304144296900461352447502076211524662597048664205981852117400551266
9595809267189528302055884346357434935625921052330459874942337757972311379613064457584799
2077802295408826859472982287377927136725752267361529134831120986828362678946917797727155
4725482803375635116237595156816078820295985301433706600553701241103940330269713452773822
2786597510911032180097485482328213886541455675641026281378340602408270155267683817792000
2292443686526982657083974766633163076613781151264688291181513117891425892102955361452500
2234299005463687707333623656893419046956635335151806646402010498737670086037191231001377
2592091013999837847217162193323076771509126791263073680388306065654469927146910415139166
5833955286979406682121223987038198227272904531569087118494808701533285436603422340993544
1719599165740943899971359387122423940551633546615004528254023234735514293153320049461311

3937601361659234664020303094390340233502593232913188635534668127824782041314756246677984
4317129337260206150873953138708412127939569722235781994153691825642062607435553950083924
2307395226145848930094950697997193110557064521525226446588857480807998890454552361883402
19175451801569162718915807701684984535717028704867994985789324304011406876159848237597
2078823238388030744083819481676016643146769815689713056294186106084933929531941028425537
3215599940584111595416319578905340537751761890997035682783443086395166459216068731414
3877520864025617561897877350740988609506039483525685810268269020567283969771261308727312
1213665807752119330538878596427837820334937157355092429587751882696978246279707112172487
9165333392601840235678038956363807740424225408565037064677853573124572090006041479276
2144757794023529242645954363401496491456002466971005962925961781889438545086240623456390
6914135291279358016559197197487663764924402927358201340601556937322314066238616583089314
117429872324215707334286617954561896352424200497916669179893064573483386482152727544986
74051643663264882592549044614931226597017729743910503489267900659436463090834663787811
5232672085465186580091270565243921746496536635167674419385600944944138088791542266160621
0447214922108046517589134260257768382118113438460578490300203049083112051401105586199
4427359976424018397202333239988917328736676938630760583880220988628788628706159703085413
28981611169154730512739251873359574164308903054395901407956101792179100853298335275199
7907786415822767660228122673223095569598657998224352084403712397642463020332039597877306
872857642144517353116638361909338724116749624336841918654487439586897706701419882693664
6165613159854855294510267478556416711586032617831051751698410283839334872656617868531272
738951878372386516436639860899380049558212834894907934174435789662847546668258978395710
86145947053419178719822158529672671038256305185499771641180951500236422482609114558463
309826606480668269846448186251844160232090583734617784711910600997548381639972971037113
7338568232094581645219961784404972033692669559543915925148248811655201479798369337729967
01318028402164313097937213315562241197232654578332127021418705983676320400003717701558
366949164781930784056396099985820996853703832248439884687926745988633152771888426546076
47329005553515682204092551878675047922199310064392931314222720959151025690450394838146
928073700585981878724345067233016732669978110980649222037563109482363604894297533701774
7667685322284772513016515049771921475188933189255859391662059905426900972479364300328276
981493336146941155106989646224248136704347004278344367461193275248604087283494227930083
884676625795547778484876206073716724559484046898498382023042679430962804012787659037479
4048559935768382418190931594551157005081410300379460423026274087272944199820025712835718
889270971074194442718349111429187222974352192071313406804515220461706560765975161037907
142788234321826511090571299696499047074408653321379128841594849870489254373102609722622
340571039870706392885043631717646444624564092468161666876004855070827284479576220731519
87288212334779016739220242126812968853866613096749378122029275946311370941602027190998
991187378979057837680347195685819038201377930245219209060307469178640056540740467166066
4271062978277884034806402675794358544320902642940855837423139335570128361090059656725065
18912172503121757538762683039107699914888699621392750232565892640492755321118156322962
394116117337088916561205205421254007664081279937535240693504743320974898048033441939425
41382174000994794898108706750513501662483007521508648244492603623232913951047549530323
4647554307220399129066574480096066100118266610812058933788821349258938878623719775257989
7497635259345456220006239888947443133921889660333624365598772552825417868093289490867431
16393171552410804322688010606482989279544934230002681812127006152252584754021674653643
64258764655958421349537777645033589614170270296162967115817580463336538178313606976449
7238660577812228633942216968451131963256102616680657504838112290631945938127465031329847
713015144143825453050931542306837732758433401988417350682144827401968112045387647356700
3803642190410259856470552133967602733747354287087752763725293811582043582740196873972
5123472033479250365095402842552154484334374434115266192840266441360058143611889658295027
239696069344436204062905043593831660893596681472040969876257302938443256842766576037917
3100572875935675060277487870828533660049006432851423412155321336056662620940379504514955
430312294840796541216886194975906144801231719437210702190895127907731132954694852366127
124863189695542135280705094860471537830205657496724432781222636368251861992025879461000
107184534642033392166040742619751874446751852154430097374437547209505054807306485878726
173561860810336952737855739519423092020366553038884188128604872388364861218602391892861
2059967604519049315947108409305171110737508763853468438394246154007191343701122475352
0894489208155649064175278720766936081040810161558989543791373290032518763230457785741477
016980415528387877779375002729608403742625446735811825550813949246290046624239856855925
3638184261461233032320736035289904858967993546429331073108843924391088936516727448595625
54239675349661494475059114733604544887937015343471943479663756081397595797999526896289
14260030827273983895596771780459193315245006212616517806418407450550297344038485834378
8135259459116313033566419639280177995107434551507064474304820340426039031612279719578684
15552559341480331751280569815111158668824200130281420685336897946196665466237756662245635
906179005241785266077701651893236916276508803351348360409532445919976620432142399109106
98508933341063991565707013430555985407035676467465647434248247417446118293012743229586
07263497551064288346423222274677943700730586522222923703014763276847364046563758957498
7682794891317116969722866053699035041939612245056249568335524123840194169929755780158575
016131915264844866793484710394357721438918229412656482913964752929979937690463124550540
998705402634734305959498788825599522495750102464936114432817862896472276327503247329506
92928589567495177084426477928094667876342657421082510385087031416735103953086851698494
504737917814401037843333169672141629156872359368199303498544018458310749934730944997592
29203059535562425203186788063700961563067779221989381996747325087778530246785277429712226
08430316827105996973292065391222222806355826191111541411488254157541570928298132997209
121003960444192221200423394092562611360013704893962306877018954456078654892699955721457
85111503301497811082482742013507374078233264330435515418943016500031747362939003557444619
58731814522194089620895732166600385084611881494757310201664311808977049939302114984293
19939490015989903183756012449250770290107969482179208714750703514440818746419994696349
545041443223895841331216652856138130543372125687168652213126890413389811898223684160247

pi to two million places

730457620586884635430762293470018348026717679450792467102581467277968692248538172340078
379298436453729795343686824030144576348771501613463800563247928756589096366096589630048
785530195395836624871155033236024758895494437249420365590463740705397766748090510460377
292127489870470299143453560537010141196958406019085467714982388228854985438956893699731
101520208183186904596105243459821774161142963879480726037449820753404358457842375729115
846066266250395145811803151013138513595495205384769407782346081450560581608871041162553
728279352967012631864770223670569081296678823539516254435650287709745249951076589719075
844645259460015237132194501365059630253906127347056523492015546063028380388555430664643
880672950016795861174886239155797201005719776736724426478981205772698596728304258088572
405909474622228404808344822727527087691590170623866329573036331719749084363121832803571
037942518086855523038989076220158468276024267829174011668050172951024059706543325042939
334586160135432967482290662271249128554825088825191245041528320195894397403504714162900
407979571378400331561520503062068934959370401710372972190066564166959447162821964071562
021054580720269528782045778663285110139631451353151329906190557114803746099949944250772
248009369295293551136590828919823346912612033447582381871047749195231223058628059003153
448117839548835984146645447033481921960343293308426229767991944701748161342655915603650
990370935254464613847324117732494891466984018838320167496736260721080630176380167268157
352937988477469983695410031744094897508282223915975817072429355943882440765874651670542
627510414553238606166503971835613462926625409145104075774978834300365089521716691184301
092328701908118866952706732713466676975735919236260334453651605737759933143021041722259
908813293093542202860195319807923915813979644907126943929370869743941098989742810040203
651371262048675609157446740660276723811144175805572340843201597299074457147436340019518
141570960885159236267416347356185814355450465910620613365946609086698532533697743789456
627248254266717012738826537532000116508385002995462971841691633251358185221576837167230
065616013394388468237638486786402047706681320227165486147409305618803780416544551921206
583099792528320295119549398143064467661170043792001164721996631961722782861795175613752
354521082067524687803096290278296999185777676317486070407884604754689365289347225238737
163959416121917368596149785492126658002366125039101916629448136872665171062091115344591
160394255512989579224893113864624887954035259932963881996343963882054967396135801506885
129553701093037604933911239416788259232250512944938176638682816349115430133985187088774
439047628106432596423317357811894474748103155439103356975525205854443482621195666391794
425746534503589972375434050522874571557237035385891552687106501100373758659555835418460
243013053047651343344522118747179066008874879689898557567497525545041726565245234072739
997676692236929979805185093853876126829995126115360747599068497820745802053510010507896
683223384622694224989254657393110150146603426986408075633082915364155032911575844570909
089329090825737215130190720366807993504340455011800158038379627868111412415509362585484
014965913355799311565413657245027440028826351988279063766916195221189910896887357610343
633025409493185952535989777002965995571902127182617133851257136692550395843062878764549
064250552200112850562881730817313339729766756874453455114396869166750603805057228162617
916200590625114541140556285637267234301421560663998136759653703984724006040824714884944
325125976170297138952333681607999040642389110022714182525494296377180890386377573292332
171517432829270805876815540757262484310640653167289953692139135341397917087218368676961
427699073723098044702448377692264873936843728635843930004506224395876836825111833277667
766844539982517327332866780084641528506455957078592656616210706585762962260341875373079
243798059877495055058001280937410916012536771865707909505437208619251564580695356973049
144842707695335519957551706788180530131878195175041928609670626827005773569300383780457
261604337147934668849290310095142969037244163158219298671040762187138885572376402291793
865057264366774618129457587129884233222230685436241682266773697265119067740721500410776
353142849308327085655569053972905681827975952004623349510924941979732498815860603222963
036809876680787752910186349169489024701447161495819875004364945531331701087202524169504
463262235671671222248078325189268847556109717708417193427747423022750226459019973672139
228025265754779872014272032679871050425138184423377259035269699199662240669521014944591
667859338866519690267779040414125297442264490745313724512460317534400949757525567603926
431859533936242463607752914077687886783469477044094155141066517236135096241918376435071
524709396010787171159494613386368799706458463789864569204733324300752051587143045750937
826272728881426998153253222403510492192203752299708997376129375331268791134022803523044
147224634480948083464867614859045181827597793675586688902257123468424627737431784204777
076936427847931183090309536787495548723589353750302735581723721770465620650477201011618
328915372060635468743352590995586861264789444149846300489528880549643182672033016656620
264051549050405778460463685796801141347836237851935895986757350287416020365485528931156
340642819096174717654450504980549394026356936574242654230029311278424030299227627342050
722185623736674872009955738682016596744107631216813358364560375480724125290568555937043
714431097559339430198038646671256816652425849623807397791811136027965344473006285243333
853339685360711945890006675404828870516178466661914617471928674938698399250100814424904
037464038225944399754306065353120672777161714218186051176789645984948034332954312612757
310928769374011947678851342816364208210489856450994537606406443502686587389298808139152
661955868421811001713504538116664273279452620839496463193446887215998250453068150677983
417895814818267308012254644021661253738754373560427280555037580579314435516753375358497
861431888936353945357925802547950057501021319433235270015484836148119061911464113392254
279727814436672836471105411963525707495041771815102247565529496976396214344421501130343
219278612011992167379051310227795279127849377037188912490215516978464393943209708053320
429250996754068711160537666034693478750473018666048713673289498947499843549473671827152
092592888044064988696008793832395425358848353148093255694909127532842075166928775982816
862267408237564728322965811633421739364376995494308891050841744884197122866442691533988
693135154076359255168723695450592906014230998532193401096087317818948960522356612430391
044791732334431513675862589569219267245289462276095440092074831832744650815915520643552
027414144598957051398857308775688580218680084003520281496874814980742740176173912663094
949721688773324093254150064099701117310624165861750318691994358069172656796287149710200

3358006031599322746334598294035145931032013741785732918025332022090957403447327566142170
0267540078869686603746290565385470205345272246011521348031203544508320577480512294547260
5209323659792313369920774075666751653167920939000626792175821867831693850095822661917870
2169945284330346356317532642567063720153856802151644263239791287500056420153069603757630
1448521494935726511926694599098297063064222066127406945397203214828474324562813263031460
2727191352043543206699894829873872435001795005304005636849027042594181357098655929533550
1783082954426854595373044440946233825658472505839432829403571725833509123633242648253310
4666419176857955420435526969928229306550533975682050056042347441594198651315056965724470
6192912039310424550698199540491827379418035241963976681980945032287052939837597179085840
9801506183446872268602929136538439157144356819319608156617764444321265638837434607945630
1290025740299064168887881571442704746809222202960238782188565111227472203669684962978030
6787794811210852515025853263077929984410787487195117758316540835477628898279770637583220
7047985751024346133827256533986891217068325156846373045436641389829664994736429420867860
0509421278297644788017093122875075233222417072099015531021345791576898834108147089570920
5248832315270036243149000477256331223665234384118688097036937483639676936107682342344970
9434234482082623593180126453488964785165657753481448616102890589957528909225384864840490
4878442808038011639541214706593832890222180571259414909469211542683098064864588585671770
7747235251307620694738669266456103119939947488365579484963610880782378948370473587227180
3537381073752364389367983288091204046227367530934332851082803701552501298416224330531120
7276109526263357042721066392604111254110648964452251457000850919950050662730684154485930
2890236867235413908943562595712543518820201907705838608514672590579039540626630066705940
8179692584371551900357406741861586997473561841479736020662091324905939150439273674912320
1470304433593577054920794408225960274630128822473857876996041996180399146659720067361710
8954179692956538022591808337043536734730462763267510368562442424474585295942151947278940
3498143668900688768440217629854457696764330640439240536825311734112705789184778282212750
3827921283050481050750925029303912585340492096661035904947375612042656397510664382145700
6531384948442243388620191176750194647364126180761165528504111861882469509050934307937430
7411151887851435552543678053723435352751917225856891312572638173522529582673191332439470
3045738021994244365931180445916983376234019463204526846539890376775925311094314543270010
1058678219946144433549366185990517912510679149475928010604688221552490476255362504665750
7418857600062366805763931246966325259017123078418630169603932200090281780181103805653140
9059521735062025720827016825351414769927448249334704243895149356113742516708258421899880
0946240496923478547948733061716713954730522918909621944150238129642570618492582114983640
6316244129849838987122331815793265449083878237857466730978884081833588006884800404808310
7220383176695718177242993501213669853434121611328860248560282777760246500340038698383330
3230842151050494164588487072323179701097934501767713755431272354409507105490008899655650
1580487761630822015121311221239363345741695823954075496346843293389765295720062887857620
9765364646991030208544544501063956245594460138605188702507264700591797326690788891951350
3708997127323194178354806307060285822835004338848763383508703018526866306970921595841890
0449079887761366431826700523884747212695763825222185250473857529735070873259378331742740
8108103245750064840907670502266738816529483797341534730756774993170425917308925902923600
7545245684528833600616911430622218763482073754438358134753498522523430538668477198616500
3280825608529094266465620216305324213755215408843883031200648716050102112346106501150730
7034214473214295774952765101974943415066439207710724349237184475304160060275333402479560
9538357934481078068378046833880428087871779627425947735332279071505656076376867963345380
2419788029468792109856700099524012730080180517305025372484611538986742273975296155333200
7942016109242118085339973121579343956284185801747088154948004772943337269997361067395900
2467147413258818707054594469432063981499052413800014836143803349306935255522790916112790
2213464160129224703779846328213676420624524099911220977090081236471179073488897967743010
5155803250633112981264115048881466116182838250473327343078218766864569702631079362399770
6131790718571069862881826382983186019672492962356536207583105585935517384170820503917910
1605631992561808007929416858364616134609250094167312737981825789904250320851974097758450
3315130198218673720686130935197616237678540888523114730950592249045887447072506403910720
9420324813291892044615917298835025313441870032167723405394008915342950156722130376547980
0074356679922285768030516068438483412110154614567641728992898006955772584198237460237620
2883368922595633476599851499743752546357741890591854477780983412295335320407208173025210
0666051327595185995330465351240759492861872504154152648319674334993787667526478751298890
2993672310829892480316613294925056542810115978166872493261944496664661788828913280375480
1023681263950957233102019419380055130089891311335526264238416609054775860592463985453130
7184369432538921110663076533837484517735421558726740236935778630007767098619169491695090
3864401071070620264717059012888658512386134288091633812713230271727918217354886988334100
4532821538150809977793768391786087009332240981678255403166242336280150002565506661287860
6418693257731833690284478977477593964777090557607042327414006369940272134578198298766910
0483309387765464867746356248628247977515993662256176316170014722863559910838346670987990
0633519593047154990369089193517355743620032611464696880970144941236703557729382248415680
2206337734562446590238548250997091591591316996770838355327985562356438782957904531893830
5320862429897311633172654844991071414287680238154636906882941289035624743275925574041490
5631418720185876387114968735045175009502536462237073494700069155317044389974071537140490
7035016419662707799293562403688754536828469440755427481746183378700420891162659989237810
5604607520731352204484745783461195641404372786593530526935954046173773560264610145749930
0825728418785257279330735268891282678392310126727297977714851022422908695329357711331050
3129553103316638791108423010541922756977464190258967805249776814687567144989508331302200
6775633631960866598487543832966408400992720719821615590735920881386501559326032743336960
6631053466992444143948627040687753610213511576508184873112471796773367512690089771984480
3019228264081242834818756297303426217072604171227971958972261697898523359654908509453940
8624694398096150115101620044327599582351763086328181282464976464720959978893024234556020
7552274648848324696035619543538809255200650466285026634563152541858799100083621558804360
2210457287821203856884499430152631530408595602007877584156108378215999306076295096159150

1217035058064900719232288137553116937052365438325297851538962966135473087020677475405812726136733585679664582824164398499111301707694659753721775683867870518315934753529986471785861260323028134593482611729465939298438005141608679374041480381620432037103537302505141674378746178761428022266773693095264451880214245773516319072302541741631429983522934935880924082833649315736488463431603834822693882096294439747153349316765061498205483894141994294969424907142364922809794330526651330794007595436763739284517136244810989654237973702841929921475527728321222506556097632760351977151867270729495633730053707188792132124305044259252441147707165817158294527402668653212680324121214585587671181451998593998025971448108825087252661592002571683442882200176442051634073309563309053751585491638090941913313849848282890613255430231222671907477284580766483202231232869322490044686142616770395633751032555693650858951662149580893456487026818547250415966018465004575295880920862217398690889153148165609959007086249055388855687316632904174278116969372441295595464140604616056299091994016674220564196282929533321821840682987650534448339278302228780564555714439488276845996284277381910671301927109785676657140227467349250637546188059869596891047377151033925460521597775237342187276009694703684527363641540222864689376588252328977434428181482281590454946574763664704841272567655234441379465074741142605112101945158633146620474170486867173237966263789072712094760057044075401790539725828401046975366093808036419777485170285317353754976503622052383970776572260069402036270360505962296245529510844440003893786503081380101445618124358513273056195601543311456717046173754928994491369617333736620825519445371179692409170371849837957579207729989517986085860058422712862349622013529171996705581851076523259845845143216124472614450157144469060971577198819870943652768574007677081410045384558893363437298249286237840657866838788729546124368958366360581932452680926341215632160624327025127970437491353453230848821749085313956471156019482505240773364043611337383707222472659143970087152830326240708958363631714436555233201491228896573708586957424661541309678100712010024661242794641985978938454907036232562377196959706560906795807380983373344740682417802947512836126806931438464139783796760399157871288230380459725473570248869217808514212825304298020820902932470767453743032441396676303065268806101541259411160167792424293082285124281587265567280130171333831750533355246697547694856118026112968781665046846525640454579622026535269251323369647476061923536995299990676136271011966745869967472871595187557208985194695063294168834046659582740790229314934114061259260776663765917428295934339842459901683005554236732160764020223404146103643774121412344156525507337891865665315721546298113464921798908074930872766048739972927261573109158678541138212835387817451310355249518750439563414824108667197159976618935093779485297899639191494409279711521807601883349112004250727971252128501107132082185322529689442603346312932992348393589783795566720093222834410103015254145288898999501299307719252518321859885359128286948297575281006204240747710791494204729977547623949929959560100079055784264424030846439918930302818232589950890993739367263550348289768289821178363046570159965675150095487275621683926842153725848285817031041839737397763497061851077591453465984535590810318835194385753613236724127538125809981873587307447703614399007802848233837464077750828126051397111080161675058792799278241827255563878915264958927898064152122763822108548526726349459094178145411090248817425053361506949418712784664790836121606371277879479230740686623214310455536165976251342048752590654845981338878151240552001277329585301775989355611551442148018769094821614214021749424233269609790857418588929929429517253483406041629477975701403940618592267841921305592823249107248651822386010944770681001754892431077874502060557156408008017205212534738345077129389362325822207344355906462669632087617014438182226940244625522224736870055175420194867109405251531358966791350018818870524038353677647196474922563620291487761632399378946838237755539744478633509057248667101578225311361943543701193892046575929260050423636625176434496343065514149213437554203544452880083773701717967822146803412414231252590986357601128605279481504854518579367761274404586892120113665918523055109845992630653807001468455935665627110422998205578499369994452114987918030167369954643655335580516043633554151540098381936920565600761063994963460580382961369643044069318698308129858214477671415738639252890445172638352304340311791900865891118150038088574646801096658919086624398620197856054335863039839498384223957903500296118228576386884206558828252840040315378060006525617044664300992225532283020450040929085079364019505365384461092417853184439506564066342898785081935314063356721613504829971824772738849231710140406951819592790782793794129424218087658100057418246347537490805498456624935353087216927748598094623072505374114414846137468568933318135872186263385433156344537781900404176881925295963507572158001113271649975311964211820527548989730027955513415575597319767345336451875011671003253739717286579519277155229918225888060488523471420694266639310759688705900982773152334560796205229644934627085906318240112352248895559598197951013157040874505848576292037018818879334217640686592439592711885786998646576869253991050091104932617298663665073783841961629648089206507809952849412930203900111134536266673507625584939032121331895686474026636839623366743136425244297147539835995198081197094763241358894164682690419416589139875015314421803961070445421330848000848804022080018132634534890377064246578049314040362955423711785771277052282496986665483753457484227520301447295156391416628389504049766936679183486366459764900237040631091306001508161651267719129473401631541000624265862990012149345611813518516792032881726297765774516256929097253611761281005016598684700931420961187077040337977031109623817605928264480676035068462806575040005227904195179323365903552638751102728270038436100420075221926333880750625028151084676951206702262412316035316705755106481673637739826103422866267410470455241216788084140755050524410498135798125600598346140185860839843067312395845575535734681980012893083373064467730899801824889622150095538772694664710342594769692139236886270885256347807834771852315817371646062092122664304670870911626449827040612985657920762765774182028006681720514234823917057765029257215610344134693445967724182347971958896463625076687422403739376590562370354578838201387682360777185539676917163320616691979351751356743981852543698905772710181186735649537799854567001576409598178337275132518454430143346290676067957794546406753529101111643173719874986863988350540833541300741601031431579398110151405952943088234998494562516496470315254218047627217795317203364527383277867047022017304574055

5793297631073327471179970754268053702481828580233267341234916657076493407413720714183389
8911157213097166506070541529071767362781030942158445921858354494004311290135677215895245
8009986109916796129611046895019134996510778904258276221597380002343935698534892518227040
0254683259759675884427654336735629251007809232410525873804900473328942535480256883249378
9774762845841629497740854257133965280993492758468647320417992850816404644611997915133984
2658585383898162919876529458599154031343163685463743395375282635924858338396219113254001
6863096102039198577553849307905719234250038684810357496664932435252476295875123007330391
9742686792010279845740044804391377703874135071785704139113952209333009792295431380941245
7998263742747189334542039605734011494037514982794217429278750351465088557428565701668045
4018543309143238766724155045940794214258707256092739856715719135092456972961314423159990
9820395116265314683936396145730187121785899647130814833406059226011241085170934867168132
2047372700640627647252899657775954241693420230265615551945420096253147150563439102663261
2879796075837336983313059025377243573536708474584090579426667143658809427355510128715350
2445099375229028027667632355178659985033417631477167198934630521970718790238335194561010
4381095168918051386958853203042652836589055777173736264589769703359583638947705468503707
4375481558507811307300256936278013780035185406313198354320574282939055940204569951613330
7812849900293320547311004950341065358124041340525921088859720749281254255101759169588401
0687232080522326126562519097770638219824963965571123738102779984379992194182395433987226
6383072946153893769362380895553281414988541962843627598003121539773316029586101061481860
5053174178272167953095047699581809806754265108834924874583802309937095265545324054855576
2580175734503175437810876978726826696188217561535358224654760031586504751898741424195515
5704777245585899001067354936536329835585878639129847349102068384215296201259110947864415
1980706912905640714978953578396682114639163301354198438120337816233003929204301417950353
6123128953778993104223156082879322902786357960886483647156495592901165931137850098045661
500562078280324341907810816430976474715282006310137485485818690828052886142639051406182979944222156211550413235056350800044568401992428331290552172096761180844373210316597468557247465892812099086767241677803757064900674149476245199462792588648315490164504856110783082626113167724290596840683885087823513352371532162101336764735498225577048518204099576948105474786169851385952052786490865118785182981701278552778034267982482998185651382102163092124678169989402352323087317453359427803866349767509968112266957604900884872475132818241541275494025924284444407457342225634748418850863933644359875069662863438744698185344310835067479218306204185070049090048399721764396033807778618585740855525239960905671222616615995879012703071934593528776329236401344886582160149029637929048504248108136317662281948077159757769447003680956777823166549715778219528600904407454772576735104724273006257218554580229022195391579232078394688613930188872629164035485115634800975271642023437559043384109691054543888187447310586213292054527355985729365975888680220434381456639996838468238939111868035555600260198440302185757970417408711171374718804874846558022094101809937697572516174940230232809604302425947637532383554420048380374107985440456050518909191989487621125925565941316217726195080254899847978987553348513481752960572904147338775873222594565591344192430140181694535301971137088815377723935692176868195957283440030905823830372429878080722331949210161527475666740208990044906625994152067089732980944586368004306977841364552979201167696551295794992789434478517731922601453339584909384908426962720585710492551877277243803335995137762839492283750134764309618622716717569153253348380010271067001895652510015454379503400339486286924829688541463101466532097162294596383857767775226081154106928466358141984686376882819945131650380011331230430498496052231607932508705971282452824930009166459724533119507657635728179869917465190724551056286040242956187812136080170254845573385161678467466709633919591011330373842741929123146518847019211558141229299037277678193622012492362161841446724867090302161728020831963667735273437769204974820526792598480601784207058520317561239133929883109297692352990768781604340305985283509664370012630975115069106140849988526062877783649816329353480108096336321673246922056464381345163031133384042092836596385624977973182100646682048563199347962642917032993123307021770691498441422509098864306356190561319874731997022599944954147833760201683898558317254437143529401866608830218802441281060798209581533833617427592669878769289169074292162435859189447051769846375635891830879569001477233456221651059615181799284362614888212747853576051044668246900979552897234910744990460992382404666849425258297938269796316490348004941858275993791994596400915417028663125002900023292492277078961658227562567147669289137053647161870559013653999529264978990414141066554850502291348350814949799190167199597267949135285968441839649771594471839272580514743744146174906248545536950925886646338120571786834356775842689831506940586474117494280391322126245533671527750950262932347191156928952488241088519382881561417526405874915320236859224901235267632130066456125403780409604327535624129366607333609095548449122625904066475450720696595267367250974867705515421770941973128157960068031549370275588085637934287278997676104632182078712893019416389971908522822843773976761752044231714509320832335162346361817418451840364557028029484756826869972039356711132980239073304878491889089083329331991432141459389676837114232046793042929875773677461349790758839639350158848581917354193625217298054188944906181388967607297125452380984837980087977313101130577771321346237142561438613294477152468256840075635430262307110922661294764040846850673769444261050329812323560638690429181079724976769444099131301246024253813091207339291141708696828490420951470528142342021759080746212769618421285778278450640077880188395437510012803288694266333177605185306695647778683897969177186177245415476953484832056274565903739330801172731438892993910141955472376789553611792612240827935982508301162530410841405073474858091321493769565001056640002272703436564898889682081784468685283063579014351954295232455958506816473390932545441912614834934214426445593912069073734102132939078685059572223789286636718673929869425618713555614255956612526029976466883305829228050168218019576871587082480442077124694307510857229894151856087732572715933091802817379902434020374474096454737728133853991562600127615451656271200753902931767191527738296922113936031917958369349496311545924830506293220045817399516369056275978162376380874562737103615082403421473805602347136579724826902514600185329753078267487706442553383530258756198810483196991780505229047307000

985391495446420881911865521553434716858422092840125889796214599701606617662957347680077
265289336531861279847409186816557950203844583850078533288171499959446258662627858682392
842424837095531730212194072727415549317270931966229533046538731011269231939923961758224
94212899929294676005664009775463068101536753338481035058157003978490307139296386716842
503701868132298738810927360173395586056138133578392466032078590686353402655984638898421
054605757257219104531405361837056295394833709342690548077688665227258112127236086466866
430999309401915777382379666076127146537966951442602703185944899072198626559837611819992
071100737651502721701854584092156078362148352218341971764423640632064885692809252099651
188234414434747045228745074244378247076422469236978453455165142341381227684496297399158
606413726302259661120572999422840517110856328003808839725858188610378931514199820030182
028606286603564310638992883785735762025863618338443736136571384143432175746918256070253
245339039802579693119439594640420045252664620215710861580875994316824066736693322658339
786029069827980094369556438130584999385357735378395664708093209213983551367787576720725
765085547549595951707783780869935758766034743143534513179602620995011333869541923226074
727674837824462748357798244991761085496298778507960424316388067817358396399576196434289
996519322135696744718539627388843756491045303514806706351634213856004419619094905680424
177663490180977879128978086716431020994931175957189028801155377415746799402067806788966
464104885346418521083612530081989381960713829981609142676533084615051049915390155625650
104738601493395392995258334280318005684080315316666551362336823593053106207269967358456
506137945133551506755206288345948756513744351048417658084990295641881102662694235754003
767036655165276200218953756887030772219469720625099083893611003328954861965114087509066
846569917036978327053767752080504574129781439726141067170431855824608431322422615821163
407437854637379441211647828098174389599839838355626355529314799371005199767083102026973
687855407351034494406149294656658843167394115459720443943735788406086567615987369190422
478261021301024964736036577696074168709220823347859283339857646314191473844362529370529
809327245567161757543208356892380946043963938883111403754127427502285054989554887813988
548120212748698085736871123769954438165776895675505871648799797011738459442608534614527
861901024816933107126469533266226688876583828248107125608920795181694688383639159419654
682266090281558178033896656997311888630089001538509885298441067140520766173963539131688
672862254935342053238476780160525661622257219932656411812822808918655306812652554692908
504030541041753431676611537206746401528738906897624695284763386219903037971333092520844
278085159192487282857550250655353188404772078809234163606658299971957363453448792498206
494442995694824113619171221241897659552565747468428643469612683459233313675894965988105
273003618004560449357457375418093000505604672774655530522756463216572242819473946465037
733002175653205347615261415557751531961937920905782739596917502519576432219350696632625
253851669482255376811560530175356109734462223337871919905756942119863707675883266180874
428376190151906303378021880614923759222987811068047510200126045621275420152549814390698
535662017515537876618784816794149745752168369100308521626018253392354314266756673433313
809434952775451772433082790636642315376741353290759831088866847633807628332561424609857
636327283949701023445832383661537471818726426002813579480593456381892974519101473990344
982510536435443777607949627868993288280255149347486154213902112125315238647570269086180
134397047957115631106176487398977762539205403136845288817144460023568406097080209656212
908411134861665637006891003906500478987362113957933246370850570742751743478691822255438
207126349095928907443875261402867348258962233775940039245236803125099307642286110727265
672453411622272367195408720120739307779299840824323977399672056108122548072893308039813
715523572500367280561682219605333605891174665581926514582437181728704892023127203047564
89324164574121296729551755567678258488837031118399274672696561126032930982233026926758940
567105216654030873811980199317384814229959971939185839736769978433718449972782847639228
716132232511403227563539929476753620801045277321400305934298954025055116993467925434383
719968844898287522659306448270980121972253650007394794011822036942284477640794389090834
574615432148323216262283022932931207886658023899124715986860498217522760422404804239090
768458930199899312467651985142237368420446410984257772187202890238993900388434471454517
733438843573208813754934776422470904457819629971449290693295863434269345011058261242944
288709943941392046391293743736135801693313320295451711695611321580310854662594904726674
824484532979051720561482588358906549396555750015985992113619437221151909605027485551945
742413561680802690454476042275730823745256553360464180507420114778635405649222525830758
008344816094412703565910260540661901599853761804407662171948887014599820752102829122664
649912018969330349560821670608620337185611303467927420467954316695645363386192307018695
930567265203259807191770435726100193888780097530910041160406611670813219619119789791070
806886717356979058701724911386863894802572897170063840952243542044214904248598644721706
050712986022084237482906395294389532803351108016127672079687037833007691736355492215655
359219416285511254595534636826649098267445247178052743283424725874502001213860143771912
966811376571118176218679784598844002735061975238620934374519161959576351961034049660861
679228757321470826912026622185862352725441375241289520155805965459413326434926759654818
631285370427692848735151704235513142175323090835634646637658091052244616266662296468361
379485936477569103007771343127349193431944725615920169913441097526705058856615990824688
0883973588041134357835976005440060571036505888555382519091443589361775884058829235984787
456676514243315869542533959029725539687960732275235462212344175989196865584257887508190
9164303832175395669372065956332379690681139884499125153807558149419484666861284537080277
783747508494333303583146689895212128016893952642544045816045343667905368616904915388785
596911597667794927561440057514009682836455917575346728232776503396733480799467122156403
403871826485472271859142629580160268330430304518707083848931742021464132398517584325440
276305707232455845946143991082546322634925408962448107790375266325800988362620486737280
488102059638843100050638629722552933023555558985680082230366277456761530464653786560129
273561696565297302458363223274803759379651235015692609191111228708692025198324565211271
149396637455499877020638162322666046142436194606519353264172614662814620995699374258553
887267928257006996360700543564797801647950892469163559398013555084008883500163192576808
361541477141846123739196024881896644862696829380909545261541482742787764267704291943373

pi to two million places

0249525858390217212596186378246765760550211191135896121811564438497390029262825425079944
4684033157107249040322349894517313877790973130920812036650210739747757461386817887895196
67967973045998419399229722665784211580727493697949812594419996271695665045113640366890449
52066588538681200248864329530153627831561871597152539610168376152293726112707027697048
8960433200242323263838947982926083217675956929642884232473377113364608402049060498102814
2975569486720967499321829182842541708079992741383136865722579713020690840384449486638071
28619669021704191931399486170596917979442302297862831893429365863329074235188300673869
83869054910173841111164898684674402309383323578014364746243372647175651977660789751730
029053227208944572220967998559591741138760118282117913340984088778391951569505619160379
7139305622324686569514190559555204159419511866732478458061603569029915601324876292041045
9799573291090549861230533340316741672625354523683223129697398210001168447557083843868344
6591006771785066360975467422927579690371328219649877878918991427571054621955331616765
1369013240780833075048439661440988318680012051336612917219276233311010874790559254521732
5610694877095091684977742398986447445782188947237291595775295033357372746947931747539427
44073392998789580541235455111176563260496495003475062864198750517076580594495125540182
096770467160577166716999854365042712502195269007149331065916634524582605879428287147024
6304912980162215721913660614315449101701912302527239977130591733481517002341451089790845
4806985737688533590246721338452459938712228227596440220819988877344977875800968480722
5959994307105156991036786676365650692597944464329477917210774349734263779110346692261221
1742469482497198767489657536774195595983488267366785496928182354337586571096374621016227
4951440251971521890816600831935848796392559857299204310228535663365843556700735592666
1955965455603034431138398055775346207604125373201725630751491584080619567967445268093027
447857136129940046470355108995872608255370459144874852338582122678623323336232630734660
516347656466295789252057685644775629938582302718227438656691880651125562159530813957592
739226594996806785963125535649361029967391331955938575964778533537805556834574323808435
236133901966795439350785092330832987775740290154456513725989538460954522658508797734956
413530848159244674780034618172912981137580284519917858694811154058602526554579825814465
964349446224556698962518772609216088193227643739818887655483229067892798877808078653784
153810970939959891479294517167166484385638245619205475596224410325536931298797699947378
5941599604540678540430117441766193799574545258961286396992056107223948067444648732774227
1608853113177938782981288721458155104438253782379827440385006926019894394520370792277
928215305639019724954558272504299436169914117320497258780627470117161868925268213965761
952873228940722945653198048493099098210098485080617070827072379474431252682649827599718
216411144710342028620002304676586092145061463339101045038565414446567610970605369228073
950006262295239898550407150399996816370998517656116666984048824937918909000821855336448
774710523720342783165517576506725475180624103487730446349434219291852929236067778383077
476669125697272857348715361726552424504396698547250989771074167714037069701936327375721
943003330364294272310847147993108866059606300529559518845004810493460280032819077405240
554530654667895086163995398801643128034996891383059051927285504775652404225033774348449
677564265826941375085093661133166008918312564352372816252784801745280861365444924678351
198154390665923723580138698242110639221467684264166797325190104689635866242185960658450
638645766178770998199052411608666083180718754628677625686989523175031235085977324189848
646673783139042187551075079192442206818110794594736337541848102657149523843965935598183
539802642719644110231469062689677411552737233685914734473627141446880813358064372581820
6247196353764481809657845339027125427785697192792799952829987160005198579772371048853671
103758082950559804500476805646443633233682453103942945499675930183143090229168246344819
682725351749338569290100005504598598306226190729296191984806803964801715289228378905375
567244703364176610167001243325108434653924047128416021140194044727979290819084334984146
216855263509467301849421038305863921151753087719174834167668986337519693814413665302910
384324753243115936528474264393706808750046563475757694198087596428676335252092910791626
307323493338994593723247394454597538493371297555259459090582869530401763136407573895848
393840420516925592193336202801020024713761692100037899931235254664109041989651593740716
626280635215200249188511380493933202693595412044119457007892409556874980561023059603232
870842387356800764205733553093687435268507101850657177697073798518484062477175147159625
353021217342657182458972261262092515701323560217513207573986815072509185683864419493733
739839983173968426108432503212993652071977555454077356016469635016720452334997937528651
240295258990481869207986034180893895278837099212337134068737451085639270006014597807482
744204530114523996315813821428163767323365688545325658277936520888525188246581996955196
089273181458380365443351288182943915073427137745968059789354163462310104316274363304839
292102098519319380869663897125288004565499260683848078343861333955609101469645351104930
940295478062307137706463977690540319274169021140970481922697043960038019180665048588451
069955575954278772442205082904609909851844374761337052109664360871436232003590600407468
006501311105104550938695666668824734385598908129797734502674177782381094792371944307215
668491887962419608908233282151373372776493299601142976453465087511622516536396620744561
405324597346656674895016790290974778069467680710657021539334661346825338425640716139203
585315084422200356450089421123279179418377919707639637722533758681081097601872987313496
961212566563501325423529703319305015358048451098685889864699457831683536081632785243212
826504885152428923582911580619904447380099879460108588532549374849994780705281903789098
555576257833427530729766855502491586633808111482011093008840837060506616334402846660419
657056070169222679445606675848784815437349058242499653895045183729118920646754737164227
058864753287808035988438189203498905850193839537527287741626915083298242519664608649799
192736646425094778637254915989275928754093410523565364096217000417532163804137820110196
717736472175641401216377603394540053874663677011590816124850785076566253231723018497652
786361877163876622079274692612934347222051985060978568460938183771391332829994048741399
175368533495986664798893354431978940771688366572811141727679658419951521640785664942048
791900405611745725210823163892498193583947039461718835878133285185809732148568375359777
478371695083760395053442687083636792830665089110740307890489533596521520347444108955640
214733323836792483710376431562414651116607094786285486841509397067561720875837902675575

5285638639328248063793980159440827244689841280351002173682629503560463318444080108550 40
7368630771849065457471048301939236993677021275919402775632116463580468421852131766203 90
6689494208523114477717289368862231639983960630823710833526431685272186540196479380113 94
3054775607259573068503904211923748186343037703873357766938412823448715255701307873365 29
9593945750430173323233225511106086031077149339784610858851457682263596088827267871761 35
7011859639607328577800826553539775079300580823912841045249232500457308960045883954853 35
7304263458895856455126200541575062514989008007667021186148376357669958763896812963854 17
0020900597186867468409696670373337924131504390019125197678985569600717552225157451535 36
4298635519809395052269121703018625039547222623846861240508492118586270906713826827408 14
0492480932984347795431706016044044359904516959066184259091868377437299472765480912473 58
8377674492534046015759276297612930054367737074246038336537789781394487582315075391118 76
3256720691988809851389232902216976665283642576878153375978337214257739085113918819984 57
7969584951346264669754039052927655647583056113859889031644096187534337529430455277094 54
1548801500290062344674655281568499385070490606655938431345289349087634863384655892820 79
6148646916080453517922457802932282442542581725363970426357438632950188815136437953094 18
7238870384636857138586967882004856460973535564003295396419793131942783455061370455064 47
7569948976145167414477471882989946073369773107059486496137438307652188644694317508559 90
7232045078503019820201121622731081934671406799046567065973540211791905228292798639536 75
8643167620009880401733414516994618474943162586614953474901935936252863702740480613740 68
3556923656900044382395306100152759503705065236550974222351996158225313039847447022165 30
3269285203993654604339965385908340021400203890526309625109456957692964046037949361879 97
3730382086997419858724096786831254260875150615921739088190667826575557451583047058152 78
7596917613092917871120768030999839438011582966804213915017213235793595619958556055559 09
0290535413669574619103478923550278409442371491842289370696173674348825422417984404669 99
7937928321645911778104237771173146525047628265989136173175053212632098446365531407201 67
6949270615732186978865047687420624695004381123672860066158691455323703714085634721263 72
7921066239303819190576158042551133196354590306608943309397332242165672086848837890432 84
5168873535278145283014759783335236422224649142427764389412255584663235409285360507412 91
5360164672271156085363125877970620766711442898720414522983583211325873563125336153585 68
3116008908489141114744281245091954517467211599709330815365764377915904100592170401480 180
3633290324539670529248393075219244184410955581220984931942280136858940332958213761161 67
0144641415353089826922296011538150858841211417914974714407429784484288630332397543119 19
3858739389995304454929918439109632402704720870491449442085805361392151593522193564160 79
7701806342239166559794307173862759967149451982129584572996194823884351303534382558794 37
0493860226593613472802823901183545243128806630830968919977627943239857704197453721730 63
6307081458373632427487910434237346373377731571254584849855682338469650064806636007995 07
7012967370709425919666288910843770987113598398272323176646708324371011340935426966248 99
5903929615008736421978455926644174816123012941443373336177449961876948808454450317598 75
1720151386855119154721192885836896548734044633905646310679325214025635505496091871509 87
7900640761552974165575465689997819253510071969943952859923341999993421298737639478583 49
9025461104114333732630304601093172516138089685109000734937070293491877087069028082550 11
1966830208514845341695198063258242734090977848976972480035682851448864338613072886692 97
1854823463817554103828569307474677201643019018538902824823523482810341966930286961645 62
6265807807879220911997584812580094971368568963994718233346816202185108610804195017033 25
6269700226820930304277132679186800806515525811729780760513982745344708428488026481542 72
3239041456516297792644157145653415512064164950527239747052811458704407227391082391938 86
8929540975128725568177269147856698073008902369810375161546555210844950234511560605394 57
5017963818052353022883993782894060468169597554727023387404764226351983317927617553436 21
1827696653022864813306578716371647969052085047518127635320726063022351953012440283552 38
0827827111130923605579097363530989232824535769511329984199004183843145950794787412012 91
4095221569164547945504848148441058766407658423349658057979028838671256966600181518256 71
7890168513643495189518673241000250827887621872633723143818892508852304911000423289101 37
3044903408428983181886684090586899182837312527824552314195254821995789759033342843728 28
0632237135536051868867036960003256553085884049230364734256657650857953788301065814073 61
4847759640759285596007126211323711100315948931639749521112319638759661221940812317259 35
3499923628051044003983869990007952220004100587557492729564792561794218575857785536003 09
9421897514093660718354262523292959533015927991422510015731453179483293797518807798430 04
5641061909863458769770108465876808639767179783484774361523298092159988500880977926111 71
6162704271466606633214130356388867464185837959539529594564189503204581791577176895684 07
5339711051624701142133673373886943989158648409692759653462396847274791291619821235280 89
5083827278880546685340847291617382901321844461075459433231674554898053122025375634619 82
4573538958845107874450840688829851988062224612591209979534375654757048245457262809173 98
5364705456645322604745321994377177302215607257454402818295289580471460872884267877595 72
6559555869759407836775587401572101296339298339544531088706008115612222682128505031705 70
6499714926170060750369497028953678747115967007244840869012932883368638368268418025521 27
6542677492183415882776321755256834062903159027080652039527570594603585247795937204479 75
2333618870509955276586509000858203951126127737075961788733020139593787604669860363654 74
5274570590552858902521623132176611577953212415296017753639258651801806830194478684071 56
2488539128275233962821020964206980236220307540864286355566446708880810396149603964487 27 88
3953819552020848191171908116653633608468803233224858673451936446137488591259802772020 67
1298766747809619699508017876599485964270262157047687058074381182693838755146484142678 60
9782872386816140488278077479273457028761302975997046135730283752965409087135916134422 01
8245576369334382526487659336881423771509440490324222665961208094651542746044532889813 74
5973446112096698182903091930059795996644250546761860129344016184843108500875556132861 54
1320407303738899518574046095187494161280831707786214785776365446076292221006527635658 85
6962025331988853472409613593282852455426154613994062659708851937922341517927625673339 90
8394875798450236198539637165387417947228698374572355957900480235585766400207060966411 23
1495356413054003182843371589007553145304009341042360302544665678602222951930662565457 25

56425201735645963043672677656507082284443046231046959310114892138338454286490980689416069771678598325255477891472719158592585369987509890109602628280594372398419469323512444362766716749760821719281578572912772760170131070662697714252543654650956406497911239788636751156944145594521697317263349923047267027546268655639559564458259705025933872915524457379057451830979745749645551154554844378688082251165576801660510894322691058633562502286343642165167105690084834297529525561418428920039294888967297115212292199740910447534514681306697215598016552650824044026908442611969681205026006101458637840807764238972691918888388952282470557580543018525876849720167621353924438982519836918422513103404461993408975527606639317961253537179983758277689548294373507896908667323735940855368471874617079709158426924400563809688618062433544072636155960236771905756230694107663569819720443783989727042031391444714924865268535935164569337213137814332425977745946321218224471660448402903902654915254037030451459351176497546607940227890214310811850373641202988864786008334745628705390194229797008092857757360982910190393879538229469008479223439459709835381917653016793133543571511436774428283688857404561886354146325801367646979510833526745057326563586256054099284349054944900097446301931953427896333456775655338716550933801747698315812948486386697013206713710196092284715199277979270888278399387546318742638008635194148647970134056989500408353466253080361899554618172337345474050252263007541015665696266062057338110970616062532192310138374804808631241068046194385924630132365060909525272778173725776602744594037087452300368576174602601420844328054861570895723509499817481998229069044978805477665742368958009587082372411354469648957759263816249132893427090221878290610068897435771376999655473722959570682872520146154990089219511821827508774157831939230322038883750396829242267477150678237312760424498619653256955678007696455556052897388802832166514810634565803549878607141901372597064401810721609624722606427903248867441158472452636149539321902558604369050300138960684226124682127333708878686477532636851498584874679150827814591005540651151322358072491968761013498784729062407866529812837858818619411778296486337936237728809412326091529382513071856310586805431709005134074725073957366989498330079933042414986929822004786084163389246456219504998062782464703388610335088988664449707657865808976865242573839421664407840777337538481613990695782553256527762754736460188350909684562329827509159375618447479159247766747854760813766823899431145366801338016180223058987202939229797702177862719369131285221380107259376992632385428937142999500586671715720725746198872190834082552316140664784460425539051769805107351847407409285523350628582359955240975536712528198209682047571676210810877220455861604215359130937016789024034652576281721961912904880896626095595499344246245973411971357141279645230059363652229397835612167692429099042273433018212696407152325239756651190536322351435939690832132468955326442727646240546546882104763266836354099502835376765309703854544265779682561109433647884196294176820230822770900124518771667241609245849796458514497509850478302493810814042218168208756287797837484177223419340958575872101525900008090448498588713146760005632178273655907953209253105619087311712816987821153320007593286472745278815209968544725473757565208077004261636922018392656027212241982768943806040338800097534443642374600567122707602373021322330282713050090991274839097695565597621707935810278670556297973264985842702298467521104764438638592093433338776544616232544488096639974924444643262712776361969275148346156912068691828274838352157640109428379165071768830582085682915462674230771569811700799175578721513784835894770734350045823570193057070298454442046167451323471491192563679302629877906317624507236377461804238979928891623696589804100168051850179104566130735287160692379770765559781773573314617243314083075696955449122990583614299286048449478189669217528796674179447455445337497669748624792215317212966244052544149628804844643963473663032758802643706021670057492182068097106640746651148322492061477085763442915372846869680794844491936853733834496843218541910338416119736269298948150696449680455570739461798986038556423302453155528597949145457619988719037730633592845921254603546020562177642024452736831059374378425732834510354754493825963329278429414219567171761588267418197672430709563605064026127365439238899841200179817891027149165001812678930534627916538468100019606987068272048773159396729451402267484428442548186218934742883484758743399346271448908368107154175356473186001482753982031226780699040760660411543360376660551071796393599033851314195964406388359507870090010784856725671396069066908394276464530160273761120788895325008871841739567273757161870013056802626824989504055026116347599421518437427454043974244499284557157187599267297878646540021009605058074744893878723848438131494174588904043684229905097487111797547047504191066506034030563768639057787208227681760906086095871130710531239420342556445964451201800516021925203514084535901914349468769095046816783510949475050238816144985847202895337136985298519727171401257221790590250173828514431047612198470637690587260531747533131165247198735324845482983622049387492794157636925033149491581541905338304385429561326999421101212562027429952713404907596846613250169655829636352612614563639837880730677354241958129120537910900909766732737734627502333073597718887795905083088711706932369834951875888634825839240885832134654815423249754814650534960443311600286227404357185153202032198132465595018819301908002513364833606618497237536967673913628188880971997525425911526653239340957640552390006538416968907601583663586245972198889458965412306972435270237258750017670662885918100213411716254105227731665004192986824169338751191401145742126471692983715938736307202330887283980456527724866620289853153077759273038083126704288036241111980880529867170821201016190337612186858212454794689749695599106789462320341161835933376091067210531309015266430653366261660398996069153644988030217897843228463869667544549577404698675535588136611619453566616797306535085187749680324072444444955590582168322521470643265185449390767663953172936089621791587565011073773626912117564253833851745136149253213586761297534493203253197851334302717988290903264170983986412223668587131746504414933180582144396635223009891250939030120941005410154094344224242360194090564124817766558637788618047088751992280433079346532050379103452448917027566296406057904114942172838444354811101754869875630748008461828888570436105268062129091713833984401940986384909659451778081414601296500898251011831480945963911179363213958388503635849904050062652314442055991431851579293884699273557478398327127715946296224174948454689447811338363899026956160328713320709525005036295627453637622075417566247665406264495034

pi to two million places

3509561644423581273606101316436359489929590120425031559048942019086685786612223845430 82
715372890279204564501142205502720780800429500876241276268124446164792218913547405201124
303266159595839548969618580417914648185569182949478096472616369184168557023625641605425
464517358173064054404752686390589224064115045382629267479921886726311163363446155125400
806717690758977483839952320591577658234820926799376172687317546419513351240264846374228
904137690770673463248710020992653721807581062213386123398057832310635343597604808790095
354511404501874291419003834630492626164128484209376171147114027634193551456893673527978
107980666197552144053728187856604694976706760201051713492610684238245116224932020431015
647871231854281286633068230205440315584454376879443839663997384415546318370352868421171
007028441205077042802663818200853306449668355260920966718560386386499033673147811438740
353788202450588831747550713924241620207909133640625330063357857880093444927932205699234
263719759430625907217109218963292240302917421229376108093435190764198861353806998977951
960694220889368734050055612500778065924185282604341171087712318242279861843628247042528
512442722958155410983431238106752698349305333963202255330519013718579071210838003122399
078146987560259573120271622013120568281268495333839885504761619921253129402271681542547
628264467731935050121328443082831114887409857776262849068247499156794714307780303576434
236834754709185347001342356138483559067228813185207280839386473771151615048080335144066
030794733111938440163372340179532830281384633549967729445683835919556742352978003109313
207408402056101557149496181576079347639817970318744832237864480911993758480307496630237
765949134922312335700283955607306452536020637144093394033880194208172235045001617396473
167793529706509123153848682126633569178851370069371880481361586583021495977245583167220
271794219664972045630121046465233605123439844605345215564088483777418359693372557603698
837452072397302803667758862744031867319908119773301324738127303314083249138978418713267
862735024295744494250837649927069962522051916844097335808852702604671443569152412022964
494210906930172442304022458448265441593685118345100866999079159536918320431037881639630
688351598600929292349268717556290592137555222063821059297965118827417731396047964764937
927729705814789055928897742488761088044730422285923791176020640278475007983028933781567
458567606038146362018179798780875438626836018539346645625814541275914125725287805865052
099932618944086009310819558120002591116674124257860999560123410487358321095189800902695
682502105825443807470784840593149649181976495528659904658201938747286411995087400561357
473495347476014347358728344254765052089908842366366200301017124010684755806797234942536
929165375145284899581772815146505002338126486670953163002073385456039497965401243301733
945884242188629576113606117918125923152909910994153646398268500170491722633207180142857
118242512388659900762333840544505543960189450309772734143508868223073744987947153032193
337014867130811415680952305789333128246167161298896188149273367589741974215737967062072
626923586110079267846707888012212248774931311591562811406861879533445363407233063280995
714463714309625189847620263259410100978301855806542899731390725139559678771628734159076
915270379758726761696405261064205525971763235632402978311229597150938726465276244866554
184215749189277260321893829445768030264973858041254113703513013173499536699173055039119
263184247556670681185023579893860560239064040510793082853789882211776217006097147820837
582458523952620733183360464180985431982848433831439117366672359086829994044809694122463
310764669230431326967854017280873970663171336371513037301702652629859745556344570005904
453338501573097245420785742975924738490835825585299229967078377090490746375295669644710
297473032447504659041277008253153747073115160597598419430636652262378809592133092496284
277655929226741368585519465204724180811298689179313400928022825801832861824148458751372
963334873995094692633443053809463064199462278863084323328085399018638754618220442782131
045092433085730164673599751741019723159634771150953745337936026585951890690519676041105
524389884121255919849843630207301574049640072512020924185994549818044474753365218180961
357493484755429388816027767393986213132107998025523210872258783621250825969405221954466
711434232841927588703841261415810803768115435382625219541154283189861019651338034281902
266426296344111832268185799462817061737430674309819396798710675722582852149555463587438
002795299640829401618043331744837261603017894231561634212490634692304991844279731182419
399377668050557152711755201513475839701836921017991006729717814032047178171917763620074
637495158216890896620501750716950050509458584478598464803174503831439637323793713555886
344742453338914594541165502029691279253260991863173256197033451869049305009404171688359
237413718441735679458735973041307361623294951935808927439443791581596994917464234812629
687840898606322729062257421926733028556315897918627600911414562396647616053622390274670
905434920166935546875336183242799862264145956686298904137307539373111756014251625172641
991471019802387692498239298308467542266816812099533624725689254377662339535417169474512
319727531687936954563400033844143153798886265072577453638221839320075956054608474890472
097570784207818901942118796399940677173022125314195426655824621030227598704610495795348
564633559875914628818399547044028524864849260827265669038419171326216664367658240171188
200665691656306393412942681715855878712823883701032712665617855558878304054427797164899
122023137315452463417508372885135146803818787356804098090738042708335065953508145963188
007289006070429185278123382010606129657602134821401240849452367870232264131276749985389
970676743965969967881725550517947269176367038368450335400121236732345706030218667138254
111784541204595338551039479669007434415390174360295529827333881161598861737439779971446
148347959177780157177181101695515827086834547019742937614580304203534428131030736768236
928807159600257450878644871358437696636725388731632290547185599895990424859095457870886
133390203108577662704843472679640651223808296919376580652111627592456749386088853528978
192897350468389279246335200746135551409658439944929109956809974383036435238040243054521
074094308233039976101272312476466020238375259747744201602338124377120807311974446724565
062057084961088000687546371902003580969616940775319356933856961150482075797930343812373
086114630604922440419315960417082515953385078442168736126021683321523575149447396521718
563816072545657383817829940925598303740774358811176316988294143946227608222367135252640
130118068457018720703474863967996208151569850320318788865899294273940933563180348259939
199124590397252439076908927735676313099058715500289484581507854738241801510269894307794
939972620734967622476914837623465446770830549488615928722235522750267228939315304388588

4799482252747838654025741754090333712848917674845269248881873283462758496801144245867 48
8346610431993770574638574370996114175343139407810293128665931984978650517169400404426 58
0829758123923016613038712867514406706251671560671199439546727620416432394053910594674 89
4866090990408028633172732897612733352037135402801805193192417620235228677326379784864 91
0284414622948195422194236021314221116651375159141548840825470014894532143805690346732 63
7659336636933157778767037694026942643918365998677122653250006180455015367428721366374 93
8055618396204075369269898631790460967841799928489956175024999535192120540767384134250 85
8401517451793218662602326172901989197600993679997367295256163233056652927033841196948 99
7819249002344046874481388460372811533423163580593743331188127279054339657330252587752 80
5574707210483192296201791138937736458568204189546361792156599326733402584611500230797 59
9737462310953773598416484726779301811867860297901011502339737836340799283636152516610 66
7003132900134560893534895437526534989635385033338345361511472086391172437711077149552 35
3049099376701686782783444546408870354033565086492904150805147094881314298002704941326 97
1755379700058300487626325230969483372581750049686158636305171607610168407845509266891 37
6431592837020961532016827033807358852263990155611366748617351058000000592775877714161 24
5754099525603494040679115151061706083557590915167532019394581864314990078405169008432 08
7561431622244572089330704768648898116609477677120682209206889790310930482285847876580 27
1837193399756198209655273025737696888211643786367412126863839084047139903659851751580 81
9282310233897918796887110165693231770726393122624870593957504260715780274387886640854 19
3212092837124879380206529274916205444559806561034330705077779485655154144954734925057 10
5149503537021939192772944481047118581457848803623844098221906145682628050969669984549 77
3665649083014853107613458766837978423560150396244506055406657121612613793927660915292 74
6681092274894436625186480891261959936214852267922525238909980399200996645521030944204 45
8999918739645283650656712556378099130770904106210463124047340485428526156184810318633 81
2830760817373201484094112444019862627203606485905036222573704137896396590195015313731 38
9886275355388265989309979859616314743687936430413504291899586167766119557228337516323 42
5131490222537494462590037473261591836188838694610773249459039987533246803835972549291 28
0027138394097672950089866985209240464821628948804088121805616583438178396272610578902 56
3584825928479281112656501817303423837646687615182635907961686323864335445820054860988 01
2016931398171040392681412747357025915960010801415422601442014350128571876712967932524 47
5049390049020621748086267809844061752356188018604359516349499326879476381000136040772 15
9663964198530803415395550823675289455787369187386733417366682859461882776286813257707 51
3737673980891938599801152353305422368501953203277230965695087195680750591749120979444 16
9111554775567539747854041216888798535385663790060414960821717825348677503579096301661 82
3129104363692372025212138089019891037904999072601670398595664903462245804995438193502 77
9331477352531464273595507808002716247139588984086921510259417985154180379431960646846 24
9099673240605556586184982722820948616971188521740022808661169907762994902705650896791 151
3438047491528004404929073997417561562201374619231193712540244909732008207525500830390 74
2724386531830978019553131869770153525390098928356890739099127737849270595174366945213 91
9485105401955606935350462432253989963833316484062432303260347637750014474969391644354 34
3707770014021585585679237199355204824422489037570311942315176502481438493595988594046 06
9391943972166082650268772780302367103824274839402924937652126185640968448360735215628 37
5720685957014270795554200476030300762749045849606095625348532751752239566347109590771 66
6230770832647882525272413972134499324850228582185005686784384436532172791952259375843 63
7570234320631037045213717997993226962624107890087176973549902281811204411986792212226 26
0171901222348204236888690325037104380138262692510612022088712979060526195701993552951 51
9671008962939810847839918461756261282355373237643047669569069792862831333052369616090 88
1911508943856896693096839034716803300580602884251952872390206092546711363084682508833 87
5612721033312435865433524249918799745151563895067870004451438934308553802619173527445 75
7129047709691136938125787302164165405142184092183151401533631683710186584024838289177 88
5762333893298791808082819058128744114225847804686509414219604999036894930861235249791 15
0670291351529129488199237639083526999594632512659962022201037216748371200002577251137 11
0552012083365878174626308391426208327053611087522963647495838083393952854349248477285 82
3509515728262984198648485467332995533843202234642406026800216341303187491455956130135 60
5123993934171097982571874372258315085216722581304479180055026999698327656402974002219 94
8771296310997317457456964953355917738786203861003658934197538358591612954624657262727 52
1288174847331157400457147216561261587906129207524746461492724286436825504810803680552 02
4685056077151930891366212231714206970393881123588539912327274414827818509994438111759 16
6934908042072129278000035351405026498276488076590774903995002042108829597984419559420 15
1257883425228971856970488080384713488248270246824618489639096624255711926814066254427 23
9594089264385636396539412166661636967453356438843622810837779624994332209436981670800 12
5422493768342200507372475175409173740396880640255164736292823107408845490184141517806 70
4526071212167526320476526808098344101304712042834153668708036150561680039795071429153 24
6252321826342119508568479250037065119118412978144666230880084803724149432593436837669 80
7465728228380274483357117714117623935939797692712837101899075147747432284661828052426 98
4064794398735544213049389412653562962790937780318567535584092110434664870157575454797 86
3232650612616244681133917306110222478498308605013498942032723269108712643778347449187 74
6001792901782554983856861896221164956864261771294678520379893487179976252782994252084 14
4067575730085697242668843640320282856300780977681327632525495277468317726709660386938 61
1771071279363306008410194669696113568926421821145119694727362923688547528522371455315 60
4328081174113254415849242453607488284789181448076833682464950963536080520130623949871 86
5252772663852259972129377803173770411901313279702799777766211810930117192191070134833 92
4880759049690418808858399086597714161508486174917722777805641542494761889137083396599 26
4168564666390336223014766172113243244087170686957715418149490392375946425543661461113 66
3617525138606350376864903230853528366625589551164535657070570406042611923692047417050 78
0180955803921828391895773290038088536440519051747513664741690039117594541760474742512 10
3151378475337322549157668954813752729861345274946973246333102028689266564184584132875 90
8617342074831823371304052518266557155254440435873930014348503921375779383449145731706 16

4627584097836548104952011378894622502639787289659102196184926343974760272719393812723666
5919563086626750893754821009964803735632512699735257273539609996409608466196588515031766
7354744917353082638302943870716738435368767288568811154008874207559206746175359928100704
1620237873698101271126193237205823096622905641934337545593211800144950520669854798688978
1772509368044625045139302927525292220179776004144328276315928084059387904578578571926133
5902039551346510051727238642607813933938188862078563796652686129254760193598580922159114
6846252299649988919214743422625152763164501024939035713405704381176095930259221963249055
6750495894714050476426397107044992846977859780760325654228348676126449557074646554038966
3489890455448046806340007060116207144242826495761514111132891897833707528071262503395797
0549337510436979181495887323186585389003109010531936440875571735194573187273163736781255
0499328963748376575772162147265345231770618203174485033727830365909870253550768458880388
6681092624211451267939709886652197376655130344497200076438454595064851929668892014401133
8732189160138772532282920170491168896894150559490337997289318544695811891299718850252822
4073660442375853712245429127717602054051085605737630523729134884443101180077766910413377
4268025730279711701303328958839236821289234850067397956166122614350312960366089242492888
1265759969389422426388548889431581167798882874557192092002576721976180339238843106555699
9885484791319350048382384417907417198873219826090704243729404415885178060862697297333399
7767778610957594596266552139999706487407541348668633662054916410364094852485883061150511
5130665958277387025117942716462833897625830061205267515150463975945598401575557297490088
1774685880708562856451344861904207964352002616299991753727699239565460450408600346701944
8391491555490344306665629420236145905159289080998415124669675655809264692609385388870533
0823431352685512738640133783933542354854202104666145692865742296959914045649572169739899
0916784544310056505807551465537637891389129918503290740463204420014324440714841600401188
0761832348450479112736186706647402725179681013735939706284983868842314735072109159135577
1325054819679921305253175427944916968162181404760787153476607990464617123917922255026755
5004391192068805570041324288614205486658301127740827406611484500897783398874715468734088
3600037131191562015781255313629109337914677187738621215582367177980847392840787398906544
4517457425612569467322273674819395557994701360057121571258402376009978979106749990299699
0970563447811120541323588941291240652037872954956836826422715504537724981810636141344466
0914416498415461549331575385407291745889125078628989221617611607162660599725167520268400
4666955373113792578958547978919307230019176121591091513315085518840702676019991630527711
4045061243216476094832517255736240541692154099991088350001940508409065795228282932684111
6169269723015815196508546436704081124791410968189133512200280173711383932150235952959911
7876687839857341769468316554886246061127618164669014727838435595400112476923834782224155
6780579254705269047454252910497231620638068103642503620193990842622680835662788167437500
7616435730400729445944700851918694434893557211800710810560867242650382975893908702432855
0620278838765635801233411587306296727591500074625487257339720327228866480188673682666911
3518999713608003671045807641302895227973086244291320158005068297472342177890376614580877
2277557177776919846276316038613278663455514648460786861280364800169349687210654260813933
8205071104731775131665292252879088281489476952340729328912892973555961118282091881792555
9383857286759873391186738684115444429021024287139665529390470010337494474194460825072777
9299664388376961424069560093449640747733363806449246967286755640688259047783797993863211
2254758869307355862592538497667566764290967010450915674432885938703363797220069460362411
1651948819407483086750602956060944971786399764084604324430555337201821919046703577851377
8936440307630062787703454392198982049145182909001638439273035920467522595544167093509166
7103836261843105265390770536626076682348083630269859787613934134071490770571340624681511
3617083690582980894409526170876484315419131858665928746265844353431187035059900300644288
9462801330459266099839385297188754916172196205953737717969859488401931806209641126540311
1057254088862472109629403978141268722114288822252461191206869774361308046576602256212299
7165377482066972339419483863809299949798575737143424672731947606412709203772383542765088
6042713344791511475785773733863985244300757297320634863674140237492887878724952239723255
7918565694923766960321362714355411000645731645779703271024035648105696410732368235115744
7973592290290666283233563526049523582803883247014076470033980679358006715736980734108799
5149904556595630523261900325170491572232656987104679867592040866510394118371972089470000
3668472513240396239063477322965782583694951571932194526049767845229792633490830714451311
7847428727363722430523288417591491346574224486459812477520050729509079349894705567599533
6973507464455266884379630207981062942505856035179566280753486358063341417809472614798099
9291315560314891762978022837889083904493834984357813903060219755080903083106800440233700
7366767322330621573042251611887124543936893974166531053520544531291073541595330893561077
6259184192574367795194743608857831807784963481716566758181908277895647294750723629754699
0842401820815618095843072216529073008614852329319745349333795604449424982918247114282399
2162261236701357550485080674167154242367503529651787371281045422705743486909930113174688
9842409765314715179993356397717875187840950099787549541813870527708200029606412099784655
2124554940097933422720078902842756881245597990772591483484127847636733727473414360540877
9565996666367538351341667990298587849836080337787210841500065200084072021875093567125555
3284694284582011110636909310047176264150264578989592026515251684103150518229697144700799
9697358877827633583765639342661099888360172156168962026689878449222067403106619714375577
5431119119055598316199373307619481476534512587916768871667126910795652009859488848711544
4523397409228208105113427223019164384830381503089781303278033498194363689693494580900833
4224170407663278163659838308257758230491316259831198978904168659487880537924927417449888
5476813101729522907887132235783467938420403150159258151295755489488345556883720224310177
8841196592941178168886009931429220027404503293368983414785112774542553696401477616472700
9678763924048383317500362650967018176878881571469969928851018419872968046201270460654611
4997234245075503002466980698621562233549293148013789002790122822404691383511412446288533
4072777249538449669967700605522823173534889569043579725607045149294836074985765562785755
0130564042354771471147833430562807255031881496925819521337465276771852039885159915976844
0978884028579787766133290236015148208573798954894632561004890726462339728233944405189600
8954843538821690557971212737011568495130977828711283256842898574771182863395150281300677

5112973990202363301285951570088270240529778635912087752176607851103982843393521223476 10
836918832771538655119244780182572340054237591629802726232147224096175139595898074061118
6320790963462536411374705150532819588054917662398864339632234314061931743673521143634 97
1731958530998949458204401423983463266210541968002582973530410979811330580128413100826 47
3357723777516039854988430711629739956083581727709077230792061933717505700231466602425 83
6593627094495509206486280120608237908877997324854076805737762742448396062232152784071 17
7835247753953062144593484716158997482249534517849592270216335732898850211626699712537 54
3717618066270347920950357173296283404331415746294836662779415396372168638787692050626 48
4459443967827130137236504948815398260698523126998035654790483077131143313845999213338 23
1113095311865671633731132897837292991963937140216228431997044802307590380337934433497 10
6036227095099726991579846124026790755792100139201963546650907743987965061097013867678 50
3373081225721493406348553956083427087635446873248357462866190139611261352202579958506 27
9016539837216414783401483332932601379992629596561510994370138751739287492017362925339 26
8349056538320658091508833849740625307988928159785478476313869463148979789361260774730 15
8939394144634301978591659291983043054971392690742407897379088971303018977287887537998 06
2256926303756570352248984359031551544789839914813927031202677579500360924098424448987 44
3483941170503824978336634917282609225549028085918085149326531267738008472159526889377 56
1422639336557132745240620466204116235218004853352375615473614651079228210175212600790 63
2097917006969820227472832957984935302989167021334565849203483522287196747015701814760 78
2756544494793312831687644543129090476069613797845121679825978003960213411101432935156 77
8891686120693424695033257401432419652489261270258175189724619267099989736123990800555 99
4333496875411261783378970311392686212773353308102043637204693063051148383149761322586 56
2900615941319547214875790994308112526515136945490391335300447665048737442283135562842 27
2563539127640446312767213795309995316083373041608403606696573460155408228856085899576 20
6151831518310923378240775450432758282671678418403474239213444589557520888652770747424 10
6329624420878507498207534963504487888055258060494194528316841850861327841038605926811 35
1294622407120916912362264390088298984519755695230744363845827897941279923515872788727 28
5800780763824949120164793957547647360361732088789475180565117223442699741415319072863 76
9518724256762119695216565941880112115462222723011256594120880732913637768821481030114 69
9720777433167452769431613920640375136843764209271829111122998193969735339532764017959 35
7353878042476640355302808137423217095011117136611113286263898771183178690345988486409 87
1898004371091654274788899744189208969911716302869194127062704695067435821936607064128 86
8426544372492733707327592947080321784515727825453852150043928659328553714027586937587 66
0808064588390296451582652759571556207648887732768797346831389326849123111390329560993 30
5644685978697077154558045831390578204932557153240210748834309973315945112940509593484 38
2290801050440918939890380143934591067474394795906994630769910893720974200427763467529 23
9546439124884542306322738510569770838669780673048086201433179607625694857437208434872 66
3795550316010007388181044280960891465733153132658383728261586708365962001035252800417 87
4124650944930830897237004685976383222094346350163686276593197636638246052415330406639 18
6116804531019903784098010894763285592767165141166748737450808775252390623323681823933 32
5945807580177239359354968908778484736006713658427067620823327344052918741825505650429 39
5084852601169469589080531772049882247266937349030385444340847025625555032873707786953 61
6987409063566130276970616149064983812775984983948742683582043360050471724265651159748 55
0941626392478194048451345923208172476440638129202656473701838904496313406882091998288 05
3142998433832148139646666647653377798416249121733304436400075107102008158288124073313 185
8836023853098373137335798016709209415082722654344663068276801489220151122992788934058 37
6604640145735064530124617804103893969195257167588236431081079819248523564865328801718 73
1829649595731818617411983094827212442000169588632772605637921898854966077807454354374 94
9950885801270128919715653147177284015053740433148175576777224465238980109655509092484 40
5499450210412613354797114423562193154010413000902961057403569574028915614982277586728 01
8192906268958654704185304401290661633331030417870772256158094065833415524323863976257 50
1389818825156209280541423214464254383028711692986311963705119738029630132229168423563 38
6444602126745177661803221881900404032214227016429614178974791147760041585164014211483 66
7169107150330215262397661024487949708574352622122998689581098665836199943465596379827 41
9961080768762371531806673424981038792107240527716809079820454124611881626429367236643 28
4752183032086806710207165317242539619159784307889641422142988554362598045965041839717 37
3856956698513299591130039165212006984999211737652703173139917358484146353164706998502 16
8455692893742544638854665930429709911345259556567035900763229745964990842752163736406 01
7602901988969947403853013975531943271928636752467049479897939888499922743499786666671 29
1347263203904717335245016403605906218908169784657185837838599126351124222750311196445 59
3221753828452077550886299317142658352991739505765470488071249814253162492339564246831 14
6503148715597591609969164409303727378099641170442450130359235107074187790519387260027 12
8510663278521641085425921175802424759791630289489223428419655673293354767936452660943 14
6892793311635630505350153446742951699536454603134888336830172413471518691056327683105 61
0013207690391311914279764645735064989835365694399231823458239035512936626264067917810 37
1017826935683476903728254571422692727808661682449446750695774538813696736997339998515 90
3948896274474428843623060789623951525606065556469379615789152158541888981079092479596 989
7957254358684882615056312315253311729010949288793482020566306697389492602384972290309 63
3753279308103452035055957335028689073837994576026530378214131346308448695727615792709 78
9465049960191428357918432977026244925367471955555834008465081389746483079952324057171 14
2818432385016853665604342304094949046373890666034607073259088306437517911098094447817 15
6261446008694465540058807208550757003812593261473047021513818694498231627728941908345 16
9150319473375666813684651202402305904881761267154530180222080419944496322972294432389 64
4638085779137589205684904761195893043955030470427197963147093931830911130575370378250 58
0093430899907487661933295328481280023517505429025724882067452756220936118003780105583 18
0617780849701079152482776192902354411583915805025031257692072910216544112034644300831 96
7623132832379816539424853020735058422040419813736372825081925394388322384826896972199 49
9316249053735800826547759494911690582251818711301033998237812431303460147838069725097 87

4437093368920427468129946383490075164731149162231325992454247971461756281259753997181050
3605299185818935012879163361995420077932622531227253949343958605862570266859546674575730
1930737560236468187942322872521985963520720219376033177346032709488256206735134056593040
1504044820318586023983127040524168612195116475552509589811518163025371927431117879771150
3787042344514289854227165628061732142260001151326881606971886114004683804862948425256050
8830514499297630195916507319798953610828275990096956161102040516997266531446961663326020
5291521005062898981496025044101392323533992428296294488688260787244934831946662746169510
9909887198805673765889222319711792654722310390001908309685412115323110680077942920856780
1667318265904020775760729666897863392751720013345446187130567884580369301950615495790180
6841210370289281400237798970524674378075524524738175682812061848462268829392040743461940
9043054810572153488623768803849134441934225757178330395950053270221949268491342729666210
1944250690998285366281752160977877137715343175447011025026246078772003440872079124287400
3569188568027367213489092912563673800641354266148048627145612595477237675942545193687500
3771883648419415417524583252490483939800527755715564172881974909653263605996430294293140
7364762591140495411937209075772620640142945487289139178434215149579877728607995807274650
7004338146962911139695475302595754152683703023550762042246434081462188718773661134123720
3491123281848173575428197874129279669398585627902539566718110935684189396507629456060120
7914168153209296570238209251774978507161222299593490528180064128474939351045978402493340
7783940903187802978807920551777061078536696399069387881855250677879331545507886759592080
0626832927605289266639281822281618099563149557924007312850181765887984880813990407469090
9084598609017720898752380634983017254408120415678329295039673583372707053324243768157250
2254104149875977241501719904945606177931565485850110689942269157461483049359256048995460
5099953792551330147400664091133102940987595142474536449727198204935735111450238887777160
9176064878744369395068074411358822471982650486139183097118450007918157773769785513155630
5841854758314176365693343230468202651447389999161104872546084311587749954333076412157360
3661224643228781081814087389277308777636861872477264733652546515477451569505278548197000
6348492262690638816962770534327250127055883979191236863199390427891853650103177628454880
3444551925723704049197554254838767730337390431212077776762038802819814628950752758087300
5090287458978538967772366560793937669728665397790851937799835877977551223001991123151120
5335899478225088369058295738275072135770496148002942897358326133845300921860326594862660
8010328434927917703428520108781825087172107546383984125475878140009507118523066034347780
2406027378605527988985821135162403367458222648102090485982114800578140070621331309893830
5930553753683280127026604055981946828219940615279147995926436234331227410558101911283580
0176125133549413921995488013984861627891622063399637942290452470585916305865648499614890
9604642593868836127615701063343535465174801537276157269438429911288400926157206191781170
4424625221378580787240445153494370834850934881399707813309074300897017938294816337934840
6372952731845741746379933670230380852671617482581570989322786355033838935194142930593100
6924622198081080442474892609655958625789461035761503716685519293200776943060086327318650
8979105223700139882428079367212885629114820935255363663921557444671643855527599364229870
3414812465173583741147472589463239803019675544914707151481470858910278655489422333966420
1188565454274366518627428731051210841869728599940176983967271392882572003896648490415740
1754653887570560864607680490280208473308926164755412939121490913531623385206218609215120
9897561782638874842483603013396420350552860989012190500161351407636297912445242629452070
9377292355424888131049728830971399999993176688420006457748673139206157949752787765430970
9279360723745435466389738142076371404884304630634150424721244582042431795870497901938580
3334306254196041525101728442234082286800114676155629573788281564952280840082989532667480
2732186681312331867460915196572095908213037185742060383245758766234854374638030068296190
6870368428944605755093058688430693501457600659666131091763909962488005593171492912972450
6629993606189642177849958414630783131819261013086122280561837411302275916906405509682530
8988508275767152519333419860352442809562459552258492373408426124415610341321472565128930
6905524451431199126517386138702698091580050658189316207623958332030476536231551283591940
7077445508811697966237618902121043727017723351125934591833530183304973601522528402946790
8399891057106405086137244609623542331424439647887642347204315223201185588691376323896000
7710797346398928751264979800635371860918069347575395878267555901349180284089124228579840
5630398685033655038173102198706075361041447233173002763743827616505585452560959308123650
2215900158400345026886081666556793950762633820683060559046083905554651019835600563778040
5601217590688697267785534445042267381536997049673140757337805770990129890109627601290990
3908367956672320067949927166551304159486359097145072672466778287357818929968964089927800
9932633302423305827622164944726902484670546865442934723662787494600818937134304993972790
7392162867280932404465529229721209476929889129235319438608177425330227594360481051499940
3955895827057833651913774070957558949740958644490084407081078794043106762634939189780490
1692553832336285921632758070595606482632116014473901217594896905374654056939022650043680
7951911634864868056538996281004216929193344798966698778301335496941690078875227165291530
4035018074101751262192647190697756994600136344829932669992625345220420637360698501321580
7045152042874576582270492955596348936479912566176616877482988340946651598642491663343360
9599661321331160633896687730097491688216529420074426147763875089982945857749551496047430
7642889133337736695385962935636876298942975183427726187954213722467751780191611240977010
8473206268661364286662367440737908274404968121296915837182788420576634203315330079761780
4299567742023336918232521520181923384249379078560051016372292432450075307804626139764180
4911836939956725515549053281998614108826660709064284108284356080698331757231454329960950
8632967998885794948627330011557496910628627026164089290050615677932972259680934749495300
2530556300087834756430424437362286917613372724362512651923046286044177782036793463910320
5984147243500677142367249109100127342262740661032938642105436326725965897428015474437510
5317145561230688023363506128460959525791059497029906406966572165808281872007034905682530
9988877425402163049153036786393244467800646228311777358026613509025715124975957220951520
0559110328184949663307729872978211958116257391460932277181810093852206573402049381773790
8717071350208880829722817841119331788971361608250999453158790428698205777711145067768610
7750814144486668488668261662631457141351210966750791350285261238294670275667320418863631

pi to two million places

6253513440173964723267249778335965796205640479165475662133388437376737752282310823934 00
2137250214320428776053285642609081221819731708985640302662380017542208015104311037084 30
2568797069846499019284963623665805170082146641532389346475473159162182946176900337841 20
7993802555347597160623083257672866270389423771457429286285568914809432522863920115739 77
2889229542637375500406402584205609017883797247436301814911089664636406226764556727095 95
7490998413124378232534815147393315682688619844017909056500355280956924114742983982749 03
3689040443743776742447121147069614507156611117780719426347216989627034331927206955877 64
9124408546257349482825208577745688984610025423049775546339169188090229211587794984927 81
1028396455773529451723107831251601597641135898993016651398732430682293420270384869409 00
2741597648124472969724879015743025148150463458760460316234267040535721267661476603680 80
2591526229421076291719813535106759218991588265112431717869508236709043374591606188697 22
4462082561110748301471863856850514043550811588032151693795670238760615572590473711319 40
4684074485699887786541934342557687411622652287623300077458339211844354586444516908670 19
1135642180265261541474238262898161035512378291229045010873027580317098175346586053469 78
2378809035387278522149165273058547193697511474020204201864068867649011950858139164892 53
8488800606583299556335369739958012268152798254492755916394923155520148407497025299887 73
0453035886862238109886994411317132776039502916166700855594332191755865015173133328753 11
9460655675483241044204436722509058751158847324373833462399673838619310641956953195246 69
1623273567571357783238836808418710108363589722011797582007732213916456711934557955480 47
3307615676143151120578708295133029704021436402944161578151832850276254270102934079435 96
3846987568359062631475839641061244169663449004177089324625230530017902557341646699577 89
1241704516129480279427011640647567257869477350487139870925291634635297519871575937620 40
7645729118225964905369029241490063192483769330802239089764986463136787459453113825829 30
7372234278073352783123053208854935274997144140672833574439526073705110944270581853549 25
1351536063204852245916475651335595718166505436869348769497358528202429621097690149345 32
8939650779652717415574498654751312187123819442821687814229966315780739716999047732673 08
5293116766255313688726266722219741598220038619746854299233034331205088586461168290927 76
7615350053131628820606598906064732676104026884647083971686909493204342547246215904053 11
0368352147923218967205583370296411374243785983244068783397977786030159331189639957647 15
4527551597708341973169486906249616069418981004403535128713333642991869201329997291761 79
3018324022528441875999521705936223129826010094347436991813123779099298699092083967979 92
4314808990080184116663597922296810352159106730444942733449166647675612860612199726250 25
5779121267044339265068286876460664688610147886213840089981530261974201449567760411280 24
5900709284305778650490893450901220318496294543643069725703488942622634251447273847522 76
5317755889759850873283044921096330168837713044725657190829833738478968460781875602991 90
2798394053551416423942576200775696321643202279842831215670941036008470631552898667323 03
2110533619410395265982994912467881383674815744041602399833666140630557263953044728890 80
9635925861772241551153395744898343122485020501889790977824384277822293985126746360396 37
9671908002824277929920352607375101316650356254987976036443856734319849950917216394431 79
8705462490997975852271994447870900539664789257759194412388380478543917324395502827974 37
5157460482042860939001061497237652081471339089354943597754970012466397340733505106922 09
0626931721491011152927162958962561394776324130932241504790003140036868744946766682320 19
5944331123582363138747064765661041001759347292725947310840586201560976606442752678732 79
9321101065324165229455562223183679941728449110540303330300747226417779306564802724502 08
5780076664241593696016454650332050562791770800913795677559854069868920395093436340771 93
2882871369265283114506722905410162241950449727258243783258643608134140162984100122351 45
4133989841552980016995125502854680904842655277348822537742752470937988474506319983265 90
6955774145538162882556866848440560473368107602567701034908204249495437364483293758323 42
7860997571835550815620861416684221982647123916508614587179929132821339403791252131901 57
5829699407648818640507680355534344332056620228039876205195586898433703009197660451636 25
9895762353156564399318439266069254698653678914635289229676107186141101874674838662889 77
9526036243120448778379821071015226818988876202913931878116943717392297381306658936974 59
4602890435520791168579870638349575709535182119788510782681946142293218182537254753014 71
3573539355225837926031477450360413238760447419211922824150743915691825726905795457984 78
9033536429830922342976017604505626839141470245619269995308461490300160615597059807655 81
9857864495326731632880866320638718869463356525603964317295194128315958578456206367718 81
4151497517253190077849618829426736311216278287130959709845276440917822472781391349415 12
8991000949512994908453013166739451774922917133204010939658972795660524943548596783116 70
3770721012375400576442624463927037449183871534911805889075214596354756702562230300146 79
9943913125844319486271702933877520078271709770520935811879575837984651726329786446560 81
0718650332019801663235107321930936688606300528180280550295948711868424512049162425366 58
6488279446266673513669679100263312422027511397545764987605367128538957583268770325288 15
8278351181895936477745996191237454977492864018133027148759495688894229489560527634474 47
4728790030127970927000577694120481886564013116589040330318391795159046482238141015609 40
6075704489827285075668835855479705626536455097200220378155186918293689301832187487883 50
6959149521927037287177505374350139183597640175833716963982494139952685427291897425297 17
7223411946497744316149214148209463984393147483886645667140690538792027627791215517220 81
6565477634383789457884354964981302334012281779330246114876283196103068680162689390440 72
6809143711510135521876998395952242132520854272882049136449162860246636950163038102827 48
5685470764890549103649697114895926119822743391573838586691100307024740127950823322385 93
3431293353145083134302322148697646594676640186926563903319155658878722489833130613574 20
0194166251216363279359507281400733378789281274628715819789629157566368683586374227154 38
7993341655748704576289586459135261996336966360158879768974952548130767690613888596993 89
3603208100754592309262744909377097943522253196357206244392487011721353311137734886276 55
9045776484205331012914770629230800601510670312034786007310701979263750026271290945503 61
4939206813528168565874012956711112923521357769709650859227065436132492441021615816405 33
1014385079242925593212727806046967415375531536662221861287366128803821250100928421146 27
5449532289742487680324415615279758306956594762689345024422506953550988231948253820260 63

pi to two million places

7860611814477638149538292667775167629639891588673014817073328720758971337020185666620424
730721218336157353976512397247823186644424443992035085556253116939899731382602851749159
847422989643092710028283731637880630960986092747881031925456588368774159105757833020388
9012289498579210930708652843966674782953630756280592761039597669306034986807013064708O1
3130297868000274847369246799033789344233438938291375483942462688284578627681114255271O1
969211059751262178040761552515403999424350145516266731865911256247375842966932043137321
356034227630097046868040289480090425357622270152454991738500652876282082844947232897746
2214065684862977291345480663160344698259473577756305161555614948782132847032242948928794
866944339632944617976682845829101181007034735179487801179223988149144397875134741624875
739095613153720413700455216803986883714081602335799444156069410887872638213020338971003
799951016335970766613394378645499920248975123249851105355532553697260376806840431030369
847624103275144680598482749833518164409364407426106383430813286821838706901670245370646
215821240801205054280908116821317559413995529540463378505961526811666007067075116809993
103232721917052235278686527253123744565398423714807131269908021498142710907984564962394
793103632503932715690214485363939535570996392613413706355283694079855691675861926761861
022206910882095003175906255558799044148134393616778698527542429084263952399371565755451
748962785200622560015667895265912090638318373362423859988343853824694259664295833241907
529342274995126034324233674418833943067302876654357385269541451369333105026535768322846
766096520368870279739778370350703717962545742445754051163217724554798081067098055175850
952219887619202813077517115927515082176311728768285936435665761256811451194170400407490
128413898373142534396718266812187402333410312253279064275345860703992770823166989754655
078646555358641595287411267944980475080107549571270721469935283375954060689577249791577
530068912962309320095761486922310891028548610054472809101465309502976500108897380505624
111300277162749787960668010985853622550375036459585607881915305682408232105734761977447
375543574363118277047855678207547357917451284528128929233019603620021968440751077409910
672529890205220867476141986504023102491528828879490997516843702839884536884782520558896
085978948619912272272442100265425242273619170560112198481877265166850985299865208474531
735375591823591271057777366205507409482064859408181593694790264152429184122421578891208
191929961882775609985860356636359069252469914184678526388777441479165457023371479473170
267499460134216425860667595778199718717717441990038089493463551331933761447158572603076
896505991191136417412284829360835010695799059001055607831543748232868730973050111899644
412652084907746022176789882108467590939734307859519683086062161338688570880778167957337
604105574448328822232366477807112396773278471223644501383798549791847487817888463161170
242845486972732738299039515633003689285102102258403151049763730937791859124236745802387
383415150174785099316522331078376158978760354156846551124328666277952861718702198668722
729283615467690868654416593353062170303791048698373076967563893515294705021327544469916
031857051856641071504874103965531180032861058780780640824227321424748062293132720099791
019392455082193819018527398366218293196302517133453360812214425341590512815206217840555
621251598839345799236917942561607027127565305803012065391861511569671811860224827977457
049153466749776919870317884868298775320079992936165622769638960056882530217749980510007
109625239664545628488327967528937682725289847695637682766796989588542844392587731814232
405154222624633448316487884689217073956997496650461487670803790811272839530625701046413
181049118384828472000395344294369187418446927247657161189515737296321205618698034409480
452322930151378661227404614838108275923511743550982307614238444892466256080744871547994
332924733034827422941928239907697984711308147339657422595031486183468415935397364916358
381312970245114734742564328808543011628160712121701195343840167941267824151917706344830
344838674833986989553883378623735886837535581021400732452526139976974030200201773593723
453393392429604637674022999917173295641451358496115485872715110679979756957211917967312
588801233594921452126730768507784660210257263118067807123994165850414444351916218206832
707476474435475243635021771387291682498141537447992130352632056642740258454244647829725
298376398037073313933734685471907621467615547586944589919407940477286113125924470358104
584665737832856247919013146502521873208118597727213601249805643545627251744772072290422
368850306213090841086629924481875548787372314464977497853754738686708209265146960181002
095391899409058463592927109528879182720553135023121127238958910123529150023735624589869
606396223581858235055578139363040675432898604647448081971321996688757725326007369075801
025710932151886714288496966744989210878815825441962285031021370219619431877579712157402
076818481104388944498331027593512663197408447378850319542251638656224131697049369947514
897426205058862349359968235667187965171052669636081131622960729293572368669774797043914
178104859912581722777863144079459945374057369324254004890738784761504898658475003647173
078144440427356694813250101911717321698784876328438857927180194526200592860132565258766
696301089807006587842893680255395136048502311067314223880788293295568432487246776927085
422947846905489063074643937315314094471341871409520467339087106627007029590295199334459
189171914773019962964526064826708089247749395547240685581130894672545365630238747868483
600538986398874434868790750940452174221175525138790488513920601465884456216623117871547
230248298036478513626596285502629212096108886336975476876306256280576324993603533884882
986531236838968854623516077694354809337149174551265036491906354283294208405553208385378
205010028718224950259176349957221719851816336805125162718991883923341976824495693634570
197209230919181128116157777228037711970974869137422592538940623672749222749700810195459
657628608079773726256392684043473283384785616985256709589244725278213837738712630652707
779501342844752394295538387062071776057301950104003718033873134796221525837482328739148
834019951170592822353935226059976931803875631460123607550360688341366357697401230519531
569025023622561793217578970910798711615056380763810966332580126936200941501263773993545
062408401394076897055010505637306881857901604037504030909383091823544926013355251008099
472794690497372937944399385070272306340260269283097750572015005342123181682509328551634
273450762959462862755370780605288127283378133557246350324824458771328547582050937466024
872006242475290408915413616124207405956140175062747442784512587186092166760491578272005
949629255429142536506047953602057230555291194143611701340381552346006032186154273703080
772848653877633102664805028786437970879599836883209363799688451420581754157675231566841

0900099568525256656416104842663107948440000584941149317485339348037870826337752806296147
2404350493616047874332799352269446584878302046868827541541222441115695692301955319190402
2761634194654865722368824529702872069965095339954869961176318195042866841131991738440234
5456298287391811340692790177693711518817189894376290332650095896507748750861095129901660
0379063936089036621437400138360263418081310852518047028151749041329199916927650831297834
5235921625661191113976950374095555803147803474014666891324971925115736096669642834218373
9268268187065079357451282741215093009789193780506013041526445859288635166058354557339126
8072230947899567060417392824049028951573448991489340595763151790565179752766311994905398
8902048286930294978715236655527170958042060630898922223596055743804886270707184051263476
5662670307902578425546537305026233131844993807878703201086707608440811021399393535669555
9080372042417935411874005687464397961590281894765411388840650875459989200321864790896939
6566869652082005771465785483853052938226131314573470940551592348436618726440186541854882
6237849010636243128813270568434764679920975224531630338019640845645288068045245244514515
0238963931302458439285669570402444844588915936113114217360569509438951026638353925707465
9096196888626941148908562395262015989882083532082605264381177500307938083282822764734068
7617438378532006252510463488508272976659458642012797309930895976053689515730977711575894
0644131654585341913414125568426145774041651288120376572771524719801805290275674678871729
9947103687669533572666281214869197529574422592670403499532519709014824593961691254189922
2508178312755677189097378935555937227850809528867510766477668128135435363228256967696634
6151087561548686976170871191396795507216568082916510301097775383319580594274470574804575
0432195644167491478952103390626219177270008587008492646046593842701730357311729181364161
6263408274079033973732715370746305174774375743971141536220171936743499522622498391862672
9744241532859806287568737935564445153701275991479765360960759877396742833566193426992333
7673554239472867029720432215633297168057599170375952788623260927573584819596066820209557
7738070857316717292205513410868329587444975212934130257432892210696738546573758184628056
5341133378804261294788480100297961397881692889165235338146188964952719386086816650995860
4764693662490041275959402723531684712860589083897662122037397530507157800899951457435213
6623061241287221625022506679677167537833049720150632713406585797546778627114451430236637
4204159536912672649931358104743791049170937185486391746440406087888730362777821702935913
6100755363382574205300295804892892015108984582910122171306239159986752090173782619952995
6729169944703629581908973306797428763243156331398344183633210987805452279641707847596783
8881802390494872128689711912489530496542248556538480214638573144081929515666286097200354
6305138139236434432412860953989337054700907195851562661424768681436522718964776562850114
5031337708233207449357034676881647033235882348938177239901810236198402232981414244653056
7368156678388730423788662801305983487556482363500166329005439063101440339575944504596888
1705350986424869215707256912363475873874462175849876794078361436408749950440603094597818
9473964713920733569611686509704677482024338359509224340296548644669246614001437823385035
3740043339155114623647520725286637588860485731709762014240859207650110013387189424969440
5847384153548849457684179768628337052088110628050286774169150401919643565116537365159392
1985721204844575756594063722523386639277985446488697683872199029157233248724622230858202
2275636304120838545454194176695946201406161729594341118420984435271498483958761560282069
0846771732946915552724233000828416382671100782272719647417141115027975237181686790444256
1499874340503870161039497365292877379924580970322216286340038497347820310456729346784178
8862355014033652993966080198718810408121378554375705800817504596307819753015754379179193
3349325031010481192947254464002733952123642040454345030121523155651860735560819131166600
4326767990129369600961807830904587797589502553802677791190074895017407407807241474544875
0086563652659154854174954872577920480624854838237508273717538016633100303057269577017355
9758370419906607262006379069733838527093020058735255986768108658294464753040575050583794
6935172307217527686493680881551323807531911288716974048627458527260806925338295592853154
1478257802509134014216277236748538575598170543719650161999223886850390410572428624303362
3306083386027018472601129990424828143778268033663627300552393622525604407186559062266514
2779715705797669575775027549597687157812272853888478787810144094990181802975806232911554
8472055090742538326009837344259715832689620521467924123147296857441065508374556810652011
9294712664936160162468556825035977972297226376316390242552423837948144809402904658613574
4772539537168042166067823778146032873979085975264510226020037165106787733345973099932507
2330318620486557291907194877109607184987623848153631778709467167601186370939137024311929
4627130854652486997234830209529671843969708746244917022503960954819159926060319373679205
7731394434676667137376839105656068886345235557700009199951983353156087691025259092691089
3300619480661625741671388846649408239350333607053504887087578797312033402284827165950213
8288837734554436445193648021672329217580423545485560464070969960378113718052474995481278
3621635651005466198476080538733218772833902001787181283713485099101193992890757328196441
6652835934630789840291360045353178422116041037404264000765826550784907481372129979739208
8755292167418199197675814797713012323517524588334805693715422216712484980648507224068590
0667597511277380058323213537008022196993793188274184997605201138453874897977558702004327
6749366048328052292931597751058389270831291128601294295383158805216778721650574030377815
9462558539666080531903553775652520091153768670120816224372415830858884934288792255919701
1521077333490655250916884295435733364307502901490921573391197275920586847928399726526444
8580105920958041543627698694994700284143766779707614490509764407170685248296695132127163
6985762873011915357619166702383700429480730582911959899181106965465008782993796136228956
9238089807126758816994391502753486607912724235123400605383338791312361459260469015193323
3052323514919185187177110469998102229664554708541634092935551146508035836503547762119632
9491999414587726626008354136437589196445072027114417289252620520903282593126069100900244
8602323159247827905118743503835535974512909851457210605208769490048361126910123246199889
8599781057908670098317725297509549268966154958048539596221471404460221622174064699884478
7097165978500174898916936139711251050756417850382706527049797708300745929252280324869808
8547191740847599947593782969088056008600529759054897365255028463120920730289994530752493
9325387579380555993720463654492118375759259285771950607818682986046855679833198378630402
6585395061059531222

5032225612863929864055106532946248142766516540509107622304423132338608462119002421315557
1370783635275177637722469102939545499341934273034734002658987481855007243264170072472 89
6663181392497974368413293492073248649875345808791715832310099446505269225943396537757 31
4289009811552426297708303967527872534760734457211799319563394323513731373770138479093 67
4533673460515426846563521806621169572707210593683648562351512930382944412353133600942 09
8347814142972374365787119118874293665161309920690894485298923256694661430201216193852 57
9951913380384118619389236779986849324697884365462269971561102701997426361659295005674 94
0653398737403388147358242532247588926743853780989861781580624998576996188959842158594 48
2403127951197983350675578966074513766089331775025950633016657035097213435222523769481 31
6683082929206391231137507920401887113895984242762810487479944576336815527736723734053 99
5508859132241569452093639461192774609816629071955633330318964102729304482767637745475 38
3940050775630105121375747589465748631705282692920447978930053699441549462911636979987 42
2492703036780074995244600192527942904514054340496225220706772417414459012775849304384 41
9169471824463589453593939083078952337880389543389035922298899744450629858915449681343 90
2883083036091668384000369513293418160373080423757387071732456232293042410972018986045 30
0317389962055357812854441828235506179232044089987741290618997312871542163821646827798 63
5420026912281461597428051200519291779221033655399002489156374867036994361212019933170 58
1508488045808028330564633328855952579588523340476137457945343567682090928879394222868 04
7782548540434679046450577120349850427778626677068882104274754870852584183136555944412 06
2419547780690567308818788094917703897813991066241291085460711296497496153293428818024 11
1921482687989241787135358182176206406195177133405387049401681725712098604187636798291 55
2622105328200847751673639162249420636920161016181878790171633393490222203157046495892 91
5660111820780330597206469937095868562677292011043899000595745476574496285533084757766 24
1675081822590306828727506987848825721700842158530486925831755822546831090486918277974 11
3104892913716636781297555208343192149972077464494451139109535345138892669390354333726 79
1851497108660607134923291714295782078096333111287299809422043134285312220555323047463 53
2150324722039448068214184338026557787770430331339161242766255908014112672276397388071 35
0929480165724265737803671936579619809240561779486152286221760799742806107862460685336 06
4007083970711459274233437403727374852715884964504424203882700398592687187852482128673 98
3534052892896389464893913128980111952245710916440143930474257190335079824129951176476 94
8514131815082633586738869611425332384486413591845479351274064862733801762071501040633 42
6243838619595191775834034308744212950952117519032259537189931950085566945149777061647 75
2985379631778891628098182218479349358008726791798662095282388987638060902226721285348 31
9317045372543422203574608825006836342018290002438468262630397781184871919736891571620 78
1247225680258617614464823051172199573937198922224913041130838543246836293647746548096 26
9491164956078096858308067229759168342890235158799056981740382471217480288150528150397 76
4681006610862782811944654397622813320444772383226041640248238079647539645234625830725 71
1759598848750796298979125520953369584198936877186947126604497187997327373491893245015 93
2454186219729259564329455131350200955933765328647961863866281243009219557257762179639 45
2769432817806571834202994652057017940680873886674438425503608032048606912448291709248 14
7871986757574183815474446428294900234455804196458392947387640106311165582157683137313 61
6226299832871546025519709507682873277232202944388929510321863133652639124594630414634 89
4189731759013884725807128078097605150396040438837482610857549018164123605387489307603 40
3229036959369649294694514257814291520595362718560672586504305103697587264140892791251 25
0124632319375128949864169520197579746546343305725667121795171573631230824010692740119 57
9021377145447117765294956028831540629720568988985876668611843835192607978937730977209 70
6337209396934119442547029441483290598332895331873776755945766108995194504218768229605 48
5089899089345368546035256900847967231897238969758673145644525198692540725327642641175 38
0155641631001265923456419031861689199827378706297400815135014671146021231597482856224 30
8710885989919055173857907826645815080230822120619972302491057809957896245297610422626 84
8085371677637152355186958833733482602206856530204771860522523250165381506144416476815 01
3957511012869624651214126280894145157569047479175182531915536425104817394584964358913 80
0961834858608864398844212985777268710547912355123750283373020162473095947612271389882 05
4366161295662458325342675520655118332156465963273125982331622965896095534354956141718 62
4126781871894688666519268360515436722888070314889636835645390280620132789955595931232 73
4809909849512032725806140560574435631010200473400823323345313206255739646087515472452 88
5399417192316954006643478902839731229608214486523986267637178428140354859462054305122 89
1398770515106964824290505167086688126723120149692447099193452230520327443007676497315 89
4771714368966384622628668978716909261187812553861271252123088640758669843061533370817 54
8460101541493356623977493620521811176341758421690263254811682023109616919219202286968 24
5072150597182115028834103131575642194014578878727971764110656283779608671347117780322 44
0251253735987500965135217548470611148517998024718823732173032798625560841813447558810 83
5371750906208384956755303849939389235710199636513419249256875462707073406590960574554 64
1251446206542917207318641479494627005420571420246216311504229196471301880607050652699 14
4511601429363159652782454334709454185941611144118579099688672078549979474686778714088 36
6683040431470694803964226218711236110921749284785436034805594327579516473619058262806 66
1886719108024163651804902451788960132554938649289030243202812567099164367382512988234 86
5150269257448093686855504681983217536990204726999829785045007841001173738549486770025 20
6884212953225981655245375140364250030177638168811985544862498383682810301740112048851 29
9177471854619174979370637733425893120298959218457070497398730814851415100426903508492 06
6736877991443254620974439962238587272173859021748093817449429311903298996733204995012 07
6088462959915772853232042176846338563113164478755605505381440490233796049294967369807 71
1838316497303944722725636665753994229170307133260072920720131067078570215314886181824 65
8576453935514087246458126880542460403341383381714520227684422747494782820981139839933 97
8232051280997699499414147039227887346295900565425739046000573445552868823251626884483 01
7084084141703800569198012653021969251563956835481180370248600160332416531355644702166 96
5153761053036823532159916011980536156243744308048915490876145930632807054397737580173 51
7852984050026391851658883457796096215577053517442885279755739267142679886594528783974 42

255967952331485016749399208245727890201848989733556863251749420704613725724635511686392
410265447620707522051223530556923842542262469082820075202593830192617815994390239132156
736988127372591213126053130776062022129449829361877569784632417086625983447647068042514
601256868794424455157791088216702384678561612672642998407249438460102519867189173850889
113785398684761957511987405603368019341587705276496809521586501438451429876660723476584
865151291103242794880816687101728573600352970077009926686733830512017337498316246318933
902027112780458975424501814904900284506344242503786069533969397720697860673703729056743
621068584343281464840001120414955668363546831283883766859316841167168394790999317218568
271857384728653957544567441395299038322361399289853088618538835132645671929328660152781
990812536771985012593298994461779414110256729478796569036323539636017631771704377350776
687941192958586777104968111542216320223477055601741204684387993359267033892345178276917
896278507654515848356670167857270809562394251797801810288005607344336420511144824956628
611879504402723204789486957916701711294894589340460774671990916055333845617577656244818
103981844322048969678307050004494702395001427148878025727431953499407001465579848669669
405899947676539645037210136977205137248500093518010286579939391759930807723978357033120
743500819002429349734065290554100970854037569511508468591803157031556814972708862629122
825373903107921812347854724662590452711430340831345611367197089771291948096542408679744
870482873775256443233211291254992673924476756666125407891548607779510146118514245894358
959288471212606367747882925054672881195999033159705896221710430393233538212727552564813
302927416351280543225086876674485235180383177565091066333061880028446542767128480254347
804630061709457812724731997123516671195522949165033398308690653279847254644479604334913
35091298264701200652831435716961259051736203338675487451251605821486751301819672555374
176337926678403192972215799007796028305526403424207843181589719496197248239664087499067
473861667197549271079660012824577002399811441883066650809939319990753519571645881385371
393920048365192976243026331532819822418267733484413510364795252212456037414770100127183
003825155734840713832128794443539501058978055708117314677733670854313950810047198859943
224568688772998154000849719486345212981252933076062874525131055554197201101500459501276
201275447689361984318444207712330812258435671167531808198994413970306906733984491787455
809254192697282160881298323493847279127025785464911676874602828079458881264960697340083
268243918893742912683680758085035658108798666330070502176158218252515579597476614657884
368882714898680254653368428716875845984258844275746897860990118501828263974528873527790
861754342615967759505477104704323920668734339623576883221651301838539392339347880682024
694548580093614403134394478999260612022858351970753666483115962022485747086496092697797
503547621456997058053831598488244157758581164220497804973862336305071306013509847049450
670796896595007694961646710924561416272824573341370014343637727857336225457824573724438
057572301273621064712826141679918675764891126577637344395061769520885707689448614892208
064202668064661014452660452475482364854453628844982477150937005524672316180429659628178
381874346490369425565450722755096280167302568420314431589434024654910031307890729021758
774338782299166290580783224076694448344714768922115999735595085870415749833019946285546
636613785465869198549179168980167105507141808269542188127465621202021971376495523081925
934779360781323019857862927933483475372210491174098676926391721051802408882103706270703
540624987561676469583175118504171823982646364758285579131483728967873319054939019034259
082582995746552034967163858538691006775558309960737265652629685271917179358900768420075
551948802103399245642689360524322401133799516798742937475985033820787770864538827243269
319639246699012454835199326877139052936624299161362923709050006300467037102129937175298
608591785164481166135189368957513966923241971500258602024042346697787732777278174740435
981161010706342634751222989699874327188619535067787826371964130903353069819179744012150
898566802274058714459591056573560299603681212828129853783171759479929689575492673450153
302479356823774313850658707807327428570393080683465501455929612941340956390408192133839
699154233144292114313557289001547326410951031661708289005237383050799055074912971339225
314337619512303726551403229547632390032904346247345136944274242579024673176748019610437
553058218480030888670975580300203944798592259512110142900450074820239305141870715820581
704315643461613395208621842385657015020760887982966137778829611045538109903889295748719
605216164863671958696195700683835338043418164250947218008599617682953197686315882324472
986924600062986350649238673925782538027900344365792476272324883584727466893527167304470
331395840846155715632853422861945853562234403535902808692220196996900684322003277151971
391627528419750823665153489695597936275135741877925855323516615638382222481264500363965
184258273245254643254298163592709275937939176872739841762606136340382594773080217636739
033231692507620814650223047848633109368210074475180252445177621983028991274319554925867
471689821313245854632220227037937013149040707460343332658552805947866676224565447602252
566929472969933695714757042662118271912419452251756415194770528689817950925998424057377
421386462334416726580255553236943228336896305571378256357397108647269076072386741424874
225158558034532830555053251312328481561140034020990869298410749622181156518970684783084
734279476008399862449137405764477050906950987748259528544245503972834945540133235392901
484940250007189315626619008377219708488109159033590629482989172228522194344422727221236
659748538766720071467310623686998596008328310293968784187836904361470125767383673819639
516433573105735545685265716682349289340218211346221727109770149525244757339885462693164
471436384296625600202983848742217124742692760446267367705620751469225244915429734539318
881140860889625068886967028758540502117338574970887989691609025677187015852875860344484
245981390335331530686028616934863449350141931227649775481476910189735515908060026107080
314020901546195867548590676224388779824556494070500093048067539383862693004053021952966
500889885655102236099419658733088736476476327940536877448297094113581707175447023337160
663458639722396590250587543662006081001366556634387485876501469551917578671233851002301
197527400425908374799394783291425360036534072469522701134279244242652622284469173396843
429731424191377080363790782326403102263732120839866316281442575578910686920004265174170
509458628854525661988864600866924305044674816951440367061744497795108994936915845708574
708576992999411498604419150054328911323011097953348299698206521621849060614558891507523
369597936307691723984499931022536478154342586071809595283003348074817865333019188244545

8856747808392875639488468288070214239773585105995276636740905808436003621909470111181672
818492879705551427514244803754775436839735775838993981395071895405920171883472980484594
376749739360665184258920668662124528016270830749944391623652768174319312025121056321242
487976480822599097387591931909820300126108898491811619888002591639896230454409961539170
364142632831927316506524676148801876215882416454683370343605568590065598046013001962324
477874860291525764819605012852773599235610142117642827127035997673194722381165835428867
871881725250392961674401013341021808357375859780685489308007359120463267716908526341413
286651000623648103220351550508071130914120462354801944427898837028152212535328454036853
640803644506514926830386229530727796909536894759213044048556819059245115582155343112578
223050011721204029552852395193365545131191969086786683874017343139099277784147683729138
231503504304382834280762654796797099221011375968182946306537693218733999732193085615004
126426840652997095410848222655654838711899958317093469875193585526122322235400306649943
902161686936959538341869103967381130394889581816964304772387136955839563033960354473549
594103169437710318772962862274931043707835188323006994668067426723329121140095429033610
009126746572895384195693022041362565631340051007707683888931709276089217452708559988300
303924626357419883381732442314159866426422690244944806987162050599254712764324305136132
570781091311627133785721977318811465243224049458625276892853375769372469311621938233457
925407528496943491261585340600247613615344122695875653124085944345231333376371704578620
523233668848535029768208376875545949547421885372254736995671462469517832531530197864254
893944867820322912527097950883150842430921180239357722095987933299945652042579536921091
356993580268490671600986920329129672122431012828561772785963975752274249498683543204413
879255573584876646527766714689294111487308527127520453842865865262416348999791179026285
870696553576917693293479767998174440357774673601328064912152433195957081894642805969940
220097416461552597792428207832406247559802922247672268000494730632159514845849063060544
493554411511447219508265072308505979664717221510626656151115516907411412686809293513876
995176083889846625594673994360873023504398068746762125767075381407009690934119916559055
441570766217773262036684979668263813129252152418835029558399823231259026813585260339230
384368842667496008734455317281501031533528936595093590203966932772054073795767690076891
429519950165896437964570477380285338348339073117092564921720624320719773734429912287597
284820642135518310274804728074284530817947159157553519826467271202960344133024285970804
876597499776244968165983199638062355564379434229835192667019685183332652528270115156970
992507634354195943316395589542823980376427278165739306019207637215790301261150424451806
766262753828119888621626010581794310260403909092375743703074793668738434405744235013582
124422250749064880959842198811647671462409445598254943462804554356345404185813236274047
086193105741395335602894149799384688028867737762189710030667162938853943343127434369206
975303177671726265630912190482316625286789518992982143301073535208181469894091604846987
172785473678751859209405393497151084975614905133319348357436523141109873223309208721605
013337256727384865751573347914315728837191030286077283981339066356086413749558383108980
623811064709690269256862194383224330649499454150905281393120275580651478286564443905910
548617936440608067367444027343242553190694538035923172994714699088304231424234265285882
375710654220286227648657881644932607781172460015037912850762791733367934376106361783422
739934107523430843122380787404353599769625386286243477478405346899480592511513026604299
757734744955798775571913234726720724614820635909480988655577202736461005613036502560203
814465294692100316155987086482304242632727283974090440461373938404573079770813316331831
031524311842185943961094716612666704959948207921026724277065913620038918528686836459533
944729200572590618434358807789680146361718258871721977588288996577958813216004068534129
634633279420905672692107885313942739093821761807584321574152776254499327605408134386010
999651150272008411909100424071071618807008210163030598251410902861771060805547971992090
514719865976116852716270860219526824415394482072385821295017622829872975163566702246365
046380562838532623239256382794883006913951899473668502669591697678513821428517114757482
290864929396368714685334928131035321101149765522917058233336354423954832236102065705311
398495461139984827581778276712221335002851174651431598682222090939473897450720402695078
320162637322169759603810472470365030934091655044825609091836940892989768097624681753022
675426790491545267231434513817128511909384268978968141656539954163665755767939393754092
302769708444734052356221353460483178045133745221756232338964672276235076882611005045096
750442144673278043090561594811006920144271008130738513075604658647551255928624777772976
680692060923293540253894700079374401884687591116529091356848750496223665847043153688055
497039883178110179499037428097762030224565216298326985250066896987765331561584288932518
772603812068838680377339388385030290146525298799193180328440673348211830625993791085057
712650648030666328189304516168414441815495837972918815212440105791775022647761871288830
207864658397876823540736617064607381726010245371805418589613077371315762914238647985640
423702839211412736138487831799107782990292088814469442193262095578200992372496532767016
017774299005809404675761818762631113081752642041142965874321687553018338820333584897177
687530533715136857788278089990918632401555052461201973537813557341038929796586388987138
354897380713932190939527816273990466944640807444153667101675297945724067360346936193105
489526486951328860064841832665644214966604884734505885662377470943300133609953305175795
166542676833035691209347599047525829220266545447234334747146922008684186172264626133908
204651016161889416363844388875268579799124034772986739085108028594883048477812106960583
594083126587270350745235179361467036244759790710044009463844196625619798849478216480799
542584561132827309527648782452888031364671296147996731316903770893728418676298707469913
153771626902242393644599682198160580219294440762308384671691341286426648516213587132527
124415772186527147676962217011325153969991794414974366018493536479219740833118501025815
992299881860097375738723567022934130785389719611433237363903116273730349507042196445256
896926452293847084281179925107621686527391024619138228693880411938254640455423213132958
746376553667695995369728207561487870842642086014045537531409475530811177911820842866865
179751514875711144417987740682341135055654125948934365373166752555211581129271157567366
032884982619834775257886393876440947234729841967733232614981662398210617216099934831537
694273046067468854914610239854853510785156003772698223505469667340079385384071712839362

6280868333831656433208136242361993726282501665963586014090745622579092390550119676198431052654496513911977912897878499182667887126207881363286060414738311798432280157776044416536985298309943565332454304217593308111153919217097005538800660581737196846478236567984470636149318269940548019581638626884953424616783509399184362420298317772751242134246280692400734021523399839327784115478653726842500269779240387944425831863855819798629317730093271623408767046200407114581858002551413446016032000929810960092941713396043822132206392622516196634648770559356152001173593989912836427795321194545180124520756256309168684462913187545506949168782391696239662400849483002657525922856646878362823518373238971914577167429955774032851695242526840313615749004685244163506250470051140603537323627869889483290834250042640129007310714215255147972107605809824295942644626962607654871049166840877120035427404021585787616466728815180209003797788141902390593613712731436736831326269620063402971487097410994695168768600341164916867021671208238572557319835147689020303495864877210466037785608165082548390341869396555330931657654275548096592900632167693137605677302273632076049243335466630337721496442877896442534167623425396983643729156135888779180218702551919903087051146651709881744458955560678692824306221093916512057714031465192002664504416864162198240294271672443716339375823848322947789161031014439393116276119048765835286661664359565606754970683869883140257354734472858340267232966581758316482171109174063218110261002971679677340313932372141166394372735321182765567197813552947035236944267223606768159555532101084416295363044683489703807902815385022995518300929090961980455112156206102176604629338410815070367450400996566566437848901065153273005873657915303393799980501010703511699555183149524749602629554210286539958877106951709626955217790361065896162639927573776022357310236857330951445040045749255905177471141954507430545870412876562628165832713460622669613290216352068377438704390102404803475597354433388875816768517431836048089082801581678846817966396560111527955742797948938323899488355544375744340525882577118818845531668215562888487828306994764821758006355139623499918675491428481186927512128264293000154137563823142147842804713190867319850263219116536633368639694460073228096557913589603097657409546743250757972866977313561268565575949971755978849719044092224826909235352771433329351875392326404892156965902735817655623909332298556682585279086641388889811556089657951602341639749340926310294176928258111013456628111281543850494157394978209214605449544417724896263606057630081793847869263936525512770818138330804771165796700172447686159964369556029800159321626021257729698438592826129253839500919972500185733944764334658453806934573476163472140463373118493560128478782399653258368219124178279032115060431362814998324409458955535435643619911331106119834935604387492091399778080068656180192778153800198531194452882409332603918403514593454874150369449295690180150433046620175205898439394027228249751196866126425208824805581596151038786204212217329236701779084976291278054334105413690343890610943838913378435663887308550086150986476677395816419263588156755387858186885651382145797606035266479066965254378770412835076929057221607856098539093969227850114695050374751848114288146541150592603594397728403907956344856341049176882165536833101729919149406760784183898958969737923921210716979533044982627140850056107540954379285476355618388129937263660410681069417387568784020105009314707228866911099252129842968299239829611755324670568379118506538781756304610704678477070577955970404718525106151213336162517223521965993268972202488727570681475328498212736759219143764367890533564770167199476610822192521089677138232650393905864358558314210978561963738198800821322625488213454865701656564138837663643953158515514293281054017445197304465700762184321393982135841365814765396500429819031900419403745695425098120401067817095396111140208829915242532665186705228499969109225580558165572108548597783708640026246922537667111330125295861958702795035596475458852848647410912004018194542282084743604873378515722558890161780963704626903179170539393688026215485568693229149453051559024141630710209054773273167503140888670339675137283440586182410398367327827695525696808929388623957574603141711764280420319128254886577075440287047403462927719211704946854034752530958476821061018129337534823030273755277975146917496594292547262473719053419107892713212985991689084322997977695615903042572603259617681763013999373735340230443500484441978907382442837870775297058455381328145669912654413192836759011421524811450281991240904023752364899730266395671987414960025182741083197570033509516000146181033741497300702485375704819316011920605889631890475971788639037907399493420802317959839516533520995972786429807520557939876120956830579092465361619956156457959923993115852791680497975594063834091470794475460299339225094634860817725347835078789940479650859469559039583164461235394728018948588446648505970098382006190362166318787402197669637792961585714886706184668722858347197230892882814557561318948786830580111596934837135156994802367349520480195911228806683534562772381377993732189105626825060057031818218979676259849744995532734934347404786415693985517697363867471119143898705681249207891624966671639083633346697871282880941013396000489703348255354781669642566747368528761027659541875674045861692730901254024099403804127415950798339034113488755638689333228426201648817274245593004856809375517045981672753152668601512531086963479371099240396324680383672899568180793728368486951017757813789966363831421610727071290550680044909943882865634136122595032087410662399448809572936909423005000147956338870302468427173113211730495350287232914738953452557192120996619957107638676562547358738813264155748073377594494690376408008154318193618664459051900517524402266814057980387100405605576873822237531800144540399413524396458188052266063013787018259901973228679002567079563510071445703513771998858435447910556929722367420641623467722570869390404718411057624636835085939899003482272456412119857682647559244657270288583804770173830628620225862617945125801767295325322964457191759013050529323394314810726550048523157158287825173397056373029924090051318122147069200638773613330569079630301563337118830696939081582759657080792555581311388902447514540466422807313334655712996019058584784747624689974120658373073066148525021016926635083811797815061481079615543351113647912503161835607338075996216196802957479555338473412398128749515984260732867732258887892059007834720716319268043522833641824703046218031664219597863131068464869040599842238330813473397266120691690125044409203755277801627243114161045218207245206740280830072433429110909187515172863885321777741646900116069117945642647382829817284527137930147369069274798664058934930018376730334544890873

6774801278642737494399073596029791065650733092117253849444758953977952557056603678655855
63704131095693691596244121943794558786461595071582979919571785796281137298382965755156
8060692045607703085295111977149374383773421429137569839446177890756804899204260858867971
9641891711969958690696231612647329918532618028521980764105878977514318527868544809125756
0114778558090447522130057638947372965789894999699006700748153732231466052326201134870636
387443221823848005171984716969639337943366089032334572415577970846553396324524312826614
209150338707801387336304627498866681206012174866045165969923800382314983620887665698349
281363440250461797778055719982737811795869587026093600982216263152306622267760762692763
074471792932302565215678005282014199613423938753103357199948509282205453073006431794383
901076813304305241519206196819304697008895768432734069453235387528317027606304811997816
9259281249439466497302638213673109972142451578236170858371083226384652357012153161744660
2557099377861763972733595199746203876484608309981086549508196039728529773904675658525
95665146129751481604219701069589139719460085119365487136884648466780592768453476090941468
75309532070680268389951109579843486150005290519372410549219269589173164075795755844345
703750518251405702811888970046259357102877487588622062400355853312118904208591461556210
9314594308042630096656220407498655604792897750914006936860095874117261298687199934107256
24736804854276439238171886528547575500381944853872654720404096939406476916154138569342
579513514628543212758167573494216457427075905793700785828013930356600449969352395344750
127951295431887193526589543589708588483636165315244257884665013693965581316487827547645
8680491181317516041599320809708351513678332556871063873234359935928538657514890186270886
249980128085842619290175151448618579935072963689399078222983206511733784786538490146551
842111656739483671581790530583066714371043603825838872222700111416180463210225265759273
485036004359214042401894756464544502668124076101908199034325671932292183177514328300534
96529351108687365959459366856039412954362320444188517346984702660264196412257870566040
2654757934251947185082440495116675620065769331529365965013819164681346645387391003730173
6008788330609067201032278019175150403475678213079717463496915111021595716778235507505
4501908562377322550801219456667066555606495354616957446685787737286765624727319931136996
5987143767852019248156884139400001192397219808905736099315131677286477238689990482564911
778061364658363248360956023837727307844931493660798924663274095452963251705262273575232
28274384831433635063593718046825306903047966341694320058469246845138128944042899595558
1893041950844098697565084479795294211999551576853601229958888583402593982025680669415428
58874608427502646227323103567437705893344782136540207749730277316554467010165488548432
868011406354621801809412622980774936680230448291330787662656283931845554680683150084056
2054987027671977096246399859359425979125636765519245276163008842423388230457981837678362
9604385239011758630371979259736008297512239542210302994130248785309784356782077558526
38780952284670968893274184533091893935490331076559757800129820005493626209088058215715617
23920080689569449583674619166261753220975857118184999793729548630609018410586941064418611
758453848374306361378287073017398747000242232565493051105334693845966326326879436884357
242408570116599489794989797416466532308313433168244523417009374289869078755908149230343
71731633710680682790340864166899901979672901119848000840645158683529279580610407386
32533112103239385861913225789672173561123653904220373429326858215370855585727546412633546
91096661598967813901640945974805246237872430580977112521806068141615817419444074908647
6034844284221399860994208565072367267886853615707329188803638833822052480918791380994776
24625880127886825592701666095809336450815067265218944204908614901901488226835126494214
90581200168242754918085567703147735312019968725929194246725835253557104498580498386194
1419097706891143900023116437003572197534128938011540986914717237160939432091577987395843
6632052982743836020243320338744338973503726521304698957709475170812890906562594460130476
59218186823513816051014747394744068013186129173653955240971071390254938866548517274874
484731321498846673279671968105611835244373972329712720016381738254101392160926877839873
53275520262462992210343518996122723077583577038792804275246960955292882708352539744757
21257778892960095653790106947217981127897196558846898886086505492789122393389764997884685
46254052075567992509247025310880898186799808018114598910281522293643929915248116215948
3598699459231910665619795950856939305045226061480664011319080915167453360921429143709980
61368944873971742109608959159690692461000568712243537216703923838917109514818175809283
265245465981181257364333046796856467417310369596921849218158933800124636354609031505729
50383326652400605683824622141672567367442779651205159310008972763291422983923703384214
1998954905785591661453206890205948138408044249258032219280418988530041278732533496083839
4538590434903958333692003438010939844516930018183663109969697796676742654643513094078049
630874612303456201013487647626736597873577676347525132016696657335034185516436742860633
7391728080433383738772914147578465327367934686484015759071274666964431341774422730282
32559352919090102965317338131633860847284176376038763522998987993998036670956078647226475
28230315117880279073456120606048309199924082241184358726289581947544330662144837011013
31842484492870732180712970277482623339299670646996123544924886848610775351770147038828
64220918614557562861821714394875794431684630845241229359179759790608813186428936653188667
0664966827249866400183099860013643521154220220647254587274774551808605122709371412124
70783862865284947625909171781412112981908470901443457260001016966119924747212311751876126
960430885825436696765611258425397434148461828181497551844706902320034510719451480629985
14732772714331524426015466397538367275319575252173057517869217087562217015777345598686
39790005293204505927197536307049552041633509866620177488145194806303504584975293225497
26745951616896550567876321377294735628650809040617254648312456276820125650722070080563756
57270553877870242114610662025923031695432988405568312235052063755550157754763820333442
32276279547125979750657455234418289662384278132006628207761749049884328291788823423870
9675095075936245495649068046459818252544460349097875256805446168030424613947849091884637
1850655099250914943905164152983432308001324552741688571396474461375357877006208428654498
71588425320201090518231634941647678063796561292307349922289188474114600354320474594473
3777954764442018952692780191414721232667526531207201988165336643816116438278444599098343
513207257618445126187836949290554899236450614002893215179331904443772017527103507880522
9762843463529988657875596344491033069669505333902417740624458495514343711016429160971

978655428139653982310541337416082235906642807026495083165562648045141673802334120888980
483967768537750091443937526414995472019287217665618085864961340575434696935073804316770
812792356685044509675802759384832865288245992667838116430246118497468291238768873801840
582354352786940433526813678263805273595673010185793089033319780687337296069069093000859
808318292570940603238443855489830159560796468462877005890540085308527637826102706541139
747263867062846812674444453182043917968871452425165885885849826822404842247255177387294
677938624133222132768740353771946647061287890934015084620877780208644338625937169217839
144917355235970232491887765523452300553169010387890838811513539348159940795336565084785
122344795907488470753888391143606858160329097717407085044927150689309384200271035314989
521118974011790786139689772113233740068971932607650117078429232281561273840997580290879
403191060842876671121135877032463514855123220273422214419763274214291267688243171082665
294490093773753487084620940863500261432479813418818813379409621414029892611269943494570
583636744872360549293082819280476332273952397176165083227376466176184522176632046274251
831492323538526631699779818345731844732510008365227170370714390978140359460699511806470
144960064546972617244792348163617738902103667613726343853672098707013730066813996095384
621788003201627743851428856614287537884224746028214749763869327549679527198010211015874
486904177427460205147259409165940668486440873731832509095170593544421014125208250006926
065419696267590181878248304772029670344610817656840342653054691459990175642634762638709
226388458830352977964753749185080483787332036306897850334737137713513270702380617442432
841917358827128680471652568489477403209002196222709863603583392781832171243231085470021
395268545350031234468911505734129121938448569387770480223928704791454061370325403324016
057104575777039019854035354382604893665700033410262005854153146934345724906058797652064
501279104287513210869557790049845758979437991051459267669033957458664667658832639303588
343860270364744637309473602119064268011550084362189712025999471271985538203073194030739
936437243479611418117588416646181579440290404428942872598987284259126873869786911225261
806630956715123805901206067712609851798943098054813921371302498087106677886883102164140
638479702068827341756646021956080782579837695257024657868940880499369776952425459263584
878263682926882833885335895850721376673414622667516590477861369761446656089901116750434
576184404555118474127765064072832880437446916128095012458655635138780524271211671889072
045579850508452072818092768891975313799952534900343862663548012283652325724684181090451
289299908897071987650429369129560390015106937167030765278019764128021381350865881218304
733144424759918863083627278579276659000852913418425635792011876682685758304652318695312
880164661105082684826544576172244855082635223397679756923045568482845703303828442747272
346204889158819224818878871016681756463461036013148643685113916183854331625430783520024
020862938567223183936698104771753319742596090522793236530000065804202701225539451082241
879513652620161611961513747551814336086634394661705380462007583233964293366885685419570
924934992335282843619410522216676816790766527523407992169683393269110095544830118753437
278109065260796566935999049538659193600258575481314066399971845110619389630901859399232
924366202182887220436671717316147420399664233950714726917978076271449376611314986850471
383027455580082308118011871597718267031039205083320534769752739089504184817117117973538
824843845553859256632842691712074499636947929106980813100160785403129972379086737012656
496981058855086548148559743239719427448395170947470158331096037020675692502330984790297
898038429703388588450091029940185895561567325358746842085486095066104214274443412905564
826159418621426796526528544638391703615488308246102504724590916416609851239781364993303
992041417347842200037873545468847478831067847160645760950787682402417587086770384920084
874593935005669664283319778689573480417982489075592641006193527321292854162539108266524
308139077775873706781020611195484739355071711094325738913591446645160178989343640262032
686254246278681414772719068625293191812905870653406476348003326867216301432884318148892
615394896756443555666411781663580422558448436226395545538821959597860053801368920406350
904110593195162534622290746473345739187146910719672164315596191009149970340407047748081
119223661414927554212246339123234527853151992410718733428347720596452540906510295997813
711119171234211437567328760111553025005562508962672378946839092838757092787126651003691
487413246468281115105992161773396172749913105240990173206027817031396913826102956062347
061936499742156319508397263275307975926726411021463661483997226841904079318991314174102
183945041844365712669129664654738204815378690694198584664462346912665973060979482516075
355355103395291361452114695638892575422427368306729757906516499905789503318825000207873
499818592394652175759918910276603418986047005242263825511043188923949326524574335592340
802646793991684430418728941150206748730891555935954175613862232497763743022556827017125
649482387889259000871457146054586958692768094039945350428070899393203766514296568497497
249831004634317214090590228280028373531865014675755057247009766345435681742998446336698
444694122348015739117781066943931397675511202109531888071360598244239802604526056287650
666836700255429983763349714696119932222354126242614686566758927542057079454522411779837
019786838703121185068449988619339654957791266206579348042280835065432336070139314463239
515658424975590752657483440312248281066111032375909099684721801760274383281869873740436
640214961141976794732364359739314725764554326616602201687960988775660616695983360597082
784788545388232830966560633694868770843818131240187135773104445719494143386514881831296
810620572960813613922605674322788640886751845398386763564938412069963307712126119981380
783746947499473585240111607554879567011702020957042027140903050479599980771560568468404
006927877919954535968850913435023753698691848831529418377014181833974696659820785368404
752896968747391142886302697380197256155199859354375617633387596829534231182945796318174
037644302214330983405918028186430372919138933317282716096025009018884178072899989041423
156230329057329905674823473673213015756605073310381614315333930479364499370995499850045
922837896302177276872397749376180124336466703257305767723774179770986889727845490097125
655155604461635235701011449686714126021708790833568374646097054283314097107518794617237
583705679513831619735834256141032828993093526412990845301213402283367371777531609479501
230162909966338446121614606978339637890038713415234822456570707309631604651392567037745
430472609409672204811577414645431522230516210670385900711201736212499815569278345791307
269996857399020317517491428262797277097335470717392667384845296260542459717825243012251

335980885602824344671544343940540731666563854665841271563266870203454748695093989093512
908362141892487912789308546054720028043456941189990352327643540279587718003055859605561
584970602609699393673295902913115699514227382869680405472877804612378630081554745154068
417537254489543991526496540971473823716642985149842316499987495860806905056463383912277
328770697000480099516606426271338462436938539958805271747614035138570268351921561652119
505772604714759446802065061117517286763490344355975730263605887125896813150637996162005
591963867351947776633412012411031324440893659554294667370114880979705458874229309648502
172704262761096294129115536439074265492954848232597889095470101805837100352220517712121
171967882011186685230391747854880428625868993974417517408202962497558078265131158189754
942091848679100112334749237104041251867738301479659108811592249876955125371287456436094
977110206686600144372131844034886481386182594455301119322608084473921173506154395619167
083103733738022664708529527069108547938819314239989655579346610847914775953201339460183
189400216581481937958409985160391935322846656428955674573336885453413559156972014559219
540561504893018340393218914113724336819630739259176247137887988571608307618982636644881
6069268182265081970205432004466668987706970509143561782139375494385506088994855116363028
977300860449271090452271829803832556043407220134052765754972896461189084010138622869602
359789630770626922644092222040835107144581056985396529276765523953342208374379435435759
401076173104116818848468370486363698850092919367165500248482234242871645792792014855261
688467965023241657198728820588704076230856856981684991657589000569750227325832635613124
806863751828316672095690349802314559217510360226175085928026341976162073312246861721520
567790482838305357924467512384028147589583232662392564108321298334963555598546798623141
357984340599459087557457872281050403186863833862985018970406813460277707685957457479428
245565247084388342022153942157711565254346853386217694114049043483332762452155627500856
904620142109484276265563240941697810324325867043608152789966850774460443678227155562955
493120215572299168457113627320784151613356869909244737467094085667660609983339575004604
129019459376353879867769337297791850967887027694818215023864111496366888302575588193477
382411587200337946172119837916087787061820484251832689439663572469135359957269653213110
092441404355137600907676247231986363545378719556706977630270741670532777370043456054657
944969747464700987472532424046451984911058158536874912145039485563170812322544245517218
332626459826724205843971612995618228952546586361814049211459787247074805929648327274035
255100517446588334905464260537244234860141197677067064376427218153293856086805237195786
002885059606193106573897205641960001195094854352327113892178950562678690839997383529899
328837050875010122164999730892443417373973245405138453694408325025629322332796933442674
371719544274744793595312770445340495505060819226909976842676746237842531517367041599326
793275269244631980348303672807975776052932087944848562672886412552024598936853347360388
695175324051593733136795042484279982532894051519456869083238588950737359531084287009324
749523158656571969812461608386610993182648692709652886808671581883006944818309574350318
392832418420837186536009785971674045677645917929742097962312950001708182890236801519397
010984405636484391681888825011023963992901508734419951097142548977458182798865490459426
107868643900401800783575810865528290134793436618496836277387041890465053069659445286713
727369898599521761504628932479282847640123664015186981007576017675324290573502120027719
352064322946788971836612837929719682490174103922545304717303599617661973411062113708932
223117459211790139444870471398692661710506988222978404888992920745425338013386637870443
720798720260247842134513091522757837659400695544877851683220682765437250622165532766314
520682325695613775798777964910816315098616028752618348210493246520757342559826918093760
867127101383760134193134868709024924016093044527070575338671814409993041418832420461096
706597116468205050149109132476617674320056621127020617893560537157189374852949238949820
502082368976418552608436175404994534596331741241251806439849480847586373521025745697850
953978273251423018707907178523213593342399588198001942945241939279975618065908309192701
175977275539140917206304099288690391418821704355249009622239346495487270919168862428197
486512112772282467923996430400327699255113283088353868876586829480981933810151125378975
682533170152062956055155733930448966831267695759703812689670244873203405536531645337087
771099347880616002605927457585959460024488679504401503120781617482987648806190221699389
250838602785877811731733541681088572166874643669634513356783147490838974901173780836214
527315692165738938717595655412416417859876006370055204961755022246433263170823209676280
765599601708606649829965243988357172164059498178509596354166140758576000291682735285862
697288606399666070492125656951324347078584368443447442895059449341434713903637282227474
745253319183806974447915853172940682642009841754184697403855940774023450041770457394847
685583155319416153887440567213624041325738090681396204064114447438982874669676692325512
112657869173936499463322312186398681888550150090839053493426138834894533809185320016920
458364102451919467686114116429686418641480747294094670574558479873201298841133361409141
031063981924800519865282579758570091576344512357129117686771559011980884373352142005120
890323957420051647864336580714662476617256825605854219396050143286319795900605892817951
383243119480433018930555603317486437439103685609007405664109006375003575106651291534432
235901208904153202609744039153301077623211571383628772303692806470119770882942461307636
035264762912736592703815521385657709489843049665129178332452592380738270943188295471198
126260073527928217132637904103290903531046530727233983926853254710626446519116805462680
430013240427349076148520068752439932838495734520290902968688683471593005709825939664087
156970613309245727554000122751450054825635890438824519085409415124903358424752954120284
719547504847239050190951298284540005897708046283227689451783926353688448253394417308689
894624918442280043510791394535408849659424042322119612354678438215042044239681458299700
172859635750221684153518435678753288237296808389414107647275108172488489706855063725143
294764254316670256652468130118074762276469945941937079860305328744059832909810260937191
751003814967673435887990635722241140259851292914795717598177726334208133074422352061866
866798532086641588279152954740779217107048775154436741193669069989126553884077543983053
921702270645601401255519867211474630473274343932890507029019354016733411056519730564318
841781829137573981024225754685912326698447863231438363403889610452336832197826755630407
670371577296390611649798716239127662382673048812886449567073669012330420885651686344439

1610840255966066104133522137002235383171496322148998718369172189339028923232756838021166
8789226374508319394667993073975641983253306472644131006543874562393917970860874817797311
0192975373761336380529321722495414622383244165371994702574798350682501400144341472996766
2920556816372170120878089441753645785113572654313504702902111769175594588508608424432199
1573994192984726225879626697742916749285012883975201108416725252122401615794730170020555
2182494040546360012074300161805327977676464730954947057012572055232920692602258061671799
8916097444676964835832244207068612431952698446557239131331643162011836666296243929790199
6829136773774129778878050280171248185825834582726176939454277973128017935460254405104455
2470214239100083032579322464474229134900733727054653275323936678947971349340160888686166
1355647860491885682051907213463419723623731670412685348631744286138704871015223230298444
8490973784644405510435201611130964023364705211290929316812930146226899879190774593434744
8754660616490713659293272077905045543863938468725659340044584817582934371658875739356299
9053410439488973663718359724555311730387695124080159584377120566004172203481047537279100
9850195174931021870512667034331007017866617638863377943577987932663136933990032781733699
9394623025550310916352661593319633370419584745732570735396588755660367960188228393036900
3736525380646368943302805782254940273167586959448019267242948814080456313268562982782200
6351382790177227371964742494249315347947679744196746497776948044237205758390215759123977
2461260820872364763765577789902138410560089360409740368052431686702905194712674349445800
8234441709926723840257439740621088275509390346202013538109099934715778013771876790710100
2534549070186654679996674608445226854419384120653731052116664978246869571025687945524911
7257854864791233179052981635566083501856815370955705986360672384065306201945454837313533
5138079610224558088322959153226240506479094570368904974937770972308190539509235048411222
7315334654855808826071345237258254657022310133800763418031676591848382113802625036233511
8123470663813342274129107562432622974640924183905999981032491371309103835344577111753211
6177883530863157687172579939516886152222227589435338647495119557852436119454244408863255
2246810554061455408245928609629883280779913806998804913378962376915033673184898076196244
5214798102786247233432932272635489598382828531711035408785393465077772126040151495656666
2289405562492272959887190432751082136881976635518262335716222542147301922538231459953122
0989346787425140251278243631298544027808836290691309051214711391357371041576241296827499
1499770664503334241719914011620374429855344176204834304983176540390470447566551068590299
5892898140093007239424417807414149313956449941263737080181224584684513538338282434137433
5750891265054174885316676685358565223226181195441449717070768428039897291838968640773511
7269995606522166399920168563474508683008436040543776246064430641569295594021399190510444
9317735531552042504803027868548308819771462845181111941477383930442609070406147662240299
2438830195337500056561392004772873396715369762965253607892425720023842420057736397431366
0327395659886124942231087350485057706139401721886423419590906027083428099887878417370911
7039099605084847242498239096429965430995835798641768434442804554604799913080970187890433
5853280031909267362837727711000621629711378610009081362578475169455596492964873283499644
4644727125589405350495013787080036076600023252160868248697375739025516908263133296711522
9101577073789358858357655107825741908164396237693067216243850958964694783393546009744200
5799167688839227081336810746926377944934621183360996694867442167803759393561275803765822
1787736092043127829617458211208512009843995663761256604964118687981040627534454145455000
4105529919970010600433054988997372098918094915929267921618115848859948432065829676337999
1495791451947667913513656969922993400095433790285014614163546294447155129372732009834255
9811288228392644412520618578455252301637854714326425680695660248871245332277654971407488
7339445315864704044495300629705317673296694520782816676512367821094690559951913825860344
5691525465324731356181386222085546865197366416126768123008807190240975399992973281747055
6161463099076616942862969718297078110199259535529376147833461609641643701897039273943011
9866638946701381688299028815185440410383977567063835329530377324896023100150257327994933
6273936132551973417434144146201431212184376537721930549712733925155020768558050577898822
0267775269899763473342333707923183983649410796576584468124097319368634996617010346534400
5101311900135062669795787848335077543724588129533736737617063363427845220693513366625022
7314311917394670491039072196057470889291544029249154248435485522340505706490349685823299
3144299135349788786498692681701160928329077122057829831394335874942380909610767341524044
9135242873529837020136596840365433178738535787610531676284423977944144808137280991235411
4754173467349688657025073758416294030488315848925585421916860383878564936556217669976100
6495471175440991169886779861342104147593441982856424416036684832604104408871206092172855
2011342306819214356594050665925262505460910546490245611052313243410630444893076197049444
7361292423589064697290344469428364996453286096040424338539407499956166537947986370600199
5472148233415936468972492881585728545247358059978310460232504346332973944962253612993455
8330136976986124964171267795711909276833089200387517775208862836240144173059944029571111
7501999266757159365663744818284802840822134452767639792959626882828708484032102638461577
9443423081038330732618372897699395700595119058707942848858433788374972540095094103471244
9731201050266872786643728084303311570060352035925092744133011753548792648369932448512544
9872625564879391180910856856490673335138022630480201891758640980105431972583465514448788
2409864555398847768818816458057502756812821961643061476438317534366828379203618877742799
2048122528179855611102601842489009383155682956586546892616533159243093120905541519782333
8748399009697293859104773185002553858122133071486005796162201908277224732618418696174722
9343810297604956673236129917070974860562148082828223559487649297478680491420445467001211
9473198969851525044554119763148630927829796887563578343431129624469783285028228517244744
0010498407151149892886299082782614881286717183630801246854960786373351611898352489989688
2927041549510342538476588111014826498785429406997565258420685489689113802362773486606644
3147637177920903932821195552168252260515227379874607905631390876978942139980395688349499
9469753649133474873525563946290542439025386692565877479839477210646627523900633685904400
8463723144863173306908344347393110781417270076636066723626360524974832660064806359423233
2950135876579093400866625033335398108276321667656557646718316059738547427966103340793699
2270857624132523462997287238629343935902684261040527486793829606828685633610476188912288
1843740570277245605666183387145575595625505723495930851559849514861265068469286491241455

pi to two million places

6764562534997559305804162658619700408076640564482316458778405383142937335602245237538864
6184402930932556527429579460342493364841633178439484576779238816303315477757120937169983
8250598514477696027227011755091784082203058661269963512486121815159898726604769051197388
4989941390327108069324530993649651722287591118784478024438443165510403854095365636474300
8416392704184150115962013622176222165854493198634529666529040661828049997108595752272522
9412987204361571042685435156629175661321878406100425491274053028523071798020151698247466
4361269502067803150640749824689473412057265410756835148762015977340359214503799576227777
9651276538488726316379656829179464867533202305497008288958658215142651807462775189559588
7230395615747045434126848277484339619044015868863127570490549180713158468926333096518633
0140949598648052521695871789949417234097668206252664332293894541519835405232536755673277
8842369773915890042387682009771951451274180903454610074689676917089601563327165928709422
3730838909575733847495801746703630479761631758461870789249560484343189295339623588752333
5345488347440445503370706904704062643635710507474798261976072207673487437006654772317400
7425244431500271200954251234267354890231889684827848509313681788401916251794377402247055
2835398673831108089842254115816746273468932726212808703971792093061009903381242139762111
0967777221928498519121325632340941189685791582993216525000705704529911042716832963610266
8065153607018191297812221146417333135705055844286877736956911237432508419355870006893144
0767459635906951898933941715759068534382867563214171115100344783761843677587172850870186
9294209081775873068574280273756601021260145773578483520852495079980432156460805380031344
0524961451874839033801294021322397054809837187286318503552088303467643590835675267209833
3177325809609033918544372497973578764715677252715715205489698523938207591597404080701633
6785982350618123454798081343385148750663918709661492310367401985346062829718606083929455
7777425013388969639115742731546737857034468208112840864869975767070524669457028539087322
6142070230414684768314181794083476235126274173417130769206233718546914261699864969656544
3989026328436748681539193929336564584669036161311529840820857092117128493566022706044299
7710127259779980180976974765597982956672474812454377317281156719876931863486419522188688
6133570508150486531927001624515390778270740873162458432298299049875459558025642558601055
8262207521782536886889561118800525307634795953529582546566018554036145935284462845530755
5663292929918106091771001014346087891439133507914275664635045195724451257980287436772711
2609870968399362991741379014055044807196809969992255619007395956672708577459873062929955
4739013554158426102143886062362929368894470127409005232370534748135196582076497084184166
2327509144269099262200684480336753804815368520282367229206940517791673411210232119531777
1448459523519422925104453067032958515635639073485444246485770082503177077075612178431400
7550629464202884937318260478333235855961927506763089819451284019016136026471979110017711
1662268844933884545789362123576701294869847283126487313162978430630208897702749294779388
0758229060181431002880048502766707118342942082066775446512862677701431810892671319712644
7325257692756497490442564058111759441624302194122158127798528429895834315253994627259622
8586132093914513827582884105647243250039282179523160506672090135768095626222007383308822
6481370091688359824018277034879896245741736082626729604558379659690233159361039136922555
4304152675003985935986781892523747904547408267926336170508793027397691474653334808140455
0281889446313467183897855255847114296968339714601557420470292887190357355586685700002400
6079169571329781908121919092205021083202861290055980958270489942972986660192916026056333
0567071998516819305946019793352269283607339967778190105336930909023780756407461595569999
5911201958420636615074452034233113101237950449353717245989960917462956386436000129601922
6087718983056751653917407540699433508196147749243481871223566681844818514155665030860666
8916496212575347856569938492323722080302624618834456319456229983755390025629299012731688
0872684839825241533158988987064667944707560435889261161513634345668964301783386498312977
8745810284601488816803995753935881578934410077023440781352342468511066179980052184983022
9161440843690369927343653904886789605817327924290597337161802306365475585068322531928422
7562568876821243651701705321575234113149981223325573146341209547541047991846853927226144
4554069135684128282335780555067821810544849702184827641442448409463980146791250481002144
5454489485752542900612275660853111439967962548501370909989132094908818667437076247772911
3304846614918669527176814310566317200074090637978702614901574902015533696910499901994866
4606754259599140859616257582511739742668878904120230451202187042377174603076181076015288
3467114617601490346451771035599823993261755310526640534215960968031170527883841504387855
8023079721889592632583773477884002507593241985459647451464554596742373435574545391322999
3036960809040466875554041670982950822533254092748101983042382203222702075972724128186188
0088884825278356213927950094256813208284853731585093814716298063387367530309701607409477
2031357843923586885071522179708832711368364763536412650272105412258049451295065917598977
7349848280437315315630753135358033133007444572318428080211512362797565782231825476414999
3158209542702525622978996894620328616598862492047692020684516183170764480180517082257111
7087131605359039344327178034743444853978577669351166858562701142178408851851983037261735
0744514704606095174669269212705470297363901472974247745468002029386373918694592644017366
0132217207231137279025262775956786225565056147387336501135356176000158713936967665910222
8047851729026911695207330179925622191276119564378750692643177114979999746070922415542933
9149130203106909071591662282397698069324715269133801616379678501622243398854383255160655
2642863495321607878906113334512289233555023564136302070001988869620552126560771069174822
0354025520943706119535820871565775098206643381427516068465977244512435181834969580858155
2491990217242084032234043041705164298620589812505032633800422941631196663701267080708722
8544879564045579993649718573797875217460499668728314673836597312695616367457975016574866
6756827313885960482964302627135038594056097830470936460889735267944770640761751909834644
1895793926515323058959974605499285751945540102519279059547083001810919103553017880338033
1231610349810050607727740613427991776729152415493944059272851990173279318522212750010622
9958639430432154391726973508865144157835335150125336086092847959822626034593293822055311
8010445730182040547209304597938728997307196829200044515572884215910014150471256757873500
4672679442551890989289249326035260916022990321582116313064303195370433428935784008168699
9111239544986061664520239997244667068492020459104514801494708368553329257339271364569955
9361235760660851100603201780416131347502114681018940065358124398077743172118714213963011

pi to two million places

2026703912536278590651267713380786628107119408156867120224068483768564571794814327538 98
8029477535221002735983841891395802286597211657090587446342347177607121875181363752132 58
5428040680095720486747053105538788465493085013357461768482512416776507463199667579740 22
4513975957422322581558444279061086946726845990397321192556726142536521077672826548540 28
9967072951222790708828213261221872861379005634532563990519018167680732514805801612316 22
6537532534307184268253708207428291151640741585283548051372694615463503788912927492435 92
6199176528944540593385024882499852573728471248516020940500513090135890946425893352158 85
0427254785377981366035741101810977506016801909509966084573294137248645716688422940840 32
6189905072748843164414542125495086657074419578600784564217066152668692862735706418999 93
6092909923491979245748499008836901942065461862701118375746554873273023717363752996413 58
2823407614781222789127656312168981885387296622311940247446053947241252738541280621538 43
5830092234310442770674879916184961054873217652622918621329008724431952107393891057541 00
8762655718597522856435116582430669460869571694885365749033213915211438516528583920143 34
2102908692761000655934631896837466934371093242517940756362317364090823502454553415293 85
7003289184022343540712366934162816365668215390579479321354452984792471259912676815293 88
4494667641170188896489905242455667576831098349048635410060714279548920485850831372024 49
9018915791305833031483396734020294647976074108290116215465953025419271184931006247207 57
1643007332201728532684526440095723539457145993197972456404632007876550234029331680605 86
5235527439861416876919407397203567832534179883352239031255300401980856607733195425617 38
9653106338568434090371751439950465290378454914070898387850125724801473577703856095030 76
4949168280588610929249750626453511446875503439772884329687921867890927195825149998454 97
1623696150205491831421697996318761422474240193740318768741754626949711122300040871093 01
0197152923981188179760100272402140681993615665140157960768657661677246031715388517464 55
4430316629173510751257025051643626350584423238156216937123494329839442084622983338132 50
4954240805455645788638991389152397137971175151122954876297055664535989648372035052538 09
6692164817133237180208818502068503985712087149728292124691548324824869069164242461148 89
7264061031688023936357026865617187418334453055289144280576977788862500373513197794417 72
5680929142940853747689065323766749512367510706721543593102507957734539439579049584551 62
6801070324123682774755214417267508054444957647468985492657107753248431565988397012382 80
9810485891602464696644608235311947551377556784227071178012307132648466308526584161185 89
0038782668364919341754544275363072816696762329901245757345353886070842457987376503421 06
8282381197045474458872486162918485567042798399221566319785400069763092996294956128059 95
7048260550834518436429220870911267773140614665325748201399431628178947332446765783514 89
6262351349932147819247171879317139180502622182003791233768645479503256034687158241595 77
0060379908200013465406045970752805687341370195291896006293327772885083305488439465563 45
8259710530482054806927559789603291679475181584394254966767124810280334002742059474681 94
7477627044164380068517358143547832754099039242610862545159286684680141724436489802196 47
7577149992355900254767307574406686277594835892168332033761695902269337934032930772146 84
8786838565223109465320214796371957626274994208181804304994953932822029898007731510524 40
4333587923814560825196069763900324376253876418421538515223948802294233091290730889599 38
4826782722116364259677037080890097529583219509110358641991992397957067984166335513347 01
1760416860767468863146114563576177571305149019681515816848068347582357007329318481443 53
4469221518746873481206024655571311149503355439972643893505691740126186027307261754834 28
9373627540712729771803007742523869123584850695871590378226423773268427831909550446665 84
8966948585979506101812321766327050887850181296270249402580542230217068579886070251197 42
5535906860412771599608448075985543585578864492366754625566393734114152975983057771677 78
1214470139922485477094212080594367701338623131377091150243845124726127806306738198423 28
6167367520562724757926877601616534138857279609143527490756445646064751009400873443980 59
0587552243886657484599292922457971676523755406170118006924289493483548811691141116937 23
1383102666294989520599303318410672225709252755085136077367113610812641782038911388558 64
1491044904096346813796406683330820490499414480322790931914348492080590677412725618552 90
4323095830977180005956305206889624916610404750690301935649066746320666009892181917227 74
7374531293258979572578632189206598814890519296792040028417057690886230245680386713505 59
3555902363093802993242226265942662161004544903542365771606838711456242255388984244150 700
1273463176410938641688592695636476886326899103595591732838497653530563354872472851622 47
0307561272404060513273547754848095846374105044804651485338675304777291940280338371770 57
5473801600446136236041565081441059349215496718692860900277999477047091496829705893262 75
1687904949233757177821954990224243254621266622924560136413158783324830123626064059965 21
3752316718790938876839053571221097389463872052613697376818010457479483540477763962857 87
7810446962899033336881222253599596916367771895568908826657378576189100033353540017588 96
4712823132823851393630872429450691827086034347962215665168837847172635173684449815600 30
7091934552710785752627926941280459704339389333856575663913812644697994257513202050062 96
8223982285160146678046023462795803371399627639743995520527565273714973076551802878861 27
1368758668986732569718470065225136239831308888031705810998990747391493143617920514744 67
5305505025360683390611663486350158101880988972049769527346863694883403428992044258373 56
3338261841313172406031892841548848512369890848432790667400044543380810520926172845144 16
3367701586082227308863498063831050100995080484851954224926058794995172580958761763442 25
4117080303822640347685198338761678124051221514751646110302956380626637142764285985711 95
2414754527715041669565435891965522651718426530539432345924331900109442814804420027417 67
2168297616976323368658118872053841320611147120522036490483906222181274712416367315210 35
6957680312348994555746005502545123496852184028235317836799184567971119742805038187636 53
1234835471448032826986727917894136340927413088552313275681524923829760362506582979900 81
3607088220470863980869462651483908716338156583606670393263109891960670615188759671215 16
9148288828247779607015774179771239768989253119060528384766157768796292933573258242142 83
7395917161280569787439978298140564282390339132408254619439981391784036475350422920228 20
2937562213193578661796549983089161394800242383183269004906035215691746525152779959026 91
1962230030001080184140356194876853297482729011819612629135478824179867130188457877248 33
5303713097215853170173977317084459983950407995910952729446760591929222370160139981481 43

6168467981323637749712384731902228775939296958536811412721852945988929445943767168341287
7132838070493327896083200803734405497613288387092863498640264005563780491309097035594l8
9814646998058555579038662123670615272453012651422866139150866557681793600388901629413l9
6471511290918498617962264156747766305327147163810195424662449034478573741009785978407597
9056521312190263942776055039954099030314824201421341132963245113169959988624090430505397
7678892854478277618121972007281328876365719571734823991876005127005156530377427216919848
8487548886472890886504996743194248579311956817426033845002166891576440486146159194776690
1561343843859036229137219817617123158589714625179125444189038715358263324369396406993448
5287121552430801774671436758679696782668224812200515544350710148040992595438642304436
2633758027214739677490853412838248542762731398660443547703582942627764494092745516353656
1127715581992320911235810581072282043090426649138781859791193154652999856368672614876795
91379970152258836409684701220159395380735214020607077557685098800965031469583982299588959
46129844449691286573274923453668823159118918522940046124575115377369649037089727054941732
71002335063581037619388629831647784180353172706963383098754421223233270859639124506
5042554636260944459902691324220875031198387960914774266588579643775067409430396242470087
91258785303602282570751341105015089291162108504901797729584824489716294531929558195477
0367484023311508102550009393012311909295411285306246998580560678080388549653244795076326
3260007816139045395394104160399314911765759757283234840063194531638375412964114642955859
82909677594855399975400473328494610579588112963866479383230422577830574103147385188567
48484105784058050362240981618670596158152925908368523231478061720756967050071380581237
85384548790523697749183223288884321021845217174634079854866134018829157907051102298952192
40899232351670475384812796508752853189534469465451050064992488327363171822141939093073
398478708961343514189744071115457854815984247757525270334904437659476509301977098668065
531161650879283628100497409416626729936344399522039808327595357661852596670029860975545
02731639478748244065720112489140507181281381474417691085553606211363172044417588897009570
1018101114688354502831152807144218262848757955561649670852742881642487439677849244405
29076034601410966447878738943938035571706676515888572455405910800664200163229610348015332
3328135341244184559130582915811121497537714597595224218138135511380259703705301555033686
723859637397386433715227001662690112374592070036174234063517087715586662172094079659410
011379000072632264604147219527078091277368828710915604414454547752741158586609984186422
754698962613509718735444601739481868290442318760193650900961870100731725493214038182162
84513462186779688523723125666358886944571886201891046361727886217808166556800210188783
828832127569195159227496666441003058309531097540003296241245720939403114331914897762045
6222230713218052905879443296582385374656134892393922136418861243997609198027387689267378
90542513851426357944780067491866820842600109487124686269093658095418800274612991970589
63243984419730067948818436876888211134960632362500549230642124180695644747256052495389
85450539582645502444069414699599716996391881190787820351842187658554063327618543600289
606208962823132911504837508561244829946052174122598683059970033661934847317573421180887
532417037215706364889080679786594463431720481693249752593667720688504368265591211178165
9128304351892318654134635340445286231339940752801204053250406527679220033547314453900643
11422896741214630462335674560377724711223086259188402880366172474223719048753852994008
6118486767869733472338235272248780814775081768735170017143810504820350727708403950255600
594257336227141700547970659703656458521851947157759202939053230156181381771954838869185
92397623463320383244822889837608012604024710001119334964352387496321548469005744129186
342215707436594161228639468255695654252541938948773738991486567368706189423002095592329
277858409923429110470687082769454174242318279026772552966524193049108226438161197052677
5742991450175798844870153083390707655104795575287640810814914758563097823758705238946594
440642837506681920680531619457992467790219677142115630901699145209587203032601874313836
66630158859482432888862875634692974768706389778307735244567564587456211122163392640805
77779625586540250455492339239029583321240637084721766232118741861765370585958170095381234
43126959247649925408913471551262292758957467486321958347457111887709347251416782454558
814947685664024663057040548587033209777427169353355224462783899384394578657277328492107
94919744533076001329860902327667597905119651934207765985502589775391119196126480577217825
048773793744917309903433434353657052303092421091355229459915455445556348841794312346659
97665578249861018481697584517669200294238257346190533408525122097066119162801979988157320
18793248835797343849317757347126190115041007381026302925182738500229964872358174943826
098910980605314318166081190872933984664430314773604306103384528471810766584288084404072
290432551267661789471753722819480094692183194141535433472225203584195682895291150874837
85991703887809386111281838222628584985091119786218887363205587205162981302256887077676182
21715322096271596645848125492305972622407123898547286396238937542149106137600562656733
1865941432375379213737440904810680005171808406219880676028467974538673517563278633382704
5113340169930134921561959355751828812588896672622115054282949548558056408874642663999
4966257849053253134617253238977733411707619956793883295942822483964234594767610196458721
1173860337448826836677233834433586830254067516291669505942522913300688139277156290968851
336202726535037595547623695296179088173670115522621223626838490428034390079826666612957
5654855596867087129294547243171520329367920584732171935061012192709397329556326990458
42290668440288595515969804941753812694702989872185370539354651569286679289199873083829193
766229728899450378553994140964317742346711143631162778219170036431920516872609174929187
72977919466093169044738205980432656601222948680197825029473599847947646188782588431955
9712067496774187837048588631439163523687461735690337066930512284061134801643898800052701
1414795985555867085234712316028860904374392869504453390592906971211307535218266270404
2434722086672464262360190581571987980781546645338005908099921442041867568287553900044948
652662782808667294086761980136130605779976480446675511692046724539050420838399123721999
693283840217815544329416484495964558014494022049200453284243548120421811693215791143101
8997191242621965388498126614833613427924008673723325326771470510779748991701926116460417
58589273725729219113670716799569304694789047787367918309511450534737029706212359700323
629225046518901130632485943293991325334667498902346917670442484478793279524409381823429
84004465399352945149060047978010052408773240709396417321200752676030276239280100674281

pi to two million places

5695347367779988989937763484271042806460752578661826981006816672655927200917234559766 77
2010303412549100896645965259207151151103421747118433486578802133426356404755152078763 33
4421892907254660986078976919951958127626337475542923566297116512457021300501876519592 28
8452630291270920977423844967492039294440478933352537521498280341783149577698023563253161
6293443099316190083640359438539847513392409778997388400218279786417788656351296184140 02
0683246981802805134167265480249337963336636821106703509616874774107936012908604664555 46
5249144361035413733325840431321409012011000130248596311672032612405558896256433301294 86
9553075877965703427368941368428088241817068042689900932187532685454192417722102313271 39
0014336583359202119237643038520532186693469445751988833160728242824032035168092750234 02
2620539493259073795858029552139309191195686403385157096001639973082961309866886471462 77
5804108455160050008115581079773524769622525786960365016026327114132953039113612458717 57
3634225221760730764126206656782935731187488833550607012053836831517455978045296779345 84
6622862902473887860413111153927562049095206164787263231244235149627518130364377800589 93
9047508979249337975822166820330690379488545535012141162423560145235620966287227610647 01
3802585478657790948257518820898463988716832074525533159305479875100964337225145010088 86
3130059962977505847758503554037112029147765657086697681710568935707127446197640825500 32
2727174272127973912391485382771956461394010223688392022829543698920883335450321556765 76
2610970772597552825321818017380563416117942544011279913545200528281040138769088168489 65
1722071888209295423823948983477448777925801057570280988755535512227157935664193321233 95
2278159261193973162224638790751747813647891795822420486458153802672781391204945750721 12
9385984934110947256603270100510315892113752286542395623783321918680067613596774979282 34
7880925543597373803671331699582481746592663244399847709173715315101489659009577152983 11
0585062241077224042716086329896380415998368526710566964411651026910412699680079913421 34
1625846579693927548421768388157275614692828478106891307107186831378574492332975256695 08
2722182396099002088087618878823709035088804892105135412063475037073302865782271172650 06
6386551164238223424214202737674086962364813229888784843544167849669048183585972773183 70
4400357556810201364660908764243455286711472319057714970616129110122508897730871735007 06
6518809570181141915399484835379168337465638461720812443968482072793628134416848208374 01
4645805541866341836956579099401640702743793779379254693708946288506465566879753452002 15
0929356067293474760913361349100769970641481653024187188733753526768671353956619668601 39
0844905821737450217518918006216809757301839654515919443652305815834138641234301098985 69
2083889791941705507080724078020816379521182537584262770848345741537747420848188386871 33
3687896308523500975845777220164869926846459459284842702777439448158663691168306735944 92
8123886840909123906385315047901744628641004764856224024692655560651573475976894071413 31
2205955610746702081572806626157493845175337610105653598369625510122789778906926046009 73
3257318803683055578215865056989202185062516296309152470635749679669413944724253966616 59
3496090705641820149186027356312826848709161232120035204679068054752600422673203450095 56
4797709609943040391967013496164334422851531073581076952657795544431380743685225249769 48
9892026114771723147544521085454354770039721453994982814202812827935533196253177265731 15
8708279951317609041851407652383433380235706363054627157229748928887898352327392849089 10
5998307302553782667282298250011580834538976838903342953017377200834247856442105737650 01
3711692821220220991322430468224732747516535623473686328640036851896095397891943006701 08
1533707862058665363959801727664609121701234045903501666509086329853892646212017380858 27
6052071900222884581769137407884985889987554345995342003315476580263859199629820894505 56
4651143750392879254918739430333678989288700804018561273912128999512658832918791340755 32
0562517330155176481961713557725648587783273760646900551648894867229421541817359015255 47
8874808379259683970117736159107901618524369098734004488547480849551978984736584048675 26
1068942174307419129199037034141624179630613881277252531344796180713865414422004915993 37
6259852265866217574170970868775485831244926586908672003809534092476114270165386685443 77
6228067106575032382048774263295630602888360792324532242314349773985635508624304904423 14
1304323949914179051513355620001897276961010798815324839918311607440980015058725511306 51
4131371449557639011319403617506910475982750636780301848191761370636072376679109982497 86
2436560635003340041131942522960032349821222329207095212161487810695593176112049368374 81
9007928808365483107665453399619835439953498167880879053238563438977315210953010157833 86
0316767758634827955723577617233393336914481868421511333185805253558786372692350031964 245
6138390093419024552459405353871964593401062991795562776219730087191836713639716089782 16
7369035958708514364207049641031188257450953919869017790720623877480361403377257089078 78
1655371470770038748852847200753145739581178705033458993888611360892126061561667568072 13
4993038956024685045922423774698521225820149061100627891291312209599713528341631169828 15
7160219031415258688891101199778412257596514905362842451606472636806506065786364467640 90
1618838198545748451614237560530885183527975122167273807365467049574775901380788504454 45
1258592059709748706927416443605789633602983588306814903239978903599767765792927067178 77
2098313496339286994741609548723100313358499107491378796819549399100369289845575521909 43
5363221808012136851753065943742592498627285119270983136826597860734728329314154486303 50
6650766600137981251106189757945840407967732441985306210413088466947707091968951520071 34
6815720714349616621848817106041881277776989137837282915726292302417620489392596211313 42
0636313915436916341299931632224735799211659957115024805709600933225792109368571644018 29
2712889796531411409121171324294773943037667293897040905339508798339764799872518537421 82
5012345570182662819992765441951041208199694401486849732904003808696781231564956055388 35
2295646068889947636291176260451822351021115604076845550226075636311662047649429961139 15
9648438315200248396682111490166065347562256822071465289346069448337917510162022098714 73
8422023967149948989898338661313699457177302833385396451847607427521796650742478340801573
4430179321364011410657114800599674306647017599513238472444580686468511546436210198499 02
3706198833497085884644158489877235695859952158053647582241726829215865049647987877631 87
7893750053842454063667537148707830172301972096008511908505312671474944705742121296098 97
5955167130210532783557138711529105925390520767215328670854770009967006493983020050413 01
6319576247915342532254802836698463155727464282333766098825940832744195704501816194483 34
9973850030658288605555786359576695430365501708336352045747944544158695475611924804231 35

4408291150195382906674328880860280581128014337102140168757806236324590113347498156391 18
7667091541245575562999567766953049470892466111575845702705964461302931219187696891428 59
1743638392575532927760931124466262487544093290770470056743118963009842533642088013321 04
6819797850267165652956264754634953239157702927456924792099015495786391947649292443358 43
3183467380756542922192699418184836937487543238512704496462114203674623360556711840119 56
4244958116413013858411178371136064046310802187070819985402746925092328317735091940116 70
1366210719189421748755952951185476438883551383547983161604425757510836874899492111482 53
9054885857726914710760844069735237526779783507070210552312195663803734886638684396577 37
9378931056746107089300966840953472122555547609981893792519490379096843473693456037072 41
1227895744208521267337686428780783126659809690274913079794473653710303598000881230603 75
1719607518214042432102088760930039737106205240406578561278067120410844330368213820981 27
1570324935119324063907227100208094449615739155562876315344946137090146081009100316860 45
9583219770130750090126561412628599116075135993634039199891184820548839892100188455903 72
6250637536471352892435282483321132053301972370670421146471211319773348096819133264309 66
9938827302570299887437704069351715775914753748554493675558019277521513412968726799050 47
4332802948403291962476334204934987664447838158583631664719319824544554086356689553908 51
3513288765417524939841472365551917882163456724269512113513337106660685621868866366508 80
3255318873652889724175657041876652272125600831566535526478488035360117637135167140851 36
2685528662630014243040503819800293140565548077103953384762259993260221743227519479892 90
4715952404873351183663563954868744115520849658305429305921626566117960340876037761720 71
5506625586581439953761631675007241107670328844321898158721874802300318212473050933680 09
5496868805040278536737034122066341106384437365304747294957502777969565475001036369428 58
4295651523794848218109497286560003388763261936440079768295740273511723747724418894271 28
8025086095744905464836079832211008460422968611267079088770936098010947743368299241180 84
5749437813347967222750495942959241441511113001488970909717491041140950242474848769552 87
7354351875556643822661396882870260079245202955888624509716128614490316478274088364201 33
0880294790789436400955656788509527193853452206696949430035714965689428412858252661383 92
8053932672536204322718048461095817165626363482301152987597393891877621371356185051317 01
9171727608801396583896272299839423779506175766168486178467177363918227101834849475810 83
7018409915315923338908352347055290405904370941199453704466114835973842811514137413950 95
2171355587271842669867463394007276306476230371500161360976472963061471276441438812934 03
9148776195057779495860116051503449393674790108078424276854046852984217261527937511200 24
3850753691929441913238511415767844315876828670302430739918330715034117504597877354057 00
7068874331311807416896613457549845440046348362378610249041010110127261351824117839668 49
4501326894432616275808025690836726592869250062688309095286204956086403543575072088581 27
3809287782972887466365570091306053388127377050914260137313157778063472393951018759171 37
0960354637411750638922929098691765404559514411353241721587531389250461625727095599294 25
9590119171031772085671021784739569851378876102249353301884920986935915317294296512054 64
1649980432761306109335085260500144507121518316209766647254152453668283513272911788010 78
3622600177936237334114393633728054574983982484520527001637995663289884928634218282305 29
9376246678574450154646235813584547756266754403410680496912363393722275823827335283524 99
1040789964453886408545012767714106603782460188211686793237037557936339558424488887642 78
6909426365091345704305916517734627415083678854513099761874164135106632607754829040881 59
8838353398440444388838928970468734236524966482736216048132204963876872346620129822289 90
5766783765931114754013105196095301192974508105552681377944004088956529559050054110833 51
8320395691459080103948325930119669471083357080667263805336803154159101689451553739518 64
0777789585390975287382602333849264918533672484632122240467164543628733769593132235980 96
7534879516067234450977973446540976031704729298598571277075847028765760635572673573934 04
0768684978518643530723632843018420134094685296802350823846936616827531275716913964179 54
0085020682466688276567793908372380023352502551493087893353482983347981099813578533189 93
9703710380091670438216101699089770920517312209778816843179710641472361071820054490594 68
6821547951521278505198267396861327707455470216049343629131937823516823646222264103568 19
4010310435492528802969604104073774894837451133313380486735311295822887760489086626881 91
7267174046198886484105412152093607331118618578177812708748201778124223812553104832081 62
7718814660813423070348079870352158221244141138353186057986834234214058529169749388456 19
6027339509127443667182002051178553795847052782325007574802661865059695954157289751487 42
5692844791350305432019681536199012932766429459463239630125461686591558844416534785298 74
3362239978990679817986356152070062155992876636084648136797565806466592959491473661001 60
7209957846399036738638146093343421724271520661187447985180491510137974811302833007806 80
6050642553129440984806470721277257290555473424509538783396333893439180004814521469566 36
5527494982810140675120959092817791271642939528296824067990312188210762782807712307784 14
9163258031538402335527097341089134266599452855774612401101940801956432140530168422529 33
7699399518908753238068008458604022225029239526789853315960195451017177233910162195063 02
5851055467777639469364126118900399173228300740289849701996172350195851302125940462571 71
1554805605455831390473331911489533078352033590499259444682580355949605216956625298953 36
6087656653543796202206523175249934464576313427203202974503273658627403400045039264251 57
1770763217659390576303870498833360467370137646823225555412286106619379044323060023451 62
3739450857433713616502002281001991553452126753050161409232236171829342929642544888307 00
8257481763237424176976809843887348623998934738138169039219745426244055446541545492871 84
2913881226935678918233133511486470127054428143250202060725597351107011247799247440207 05
6857314759682780066073018811431330492135072097247542695449083921071971340824537910048 42
8104385022559677723739439691148039550242704332509730134035469530713634167695466213781 89
3943553764186184467678036392408085683802934328502884390439929956342331799487622263868 71
7856896700893468725716307477791004347877889195516884065143271128621149943965823114187 57
6032559125960984546032187302184554351414136449962730326337221462987900561331123121805 37
5779634727322040979944276883328440034338774407404204031589404964271705548810997218084 28
6869739076660803010869416747481452200091719129415246130475313512637251616331652602598 56
8260799615224609845497357288434506307332301447940033511771936218489614870951947842347 30

7671563605896312221586848142952152582003567287964368570767927253478058894005283994565 82
8767988447154815032357964032330489444660248266532500258670661427484935867368406027367 61
1998099685435642430359055687088278797060306912357088257208825200000422019753902549082 13
2531760928707692838417970442558752506494317042002504743706328652930122103761281292389 95
2693741468755317722693340262051436245947128321140712096468993417710195868202755950847 03
5869245454060918314043252448176213303344690545288964701666808887411327729672916692474 89
7342645248559715042579111004603451985908156250091148693942736284495476981979222919353 27
3739243896745087933454334528897252731868805246540617667433706369160408610030108024948 34
2272870217626486730169672548967268270766922690266677861568167642966005489990615581258 65
5710393502340501929192180096496000972515301070469001962361684026392103407548601860215 92
1678173555397589995280601378013549942911771951285383042790634690936467587052009805211 27
6406916204269346376296863974315894309038752737639538554006527570821745220331763906532 87
9139867208165612471013922890859016313239653878326814575759044664139806675964261490083 33
8681529727269565092555095384804880830398126651188272291655591438149506968256452102398 39
7368389675681534690233636588096504270468634899252700356342405946737515286225428626299 95
9134383901531611291470172036783683828465194398496655285627903848911197841190803490757 78
3670364646304222797120044567560289547363491587995655130729868343898443811051835931539 99
4856741203485445528191747945943768924253168629254938022256621653469466384874810521578 81
3192435621922921448323987483333108004701505417860214865425888216155605230547142996878 28
6879751503972666149742650066196118708454903395631112528635333423864151609730554272818 21
5914721929091253113532012236066388437324460309123971524002378383047291687409258353676 58
0109411378994195214847800271096026557761280419188357202525637729409969536937505990393 04
6972537004141348928315974794916265341766604426314206575130421883392616843074535285606 63
5313241933857195972204427082182936549939347657996213073690321923094012413912701813011 99
4765037975698636095667468138815134062829379709124701650733820500049868277391947802497 36
9119534228864321564196079538166071988703912356037304470833545016216436375575925994810 73
7974392223319336935371175966484634277942603246586078292233306557180314670861359139050 04
0906388487929751844791473361250555607252240786681606740000763055084839022131270544586 21
7987783263554971557210750183989719793167793256867129178949078153690278433324824899318 73
3907938059425647033196317721813233157474965697817072978777014848715192721503233501637 58
9498573347917745045476128237246266727252655225499568052215065015446331501373314515514 98
1618184453128033115413547323278805664137731293401085028344620498569436625911620875011 55
3447742974588383573192465418658738879393014711300367518556353921453962472081809399642 44
2976718102617968683877919001794935413839218932060432087403590166629906085731764933709 43
1272430504036437434879300223245762823158871143422876820129862394378067689187285291911 60
8408610556087783327803514968305651600163498133159543101881163212278532652032994596881 56
0437571859657103577932821715750110405721237504504735756995813962844045941083604638678 25
3618692334442560877834823453013745240429263793485698991544826318407805874151037859665 21
7047870631282402675575033541443080742251492445567288130820685806445096187814676310697 72
0435206860220030086333930678460952915737528002193349431680806664547260567969521974106 05
1935728111077190207986286383417790414689521199581599718334770111569096346749476107444 53
3476695274750093080990953259144674591277430317218272825058808135452843884536989816880 30
4184973878549751859921083753586469754574457806838014055634005874635472855018219097732 63
0164827560410987412976124272119992625959988656857411532828614672978011006818650259889 07
3486731365748500934155878849561708551073374680599126956423054265529108623526558074001 09
6072252381836486349080592799188406930113629570329395443624962285051890764054080327155 47
0616385756835604474887590849545557503710789509441082475702734011199934801958092639463 821
9570211309003202753382722427863641342223482817249099360868553181725612323072278301265 63
8129206633246379466910093425995523055097304208480168858781258887661604428416858539531 11
6380398710653803536100357906057311953078332194093205691669394100940389471190499582398 73
2818656798629532063298824121908999331438847303343513590335974803879463354137729999301 99
7987696903468044069815289207736700311798344354741934277296512971220376732999418630174 34
7007606098058046304318706872593575944587310227657755239044225082818091870111105182784 16
5371271460124261639593453720316188030537279843029167275139752020316952878268657835999 95
5581312725252923739690609487289173357180463287733480678226661743340027622688457401492 99
2456446744664546955708420517638298827120957019524513302627536911551627111825721280628 71
2731785944495348834786403088304066505454210493193296351175782153830478353785007505528 50
0556707202505289600274128858907173367142938718699532277316262690515159821555532097513 03
0803320397860444792057233144184073803586091686068826432567018502630536870440024272887 93
3547272226395098940150971997311864374644137697442787228710763284087383956479659921870 39
8481895828273082879889751079703379172250120705693975001319817825477504541812943389937 08
3444512728772502682176477426755119319393447116996252644635863296912140153275100058449 93
8948155775003867813519152858884739860651923975881372857787790763599947399723084703772 24
4643108133720423636365076999084160699352616327799041717499451495190545224334293448097 92
4732780805642223651153109400416593403480757275067120442458468001368414384380151564718 75
8336763442615714186252417912020384234798016242711550557913646436445305202661774916921 42
5416119399588463855015585114613138847312588836087217954053968793860043027129571139025 63
7978314935320455463419793231413492713012208621163707613560260951821797562365290740256 70
8239174958981288287871086854544532540209697084791868842052292001705072972967932658463 06
8562833147042334538695572625766051042724373114325266198975030256805540836952393034984 94
9986186101863548178496989647464654054533455224782232570519452084033813008323241495092 62
8449991373598119185135455949778131853669837284570956491319112710511762335104524937646 61
2828754929785503570898291835506559099940645129134424297625106875750141222666727820990 53
2920640329394677161574371892037516463974807411927032461468405352164597848596950169557 36
3817251928304149447007216807199153178892731694153798382602763522845321274320961526897 91
9637530818073455503823451324567226375799306360624684198339046351275512463995911227060 86
9207723911630509999543713417027236841114621396567191184691387084972968691246674552241 63
8084178017972198928100786754359856425423531798447159283299523221424673437210888007313 21

2052824367911871699020275408746229526719625693091211379152065577003007139459731877788183
3004368294293826784038446732083740946165303211947753515977276501000496030095271339087425
0022270783413892730119908963302533138402252429502759955753930704498915485529663547271355
2016719608790229812002592642917721367296466842995340328864450160845220622752780710166295
5344609093440516096610964733422590564818090006423609764993540420647962466952017425811734
4555545428018573245004966342366895617972281895367986821197049490516267301320464817404275
4680522363999699975031263043580856913855945411810199905352153980503074181665827236126396
5726814009781823760129907469028532969192123397209243944432302866049012206502385494231427
3794991209412593284162584844845852837905796233813729524655023598948004906157721703928718
6886163519856261407826385323050136297324487498993154023937429769026833397322153192975029
5994345917156770541415894535292880031835036500892821275532997717377081187575559295669331
9224591851300746694770160710002481788310100837315602899783810922951754329636118176429332
7157624873181475593432875218936543906028014758427229981307989530295196315340211643238473
9065990327656478076199715127587748253998520694903013977581273478921347880101819942629164
5714600012025737110311267009013255317445178535649611148687269544806994771503581519053025
9187646391750256633370044964743848761795128950343426437521040926737252705603484551340496
0118192964649233393082929351351981383938586950868951952249492490636077368055062425404657
1878344360247149375580726130569743795530451090937109961535188564045368507647709614005658
2229655383445122437768087299078121610177161592548065516432893118861766315943408491668958
6802842459905129067546557135568153302939489956588200533779217351618430748642880856904289
1679625423312477710106625662722276781819300352018505158930726173141719443975668527106981
7687419728233164156082905058852075204410788245904324878555449250562214365244858189517582
9450735558652133352018412518772322635431652846678466833867441370421184064270282860245162
4902974322130777077572078834869304346327575987400471995511553150056302285204940577370722
0023968331276509540247364712825501886063820005474317897553335592511137710269907056318988
8809596858742902619750602153900075689201998808662144297220703194824160560390489099665626
1260821066773806603230735034691160751748887785963735808110628576477428694524565066294571
6380450871434030527045154581914531965140465132091458834528553405523903999039928116346052
6643979416510489961074209416634928240166224554495032910418342848788886452083947763661307
6382272116715660994169092878429923436161612469962483647976764048627063753219322296892698
6660143883238535552325827406312011452489729692395688734335024519119573028467089261150664
8650060739817975903610217744783249173027774628608917635228074609735212238170095685471311
2645355739997758791484040081448728084958416300660030583304512659654291868749588803185833
9149270705125979716939488141899343956724403607883931297567237576204454530936898982359523
1402676956829972481902658120334346350005435589196211707580014021808115138782466382933366
6272037518349937214079365300339365056101170016715169568865228197507551423361351377326597
3694435470645114447567178277785109708059416786452103805885857572926263677345066397774127
4755525993081128912124314122467603911203920332192225940754017866760068536697086046478639
9339839157716658318660682574440640805623849068098120834656263118480992369810729964463233
3826720364093242097705778971141013457796412039792017003365179043268935882406522479830742
2092600437689453583686409637807887487364275873223229424121140710595126575784176979693660
0463065802638125355467413642784161285086328573228120621673078952601517218329333428205637
7782480910734132459187275695105884120173568416073092356796115837204487479896650078354545
1527639675252957687531912589914664678263831523308617702281185662487751756374297129717268
8716430754826741898731702776193391842587270000991972812592573573987986784535063536209097
7881504442460467489380383146753777104386953683776112171195232136335184601818915461287141
1790353352302460563132330388934955260175937259264283398800733298750167138462546932978164
4837190092862452624019925988905931154175404789570226471933089876863087151379057029462639
9226885475541148960946173947913004167650252850399401258625729147883843323267952384405299
9112268844281934236124890913749609744124707241445113438474639356985998236560555964817261
1162824113886489539264169040092479762559380566717000538364297558108974704388505979410977
7385039209523054402048310046866613473381083443033726682553562544792331257852504543549694
4070041966014154553570284399520715005356335153779165000637950601989758521326956299390368
8466976085308639556824300302172194498818589399208508746696162697178267273506891965659738
8492696624040540681245457212193504439594114887872779410201196571346434119471025846286399
9448965962207828891908458931378402572078059036769566594378801470942356018688987926879396
6094714711735260313680024773152114734169643751483528047995813510777772539836589349083246
6691885217805197849240891941410476294027252381830294909569144717993754950500134971978268
8002930167552295302149408500411127694965703099512908701453400216295615761363737676139365
5882445803998835330167623269076046383969055299646672788670850594676011495075028933635875
5406137392586058815927084435893142498878204640063428605466116741289724119666336845101890
0244971060074292213016376392612492890430302748431042638539584788565852473215998535672237
7450989491232623025362439613781151319137999285890991809844876287360602477370126415763500
0935583053177016271723883595316125467137830079070540597163878726303914387010438732932096
6797332918220495230762931108733609207725277077237463889387257668116356274977663099981454
4562183185087283234336065377846204883832301333641592011640141301706890366252892222975731
1863308583649800986429278141772790761608573556616897227545111082076138726711253024919780
0317913635039512597164080540080867490782233386482987519078212150479773578943307525973129
9489798839543335848987026338475972596039557374930995042720074029654351137666025767133518
8721910807231805906117407804673242255985747464837941436347171623414398078081836071469681
1656865997808738213462511305732434262038497043800321547898630324853136836599931501961409
9344344640255984949662416500428435521506302714326967558409643466452977839293505943657684
4494836412718117742602646422636544048368890129432253308939758869189989739684328120401890
0055719029941449000982095699506835904133507490725984895081442619307470736395901161388543
3474397773732200919463502642895422375807421892163584749773169884877371723191532443596408
8773532940249350864199110364605698366798128792782259797621306668768255916296379874392307
7453119556460909806717435360801283532139791198042725712086962688878065038234384623486353
0799971016983019964803600741870216126199578924442995892178854297333311008867072215166

pi to two million places

8335762492009010311009976090980665252668644423844360766963859177478915545584020795317750908514445278513237216299637448382505029630324993722882359346853526869221215924140393019846369925873307535506494682874918013752324158150202284358292260562114956355740887409290482741136799606171555958444370951770805462539281514382463620364152941036627107381598069667565831358303265265873952775474579068017107396422361952821997921238565709822653718664175929739062965443517651450456535311170551917772236905999512092895459369385043771890083405436321326502790870136700137979498250205731747110470809882353049933341189832923500980217128133576710403608443157409781630642796584166729369000992942296513870287218933808446775579689082317060020286413437769907522634067146042257895238967752243167851332688923857695967024166921242002257839357347073636648282522709983766617598497000262322746868821625811903903796371294540158053140694425595306597684989996491339634750875021799075292429554159989967980386846670537437791372320047340481723877009632430288636001475123687400449802812379561212954427779610501966039746177863488749999214496262911576362323842484377742333145953477189810641701054204022039126124978066279631349595857354798232841749895463986074052063457608076331040997974605551419591441656305749158131433462476078236134202851213974151039718022960719974292445974503767832844774925115530471772864381576516416978332765351889005324499233883587798585785799511654211027728412901805051138076515027191432575007985838143649003976017533986602879688899414915866071265204370969410382719718507246502487395694974028957480303307368236442761495313955040642763960333304537937666178287844462706361063634540758895695296914651549978836805091833098901122147429330884266916446617942094246390920832358461376811701866879521302963763642951792802454258806597811429068218997214045799354069439352786565875643221869326769301291916805536723179528482066339473120901823684316724109090183209486917205713293347184883798199152157522223999850828937114069401164946665866574292713380571248865401112843215444016157638367474526198326213060481923415890885780604563543149196205894820039406577897668913485016442834916328199981571237193237334667268173442642295400761546514900935380972814486586029144945034409364312285242282715320715489332561425803365976222603062400682146535872818912552412428749834180846658124690205856941002826742845237106679505135444450508475704203325106189065530185450648542348634765841680405042497172672586938510161987057592986877017434955242275428899646051462174951754360104163106481620779830946914067692508669924819655700097016987132992740435400693307197653607951235271219855197597592586832219597999842043377794390796718982099716369971360059554570070482563925322024202253703277948425807110946160303302640998648738079707526581332992684913153731759571319772627636701126319957477655537441989279915981222403701867180563280451455644379573415064733122440354479644098389467719564515140834811208010961936353116931766476983351321258307018974365835628419755801127238662324536982269889450832431747272282468638111728386003408756357338265896053910796010517471595353817231181968035720526598986363808759857553668673381364060703412955867006013957613092913318519621924845950374659141095447582664578402990821064801881078629969629212030696813642491571734892627963242246599481674755189542214953076863524945601267070765196612433933072579398557462843829648201272383120090454623373954912410617288863008875674571369442350473431981486341267148467441368333953004133010405298330356095626197046658729597456972353953004345020616216371114662775061554462852380245423522584657460138352340713941302161373462278452833251895410676638336062815622657812492325951730515377903413161150354248764649981401989459213702888302579451504593398381508677581835086885953680315978322130052248089528495040995854881228645819245214537307169632466632625256358192249780161995147341596653170252071669287901868944710127913188620863137053225300187674887214671375236510526714169739623848466498413964695601300104379127674249895596852390386297882338958812745644978020790330648989255566052139581550984399252044420055599537993947281024008824735280009750667240789309900475472545197096283213522235440963730117768320754744921675580572558803527378383640791500064203600907654373835155092217896679782972558423793996167163621030460946649329239207196550105029102338128244352244061176369194702694195205522106663836290681595556748631026451327181852602799355909752989677394347757627791597649496432925704953208383569733223362871607563151068565932241420954305523997757279678962042258064324571762713856944342681341299763795092230786989656036876519219949802416497956433058013628668303022192259043031088533518666998868974659713140191860159836503470452011384131314084441564639345043614400603648107963105333582048004252368141609258227494824280132338014366159664050116729114647050324746678058895225301430379816485985099620300239560420245007679066161312557713689399461003247277162091102345988988259345791963993992783963065875800219877424585287264861591191253936881656058971502744914963296582671450636621053688454555634051763762905698215851139545676155977734356408232668335362375968718885145786072771026383527800333008562478354275331511797820889953426686226868888626373649532530738522888984477612040424848579515315377216535054651258730994462444376647425233461112362039275095970939223504370991630255731943552128088412365866558049839266675468036864904472588182017587045811684859268208771427888962904841897719923823178081430690262544411351627084867954073723118004151578052428696551682663969695101155411137971096019572631955333901475623027703281961961673959805569694971974324736451733216999378041000469725909205542500616427926748160388620409449873642971099457369720288998647349847072956855292742343245953292087359712568933406934694464652876093705954622046553819640754725399027039905933106417792453034689400126988065805733760536811515786966208447060220708662946323948280395593129789021212883891711804938479655148315344857990684599355760056760005938717010802646389648701090143973633445535550205709160058165835889872182240164935198531823397354716174563584641128449959494432531281396441454205083859915727086502350282131240684366646886933082293197929721709884100170825680526787171853488140779097001605382308118504613385927588837373776825471713339734496728957101159251241834214422855585524000856176444723714831333673612418040552398251681545646381737699854950560327787702692694657094172899539809293596578692339305452843243065283208506083587484943467391432727253967219272443702371134507750768456279740048153420065817414821521991163015602771552956240115264627064398532088247542604581340879191849396896813392910165892692403445606084097770434511329458547736701164779796529814230304276504332466387448190269911437412335322975057890774

6062443084345209128133078144970081700404838944210929935109846762481049587363002158243O4
7652840790672106063248660616872596146470307950911427294824069959783240264066471572539O3
5219323302575670144078821134820500394028633617957921331498404348004722420253895397792 77
7043757610413580129539377089303048684679723283510913029037050685130393462774364267409 72
8717526582583759439421494122420615768640982475431284162969640178661730852682555404297 79
6734317446858146538276463035168609572052671007625113094409423223741552689925852804120 52
1132150553439055144495407649738541466780136399830050098266715740453186288435953080125 97
3024985192779752569495432393595393258587235367601096814123511403656440502037745483326 97
8423675634948324516101576975175934965318164917195872594182485019826976657996508877028 86
8584122022193994739662682348999591244119451983426808651354864539172096848529223826573 63
3274899092673419408197068196342040300693179496756714032347291719068721279253069939962 78
1242682760502085146310573243863254662787271307025245976501139958031275323814332947048 99
7802687259554406464487765841060729503512981744935490255187441358690444710528827989728 24
0620895130446374591055008736422475309825556697932268930127615804099906940246160936266 30
5723213154895292401944446467039926913133684220807214954727292895582203537622435118874 18
1262406246155707973055562723115024788276057101354713325638964118191813556232397546587 25
9511370331596450892405328550963496338730686571247706417983579313270353081389299488868 92
0067133309927423863771100403755143053830836863793006757603329719203023990281261576815 43
3800373203459314545696890785766893508557576647501856653724370080954020641847572904978 36
4006788708311389148886932641592121851721463867041588565610820315916453296809078343604 35
0770126013712935581200500161522831649557894273731623083777580395267278946455425298790 86
3327835463908163854898436501827411419181669353324684858054586031219415225018388910885 85
8273370031543978358488247970298635798132962537728841757228069166778790028597498377490 20
6051831823713016333454411959340377323750266199070578531414077788017259056455591375850 34
0397816948761767993917160428408694006576639780804535250349589134413229776342541893783 12
3423854767687643731459769587833348596936912023547757281273053256343086168616744611989 15
0558285399312460928270397335281893733597067212456822383289559723910485536652897573993 12
2180733017728153130157546842105629150197309179417745503327391776540853359751695939659 85
2281874088719339000570731792622594419902051556019308192544228287759785102394078346390 42
5096892042208425650955061211051822456355176775791262547126915640016913330215809979587 28
1570101984530916775572416236060109821010300273941917378600914669585714659190513318666 63
6214608939836775003157204264302891805860761018282017891365529110573900161099615740678 21
4967506848605810950103932527047498474562589348951916720448262209787621496186565507347 77
8348777245278873978491472126029161694317697150551050491919455981063931508035600753736 58
4182420721114797128218001900346348344023939268955874689236880089123770520718409592583 66
0835920555953895647672231371353400922641035331702686046771221284192001449854140957063 10
2407189356080303115911611605084535722431724545178478679346181229321201923124751185608 37
5788946854501955234729967879028748922407313185945837101878015439081185771875259967525 36
3588788544889430082596599907593518572238525036169746840295491186498650261573341080085 62
1548420840780312996932020212250619096081831360346165173073940477390906452841229274626 84
5913699420727875490256643737119312667076849362378463810157844172758778412243607182766 72
5105924528599558065889826882125390299649916766964956977994506562594746895273228922021 43
7901358957097990376810596631864246097216349116158794051245173839049729442477651226337 09
4604853604866535207902889531131122455426744665158434759313768897133386762642366573669 79
7657336359688570007296194358434936837889184599290751415317976949580494697564736327998 66
0262570914383517423201623422944706728647655711221548608147201952637372891710884420496 94
2290477555559018969496850949980916935997764789604506996537337597407311938552795329969 61
8930010758835334760200314110562262628157920965556321989635256514923468323750744765550 54
4809589075458530956553446205441683717542220287986003761580958237845395940428247506557 24
5006694362123461416235731469303670887587958875837406075120582553425944785580797818069 10
4612841905112236276602140228990343919784153044132971690534735418773667884013932398721 77
9961916869645910145985955465539150112726556753302539404795084807192999560827512041802 13
6925022437416911878246776406492213562451011579003562017495204793834798372793081899551 82
3228136133315966860823825736466901668794276708698016187647915621441511794177752503676 70
2559554051620709498021323464648213413706500261120449993210714106167091276840981268676 71
0075441637367343853127277565859618291277966809554285020663614811753359323868574908739 86
8915779054766257987190384076143799858125430315562877378641092008299691451612478588629 67
6703978456527162943027940854046200037348768753973979052142164109710928324069064518210 30
3249643238001081572620755246374885779097742547335913379335328120991078258024076502512 49
4267012561546712726755215043682674162356361828598796281150782216200721683910021961766 38
1377432479337621046305280762808772973883110231408239019284931213034843947514971584267 75
2800873213355449945154823377373608465292149530744368027901303024841087187322658932387 88
1664336874044405270458761321449829549959726942500358494917016297805570225616749090934 92
3319790038121194241402331582468111991505833708495771204132014706829087083801101590876 36
6050021953681685309692339412353670666906220759826609590021700853079804265833278014142 75
5278599156992517012049128553861666199248072048190985765805617004654917248569523707983 81
6284801985460788105661967436921977464961979380539696345965887301987727349889569937267 37
9477435081242392569239408213412701092023546929064312500648221994550006756631887050708 07
2501933826760744941822225176833170748179897171069954456450723294741900471969959117757 65
3110292478258540048283945450630052395171766098804442295770853868322357959642406075477 74
8269244357821540690035329045823686402544703988394295413001855732215977638440904030308 57
0900456372121063067007081615817989138852158799802580781408279645000620039353485662660 46
7865190967459592498548358808844568563609073175197658760501385203721564456684381460512 18
4142319785529395494811894434667441237305396481665962299807814968580917629551204356807 72
8694623618297310802986365403240812457564745101429796302023020311825338361193429584363 77
2087838203953247856451803666333243333826959374050072203154183092584141859362138473083 69
8289810879273824824459670531093480177048859131097874753455381709082143374170498642601 13
5023843486843872105359841873507807067680415815642944171981214009223478743950581031848 98

8706168172767116946863328795087374076787633819191673875694293934638383121430534268382134
4409140062053455543421961011422859429678688417059270583605540532941111655479558520979865
5548720699012146304961731903088084720272956452257255689762473387326292727232301155867387
7445776940297182470306416038372839499742585900356679589293386178716133729481143673913934
4434014238498348294773709144030833420609747896242073415460245891398533490416049670728057
3732654709534055134824701054572334423602490658850103689007435190220631073951748236828149
9434616934394064527426779601203417836990109828559630572690383401580754997616397518038874
0429874852046293516824056154863263134968535740605900036540215422131951072692964189479854
5417604588840040232440573405386669278020095219691009098941458725213150384072764988224217
7162271991966669052794158793767557147206232531904164664255105403129120898407816595945330
0343837098528539935754641046873344107243415430168529730782840814404868706189473158609610
0223985525129129438428217510467766178167209692652351240806595574130164063434668746337511
1516060585105089494201531548192610949223659801941625499035173063955175655055707026962068
6419969464288004380942508357639887408997929511629782218276480907519305456485030623273286
8092106899048683433565117227835777534310788432148336503831920237199347322039998881151
6889150208030124763628439834207167958634076255765265151755356401233510127326167284598605
7362325949812733033808615093112039861290653748446075011904913811584744185291943816251280
6985061395355277287036417149025736999903254505852716784696471920067375975552864675220
2939525138355782832563192924116987682988542004854186257340670458001865670967023143659871
4280851180117682110319152105902132598264977208543304399197392201562129298144206920295241
3539867641947331410168290159259434541720586509556666485855763118998541552351937056959
5484696709750686415880248795091764785699679232958533617500971543698305927579522631737061
4808248627574677498036187392616073753721545297190606136092101437714645999023120220118389
1511554850096468456340881715935905404441185718284289447593409868978680007809664333655
4954335143315174921085234360326134559340205311476628092289416330973757682439036047139523
9707561489025153340356933556240042248031285543159327048712843154979277779211646154033812
2070009107093502403849914095427192350172255811592289630348327640575227854865984098290
6182361583522118289737543668275679100962584180567286109412183302385811928785590250173843
2831769407310706448587447609474985213969577631932155430428483702990671588545463574144053
8124175176919124498344137480673355665813479365091229887360410839318863577680865355361
5447842447866291169846790592056310869493287751852391188770873357707788758346140521168817
7966891106057224550057306488528457911184406798924691166466853573934854757985208427112539
1488936479746313795413644766266522016034013495204021449262250441077572716441540082197
5552073584185709502584365140648849658632944233786731241182488226839475766558569968287854
1826624288084686736168135823437362019529785050162021708729867000254076932854979933495851
8775512396048284119500490246706405099021472738824034191090520251140026777993345392353
0742901760852742368221906446187428375923828629679596904800495613387610168312456445678090
8647792459929217251063671031217818623155408566326628162806081698284661165969079788801203
6398320498995393823782593208299879084277372170589619942375170104806533632470470018155
4788961949005682323492343680498124595686974891832720129578863716392863674212969228922621
0972578571700387957048184079779974812856854055552246135329033828748479794338897300848018
4853675192909498586293187007834992055433231287330281192381754563624194642615953642055
3188722067269744054461771197930440030160234617193899866193658480957955377294192159083244
4357100284874555945484652626646403412962539986473437324502038187918207202204349746607655
9948974678074612529810916840139790356392744283382805662001738036303760907200329013429
2867435057613732452233987166865019708572614371241676233256100947495974122486239678253427
5851861175483034840965200599567419264860005978975193496066120164854966973188879807910363
4207604563746326627611818211673187158558684280242866637020811962946411074898767551102
6685619729297838327589622699283633961823503953049277054584509354791494195654368812800021
0390672211006760695170462658022402527514963698986200686349489095297025558574500753474145
0266251993375014966531718132406860133069529999025604876236376186227688557654905065782565
3628020421485945517677117983705239990295704998372275123887498490034366559809585113040830
3127378711287692226522101671571012739555358231352243591630936007233569572041044983
9588370196173757083992005852855091999512876993984323716378520901964810853288867605059869
4182635759348944442715684340784775941475966095958070808965916700945868882982610602170515
8225096743803037241490221521133670885232786062747359053235901146463682157696661192511
3044918741552422520546850343533499725343463713765619756115654517827039178608674974339236
5758554482574916504221856081989703459403075660992362812875290806091307597119090919771613
2799595845943040152053746992696829536242951163843285498059062970439752414523975071724
1085032588784087605331645731223622289779185940301210556585842632092374271207443657665395
2341863221350340631338840693283146000589494222125120201218499416640968897232695826051117
7443000940419977792815263713673710726330604946973857996243709883499409933343565187073
0080118425822352711006061299565818796254919816244788738964841749010677835481492536127698
2888405486755634604292013919351773781730452110771884612901445224520683844511684830340426
6879300999357459211764469587896093238662187411093989390920074978927706206093551194604929
4414430973683670478899298986482519618587523807022116966088957041984146466864715083934
0402564735616001063144618262100584227489339484378215064080471414149877478590244573525358
5439469310725437910045267329748974305898521382040464007942685408639332490425695616183259
7352133741127724653833977524273098718968562181502100398670364180442493458584624325216
4970932436260154842504786966605124227241413227839589204640825134307177109326437937552591
0351794293943209120286640716959560566922139594093179691319311197918807175959717837461708
3784582266718641757472739505877607651170564192035855730789951299772691786153179510367
6915860102166873065645694180095711026989571433229955753302160036636283250819108414240310
5149276596842878406292268970234455656644538703439035269686947148292859189723001675329668
5327646887648945006403374203544896667029111803924411530335206266626206152571178255466
3731044275271760616841863559308758113353437580489286834579140901994366837452386828736313
7098253566148258783487444094152445882630620596472368030815094075603977228386958334403767
0805113404206290277271029420583683872592451357411244118087911969923402319551714020566

483785453129104187689817452273665436784965265430670625682944303687601751410402738405849
242130846431218681670178076833115046689114994223977891829408933618457541110172900300575
424498408854626521102859155245114427761088341369147465021625196484362829321539008789356
430956462496394644371108788728690892068385921832594227850099114671759808838948537413445
638404647308174937876343869973523573205755974077331703156658216285601001838265895084136
682979124331982111620849009868532838669595909206313205849359378956860269585837132396496
463461682022764518113407436698307854784645728279932853002766203359461173164434001106586
360409790175024860578173110046385145691258770347570273061914319433992927318333650356515
495395851232066037114057905422242938674071194552986236097248670343871747710462259015901
530087452839262024207757733507461099956394246004267946651710992373947097895797959981459
405260783264128417421486988697853001835155705559941604815952264907317407433589469379894
745637323074245055323203477266319547345978199017901940231883181371580507495932696891870
480539874677121483738599030036653927841318154339429665270627671816166491787283054908803
470836973442509301133563156995584867574547132408436950703563315066724067577355528109187
427321640711194903730530423748120237050009768797552688366841458072107717898311713907244
461116436996655708976003612591264945781728422255858244375977648016443550622848387960746
919515752057247257442123314625828237530706735184120110776383842877195542942675237280111
837471579064171479813493584556196805323728236278584373433870878038995303974902851921152
674958717705369808911544126218206537802423970448936143899253465826690067866922510535711
846993727915470313626028796252012993880087324879208749286665064265127273486379886161739
243824914860345669416348973301878770258866370332284935401157588094101469590727918515152
947227520969806340143912438396977161695762263277703100642700493244345540646915282902070
510375940715383243465051422358355499154171985815897045011822527406307832198823126713400
5760893634229664927217962703403145383351915274237861446767776855520854806191395508145983
545076848291112415608503458063185735924074822232620693813478398784091975182942305043933
325395843340305247264393917809390704384351389345213500176967812453766852980525276178733
494814245111189583124523068546193215476398839498142929243528809828188366614268731891802
050866328480014848810169248411304936947524914433191799560957461988532197925874704824891
556015833811089240846382065403636629346141461341730708445166150389166680221839012879137
871656606809543342457762879991640779442215059737836090389375081579016186113609748770811
049444230479014038544272965937140727188613483228719341898783436249523616053168672401250
516002088545107535618824172013791850109484059802128659634387753932220028861543179779772
305151254531408849669098778107392006670138937349004681281395465809828380759688414054404
707163564309998615957637196792733538466398239773819756142401951120654053354528476529594
782453526611304502198268397550360014222124523129146132003176108588145343642365335756462
369497435619460041519225015719840478810152462270814824795873797077191961132013089124647
143686327710389291366811108756944532843472534843022633755345746946811208640735581212857
126394760377908881599077479634605052698357090651427311629868630883018816849018106572383
017977801938939451317186265365638718839382011056016198391723039019647545587016970017294
217784356924359620637635547371759780776267106828695597227103313595321347860665657728259
157190400178151904236985466970368444390033688655759978130456407232936288792397359687172
468581321049206901055249963902601944560212841530954880852007999259135115717060011641239
491700853281033392324830842915262668900902138262921386644065213690516503832265630559436
608262577100282580411455514372223475269562835572274034338359879800113203888713369668139
114729907063178057276321846386276022624624823618840168096630231487693544668164190226665
544124845739571267755503727271431358654519363916270566004163035241614114906237667736959
698849216336930526965363783555188747664054148948192155339129547878781923520847472134141
974448080930463677647825215581150320240971376012634645072882451510165564406853137525261
196734675156779779681941110135392540105434858685840126370366971720746582951797422212235
398551316014847139409373559541840624878162059703278546055942831590279285787934372777395
790274094456130832603725972861242162604300483253687349655093734426030557194560705912475
149752219275741588358415347464680650865226913049331118542345836791977683139970770443146
167646802474121637255152966782752663869481160324699102355845809193789262226305460680382
109361575725175297048569871319907541166601929409694900546676431322329889547622014033466
819952081354422950285305903781967917951707561428196156889987867062155191998010997419509
727533469313855230422002628300068855819908819833299623408155023687470969156760486058934
314761462888404538775033180125618710599842632123049806123075347761021170050034307299761
955465438718953648163727738171200444905524388694103097515725535870038881138312413420557
783484417674868466405980539468129193325075874258615743782608623221389887404695863693935
591823505582571561628361672209446056025288192524759518440749194887994353304018969354741
675673563352991983626517340735658716032578054208018675636248401335411918364152323968859
735744099719270997935063477864511233346318229037521682204874114270791544001756799265718
284711803778810511210025120785427029031976214059955544874107308975456238175173281896318
105186996425959991458969981143417710430064796372451776436792681647082582257824955425808
804641578631534307079171010044260011177302294109950414326445621115567536603778149970481
018841066696902349062340132154858104595820172205510528095549765182819394313318080495635
096549542624367479457127198541521804467657814338763617872452528815288263386421826949947
962565052446766344120278453394482844725436698922381984680503326008775954382564976465006
2456741751417592035292308909804533253621806145164052245028278333134740396934846548290363
186533521647268045983689253542320036607578811275627712406119152385791788110663990113418
128450783839841812823367118376470800015724269167905520883205103394037913478191136247129
385537688924156809270247686197706364490766564017474227271696923367335870264708414848490
393539765720229774668426407028219190847982241247945316614310054980059052392887667058336
885124822061495347109840810441862825548055322346934748149596133427068041169121053360630
277921842498216420331065436101517853284471538459027247166289511861993040423234597264399
090213531154152325460832991224204247775219478162269037424047748934439810630357712838009
549153080840870803663837350997206705454441213029591917742629939514995307237270170415364
196528927203728336860329061481527742896236191048328350414062420407000249939225745024617

pi to two million places

1793294082079283531936652694072559445990998069688288270508443992466325749946007920429864494653239738612041795052124273217380630600117853767871690995689966747341307526431677601722337977244988154341024114137565109817995605466773246614749217310370303939838772270261165548783462578741333545690827385304805327381455856313999012947990806051240096850286470677304592093852239392125935558028050420317822161808740658069730589583588998934463507612474166174975562415579901041030856355634394540462651808981519987249969607807018299918085627715482759321826665405971152225274306977589201565352480641613494845071130273754064195830231773206458223223620853240066718642598656259790392387998018943333654872833174577448069368385731267768440595744709696386596516757494819256737925970097104393452223961584944125667615750177592337084182732297736148715913662958905855903764111538825344391673731985982194663706780179518514116029041960764455470034525143230264792914949486714927007714171177974324896072504843263512081573503575823559391894560611472942827396896938048783016110990218940902027408851588085536576456795140751876863213115009985389056127901440785141837883969799525899524657046806407839328150557151722333976935965845275062384841492677767870398260459268691266896358224532947039146622568228028515614781934535819489451037674620279837317420710085587725985978634035344660416210235720186014310372753074468413012629933508685097091927266734836791840414625178258520417704476129024053361235846599390815227700436562256900427095161805552571921484337895074558983692855087588330341713376251146142907521790170964780818001936818079111426299247288212427320298839301844769076139335259040773283998124400806893900303051795208450494217073756743383108221111105798772971078172509390762116919928587844757451314437465702932713500775691594495595619249896155774434352980642102738555454228973550160042230199079059834770260312326743054435128975807847935694771711712099933678295626026670901461301468153899786834554952224077877567437994233136610052847502811624563293509946084212401695871899861793363457417338654616595676528250272556068729106828998874866533358714347907215554260024009634183502707281918158909406842080934128035109653790142151455912960306023624426067981290176758826998967089472406902109780447381983925751068605694954266468567390799062964527790529898994068972785530461507394895685379582431192073588366251916032517109159237880369200880308711452904348799994816038757056861460508893593631957700469187435558165055256567576960758152571455617427042933864392412845527625283522937410224083086925277261582163552843318812457170535563348195387039344813398865378439114825329456246208404919559703732584302023763275056042105532959397113296352734062815851476328611810541050806641717720894702618401160241604644168955755026295170989584093460438010741648497794547576830255008566249740912220904091272744365982663056479916582656086657826935847565880365205027113951023945009044169044921452981721312759950146895168832485878826626466625727383809995909013382167690879045502215140429850901770394073896628872803234961445274047509385375284900517625529048770748613439087070426923773444136336408014307049899147909822412680098799347563799952814493598024339295285920502595216137212334981113321685049523840848654904540227629044002331195365320970490311108838252614826661229259460812612786311212066740138468073747385942644839578919526646790622030864277330423204355035247178336246350884078753316915656038432373384339124348491882711655958898062113609478038350105173132503835152381757442849359940804918154498708977966652044956769020263469747167784445952471085204961141666990280936381160093690702884275175672550179162666630613015153214506717600325882151824590253417114229755754182949010938030766207971073346439456404248165265901578529081034058686940050247611312229304847150088743938402446680940138322917778751385613070512325734857189851674583822835593806557717929565223444330439896883333264153411574447990080213927599005188660717665167330588760386702357537008109611396841111867676552704300353827753029948522102931363848369908894274631642533087752872614110911908338302678696061835834304738196623213267356175998351813753026203786794179701281387673024921475248757626418700145341476634226101897310657776173089499699088320514839714492551256083472041322731937943764837165615394155959092012941491686144806310708740402385988034973067873134040563961752863935218724297427246023641126996521768978574040847458816526711953473379492717079903080539035238782526108167857841167296506619886237810999388630959492105611275536481936688457969615646771298602214771720791797341781092165286527497275916937072234235160282265783892383337550225425409707981844708857895521315009096802056465389513969139508657595621907109459064207461541338450484837633831729032552788806468316582564823905206771819026022300832112046908234303167348657043031074814546936712614022978913084700345349618129327221473659630733815288335598959818632269483194201444985968686842250037916208152966777616340177891067586099813622257999843933732890363830694739653996882250096917038403914227115819918160808899568056062674929155882566746659901093272022813463335699944133483094651460860657719581486605377431595567472594957897048465655237392419086964156190715920754765808285688810902093158471847095039776170891545453835136896041638117123779830898884549181879958053283126899609075151619571611382711413310339017481105416766362341746296205723980765564299619819417344578728953393240620209408593693984599159719224217709675141407296769430422145076865686780277545734289197991884946081164360584461923156794888627149414154262837492784947779961026504073919496557412987534374219475414863024474551459665274992057295883521561596431189541681492595699032944841410727963613986421310637639812590490324094980384223808051051305655596548685487810851892940992814419469970695872339226984996192986380998978964659730501380948918228816446664009048528315868437197528977503266153253853492210378164318224791565844870044750552405805403935508387191133532612769582363443121704953545983000916383163557177923138637157409138604351645146468094965923975007335161771358550034453848077869341617191062611367846793195556531189565736895437128586603474681819069431551465260817262313031424299002065097676347508651504535210467209433704811631643669362241245452790293847979933564676901622677802949234274101879416294890193725996941174500184610907852942425397985063835745460701605268604329778043581679366178262110225273614044289955510253733447790106987554947124500439617353514003809637002332126819997294002931607537513818065205769560449835415848485806062310331902720540679806763449316957481750509833251783371522848714641945211225622428204591544241353188249033128085801758543976968631827764942423506222337791638498850562636068110523709322088028786183391231243618508222196324522095

pi to two million places

3471161338275056582792548717048003453834074103740000648403139662266781048026318700486610951637952039605576961772228522387163270888991259699482764249430844810281012985407286010769889233194315193635427908658424859255789860195309629233840821010250400665054854194593019769503699102554123252672507979902753816457650962747392543291700977634195342572404665417094153193610311066691360074753291201235886897175974505658077457385911477112446378099811248634933261056646411158782190940696875451098440801371441905507663060267691194317546832689977117015666147285464153345621555343225014884651423977159161753063706426493592325850186431363810520212064656767676222556142507404495323218500966064080406573588030783039614442244782665001410643934488582775365713536838158567552951657896730870596754071439874248981463765967726925425278613311895263001407323521619391128601672078136651285930209512437250688985402083991602670070529631904389664399868561990967245608199043175785266946738413781911925274110609796604169388491558905717133574916918530794179817659547167104235506749346035575679261546980194211627129322512131860668139640419054814735856351708273531991197979046460452706631989074522137420364828053744958686283371483541771332674101946959601335948560994379510253138773920502993139220082739025758139144454648179722269089089903461086197883088451891718920350009262013278202583266794186019096732053551727113956436606071711386227354156983555862687167875077968444707469662440340224829946283210349232997527973806853049798357661302427003304872469157477058031755088783527829658779401347069782910653622174793741254849352935665762767209530156982562841213944347044152776401659848324506717070926928575794511461995424954047204142815474600066249143517249544334352870804543756381816097269543766917767385598203450523212716126297894826079346583005479098893631784055718886301881618599758487628046580334399594582017239151592838838794269244672055696851409267003983064213866355837875523315917916613075319188632450573731406559894541617181246703084518985110142258763149952308549828786925608933007695716702026534319776408194360332937225160789741998509542622005438312622100554949635171970934449361986305698362532584500907342945858114003945155576071086091704857740856183534621087703392521233434370133424119552066402407925354823715679218353537115220107588411672620106691972303488400254482165291177508116265046296819988326661346850153044385132055787844478370178546147040546904301889058912760256919505561258278875186847250097623548708481074396039321983834524196428408409317393080276797841773030175962798291761601954465669650850386612359772189807228882405916184848418300710380467486809598296355648357799427179512691983597842176480543811772186783801176233654133188656064713109793938075148838721809910098426230726582014931176652512571004218346400449599050851204240990011874417213655229792888434338842084307723707010716280617749546157624728717174038162500053504060090681053087986634618278179457362934451223939192817004288285758485732239832977352296841651662403883931023670978745376540104355075816513421207952282536560381234188861021702211074713442639967236348671463649589653421137077843669491999624764863608214444549077949720318264902062278980941713150458794817075804141683959031582907299314216270417504145006073517307881190054121794837941052700839465792082480629789226354782831680894955591095895452841537518793536779217212644205484439632350085336096960616984279645619096027894957762232438202885459373855204179161451244877079221705053447742685474896476104516745010784063080915681212832032703001578148986910682910185045747350271343877548578304782891626052503554376781358066306614074952192943429702520915240503117955763244944115411163891626885071185213408018133521057911914527804166878949324517483935719738518100188096238475924888832326892627849767166375539836561387163864762964461200153308154260864927001108405070476942415424163738849073811804357534975064622601237770440238022441318985855325323523163605956260141181746688874144666351225163365868939007846719185982013720688940050560910259614001752418207786856857919623460744379616199718119747235688441584126476907407003101125837270777811924744524155986651343801015606584942821629974269384543004322136336545965641225894899228909234649794470382417756079662127129255424687364167371924226330700144509734915906721721849831738879260727592752357328604716093572902230017379032138145558223229394055775449753064550562008300264895543239861882517255547936306603108772707676456721816425712207659386426228914042973574911732873025849326050598093142262097125968943380457914871038062407672328312729159423949614366927674782845922407114866903841764238273968702917235214914423403377561853214816900512035950919554875852580908211012109720773989747212199047879679487800082431869095350323364908278969116223932190633318546070347267079116016771071503520040262380136305691651916962110405743316887789310776895391289783161431613964459459069379398044471294735029571958429549299355270960498773980241780481208557089258390099780736920334031145972238320480555564543909199498480627054493553336621219300804451558072272871573697138912292574414787384953822172617265870882214720182006364087307744500492619492409374899121109468242759852677804023056243113613374002423200106783941232089798241997573170556522466481180768813199626736986166640179432359990419931087787422371726815206726903940850174752375603972553058042449903739074598344692231049861772313941838071345413834954565141599606384217557088785210286478161642433846160410200960596868239894173678946834663598605481593469605137418127748981342292848886168461776206398080124503486775320517662703897798592536115146153106578340087672191169694471736864230447776278699355587101845999124680708497503641911557880305388830545767004366766389610511736802591617569453692560805687297985118287432244889454848835746494987952124102660873513960191094452660865779294404323384393169016481295034053483651803814889321767904077478050192809907604546213483556743588674965418272238536716094299172888495492802328412251963832462594638196667875450866112799847385633482676329278271184852094021580612862018159168878822928190311807419301301441155028549579667321803184741856558672183023561943111177581303165289124520514325966923899898902451549936287369991698065163667758876478451791140836354063847457470059736105174683499077795282319415059766445601775609878691809978407351225831957573582676101327937365778841093117624355106684231634859179744731806390074952867449036102253025083113995473087893839509674341653712327039363151667629912250314140120034413322580876740770724469276806592699681367911973423281643018081973501498660540832327871517641151816915778798607107638082924982746189017394347708728333105806580877255649309881532127497435161388546736165786830859780091755967853604285071982507434977950396421575971455196834

979296607059550825540546157334344060299937007516712781676909189197850938136687947050507
120236108507592245045239758411125936821812143368196131842914929855048302737676072560946
988985762766843069258908819635104262091134346526778537388758566066508292909601872883657
909037816181010455809924548032153515664840858425467507129896902720853806994583114420328
448158158765046632345676949989727124478275013040474670689810593744587859257839014551004
336481587567809594762293283034489233972378085879769095734425858370955845957296283881282
605625338116422633902987649285352848149685516132386716652476901772946080526661628272920
683813698946391137242894254062205986700338599866673862686363390920244042918870117815433
163616384257234178460865803372939091428634056072575619716659883339522354537259556363331
734114860386697640303734266495470302120035055170392780909639566469829733078634698295066
802795219628732545081764442691314632972128749772391506353872112282597396414507619010435
446723961988971383070520084792059056902491153495144825526441778359540914454855989907047
458613696579285016037456291046236116573066764570440250770410751855216155677070975753490
865286496984655751461263451632323697334650433395017569584940223115545606643420585359479
849141696387549905549615494061878025958141277670520873050209472820618271344371769241420
798003584495413558554644205336977166457513596078993767717747323781189113109706568508040
143895995249541057661876354865383405790055159728007918919332545246737737505308728350215
357540339331259192958998096456452118208526344399307955873111696147212597175736940263688
220233048982638597227391294683235496609622030277788006564558523990144116391157988363121
485765469605253692369802293744683661083469289759487414869585493544441724611792455832644
627398971315532320657304311381630113341663583807805759343465467445625701602773135061497
538975524713113828391038387470221899298143814834151502276738355572749747034475765258950
012526107681972947337518102321507205622192789876581330190341848578427718690455814666813
745402369485554213809784350576376916588580362618188401725574228735883503217898370449388
822500705963555283160368121063262572740169831236045173986472116839706824763354492360279
409554679179158893354915742789859944511513179044711707372694009724355581261606821784244
460636408002519089350375677727988107819448097943716586449450297192146587787399840028841
203280949827083087772151965859643925793731630159141853083437636896470749411291899554656
622547588206226953114881603904316565833098756987915255520632038501098768034046334630304
397161765101868940945225582863651189422338790087619706944929695675247131869710885061432
362239522326718621716046923032075990165538719688264690946667188574240800874529163057952
984328141950744352842601404174961985254939545917863064084944343858282573362327323090033
396559293807339326653816863351114254462890095505278580314750685865257906138667223187289
790028336381704756697110788329667000489709699267876072628655090651729263799959291557953
947463327056765083683157009416697358344928214379096476377564116842694113000037450365437
707227712339837808671247142159731023458370652189087184800724453107204016569388346496384
135591401571505938295887791048687184763009120638352591873392590492299703328060313169596
595567362611136181719787072434947655152093227735082515232717614508720402136617243278191
753051362173214977778295391126153742629220853320300417883442622609455405852224660358775
327356653693994174934984642190627988307626219334234639267337314717271404536401296910218
851131602736667171899360285193579759142181708690561891830360358312462499976865002367842
676274481124306005293244285853168047710338995110952482981192561767649740515605867228818
294264934696927459329904485825351819683302529448099991390059710123134736512681217699455
336105563743394633631714324036996658414647736621950290428580658521741862825999301521460
843891107730017678424820263651271129454112130443063040706997189872182225698064178783731
140818483164347865834700895106158325984553982060018111711376496507579618609234670539805
761668898293313822522035273606796639485108768854795216307094880342598713793421583791474
012502338989107952623406236332065471277819442794028366828693126846093338599423861255685
229528786801737681858309382644967946301673709967989593187593106860568605400014243257435
792505415100286633469066914729927720124522696413335386165270177216124792446103033803752
427298901531233930397458230043469868753890524002887487412245964811428540290437323170365
175161604591284435178321910772301470743623310715919594595805854040884257931188785740641
032099380757862566745164377827178590770043335145439919761542679634921532303305062532440
681740132264523868437463280419079823569354698739092825047002969734553717760858725815620
118989566381420513760314111152717472639204793890508814242033653603643254359515958572498
877773830254802993756508844860125719146654556258101287675840666018571909647269339631082
341449210969575231790190710293864212706173065012705765672424207844501396119812620199718
291713551750256256259490846542282844835327768990079039676424017305442958357961105364237
526815209122981844566774081942596640596217314295724961222464471794208106146544048457641
085494581563075669750412959940346500062985067413497920872329584259068487719978191682427
966850533943139093561590996382209000862243457617965909998562096817753239052860432006401
028761347343828489471826616528867073408363717772497874327018579156431681788304102404691
749112338311265691276530694548825695895248188131886027224227899344422713094326819058630
027840048741356926033722740889360256522941011382537469595727828497143246556836845762892
134674952990123319112512677107582746086594533185468884746766964687629387523242500951927
389495794750677163896042130111237739960419125785891607909976165869149439517837425471277
41333924120947683017086402104058359640889718364385200027523757891432618988941003084155
699519808125609349862022722853406435375675690723136312121208840945104412337814468432852
517366429156697466522740959216632750190511643330684463732478315175835865990751887241991
784130762536941137509136613820158121029115021436521611626089598904896413038914430546647
838759136403582147218643464917855894164206994156605950560673912091702689005561693892545
591234129100437198178628709915829947193085661647773163214184989901397164887866909270091
609781282540677895299294664117890238256172050252789378029735023901586724260707193848006
512383264104491993132076673159222837811053649567936783978648350891619169454051905035027
167422014966612222653952015055933140998119828844520389924918452173561199935281064415865
586073347537556746834316001043204602245206571289894315317834560687888819109328289039616
743681181531780910310225542230317553514132678264082440094263048098023273463543790124638
807653810804330514301658129804764482806989444712169027067198097183665621967942977786033

9943720159200836515447415950096278910846167763203691173610901599705692068267468915339O9
2653974859254580447089516741914090761019847012565139697482914755238336025477113514827O8
6517452193103400681624463974688410932435768030404625136436288299599402305957422200448895
6847075148897368275534229962597473449590717962628174624206058868443854768466457151989996
7303974168036453934957670897169531092348681621708078804747475868799276593521745893171158
1834878999066792475706665317308122340426863912238699807336870838906264005341566904106618
3020374941550168990446377356371293992335370576860290831232461173470045209128875527359904
0964860841349100186198599724454376435803976740052139585207040476175107323131031201295506
7848861533325038625620303862085612569104719598633147612483428611064065984355909698279949
3617126552250626304546516736933946757245300044308084314745395517508169972963322657801142
8610564210392915989736284862035786993101970141815667642396105658160518466414201909325568
3433213359816864378980172363697214938990823956460697982850074006485780044896630306926689
8559320818784257847166196096764604664358604987197232350925353263870472089229920078000048
7552527802254968944256092206259635245248299278578496628633345149561300578795162149284479
4479422878926826675507951169577235231466813583197348014179482090251129574305156126520070
2927243972151050836341006468508273507170011838610993118976214573111170983297054706145574
6965394927650280637720244980800340847670034137001442671147740746832119934318882187988859
4164734938232474073131229487788552256915092957280541877992715003221889586090758586074425
2792022569760325710329453193094060914868153937049838291959217349422542752901831092354498
4833928483169835716967578538156265659100339246768537448014575928008700855603558812779927
1853946952643671631985987322763397606695817620374997182119437915564325714125918277952236
9635666136571748328030898694825503025274033890031018080072593036570290346238378123792270
4049190373760634676481541772569316024289110174045031808353913453413209880921037342704430
0923690464988304960666932784835889161500513370991556495754197421723800345307528319825531
5052029622973346152601664516320355742236173083278031533251557966788808924507846388304697
6751436639707383918942564209846167836486902033305465988738002137912164453019097060444477
5964406855319811107760135314992802046896094961512133817856905686622980127671525992393329
7433401586800384311097249479104498278529856937808095336753535822180457064514797410588842
0119003231726904456942216232608860486518290179750966172739244672585624409428408963188807
0858807720402141900930106575661788781758263081452086397475777812347134361844239310627764
1422946602148562434501319358497350724624426675203987959917785475288556922080615847521111
8596623312118679093196348159577638668245060772268559865932947224947090319547759391161128
3860597883730100092832523990829240705229525133313197582597889610585627574505288625927712
6574008035182093977332194013383043878101054100130063857059073842061422507762531729725567
3737178307086384805375089618315911474383017441106447744282158176525278776059391631106661
5448395555592919264955422589032882447360123627721872587400242070600882973683294846055532
8837062379894462600321074390036528675954061905906070365444646416019384050832316140842203
2847844379263697787413859179467517115160214491841953126024526792592450093236688559544429
8432042036210194955658859072844272550719142964748417059143920521564739761034749130427755
1510068962678521324825216760744986033554290164583918151556854810432488826540955520353369
3262312004990982101753973445017910958130146878155982163346503101660240258638841327173380
7784591801488960607032610317418932140858334487979725677503706542499175915743726395448884
3731425548084559012214332195460981988339794134668726597779440857471202980732982546709940
3075548279765578249930451460918350901419372463733213665629021972212642264797020783106611
7438295416257140816082013919170484278800526774530779333688876442713441166852271447340073
5610310795727954935432325105512261531790359301045345471237243164293767004022534403929998
6654053037286438553180740931469511514009999477980240026215843792032162978113824817483330
4786591517793532178505772777674570863104412785773215125555489867123494972863343716626663
6797252396809426061687182013768419767747989382201299201008238734370299311126880392784478
1019495017155061297607463772587450540314503300049682903385583708458584291280191081817714
5994609035589521118722385963096549029516840763958366684215076074516830774074152091637762
8778205938934609692167428724710338703651328925667966345217325407557440717076556158286692
1043409968809122774437737719504641020375083866204544891541291590571010816942596001129941
3553993553071760883742420594843842259156429157621311817622188423613496240531712898102248
7702614783842556874299519993732739297273799562476505618607860508685828586679481966167755
2201977795187233778434073008636881430984303706568930462639044081479776187728304589359913
7317267182187196205500282416487105719010771364346649781407506435796913764364422201015575
5784587286805884182775425087146053622029173410291661862226670172036403015640376613817757
5713470840030149678761260871669033825029844339842052403572369912256284145720011258216667
0357473674407629231845999255315602281072892805929578394310000469030402377842771673662257
5247131401978431383815348919898214443899611961801786086339770898165789410080782871413358
5923122535470683377098696704365183708428805204391083693821746313198242877984201437196672
1331588095295515258890547482681257941259872896029631696217569374152319717041386917111119
3116277339536225613804891571020748728296149845757923152371071457673264761288808862620067
0942772520015078789662407674333782725015209353887443294401556552599627070774844983675506
0361435359859801040078159399997457426590721664391358648831240596432447543349139617133338
5021381755818440922356660520759097405430839999537598039780141638131060658012215909321119
3736126445030178748246446123983216579696644400737938662050371648175445293356469130422253
9751029837420419453378130004885984551635738091690011159691048635574069818583732729877792
5987650273021138574483186854747596891235744236575820389078216570541717184035300204621174
9100899005805995029244834075446426306030792093271632774951732902767784464607234675238443
6955618849437869680285803394997194123279371131807725414792324494288461997350388113199925
4254841616342159533056714071725696377620278654174208447508690784161076304493710126823308
5569176191330695401942545556892485068851630616615352594729831676735958833567833941666666
9635041906544266303975762945115968225328490732726355958897903699281333072589116424163317
2476980308904637262708235722627282918410523108888166618787374877295432236615174859163323
6459594595584916392378269507812531260013191584237817030877874323187908592164817964313363
9922482314117037023599102997715762116867446580752887252469358684254572426687947372091168

2785577202040274826777334242216194090529903056147254990678340354907380390911634019837333
4561054188974466294705583894731052784286980806490269641423706461119618624849934608521065
7302831252702680532944886061640230175394086190655591251236608099393123246635195109708885
4373099453467899045917965323704719824858228996599066971198420753174820370664618031208975
2160307917512954145856757803885885672655008012939822146026846809211475719390484223533825
6485742932291055152181323140086734039615637026519043446149901508149133301303609367558605
9916875905673881055515488822070681890415107340231980835324730641023014975172736481005655
5862735637825250087175552215440668109444797766815227102627547039429935865510629426627005
0895160338438409386830969136862513621957749005928809429737152660139355249594691330832715
3703101807787385962742634413808751837537409707886698013844306861450500655500168287124115
8510586424639257096032012713633076219297658796823738392963342542675902077681425671453365
1557072077938668075821695682861302381579894057166848090864649343980801435775878339643125
6857236868990525504635020709613404433758462075207146447245285650459343852399469276535505
3187152732664077304944085591722825872713213952920169596948408927479096026447881214210525
8910449375948286287558090098909104208359896072700909237312046924873117524669597110848375
4135411676937319928546048202045333587003180744884684095923264812087238016881239451119885
4482829557246430900396944584192711381349353502316871920553686668433863077559731697104375
7913089886990950326488942682311955751088365388709915006829292299586358455812729592514985
0560020920916291008018647040791617393906103253049829514167136584651038110581751841788705
1863859768547523329120867653056236913222834646056663440422159022723345457377656887029925
1229305325397089012741159786475274360849470192699252789958678464622807728719960180441225
7095567673509976546545334328639881026136439340762644101383044385614568988516470218622885
8117892256956035957580002999286624298674621088928401879148161341585486691023317406502605
5159724762327205354569663877306603291604339087341729146366779107833946329160250672916775
5832840521982108912707682086585876434681500749225191450089488840046900791426702594599095
4159856250396167728279713482991739960990884565078713244492610754616013418520534358748725
6616277556904105595526879407013922247575940635807981868303367698879150273156911036892995
8406487350102559238160913381577482259355969264629066452455878136930122341102556455411715
7980426181995365847463466476970411587791868622097141078263698862702925686668348110364805
7908902088283312318751985365507377698722457915639830441958324832398192823110200887796645
2255363869797187866744841753456639104373741630427787162858621528421572064652606478955585
3687256400535516953508574023118502566927869178966906738021390834815582474914979410606215
0580764193749237004271310782301191695883349905173015751975353732945434641319714604888105
8020193369821360086825889695700387505431646652356169363156398284604434710047587069280695
5430657970564769952170353353821651582087684869551954991250996762060079543935873474641735
9219377713029770449577958277960287400782922920305528694533465940100917490929777376715925
7515669118368254655135858084345129378992129519445397208176656707297570191516390075381525
1804338049901054846257344310968479864812729387828799664722091982419124805149481698614245
5886667745547312652139502973221067565519885079809186839925999533605784896842558429167705
8480993929703524002006993272615427016460417119775276306432511674910259738388109776821915
3300393172313434572117904843484447989883130802896601669710329398931831343789146360577015
0656683929369835796472785038602435884241514014358952954350291584572143682032128178573755
0038070385146284991729283743842324381989906648385607078924503781840812777307865179741445
0698155321473763601867846020525872308738598963520314762753451911693078713640708034949395
0917609698693522176794455314632371203646274101238936180029883407823536104487928535815245
3638084477908599143472943012250541984907903289974881415212232869520237804013635234554385
4647449000675116597635036285006956480501748844025607018554473903056631950202720190691015
8516125139807962135149167328768219847795042472019379484926814473796598432308531317606145
1580382048566768238456446879019367588214775729650583894180597011286883241611059511195165
1009694188907118154692167138259691752785138561641835848097844262867916708491159922056595
0811941351804138410331001741536852360406213686267927496310894147783650125346368214943875
8122003638091469446976215851285892695493927429283514522149294470118376904994187334867065
0211225161293513590740294210433028874240723145483735242500857851065780354216782841654775
4139332769481286589641561133385093646112977830937142737613910647152233370812019516006945
5554514780056520035351576272823767171348479055329542561767426711258003271351883897870165
2733424182143316113747580456559410833409470287574102983572089196440934007244881280753985
1211347760095738890813766215014468544424237684663559469191243583293070969730434458703775
6646751418161944726642656215671861161538608980942412666500171961812964658746950513787935
0649857785496152387018985471934733281848024520918953909885991388099727154861481147732665
9186421789761509441372170199102368935423555034419596893210633283123428611650673733788745
8408020574485897669862268312406167579178681366223290851427814126930086709288549999077365
6160467558880925411396887638299263479863427065733965321567483276397760516027501358003665
8426991585909631393993698622763545727913830421147330512994332855571941592177668491908445
4809567577808212738724104209463208427293225476289557579303186480636282798001916624301505
5164836492435944157020482370253453894165155812608196207500957596062427483486705402583715
0122424311475772064099508490430235056123641716176023450594128960326987429429881826961365
1219718141339111892077276330943347645418256504788481193272600810060303864775738621784305
7461520653400690334251507660489080805529375599338974948564694766747783365448724498117125
7969183691367062647275013647027107347694462437693758245995486313430949524247708598610175
8968717976586602883626647183737285203596362198141449597422839595773152927056367153586785
9631817045225695510749082357568626376132980643220099293233929159644044189610772188164505
2096447146491719433800858274145999767496719086238348573501749755801758117382129537131165
3469487279626734681881909574821480765723798179886830513689277971586161972984431381016495
4297026387083409094246187885354637430802966032574618405872560710826972504206478367118805
3737936617823389016847738719249119534244510056760360655857575230315967490345572678690225
2427589618061537015202983068795547231779245187463092772096726313537170116336921204955045
2721584334607023260078746057229055778047152528716069158536361323371141304996180959873005
7534069370517968852006744009721020945093482056481226966458485068556706199458399900000295

5902221433032352181714057755517964659823083382359526829803446920337912627333995935654193658904867949226839038337070732095514731277632365571567571036238855003594058039339433098168098901902061839047863125919742060553039425017147294871132809631230302218743711195571189615674528734244491496872785262489984542948755658884670641064014340893465643367462033273383797677100378655902885778906401338648109657136583185923040431468760342344836487688908809069747861366085997295926438233776792645942289801085427122238373841027785502669860175089763403984838024024962504985667019092598590207890117845327113774952881749513648361153194454633553156795215061339009475258891824493822185960025379893649519903635355660615435641023476716514384292165783435592533578498766756603963422079058780945711931112306487159771871523319266950762773993780847269744626310014811704247127561875372587639389947583062018931856266693329546015523290318320521388169002223965145774769216538650502590034319214766473021380300724610469862177592041227694321461638306326557376845852446778427785811026089762871143207984463377281525270414363543548853787443359516946984195836000665950211909195133225896638825937096785289074396008077463950131519709406845265237126161832357310428578906670980572233783001608227074931945928378627150690214038470618201454705598428312024002805960482514009752381172937694047265490243164341234759599141413736268970591718927272790307420214118365933821861839660733426113160833279876165616822585756931008433896615329807940396114020287169077058602105556152368001781440889556650846427151725860380144224109806263521813734909618626323679934722087199758046409673171356485997705678840060598745119943695776015135439557656501201596577947818942050441403309071454937026667249165006700153737391142780550313333907137835088196623639827724588088060407511301060433490435611814096008890166577789241378522942412461156952757754721320474665106011809212901316232976030510401798527741164272431669861252316624834454291989833086429080970498968502855294811910360692895901086296221516251600033863600794255082009998902262082165974088680849726201263637558624281742957879185070131360610331700123538061925042127930079906840157300823734646398977354138357117859636644703711572998422965213137683169196065607273217825306926433596069380929606041965730438200410734033370399851932294994221860158660603439537843831521288740472880295517489198276299571987220613915298972504544375656441703853933047591164873656986741807735653482115906205620591731600375675769791779139161464026587974368815217899047827025727072892976868566262537891820106569026932403117312521818656520450458540943802467213152099041859181292133606634263684721584961679338411929846138718317210992737068037978207540211933160990463910894770760466331077445456840965733785186430790922373136296265024776575325623368759276045772539787301088470254556912631651400745733336495560866950360302961572781015371753127413778505483551036936800521398820807350546986609833643159395538882357581167721343089635244999290367323427849387970495593165056053394037246770363348401237529133313877350094566850620377973006113139269318222614340917404881602226966046217210159302582498345976390260159658681362841387281209248256891911083443952134978583891423275997167844216530177778105442280930214516783614821708316311186013078146558518105271582598325376632246524863414781730893247474744749447765726601663987651421331167722658452679502717036292286961399205868248605760699864929827838754057048139655838406568170651623930334668569018977005780553595958923822343887432529050918558439708758624329851200478061588972224915766964585589840387709510555916047621638678553905606458734356774312616792478737933651428573187967017304355278937816761846329848721426799043355515200131521232810305848923327389683288075559493426215759988923567023534462849367136331030762065879239536323426541971037983812010910537743645385618505647776431807769624154014915434045180989233289187214715547406539780606590419084681294647963892460326444260039131498076905441331580771689756366809756098020960101986230107340217133669548388280445940781218592982431713689313624409107801805225708826696215421250664212920406160527302828266178905688753325647852630720017234150161467447792728336417097399819876018213971553686328269152583295998370343539800913524188713546615814991232940460965540554079931076824475822995348029434761046199488305588523534173906070786560731669828535563560135300330320134145313666177841619686994215794739250689574655686341890888739323031919192356168719819102091661150843812844530596209266422807617624007750363050266958227066486126059439991456366170272878713168275921786037000395388159070040736818681962857937250656190524458097387357783539417477962864000714722008740580972152012178057257971319834206980270037937439372527184974852742706862151530786353472823519119818079901099367368405333961902804699151231747564395792615944044561083079144863138913090400874184243113321252744929177230079903739477387115435854108622504249772458467184111679839087219073751191650021612629514614987485589713945512788497934789462449403441316949119499593039072975023103641314895176435972040376397414842093321021098863426359819154868442484340622615972338171425717260148386925021328959983092837136874907946331560931586093133408219390906082077177398771307367880021636823848061394868745465758561738341809044199591584679611352763370128523804246943076576905537683676837907305051135874834537081552854050151821438122185838651616121676930587962686423312012562224138355677741598649225231074349061241270305809006004954122875562915776567221108161183885964884562440638880923279476306011996297280511702746036584284470182650173151416721190168526987986012029324333531370194863045600526557489277783387001881160016813552165101516067500938156633852506623135595566795576003377936640048336515730220064543682259398900445033876689098614582563181340541591134321455933002428795689699159878970442882048272233655476522210599278850669912096763532132314339016332268241485293965052303901570004533711882938709388474609655327710620503385870734268635890508908410247228640926431520482493618323423957620556940020257315691743421535209678647874687233993317527043059417291721932596053789744247577019655472968797244391686720021050333604144760708500507747661951000511511269934335053159452026705263109053905530884254501520656106265594065486508674438667245175487412750272077120446111714244357235047725317164018266315690235621857457744734878869072355986312033657738002649229314559645435317130063792190837144178774434490640723009088761776820465913917201035078613483671320225928811927917992265630843941186165869867993280832277978132798640565000709418440843264110550513099586130753144258749956676854097861776843479376571003520052155230186651514530494650611437659273598576331201859580443731045506546540228425363133885457891419471

1822188993798708971841742560491892688159375352102721391998956669653300703606091818 48321
9473151928797854800244358815034832116016185339365647939509538384332519604670718474 35180
3863659947775480591521340285603497617421008050126620917076190191927608158581798608 21294
0236144942223645890059871137434337485790057655671183154133809612123799393884024976 5178
0779292122879383793588151557420762589293761135322432835084989576119139593515632812 93739
7446179935946835973855459994946524941616190176435256648190671827897544291626615561 27137
2248592753521098186839045616792052212678752385927794873844468582550741146971791870 48488
5578169503767439113663446480342695339740409166871360963932877974105245112501779778 18956
3025357363362371445001826760901588083635636528807776232522852819565209714749329761 23480
1493467946777860562173347414238520769930305901768533310302261427361788410477949817 19032
8751753020733899675291279262222393995850816675892593828992328134312611134624086704 80841
6954137124403155201033927290359540904448430807185160213170460718932735617273966906 51527
3393917233352611365407469726126525627657455782466578719116060805570371030857898082 96287
9784207701770591705948793668877267826861137704208897381837081346603397392426525817 16002
3424579513885088194663251456636034164473791110123267475040575172002805211576562580 24592
8684942287098899578856471023999470489424986182293092259980920299536585827163845864 15000
5896662763893613463497157725169988298557013790562335928473464521445263535513425215 07740
6498245091385388779646976201270640659703117575969626867485634264225242735066534611 96777
8077861266712269765584033665232781730604309829306218419162855397452335790278169141 42011
1626482241336458048062141605681892443816526979401446698300156501369211604995654991 10498
7936530853531333941051596738019112885075386748117531170631763597278273966759782289 74529
0929836667344283565299487240515384489239125875911717067547345985960595170469838549 89775
4466421627636826037175526826217987441415874174726634310513958760222901836073040878 76331
1886731991666292119676487093402142840333804181366316828260072378739097930301245285 38166
2356415886995855371185021219211186053190141758122541491680916689126806831123666118 38382
3938478838952988462978226401763317509267097172348654431762050930814279943452268453 00008
1791145813329126628697794209839930222498931025871411672270177897645860379032312756 62409
2224675753428856533510026118874998975935840629019541162729672526434947661276506381 80239
6375380029320646745395130442240955738207533242731077671189645850977567240702631400 48470
5120798580213079700220167305530965813744674063311147078824040935713640192484396144 99233
6156257316791801862519516783407065198192381611579533909383951039457822828150127146 44848
4249194797964708781416778080293883061277618340418321611460996082677339803037544611 31307
1816625742991800103119934389966138333564554508118313099462912766950203737639143705 80873
4486287945074165843956109874192416920858057206163707062357870735546624174289242240 14158
4317654618901629972412845685301972946754433431613679021894090209794852417631993614 39377
4555518554741106400787152805457231257119809538526292357864967980599837459646758845 50338
3962559201142363740684063493746084864395244334342157155102255006985717040131843575 21653
6445998270299879223351599292612063195427311773101531154023769338600035111488291189 43062
3364515352364147856530283657341806636435874672067458757391977753432845441290529504 66176
2320862577511155069140034076443792165131913137162464010087950219214407854730171315 29563
0077671189792893041830719924876571337632289497886545255586330707330373070584352101 29860
8902442784805623915949687311468723707586531775145485541861404290241346149453767803 65082
3169521650360535239749489200133210656051666597268587465662196405124101526940030532 19750
0516211201670338671950271625900292541900876789817926459776942498321822875928922931 49632
4446464276658563470765544347114017384689732685027775472499553479183056072972236643 35833
1272204136322887821461755638916862070579136417989523249472062256202403298503786879 25203
5757845185291308322346432383723676453913825188522020842526541139028074843105919565 25539
9498334885265295968580861010021318283582495692627719428548638557726977063050322606 66756
5378309169367862103968932730778899026062410759906025961030467304640169815405535362 90689
4680680728652001486761210295178369530282700420802347291485713481111337103502617097 43787
8567144280150293363716307167697349202477189899756053544597584530799711628528447755 98660
8291232852910059560749677912565563936722498318753236359165408554224314481279208023 05633
3574921899694436346913193823613944532501213042977700318766645843403164295964766106 85080
4930458959230321984080827815720272956525123000300667687811664965058632956940842594 45851
6285737392335651925850357235238448826435553596074300045646515842353836068106514467 41441
0018896689589625223573001183170066194168556046760505006068577713665034540769695767 46253
0997128498403018746857930227053366059934469527075560280304319429805231064942356752 70598
0818695692085008073959469024649272077186412319040857902755610448135404817372025332 93102
5309075445911816626015521365761928038934315405658647190726138468676529602195783053 48467
3511934770660240062076006544192848056139507901633352596522929210239665447776601990 32553
4997530657535878595369048515438239295907267922835215460559669340630334227066091317 61854
8630029147548617498066801552105210731203826655306078844198933288877997222031341489 86517
2143802455359161464112014952168697910023507004131262638340927714090972304785839625 08176
0765694040139558355153881365152148433878193466231511946329939204838110065954794115 03294
3309269456260256623449122134247394951812737018470123089237714644409232043790250664 72702
1389722820277612389208781236052428750168400456718639493441009788002412486802235233 71369
6220484390437831000616930782548879023218151594259204873156986198836461927671524680 86976
0879816228727289314807130677813117524421181922540990736236123721029438760479238762 26394
0701307006179953496145943120530223567230685973295898142895043212371815983523669064 45889
4273429155469393501222224399082393164248655706784213150228657739202336052836928805 82902
1453697644109665205066934685919773491664728422716719500286825029335715031183616497 59861
0357806556648430281622777682882284680501843548370786203408230900697442113302625009 31579
9778268259870606060218416527394756309496944233397358793800198415225275303549704290 72641
8275067925021312295468789364039582979442469559118988918125269205092265698383341210 92503
1616446154468669210648482403466670226461366577228827994929800613947898598055670121 05239
7376021791499007134795808596546605642535224020606785407484367571924951230954599640 72587
7240307476089128899445483377266453521413288093156552913436664244052317239038591018 81285
8480896642265710574595182164012019268020109027092477298921289773049396949697367383 90532

pi to two million places

536811425061719119769579312029449746261422766741179968146986940698079148295709210154612
217744362444245615290482819882454939733835658679405858910062852165241994666516354666217
450109249976451496332947631392057959185482772804334672815067867238501699073967911367915
434521830879818586151270660426806272430537158935339997913133923266065703088710398725993
650056776364664911383590735782513632517615442636133974040170856839761472931963770820679
532764350352618356805109223195010442449042359449193402837612637113934002722543291080612
452731499459088370513803410702883609200556876622447401245399182697798623006457376652636
203836719848806942794704205996048755173123273441540077340145132122752937032285475100909
560925511631125760428478490535808747118477710517639295442711988024494519686799122457248
646951281913961533674710469363056736261414462876678279669111452456658129896931003581253
702438768495433852989364316306107683205182175565682979943095677094402641182250354130288
756781165428692841205478600433735553732938677548033665565789569708710700248212963420727
006776825445079540220013652474014712458381082957275616833100372701873259386672788185015
082769149702195402407971457535977211161821544782781019146617388702374933823349642936487
338236599077207064527153393790677700854751124391165117816594956188128102127853397341082
963050727759725665073617984875294404755220237998024505795851645954918347709785045532336
614119775836964855500677822623416467894114064589766666823424588278293554735975790597966
968460798593828579038354088546891638516347255201821794433143627783176826094572567406526
806262136586303984941938302426507780359161062158427955917128460812774246478276312134706
059220742101454923518044074246617803980613577890134396485358808526503297946106739660108
478210500395034103950611901555522318416970360911866447085000413717558022596810292005015
474569078183724180794761416828084262377365333765247259510386077678882653425358713560805
826718866230621987156031135995692812221227863011358849818895285470611622408635143723471
919845064916367258608343611910774145468495208272702577314161221438192751817295334070771
480436955619643617668099768089997858089064777534987165675024996259010652450651903572350
578309545539331437828953931330700832195812884065312962728417737899485016834495765609894
934313703407205236645771059984253362654084785865368566543896126249146638769659745758854
345700703193203404762328685897638561368409641510005230763333529911133437512428938598465
177637002541457522670191035034190701618027208375766974970589717728962453167393014683792
433668009439053002646964451504964624836722449932486004568754615711786203031370671521901
374798616900298085961380159478959767590860939901387812807607711162778415508352821668414
143906033431900433198046669247977158249123915939079709473977312659160956625471495068482
493181819436983040092166580651670234873335344431251548633407081455279391313946902999093
784225824172439794274748951007943163271673922205281804182842905283466364526803532021064
351095983986720937857241740934644161271642013906157709049497451514987118907771463332121
575795786906251980236866372735458752771331189316589316444682750758448009182687367775819
177135676561543906797427503207060743121359532109627323931928244932944136251002230904999
122762304426983663839882798992280570698922660535681573773633766990736130777619706444113
435831273209581404203919580959953708915195797046451889505623291975596206857017550873719
308701399186459170484187414979209257709944677409777906333660009497066148207022097007480
928538632686209261746121744831591813071900707843752401915587260521318046580406747102610
191904771146769525983920252788939852596329837799004719975128554316835747863772082361811
483077074926043731449986093797101583281008911229685531024694483350102890428385437909207
307216176958447116629164402518168786636350780497118838156910131713493652174288732547140
470328045286421694800840178819156738338572441725118661619553380414893420616053237494360
669228862972415326897621004254599022698969895703655558839331729054497857369445111461712
139778167128675948735602470520996488512983589669981287942685293857971332226149425171811
226665373208542047067194930343413015492364718418210010489140452514096466594540789058835
372816066006269246561849853906808485483044652082843259636990678009737921142559101538130
535798033791221600622478928351430294566974226338659292578574797154462886435286573800636
250519410527201753182334809624245524294116144374574625161145938764035874932671520506798
453919912838261353663430440244167408509157693915125683303216718767108159384344399860348
931974866987443700289743208561568416816750607266305096955761580280610582244273330366998
179702946805442219397541567816178056759310625936576378412915789509766491959324149161953
710794300114238710900052201811623571127525109820309139639714020908237665579387586326857
057820833562046202542998117799074654810963394569501182024749346392110793740293871727435
340720923977612566788093647676250893949837170484076926622557141413649519627495422325055
780768275989029451889315437507923615612504808960550805165519480194468965243953134729606
018341873976535332939099856718239377519298297641916533330272578076564944517881861129412
277722593695007062610655684609405038659119028109249801950834718426339614429619214745269
458810302974959085514398482946521469169128940943578123617100303242812497286459761363578
531026403758949709762662122824148452950666389101277908667742944535096892659769741286762
657097265975959957892262847299918734677906115671802879023081030202231777733475880879693
863135113417598071120572891686475360303888304421977567517277229679837256776915788670804
920758056467613704486661325817708701083484319002288344751594854024931782056414003100828
091160521729263874208371081940735618017929891538611748742390851598229807986415727434981
419426641810460660798260632776758831903935947442866178046272784135525602633439916389872
416115152115408626327856678721328640219153754724406676950321965922168467352526889072757
931542123068545623645769959957935242086408196573653893682063855295378838829972338630639
195451419874649630359689997656339393529821781330967764550364013274639214987178935008139
018299541318709036594417514012426971880756666091708257584137499800026356955684172550508
091314624784276912959337504221430454181269678632991906354771341600339139880340161053706
096783262961821087365734424215488641540813869845002504548048553506686360659814071313897
277926471051795647996957496189904973820368769369260196388465699506781147753371150872588
287643166790027435214758439117028593313337048043652412061403805690141848712709582342903
213763761279398076475544726754265765350309335209273524934170505355619085841695880867533
308374361552498668329699846299173122173364433946533930251904456090953336204274863448273
397945427402729857906135393300633489463672224835064941414368314426157412313239447066808

5215968002999173414924166971674312390085936720261039190672560730454895924052629907422676681297632839907251124761424153262437455905416375476039670007660138553799113803664358589743094800772465866595731713653555409871938613634390373369834700624025550670058397082304841135713898843405016603030154663961072709397076029769454524118525089234592325311192900630124627277856013378715312483986752944943485008396705036125396288637441760653566979827505959915652036183468415439425992555992475106535735739646453149401228240471339762472506508308035830041806737182096700540200578621783264457319047003300818494087591738233951176576955342332443066867692551186862238441091101703606250406548347345927878296827412822514673692311779946542482163881812850204534048879290411124487692534510453402459955827392357040249419067034178130784701220141385621779156102588388659410449262014329959009659422519695360438228054148476254243010615661459214456417942766374114686662375593811308812987658618155226807134848192432846175250197078705933990481451355847257370606528499547956240015651948839166383112085459392519859324336561479215085515617612083802948427404179116043179236494965684832940771065830722120983674396607152950745639625597754279092277006091742467351575028531616276235239256054843594869861959424851686125036936764519317166856498545217518391248155364035223954233292649467335315794562416779608438773922194930167598662761944040465083777473076512395679222293072577538529242207710425437163069122590319814395704094275063529617620583457733210344580832481957376415732041564145484793156491636598157662399666915280923960307630786605112309074398150497900124717874652511492718292098256447216377811304945825123044221465655775344925067274743829459688560993982463053741261658117630667354154079997183685764895836574137122351948693149631709007661977224874079218345091700898501036925035241885757138489900575498558899718945806256323885414338945977999426256258687649688837985152101434159783056602555535982225088631700073184977500566316564575564410103799479429373959154594764447097514806762233849807469974977854052275685157571895060821029769547117403528192224745174193616940315494621209745949585613214670176071921797016927263309813144887606994503465262052734181710696454141888596833764480437138106698242020734132298075073844391526846234076139363568622073551570860885096456490221845847970074786936065369664231199903984580386452978465142265561155891235194925212428585383579209896993111152438502407696832554399819965919445017049242797118635181034507996761632979477060471565330122418354248810922291352720210119353135031899072752267861435480767348385199928813734510280255636404997597108022519302213694326111747566703194314320910627816064069178252234404664359399348723411664468316785554726289274078597213988560848211815248400771118297764202748207168604338600673757472946816951017261063808475791489031684966333858020394321515636965470150532871836878854496162219449602730679699558375737842669756759492204168932935012548413146427346547394123986031124754420184677773277115141634825524037048619361099122662328653908043673745874977491412161592218916197980511574422151730806578423135212138345606318828125084579666994735705397959622937981517464090715863136690627543049034287603813282858414342206947162490574006132060381714783057261751129051656302711083269924773187887075454766339466299139964574077538761839956276691398617485752948163941820337092218607282178323093642130511248434364171873415067106984583537138755000493539435483109920809356166949439735388468452981655481172630059832183900983777372687737203798799988861116594995784688634784025128964204982993433717325252828241240392028869681545854726259853669802768344245068736968800909601141431620456691817116238575809743707234902073155509757070536056685307916472412650725481765389846591314154009837043232650492142895729999640025794908328480593774368329348050064365075843348505543889775108866402352038691143591595648554979237139712368738075917310983365983398187780053011000942841687001289568298298451035995731622108010097782732461627803187396054540906913979469990797319486155117768237939508920914677045468447855572086897662268657201945783496199190230935187744840921693617922960598022075820618779652401411968866589334256919765285906979728908441154131939457443195376731093604329976418435657520666959296510257736793261286016077229746671058317564573601315333979571308929509565541945548230369192999235168201619474588629005464138847270278244519790348229304124693054285597026552302641193268185662166922073160955636854032654830072062308278374627858738156117473578849190587191612412726617998539786349549167918386695601723370717568271519538797229694640330108073759481039845441014273432804353884052574417485756674671809650655171693113758641899938516054718558617030160311496771127941886204336095759411832528795861943440296074737642049078787663209859416511074702575122315556577060959975218017489656692851131488787823855284315204102171717838719069821380394263659518374951627833770395541955723273746182926876878348823582513792637641200651691829899737760223807477567405781758897572068402758238079975959746305620974968255875429258637545243824701570920789544137963280688255565545033979287050840918293585058705565394331059532114974600654359005780343258655773890121364373438803618956819773375247469979569555145364497838602040298653689363783537806126130722724348324932365347234841767386852445407402713420068053877529995238843104859761470473527518265689505023333951615065410987171507503143369755195009982111021649429663433832001834584536074163078189429034821467802220260641454679175894378230178979644323899780710518427195071419515573162957955719785686415814644376947331877542603210649944699821127958729755498406088042269898714949942871510450537242156569240189654609633117429986992541477594125238722861974738364816775928265667197267937910558279294291061607721097902185894151474077390416033724586793605371870104047205180284074662030919321040086441209302635862139936496915422346654228356278791332668645750283057475042139324365471671310478159419653498167141031666603998498576766422268406858831413868765425777727342629539870092521131620288666543532317832139100278633402588675821595338767675817869662651597762777581740185656059730895041110527648982537203002944856907849057956312565814781925956071815290990996377668731628985065576874775660355956031121660922776786007725431791733034343650497731651683694532615351964777863635110195309390854032020640245059830147365427318736350108894077694123764559615097983651155876807621761253400940146675648440984889371058717085084619667992625113858670659802710579841891058981435055441592479794679644038091198018002127713224439846733581973388437448916777277253221422063713206041103011289753248116876320627917618564524097494284159338841449908555631852733620628722492242251271226589963201834227490000

pi to two million places

9291994226774982759016145623833843457025812398474774862907162322780563021369412799395744
4639494317958639135916618191712246886588935063596193009135936021324626796541349839602122
4769039508013164055767053784942614707833674069615378499326696049322025871676892888107111
1087696342521598832335567065992207721453041708807723327393363305639287576710162394711555
5472186701400394874665286821866944078819523281508728512608894061495867142163050984113622
4846555859983097493774670522420081833233804941108600348541513474511331806960793774213477
1658424970819254507549917183735435240930171668201286686066619982997410111795945916808677
5360236859597226729163116342955886057740404899434761540419629828401696380294841376377955
3844750838193029276556945440462260510785886195502988952450654035627284968134368975607244
6549534579828451649819108246640152164714405933866469660803281056919420802794299400215699
5386841483175378922957509124246668565938591927751087409511201506518009587320765830406388
4422539454365446189616171458741031458902887799673354246520173325973774474023098604270333
8270261588392949554189355706642109531441765841299854430877999015174258305108596444013788
6504340166079492079634239626949809409064060915422690625751338513351679061183831768219844
0879001953556948407768712834839651968427433369175345616157016940305891005795199513338133
6022406961478372244150747443730992020925446774937524159519335203365629714474144719696366
0681659804155385975948390542467247421014061804809799119108051301544942858426021551211088
6935100211133517878085206385936600182563052147278979096024266926598129865754900291549099
8722974902730331739483213456236735629091923795554203592065154970295396731324186547668700
2144292781204996509197630527448393095077573645380884738951681469347417301387073999176844
6527377931803580908667740878118067121342619353732703629631199743260175146588094443006311
5230124188763360923743106885487439592452619168954136730722777888419275272174205200520944
0264305569729946524072454150090466591540842928612818933973653206485578478732900000725311
0304967559985157073594449422875502250003329350038977190258581076089795709840818847962833
3414931136502887460199109773510483859421586659110821245375685177071956007773186184214044
7438573977313218104423015854445075327467046247730556582207058057033901257484943956983700
8318640454834799035659271686062615608905969965228372813623091659011946638957051795109166
1080771191832169480245060988906703885035143643983154129357806116479729067889969270858966
7935773144175257027747881005443672272425019593388108743553325517267662566813878491912800
5703153601493626958536403438283443287487203022329665133970937514870599738351642337469700
0990122675434415994507886086677953715739963779303296496112053737323484051323830247956188
0410592568640477276402817449688972308832564430543697865078707161863104293492051483379900
6182105640947452079867983933673826406008392496618070816934967462576138216574480562001277
3446840532882409958918154020522999067683524158135923930438945112707927928731541510540511
9091326151975781069524727215064263302821837451344557439634481562571091947239274946896511
6984119116845839510008874207996866156088388078885436092330873689291440384949967769804522
1881287506062398137017261252169419115810894429840530477497238391545488170429646443433233
5757151687454361543268886589360631797929438282022617264984659608774214339216642472557599
7355319349446435224094706865549059766313064901333828759417255852592219608921811761522266
4849292881580618384298835656093969480074501761122990342337611376513876242359077032875699
1603256164851131385802303090375719536663558287393162216613808644739051237222980297963733
8437800623014155533147283270378443773617386008998462040262390260051509232816945244723133
6701968109913266900673573077864054754714628416410814887549579304065002574897958631312755
2941332224740533527252416322717101959749375195646147640503454587687156616399598842157822
4945178962851814081919251319309531190716259450593755138749533554043252450849093059821522
1548685302726783823504023230074103504977155220047622344004976796126622611370628252982511
4239291287054954507016171846895005147462293692987153475011527728713176767399037079736822
3980761343813419137156680276796837396983034024576668100194655257710678492197634229030844
1933903374529096090098283134302273003959720509861684971413359579489133557352667901724888
1311611036139784229579591634031071133034720757449411076603290357073382823262667284350700
4783703183120991425958221920197519117481985345997012635925227305720004400805544344910988
1950603825719431612982265235033650128340674967637769123418468760988248203642880110959111
4217553836901155078553353423012179097553438440615345077012216580642841524633987118071644
1397773080904649993483091667737560626652657595133473824480216704921151929510712965081788
7137077042469602660609884593794452121176165299480180121823010719576313513101565441610744
3614266612401404295756070512887263646510081568564884896789390066258906107488013947487155
7042738004424455049624935129394273386166804868932285597326051381678385274411717408896611
4992670277243358401999489971647110582034378634511031137134322268915603068565169306733311
4377877525674349671103690638675262682851154948154129039364462233889023009195459253257133
2454025029574902827484273197187251118515470058851841927330943857723874228514328949985144
8591485094142347057958363110519736571767515552981269186264034492920217245204503141865111
2837172013915372308792250803003080318968720198592478234076129222746858882513643595220300
9774104002757733348395573707828940790839292779796466742363052898092443359925514124040255
2877838716280266382538949949902350513632482919520862390965172531435974560552255153852555
7045464064544199923659746773253512629186426937857584353121104210372993485687385399944533
2428023806501574419517105325655854950483883541058772955178791806976533640547989506577044
6027101837587399521707109989735082558417362343720953528834734691466640352802371474628422
8042792722924845634976839868217080300056671916222950028647099063604104399194052325854233
0639737466333374676725195405246094654979049114573453557698626869593724361488952655784511
1017427388953650558860778547914546444683899263803597775792357868767715903677659573837288
7229241188281068104211511179891595052776135126384895072807951373383703991979804675818011
4940356530604365374971057107353973430220768307364048182022768161656862281779290298659755
9910303897240960562120834175825335331613351273559726504292198648147955414253307013799677
5359576480302125440812676406780413827612196942504510164277244924434085297674171901606811
7088293389636289100662893440089766639379549450233612234332534596917105449844201140628199
9926803703085860364588078936417992459945005686784414380971994910120390253662674470840222
0532733373080598522631138541036957206855344618623013222519868543383022257803036893614944
8376183084519583300581919657613884706355204953212089769506852278220189630312748304751488

pi to two million places

2073787006319576543785969555888937330582397422051623241616091798349680912026495431909677
3786608100673829100194510728125603327434898929091731615206874633767681514384728528247 29
7769837052913716620793004841791576603519366048571492413589964207351992852670099155432 49
0017304927507445261790422491923573083423146742116896512068090253921867257008659512224 99
8678701908012248236630076981480932853686974152252034319298562477921046406655170584626 45
3697450450326492569677606853314689757027492947030214301514506367823512150435131561443 21
8191407790592892039018945674361203349556651881290090823351357554954402149955893987036 08
7872382267160941324615055361428143336997319437146949717225320276146608375460475196826 95
8966816021060640737941547548424973787520914835390257609799652198323892051940372088561 95
7012528027375944515631888104489782150705039503700766421848347584251504960595156106416 94
3943080350821515555279646895965687169907344133943233247123811117194261710851850635379 4
0565388478135343184614328913754726921395257297186305955088686431369913215361284294410 63
4055789691965192578506599072833609606436438891837576060669199294111602626725506364590 50
0483533192660812231596236600909355423118403578140096538456430643033735515761473126762 27
9351978775254489709184222069842350721993916586011942152955304274939265482686262505865 47
7108500752549856105923665650677321813156372802955901773763388493610593509654555702340 37
5931786176929568196719198928185267321004533114032419089576537996886016310389747956657 79
5195346752562322151164018887568610660251638816886676137477729274906352007675189978972 04
1473450695728064292601017951787765326697649325578120205798000043929971509082389972566 38
6193520081095198882895971858147576720464949957344122981347756007450510999826484522113 24
1669921415070656052698051282362262376942296553557987787667759639468722683930550020499 75
5770611804678188555391378447281131066671075552677461015733901984044007175658367497688 51
0276641272474477016274991676830900838362709334928754840908004166474114486700582328196 27
8724141874929764959428671563314690648240078673318314862849005482098334413867617983386 67
3185773418481785407797234324007915185670230413153697941862799501592738952997427822558 83
7250064918633957141882535311486942489842791768061186280090697801653358925160782402814 39
4274727886270429144105203020975265332054305568395517813012990460659406577089154561728 43
0251634590195527302578015054212474133451309995799459254411728729626555655932864654058 946
7554654795288819846978586869999286104661550499781042057532443472936773281374346432640 46
5507741400735978242795147119977044792897925942062855566364973003337675840920313465590 71
5550062761657014197808773180007374664670134110495134682788925928197700641749931112028 92
8089450836760646260844777775200694674902457492874139611053844422782056684617754664606 512
6138482378290670504770994809798160807917182275343368869415971446758582307527745227424 99
3360348662925901898068701177396046147302297506976535643083302723237784992129942748038 19
4629220553490551844091256069851799459218129300739619446438427145260066413903470212849 90
0652134507948838028815465632238816627454086903704144089151434537028787474645949264397 06
7874577114186507302441834665952567262138946699339686608012888391011926492552785014493 17
6054225954470096227652135762921208073786224855597659312104309995437695852583470532407 99
8353033589639351938703746390593125562242657432534304179299785203208818790540374500299 92
9608445071003243795735264159797428054958470365316342204853874628339026737451490038901 35
5577210904546381269439361260999111517175907178253690017229648307359900147829054581195 1
9436589597347963629279196648979047560232717100088457869237708197968626174850594504159 22
4639423648631662906700026712199161254212464006523152778807371486173430861483739309080 97
0117883781059829818890632889920061343103291389669189676504187912648939542782129850532 66
0322411943113948384003441677863368893779740325518108194118031780638331676754080114286 22
8511322511301925518454640648197263576558802216006210930475283482365710600567780230850 46
0645076902590492536583626489982652476590292223468620269162852765362504277535551497359 55
4914212445008366765914420657903157315358138365108059991782124067366362302679080728080 29
7374870720374185433670481210936171383255623769858325698430211067564902798638323462156 02
2398892717827319867447988908510159110704277056093472704170296044384473816343908894845 09
9092382809924516100145912918131508718918833824589453565986596545208488323465088343699 82
1438559758472776924514013730666449018321182864552984641671452778971997436774234741545 57
9427415883958380417958636704295920072058489995152875038060250033111127760098137315029 90
9808261525959556019086537604720147225572680449978212955808626171962944626152102020998 25
6057370573772549092104580336076900143824417253806264803223117927788653272479887825430 19
3729091707012548980549894911648375964260713285304567362574940772476278944131074424081 63
0681167741611730273361943774348203889367577944775650177429076780431733404068567393759 47
0256562383046201290347294443310915214300017883081476724414299767050796598225199104770 06
5877676764784437293267361902997679582631765076224881898450404589837577624378853824490 62
4490836831835686494951765130603765096345716652039410850777202346081985362408812326316 30
5125108798974459035590615063377795081903554819539332899910655787094667541296788134008 93
2741122591565148373816760355652685824985901446048899448164539587940248890529572571497 48
1556459500411636432248876935023537200792930713879939546051813695815021074200366376042 35
9948428960259067928301607147855907454065428702372251248589002595072102580366366808412 55
8543378059993689258453488654269859865935254912206763713724087395538748904405761228149 29
0048828554936491249492439757145404321677565979666517885361103277997978150604847387537 58
8787909613974161569562546420156474063606597582906139522632581393607912426527937312532 93
3414213842929804773460519110698691695963743254336956881840336147832753900496018429844 80
6147488760139518082391724557136508078479470802289384463192218925708102368051111784011 93
1927221941030562498582526636580062484298520092629101870520927628971621205779000013229 02
6901750628643401689438630916189851379067588868095347408459700023743470076953662876150 26
0165246817153479087304648508255273309781001716536427580385380154235719454962329953943 85
8215832502469412989613547187623163217432634184578597698395080971507901443217056228826 93
0219811895097321437671726757533916182581305507540512122236938000379423210633054371335 18
1715699654810548082963723603180455090011398582867299481575125971416195826285447724824 36
7557881149470075372143949059331230639998981954403022608078790206234076002243184488999 05
4186867484853082402536232479562407933979762660362971180342680896418449325463304917969 11
0613763594422303665739075670709306588491446501101414886205958134879066644711977950729 21

0993494804149262651364002820810839982262027638056744107302780146385991434449949862745995
311456033373630280991171732762706722641591914913635265438405719467667366749583602104244
951464866826052185751895076722909243710214818883994697680252870132538798159720597874056
172703310330663188216385289542431398849884219224268823941537136569251815519535011165666
118807926284046466487787689266766710229206031242470449444819826818546530688105331324068
012372323110847246798117748571925592106622215837292449172367902042223431790041810128981
016020270580221589647928000739239700640067958480559057419673302259986052649438851103804
100859677976177716269897894774304168232482149343481199237565157758185521046654188841812
646711312318267842886922883973450586457605900815437218669999870609785550584038808657079
656676271923372872155767986979154854717842164904431783302673627810136052739214585226321
562523721460572783300009705309314123874112699476911730904106334576728017516184556062044
975715537269161229826118935075940455226377209333956073606246881882200664298331010880287
494704277952043473259127914304707723672215928803277824072295322441497585522442930608930
682398486649264438911114663132620380779329431885753471536405942754856205818744766426735
376562641431513231288803840299611026215558589140924191892311936021776594982239703696500
3741196622481093203950911828372968295190372665621491038432236373047508942005354811744398
778800017117557249866382539768829091974386879827577269311769602527343337443077990739488
106494161202047823395578525786948259483984205454227478994381155324159800893815581004951
458378633918652322671707967010093491258548397235979362797681264070195860521413277462362
358314654106986566798433315233830447218820522763351304995821648707234599835254115664542
921964916191542900669965886672047895955697554300769589360388400675387010152848614014939
672512410846836266343122043262655206993398159590063826064616693445406351388467734058255
519952121802978557157752991373267215201215851314345239362970136317415612733761392267281
822798074464165283686265464923334149141840545054634555081267199104428277227575192028709
356418897033452421261028241273202276786860322305423139553962179716343467616131745503826
039831281119492434306216284572475496576891115201842133635355715547243795116352021725173
249584996208708731651769121421619653027649424975681440709438033841665721153580486582426
319936501757333760676142401159507715912926766387196827158122518092128490081733499922795
254010420151349577358021053002775986920431185501790314544334886462736336990424474450048
019401238438457502355233285083330776332962315633247798733088263657325194704274504471366
555985374502488669419094249571679906807660538289386236972001259703186311181323412156728
467443147319062428433759574290321304538046296018975646768910955453397242053408962985039
391173688310996821036496599183296239027517311186110326638057140876648625475532725295828
048602921973275364490557567983372878179761584131647257950935598837219061082339524442719
542423494085131224880529134929600752647680581191263577225681517584633809329134045114407
063233088925488811184983890281313874572487668679515723526894126295814493773572183154622
329989037401961273877161474622857590694650654409456453029665409184466277567013898563051
794368979776220881938298616095554900122267811327853578004152262352455496707813893552763
432841029187253320593030326599762914753673591034441509507162269745450679231295686921965
703301151094523727004765844406146353444062032949412580749291503038114369350081744708277
227525374897437940222471051471523047504066866644413131603418157288990595471261045072332
2763227461901480909608522566571461440511776821263022068772085338277206869346161336431612
825656944556110395781807329846468119791024492179150668449782783121550935522946930736130
562487686064533435847764240887391975958861307810884545231167661231971772447458757466382
562784672936132175531332669123037057255126905062364095907945810467031954362754364810488
527824060205671247617270453183047139658572301891508491381167006168667235029626900983122
548835473332171552444184889316715072263786838842299662146201539915156985205166620507018
024099543291240563130499356499769672154641185490098068987079638518744612039520189089456
624130526725938061514554470714390274742451485298951834468777980718576747056043945485274
796074319580253621205064534234329545521611147822294491079401474504503688216580920931022
915060864745842746726905647737172621145664385934457631483461394268184378683989262608864
359957956862304982960517567219409743699432358568712608865357253005948069655014003341428
267766578345773616861920929519998996301740073887974969172868812283814922541859661088098
808071127716019361077933348161188026609935598130139647687698981300733075181257745199670
830412243261767911675240813112012423844846367789169259495567189356452567014255287066068
015375733356752268338517629666931242990766339870435227539540360704172994604535187488359
345964408568864262953552664042648693980771622521566234456013187571267801814922526915434
351388550494401338252313419510829494274396813976389868046059347172207104756740739172532
465405407537858250938302317814475161192019648934401816659536372862177588766334469956393
524497076413430110165084307255149278116129418185385784609424719000997683792886429294022
559817539860621806772885267160729123602033882505482703733281773475896095962693070629637
065835695967253469081518838234456279620337512333181598880732570097347345915672226544416
800347006615272419739741753053872650386573381235648109619525143422590106177882993142813
321022872573148938091217127268494498832137031350639104757066361189504386454965820580839
714630523337756264202547363996980694945972764218052269511428814028840150698953376580733
143686604910365454536319820283378367036476875148326741217117465917395940606538018798173
561745145261927217065947814910616078854488427088457673122291181675769061546845928489389
872211507108177313167970523605712852648901702779882656036036425946100654312515284318121
069148124514848319984745688200124725257712909389527432874308306226549228088186224009796
397389890817162174687081964220899919329516341692665094836861893743911571380315558426515
795243388380664875445509788424351665771616343682605144443309398192494166703495522879935
934587134996964886146836811790956119481408503459292925286547810684815783328358183615613
081083914297207408323243453943183602941742830762500421539366458222635109368920549264409
242536259304391529761921742593696600279911879143155023175407878364963349828732337301345
092071396635851347236636050080690548703517397770662565002788040473081342587384336644094
690700805580771337938573095428710602428662306875592703292094844585787908833103013981331
304435265055423268087818943843457575740910258466063589841060047604402396670777900307600
2607535793557570572462736208054863917612453226099325328088302850515707474590453383116 1

2312440829280071049227775204595879379716801168819506978225860464265452862565106896527952
2168512486083347947285256902221637767180365258967057383781326645311614634579633785474257
7288680072050497017421784982547653092309689569662550516285085377084098442083087733517480
0602221085547872613937262799107642657774785862278554973921139810935556347573844161491909
3229283225012432402241860928852326643409350403658249118305607348190230947773849582529322
1323463318701445918057276923280084378916669445748422911182985269593225654182290003608
4344162810462153415173262827991783980169473213575302789632128738466265084554867177255451
6512396403889019767238834107504385688763028493589341194517615641874580331364811916048740
4555199522686227282577947877509709050765843794918815242655489381200376226412574390346
3710491950386003124602727777822934795951357133196486605612273685194289369894749269974900
5485930740598073853684149538460475244838895242017880695935007159198845026764211288516555
5880247939630267424743052201708700243039836108086130684547250521397827440584441878192
5669501713798883535235885297460717107779632674477507292772871491501065639470917804609540
2454318751435845078282022514728582082025168606292370048695240573278165479339677067496452
0066050594883375333418336384283967671496309819061080778792733183270464109295089155736
1955919296815876941623470610769781208417319041527070885528480333751947413302427900597328
3904830120008716178326332734615209822426815282302459749656982526128420001366754367150402
1612718446212024473182724830267985298975105348388807788763386570514841986642083243111
5740834328293726761364493396239722772356323278889009738210120443555015028313261048405384
1555003297158112982560286811091827469425219961659888617954393071023058863656832320179740
6555608858388766642068406282691809464338859277221268958632160956780237634599840260040017
9841686173076809633482119953168840262687003181796985253134656507584019743474051540037
0198785106060874599446695014540352570031546334318165231198110270083200993819460961574081
1912167380768981505991052410671914746837566627126007296927282268656536524699761728602528
7726271640344125577393506273522374772250014287765897217988093319245907440227980569021
4264394254810861687068347816252173850576491072683937870148769982155509314165618595245007
1025756726282595177928405628829700011546419865595352544996834543991977275367577600211083
0150881819854147376943459386561069320736797076471298493551694546652267090472942955083
6157240280482005825739183752952526736506496453282054508276576046788697632970374445563775
6378561084672717707580825514964501369216617785765432841502811835031591124044461803366199
0800521360880357385128406547882769058326681835881607001362946695350710316558569704192
7786005935078551589529123382046304921814494052053499713106494761578604400320885602889206
9615344860233155094269047028980977793537885796495500203176504627651740571540617060556810
5134056391604287655944926063873322899971410839137121503436133645142259986424829208305
6485897581603256383128161380960007908935734957310889631462409528014068190931239998988177
1763898555621844678030169346353090305927962342402106449294831020850645868559433312539513
9524838688898062346679486890650903137855452250728662262683870750011352660291818955209
9262967569204845943262476938529731267105064168639505273790101943292694527648904787890761
0126370136794369716213466888141891993363958601800350752311880372725257289666607166217142
4440801983478847012145344728599155557751206583279140053175781538856847128680277326385785
7441328392042729527926504880137401784521928995686186829357221339510448464882821735899
5766512964725636335975712145877833349572515461398601647453999648539769311673287373055190
2588835238311357879916801903326522714816421732506704342138609935553093194099560262759136
8697415261669351800733010073347712770141982593018746435630965905683456340475293323748
1009237396165750138276831763589624328152472882977893610469543915503502826559157739847632
6430902916874343280159236172941536736171330029017371318443127049688675611298969326529659
4465499667471535768696660558296310995106175792717578161573125856285897128276426779171
6893165800057218282357750004059567044091096667357796596052506050666833639190204975710634
4872344986348671414667822971457070330926915123679743067387581968121298642723541642698069
8879790931784518059159901930243208904155530493648654898714577652218689894805439176308
4645259007275238629236074720701585470755405210996509698566409643722870152885339235626392
7312046559262864788402838720589249891362014140116438701954364134908633792812816479946379
4627588889088646470966993849937491719882603931726929203942777681983683313383799583742
7681836216679871214211905011989208781938393776488628449485561805006089077650751586292770
1959259932442619273269727282349591170153098043676413856262557332174809638028686850431393
7916168857634951731734859223470560443030172884025800753908911364147726027748745994192
5097705086980709364105512043219349034955789940741852713087627118882430783202381903007690
7422083327977796181370663373043271297538919350051789161968135373404265563118960258640067
0148317603541154297223781992199556023858024503661467838605799056986863555406479548954
7204056394333390506496791267481311140510732501540805616465900806315159221174402378865157
8392189829429175508276709326894290315493806998437933333869110527761985924525647789587848
2052188521414090323051816901214574000009966442659876813595238437727895869528100085711
2561862573652015981825620202163573083792092430786700079613736910219151054757357051501146
4338961283872425467300505177032344238412357186969741080593350583771270719869435995079630
4714577450582336507964044785917607165988886808265296963837809706113313717750364278013
1010795150616285422063203446284663263327454831544403946672431331160325082494490505332384
1032493463922021870727593349291267258629334809932531823747239254168211340581999954132751
7016320636278222271386691383348995336712136091150015173132961707402084824416264549590
4984441129545512925447785182211991094032878118886905714686177128860539799698185577922155
3925671787348163160922482776312660684745835366035159779909896769595855732036129228443292
5853647979609412054158932642949070141158929111517475146758281987892607050622094077745
2074663217064155852661707885427500925795160755955935869473424959095581322204412566794346
7753504476532650197123737317571383329927089740755597885128351304598455284883725505348963
8239935222169298830219801658596549312480683722635593267928851785450284402277438542102
2824903257802006948502771621276658064074604429517541349195216416007268506509260553378969
4772396579050461565031554592083799941198404765328113176568076722952904454592704054640301
5287772775804810676349293171340851049207675359972862966146157420220566537646982643170
5138047340697798725308838911494385723374011860751602326020277648969076985433651473082275

433882738695101099833670024128727413565503727405947236565447771399796685135931621234337
544851903148567345541814753513518242417209118014685847208000730862144747606523813023525
783587455301825281717171171015718862566712633423725220570050406098467854483788176358108
948533115971540207042303614525820590385294533054215525297530558453819167532187528598516
214901327719024316381694861068173772189029429012572557985990684019686316410212246831985
853112323567481945936102796550518939314502487518074948177301109341947016827082389900261
470702361643877766080942041049054468635485493122305933781024330895127082273531440571655
947110323005705951169588095477233931840339027767987911592480395689850226115275444158320
618765140222224029502054559218094927045766933472596857623787924529781560787343211974293
958396751953897593814876988250625385358777662724507546902465810155429989048070089258311
713170637950445268840612738875901584142654170475607523030375143208733562082592850178264
308775160926789820001390613505871376592462332190404440055410942398267905897885139689547
254922622137334991178564280498375609668141840048848730123001296710747784360661237285418
203866476349827447715198814305491911669585617410233435433061542496201170954475619736192
403103667583443494644378941368106444501687764504945634650421272848731047695308709883688
646249809554494260241871226579845785739308975392230739226143943963612127146941410229439
763311559861076503823370849856086741013044734018270146996320532480321622612198722761063
082967359523544105278378892129976177519117764465881608000653371191361296724344282269125
266390925412447450210926065513724173419891767557150588596357598793490197538783563658367
697267863568522582833163097474123481253074073733073609998194262287404434549650747294904
633417965646870617383984091293859352818225975882143718052913295447657845860573539077705
255793937971890446253320489638960187738469072011853132503077041069244186558093130464544
826576035620897490872484742374228448213547948834652793400540354495932568472177740979905
736395866975286081479810254140777498243821767156774786614116483981123794791015664358048
572783671091600862419885755080132281753843595684111703726561147141239651878899697255163
722793292460284405767699251361815879579964427553237302394204383780437233958958654624999
575196742575521197581872950935410675515939829373521611154333013101218397606197393639476
252192033417790399006033197368283958253244627168591687355022612585968559972933286348624
787364150024131118383245202231890129755644954732795227621021239612213463903158915659464
565170028931151115838995840891206796472853033188981376388587555107786886133363310754707
022741418500362131310361840406953168805178895472784920094230420492712867131644669700850
727102018463616874856588970410659555414626982879264178083593342725702652032180950707546
744984594470135607865478451604595391069748769298974155523133964422645501366382166517787
242630974178641994515224617209757080825005108058544470468982995529324944355070541728582
642914040967145656221243709004095792166706511165008760646071273762433734212149082582468
986038099644176184618689572564502735402238330955425104263133356752324019357512695127790
299018595816255206052521411701352429193354046859210129414428254516897691209514436280639
457148833982708816842955582352399731139179320095324951789456655108842397799318241086990
011684178564668651721746650762843567633742156096164220261201612394682596404506621745643
494583273780666610519752481871694062466609695309599806955267300116765997688403728673087
740144661789276382246666333464390095509747252303942650540425308932280146775346431203610
557794395524480251412231848215988262028259644259967619867396570658675918183077336559266
491516066778791051898706158876319019072007999420011210607273526635817597141241824062366
925376555868441202675680559817657256665332070202064686575511021574253755022247021517925
993343302541345605489426826333314549024487971506727720282217255867715657878582050820252
516940114359515637451652608574994595025434389527768003520264717409086892423514519089400
530971851647234523173506926172668878267171063688173881578897796060631452183858778693708
734540094966693003360135869026540427097270623600492292296081310863820623632371577467779
415797402484532798082450639053134742445006798450835594480744769945415906439499475775872
651671793542742497006319636964300292292105200671567613966611866628297293707079875606803
365449055456174068804328975947443342155278698562973267576655053715029185137102846357706
980827329232531502006911852427392874443494647938213496398006129780887756245049988455372
403877702348309998608366999859756737106958886271914691941912732072570723773540909339666
469882967037862462373727420973226738742059575980803876288431602697568771412581404183550
497490594915799545865238939194742639325151091858556902894393469505573066259692265689160
715818537236650471672940961573025352126600539121701451300622014702523493475260476741639
590659257143976196544752993213729694867219520214964209756903685163739345305181783466573
867095348795181729514491253484081074026569849939436800822125995597654213257352321038454
869578436312112594619326722053090046149457050825276388969813510186661165397142404647174
358655196291359358309791076679204593365410608538223802648413571872993380233429856651508
502649340352807197155910794745390065517046281762467044558322562717521852255957354599497
783140787677153792536566105478976853462743996130611415112132548466122660618788735716028
123226747181921528888523263669210346564873161965422985362182045726235421477498978612133
979025337473446140068063972589239787944085099937220744000908744807063865971024146227630
512290620085496190713583799561482945690351035198342287315755934262484661615189454945582
022679624435043712298667900943181569004169208839726031268939689566193121833197827842142
259055221401751680376956490743389418715932175684929262354393738598185338071320516816309
514199459009694979363545894808328547660132882941444225114449681992730153750546645843363
598289364660036578319172432850390018182148487616941457628961538382188701890863412093856
427485936874149446875827103451741492004319567298678413978884786649045991569612881913618
107480949910415708977401262815101242861452865802981066090455250499385987919041976120841
978698760392527759782707450113483061035018278792965991362851752723479032019734473147292
709622215939424347814352262747627307677104696840017100224038852950672410051667635212376
481720052110679446599562233491873987483954834823498968805230934707587477385232877924798
127896917631344517933729971274456248609576673293825568186446636406996989126898126005869
070083209760165167026819637887897604890997230557171338375464999624068723855694875068938
295763258938024541153299157039796422302399545609544464111391361006881586731846878265101
696570049228345825139404655153170197076456238618238807937131665209078625310811533626689

4270612201072180413764483243032874773141202406831292014523261195403727015190784869701521
1814381043544416171734139163184856547376681110702204510818209221280723044332929124638244
4336301743384828584233796212323477117793895616687237809242519081296027936424193628184974
8940990208983798935830032421390456267126734572816186366589306248014609505597734735455475
0139602173216540096720666290583981806131688670038187059378106385774367863337205955751842
7694758327773759194631398427243936334188376310834644995756352932716750417168969882165
0648076245182062500282624495681042643032003879171181952919853127293670417047963158380327
9365473939857208557192035086657551402004214770234830389426903212432500258859910883941811
6497721384616412941855176171449142691293003597079831727614858632501618285460862041791
4485392820640005454345597304158211690285985682984105046028612965083285080329967865104009
5648664867159033719334677194350891065731490678039563888856636120160050805921745137924632
7838665734472862863030077297595047516604545502407028830628040351382168525730634149438
7068515956395936526814496968956145643864880494504349902255671166299639240221440997689487
0875671432390355141359521037244441846307985416821904504338965724640214257044953562645495
3662256500666967608460937064825127434811013467273425456479755805739750593124835487637
1580056028471346601081062546627223334637334300500962105194636726949047877374912405420699
8875797422158887968385177321406209871182774315762418171847869267900415587507121433601085
4001360436143232895302095723593206714156110714319271992678558574157594257706146494556
0999552401001930588972927266859001528748754661687092245613133185130601755513126079433440
2309229537907376633122585990420900946722670259870799277615990000931776653161528804416226
9427921689762984424627086362210910933759848947170683950956349697338558358260813169567
6215749333254707734883220480079033943623108085790529491724423069011301113657174136729203
1305878555105039631094535030952218312738609909579418876434542217174163341898636399966114
987157781219546407977557588846403937077193469087088965536912400030650096894919810910052
9774388490631250902737856046776237721578555870445997612116416145716077093607546517425
9953259347107577544350275053673134131142151523230609938392536913858845846558330032898178
3740081818750715230806200313004540211260868774528210801714080209025019302904190835899329
1865938452890400184964371422767592128913492688625668113775493673103222674200933922950
3286051501899317271026717013773777478068198009204028077860055452148757552940990287324819
4013610819428190470689114022142526974181046535145505083882265139194530755489780807554244
5696932196711005718521638648159809872679331664390265225713819130934736312569789201105
3960924282364559170477285341811402639321945703406017953885618436295894606499456445688453
4260278605126312521700825881938297030445738554413388360308065285703284651271685212444238
1834142017783679437013947909311434814254859085530727882948450869499354859854127044459
6159546603704388620071446618750901304265565906955857280800743670206816972183517827428721
2890200809069634003777089145190128393774671339739619263865304770679339560519270510672331
0309545779961252059096069452429750630455911318666948313642589452443235754979677412397
6479466881595764173556869836143273170029032516182522700301594928061770511192589115371291
2448237552457169647267653944748610995241519561470596988721410151580233901239761083088219
9117404322356008911792084863059073388029852774767535322551612428367376979180054434711
7232334520091246908002164559179524174267829385153189306909293266152425660581889441967434
3456022709727360342662663293332460147060480984677870815939354282100342807812540810675593
1069647392948262847057518124302884132653959725874119645501633325305025383502914686146
1617478562135574543855998553680934633421813599249185751302585476037030128817609364336840
7829046643979074260657889875070232936409310221412174671651380240699705434985820234902593
9888642481855526553925237195681271750701842124791424929017032147007243357015970121197
4487005523405020773257002803879932970475431199312365706828079352381774539205343288988952
5729924276130454542719431733329906499702589876418489167682854083061104620934122171986713
1576864161549942765384099159560690953138315503090920169456133454175784758520633312043
5801758720939485541655086915180687797962382252155342503221388405441873046959976877030507
8952437731941691159988820522504997829406594753457651159100787163985415613456743080916113
1379633369817885586717413414135624336005447938745116496284313986194980906009398323090
0905816829472262572150135372388793937812208630799023509730205123922445632890228249256304
1374965855818486513730849341492435708684855135159799902503246106020272290938885380261363
11393995424616487801827937536625047070069147397331706904684960315105523973517573203926741
5391027318967677120090638872888536869933205207556875837563603347508741564417601081667302
9198608400088033630287086900996817485848932184601087058586682486371844389227665524
5767941640438831832423498299037654083476895377032201866803971010714390473868140216052675
6924847278468343160998429655272074857983721769597877566965816762265514263138851500029463
7781009953999484287585441946943472649341651277938308626649058850081322768050075865491
7845129274926312820890220215326064985899909536153454281456664988465333547576205127922400
7840306588179317349691385419030792871411618904735766024650801839335389171063110127286667
1634982962423289411483310313279065817329196766494850076768129269935387992768214746059
2952748874993037262597929616656587786392486389524151174329486217752054176175690931085588
8968473494209323665569771317681401250124226846482228750429231836913579596932997917249951
8621868441296078646960289821200589293098838773815520117565114199952797094114378817494
4968213742860156220391799236686004854434589078533611910302737808989335252475535230843824
5867756621956044928479629946792772135435892988188021766783374771512365742650944277878804
3350574747406237802259008494111884213829244582144873346591901591205179054776770371230
4249172116265913399037622083284442738767047837414547232187942331243607588488882771504586
1543510939002643601682992803197506809366470666203205552416310289845038254621159578764756
7302752686929159669618549619044404044662199537289824109617707988366088887654648126634
5383532784505254198618337678646150550393790260681921141972384146128642458433235290322236
0905277755004712298512295034462883905569975981720603891611238060131318105122494328359577
3306386309259851103472239464574086266531490001290366504156968191023297969769847290863
2298673625985784801676225265108112016335490215471442937431141975634586560157182991448118
6126877183400623878548437978651785682912432953988141623754485668630773766103746323586821
3275665314854464506820125484535927360936186264766196231991833539147317185822298570171

1167198114321090854434250995031766794195448597299778958568412400787726618629144166398690
0080112196831336509690388607660128461596797832925880349042063398827207906992806790336470
3776817406000584415438581555772772295649752568643868949261549289242281103450161202814780
6213506165074631824647369975188202096435878027594649183662978213194771411452163821127470
7858521073676592391306086899850625921114152889719327870137825427134275634294807801331650
9762642872460027307094590110964339848745855569740924598089885435454313053664774919881500
6892897687779052390542681934357738634441513341862757636310330922195956203215290961611150
3685739850246058647010516677093232103064901091218555140080216601473625469990757789341660
7924480798967324179992503579728065637862421429843419725017884357176949576992398474525130
7124601847069314815508299698961172058217191105416766532433095673499951532031878452325760
1328623158175219882963020982287494311476536039937261066354950778158188580305550542662450
1793842510847262774840941625543412354383260941951590947168093052618580653810494509778580
8082214107635639475580905772471530722819487355835346543575188735512400918724641670434700
8737158493158821548434544930428082014655351469522808478305357196952737532379168519497290
6531354042949046604782885591851482763239489953795565249201148969394451505727001755906800
4999040650039473642420505467130050192262591364234739587949647039384936419098098530512190
1621558370045577310419662662507783512441077625625285738988589660214128348956074093254650
1181353243587604095375521215643364986047106280265202770977319515535396436757051151166920
8929792201097165623859632518175615932409773122534533106080717911505072302717148271801460
8963237829891120610276527178988163345417514441473064365430351731621308486764036914376840
4320016555987766873301871634534011417135689071831702710641445210584948829911531164976120
6784909397348106854900394978740356320605030736273570324159687089620643717065436397290460
9122216716427888535343372115831471982402305671616657769170178204752905773443970412847340
9545208960185542153136280719740786321231653978301297526012731126414407421906106695505250
9866813551495083840347231036870312555137259384585149927005013275934627774401011186284250
5256816329808614988780665529754260860424160682983073381812311614969179314060777202474200
5425282522310772705986412611941417314411684087279215303951794379325038950350379516997920
5183962279012532046316109716981397061452327226279366007086271394457758796426229914497920
2545209960976664477076992153613686025238118919227869145484714553128960400300537363318190
6033294140686543742207734030046033858962622624914352407456457049225044993493360293930870
5293349273129600917095224860995537482710547859161005138392047644788797683638846571330500
9329165820640169469425752029600556619578019276742934566777291559536990206106989865429550
3785598799106190704558262023852154327303194037729871845191269514943207628695325951711510
0077776832820274969017179256898545675956507417512128762396724736951674199195036110049460
7991355288402269400931838195646088131692112649656237413831682406239087055671539739714320
3181967293239689263514314925312245436289062441606585396768823576764744363980140572614250
9822342561074243161625488636718123532868778190505930069407512921348443987792577627269630
5445371307224686145131563149621020138838593714401889477921529277658454377345516343893370
1988738468811105486723356932794005440921219609343490675983897554524805054520800498152740
9200747877396284199965065331282696188677043348063691780799646256857245519628823197948740
8965577021479642894521512883360158751915953810370850639797299735689690584007788429708990
2608484570355183157900890348637184875416685383234608639198278087861239701947102652931880
5737214809046707762474014321682502610112667613742170898136751231083191764166934869984890
2029180277851355815010066744310597343833944735567725117836458046464615451179110351240930
4984918551593866892665938136961130354024067298961219762427109325385869446359423002541750
3321565657250243603303884613932993937865959279678964044588046078153059477510151153662450
4418898329892357775660931477161612821804669206134459799293679366502420179666922897975370
3073644818898314668245552307321257581535039488127264115042317215482443561244182528835000
3580332786706813879504627906870490116712376071643339855591302403429998614177471692672860
8803266255752108847842167745295910245506769303041791375245768802523589809748645024641880
7666375395553662635915381640596219825235870966413773128739336765790120030073059184096330
8208230332082989535349363242614029014846162918618765224111053776952694561212046187074310
2341950219314483236383088200275662986843078418065621236090959500784650336417548080985020
1354733949247758495304548113702074026329982580740968256127916348588516591447588799029210
3724484727064461190992366647575211498364865867032110956454715787915092174702627243733690
6974862807417258185141822083551705270634850960040443068436635163061826437466064575132650
7060745205037211201515169934911278392933911922239625437654470861056645742725569231465180
5535254260011608321678794429779424962733813576864466943336000352414466649007679870013156
0263492047563503495247919804282222185278059569305249902499544309392692436127643824686110
4268197818083502056125433725145086928036505350298432622855506042603199315127938741173180
6218796986997634477503886909806836040445743722862880570942919176933948988765809838737970
2692322359662338596934965123374347476991436318677827011482214016647547365610392372750410
7481769037539367334032467534153627730115501020934272183689094155910864400019096530153950
4101206901250603932269901811547829502282032233294005183868364508051319548143727208064370
5319322296989123492481038512843617560207402226558285085119469948842512588869691454581590
2065654877912337542707098451747937039956902676262014075064802750327238065670591321708270
6986091864294475068937683826645445595483089486043222574579380028503186276579406961473160
2813569174441598027998154355924068114906680946709929845745987015484908759246053742290640
0373604467742854060157617053258755713356838165740712642593248051691370447735683718307120
0813930352821429444752842340289752259179215160740108042365426285710974864466763571440820
0724008145338160923262035279814699967495723236578267332151091118022636939916018147122970
2023804757483953564553196945408002264803425552680905758206112444701757716058269257018510
4413944688490536072043610096757212802724052092407023610606881871382476516361742619133970
7634561383181365012560911047663183377228720637361595456472374147479748983593518162651430
7669712003413553498946837084893797073850247600099115866174299161707233784893531177170770
1523957767403083393142305027292630631081042705651302864846625392029885328796316776230380
2299372472723880411875447697652523507829505615255394759405824782421806718582457046994070
7394984720663856205573233445674345660458467837742136922589073101855995841068834622503520

353728534963319528370625094250743576997497027094055747046645219762536141796662444272667
694189303906676342859230048152445224261681133173758632783786881621047591537258975966869
644976142248371194374904086560255007219893502650158163467368846512877064554251661078288
560318969998871266451468410214402319771058241497970439246126604419241140601468338116122
742268315690541247152897991464509522966270856875012701926725596500194500691612906696802
810447518194292025511547494248704402631188754574476100184791278186581739976927180810007
289603084898604618682819856309424121676316136237767714056231946113383280742418495053461
686110194604382531930977109152591736005534399731333135849501017289547184952317906566250
985046174665577863148875458623872026102194823242877652986385229004252619718067042541172
562236071642639707822048598594167952360852817687342773958830828691056331069486623609543
505565751216514943726300699722505678302211912865501271465487084140225223153149223497227
991353773065966959998750399937809447331906601103781235090389018232562430634496575814744
527861854559736247006319403745213472898121526972478109128556522537963473738871671260329
930685488860197364027169827259109531924076555588844508157855245841112044490290390695423
912848842519042245588371170888162750746147976246189235606500817934947717429070123897987
807826051352689926894698419053814626163305900244704715876550365950178939076581413767687
372513056774981449385392942166782820255779041732656058322935494834190658600928961029646
683578155296421392104059327595406060943826670317269638320413335494679697700418397854342
046766321555194844828832254976530807859106030120865055406042113390401326884294830908988
458874302308227993023592854783522505718491063812864764548455171809907731790147058800482
920620150442067845397717741908020153668160752550314373827952670837612432049105623462576
896289194352666108980139562450209518150431998234087432113171069061682401787111226713659
666397920101981704274586664036897732827327834824354272650977160431804192252309441298884
469731083473896367381090190899526718997193921451058110812718546816370155095825516707124
064299341476670161150355441441194041958291773485107112125662979098619999996540849517043
900368922001042267078694579723196923585063115708693845969618420301450551502590120510826
393703997436549880619703741739219144469416514106636079456974844934847539039281386745821
183010667515431840484689692468261959399861571628430139644231095482686990322371644747157
274936323184407429819539857799561475922984017440485153041766797849006369942693660002844
979797785554026051121305234955737062675825799714395991788544988331066551623665856123359
253271839646618237644829731889332984235825957793217130141217875919415974402076397295715
318923688230819364762067147790956749070154985003751346191143098492770770798138818879959
480224164865802702510746794094352044953103856776649587931702398251039066021471108615831
162126071865990081911399381598101920698000241127350474628896671903790748563321428342740
092168205493805627071218082873092264983066575219118534647501850135320606417641227128460
558544761739453056560746356686442454462741648375569051412067351406096305126451667489406
638299294383654360746506143950704968071826849163709545576424848386926818411912110924906
610206884689715076660327450577773343041627637106097519790330632956817842597464262878940
878385168455964410085119530978589644726690699998440632186762055303418051862939174406365
332926189220856314319780999680657067623053407591615808200824362073932301142238256300716
736582771093991522624799325280011454563312695691324690032873899391133992497324480045876
5877136821489766160972699382570765115866631072108284970496493655241292224044583191702848
898207178884854389814512916675879078384962096048603692963291475594980370318139786797552
893736137354696276157766296220771905122829174297094723387017653443099578431063479542353
488058881231657678707058508246500048324511067253389916231431901305221543679126128662126
202525285217912171349692406692095581130069997949230374601284527711387700629707699711438
037455115057984111004434208922444051637714507618625054223567039996225197559236773789874
643199570161360532291164093609531373490049969012995668979555332711001385841753033461134
021084582654751012860004887609682955219284748870722926282904573311985604399900510484707
663102785911891124646255931909842198789269785184378679504406465847572281723156769562991
757576116072033804652994875879330654487129164307524196509275793462442419626650689799486
103279273765620052305756620792920167880496283871511510033649592719187955925875040744753
907196864979388197621845175826254789534385133481989809513250776516032189367643935789874
257182383956741818445356995453150082542133306517010308732535611610693267909309202869922
372785468521701628067475994913914993704927888567167506237662414914706279647071773520852
781017167335998848281371796097777952941722426332318518355317167332029976303241877644623
541759688590251447822207258221934007218347777608119640621380414982008082501255000941766
539288008617917781449591615434577247406519542133708763636682423973656328189532210831624
585975166149030189591823199426329955963185619311559853962760549639977759717446047517551
697607483562759819357200082297757428277619802497403878260857717125856049294600640610936
731278999492691310345764588448614990440208426302703669579307438588443140609014361706015
410377588673856929793776645755861239978935400010928574449580222935467498754378309709514
101261834854838480430241381337637000454392225853714037714479457358529388458142893720019
798801308454781892628122220191204628353506938014819743955584677564588334496279172636082
5927611248659200599989832184060109355402376630907822491629213847591024241141382835251602
625353264435600570070840451193106652690687989966771893249269469688345264273090749863838
772405283366781231370455336979991957829232134682137696312863543658225069243913089568686
769743433408444751908668832572786176711612207847218246948932532339566698857067024691758
606989056945060356554952953970519099492885848945584654061082743735606094125813712294849
204474005805187302300883740866929461498442748676303167772462752402015402800923415276891
706235714025378508790372935560072864595428497327769753982087961076731134800698144794254
862088170466743965487214385451904279579068136043166348631745694327734852881015985109045
052789531469123475508715627103917239578792875976672215167624262901074540331212909101122
408326517693204979603234477416252168222097499159489203371376609450830034334820944250251
080004208805754692212049324914127403429149124538021483411683675384194160195388927703325
746632749339705687372544875506221412737504640649213672165369431114748332224980621548316
159465443908989830964153306958486278581514770351522551461015648556801551451579171818206
935769917187560879264686555598801663358261869069835490808352983928611931266539140271238

pi to two million places

2355315960381577860986824167842882888906935426095107932400386980631122798817199112860052377045295945646679433112929911631026904499920806724716918121994414672327761362922962894963354828582248486035627995208422410591444838335198652172331478016424432466805977368595377413430066551651461749035324251190070249576024735054695376433532523849388045847442279832393927585805953902396893823771283293385537873048442390783705836674196657066165500536645168202374503910749117674459632075945204954674341646346117978453363714430103902256508271911023615932369578312408093209988951551222044702519521768909200457761299388993167263916545073537102374111797467742589229534419022297588846887428396795284431983203184707111946846841412444186749705085618158902163171987758989525771349462047318983190684935059056748665606636149416187281484976653042222244137565176277436844536195895000602498130043450987713762463275355622902259700447693569386717575223737336763902387807508478999304666167525689085346755244694032897011670909700195343921033297006319642674010577008035148829771798020899951155435069771918312921014295000412547369839690092625757617418217285451418090261417937586289984055781829869216488669866311027767601314870574595534063602210493845371086798753552300415084475554328382345340505167636737734435740356769360532775639779430015655396644317292414282220372659382814494564574118958257396763607403875678468919134116457365526844823195071686209998991854740816747696526459352447205019157770113454318960866629832780076951704649374183976247833709791438522418547490359977792730018563503069417517669898828337921384251067140264505840795093431796344345213812045700863478946499632502695806810343693987940006364358499908160484536383345938466846043068215257288582616984964194825384992520628935004069880794079831963839236443883299605818255130768306655909310376848767243320389451282346340701247214637967674435270796736390517498369552055472923054379993992466802982060562655867718365656458262401341613122993110075317943005543075673479011030813373018187461428282709488283728721491596665338305498156675523181861159980493764704670352243891506789090516032939152322736117095440753371603747714232688110538263838872461673616621815178052763810667586619862675771321663375058691996181808885809030453416942522607205174284438592720627563137514729685950742803095197971417263317393680104061066136833861297158511652364564978097390482367021566542483079034459666444977332720654019922894976588505651500314521119049014317705778615282616345558170992060355620562170336671680274327398863218900503710414661592939203416916939732069928176641726073514819548151202908054146724975545371815424014803824880408642797743589891612170354505181704336790744931901216388147668829549428288766084405196380615447100601816760416605967848831453287181420187890959751576296604537901756653150243739237873375877180871097275368939187382870095733925493173497022366675478129783386866172310589523041094067484693935491640219722475147358393798945005430783049328297619362432803192064663919348595931187492120232849767501704144656272937695334614928564790566159838544254520906216522064242643169929850788049603562205892264441183814540887940456152604642646081518100790786253808888459164460749290792014776061791654775036456543063719126897051107129955175798864503225396039477573596160651333709828760989674961481092550416348907172083484913039395856156153788511253058456413850893912516842332977169998426559734568861671652681581744202400647635087836128182161462457641325948035674231361399643436393913410657749005471861745992859391203002466203085594304274032156961815081290853233267827251854624651877808066156335394236175098685775305577260345500424287628816005466348435835950600399040483742325491792279165005443845073652265177248905509237211882508520359061511601291080058326348787823915519188553688458109520277816821133577153943866050679060657587334366500667763995212155672480632609841461142493709846584692378680473652301703196317530467860944697406701975450030747586671247149330060722978454298265922326509871354876302974332536514304271108149700160998187087193133814629764484545952202409628845071163482774899350680761521647309162659576733434014974508877788663280912709006884916752205621356451448350554199154306444803541042761871423250277765444311899113675814256648224546762357352610651197919381145853116206701867530637021097360878225320120185544646738782583791212967315057226991898272029246060901000445464923922695823807702261434629979440345501362860138448170567978194078867018752402454298938292784803208503008468921776878082862044177766867301770171858710422446710121241586685243731768336375743574922209717353839850496265914139177476853263434542002271379751534515291235804892956215632268222240045550198283993360031757198544695120178194983129576144008630412328156550541801608079631629481404942889792516842151297648241816041732667771491529188321821595862113772571777526038535830259431341104031281840309708735307744033193170234083162390284816668614438282059884693113666082217582778330477618575558743387072767926492966776452756343709727152451400396951464729480894427190494422165878796289923198080578982950849044849702744472815094507452714636045491812704068971130947259826384699166909858284463485159645719574745789444078030569803419742265147581880599782099727709997065988629466708851071806388515362808704646023304407120251598840781471319249746053671937097947084746809249429675999239747777201274248256224834207424205379522783077591711853020470906159109668076709677560768368666486313302405564400786020583051395920013363422305560406259550977865826117875070024034614149367203036528048459467371603346158077668739478840905319311591307243303519238826829211733360476563764554735708637300904704840919990453105078103080979122610735496844118065728267630502945319617542780473412220721813553388758185695327243589416277726644519150561198823919773017584364638864658378871929159351096829409773997029127474772683172038758984845969164620740498549151108600918849091590639766055353337304249451463974913939406967815962331132313900119095789784705620482550134281175557040319455983679651476063971480414260589730191943638978862757551897515425366279836398645904175237303638915674362383824400576023669321878385256897351215801981834444124818684527273892114974665005830424684202625064864025418782191399647787844769164152061376928396815195566074409531701703994748672736601894695988293828632300474370013845796836422762887202510115586290890656691360663069899485101816380520896900571336339970135427187849723967099548709141264336984718651136987737348811512512412583717999584834852494121672340104873304918215109222609938976122320524936161608505389819426343103514536904444996358869294552888075162052055478949444034720622572990304307297951148304344028898949576356355002448587462283582061144774727653380499362329080612415452406386788488

8066034190642382510392065802435391681445326133690309558269124625547377092193942419997366
7237427964870665573467507064006070702431362344374577410535083062205937499206939694968222
9570830059215398620203416599376959055913328879786256920925673660924927775094370519399622
5770714869098747312889527659466767235124208758141018939572185813380477329991292659403766
0495554824830207878012863593305259979657034333232007305519107967323645177791154535038155
0642449857830819137960436085325658310504907898016519681113598867584484209456644032472333
4822748354941456194544641598625123501474492022668858016247484276017798094418670448700244
6812333177774369756596789072654792944394046910761910347011886314350557388989453763899855
5589010759476583690185474808022375062452550576359470476174391616344837883575336852910322
5698715642394667429201428282747163676676396083403322727388870888451103923746783680369233
0512303497807584975251183105572774486705376659473404247917253573675581937572559628844900
8011066444167543854722892303330367245085576029869857601107914854416734966385833310448433
5101802565021023352945476315173229225137998040680350124017179778045795935289355746298899
1801117923156335080513558896958851594669256881504863637328047836680876883571335909082488
7423589569225121853432135749969499674489836421458127316999050262158939736992904023185388
0762305205410977177204729664512654057590236729896110236961465974989670909548921062506499
5528265961274385259110241012307803149791757107269526728131987179854553178239940885772166
1453941523730526577673920866810912462629410636258348318586059461246737991187692515733722
3949855492734735701873595725512682224542016976958607207287746093418219090093926465385633
8647752874146177299611493925900637453562530308317498843293581281780510432229971122444155
1120826976816742625912678645277273777527399854556103047251453546760224077180816094532266
2316977299962731440506393549361430032900037022216250885221870794350300748079667245111700
6194231601304267219633741434784999771550990774233422365508714741255647628863843831659155
1499036237458668967214844731356687079939016423811183996217479683216082443595649516901455
8024925520775767313601173851325863232419886342247342649198000048119723475032561582830800
9134331665055224443523871600822803540940386921426648110216927954111388181273220199913766
4643952008020356106747475896884625547624736347420507112698570308417824681153051680155166
5491291863961412924182699224706991571658001340717513650333188771429214055748695041211599
2792339261683797587686838947302747896147486167629768582995622801017465451017011215612544
9413019015860351723464733667168301349924801742251392905122942094455992033492760340894600
8464033876095367596380190919417586559312873960279125105965404496212151770209578971066555
25923697193382282267491322907174473589256504616637356323687106519145729790 9...

pi to two million places

www.ingramcontent.com/pod-product-compliance
Lightning Source LLC
Chambersburg PA
CBHW031824170526
45157CB00001B/179